Trends in Teaching and Learning of Mathematical Modelling

International Perspectives on the Teaching and Learning of Mathematical Modelling

VOLUME 1

Editorial Board IPTL

Editors
Gabriele Kaiser, *University of Hamburg, Germany*
Gloria Stillman, *Australian Catholic University, Australia*

Editorial Board
Maria Salett Biembengut, *Universidade Regional de Blumenau, Brazil*
Werner Blum, *University of Kassel, Germany*
Helen Doerr, *Syracuse University, USA*
Peter Galbraith, *University of Queensland, Australia*
Toskikazu Ikeda, *Yokohoma National University, Japan*
Mogens Niss, *Roskilde University, Denmark*
Jinxing Xie, *Tsinghua University, China*

For further volumes:
http://www.springer.com/series/10093

Gabriele Kaiser • Werner Blum
Rita Borromeo Ferri • Gloria Stillman
Editors

Trends in Teaching and Learning of Mathematical Modelling

ICTMA 14

Editors
Prof. Dr. Gabriele Kaiser
University of Hamburg
Hamburg
Germany
gabriele.kaiser@uni-hamburg.de

Prof. Dr. Werner Blum
University of Kassel
Kassel
Germany
blum@mathematik.uni-kassel.de

Prof. Dr. Rita Borromeo Ferri
University of Kassel
Kassel
Germany
borromeo@mathematik.uni-kassel.de

A/Prof. Dr. Gloria Stillman
Australian Catholic University
Ballarat, Victoria
Australia
gloria.stillman@acu.edu.au

ISBN 978-94-007-0909-6   e-ISBN 978-94-007-0910-2
DOI 10.1007/978-94-007-0910-2
Springer Dordrecht Heidelberg London New York

Library of Congress Control Number: 2011931897

© Springer Science+Business Media B.V. 2011
No part of this work may be reproduced, stored in a retrieval system, or transmitted in any form or by any means, electronic, mechanical, photocopying, microfilming, recording or otherwise, without written permission from the Publisher, with the exception of any material supplied specifically for the purpose of being entered and executed on a computer system, for exclusive use by the purchaser of the work.

Printed on acid-free paper

Springer is part of Springer Science+Business Media (www.springer.com)

Series Preface

Applications and modelling and their learning and teaching in school and university have become a prominent topic in the last decades in view of the growing worldwide relevance of the usage of mathematics in science, technology and everyday life. However, although it is consensus that modelling shall play an important role in mathematics education, the situation in school and university is not satisfactory. Given the worldwide impending shortage of youngsters who are interested in mathematics and science it is highly necessary to discuss possible changes of mathematics education in school and tertiary education towards the inclusion of real world examples and the competencies to use mathematics to solve real world problems.

This situation is the starting point of this new and innovative book series established by Springer, *"International Perspectives on the Teaching and Learning of Mathematical Modelling"*, where the aim is to promote academic discussion on the teaching and learning of mathematical modelling at various educational levels all over the world. The series will publish books from various theoretical perspectives around the world dealing with teaching and learning of mathematical modelling at secondary and tertiary level. This series will enable the *International Community of Teachers of Mathematical Modelling and Applications* (ICTMA), an ICMI affiliated Study Group, to publish books coming out of its biennial conference series. ICTMA is a worldwide unique group, in which not only mathematics educators aiming for education at school level are included, but also applied mathematicians interested in teaching and learning modelling at tertiary level are represented.

The planned books will display the worldwide state of the art in this field, most recent research results and new theoretical developments, and will be of interest for a wide audience. Themes dealt with in the books will be teaching and learning of mathematical modelling at primary, secondary and tertiary level including the usage of technology in modelling, psychological aspects of modelling and its teaching, modelling competencies, curricular aspects, modelling examples and courses, teacher education and teacher education courses. Herewith the book aims to support the discussion on mathematical modelling and its teaching internationally and will promote the teaching and learning of mathematical modelling all over the world in schools and universities.

The series is supported by an editorial board of internationally well-known scholars, who bring in their long experience in the field as well as their expertise to this series. The members of the editorial board are: Maria Salett Biembengut (Brazil), Werner Blum (Germany), Helen Doerr (USA), Peter Galbraith (Australia), Toshikazu Ikeda (Japan), Mogens Niss (Denmark), Jinxing Xie (China).

We hope this new book series will inspire readers in the present and the future to promote the teaching and learning of mathematical modelling all over the world.

Gabriele Kaiser
Gloria Stillman

Series Editors

Contents

1 **Trends in Teaching and Learning of Mathematical Modelling — Preface** 1
Gabriele Kaiser, Werner Blum, Rita Borromeo Ferri, and Gloria Stillman

Part I Modelling from Primary to Upper Secondary School: Findings of Empirical Research

2 **Modelling from Primary to Upper Secondary School: Findings of Empirical Research – Overview** 9
Thomas Lingefjärd

3 **Can Modelling Be Taught and Learnt? Some Answers from Empirical Research** 15
Werner Blum

4 **Can Modelling Be Taught and Learnt? – A Commentary** 31
Marcelo C. Borba

5 **Upper Secondary Students' Handling of Real-World Contexts** 37
Andreas Busse

6 **Word Problem Classification: A Promising Modelling Task at the Elementary Level** 47
Wim Van Dooren, Dirk de Bock, Kim Vleugels, and Lieven Verschaffel

7 **Understanding and Promoting Mathematical Modelling Competencies: An Applied Perspective** 57
George Ekol

8 **Secondary Teachers' Beliefs About Teaching Applications – Design and Selected Results of a Qualitative Case Study** 65
Frank Förster

9 **Secondary Teachers' Beliefs on Modelling in Geometry and Stochastics** 75
Boris Girnat and Andreas Eichler

10 **Examining Mathematising Activities in Modelling Tasks with a Hidden Mathematical Character** 85
Roxana Grigoraş, Fco. Javier García, and Stefan Halverscheid

11 **The Sun Hour Project** 97
Thomas Lingefjärd and Stephanie Meier

12 **Mathematical Knowledge Application and Student Difficulties in a Design-Based Interdisciplinary Project** 107
Kit Ee Dawn Ng

13 **Evaluation of Teaching Activities with Multi-Variable Functions in Context** 117
Yoshiki Nisawa, and Seiji Moriya

14 **Mathematical Modelling in Secondary Education: A Case Study** 127
José Ortiz and Aldora Dos Santos

15 **Students Overcoming Blockages While Building A Mathematical Model: Exploring A Framework** 137
Sanne Schaap, Pauline Vos, and Martin Goedhart

16 **What Did Taiwan Mathematics Teachers Think of Model-Eliciting Activities And Modelling Teaching?** 147
Shih-Yi Yu and Ching-Kuch Chang

Part II Looking Deeper into Modelling Processes: Studies with a Cognitive Perspective

17 **Looking Deeper into Modelling Processes: Studies with a Cognitive Perspective – Overview** 159
Susana Carreira

18 **Applying Metacognitive Knowledge and Strategies in Applications and Modelling Tasks at Secondary School** 165
Gloria Stillman

19 **Effective Mathematical Modelling without Blockages – A Commentary** 181
Rita Borromeo Ferri

20	**Modelling Tasks: Insight into Mathematical Understanding** .... Jill P. Brown and Ian Edwards	187
21	**Mathematical Modelling of Daily Life in Adult Education: Focusing on the Notion of Knowledge** .... Susana Carreira, Nélia Amado, and Filipa Lecoq	199
22	**Students' Modelling Routes in the Context of Object Manipulation and Experimentation in Mathematics** .... Susana Carreira and Ana Margarida Baioa	211
23	**Engineering Model Eliciting Activities for Elementary School Students** .... Nicolas G. Mousoulides and Lyn D. English	221
24	**Project Modelling Routes in 12 to 16-Year-Old Pupils** .... Manuel Sol, Joaquin Giménez, and Núria Rosich	231

Part III Mathematical Modelling in Teacher Education

25	**Mathematical Modelling in Teacher Education – Overview**............... Jill P. Brown	243
26	**Models and Modelling Perspectives on Teaching and Learning Mathematics in the Twenty-First Century** .... Helen M. Doerr and Richard Lesh	247
27	**Mathematical Modelling in a Distance Course for Teachers** .... Maria Salett Biembengut and Thaís Mariane Biembengut Faria	269
28	**In-Service and Prospective Teachers' Views About Modelling Tasks in the Mathematics Classroom – Results of a Quantitative Empirical Study** .... Sebastian Kuntze	279
29	**Pre-service Secondary Mathematics Teachers' Affinity with Using Modelling Tasks in Teaching Years 8–10** .... Gloria Stillman and Jill P. Brown	289

Part IV Using Technologies: New Possibilities of Teaching and Learning Modelling

30	**Using Technologies: New Possibilities of Teaching and Learning Modelling – Overview** .... Gilbert Greefrath	301

31 Factors Affecting Teachers' Adoption of Innovative Practices
    with Technology and Mathematical Modelling 305
    Vince Geiger

32 Modelling Considering the Influence of Technology 315
    Gilbert Greefrath, Hans-Stefan Siller, and Jens Weitendorf

33 Improving Learning in Science and Mathematics with
    Exploratory and Interactive Computational Modelling 331
    Rui Gomes Neves, Jorge Carvalho Silva, and
    Vítor Duarte Teodoro

Part V Modelling Competency: Teaching, Learning and
        Assessing Competencies

34 Modelling Competency: Teaching, Learning and Assessing
    Competencies – Overview 343
    Morten Blomhøj

35 Drivers for Mathematical Modelling: Pragmatism in Practice 349
    Christopher Haines

36 Identifying Drivers for Mathematical Modelling–
    A Commentary 367
    Katja Maaß

37 Documenting the Development of Modelling Competencies
    of Grade 7 Mathematics Students 375
    Piera Biccard and Dirk C.J. Wessels

38 Students' Reflections in Mathematical Modelling Projects 385
    Morten Blomhøj and Tinne Hoff Kjeldsen

39 From Data to Functions: Connecting Modelling Competencies
    and Statistical Literacy 397
    Joachim Engel and Sebastian Kuntze

40 First Results from a Study Investigating Swedish Upper
    Secondary Students' Mathematical Modelling Competencies 407
    Peter Frejd and Jonas Bergman Ärlebäck

41 Why Cats Happen to Fall from the Sky or On Good
    and Bad Models 417
    Hans-Wolfgang Henn

Contents

42 **Assessing Modelling Competencies Using a Multidimensional IRT-Approach** 427
Luzia Zöttl, Stephan Ufer, and Kristina Reiss

Part VI Modelling in Tertiary Education

43 **Modelling in Tertiary Education – Overview** 441
Peter Galbraith

44 **The Mathematical Expertise of Mechanical Engineers: Taking and Processing Measurements** 445
Burkhard Alpers

45 **Mathematical Modelling Skills and Creative Thinking Levels: An Experimental Study** 457
Qi Dan and Jinxing Xie

46 **Modelling the Evolution of the Belgian Population Using Matrices, Eigenvalues and Eigenvectors** 467
Johan Deprez

47 **Modelling and the Educational Challenge in Industrial Mathematics** 479
Matti Heilio

48 **Modelling of Infectious Disease with Biomathematics: Implications for Teaching and Research** 489
Sergiy Klymchuk, Ajit Narayanan, Norbert Gruenwald, Gabriele Sauerbier, and Tatyana Zverkova

49 **Using Response Analysis Mapping to Display Modellers' Mathematical Modelling Progress** 499
Akio Matsuzaki

Part VII Modelling Examples and Modelling Projects: Concrete Cases

50 **Modelling Examples and Modelling Projects – Overview** 511
Hugh Burkhardt

51 **Modelling Chemical Equilibrium in School Mathematics with Technology** 519
Mette Andresen and Asbjoern Petersen

52	**Real-World Modelling in Regular Lessons: A Long-Term Experiment** 529
	Martin Bracke and Andreas Geiger

53	**Modelling Tasks at the Internet Portal "Program for Gifted"** 551
	Matthias Brandl

54	**Modelling at Primary School Through a French-German Comparison of Curricula and Textbooks** 559
	Richard Cabassut and Anke Wagner

55	**Modifying Teachers' Practices: The Case of a European Training Course on Modelling and Applications** 569
	Fco. Javier García and Luisa Ruiz-Higueras

56	**Google's PageRank - A Present: Day Application of Mathematics in Classroom** 579
	Hans Humenberger

57	**Authentic Modelling Problems in Mathematics Education** 591
	Gabriele Kaiser, Björn Schwarz, and Nils Buchholtz

58	**Using Modelling Experiences to Develop Japanese Senior High School Students' Awareness of the Interrelations Between Mathematics and Science** 603
	Tetsushi Kawasaki and Seiji Moriya

59	**Stochastic Case Problems for the Secondary Classroom with Reliability Theory** 617
	Usha Kotelawala

60	**LEMA – Professional Development of Teachers in Relation to Mathematical Modelling** 629
	Katja Maaß and Johannes Gurlitt

61	**Modelling in the Classroom: Obstacles from the Teacher's Perspective** 641
	Barbara Schmidt

62	**Teachers' Professional Learning: Modelling at the Boundaries** 653
	Geoff D. Wake

Part VIII Theoretical and Curricular Reflections on Mathematical Modelling

63 Theoretical and Curricular Reflections on Mathematical Modelling – Overview 665
Pauline Vos

64 Making Connections Between Modelling and Constructing Mathematics Knowledge: An Historical Perspective 669
Toshikazu Ikeda and Max Stephens

65 Practical Knowledge of Research Mathematicians, Scientists and Engineers About the Teaching of Modelling 679
Jeroen Spandaw

66 Evolution of Applications and Modelling in a Senior Secondary Curriculum 689
Gloria Stillman and Peter Galbraith

67 Sense of Reality Through Mathematical Modelling 701
Jhony Alexander Villa-Ochoa and Carlos Maria Jaramillo López

68 What Is 'Authentic' in the Teaching and Learning of Mathematical Modelling? 713
Pauline Vos

Corresponding Authors 723

Index 727

Chapter 1
Trends in Teaching and Learning of Mathematical Modelling – Preface

Gabriele Kaiser, Werner Blum, Rita Borromeo Ferri, and Gloria Stillman

The 14th International Conference on the Teaching of Mathematical Modelling and Applications (ICTMA14) took place at the University of Hamburg from July 27–31, 2009, welcomed by the Senate of the Free and Hanseatic City of Hamburg. One hundred and fifty participants from 30 countries participated in the conference, the majority coming from Germany, but with strong groups from China and Japan, from Scandinavia, Australia, Brazil, and South Africa. Furthermore, ICTMA14 attracted researchers from many countries all over the world, such as Singapore, Taiwan, Turkey, Nigeria, or Cyprus. The attendance of so many researchers from all parts of the world shows that the discussion on the teaching of applications and modelling has been established and is now an important topic all over the world.

Furthermore, the attendance of researchers from all over the world, coming from both mathematics and mathematics education, shows the special flavour of this conference, namely, that it attracts researchers who work as mathematicians and are interested in the teaching of mathematical modelling and applications at tertiary level as well as mathematics educators who carry out research in the teaching and learning of mathematics, especially concerning mathematical modelling and applications at school level.

The reason for this interesting characteristic lies in the origin of ICTMA: When this conference series began in 1983 at Exeter University in England, initiated by David Burghes, together with John Berry, Ian Huntley and Alfredo Moscardini, its

---

G. Kaiser (✉)
Faculty for Education, Psychology, Human Movement,
University of Hamburg, Hamburg, Germany
e-mail: gabriele.kaiser@uni-hamburg.de;

W. Blum and R. Borromeo Ferri
Department of Mathematics, University of Kassel, Germany
e-mail: blum@mathematik.uni-kassel.de; borromeo@mathematik.uni-kassel.de

G. Stillman
Australian Catholic University (Ballarat), Ballarat, Victoria, Australia
e-mail: gloria.stillman@acu.edu.au

main focus lay in the teaching of modelling at tertiary level. The founders of ICTMA, although the conference series was not named ICTMA yet at that time, came from British polytechnics, had introduced new courses in their curriculum called modelling courses, and intended to start a scientific debate on the evaluation of these courses. The introduction of these new courses was an answer to the requirements of industry, which needed better trained engineers, technicians, and mathematicians who were able to meet the challenges of the technological development, that is, to use mathematical models in order to solve real-world problems. The success of this conference series might seem quite unexpected, although already at the first conference 125 delegates from 23 different countries were present.

Chris Haines was one of these very first researchers at the ICTMA conferences and his plenary paper together with a reaction by Katja Maaß is printed in this book in Part 5. At this first conference, Henry Pollak gave the opening speech and his influence on this debate is still present and can be found in many papers of this book. The debate in this area continued and led to the organisation of the second conference on this theme which took place in 1985, once more in Exeter, with the majority of the participants still from England. ICTMA-3 then moved to Kassel in Germany, with Werner Blum as the chair of this conference and Gabriele Kaiser as co-chair. So, with ICTMA-14, the conference series has returned to Germany after 22 years. Werner Blum presented his recent research at ICTMA-14 on the teaching and learning of applications and modelling by a plenary lecture, which can be found in Part 1 of this book followed by the reaction of Marcelo Borba.

At the Kassel conference in 1987, the acronym ICTMA was introduced by adding the phrase "and applications" to the original conference title. This third conference was well attended by more than 200 participants, many of them school teachers for whom a special "teacher's day" was offered. At this time, the emphasis of the conference moved from tertiary education to the teaching of modelling and applications at all levels, especially at secondary school level. Many projects especially from England were present, such as the challenging materials by the Shell Centre developed by Hugh Burkhardt and his group, as well as materials from the Spode group and the Enterprising mathematics course, all developed by David Burghes and his group who were highly active at this time. From the USA, writers from the huge project COMAP attended. It was at this time that Werner Blum started to work as continuing editor of ICTMA, and he has been involved in most of the proceedings ever since.

The conference was still strongly European based, so consequently ICTMA-4 took place in Roskilde (Denmark) organised by Mogens Niss who has contributed with his work strongly to several conferences of this series. Mogens Niss emphasised working groups as a special working form into the conference and invited plenary speakers who introduced more philosophical aspects into the debate. The discussion on modelling competencies, which is a highly important theme until now, began to evolve at this conference.

In 1991, ICTMA-5 was organised by Jan De Lange from the Freudenthal Institute and took place in Noordwijkerhout (Netherlands). Societal aspects and the connection of modelling and technology were important topics at this conference.

Also, contributions from the primary school level were more prominent at this conference than ever before. After that conference, it was consensus that ICTMA should become more international and so ICTMA-6 took place at the University of Delaware in Newark (USA), organised by Cliff Sloyer. New researchers came in, for example, Peter Galbraith from Australia who later served as the second president of ICTMA.

ICTMA-7 was organised by Ken Houston, the first president of ICTMA, and hence went back to Europe, to the University of Ulster in Northern Ireland. At this conference, assessment became an important topic and the development of modelling tests started, strongly influenced by Ken Houston, Chris Haines, Rosalinde Crouch, John Izard, and many other Australian colleagues. With such a strong Australian presence, it was quite natural to run the next conference in Australia, so ICTMA-8 was organised by Peter Galbraith in Brisbane and brought in many more Australian researchers into this conference series, for example Gloria Stillman, a former PhD student of Peter Galbraith.

With ICTMA-9, the conference returned to Europe. Joao Filipe Matos organised it in Lisbon with Susana Carreira as co-chair, who presents her latest research results in this book in Part 2. At this time, the discussion on psychological and cognitive aspects became much more intense, strongly influenced by the Portuguese researchers, what was continued at ICTMA-14 with a plenary talk by Gloria Stillman and a reaction by Rita Borromeo Ferri, which can be found in Part 2 of this book.

In 2001, ICTMA made a big jump to China and was organised by Qi-Xiao Ye and Qi-Juan Jing in Beijing. ICTMA-10 was attended by 150 participants with the majority coming from China. This move has strengthened the visibility of the Chinese researchers within the international debate on the teaching and learning of mathematical modelling and applications, and several papers of Chinese researchers are present in this book in which they report their ongoing research to establish modelling courses at different age levels, with modelling contests being an important part of their activities.

In 2003, ICTMA returned to the USA, organised by Sue Lamon in Milwaukee, Wisconsin, and brought in several new American researchers. It was followed by ICTMA-12, which took place in Europe another time in 2005, being organised by Chris Haines in London. This conference stressed in several plenary talks the relation between mathematics and industry and related strongly to the origin of ICTMA, that is to modelling courses at polytechnics.

In 2007, Richard Lesh hosted ICTMA-13 at Indiana University in Bloomington (USA), bringing in many new American participants with various research backgrounds. That led subsequently to a broad thematic coverage of many aspects related to modelling with a strong emphasis on the relation between industrial applications and modelling. Recent work of Richard Lesh was presented as a plenary lecture at ICTMA-14 jointly with Helen Doerr and their contribution can be found in Part 3 of this book. An accompanying satellite conference took place in the same year in Kathmandu organised by Bhadra Tuladar (for details of the historical development of ICTMA, see also Houston et al. 2008).

When summarizing the historical development of ICTMA, we can see a growing interest in the teaching and learning of mathematical modelling and applications in school and university. This development has culminated in an ICMI study on this theme which reflects the current state of the art of the educational debate (see Blum et al. 2007).

One striking feature of the current debate on the teaching and learning of applications and modelling is its diversity which has been developed over the last 30 years. Many different research perspectives have been developed with different foci, different aims, and different methods, so not only the development of new teaching courses or learning environments and their evaluation are important, but also various theoretical aspects are prominent in the debate as well. This diversity has led to the development of different research perspectives which are characterised by different goals and aims connected with the teaching and learning of mathematical modelling and applications, often on the basis of different theoretical orientations. These differences led, in particular, by the discussions at the Congresses of the European Society for Research in Mathematics Education (CERME) to the development of a framework of different research perspectives which was structuring these ICTMA conferences and consequently also this book. The classification developed by Kaiser and Sriraman (2006) distinguishes, amongst others, a so-called realistic or applied modelling perspective which aims to promote the understanding and solving of real-world problems. Closely related to this approach is the model-eliciting perspective in which students are to learn to elicit new models and new situations within the modelling process. Many researchers emphasise pedagogical goals, that is modelling promotes, apart from modelling competencies, more subject-bound goals like the introduction of new mathematical concepts and the structuring of teaching and learning processes. Another perspective emphasises more general goals, for example, the critical understanding of the surrounding world. Quite distinct from these perspectives, a meta-perspective can be discriminated, which is more orientated towards the analysis of cognitive processes within modelling activities.

As became obvious in the description of the historical development of the ICTMA conferences and their thematic structures, these foci have emerged over the last three decades and they are all well represented in the four plenary contributions with three commentaries by discussants and the 52 selected peer-reviewed research papers published in this volume. This research diversity is reflected in the structure of this book with eight parts, which all contain specific overviews leading into the theme of the particular part of the book.

The first part on "Modelling from primary to upper secondary school: Findings of empirical research" is introduced by Thomas Lingefjärd. Empirical analyses in its 12 papers, together with the plenary paper by Werner Blum and a reaction by Marcelo Borba, address various possibilities for introducing modelling into mathematical instruction at several age levels. The second part, "Looking deeper into modelling processes: Studies with a cognitive perspective", starts with an overview by Susana Carreira and concentrates with its five papers, jointly with the plenary paper by Gloria Stillman and a reaction by Rita Borromeo Ferri, on the analysis of the cognitive processes, especially meta-cognition, that take place during modelling. The third part "Modelling in teacher education" contains, together with the plenary paper by Richard Lesh and Helen Doerr, three chapters with an introduction by Jill Brown and centres its debates on the necessity and possible ways to introduce modelling in teacher education. Part four on "Using technologies: New possible ways of learning modelling" is introduced by Gilbert Greefrath and explores in its three papers the challenges of the introduction of new technologies into the teaching and learning processes of modelling. The fifth part on "Modelling competency: Learning, applying and developing competencies" with an introduction by Morten Blomhøj displays, together with the plenary paper by Chris Haines and a reaction by Katja Maaß, in its six papers the current debate on ways to learn, support, and measure modelling competencies. Part six on "Modelling in tertiary education" contains an overview by Peter Galbraith and presents in its six papers the current state of the art on teaching and learning modelling at tertiary level. In part seven on "Modelling examples and modelling projects: Concrete cases," Hugh Burkhardt introduces the theme, and the 12 papers describe various teaching courses and projects from all over the world. Finally, part eight on "Theoretical and curricular reflections on modelling" is introduced by Pauline Vos and explores in its five chapters various aspects of modelling from a theoretical and curricular level.

International cooperation is the motor and the heart of scientific development. So we hope this new book will contribute to the promoting and fostering of the teaching and learning of mathematical modelling and applications all over the world.

The present book is the first book in a newly established series *International Perspectives on the Teaching and Learning of Mathematical Modelling,* the aim of which is to promote academic discussion on the teaching and learning of mathematical modelling at various educational levels all over the world. This series will provide the reader on a regular basis with insight into the development of the state of the art on research in mathematical modelling and its teaching and learning.

References

Blum, W., Galbraith, P., Henn, H.-W., & Niss, M. (Eds.). (2007). *Modelling and applications in mathematics education. The 14th ICMI study.* New York: Springer.

Houston, K., Galbraith, P., & Kaiser, G. (2008). ICTMA. The International Community of Teachers of Mathematical Modelling and Applications. The first twenty-five years. In *The First Century of the International Commission on Mathematical Instruction (1908–2008).* http://www.icmihistory.unito.it/ictma.php. Last acessed 20 November 2010.

Kaiser, G., & Sriraman, B. (2006). A global survey of international perspectives on modelling in mathematics education. *Zentralblatt für Didaktik der Mathematik, 38*(3), 302–310.

Part I
Modelling from Primary to Upper Secondary School: Findings of Empirical Research

Chapter 2
Modelling from Primary to Upper Secondary School: Findings of Empirical Research – Overview

Thomas Lingefjärd

The study of any phenomenon is enhanced by attention to the various habitats in which it can be situated, and that is certainly true of mathematical modelling as an educational phenomenon. Activities in mathematical modelling take strikingly different forms as they are institutionalised across the world. This chapter will offer challenging insights into mathematical modelling activities around the world and at different levels of educational systems. Several different research paradigms and/or theoretical frameworks are discussed and used to present and discuss different findings and results.

Some of the papers use a variety of theories connected to the mathematical modelling cycle as frameworks for illustrating quite different aspects. In this classification, we find papers by Blum, Girnat, and Eichler, and Schaap, Vos, and Goedhart. The fact that these papers are quite different underlines that the mathematical modelling cycle may very well be, and obviously also is, used in many different ways. Schaap, Vos, and Goedhart also ask for an extension of the current view of the modelling cycle.

Werner Blum uses several theoretical frameworks when outlining stages of the mathematical modelling cycle pointing out how easy it is for a novice modeller to make mistakes at every stage. Blum underlines and emphasises how complicated the question of what a mathematical modelling competency is and the many different sub-competencies one needs to be good at mathematical modelling. He argues that meta-cognitive activities are not only helpful but also necessary for the development of modelling competency. In the DISUM project, different ways to help teachers to teach mathematical modelling have been tested, and Blum reports on successful and unsuccessful approaches.

Boris Girnat and Andreas Eichler used a classification developed by Kaiser in the 1980s taking a pragmatic approach or scientific-humanist approach to mathematical modelling in conjunction with views of mathematics as static or dynamic

T. Lingefjärd (✉)
University of Gothenburg, Gothenburg, Sweden
e-mail: Thomas.Lingefjard@gu.se

to provide a framework to classify the views of German teachers in their study. They found that teachers' beliefs about modelling were dependent on the relevant content area of mathematics. Apparently, key elements of the modelling cycle must be rejected when the learning focus is geometry and a static view of mathematics dominates. Nevertheless, the teaching and learning of geometry can occur in a context where, even when deductive approaches dominate, applications of geometry to the real world and solution of realistic tasks play an important role. In contrast, in analytical geometry and stochastics, the teachers appeared to lack any rationale for emphasising applications over modelling. The authors question whether these responses are a result of beliefs about the relationship between mathematics and modelling or due to a lack of convincing modelling examples across mathematical areas.

Sanne Schaap, Pauline Vos, and Martin Goedhart report the creation of a framework based on both opportunities and blockages that students experience in the mathematical modelling process. The framework builds on work developed within the modelling debate in the last decade. Six students were paired into three groups and five different mathematical modelling problems were distributed so each group worked on four different modelling problems. Two of the problems, The Swimming Pool Task and The Horizon Task, are discussed in this chapter.

The following blockages were identified: lacking algebraic skills, overlooking essential elements in the problem text, impeding formulation of the problem text, and blockage in communication. The following opportunities were found: rereading the problem text, subconsciously simplifying the situation, slowing down the process by taking one's time, verifying the model by estimation, and assuming that a solution is valid. Some of the factors are not identifiable by the modelling cycle and the authors suggest that a framework for blockages and opportunities in the mathematical modelling process should be extended to also include metacognitive competencies that overarch the modelling cycle.

Some of the papers relate to the use of so-called real-world problems. The studies by Andreas Busse and Thomas Lingefjärd and Stephanie Meier, are different but similar in some ways. While both papers examine the use of real-world problems in specific secondary school classrooms, the reports from the discussion in the classrooms are quite different.

Andreas Busse describes how four secondary school students in a German higher track school handled three different real-world mathematical modelling tasks. Real-world problems always have contextual properties, which will be interpreted differently by different users. Busse uses frameworks from cognitive science as well as from the modelling debate in order to examine his tasks.

Andreas Busse posits "Ideal Types" for dealing with real-world mathematical modelling problems as: Reality bounded (the problem is fully characterised by the real-world context); Mathematically bounded (the real-world appearance is a decoration and the problem is solved exclusively by mathematical methods); Integrated (personal contextual knowledge interacts and develops through the given information leading to a solution process invoking mathematical methods); Ambivalent (there is an ambivalence concerning how the problem should be solved in order to be legitimate). Neither mathematical modelling problems nor students' views or strategies for solving real-world problems fall into just one category. Nevertheless, it is important for any teacher to realise that students' contextual experiences and understanding might be very different from each other's and from their own ideas and experience.

Thomas Lingefjärd and Stefanie Meier illustrate how two upper secondary teachers, one in Germany and one in Sweden, taught the same mathematical modelling project in their classes. Both teachers wanted to teach the same modelling problem and both had classes with students aged 18–19 years old. The teachers' idea was to either illustrate similarities or differences in the teaching and learning of this mathematical modelling project, which was about the phenomenon of sun hours during a day depending on location on the globe and time of year. Although this is a well-known phenomenon for most youths and grown-ups, it is also rather difficult to really understand.

Lingefjärd and Meier examine the dialogue between the students in the two classrooms and analyse how the students attempt to understand the complexity of the problem. By using Goffman's theory of framing, it is evident that the students, in order to manage the handling of the analysis of the complex sun hour phenomenon, try to simplify and take shortcuts at the expense of an acceptable solution. It is concluded by the authors that the more open a mathematical modelling process becomes, the harder it is for the teachers to control the work of the students. The students are after all adults who are rational and will try to find a way out of a problematic situation, or from a problematic framing.

One of the papers, namely, the one by Van Doreen, De Bock, Vleugels, and Verschaffel, describes the correlation between how 6th graders classify word problems and how they solve word problems.

Wim Van Doreen, Dirk De Bock, Kim Vleugels, and Lieven Verschaffel describe how seventy-five 6th graders from five classes in three different schools, both performed a classification and a solution task regarding word problems. The students were given nine cards with different word problems, nine envelopes, and a pencil. Three of the word problems were proportional, three were additive, and three were constant. The students were asked not to solve the problems, but to figure out which word problems belonged together and had something in common. Subsequently, the students were asked to solve nine different word problems. Half of the students did this exercise in reverse order.

The result shows that the students who did the classification task before they solved the solution task performed significantly better than the students who did this in reverse order. Most likely, the classification task made the students more aware of differences among the word problems, an awareness they then managed to transfer to the solution task. It is concluded that classroom attention to discussion of similarities and differences in mathematical problems is probably essential when learning mathematics.

Four of the papers discuss the relation between mathematical competencies or mathematical knowledge needed to work with mathematical modelling. Whilst this is also part of the mathematical modelling cycle, the issue is a more direct focus in these papers. George Ekol's paper examines how mathematicians view mathematical modelling, whilst Dawn Ng discusses this matter related to 7th graders mathematisation of mathematical modelling. Roxana Grigoras, Fco. Javier García, and Stefan Halverscheid investigated how the formulation of a mathematical modelling task triggers the depth of mathematics preservice mathematics teachers and 8th graders used in the modelling process. Yoshiki Nisawa and Seiji Moriya describe how a multivariable function course was conducted with 30 university students, aiming at becoming mathematics teachers in high school. The purpose of the multivariable course was to help the prospective teachers to understand the general concept of function better.

George Ekol discusses how four applied mathematicians view mathematical modelling. If mathematical modelling in the classroom is to link with the real world, then there has to be an enculturation process where students, teachers, researchers, and educators share a language and practices, and develop knowledge through communication. The findings from interviews with the mathematicians were that they viewed four themes as significant for mathematical modelling: Finding similar examples or phenomena, connecting physical phenomena with abstract concepts, building a model from the ground up, and communicating a mathematical model solution. One of the applied mathematicians also expressed the importance of "play," as in exploring concepts through guessing, estimating, simulating, checking, and cross-checking and in a playful way using tools at hand.

Dawn Ng explains how interdisciplinary project work was introduced as an educational initiative at all school levels in Singapore. Dawn's study aimed at answering: What are the types of mathematical knowledge application and student difficulties faced during participation in a design-based interdisciplinary project?

Research studies on interdisciplinary project work involving the nature of mathematics application have seldom been carried out. In her study, Dawn Ng observed a cohort of 617 students (aged 13–14) across three Singapore government secondary schools and tracked the progress of 10 case-study groups. She found that students did not apply all the expected mathematical knowledge and skills. Furthermore, limited activation of real-world knowledge in mathematical application and decision making by the majority of the groups indicated that students may learn mathematics in isolation of its use in the real-world and form certain beliefs about the nature of mathematics. Effective groups spent more time on attaining project requirements, had members who added value to discussions, worked together for more accurate and appropriate mathematical knowledge application, and engaged more mathematically with the tasks. Nevertheless, even effective groups had difficulty sustaining their interest in the project and maintaining quality mathematical outcomes toward the end of the project. These findings underline how difficult it is to change the teaching of mathematics into a more explorative subject.

Roxana Grigoras, Fco. Javier García, and Stefan Halverscheid ask the question: How does the formulation of a modelling task without numbers influence the degree of mathematics the students will use to handle the problem? A task was given in three different versions to both preservice mathematics teachers and 8th graders. In order to analyse students' utterances, the authors used the theory that modelling is mainly a cyclic process where continuous transitions between mathematics and the rest of the world occur as described by Niss, Blum, and Galbraith. The Anthropological Theory of Didactics by Chevallard was used to understand the performed experiments from an institutional perspective. The results show that younger students made few and only tacit assumptions, whilst the preservice teachers made more assumptions with the majority explicitly stated. ATD showed that the two teams differed in the techniques and technologies they made use of; however, no real difference in terms of complexity of the mathematical level was noticed between the teams.

Yoshiki Nisawa and Seiji Moriya report from a study in which the authors wanted to introduce mathematical modelling containing multivariable functions at the high school level in order to solve the problem of university students not understanding the concept of functions due to their previous mathematical education and teaching and learning methods. A multivariable function course was given to 30 university students, aiming at becoming mathematics teachers in high school. Students completed a questionnaire after the course and the results showed that students understood single variable function better when they had studied multivariable functions. In particular, students understood through mathematical modelling that the events in their surroundings are not decided by one factor but by two or more factors, and, therefore, these can be modelled using multivariable functions. Another result was that a change was noticed in the students' attitude toward functions.

Finally, we have two papers mainly concerned with beliefs or attitudes, as manifested in the papers by Shih-Yi Yu and Ching-Kuch Chang and by Förster. Both these papers address attitudes of secondary school mathematics teachers.

Shih-Yi Yu and Ching-Kuch Chang present how 16 secondary mathematics teachers were engaged in working with different MEAs (Modelling-Eliciting Activities). The concept of MEAs was mainly developed by Lesh and Doerr (see this volume). The teachers worked in smaller groups and wrote reports where they reflected on the activities. During this 9-week course, the teachers revealed many opinions about MEAs and about the teaching of mathematical modelling. Despite the teachers agreeing that MEAs probably would enhance their students' problem-solving abilities, they still thought that there were too many obstacles to implement MEAs in their mathematical classes. The researchers also had difficulties to engage teachers in reading the research literature about different modelling perspectives.

Frank Förster reports from a study involving questioning eight in-service teachers teaching grade 7–13 of upper secondary schools in Northern Germany with several years of teaching experience (2 up to 20 years). Building on results of research on beliefs (the author sought the teachers' beliefs on "global instructional goals," the "picture of mathematics," and the "reasons for or against applications," as well as the connections between them. The researcher concludes that there was an unexpected high consistency between the picture of mathematics, the global goals concerning teaching and the selection, and reasons for or against applications and that application-examples often come from the teachers' own university education. Also, the German traineeship for teachers has effects, because frequently the trainees are encouraged by the instructors to teach applications. However, if the closer contact with applications does not start until this traineeship, this "retrofitting" of competencies in applications is considered by the teachers as amateurish (dilettantish) and after their traineeship they do not teach applications any further. A positive attitude to applications and knowledge about applications and modelling is essential to be established in school or when training in university study at the latest.

Together, these 12 papers illustrate how widespread, international, and rich the world of mathematical modelling is. They also provide an interesting overview of how many theoretical frameworks may be used in different ways in order to investigate different aspects of learning or teaching mathematical modelling.

Chapter 3
Can Modelling Be Taught and Learnt? Some Answers from Empirical Research

Werner Blum

Abstract This chapter deals with empirical findings on the teaching and learning of mathematical modelling, with a focus on grades 8–10, that is, 14–16-year-old students. The emphasis lies on the actual behaviour of students and teachers in learning environments with modelling tasks. Most examples in this chapter are taken from our own empirical investigations in the context of the project DISUM. In the first section, the terms used in this chapter are recollected from a cognitive point of view by means of examples, and reasons are summarised why modelling is an important and also demanding activity for students and teachers. In the second section, examples are given of students' difficulties when solving modelling tasks, and some important findings concerning students dealing with modelling tasks are presented. The third section concentrates on teachers; examples of successful interventions are given, as well as some findings concerning teachers treating modelling examples in the classroom. In the fourth section, some implications for teaching modelling are summarised, and some encouraging (though not yet fully satisfying) results on the advancement of modelling competency are presented.

1 A Cognitive View on Mathematical Modelling

In this chapter, the actual dealing of students and teachers with modelling tasks is to be investigated. In order to describe, interpret and explain what is happening not only on the surface but also in teachers' and students' minds, a cognitive view on modelling is necessary. Hence, when clarifying some basic notions in this first section, this is done from a cognitive point of view.

W. Blum (✉)
Department of Mathematics, University of Kassel, Germany
e-mail: blum@mathematik.uni-kassel.de

The following first example is meant to set the scene:

Example 1: "Giant's shoes"

In a sports centre on the Philippines, Florentino Anonuevo Jr. polishes a pair of shoes. They are, according to the Guinness Book of Records, the world's biggest, with a width of 2.37 m and a length of 5.29 m.

  Approximately how tall would a giant be for these shoes to fit? Explain your solution.

This is a mathematical *modelling* task since the essential demand of the task is to *translate* between reality and mathematics (make assumptions on how the height of a man is related to the size of his shoes, establish appropriate mathematical relationships, interpret results of calculations and check the validity of these results). *Reality* means the "rest of the world" (Pollak 1979) outside mathematics, including nature, society, other scientific disciplines or everyday life.

That such modelling tasks are very difficult for many students is shown by a solution of "Giant's shoes" obtained by a pair of grade 9 students in a laboratory session. They multiplied width and length and thus reached the answer "The giant would be 12.53 m tall", as shown in Fig. 3.1.

This kind of solution is rather common and was observed many times in our investigations (see also Sect. 2), not only with weaker students but also with students from Gymnasium (the high ability track in the German school system). So, a pair of grade 9 Gymnasium students applied the Pythagorean Theorem in "Giant's shoes" and thus got to the answer 33.6 m. Also in this solution, no check was carried out concerning units (in both cases, the unit of the calculated result would have been m² instead of m).

This example and all the following ones are taken from the project *DISUM* ("*D*idaktische *I*nterventionsformen für einen *s*elbständigkeitsorientierten aufga-bengesteuerten *U*nterricht am Beispiel *M*athematik", in English "Didactical intervention modes for mathematics teaching oriented towards self-regulation and directed by tasks"; see Blum and Leiß 2008 for a description of this project). DISUM is an interdisciplinary project between mathematics education (W. Blum), pedagogy

$$2{,}37m \cdot 5{,}29m = 12{,}5373\,m$$

Antwort: Der Mensch wäre 12,53 m groß.

Fig. 3.1 Students' solution of task "Giant's shoes"

(R. Messner, both University of Kassel) and educational psychology (R. Pekrun, University of Munich), which aims at investigating how students and teachers deal with cognitively demanding modelling tasks and what effects various learning environments for modelling have on students' competency development. The focus in DISUM is on grades 8–10 (14–16-year-olds), which will also be the focus of this chapter.

The DISUM examples are all "medium-size" modelling tasks which can be solved within one lesson. The spectrum of tasks suitable for teaching is, of course, much bigger, reaching from straightforward standard applications to authentic modelling problems or complex modelling projects where the data collection alone takes several hours or days (compare, for instance, the "modelling weeks" presented in Kaiser and Schwarz 2006).

Why is modelling so difficult for students? In particular, because of the cognitive demands of modelling tasks; modelling involves translating between mathematics and reality in both directions, and for that, appropriate mathematical ideas ("Grundvorstellungen", see Blum 1998; Hofe 1998) as well as real-world knowledge are necessary. In addition, modelling is inseparably linked with other mathematical competencies (Blomhøj and Jensen 2007; Niss 2003), in particular designing and applying problem solving strategies, reading texts as well as working mathematically (reasoning, calculating, …). Helpful for cognitive analyses of modelling tasks are models of the "modelling cycle" which show typical ways of solving such tasks. In literature, there is a considerable variety of such models (see Borromeo Ferri 2006 for an overview). In the DISUM project, a seven-step model proved particularly helpful (Fig. 3.2, taken from Blum and Leiß 2007).

The following example (Blum and Leiß 2006) is meant to illustrate this model in some more detail.

Example 2: "Filling up"

Mrs. Stone lives in Trier, 20 km away from the border of Luxemburg. To fill up her VW Golf she drives to Luxemburg where immediately behind the border there is a petrol station. There you have to pay 1.10 Euro for one litre of petrol whereas in Trier you have to pay 1.35 Euro.

Is it worthwhile for Mrs. Stone to drive to Luxemburg? Give reasons for your answer.

The first step is to understand the given problem situation, that is the problem solver has to construct a *situation model* which here involves at least two gas stations

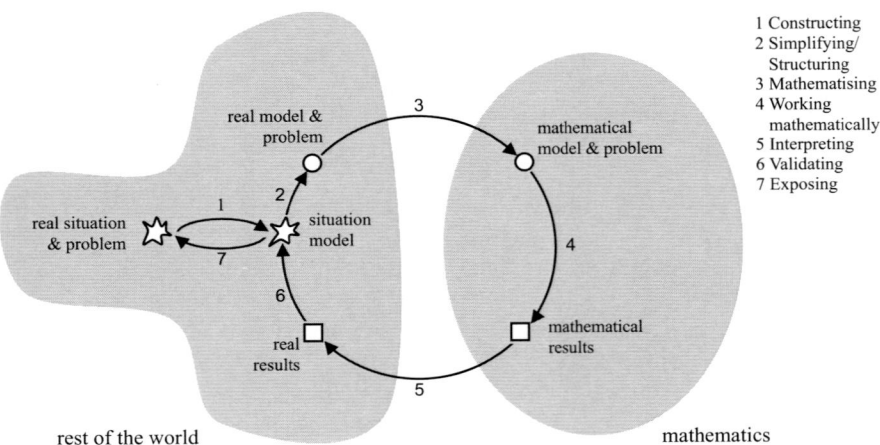

Fig. 3.2 DISUM model of the modelling process

and the 20 km connection. The second step is to structure the situation by bringing certain variables into play, especially tank volume and consumption rate of the Golf, and to simplify the situation by defining what "worthwhile" should mean, leading to a *real model* of the situation. In the standard model, "worthwhile" means only "minimising the costs of filling up and driving". Mathematisation, the third step, transforms the real model into a *mathematical model* which consists here of certain equations, perhaps with variables. The fourth step is working mathematically (calculating etc.), which yields *mathematical results*. In step five, these are interpreted in the real world as *real results,* ending up in a recommendation for Mrs. Stone of what to do. A validation of these results, step six, may show that it is appropriate or necessary to go round the loop a second time, for instance in order to take into account more factors such as time or air pollution. Dependent on which factors have been chosen, the recommendations for Mrs. Stone might be quite different. The seventh and final step is an exposure of the final solution.

This particular model of the modelling process comes from two sources. The notion of "situation model" has its origin in the research on texts (Kintsch and Greeno 1985; Staub and Reusser 1995; Verschaffel et al. 2000), whereas the other components stem from applied mathematical problem solving (Burghes 1986; Burkhardt 2006; Pollak 1979). There are several advantages of this model: step one – a particularly individual construction process, the first cognitive barrier for students – is separated, and all other steps are also essential stages in students' actual modelling processes and potential cognitive barriers, though generally not in linear order (for more details see Sect. 2).

With this model as a background, *modelling competency* can be defined (see Niss et al. 2007) as the ability to construct and to use mathematical models by carrying out those various modelling steps appropriately as well as to analyse or to compare given models. It is a natural hypothesis that these modelling steps correspond to *sub-competencies* (Kaiser 2007; Maaß 2006) of modelling. The main goal of teaching is that students develop modelling competency with – using the notions of Niss et al. (see Blomhøj and Jensen 2007; Jensen 2007; Niss 2003) – a degree of coverage, a radius of action and a technical level as extensive as possible.

Why is modelling so important for students? Mathematical models and modelling are everywhere around us, often in connection with powerful technological tools. Preparing students for responsible citizenship and for participation in societal developments presupposes modelling competency. More precisely (compare Blum and Niss 1991), mathematical modelling is meant to:

- Help students' to better understand the world.
- Support mathematics learning (motivation, concept formation, comprehension, retaining).
- Contribute to the development of various mathematical competencies and appropriate attitudes.
- Contribute to an adequate picture of mathematics.

By modelling, mathematics becomes more meaningful for learners (this is, of course, not the only possibility for that). Underlying all these justifications of modelling are the main goals of mathematics teaching in secondary schools (Niss 1996). The goals correspond to different *perspectives* on modelling in the sense of Kaiser et al. (2006). For realising these goals and, in particular, developing modelling competency with students, a large variety of modelling tasks has to be treated.

There is a tendency in several countries to include more mathematical modelling in the curriculum. In Germany, for instance, mathematical modelling is one of six compulsory competencies in the new national "Education Standards" for mathematics. However, in everyday mathematics teaching in most countries, there is still only little modelling. Mostly "word problems" are treated where, after "undressing" the given context, the essential aim is exercising mathematics. For competency development and for learning support also word problems are legitimate and helpful; it is only important to be honest about the true nature of reality-oriented tasks and problems. However, word problems are not at all sufficient for fulfilling all goals intended with modelling. Why is the situation in schools like this, why are there only so few modelling examples in everyday classrooms, why do we find such a gap between the educational debate (and even official curricula), on the one hand, and classroom practice, on the other hand? The main reason is certainly that modelling is difficult also for teachers; as real-world knowledge is needed, teaching becomes more open and less predictable, and all the competencies required from students have, of course, to be acquired by the teachers themselves (see, e.g., Burkhardt 2004; DeLange 1987; Freudenthal 1973; Ikeda 2007; Pollak 1979).

2 How Do Students Deal with Modelling Tasks?

Studies such as PISA (see, e.g., OECD 2005, 2007) have shown several times: modelling tasks are difficult for students all around the world. Analyses carried out by the PISA Mathematics Expert Group (see Turner et al. in press) have shown that the difficulty of modelling tasks can be substantially explained by the inherent cognitive complexity of these tasks, measured by the necessary competencies.

All potential cognitive barriers are empirically observable, specific for individual tasks and individual students (see also, e.g., Galbraith and Stillman 2006). In the following, I will show some typical examples of *students' difficulties* with modelling tasks, taken from DISUM studies.

- *Step 1 constructing*: See the introductory example "Giant's shoes"; this is an instance of the well-known superficial solution strategy "Ignore the context, just extract all data from the text and do something with these according to a familiar schema" which in everyday classrooms is very often successful for solving word problems (for impressive examples of this strategy, see Baruk 1985 or Verschaffel et al. 2000).
- *Step 2 simplifying*: This is an authentic solution of example 2 "Filling up": "*You cannot know if it is worthwhile since you don't know what the Golf consumes. You also don't know how much she wants to fill up*". Obviously, the student has constructed an appropriate situation model, but he is not used to making assumptions.

The next few examples of difficulties relate to a third modelling example.

Example 3: "Fire-brigade"

Die Münchner Feuerwehr hat sich im Jahr 2004 ein neues Drehleiter-Fahrzeug angeschafft. Mit diesem kann man über einem am Ende der Leiter angebrachten Korb Personen aus großen Höhen retten. Dabei muss das Feuerwehrauto laut einer Vorschrift 12 m Mindestabstand vom brennenden Haus einhalten.

Technical data of the engine:

Fahrzeugtyp: Daimler Chrysler AG Econic 18/28 LL - Diesel
Baujahr: 2004
Leistung: 205kw (279 PS)
Hubraum: 6374 cm³
Maße des Fahrzeug: Länge 10m Breite 2,5m Höhe 3,19m
Maße der Leiter: 30m Länge
Leergewicht: 15540kg
Gesamtgewicht: 18000 kg

From which maximal height can the Munich fire-brigade rescue persons with this engine?

- *Step 3 mathematising*: Often, after a successful construction of a real model of the problem situation in "fire-brigade", students forget to include the height of the engine into their model.
- Step 4, the *intra-mathematical part*, may, of course, be arbitrarily difficult. Step 5 is usually less difficult; here is an example:
- *Step 5 interpreting*: After correctly carrying out the first three modelling steps and successfully applying Pythagoras' theorem, a student's final answer was "*The ladder is 27.49 m long if it is extended*". Apart from the meaningless accuracy and the usual mistake of ignoring the engine's height, the student has obviously forgotten what his calculation actually meant.
- *Step 6 validating*: The introductory example "Giant's shoes" also provides an example of a missing validation since it is obvious that someone has to be more than only two-and-a-half times as tall as his shoe length (or can giants look like this?).

Particularly interesting are students' individual modelling routes during the process of solving modelling tasks. The notion of *modelling route* (Borromeo Ferri 2007) is used to describe a specific modelling process in detail, referring to the various steps of the modelling cycle (with the above model of the modelling cycle as a powerful analytical instrument). As Borromeo Ferri's analyses have shown, all these steps can actually be observed, though generally not in the same linear order (for detailed analyses of modelling processes, see also Leiß 2007 and Matos and Carreira 1997). There seem to be preferences of students for working more within mathematics or more within reality, depending on the individual *thinking styles* (for this notion, see Borromeo Ferri 2004); details are reported in Borromeo Ferri and Blum (2010).

Seeing students successfully performing certain modelling steps and having difficulties with other steps points again to the supposition that these steps correspond to sub-competencies of a global modelling competency. It is a particularly challenging open research question to establish a theoretically and empirically based *competence model* for mathematical modelling. Essential parts of such a model will be to identify distinct sub-competencies, to differentiate between various cognitive levels of such sub-competencies, and to set up connections between sub-competencies, modelling competency as a whole and other competencies such as reading. The proficiency levels identified in the context of PISA mathematics can be interpreted as a first attempt towards such a competence model (see OECD 2005, p. 260 ff). Another attempt was made in the context of the German Education Standards. Roughly speaking, the following five levels were identified:

- Applying simple standard models.
- Direct modelling from familiar contexts.
- Few-step modelling.
- Multi-step modelling.
- Complex modelling or evaluating models.

In the following, I will mention some more empirical findings concerning students' dealing with modelling tasks. An important observation is related to strategies. In most cases, there is no conscious use of *problem-solving strategies* by students.

This explains many of the observed difficulties since it is known from several studies that strategies (meta-cognitive activities) are helpful also for modelling (Burkhardt and Pollak 2006; Kramarski et al. 2002; Matos and Carreira 1997; Schoenfeld 1994; Stillman and Galbraith 1998; Tanner and Jones 1993) for an overview see Greer and Verschaffel in Blum and Leiß 2007). To put it more sharply: There are many indications that meta-cognitive activities are not only helpful but even necessary for the development of modelling competency. Indispensable for this to happen is an appropriate support by the teacher (see Sect. 3).

Another important result concerns the transfer of knowledge. We know from several studies in the frame of *situated cognition* that learning is always dependent on the specific learning context, and hence a simple transfer from one situation to others cannot be expected (Brown et al. 1989; De Corte et al. 1996; Niss 1999). This holds for the learning of mathematical modelling in particular, so modelling has to be learnt specifically. Therefore, a sufficiently broad variation of contexts (real-world situations as well as mathematical domains) by the teacher is necessary, as well as making transfers between situations and domains explicitly conscious for students.

A global remark: Several studies have shown that mathematical modelling *can* be learnt in certain environments, in spite of all the difficulties associated with the teaching and learning of modelling (Abrantes 1993; Galbraith and Clathworthy 1990; Kaiser-Messmer 1987; Maaß 2007; see also Sect. 4). The decisive variable for successful teaching seems to be "quality teaching." This will be addressed in the next section.

3 How Do Teachers Treat Modelling in the Classroom?

Concerning mathematics teaching and learning, the perhaps most important finding is one that may sound rather trivial but is not at all trivial (Antonius et al. 2007; Pauli and Reusser 2000): Teachers are indispensable, there is a fundamental distinction between students working independently with teacher's support and students working alone. Meta-analyses (e.g., Lipowsky 2006) have shown that teachers really matter a lot for students' mathematics learning, more than other variables such as class size or type of school. What makes the difference is, of course, the way of teaching. There is extensive empirical evidence that teaching effects can at most be expected on the basis of *quality mathematics teaching*. What could that mean? Here is the working definition we use in DISUM (compare, e.g., Blum and Leiß 2008):

- A *demanding orchestration* of teaching the mathematical subject matter (by giving students vast opportunities to acquire mathematical competencies and making connections within and outside mathematics).
- Permanent *cognitive activation* of the learners (by stimulating cognitive and meta-cognitive activities, fostering students' independence and handling mistakes constructively).
- An effective and learner-oriented *classroom management* (by varying methods flexibly, using time effectively, separating learning and assessment, etc.).

For quality teaching, it is crucial that a permanent balance between (minimal) teacher guidance and (maximal) students' independence is maintained, according to Maria Montessori's famous hundred-year-old maxim: "Help me to do it by myself" (see the "principle of minimal support", Aebli 1985). In particular, when students are dealing with mathematical tasks, this balance can be achieved best by individual, adaptive, independence-preserving teacher interventions. In a modelling context, often strategic interventions are most adequate, that means interventions which give hints to students on a meta-level ("Imagine the real situation clearly!", "Make a sketch!", "What do you aim at?", "How far have you got?", "What is still missing?", "Does this result fit to the real situation?", etc.). In everyday mathematics teaching, those quality criteria are often violated. In particular, teachers' interventions are mostly not independence-preserving, and there is nearly no stimulation of students' solution strategies.

Learning environments for modelling are generated by appropriate modelling tasks in a general sense. Here are a few well-tried proposals from literature:

- "Sense-making by meaningful tasks" (Freudenthal 1973; Verschaffel et al. 2000).
- "Model-eliciting activities" by challenging tasks (Lesh and Doerr 2003).
- "Authentic tasks" (Kaiser and Schwarz 2010; Palm 2007).

And more generally (in the words of Alsina 2007): "Less chalk, less words, less symbols – more objects, more context, more actions". Often helpful in such modelling contexts are suitable technological aids (Henn 2007).

Classroom observations (see, e.g., Leikin and Levav-Waynberg 2007) show that the teacher's own favourite solution of a given task is often imposed on the students through his interventions, mostly without even noticing it, also due to an insufficient knowledge of the richness of the "task space" on the teacher's side. However, we know that it is important to encourage various individual solutions (Hiebert and Carpenter 1992; Krainer 1993; Schoenfeld 1988), also to match different thinking styles of students, and particularly as a basis for retrospective reflections after the students' presentations. To this end, it is necessary for teachers to have an intimate knowledge of the cognitive demands of given tasks. In the project COACTIV (see Krauss et al. 2008), we have found that the teacher's ability to produce multiple solutions of tasks is one significant predictor of his students' achievement gains.

More generally, the following elements are necessary for teachers to treat modelling adequately:

- Knowledge of task spaces of modelling tasks (including cognitive demands of tasks and own preferences for special solutions).
- Knowledge of a broad spectrum of tasks, also for assessment purposes (concerning assessment see, e.g., Haines and Crouch 2001; Houston 2007; Niss 1993; Vos 2007).
- Ability to diagnose students' difficulties during modelling processes.
- Knowledge of a broad spectrum of intervention modes (Leiß 2007) and ability to use appropriate interventions.
- Appropriate beliefs (Kaiser and Maaß 2007).

(compare also Doerr 2007). It is an interesting open research question in which elements of *teachers' competencies* precisely are necessary and how these elements contribute to successful teaching.

4 Some Ideas for Teaching Modelling

There is, of course, no general "king's route" for teaching modelling. However, some implications of the findings reported in Sects. 2 and 3 are plausible (not spectacular but not at all trivial!).

Implication 1: The criteria for quality teaching (see Sect. 3) have to be considered also for teaching modelling; teachers ought to realise a permanent balance between students' independence and their guidance, in particular by their flexible and adaptive interventions.

Implication 2: In order to reach the goals associated with modelling, a broad spectrum of tasks ought to be used for teaching and for assessment, covering various topics, contexts, (sub-)competencies and cognitive levels.

Implication 3: Teachers ought to support students' individual modelling routes and encourage multiple solutions.

Implication 4: Teachers ought to foster adequate student strategies for solving modelling tasks and stimulate various meta-cognitive activities, especially reflections on solution processes and on similarities between different situations and contexts.

A few more reflections on Implication 4. For modelling tasks, a specific strategic tool is fortunately available, the modelling cycle. The seven-step schema (presented in Sect. 1) is appropriate and even indispensable for research and teaching purposes. For students, the following four-step schema (developed in the DISUM project) called *Solution Plan* is certainly more appropriate:

Step 1. *Understanding task* (Read the text precisely and imagine the situation clearly! What is required from you? Make a sketch!).

Step 2. *Searching mathematics* (Look for the data you need; if necessary, make assumptions! Look for mathematical relations!).

Step 3. *Using mathematics* (Use appropriate mathematical procedures!).

Step 4. *Explaining result* (Round off and link the result to the task! Is your result reasonable? If not, go back to 1! If yes, write down your final answer!).

As can be seen, steps 2 and 3 from the seven-step schema (Fig. 3.1) are united to one step here (step 2), and the same holds for steps 5, 6 and 7 of the seven-step schema (step 4 here). There are some structural similarities of this "Solution Plan" for modelling tasks to George Polya's famous general problem-solving cycle (compare Polya 1957), but this plan is more specific because it is conceived only for modelling tasks. The Solution Plan is not meant as a schema that has to be used by students but as an aid for difficulties that might occur in the course of the solution process. The goal is that students learn to use this plan independently whenever appropriate. Recent experiences have shown that a careful and stepwise introduction of this plan is necessary, as well as repeated exercises in how to use it. If this is taken into account, even students from Hauptschule (the low ability track in the German school system) are able to successfully handle this plan. However, a systematic study into the effects of the Solution Plan is still to be carried out (and is planned for 2011). A related approach is the use of "Worked-out Examples" (for details, see Zöttl et al. this volume).

Finally, I will present some more encouraging empirical results from the DISUM project. We have developed a so-called *operative-strategic* teaching unit for modelling (for grades 8/9, embedded in the unit on the Pythagorean Theorem). The essential guiding principles for this teaching unit were:

- Teaching aiming at students' active and independent knowledge construction (realising the balance between teacher's guidance and students' independence).
- Systematic change between independent work in groups (coached by the teacher) and whole-class activities (especially for comparison of different solutions and retrospective reflections).
- Teacher's coaching based on concrete four-step solutions for all tasks and on individual diagnoses (students did *not* have the Solution Plan, in order to keep the number of variables small enough).

In autumn 2006 (4 classes) and in autumn 2007 (21 classes), we have compared the effects of this "operative-strategic" teaching with a so-called *directive* teaching and with students working totally *alone*, both concerning students' achievement and attitudes. The most important guiding principles for "directive" teaching were:

- Development of common solution patterns by the teacher.
- Systematic change between whole-class teaching, oriented towards a fictive "average student", and students' individual work in exercises.

The students working alone came from those 18 classes that were reduced to 16 learners in advance by means of a standardised mathematical ability test, in order to homogenise the classes for better comparability. Both "operative-strategic" and "directive" teaching were conceived as optimised teaching styles and realised by experienced teachers from a reform project ("SINUS", see Blum and Leiß 2008). All teachers were particularly trained for this purpose. All classes came from Realschule (the medium ability track in Germany). Our study had a classical design (see Fig. 3.3):

Ability test/Pre-test/Treatment (10 lessons with various modelling tasks, including "Filling up" and "Fire-brigade") with accompanying questionnaires/Post-test/ Follow-up test (3 months later).

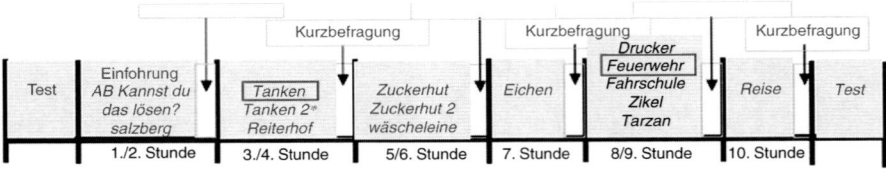

Fig. 3.3 Course of the teaching unit with modelling tasks

In all groups and teaching styles, the same modelling tasks were treated in the same order (see Fig. 3.3).

The tests comprised both modelling tasks and classical mathematical tasks close to the curriculum. According to our knowledge, this study was unique insofar as it was a quasi-experimental study with more than 600 students yielding both quantitative (tests and questionnaires) and qualitative (videos) data. Since two optimised teaching styles were implemented, one could possibly expect no differences between the two treatments concerning students' achievement and attitudes. However, there were remarkable differences. In the following, some important results are reported (more details will be presented in another paper).

Most remarkably: Both students' in "operative-strategic" and in "directive" classes made significant progress (.45 resp. .25 SD), but not so students working alone. The difference in progress was also significant, in favour of the more independence-oriented teaching style, and the progress of these classes was also more enduring than the progress of "directive" classes. The progress of "directive" students was essentially due to their progress in the technical "Pythagorean" tasks. Only "operative-strategic" students made significant progress in their modelling competency. The best results were achieved in those classes where, according to our ratings, the balance between students' independence and teacher's guidance was realised best, with a mixture of different kinds of adaptive interventions and, most importantly to note, with a clear emphasis on meta-cognitive activities (according to Implication 4 above).

However, from a normative point of view, these results are still rather disappointing: The progress after ten hours of teachers' big efforts to train students in modelling is only less than half one standard deviation. In fact, there is a big potential for improving the design:

- Solution Plan for students as well.
- Directive phases also as part of the independence-oriented design, especially in the beginning (teacher as a "model modeller" according to "cognitive apprenticeship").
- More time for practising sub-competencies.

It is the intention of future phases of the DISUM project to investigate these aspects in more detail.

What do these results tell us about the question in the title of this chapter: Can modelling be taught and learnt? The global answer is: There are several indications that modelling *can* be taught and learnt, provided some basic quality principles are fulfilled. Although the teaching units designed so far worldwide can certainly still be improved considerably, we should not wait for future studies before we begin to implement the reported insights into everyday classrooms as well as into teacher education (Lingefjaerd 2007). At the same time, there should be more research since there are still a lot of open questions (compare the lists of research questions in Blum et al. 2002; DaPonte 1993; Niss 2001), among many others the following:

- How can technological devices be appropriately used for developing modelling competency?
- What do competence models for modelling look like?
- Modelling competency has to be built up in long-term learning processes. What is actually achievable regarding long-term competency development?
- How can the interplay between modelling and other competencies be advanced systematically?

Particularly, the final question points to the ultimate goal of mathematics teaching: a comprehensive mathematical education of all students.

References

Abrantes, P. (1993). Project work in school mathematics. In J. De Lange et al. (Eds.), *Innovation in maths education by modelling and applications* (pp. 355–364). Chichester: Horwood.
Aebli, H. (1985). *Zwölf Grundformen des Lehrens*. Stuttgart: Klett-Cotta.
Alsina, C. (2007). Less chalk, less words, less symbols ... More objects, more context, more actions. In W. Blum et al. (Eds.), *Modelling and applications in mathematics education* (pp. 35–44). New York: Springer.
Antonius, S., et al. (2007). Classroom activities and the teacher. In W. Blum et al. (Eds.), *Modelling and applications in mathematics education* (pp. 295–308). New York: Springer.
Baruk, S. (1985). *L'age du capitaine. De l'erreur en mathematiques*. Paris: Seuil.
Blomhøj, M., & Jensen, T. H. (2007). What's all the fuss about competencies? In W. Blum et al. (Eds.), *Modelling and applications in mathematics education* (pp. 45–56). New York: Springer.
Blum, W. (1998). On the role of "Grundvorstellungen" for reality-related proofs – Examples and reflections. In P. Galbraith et al. (Eds.), *Mathematical modelling – Teaching and assessment in a technology-rich world* (pp. 63–74). Chichester: Horwood.
Blum, W., & Leiß, D. (2006). Filling up – In the problem of independence-preserving teacher interventions in lessons with demanding modelling tasks. In M. Bosch (Ed.), *CERME-4 – Proceedings of the Fourth Conference of the European Society for Research in Mathematics Education*, Guixol.
Blum, W., & Leiß, D. (2007). How do students and teachers deal with modelling problems? In C. Haines et al. (Eds.), *Mathematical modelling: Education, engineering and economic* (pp. 222–231). Chichester: Horwood.
Blum, W., & Leiß, D. (2008). Investigating quality mathematics teaching – The DISUM project. In C. Bergsten et al. (Eds.), *Proceedings of MADIF-5*, Malmö.
Blum, W., & Niss, M. (1991). Applied mathematical problem solving, modelling, applications, and links to other subjects – State, trends and issues in mathematics instruction. *Educational Studies in Mathematics, 22*(1), 37–68.

Blum, W., et al. (2002). ICMI study 14: Applications and modelling in mathematics education – Discussion document. *Educational Studies in Mathematics, 51*(1/2), 149–171.

Borromeo Ferri, R. (2004). *Mathematische Denkstile. Ergebnisse einer empirischen Studie.* Hildesheim: Franzbecker.

Borromeo Ferri, R. (2006). Theoretical and empirical differentiations of phases in the modelling process. *Zentralblatt für Didaktik der Mathematik, 38*(2), 86–95.

Borromeo Ferri, R. (2007). Modelling problems from a cognitive perspective. In C. Haines et al. (Eds.), *Mathematical modelling: Education, engineering and economics* (pp. 260–270). Chichester: Horwood.

Borromeo Ferri, R., & Blum, W. (2010). Insights into teachers' unconscious behaviour in modeling contexts. In R. Lesh et al. (Eds.), *Modeling students' mathematical modeling competencies* (pp. 423–432). New York: Springer.

Brown, J. S., Collins, A., & Duguid, P. (1989). Situated cognition and the culture of learning. *Educational Researcher, 18*, 32–42.

Burghes, D. (1986). Mathematical modelling – Are we heading in the right direction? In J. Berry et al. (Eds.), *Mathematical modelling methodology, models and micros* (pp. 11–23). Chichester: Horwood.

Burkhardt, H. (2004). Establishing modelling in the curriculum: Barriers and levers. In H. W. Henn & W. Blum (Eds.), *ICMI Study 14: Applications and Modelling in Mathematics Education Pre-Conference Volume* (pp. 53–58). Dortmund: University of Dortmund.

Burkhardt, H. (2006). Functional mathematics and teaching modelling. In C. Haines et al. (Eds.), *Mathematical modelling: Education, engineering and economics* (pp. 177–186). Chichester: Horwood.

Burkhardt, H., & Pollak, H. O. (2006). Modelling in mathematics classrooms: Reflections on past developments and the future. *Zentralblatt für Didaktik der Mathematik, 38*(2), 178–195.

DaPonte, J. P. (1993). Necessary research in mathematical modelling and applications. In T. Breiteig et al. (Eds.), *Teaching and learning mathematics in context* (pp. 219–227). Chichester: Horwoood.

De Corte, E., Greer, B., & Verschaffel, L. (1996). Mathematics teaching and learning. In D. C. Berliner & R. C. Calfee (Eds.), *Handbook of educational psychology* (pp. 491–549). New York: Macmillan.

DeLange, J. (1987). *Mathematics, insight and meaning.* Utrecht: CD-Press.

Doerr, H. (2007). What knowledge do teachers need for teaching mathematics through applications and modelling? In W. Blum et al. (Eds.), *Modelling and applications in mathematics education* (pp. 69–78). New York: Springer.

Freudenthal, H. (1973). *Mathematics as an educational task.* Dordrecht: Reidel.

Galbraith, P., & Clathworthy, N. (1990). Beyond standard models – Meeting the challenge of modelling. *Educational Studies in Mathematics, 21*(2), 137–163.

Galbraith, P., & Stillman, G. (2006). A framework for identifying student blockages during transitions in the modelling process. *Zentralblatt für Didaktik der Mathematik, 38*(2), 143–162.

Haines, C., & Crouch, R. (2001). Recognizing constructs within mathematical modelling. *Teaching Mathematics and Its Applications, 20*(3), 129–138.

Henn, H.-W. (2007). Modelling pedagogy – Overview. In W. Blum et al. (Eds.), *Modelling and applications in mathematics education* (pp. 321–324). New York: Springer.

Hiebert, J., & Carpenter, T. P. (1992). Learning and teaching with understanding. In D. A. Grouws (Ed.), *Handbook of research on mathematics teaching and learning* (pp. 65–97). New York: Macmillan.

Hofe, R. V. (1998). On the generation of basic ideas and individual images: Normative, descriptive and constructive aspects. In J. Kilpatrick & A. Sierpinska (Eds.), *Mathematics education as a research domain: A search for identity* (pp. 317–331). Dordrecht: Kluwer.

Houston, K. (2007). Assessing the "phases" of mathematical modelling. In W. Blum et al. (Eds.), *Modelling and applications in mathematics education* (pp. 249–256). New York: Springer.

Ikeda, T. (2007). Possibilities for, and obstacles to teaching applications and modelling in the lower secondary levels. In W. Blum et al. (Eds.), *Modelling and applications in mathematics education* (pp. 457–462). New York: Springer.

Jensen, T. H. (2007). Assessing mathematical modelling competencies. In C. Haines et al. (Eds.), *Mathematical modelling: Education, engineering and economics* (pp. 141–148). Chichester: Horwood.

Kaiser, G. (2007). Modelling and modelling competencies in school. In C. Haines et al. (Eds.), *Mathematical modelling: Education, engineering and economics* (pp. 110–119). Chichester: Horwood.

Kaiser, G., & Maaß, K. (2007). Modelling in lower secondary mathematics classroom – Problems and opportunities. In W. Blum et al. (Eds.), *Modelling and applications in mathematics education* (pp. 99–108). New York: Springer.

Kaiser, G., & Schwarz, B. (2006). Mathematical modelling as bridge between school and university. *Zentralblatt für Didaktik der Mathematik, 38*(2), 196–208.

Kaiser, G., & Schwarz, B. (2010). Authentic modelling problems in mathematics education – Examples and experiences. *Journal für Mathematik-Didaktik, 31*, 51–76.

Kaiser, G., Blomhøj, M., & Sriraman, B. (2006). Mathematical modelling and applications: Empirical and theoretical perspectives. *Zentralblatt für Didaktik der Mathematik, 38*(2), 178–195.

Kaiser-Messmer, G. (1987). Application-oriented mathematics teaching. In W. Blum et al. (Eds.), *Applications and modelling in learning and teaching mathematics* (pp. 66–72). Chichester: Horwood.

Kintsch, W., & Greeno, J. (1985). Understanding word arithmetic problems. *Psychological Review, 92*(1), 109–129.

Krainer, K. (1993). Powerful tasks: A contribution to a high level of acting and reflecting in mathematics instruction. *Educational Studies in Mathematics, 24*, 65–93.

Kramarski, B., Mevarech, Z. R., & Arami, V. (2002). The effects of metacognitive instruction on solving mathematical authentic tasks. *Educational Studies in Mathematics, 49*(2), 225–250.

Krauss, S., Baumert, J., & Blum, W. (2008). Secondary mathematics teachers' pedagogical content knowledge and content knowledge: Validation of the COACTIV constructs. *Zentralblatt für Didaktik der Mathematik, 40*(5), 873–892.

Leiß, D. (2007). *Lehrerinterventionen im selbständigkeitsorientierten Prozess der Lösung einer mathematischen Modellierungsaufgabe*. Hildesheim: Franzbecker.

Leikin, R., & Levav-Waynberg, A. (2007). Exploring mathematics teacher knowledge to explain the gap between theory-based recommendations and school practice in the use of connecting tasks. *Educational Studies in Mathematics, 66*, 349–371.

Lesh, R. A., & Doerr, H. M. (2003). *Beyond constructivism: A models and modelling perspective on teaching, learning, and problem solving in mathematics education*. Mahwah: Lawrence Erlbaum.

Lingefjaerd, T. (2007). Modelling in teacher education. In W. Blum et al. (Eds.), *Modelling and applications in mathematics education* (pp. 475–482). New York: Springer.

Lipowsky, F. (2006). Auf den Lehrer kommt es an. In *Beiheft der Zeitschrift für Pädagogik, 51* (pp. 47–70). *Beiheft*. Weinheim: Beltz.

Maaß, K. (2006). What are modelling competencies? *Zentralblatt für Didaktik der Mathematik, 38*(2), 113–142.

Maaß, K. (2007). Modelling in class: What do we want the students to learn? In C. Haines et al. (Eds.), *Mathematical modelling: Education, engineering and economics* (pp. 63–78). Chichester: Horwood.

Matos, J. F., & Carreira, S. (1997). The quest for meaning in students' mathematical modelling activity. In S. K. Houston et al. (Eds.), *Teaching & leaning mathematical modelling* (pp. 63–75). Chichester: Horwood.

Niss, M. (Ed.). (1993). *Investigations into assessment in mathematics education*. Dordrecht: Kluwer.

Niss, M. (1996). Goals of mathematics teaching. In A. Bishop et al. (Eds.), *International handbook of mathematical education* (pp. 11–47). Dordrecht: Kluwer.

Niss, M. (1999). Aspects of the nature and state of research in mathematics education. *Educational Studies in Mathematics, 40*, 1–24.

Niss, M. (2001). Issues and problems of research on the teaching and learning of applications and modelling. In J. F. Matos et al. (Eds.), *Modelling and mathematics education: ICTMA-9* (pp. 72–88). Chichester: Ellis Horwood.

Niss, M. (2003). Mathematical competencies and the learning of mathematics: The Danish KOM project. In A. Gagatsis & S. Papastavridis (Eds.), *3rd Mediterranean Conference on Mathematical Education* (pp. 115–124). Athens: The Hellenic Mathematical Society.

Niss, M., Blum, W., & Galbraith, P. (2007). Introduction. In W. Blum et al. (Eds.), *Modelling and applications in mathematics education* (pp. 3–32). New York: Springer.

OECD. (2005). *PISA 2003 technical report*. Paris: OECD.

OECD. (2007). *PISA 2006 – Science competencies for tomorrow's world* (Vols. 1&2). Paris: OECD.

Palm, T. (2007). Features and impact of the authenticity of applied mathematical school tasks. In W. Blum et al. (Eds.), *Modelling and applications in mathematics education* (pp. 201–208). New York: Springer.

Pauli, C., & Reusser, K. (2000). Zur Rolle der Lehrperson beim kooperativen Lernen. *Schweizerische Zeitschrift für Bildungswissenschaften, 3*, 421–441.

Pollak, H. O. (1979). The interaction between mathematics and other school subjects. In UNESCO (Ed.), *New trends in mathematics teaching* (Vol. IV pp. 232–248). UNESCO: Paris.

Polya, G. (1957). *How to solve it*. Princeton: Princeton University Press.

Schoenfeld, A. H. (1988). When good teaching leads to bad results: The disasters of "well-taught" mathematics courses. *Educational Psychologist, 23*, 145–166.

Schoenfeld, A. H. (1994). *Mathematical thinking and problem solving*. Hillsdale: Erlbaum.

Staub, F. C., & Reusser, K. (1995). The role of presentational structures in understanding and solving mathematical word problems. In C. A. Weaver, S. Mannes, & C. R. Fletcher (Eds.), *Discourse comprehension. Essays in honor of Walter Kintsch* (pp. 285–305). Hillsdale: Lawrence Erlbaum.

Stillman, G., & Galbraith, P. (1998). Applying mathematics with real world connections: Metacognitive characteristic of secondary students. *Educational Studies in Mathematics, 36*(2), 157–195.

Tanner, H., & Jones, S. (1993). Developing metacognition through peer and self assessment. In T. Breiteig et al. (Eds.), *Teaching and learning mathematics in context* (pp. 228–240). Chichester: Horwoood.

Turner, R. et al. (in press). Using mathematical competencies to predict item difficulty in PISA: A MEG study 2003–2009. To appear in *Proceedings of the PISA Research Conference, Kiel, 2009*.

Verschaffel, L., Greer, B., & DeCorte, E. (2000). *Making sense of word problems*. Lisse: Swets & Zeitlinger.

Vos, P. (2007). Assessment of applied mathematics and modelling: Using a laboratory-like environment. In W. Blum et al. (Eds.), *Modelling and applications in mathematics education* (pp. 441–448). New York: Springer.

Zöttl, L., Ufer, S., & Reiss, K. (this volume). Assessing modelling competencies using a multidimensional IRT approach.

Chapter 4
"Can Modelling Be Taught and Learnt?" – A Commentary

Marcelo C. Borba

1 Introduction

The questions, whether mathematical modelling can be learnt and what we know from empirical research, are highly relevant not only for the current research on modelling, but they are as well essential for curricular changes introducing modelling into schools. Werner Blum tackles these questions in his paper from several perspectives and shows important research regarding the possibilities of students and teachers learning and teaching modelling, respectively. Blum dedicates his research to cognitive analysis of students who are involved in developing modelling tasks with the help of teachers.

Building on some of the important research developed in the area, he sets the stage for his argument defining *modelling competency*

> as the ability to construct and to use mathematical models by carrying out those various modelling steps appropriately as well as to analyse or to compare given models. It is a natural hypothesis that these modelling steps correspond to *sub-competencies* (Kaiser 2007; Maaß 2006) of modelling. The main goal of teaching is that students develop modelling competency with – using the notions of Niss et al. (see Blomhøj and Jensen 2007; Jensen 2007; Niss 2003) – a degree of coverage, a radius of action under a technical level as extensive as possible.

Blum claims that these sub-competencies are related to the steps students take to solve modelling tasks: constructing, simplifying, mathematizing, intra-mathematical tasks, interpreting, validating. He uses two problems to help the readers understand his ideas, the Giant problem and Fuel problem. The first one is more of an open problem and the second more related to the optimization problems that can be found in textbooks. He also refers to other problems in order to make his ideas clearer.

M.C. Borba (✉)
UNESP-Universidade Estadual Paulista Júlio de Mesquita Filho, São Paulo, Brazil
e-mail: mborba@rc.unesp.br

He then presents the difficulties faced by teachers to orchestrate the classroom that is involved in solving modelling problems and discusses the steps that students use to solve problems. His chapter is anchored on strong reference to research developed in the field. My comments should then be understood as a means of expanding the very important ideas presented by Blum in the preceding chapter.

2 View of Modelling

The very first idea that the reader may want to consider is what view of modelling is embedded in Blum's chapter. Blum's view is one in which the problem is presented to students by the teacher as a story that is connected to other realms of the students' experiences. It is similar to problem solving in the sense that the teacher maintains the role of presenting a problem and the student responds to the teacher's input. As we know there are different views of modelling when we consider the role of teacher and students. One possibility is to have students choosing the theme to be studied as is done in some of the Danish and Brazilian traditions (see Borba and Villarreal 2005). We propose to enrich the modelling approach by such a competency designing a problem within a chosen theme, because it seems to be essential in order to promote autonomous learning processes. Alternatively, in some cases, the students may decide not to design a problem or a "research" question and opt instead to study and discuss the theme of their choice using mathematics and other established fields of knowledge. There is extensive research in Brazil illustrating this. In one study, for example, students enrolled in a first-year biology course chose mad cow disease as their topic of study and ended up merging biology and mathematics. Others chose photosynthesis and came up with topics, such as the logistic curve, which was not included in the syllabus of the calculus discipline they were enrolled in, as Fig. 4.1 illustrates.

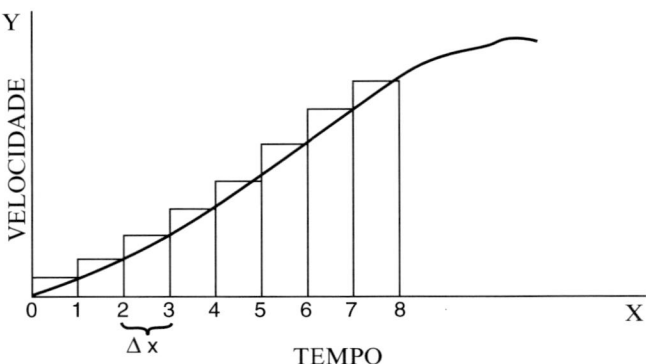

Fig. 4.1 Speed of photosynthesis increases as pollution increases over time

On the other hand, some students chose themes that did not develop into a problem. It is the case of the students who chose Nietzsche as a theme. They studied the life of the philosopher and elaborated a project, with the help of the teacher that included the view of mathematics expressed by the famous thinker. This never became a problem in any usual sense of the word. Is this problematic for schooling? Must students be solving problems or demonstrating in mathematics classes?

Answers to these questions are beyond the scope of both this commentary and Blum's chapter, but it does help to show that there are other possibilities of modelling to be investigated using Blum's analysis of students' cognition and teachers' competencies. On the other hand, little of the kind of research conducted by Blum has been done in classrooms where modelling is seen as a practice that emphasize students' choice of the theme. Both kinds of research could benefit from one another.

In the Brazilian tradition, recently Herminio and Borba (2010) studied what drives students' interests and how this drive changes over time as a group chooses a theme to study. Anchored on the discussion carried out by Dewey (1978), we make a critical analysis of the Brazilian tradition of modelling in which much importance is given to the students' choice of theme of interest. It would be interesting if cognitive analysis like the one proposed by Blum was incorporated into Brazilian studies to consider aspects such as goal orientation and motives of the ones involved in modelling projects.

3 Technology and Modelling

The research presented in this book by different authors and studies described in Borba and Villarreal (2005) show many examples of problems and modelling activities that were designed to be solved using software or the Internet. It seems to be necessary in the future to include technology in modelling activities. The examples by Blum mostly do not make use of technology. It would be interesting to find out what happen if studies like Werner Blum's were developed within a modelling approach in which problems were designed to be explored using information and communication technology (ICT). Not much has been done in terms of cognitive analysis in this kind of modelling project.

4 Critical Perspective and Cognition

In Brazil, a perspective on modelling is prominent, which is described by many authors, such as Kaiser and Sriraman (2006), as the social-critical perspective. In this perspective, learning of mathematics is connected to a critical reading of society as well. In this sense, it is fairly reasonable to say that connecting learning mathematics to a critical reading of the world (Freire 1976) may demand new "cognitive activities," a terminology used by Blum. In this regard, it may not be fruitful, as is the case in some areas of mathematical education, to see critical education and cognitive perspectives as a dichotomy. It would be interesting to see the design of the study developed by Blum and others incorporated into a task that engages students in a critical perspective as well.

5 Teachers and Modelling

Last, but definitely not least, Blum's paper pointed to competencies that teachers should have to teach modelling. He suggests that orchestrating a class with different groups developing different solutions at different paces is something that may bring uneasiness to teachers who engage in modelling activities.

Werner Blum claims that for teaching modelling you may have in mind some principles:

> The criteria for quality teaching (see Sect. 3) have to be considered also for teaching modelling; teachers ought to realise a permanent balance between students' independence and their guidance, in particular by their flexible and adaptive interventions …. In order to reach the goals associated with modelling, a broad spectrum of tasks ought to be used for teaching and for assessment, covering various topics, contexts, (sub-)competencies, and cognitive levels …. Teachers ought to support students individual modelling routes and encourage multiple solution …. Teachers ought to foster adequate student strategies for solving modelling tasks and stimulate various meta-cognitive activities, especially reflections on solution processes and on similarities between different situations and contexts.

There is much to be unpacked in the principles brought by Blum for those who would like to be involved in teaching modelling. This unpacking is a task for the community, and I would encourage them to consider the following questions, as well: How would one consider these principles if other perspectives of modelling are considered as well? How about if we consider technology in the classroom? Or if we consider it part of our task to teach students to be critical of the world they live in as they learn mathematics? These questions may contribute to making an already important agenda outlined by Blum regarding the teaching of modelling even more complex. I hope we can tackle some of these issues in the next ICTMA.

6 Final Considerations

Werner Blum's work connects cognitive analysis and modelling. By presenting a careful analysis of cognitive processes of teachers and students involved in modelling activities, he invites researchers who are focusing on other domains of modelling, or other conceptions to do the same. It also invites researchers to consider other cognitive approaches to analyze students' and teachers' actions. For instance, one could use, as was done in previous ICTMA proceedings, activity theory, a perspective that focuses on students' motives as they are involved in a task.

In summary, I believe that Blum's chapter brings new ideas to the field and inspires new research as we try to change curriculum structure in school through our research on modelling.

References

Blomhøj, M., & Jensen, T. H. (2007). What's all the fuss about competencies? In W. Blum et al. (Eds.), *Modelling and applications in mathematics education* (pp. 45–56). New York: Springer.

Borba, M. C., & Villarreal, M. E. (2005). *Humans-with-media and the reorganization of mathematical thinking: Information and communication technologies, modelling, experimentation and visualization*. New York: Springer.

Dewey, J. (1978). *Vida e Educação*. São Paulo: Melhoramentos; [Rio de Janeiro]: Fundação Nacional de Material Escolar. Tradução de TEIXEIRA, A.

Freire, P. (1976). *Education: The practice of freedom*. London: Writers and Readers.

Herminio, M. H. G. B., & Borba, M. C. (2010). A noção de Interesse em Projetos de Modelagem Matemática. *Educação Matemática Pesquisa, 12*(1), 111–127.

Jensen, T. H. (2007). Assessing mathematical modelling competencies. In C. Haines et al. (Eds.), *Mathematical modelling: Education, engineering and economics* (pp. 141–148). Chichester: Horwood.

Kaiser, G. (2007). Modelling and modelling competencies in school. In C. Haines et al. (Eds.), *Mathematical modelling: Education, engineering and economics* (pp. 110–119). Chichester: Horwood.

Kaiser, G., & Sriraman, B. (2006). A global survey of international perspectives on modelling in mathematics education. *Zentralblatt für Didaktik der Mathematik, 38*(3), 302–310.

Maaß, K. (2006). What are modelling competencies? *Zentralblatt für Didaktik der Mathematik 38*(2), 113–142.

Niss, M. (2003). Mathematical competencies and the learning of mathematics: The Danish KOM Project. In A. Gagatsis & S. Papastavridis (Eds.), *3rd Mediterranean conference on mathematical education* (pp. 115–124). Athens: The Helleniic Mathematical Society.

Chapter 5
Upper Secondary Students' Handling of Real-World Contexts*

Andreas Busse

Abstract Using a triangulation of methods by applying a three-step design consisting of observation, stimulated recall and interview, upper secondary students' handling of real-world contexts was investigated. It was found that a real-world context given in a task is not only interpreted very individually but is also dynamic in a sense that the contextual ideas change and develop during the process of working on the task. Furthermore, data analysis led to four different ideal types of dealing with the real-word context: reality bound, integrating, mathematics bound, ambivalent. Based on the theoretical background of situated learning, these ideal types can be understood as effects of – often implicitly given – sociomathematical norms concerning the permissible amount of extramathematical reasoning when working on a mathematical problem.

1 Introduction

A couple of years ago, some enthusiasm among many teachers could be observed. A lot of expectations and hopes were associated with a real-world-orientated mathematics classroom. It was especially expected that students would be highly motivated and would find easier access to mathematics. Burkhardt (1981, p. iv; emphasis in original) optimistically wrote:

> However, realistic situations are easier to tackle than purely mathematical topics in that here 'commonsense' provides essential and helpful guidance, and because there are no right answers that must be found but only some answers which are better than others.

Reality in the mathematics classroom has been different: real-world problems alone neither motivate students nor do they make the learning of mathematics easier.

---

*This is a summarised version of a PhD-thesis (Busse 2009), intermediate results with different foci were published previously (Busse 2005; Busse and Kaiser 2003).

A. Busse (✉)
Faculty for Education, Psychology, Human Movement, University of Hamburg,
Ida Ehre Schule, Hamburg, Germany
e-mail: andreas.busse@uni-hamburg.de

Students sometimes see the absence of a unique solution and the need to include commonsense as an additional barrier.

Another hope is closely associated with the extramathematical field which a task is embedded in: the real-world context. Quite often, it is assumed that a suitable real-world context makes the approach to a problem easier. Another aspect of real-world contexts refers to gender roles in task texts: Depending on the perspective either real-world contexts that are assumed to be close to girls are used, or, in the opposite, especially those contexts are preferred that neglect traditional gender roles (e.g., Niederdrenk-Felgner 1995). In any case, real-world contexts seem to be important. However, students might look on this topic differently: When asked about the importance of a well-balanced appearance of males and females in tasks, students answered that it was all the same for them; the real-world contexts of most tasks were very artificial anyway so that the contexts do not have any meaning for real life (Niederdrenk-Felgner 1995, p. 54). This comment suggests that analysing the role of real-world contexts might be more difficult than expected in some circles.

In the following, some aspects of the current discussion are presented. After that, the research question is formulated, followed by considerations on methodology and methods. Afterwards, results of this empirical investigation are given and embedded in a broader theoretical context.

2 Some Aspects of the Current Discussion and Research Question

Although the real-world context might play an important role when discussing an application and modelling classroom, no (or at least no standardised) definition does exist, even different names can be found, for example, *situational context* (Stern and Lehrndorfer 1992), *task context* (Stillman 2000) and *real-world context* (Stillman et al. 2008). A comprehensive definition will be proposed later.

Several researchers claim a fostering effect of familiar real-world contexts on the learning of mathematics (among many others, e.g., Wiest 2002). On the other hand, there are strong hints that the familiarity of a real-world context might have an opposite effect: It can be a barrier to the successful solution of the task (among others Boaler 1993). Further analyses show that a positive effect of familiar real-world contexts can often be observed with *primary* school children (most research has been done in this age group) whilst opposite or more complex effects are related to *older* students.[1]

Another aspect of real-world contexts is based on observations that not everybody seems to perceive the real-world context of a given task in the same way, obviously there is an individual factor (among others Boaler 1993).

So, there are two areas where the research seems to be vague so far:

- How do *secondary* school students deal with the real-world context?
- What role does the *individual perception* of a real-world context offered in a task play?

---

[1] For more details see Busse (2009).

To investigate these questions, a definition of real-world context is needed, which also includes individual aspects. For this reason, the following comprehensive definition is used in this study:

> *The real-world context of a realistic task comprises all aspects of the verbally or nonverbally, implicitly or explicitly offered extra-mathematical surrounding in which the task is embedded, as well as its individual interpretation by the person who works on the task.*

3 Methodology and Methods

3.1 *Methodological Remarks*

The explorative character of the research question suggests a methodological embedding, which emphasises in-depth insights. For this reason, a qualitative approach was chosen. In contrast to the quantitative paradigm where the selection of cases is based on the idea of statistical representativity, in a qualitative study, the cases are supposed to mirror the range of possible phenomena ("representativeness of concepts", Strauss and Corbin 1990).

When investigating complex questions, the approach of *triangulation* has become a powerful tool. According to Denzin (1970, p. 297) triangulation means "… the combination of methodologies in the study of the same phenomena." While some years ago, triangulation used to be considered mainly as a tool of validation, more recently it is seen from a different angle. This change of view is based on the insight, that "What goes on in one setting is not a simple corrective to what happens elsewhere – each must be understood in its own terms." (Silverman 1985, p. 21). The aim of triangulation

> …should be less to achieve convergences in the sense of a confirmation of aspects already found. The triangulation of methods and perspectives is instructive especially when *divergent perspectives* can be clarified, (…). In this case a new perspective emerges that requires theoretical explanations. (Flick 2000, p. 318, emphasis in original, translation by the author.)

In order to reduce the complexity of the analyses, the Weberian notion of *ideal types* (*Idealtypen,* Weber 1922/1985) is used. By unilateral exaggeration of some and fusion of other aspects, an essential structure becomes apparent. The purpose of creating ideal types is not exclusively to categorise facts, but to emphasise the characteristics of the real case by contrasting it to an ideal type.

3.2 *Methods*

Four pairs of 16–17-year-old students were chosen. They came from four different schools. In addition, both sexes as well as different mathematical abilities were equally represented. These eight students were asked to solve three different tasks in pairs, so 24 cases can be distinguished. The tasks differed in their real-world contexts and their degrees of open-endedness. So, the tasks as well as the choice of participants contributed to a broad range of possible phenomena (see above).

As a first step, the students were videotaped while working in pairs. Secondly, they watched individually (together with the researcher) the video record. The playback was interrupted at certain moments in order to provide the student with the opportunity to comment on his or her thoughts about the real-world context that had occurred while working on the task (*stimulated recall*). In a third step (*interview*), the interviewee was asked more detailed questions about these statements. This three-step design enables the researcher to reconstruct different levels of action separately although they have taken place simultaneously.

By this methodical approach, a set of data – containing three different kinds of data – is created. Due to the three different conditions in which the data are collected, each kind of data has certain characteristics, for example, relating to the role of the researcher, the time which has elapsed since working on the task, or the means of collection. The three steps can be considered as three different perspectives on the research question, thus a triangulation of methods is realised. Consequently, data analysis had to take this into consideration: First, the different kinds of data were analysed separately. After that, the three partial analyses were brought together to a comprehensive case analysis. These 24 case analyses were compared and contrasted with each other, and finally clustered. These clusters were – according to Weber (1922/1985) – idealised to ideal types.[2]

3.3 Tasks

Three tasks were given to the students. These tasks follow in Figs. 5.1–5.3. The first task is Home for Aged People (see Fig. 5.1). Since no criterion for an optimal

In a little wood a home for aged people has been built. In the figure the seven residential buildings are marked by black dots. There are paths in the wood so that the aged people do not have to walk through the undergrowth. The paths are marked by bold lines. On the path between the two crossings (marked by a dotted line) a common house is planned to be placed. This common house is meant to serve for afternoon coffee and evening events. The question is where exactly on this path the common house is to be built.

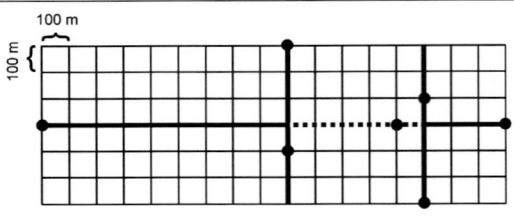

Fig. 5.1 Home for Aged People Task

---

[2]A more comprehensive discussion of methodological and methodical aspects can be found in Busse and Borromeo Ferri (2003).

A transmission changes the rotation speed from one wheel to another. That is known for example from a bicycle: In the front there is a large cogwheel, at the back a small one. When the large cogwheel in the front rotates *once* the small cogwheel at the back rotates *several* times.

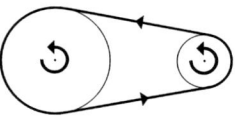

If for example the radius of the large cogwheel in the front is *twice* as large as the radius of the small one behind, the latter rotates *twice* when the large one in the front rotates *once*. A gear transmission changes the radius of one cogwheel or the radii of both cogwheels to change the transmission ratio.

Look at the following *continuously adjustable* gear transmission.

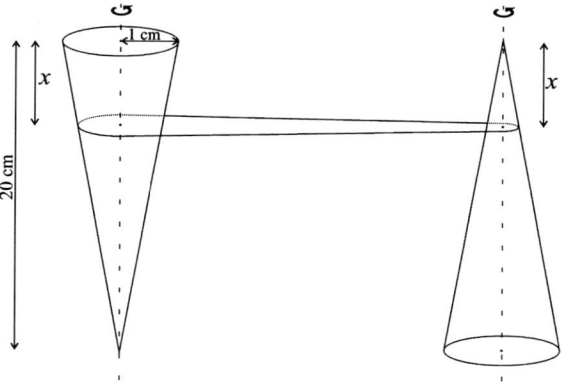

Between two cones of the same kind a drive belt is placed horizontally. When the left cone rotates, the right cone is driven by the belt. The belt can be moved up and down while remaining horizontal. According to the position of the belt, the transmission ratio varies. The position of the belt is given by the variable x. Height and radius of the cones are to be taken from the illustration.

It is assumed that the left cone rotates uniformly exactly once a second.

a) Where must the belt be positioned so that the right cone rotates exactly three times a second?

b) Is it possible to achieve any rotation speed of the right cone by varying the position x of the belt? Give reasons.

Fig. 5.2 Transmission Task

position is explicitly given in this task, one has to deal with a certain openness in order to solve it. A criterion has to be found on one's own, possibly considering contextual reflections. There is more than just one possible answer, so the students have to give reasons for their choices. The real-world context offered in the Home for Aged People Task lies in the field of social problems.

The second task, the Transmission Task (Fig. 5.2), is contextually strongly associated with physics and – depending on how it is tackled – there might be a unique

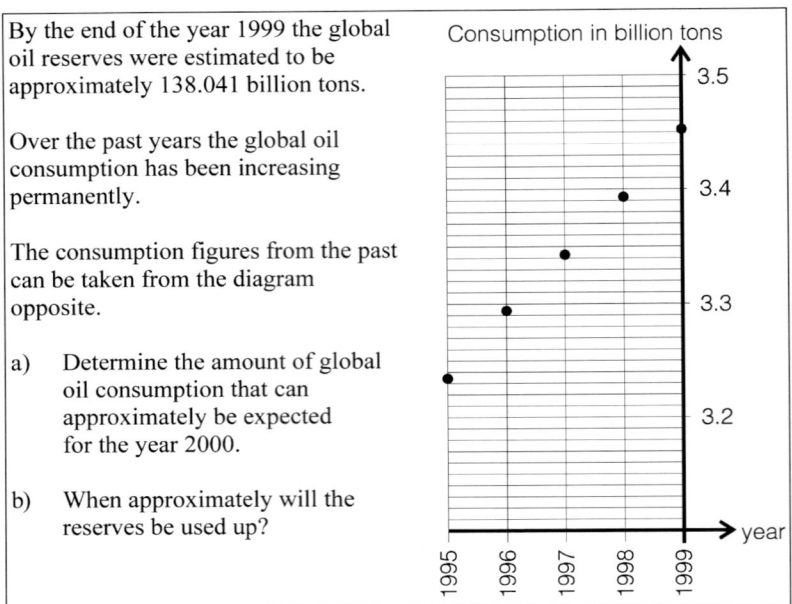

Fig. 5.3 Oil Task

solution. However, contextual aspects could be involved, for example, the role of the dimensions of the belt.

The third task, the Oil Task (Fig. 5.3), offers a wide range of possible real-world aspects. Depending on the modelling assumptions, how the consumption develops in the future different solutions are possible.

4 Results

4.1 General Results

It could be confirmed that real-world contexts are interpreted very individually depending on different previous personal experiences. Usually different aspects are chosen from the task to form an individual context; for example, while one student embedded the Oil Task in the scientific context of chemistry another student emphasised aspects of personal responsibility for the natural environment (see Busse and Kaiser 2003). Thus, it seems sensible not to talk about *the* real-world context of a task but to use the notion of *contextual idea* ("sachkontextuale Vorstellung", Busse 2009) to indicate the mental representation of the real-world context *offered* in the task. In addition, it was found that contextual ideas are *dynamic*. They do not appear at the beginning of a task and remain unchanged throughout; rather they come into being, develop and change in the course of working on the task.

So, the idea that an attractive real-world context can serve as a special starting motivation has to be qualified: It cannot be known for sure in advance *which* contextual ideas an individual develops and *when* in the course of the solution these ideas appear.

4.2   Ideal Types

The analysis of the 24 empirical cases led to four theoretical ideal types that describe different ways of dealing with the real-world context (cf. Busse 2005):

Reality bound: The task is fully characterised by the real problem described in the task. Only extramathematical concepts and methods, no mathematisations are applied.

Mathematics bound: The real-world context is a mere decoration. Only contextual information explicitly given in the task text is used, no additional personal contextual knowledge is applied. The task is solved exclusively by mathematical methods.

Integrating: Personal contextual knowledge, which exceeds the contextual information given in the task text, is used in order to mathematise the problem and to validate the solution. During the solution process, mathematical methods are applied.

Ambivalent: There is an ambivalence concerning the legitimacy of the way the task is supposed to be solved: Internally, a contextually accentuated reasoning is preferred while externally a mathematical reasoning is chosen. These two ways of reasoning coexist without forming a coherent whole.

In Fig. 5.4, these four ideal types are presented graphically. The lower two arrows indicate how the two extremes form a new quality; the upper two arrows illustrate how the type *ambivalent* is torn between the two extremes.

Fig. 5.4 Ideal types of dealing with the real-world context

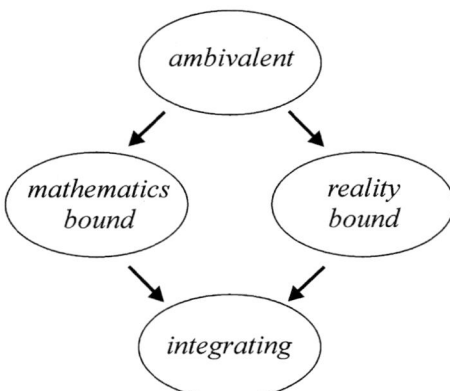

In order to create the ideal type *ambivalent*, the above-mentioned approach of triangulation followed by a separate analysis played an important role. When analysing the data, there were students whose reasoning *sounded* mathematical during the solving process; but later during the stimulated recall and the interview, it appeared that the actual solving process was in fact reality based. So, these students translated their reality-based solving process into mathematical language. This phenomenon can be explained by compliance with certain *sociomathematical norms* (e.g., Yackel and Cobb 1996). It is assumed that the norms in question do *not* permit contextual reasoning in a classroom-like situation (like the first step of the three-step design), but *do* allow this if the situation becomes more distant from the mathematics classroom (like in the two other steps of the three-step design). In other words, sociomathematical norms are considered as *situated* (cf. Lave and Wenger 1991). The existence of the ideal type *ambivalent* underlines the important role of social-mathematical norms and their situatedness in the field of application and modelling.

4.3 Case Synopsis

The total of 24 cases can be described by the four ideal types. Some cases are described by more than one ideal type. A synopsis of 6 of the 24 cases is presented in Table 5.1.

Table 5.1 Synopsis of 6 of the 24 cases

Student	Task		
	Home for Aged People	Transmission	Oil
Karla	mathematics bound	mathematics bound/integrating	ambivalent
Evelyn	ambivalent/reality bound	mathematics bound	mathematics bound

On the basis of the results from these two students, it is evident that neither a person nor a task is permanently linked to a certain ideal type, but there are hints for preferences. Considering all 24 cases, the Transmission Task is more often associated with the ideal type *mathematics bound* than the other tasks. On the other hand, Table 5.1 suggests that there might be personal preferences for certain ideal types.

Deeper analyses of the data showed that an emotional involvement or a special interest in a certain real-world context is often linked to aspects of contextual reasoning. For example, Karla, who might generally prefer a more mathematical manner of reasoning, included contextual aspects when it came to the Oil Task. Karla embedded the Oil Task in the context of responsibility for the natural environment, a topic that worried her very much. In a similar way, Evelyn included – in contrast to her actions in the other tasks – contextual reasoning when solving the

Home for Aged People Task. This change of focus might have been due to Evelyn being very committed to social problems and her wish to work as a volunteer with elderly people.

5 Final Remarks

The results of this study show, to a high degree, the important role of individual aspects when dealing with the real-world context of a task. Although many questions are still unanswered it becomes clear that teachers as well as researchers have to take this individuality into account. In school, teachers have to be aware that a student's contextual ideas might differ from theirs. Also, the way in which real-world contexts are used varies from person to person. The system of ideal types might help analysing learning problems concerning this matter.

Another important aspect is the observation that sociomathematical norms concerning the permitted use of real-world contexts are often implicit which might cause some irritation among students and can lead to *ambivalent* behaviour. A more explicit teaching of these norms must take place. This can be realised by including the teaching of a modelling cycle in the mathematics classroom.

References

Boaler, J. (1993). Encouraging the transfer of "school" mathematics to the "real world" through the integration of process and content, context and culture. *Educational Studies in Mathematics, 25*, 341–373.

Burkhardt, H. (1981). *The real world and mathematics*. Glasgow: Blackie.

Busse, A. (2005). Individual ways of dealing with the context of realistic tasks – First steps towards a typology. *Zentralblatt für Didaktik der Mathematik, 37*(5), 354–360.

Busse, A. (2009). *Umgang Jugendlicher mit dem Sachkontext realitätsbezogener Mathematikaufgaben. Ergebnisse einer empirischen Studie*. Hildesheim: Franzbecker.

Busse, A., & Borromeo Ferri, R. (2003). Methodological reflections on a three-step-design combining observation, stimulated recall and interview. *Zentralblatt für Didaktik der Mathematik, 35*(6), 257–264.

Busse, A., & Kaiser, G. (2003). Context in application and modelling – An empirical approach. In Q. Ye, W. Blum, S. K. Houston, & Q. Jiang (Eds.), *Mathematical modelling in education and culture: ICTMA 10* (pp. 3–15). Chichester: Horwood.

Denzin, N. K. (1970). *The research act*. Chicago: Aldine.

Flick, U. (2000). Triangulation in der qualitativen Forschung. In U. Flick et al. (Eds.), *Qualitative Forschung* (pp. 309–318). Reinbek: Rowohlt.

Lave, J., & Wenger, E. (1991). *Situated learning. Legitimate peripheral participation*. Cambridge: Cambridge University Press.

Niederdrenk-Felgner, C. (1995). Textaufgaben für Mädchen – Textaufgaben für Jungen? *Mathematik Lehren, 68*, 54–59.

Silverman, D. (1985). *Qualitative methodology and sociology*. Aldershot: Gower.

Stern, E., & Lehrndorfer, A. (1992). The role of situational context in solving word problems. *Cognitive Development, 7*, 259–268.

Stillman, G. (2000). Impact of prior knowledge of task context on approaches to application tasks. *The Journal of Mathematical Behavior, 19*(3), 333–361.

Stillman, G., Brown, J., & Galbraith, P. (2008). Research into the teaching and learning of applications and modelling in Australasia. In H. Forgasz et al. (Eds.), *Research in mathematics education in Australasia 2004–2007* (pp. 141–164). Rotterdam: Sense Publishers.

Strauss, A., & Corbin, J. (1990). *Basics of qualitative research: Grounded theory procedures and techniques*. Newbury Park: Sage.

Weber, M. (1922/1985). *Wissenschaftslehre. Gesammelte Aufsätze*. Tübingen: J.C.B. Mohr.

Wiest, L. R. (2002). Aspects of word-problem context that influence children's problem-solving performance. *Focus on Learning Problems in Mathematics, 24*(2), 38–52.

Yackel, E., & Cobb, P. (1996). Sociomathematical norms, argumentation, and autonomy in mathematics. *Journal for Research in Mathematics Education, 27*(4), 458–477.

Chapter 6
Word Problem Classification: A Promising Modelling Task at the Elementary Level

Wim Van Dooren, Dirk De Bock, Kim Vleugels, and Lieven Verschaffel

Abstract Upper primary school children often routinely apply proportional methods to missing-value problems, also when this is inappropriate. We tested whether this tendency can be broken if children would pay more attention to the initial phases of the modelling process. Seventy-five 6th graders were asked to classify nine word problems with different underlying mathematical models and to solve a parallel version of these problems. Half of the children first did the solution and then the classification task, for the others the order was opposite. The results suggest a small positive impact of a preceding classification task on students' later solutions, while solving the word problems first proved to negatively affect later classifications.

1 Theoretical and Empirical Background

Contemporary reform documents and curricula in most countries more or less explicitly assume that one of the most important goals of mathematics education is that students gain the competence to make sense of everyday-life situations and complex systems stemming from our modern society, which can be called "modelling competencies" (Blum et al. 2002; Lesh and Lehrer 2003). Although mathematical modelling is generally associated with courses at the tertiary or, to an increasing extent, secondary level of education, an early exposure to essential modelling ideas

---

W. Van Dooren, K. Vleugels, and L. Verschaffel
Centre for Instructional Psychology and Technology, Katholieke Universiteit Leuven,
Leuven, Belgium
e-mail: wim.vandooren@ped.kuleuven.be

D. De Bock (✉)
Faculty of Economics and Management, Hogeschool-Universiteit Brussel, Stormstraat 2,
B-1000 Brussels, Belgium
and
Center for Instructional Psychology and Technology, Katholieke Universiteit Leuven,
Leuven, Belgium
e-mail: dirk.debock@hubrussel.be

can provide a solid base for competently applying mathematics even at the primary school level (Usiskin 2007, 2008). In this chapter, we report a study on word problem classification that proved to be a promising modelling task at the primary level, having a relatively profound effect in breaking pupils' well-documented tendency to overuse the linear or proportional model (De Bock et al. 2007).

Proportionality is recognised as an important mathematical topic receiving much attention throughout primary and secondary mathematics education. The reason lies in the fact that proportional relationships are the underlying model for approaching numerous practical and theoretical problem situations within mathematics and science. However, numerous documents and research reports on a wide variety of mathematical domains, and dealing with students of diverse ages, mention pupils' tendency to apply the proportional model irrespective of the mathematical model(s) underlying the problem situation (Van Dooren et al. 2008). In a mature mathematical modelling approach (see, e.g. Verschaffel et al. 2000), essential steps would be: (1) understanding the problem, (2) selecting relevant relations and translating them into mathematical statements, (3) conducting the necessary calculations, (4) interpreting and evaluating the result. A study using in-depth interviews (De Bock et al. 2002), however, revealed that pupils almost completely bypass all steps except step 3. Their decision on the mathematical operations mainly was based on routinely recognizing the problem type, the actual calculating work received most time and attention, and after checking for basic calculation errors, the result was immediately communicated as the answer.

In line with the above interpretation, the overuse of proportionality might be broken if pupils pay more attention to the initial steps of the modelling cycle, that is, the understanding of the relevant aspects of the problem situation and their translation in mathematical terms. So, when pupils are engaged in a task with proportional and non-proportional word problems without the need to actually produce computational answers, they might be stimulated to engage in a qualitatively different kind of mathematical thinking, and develop a disposition toward differentiating proportional and non-proportional problems. This assumption was tested by administering a type of task that is rather uncommon in the mathematics classroom: the classification of a set of word problems.

Interest in the value of problem classification and reflection on the relatedness of problems is rather old. Polya (1957) indicated that when devising a plan to solve a mathematical problem, a useful heuristic is to think of related problems. Seminal work was also done by Krutetskii (1976), who indicated that high-ability students differ from low-ability students on their skills to distinguish relevant information (related to mathematical structure) from irrelevant information (contextual details), to perceive rapidly and accurately the mathematical structure of problems, and to generalise across a wider range of mathematically similar problems. Studies that actually used problem classification tasks, however, are rare.

The study by Silver (1979) is well-known. Silver asked 8th graders to classify word problems according to their mathematical relatedness. Afterward, he did a didactical intervention in which the problems were solved and correct solutions were presented and discussed. Then, again, he offered the classification task. In analysing pupils' classifications and the criteria they had used, Silver distinguished classifications based on mathematical structure, contextual details, question form, and pseudostructure (e.g. relating to the kind of quantity measured: speed, price, weight, …). Silver found strong correlations between the quality of pupils' classifications and their problem solving performance. Also, classifications were more relating to the mathematical structure after the intervention than before, but the pseudostructure of word problems remained an important criterion for pupils' classifications.

2 Method

2.1 Subjects, Tasks and Procedure

Seventy-five 6th graders – belonging to five classes in three different primary schools in a middle-sized Flemish city – completed a classification task and a solution task.

In the *solution task*, pupils were given a traditional paper-and-pencil word problem test, containing nine experimental word problems with different underlying mathematical models: three proportional, three additive, and three constant ones. These different types of word problems had already been used and validated in previous research (Van Dooren et al. 2005). *Proportional* problems are characterised by a multiplicative relationship between the variables, implying that a proportional strategy leads to the correct answer (e.g. *Johan and Herman both bought some roses. All roses are equally expensive, but Johan bought fewer roses. Johan bought 4 roses while Herman bought 20 roses. When you know that Johan had to pay 16 Euro, how much did Herman have to pay?*). *Additive* problems have a constant difference between the two variables, so a correct approach is to add this difference to a third value (e.g. *Ellen and Kim are running around a track. They run equally fast, but Kim began earlier. When Ellen has run 5 laps, Kim has run 15 laps. When Ellen has run 30 laps, how many has Kim run?*). *Constant* problems have no relationship at all between the two variables. The value of the second variable does not change, so the correct answer is mentioned in the word problem (e.g. *Jan and Tom are planting tulips. They use the same kind of tulip bulbs, but Jan plants fewer tulips. Jan plants 6 tulips while Tom plants 18 tulips. When you know that Jan's tulips bloom after 24 weeks, how long will it take Tom's tulips to bloom?*). The word problems appeared in random order in the booklets, but a booklet never started with a proportional word problem to avoid that – from the start – pupils would expect the test to be about proportional reasoning. For the same reason, we also included six buffer items in the test.

For the *classification task*, pupils were given a box containing an instruction sheet, a set of nine cards (each containing one word problem), nine envelopes, and a pencil. Again, three of the word problems were proportional, three additive, and three constant. The instructions for pupils were kept somewhat vague because we wanted to see which criteria pupils would use spontaneously while classifying: "*This box contains 9 cards with word problems. You don't need to solve them. Rather, you need to figure out which word problems belong together. Try to make groups of problems that – in your view – have something in common. Put each group in an envelope, and write on the envelope what the word problems have in common. Use as many envelopes as necessary.*"

Both tasks were administered immediately after each other, but their order was manipulated. Half of the pupils got the solution task before the classification task (SC-condition, $n = 38$), the other half got the solution task after the classification task (CS-condition, $n = 37$). Because both tasks relied on nine experimental word problems, two parallel problem sets were constructed, each containing three proportional, three additive, and three constant word problems. In both conditions, pupils who got Set I in the classification task got Set II in the solution task and vice versa, so that in principle, differences between both sets would be cancelled out.

2.2 Analysis

Pupils' responses to the problems in the *solution task* were classified as *correct* (C, correct answer was given), *proportional error* (P, proportional strategy applied to an additive or constant word problem), or *other error* (O, another solution procedure was followed). Obviously, for proportional problems, only two categories (C- and O-answers) were used.

For the *classification task*, the data are more complex. Two aspects of pupils' classifications were analysed. The first aspect concerns the quality of the classifications, the second the kind of justifications pupils provided (as written on their envelopes).

The first aspect involves the extent to which pupils' classifications took into account the different mathematical models underlying the word problems. For each pupil, scores were calculated using the following rules:

- First, the group with the largest number of proportional problems ("P-group") was identified. It acted as a reference group: If children would experience difficulties distinguishing proportional and non-proportional problems, they would probably consider some non-proportional problems as proportional, and thus include non-proportional problems in the P-group.
- Next, among the remaining problems, the "A-group" and "C-group" were identified (the groups with the largest number of additive and constant problems, respectively). When more than one group could be labelled as A- or C-group, the group having the highest score (see next point) was chosen.
- Every group (P, A, and C) got two scores: An uncorrected and a corrected score. We explain these for the P-group. (It is completely parallel for the A- and C-group.) The uncorrected score for the P-group (Pu) is the number of proportional problems in the P-group. The corrected score (Pc) is Pu minus the number of other problems in that group. If no A- or C-group could be distinguished, these scores were set to 0.

The second aspect was the quality of the justifications given by pupils. The justifications for the P-, A-, and C-group of every pupil were labelled using the following distinctions:

- Superficial: Referring to aspects unrelated to the mathematical model: problem contexts (e.g. "these are about plants – tulips and roses", "they all deal with

cooking"), common words (e.g. "they both have the word *when*"), or numbers (e.g. "there is a 4 in the problems", "all numbers are even").
- Implicit: Referring to the mathematical model in the problems, but not unequivocally or explicitly to one particular model. For example, "the more pies – the more apples, the more you buy – the more you pay" does not per se refer to proportional situations, and "they all relate to the speed with which activities are done" does not grasp the additive character of situations.
- Explicit: Referring clearly and unambiguously to the (proportional, additive, or constant) mathematical model underlying the problems (e.g. referring to a proportional model "three times this so three times that, and in the other problem both things are doubled", or referring to the additive model "one person has more than the other, but the difference stays the same" or for the constant problems "these are tricky questions: nothing changes").
- Rest: There is no justification written, or it is totally incomprehensible. This label is also assigned when the particular group does not exist.

Our classification of justifications is similar to that of Silver (1979, see above), with the exception of the "question form" category (because this was controlled in our set of problems). Also, our "implicit" category was more inclusive than the "pseudostructure" category of Silver.

3 Results

3.1 Solution Task

Table 6.1 presents the answers to the solution task. A first observation is that the proportional problems elicited many more correct answers (2.68 out of 3 problems, on average) than the additive (0.88) and constant (0.61) problems. A repeated measures logistic regression analysis indicated that this difference was significant, $\chi^2 (2) = 23.87$, $p < 0.0001$. For the additive and constant problems, almost two out of three answers were proportional, indicating that pupils strongly tended to apply proportional calculations to the two types of non-proportional problems.

More importantly, pupils in the CS-condition performed significantly better than pupils in the SC-condition, $\chi^2 (1) = 10.72$, $p = 0.0011$. Even though the Problem

Table 6.1 Mean numbers of correct (C), proportional (P), and other (O) answers on the three proportional, additive and constant problems

	Proportional problems		Additive problems			Constant problems		
	C	O	C	P	O	C	P	O
SC-condition	2.61	0.39	0.65	2.08	0.26	0.24	2.08	0.68
CS-condition	2.76	0.24	1.11	1.86	0.03	1.00	1.70	0.30
Total	2.68	0.32	0.88	1.97	0.15	0.61	1.89	0.49

Type × Condition interaction effect was not significant, χ^2 (2) = 2.76, p = 0.2514, the most pronounced differences occurred for the additive problems (1.11 correct answers in the CS-condition vs. 0.65 in the SC-condition) and constant problems (1.00 vs. 0.24), whilst the difference was much smaller for the proportional problems (2.76 vs. 2.61).

Table 6.1 further reveals two explanations for these better performances: First, CS-condition pupils applied fewer proportional strategies than SC-condition pupils (1.70 vs. 2.08 and 1.86 vs. 2.08 proportional errors for the constant and additive problems, respectively), χ^2 (1) = 4.73, p = 0.0297. Second, CS-condition pupils also made significantly fewer other errors than SC-condition pupils (0.30 vs. 0.68 and 0.03 vs. 0.26 for the constant and additive problems, respectively), χ^2 (1) = 8.05, p = 0.0045.

3.2 Classification Task

Table 6.2 provides an overview of the different scores regarding the *quality of pupils' classifications*. First of all, this table reveals a high mean *Pu*-score of 2.37 (on a total of 3). Most pupils put at least 2 – many even all 3 – proportional problems in one single group. In contrast with the high *Pu*-score, the mean *Pc*-score is only 0.40, indicating that pupils frequently also put some (on average almost 2) additive and/or constant problems in the P-group, instead of putting them in separate groups.

For the additive and constant problems, the uncorrected scores (*Au* and *Cu*) are 1.73 and 1.71, respectively. These scores are lower than the one for the proportional problems. This is inherent to our scoring rules, because we first determined a P-group (which often also included some additive and constant problems) so that, on average, less than three additive and constant problems were left to create A- and C-groups. But still, the size of the *Au*- and *Cu*-values indicates that many pupils did make separate groups for the additive and constant problems. As was the case for the proportional problems, also for the non-proportional problems the corrected scores (*Ac* and *Cc*) are somewhat lower than the uncorrected ones (*Au* and *Cu*), but the difference is not as pronounced as for the proportional problems (1.73 vs. 1.35 for the additive problems, and 1.71 vs. 1.27 for the constant problems, respectively). So, even though other problems were sometimes included in the A- and C-groups (i.e. on average about 0.40 word problems in each group), this happened less often than for the P-group (on average almost two word problems).

In sum, the results presented so far point out that most pupils created a group containing proportional word problems, but often also some additive and/or constant problems were included in this group, suggesting that not only in their problem solving but also in their classification activities, pupils had difficulties to distinguish all non-proportional word problems from the proportional ones. Nevertheless, there was evidence that pupils distinguished some non-proportional problems, and made separate groups of proportional, additive, and constant word problems, even though their classifications were often imperfect.

6 Word Problem Classification: A Promising Modelling Task at the Elementary Level

Table 6.2 Mean uncorrected (Pu, Au, Cu) and corrected (Pc, Ac, Cc) scores for the classification task

	P-group		A-group		C-group	
	Pu	Pc	Au	Ac	Cu	Cc
SC-condition	2.34	0.42	1.76	1.18	1.58	1.05
CS-condition	2.41	0.38	1.70	1.51	1.84	1.49
Total	2.37	0.40	1.73	1.35	1.71	1.27

Table 6.3 Number of superficial (S), implicit (I), explicit (E), and other (R) justifications given by pupils to the P-, A-, and C-groups

	P-group				A-group				C-group			
	S	I	E	R	S	I	E	R	S	I	E	R
SC-condition	19	13	2	4	12	18	3	5	12	19	3	4
CS-condition	16	14	3	4	14	17	4	2	7	22	4	4
Total	35	27	5	8	26	35	7	7	19	41	7	8

We also compared the classifications in the two conditions. As can be seen in Table 6.2, the scores for the proportional problems (Pu and Pc) hardly differ for the SC- and CS-condition (a finding that parallels what was found for the solution task). However, classification scores for the additive and constant problems are somewhat higher in the CS-condition than in the SC-condition (except for the Au-scores, which are approximately equal). So, whereas doing the classification task first has a beneficial impact on performance on the solution task (particularly on non-proportional problems), the reverse is not the case: Doing the solution task first does not improve performance on the classification task. On the contrary, it has a slightly negative impact on children's classifications.

With respect to *quality of justifications*, Table 6.3 gives an overview of the various justifications for the P-, A-, and C-groups (for the explanation of the different labels, see the Analysis part in the Method section). A first observation is that explicit justifications are very rare for all three groups of word problems. They occurred in a maximum of 7 out of 75 cases. Second, many of the justifications are implicit, particularly for the C-group, but also for the other two groups. Third, also many superficial justifications were observed in all three groups. Of course, this does not necessarily imply that pupils actually *used* these superficial criteria while classifying. Their classifications were often in accordance with the underlying mathematical models, so children might have used criteria that acted tacitly, with superficial justifications occurring *post hoc*, in response to the instruction to provide justification for their classification. And fourth, the kinds of justifications are very comparable for the SC- and CS-condition. So, even though many pupils made appropriate – sometimes even perfect – classifications of the nine word problems in terms of their underlying mathematical models, they were rarely able to justify their classifications explicitly.

4 Conclusions and Discussion

Previous studies (e.g. Van Dooren et al. 2005) have shown that pupils strongly tend to use proportional solution methods for missing-value word problems, even when this is inappropriate. It was also suggested that pupils' immature and even distorted disposition toward mathematical modelling plays an important role: After a reflex-like recognition of the type of word problem, pupils quickly jump to the actual calculating work, and afterward, the result is immediately communicated without any further interpretation or critical reflection. The current study assumed that – if pupils would work on an unfamiliar task not focused on producing computational answers but on reflecting on commonalities and differences within a set of word problems – they might engage in a deeper kind of mathematical thinking, and distinguish more easily between proportional and non-proportional problems, which, in its turn, might have a beneficial effect on their problem solving skills.

Taken as a whole, the results supported this assumption. On the solution task, pupils were prone to the overuse of proportional methods: Performances on proportional problems were very good, but almost 4 out of 6 non-proportional problems were solved proportionally, as observed in previous studies (De Bock et al. 2007). As expected, pupils' behaviour on the classification task, however, was different. Nearly all pupils classified the proportional problems in one group, but they typically also included a few non-proportional (additive and constant) word problems. Many pupils also made a group of additive problems and another group of constant problems. Most often, pupils did not provide adequate explicit justifications for their groupings, but justified them implicitly.

The difference between the two conditions provided convincing evidence for the potentially positive effect of the classification task. Pupils who received the solution task after the classification task performed significantly better on the solution task than those who immediately started with the solution task, suggesting that the classification task made them more aware of differences among the word problems, which pupils transferred to the solution task. This observation is remarkable, considering that pupils' overuse of proportionality is deeply rooted (De Bock et al. 2007), while the classification task was a rather subtle and limited intervention, especially for pupils as young as 6th graders: No classification criteria were provided, no feedback was given, and the usefulness of the classifications for the subsequent solution task was never mentioned.

The positive results on the classification task also have implications for educational practice. They support the assumption that the overuse of proportionality is to a large extent due to pupils' superficial approach to word problems – jumping too quickly to the calculating work and immediately reporting the outcome – rather than to being really unable to distinguish proportional from non-proportional word problems. As such, explicit classroom attention to discussing similarities and dissimilarities between word problems (both in terms of superficial contextual features and in terms of deeper underlying structures) seems a very promising approach in order to eradicate the overuse of proportional methods.

References

Blum, W., et al. (2002). ICMI study 14: Applications and modelling in mathematics education – Discussion document. *Educational Studies in Mathematics, 51*, 149–171.
De Bock, D., Van Dooren, W., Janssens, D., & Verschaffel, L. (2002). Improper use of linear reasoning: An in-depth study of the nature and the irresistibility of secondary school students' errors. *Educational Studies in Mathematics, 50*, 311–334.
De Bock, D., Van Dooren, W., Janssens, D., & Verschaffel, L. (2007). *The illusion of linearity: From analysis to improvement* (Mathematics Education Library). New York: Springer.
Krutetskii, V. A. (1976). *The psychology of mathematical abilities in school children*. Chicago: University of Chicago Press.
Lesh, R., & Lehrer, R. (2003). Models and modeling perspectives on the development of students and teachers. *Mathematical Thinking and Learning, 5*, 109–129.
Polya, G. (1957). *How to solve it*. Princeton: Princeton University Press.
Silver, E. A. (1979). Student perceptions of relatedness among mathematical verbal problems. *Journal for Research in Mathematics Education, 10*, 195–210.
Usiskin, Z. (2007). The arithmetic operations as mathematical models. In W. Blum, P. Galbraith, H.-W. Henn, & M. Niss (Eds.), *Modelling and applications in mathematics education. The 14th ICMI study* (pp. 257–264). New York: Springer.
Usiskin, Z. (2008). The arithmetic curriculum and the real world. In D. De Bock, B. D. Søndergaard, B. G. Alfonso, & C. C. L. Cheng (Eds.), *Proceedings of ICMI-11 – Topic Study Group 10: Research and Development in the Teaching and Learning of Number Systems and Arithmetic* (pp. 125–130). Mexico: Monterrey.
Van Dooren, W., De Bock, D., Hessels, A., Janssens, D., & Verschaffel, L. (2005). Not everything is proportional: Effects of age and problem type on propensities for overgeneralization. *Cognition and Instruction, 23*(1), 57–86.
Van Dooren, W., De Bock, D., Janssens, D., & Verschaffel, L. (2008). The linear imperative: An inventory and conceptual analysis of students' over-use of linearity. *Journal for Research in Mathematics Education, 39*, 311–342.
Verschaffel, L., Greer, B., & De Corte, E. (2000). *Making sense of word problems*. Lisse: Swets & Zeitlinger.

Chapter 7
Understanding and Promoting Mathematical Modelling Competencies: An Applied Perspective

George Ekol

Abstract What is it that the applied mathematicians actually do in applications and modelling at the undergraduate level, and what might we learn from those experiences? A qualitative study was designed to answer this question. Mathematicians involved in applications and modelling from a university department were interviewed over a period of 6 months. Part of the interview involved interacting with dynamic conceptual models designed on Dynamic Geometry Sketchpad software. We anticipated that these applied mathematicians would favor the use of dynamic models in their teaching. What we found out was that there is a strong advocacy supporting "play" in modelling, because apart from the fun and the interest it generates, it might also lead to discovery and a sense of wonder. We identified four other themes from the interviews with respect to modelling and application: finding similar examples or phenomena; connecting physical phenomena with visual concepts; building models from the ground up; and communicating broader context of a modelling solution. These categories not only add to the list of competencies already identified in other studies, but they show a strong need for multidisciplinary collaboration in modelling and application.

1 Introduction

Mathematical modelling and applications is a central theme in mathematics education, evidenced by the many publications in journals, conference proceedings, and programs of the International Community of Teachers of Modelling and Applications (ICTMA), International Congress on Mathematical Education (ICME), and the International Commission on Mathematical Instruction (ICMI). In teaching and learning, modelling is variously covered from elementary school to tertiary

G. Ekol (✉)
Simon Fraser University, Burnaby, BC, Canada
e-mail: george_ekol@sfu.ca

education (Greer et al. 2007). This study reports on modelling from the perspective of applied mathematicians actively involved in undergraduate teaching and research. We take the position that the learning of mathematics will help develop competencies for extra-mathematical purposes. Extra-mathematical worlds are other domains outside mathematics, but which are in many ways served by mathematical applications (Blum et al. 2007).

With respect to the application of mathematics in the extra-mathematical or real world, Burkhardt (2006) reflects:

> ...there is no point in educating human automata; they are losing their jobs all over the world. Society now needs thinkers, who can use their mathematics for their own and for their society's purposes. Mathematics education needs to focus on developing these capabilities (p. 183).

Burkhardt's reflection in essence points to the urgency for mathematics education to take its rightful place in society, and play the role of linking mathematics learning with application, especially in tackling problems encountered outside mathematics itself.

Much research has been done on what kinds of competencies students need in order to engage in modelling. (Blomhøj and Jensen 2007; De Bock et al. 2007; Greer and Verschaffel 2007; Henning and Keune 2007; Houston 2007; Singer 2007). These studies describe a range of modelling skills, which serve to guide the teaching and learning, and assessment of modelling as a discipline. For example, the studies propose that modelling should be properly incorporated into the curricula, and should start in the early years of school, taking into account the appropriate mathematical disposition of the students. This is a crucial point, given that in general, modelling is not taught on its own, but within mathematics, making its status in the curricula unclear. A challenge for the teacher is then to incorporate appropriate models that help the students relate what they are doing in a mathematics modelling class to extra-mathematical world problems, without over simplifying the mathematics (see for example Greer and Verschaffel 2007, p. 220).

We propose in this chapter that if mathematical modelling in the classroom is to link with the real world, then there has to be some enculturation process where students, teachers, researchers, and educators, share a language and practices, and develop knowledge through communication (Lerman 1996; Nardi 2008; Sierpinska 1994; Vygotsky 1962; Wenger 1998). In that respect, our study focuses on one group in this shared community: the applied mathematicians. As part of the enculturation process, students need to learn what applied mathematicians do, what tools and language they use in the modelling processes. For their part, applied mathematicians have to understand the needs of the students by putting together learning activities and programs that build their competencies. In this study, applied mathematicians also assume the roles of researchers and teachers.

We seek to find out what modelling experiences applied mathematicians would give to their undergraduate students, by addressing the following specific questions:

1. From the perspective of applied mathematicians, what mathematical modelling experiences are needed at undergraduate level?
2. What might we learn from those experiences to inform teaching?

The purpose of the study is to record the modelling experiences of the applied mathematicians that will inform the teaching and the practice at that level. Our data comprise the narrative gathered from applied mathematicians, paying attention to the use of language and practices that are taken for granted in that community.

2 Methodology

Applied mathematicians in a university department of mathematics were asked by email if they would agree to be interviewed. Initially, it was not easy to find convenient time as many had commitments. The interviews were scheduled on an individual basis, convenient to each member contacted; the resulting interview period stretched over 6 months. There were ten respondents, all were interviewed, but in this chapter, only four interviews with applied mathematicians are presented because of space. The interviews took place in the mathematicians' offices, to cut down on their time of moving to another location, but this also gave the interviewers an opportunity to see the work space of the interviewees, for instance, the tools, equipment, materials, and resources they mostly use in their teaching and research.

The interviews followed a qualitative design (Creswell 2008), with open questions. The initial responses were often followed by probes to get further clarification on what the interviewees meant. At the beginning of the interview, the interviewees briefly described their areas of teaching and research, and the problems they were working on. Later on, they described their mode of work on the problems: For instance, did they use computers, drawings, paper, and pencil, and if so, on what kinds of problems were these tools used? Did they use mathematical concepts differently from their colleagues in pure mathematics? In the final phase of the interview, they were introduced to dynamic conceptual models (dynamic number line, and matrix transformations) designed on Dynamic Geometry Sketchpad (DGS) software. We conjectured that applied mathematicians would support the use of dynamic models; such models help students focus on the behaviour of things that are moving, and hence the concepts involved. After interacting with these models for about 30 min, they gave their feedback, relating it to their own practice and experiences. In the analysis of the video transcript, we looked for themes that related to the practices and language that are taken for granted in the community of applied mathematicians.

3 Results

The original data comprised video recordings. After transcribing the data, four major themes stood out from the applied mathematicians with respect to the modelling competencies they would like their students to have:

1. Finding similar examples or phenomena.
2. Connecting physical phenomena with abstract concepts.

3. Building models from the ground up.
4. Communicating broader context of a modelling solution.

We briefly discuss each of the four themes in the following paragraphs. We note that Joan, Jeff, Bob, and John (all pseudonyms) are the applied mathematicians, and Nathalie is the interviewer. The response to questions does not follow the order of names.

3.1 Finding Similar Examples or Phenomena

01. Nathalie: Can you talk a little bit about the mathematics ... what is needed to mobilize the math in modelling ..., what is it that you are good at?
02. Jeff: Part of it is having an encyclopedic collection of things that are important. You know that this thing has been done for these sorts of problem so you could extract things that are similar.

Here, in Jeff's response we observe that modelling requires an open approach to dealing with problems, but it also demands some degree of preparedness to handle the problems. On the other hand, transfer of experience from one area to another is essential. That implies that students can draw on examples from other subjects such as physics, chemistry, biology, and bring them to enhance their modelling competences.

03. Nathalie: What else does the "encyclopedic" collection include?
04. Jeff: Having a wider understanding. In this process they have to identify what the tools are, and then also they have to know how to use the tools once they have identified the problem.

Use of "tools" is important in modelling. The tools mentioned by Jeff could be physical such as pencil and paper, or use of diagrams, computer simulations, but they could also be nonphysical tools such as a procedure or algorithm that one uses to solve a modelling problem. All these imply some knowledge of the problem area one is working on, and related information, or experience that have been built over time.

In the next interview, Joan described a modelling problem she had to deal with in biology.

05. Nathalie: So where did you start ... when you wanted to model that, did you think of an equation, did you think of something more geometric?
06. Joan: The data was very suggestive of things that we'd seen in other contexts, so we had to think about what types of mathematical objects would give rise to such pictures.
07. Joan: That was much harder and I have to confess that the model was cobbled with different terms that each individually described certain aspects.

What are those "mathematical objects" Joan refers to in [06]? Sfard (1994) brings in the theory of reification – a transition from an operational to a structural mode of thinking in the formation of a mathematical concept. Although reification is beyond the scope of this chapter, it is worth mentioning here, just in case there might be any parallels to what Joan went through in [06]. What comes out clearly from Joan, though is that she draws from her experience to make connections to similar situations she had seen in other contexts.

3.2 Connecting Physical Phenomena with Abstract Concepts

08. Nathalie: About the concepts that you use in modelling do you feel that they differ at all from what somebody doing pure mathematics might use?
09. Jeff: I think the concepts are not any different; it's just a question of what types of problem you are interested in. The basic tools are, how can I write something down which describes certain aspects of what I'm interested in, in some sense?
10. Jeff: In terms of someone from pure mathematics, the questions that they are interested in are slightly different but the basic ideas are being able to make some abstract representation of what it is that they are thinking about. That's the starting point and that's always the same.

Bob's response to the same question does not differ very much from Jeff's.

11. Bob: Well, in modelling you have to have some physical model of what the object is, its structure, shape. So you have to write down the equations that actually describe that.

Having a physical model and abstracting this to a mathematical model is a key component of modelling. Students should not ignore the structure and shape of objects in modelling situations, because these can provide some hints for formulating mathematical solutions.

3.3 Building Modelling from the Ground up

12. Nathalie: Ok, so all of these [dynamic models] that you have looked at … we think of them as ways that help students focus on the behavior of things that are moving, …
13. Nathalie: …and our hypothesis is that the dynamic models might help them with modelling or applied situation. What's your reaction to that?
14. John: When I look at a tool like this one [dynamic models] my first question is what would I use it for? I'm always keen on anything that gets people to play, anything that brings a sense of discovery and wonder, the fun thing.

15. John: Well, there is certainly a lot of test cases and playing around, prototyping small cases, "what if I had to fit a square into a circle, what would really happen?"

We observe that John does not object to using dynamic models. This also supports our original conjecture that dynamic models might help students develop competencies in applied situations.

Jeff, still referring to the same question [12.13] says:

16. Jeff: Ah, you think of adjusting a parameter and seeing the consequences, for example in a dynamical system, you have some function that has a parameter, and as you adjust the parameter, the function is going to change.

We observe unanimity among the applied mathematicians about the use of dynamic tools or technology for experimentation purposes (what if?), prototyping, play, wonder, fun, and discovery. Modelling should incorporate all these attributes. Evidence from research also supports use of technology for exploration purposes. Papert (1980) has very good examples.

3.4 Communicating Broader Context of a Modelling Solution

17. Nathalie: Is there a different culture of writing in applied mathematics?
18. John: In applied math, there is always a lot of communication between the people who are working on the problem together, so a lot of writing, starting right away with something broader, with a lot more background, in some sense.
19. John: [With respect to students' modelling] after working their solutions, some will say, ok, that is the solution, even if they made a mistake. You look at the solution and it can't be anything physical. The physical world doesn't work that way.
20. John: But they are resistant that they might be able to apply their intuition about the real world to the solution of the model.

Communication in general is arguably the most challenging aspect of modelling. Students tend to ignore it as not the main part of modelling but indeed it is a very significant part of modelling. Clear communication is important because without it, the process is incomplete. So students need to develop the skill of communicating their results, and the best way is through practice, through individual and group projects.

Modelling solutions should be realistic and amenable to the extra-mathematical world or physical reality (Blum et al. 2007). In communicating this solution, some background information is necessary to inform nonexperts in the subject area what the solutions to the problems are and some implications of the solution.

4 Discussion and Summary

The study probed the modelling experiences of four applied mathematicians, who are teachers at undergraduate level as well as being researchers. Four major themes emerged from the interviews:

- *Finding similar examples or phenomena*: the importance of drawing examples from one's experiences and using them in the modelling situation.
- *Connecting physical phenomena with abstract concepts*: Moving from a physical model to a mathematical model, and [after solving], interpreting and communicating the solution in the real setting outside mathematics.
- *Building a model from the ground up*: starting with a simple idea about a problem and experimenting, using appropriate tools, until the problem becomes clearer.
- *Communicating a mathematical modelling solution*: recognizing the importance of clear communication of results that accommodates a wider audience, other than the expert audience themselves.

These four themes address the question of competencies needed in undergraduate mathematical modelling, although their application extends outside education. We agree with Burkhardt that mathematics education should contribute to developing the capabilities of students for their own benefit, and for the benefit of society as a whole. The other insight we get from this study is the notion of "play" in modelling.

> I'm always keen on anything that gets people to play, anything that brings a sense of discovery and wonder, the fun thing (John, line 13).

This remark is rather surprising because at undergraduate level, play is not usually considered part of serious learning. On the other hand, John is probably advocating a different game, perhaps a more serious game than just "play." A game where students explore concepts through testing, guessing, estimating, simulating, checking and cross-checking, in a "playful" way using tools that they have.

Furthermore, modelling demands adaptive expertise in nature and is a social activity which should be properly supported by good curricula (Greer and Verschaffel 2007). In the context of our study, adaptive expertise means an ability to interpret the context, environment, of a modelling problem, and to apply the requisite mathematical tools in resolving the task.

> Having a wider understanding. In this process they have to identify what the tools are, and then also [they] have to know how to use the tools once they have identified the problem [10].

We have argued in this chapter that modelling is also a multidisciplinary undertaking because one must draw from many areas to formulate a mathematical problem. "Wider understanding" [10], also implies that students are being challenged to look beyond their subject areas, draw from different areas of learning, and also from their personal experiences.

The data was very suggestive of things that we'd seen in other contexts, so we had to think about what types of mathematical objects would give rise to such pictures [06].

This clearly demonstrates a multidisciplinary approach to mathematical modelling and application. Overall, we believe that our conjectures have been well supported by the data that we collected.

References

Blomhøj, M., & Jensen, T. H. (2007). What's all the fuss about competencies? In W. Blum et al. (Eds.), *Modelling and applications in mathematics education* (pp. 45–56). New York: Springer.

Blum, W., Galbraith, P. L., Henn, H.-W., & Niss, M. (2007). *Modelling and applications in mathematics education*. New York: Springer.

Burkhardt, H. (2006). Modelling in mathematics classrooms: Reflections on past development and future. *Zentralblatt für Didaktik der Mathematik, 38*(2), 178–195.

Creswell, J. W. (2008). *Educational research: Planning, conducting, and evaluating quantitative and qualitative research* (pp. 212–242). Uppler Saddle River: Pearson Prentice hall.

De Bock, D., Van Dooren, W., & Janssens, D. (2007). Studying and remedying students' modelling competencies: Routine behavior or adaptive expertise. In W. Blum et al. (Eds.), *Modelling and applications in mathematics education* (pp. 241–248). New York: Springer.

Greer, B., & Verschaffel, L. (2007). Modelling competencies – Overview. In W. Blum et al. (Eds.), *Modelling and applications in mathematics education* (pp. 220–224). New York: Springer.

Greer, B., Verschaffel, L., & Mukhopadhyay, S. (2007). Modelling for life and children's experience. In W. Blum et al. (Eds.), *Modelling and applications in mathematics education* (pp. 88–98). New York: Springer.

Henning, H., & Keune, M. (2007). Levels of modelling competencies. In W. Blum et al. (Eds.), *Modelling and applications in mathematics education* (pp. 225–232). New York: Springer.

Houston, K. (2007). Assessing the phases of mathematical modelling. Levels of modelling competencies. In W. Blum et al. (Eds.), *Modelling and applications in mathematics education* (pp. 249–256). New York: Springer.

Lerman, S. (1996). Intersubjectivity in mathematics learning: A challenge to the radical constructivist paradigm? *Journal for Research in Mathematics Education, 27*, 133–150.

Nardi, E. (2008). *Amongst mathematicians: Teaching and learning mathematics at university level*. New York: Springer.

Papert, S. (1980). *Mindstorms: Children, computers, and powerful ideas*. New York: Basic Book.

Sfard, A. (1994). Reification as the birth of metaphor. *For the Learning of Mathematics, 14*(1), 44–55.

Sierpinska, A. (1994). *Understanding in mathematics*. London/Washington: The Falmer Press.

Singer, M. (2007). Modelling both complexity and abstraction: A paradox? In W. Blum et al. (Eds.), *Modelling and applications in mathematics education* (pp. 233–240). New York: Springer.

Vygotsky, L. S. (1962). *Thought and language*. Cambridge: Harvard University Press.

Wenger, E. (1998). *Communities of practice: Learning, meaning and identity*. Cambridge: Cambridge University Press.

Chapter 8
Secondary Teachers' Beliefs About Teaching Applications – Design and Selected Results of a Qualitative Case Study

Frank Förster

Abstract As a result of the standard-based curricula, in several countries secondary teachers' beliefs about applications or modelling have developed in the scope of mathematics education. In contrast, German secondary teachers rarely integrate applications or modelling into their instructional practice. This research report is focused on teachers' beliefs that hinder or promote integrating applications or modelling into their teaching practice. The objective of this approach was to reconstruct the teachers' belief systems concerning applications. The undefined term "beliefs" is specified by the psychological construct of "subjective theories." In this chapter, results with reference to the subjective theories of teachers with respect to modelling will be presented. Furthermore, some recommendations concerning teacher training will be sketched.

1 The Call for Applications in Mathematics Curriculum is Quite Old!

Everything flows. The way of teaching mathematics undergoes steady changes. To be more precise, there are two aspects in particular that have been altered over the years at a slow but steady pace. Firstly, one strives to simplify the teaching subjects. ... Secondly, the approach to teaching mathematics increasingly seeks to adapt the needs of everyday life. This effort is reflected especially in the selection and status of the so-called real-world problems. (Heinrich Kempinsky 1928, p. 9. Translation by author).

Today, applications are in vogue: Even if there were always oscillations between utilitarian periods and puristic periods in teaching mathematics over the last 100 years (Kaiser 1995; Niss 2000) and, after a "weaker period" in the 1990s, today applications are brought back into focus by the TIMSS and PISA discussion – as one can see by newer German schoolbooks or in recent German curricula and didactical journals.

F. Förster (✉)
Technische Universität Braunschweig, Braunschweig, Germany
e-mail: f.foerster@tu-braunschweig.de

But applications still miss out in the mainstream classroom at least in German secondary schools: For years we asked freshmen at the Technical University of Braunschweig: "What applications of mathematics do you remember from school?" The most common answer was: "None." Often applications were mentioned as *pure mathematical applications* like curve sketching as an "application of differential calculus." There are still only a few research studies to support this thesis, but a lot of direct or indirect evidence by many studies (e.g., BIQUA 2007; Hiebert et al. 2003; Hugener 2008; Kaiser 1999; Neubrand 2004; Stigler and Hiebert 1999).

The burning issue is: Will there be changes made by the new German standards-based curricula (i.e., Bildungsstandards, KMK 2004; the Kerncurricula e.g., for Niedersachen, Niedersächsisches Kultusministerium 2006)? Not necessarily, we think, and particularly not automatically, as the following analysis will show.

2 A First Root Cause Analysis: Focusing on Teachers

> How teachers make sense of their professional world, the knowledge and beliefs they bring with them to the task, and how teachers' understanding of teaching, learning, children, and the subject matter informs their everyday practice are important questions that need an investigation of the cognitive and affective aspects of teachers' professional lives (Calderhead 1996, p. 709).

Is it possible to identify issues that can explain the gap between the educational demands for applications and modelling and the instructional practice in the classroom? We use the following didactic triangle (see Fig. 8.1) as a simple model for teachers' actions (Tietze et al. 1997, pp. 74 f.). The model incorporates the mathematical subject, students, and teacher and, as indicated by the arrows, the interactions between teacher and students, and within the group of students and the involvement of the teacher and the student with the subject. From the qualitative point of view, the dialogue between the persons involved and the subject matter is a dialogue that can change both person and subject matter. Last but not least, we take into account the conditions that frame school teaching.

The key persons in changing or reforming mathematics education and to apply new curricula are teachers (e.g., Fernandes 1995; Wilson and Cooney 2002).

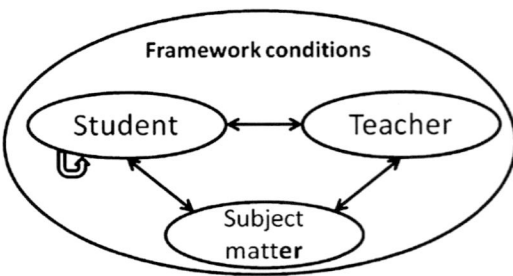

Fig. 8.1 Didactic triangle

Focusing on teachers as the main actors for planning and performing teaching in school is just *one* possible perspective. One could also focus on the *framing* by investigating the effect of curriculum, assessment, or even social expectations (of parents, society, or even the business world) on mathematical teaching. One could focus on *subject matter* as the context of schoolbooks or learning material for teaching applications that also have a significant influence. Alternatively, one could focus on the *students*, including their mathematical competencies, their attitudes to applications and modelling, and the students' expectations of their mathematical teaching as the research of Maaß (2004) has shown, a factor, that should not be underestimated.

For now, the focus will be on *teachers* as the main actors in the educational process. Two, at first glance, very simple questions arise: *Do not teachers want to teach applications and modelling in the classroom?* Which would correspond with the teachers' motives? Or: *Cannot they teach applications and modelling?* Taking into account the application competencies of the teachers and/or the objective or subjectively felt barriers that hinder them from teaching applications.

3 Why a Qualitative Case Study?: Methodology and Methods

Questionnaire-based quantitative studies of teachers' cognitions and attitudes to applications and modelling show that the great majority have a quite positive attitude to applications and want to increase the number of applications but see a lot of barriers connected with applications in the classroom and even self-distrust of their own application competencies. Barriers mentioned are, for example, "too few materials for teaching applications," "applications are hard stuff and therefore only relevant for high-performers," "applications are difficult to assess" and most of all: "There is not enough teaching time for applications" (e.g., Grigutsch et al. 1998; Humenberger 1997; Tietze 1990, 1992; Zimmermann 2002).

From this quantitative research, a lot of questions remain open, especially: *If teachers want to teach applications and modelling in the classroom, why do they not create the framing conditions to reach their goals?* And that's where the qualitative study starts with the following research questions: (1) What are the teachers' reasons to integrate or to ignore modelling in their teaching practice? (2) Is it possible to identify issues that can explain the gap between the educational demands of modelling and the instructional practice?

3.1 Theoretical Constructs

The research is based on the following constructs: Teaching and planning of teaching are actions (and not behaviour) of teachers (*Theory of action*, Hofer 1986). Sources and reasons for actions are not observable, but have to be reconstructed by interpretation (*Interpretative paradigm*, Wilson 1973). We have an *epistemological*

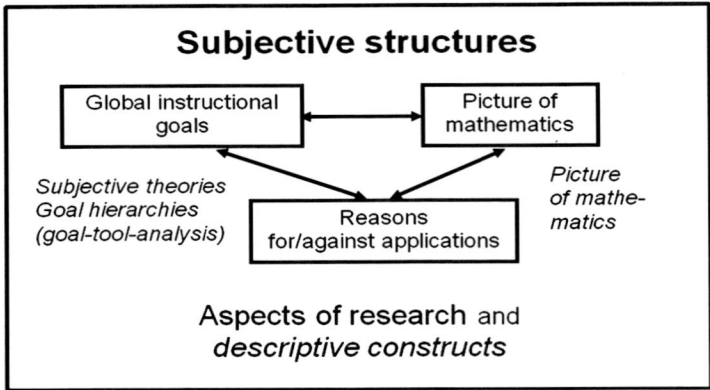

Fig. 8.2 Subjective structures

conception of man (psychology of the *reflexive subject*, Groeben and Scheele 1977, 2000), which means especially that researcher and researched are structurally equal, so it is possible to communicate about the reasons and intentions of the teachers – and to validate these reconstructions.

Building on results of research on beliefs (Leder et al. 2002; Thompson 1984, 1992), the *aspects of research* (see Fig. 8.2) are the following: We look for "global instructional goals," the "picture of mathematics," and the "reasons for or against applications," as well as the connections between them. To describe these aspects, we use the following *descriptive constructs*, which are explained below: "Subjective theories" as a background theory for beliefs, "goal hierarchies" and again the "picture of mathematics" as a descriptive tool. All together, we call it the *subjective structure* of a teacher.

Subjective theories resemble scientific theories in structure as well as in function (explanation, prediction, technology) but are, in comparison to scientific theories, less coherent and consistent, are usually implicit and not explicit, and have an important function of orientation for the teachers (Groeben and Scheele 1977, 2000). Furthermore, most importantly, they are *subjective* and not *objective* theories.

To structure the subjective theories we use *goal hierarchies*, which we gain by *goal-tool-argumentation* ("Ziel-Mittel-Argumentation" according to König 1975; q.v. Scheele and Groeben 1988). Added are specific assumptions about the structure and the function of subjective theories (Groeben et al. 1988) which we use but will not mention any further in this chapter.

A brief example is given. From an interview there is the statement: "Applications motivate the students." This statement is now transferred into a descriptive sentence: "If one wants to motivate students one can teach applications." From this description, two prescriptive sentences can be derived: "One should motivate students" and "One should teach applications" as shown in schema in Fig. 8.3.

So, we have "motivation" as a *goal* and the "teaching of applications" as a *tool* to achieve "motivation" in this hierarchy. By adding further information from the

	Goal Level	Goal/Tool-Level
Prescriptive sentence	One should *motivate* students.	
Descriptive sentence	If one wants to *motivate* students …	… one can *teach applications*.
Prescriptive sentence		One should *teach applications*.

Fig. 8.3 An example of goal-tool-argumentation

interview, we can develop a chain of arguments and a more or less complex *goal-tool-structure* – either in search of higher goals (*goal-perspective*) or in search of suitable tools (*tool-perspective*).

To sum up with the example, the teacher might have also said: "Applications convey a representative picture of mathematics". There are at least two ways to interpret this and you have to reconstruct from the data, which interpretation is adequate. (1) To teach a representative picture of mathematics and to motivate the students are goals at the same level and for both goals the teaching of applications is a suitable *tool*. (2) To teach a representative picture of mathematics is the main goal. The tool for this goal (i.e., applications) is now becoming a *goal for the next level*. The first interpretation of the sentence "Applications motivate the students" is altered to "If you teach applications you motivate students" which means application leads to motivation as a consequence. The important difference between these interpretations is that in the first case we can see applications as a *tool* and in the second one as a *goal* in the hierarchy.

Note: Due to space restriction, we leave out an explication of the construct "picture of mathematics" in this chapter – roughly speaking, it is a mixture of *beliefs about mathematics* and *content knowledge* of applications (see Förster 2008).

3.2 Study Design

Understanding action as an inner process depending on situations determines an inquiry in the form of case studies (Stake 2000). The definition of the cases is according to theoretical sampling (Glaser and Strauss 1967). The main study involved the questioning of eight *in-service teachers* grade 7–13 of secondary schools (A-level) in Northern Germany with several years of teaching experience (at least 2, up to 20 years). Data were mainly collected by *(in-depth) interviews* up to 4 h with open and semi-structured parts. The interviews were prepared by evaluating a *standardized questionnaire* (mainly statistical data such as age of the teacher, type of school, and also schoolbooks and teaching material used).

Interpretation was based on an adoption of *qualitative content analysis* (Mayring 1995) and methods of *qualitative teaching research* (e.g., Corbin and Strauss 2008; Jungwirth and Krummheuer 2008) and went through the following steps: transcription of the interview, sequential interpretation of the transcription (to gain the subjective theories), summing up global analysis (to gain patterns in the goal hierarchy), intrapersonal description of the teachers (to gain the subjective structure of the teacher), and interpersonal analysis (to gain different types of subjective structures).

4 Discussion and Some Selected Results

The following discussion focuses on one main result of the study. The process of interpretation of the interviews is not outlined. While primarily one case will be discussed, some results of other cases are used to complete the case description.

Figure 8.4 shows an "overview map" derived from the goal-tool-structure of Teacher A representing the global instructional goals, as part of the subjective structure. Omitted are subsidiary goals, the connections between the two excerpts, as well as most of the substructure of the connections within the excerpts.

Teacher A's main instructional goals are *significance for the future* and *fulfillment of curriculum plan*. Significance for the future means *university and vocational preparation* and *school attainment*. Firstly, students have to learn *logical thinking*. This goes with the transfer hypothesis that logical thinking in mathematics leads to logical thinking towards everyday-life problems. That of course is objectively not true – but that does not matter to Teacher A as long as his subjective structure is coherent! (An explanation of why in the subjective structure of Teacher A's logical thinking can be achieved by *accuracy in working* would take too long to explain here, but it "makes sense" within the subjective structure (see Förster 2008).

Looking at the *School attainment*: For Teacher A "no one is left behind" in his teaching – to ensure students pass the examinations means *repetition and exercising* and *training of techniques* as a *stockpiling of knowledge*. This altogether needs a lot of *teaching time* – which of course is not an explicit goal, but a *necessary tool* in the goal-tool hierarchy – and the needed time is missing for other goals.

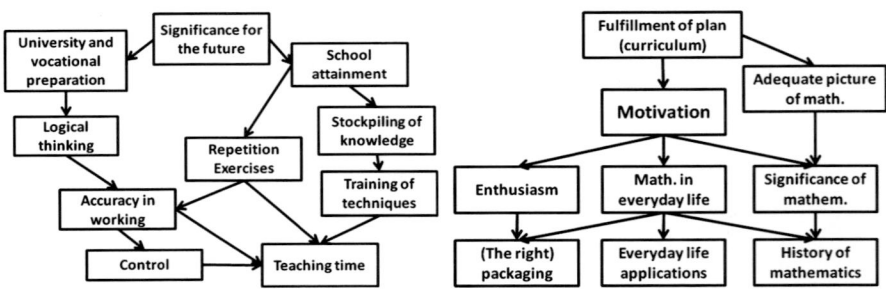

Fig. 8.4 Global goals of Teacher A (excerpts)

Where are the applications? We look at the right side of Fig. 8.4: *Fulfillment of plan* goes hand in hand with the goal (!) to *motivate the students* for the content they have to learn at any rate. One way to motivate the students is to teach simple *everyday-life applications* – simple enough to motivate. However, Teacher A can also show the *significance of mathematics* to be motivated for example by episodes of *mathematical history*, which can be told to the students "in passing." Last but not least, the *enthusiasm* of the teacher can motivate by his choosing the *right packaging* for mathematical content. We will come back to these aspects, but first have a look at the type of applications Teacher A uses in his teaching: *Everyday life applications* have to be simple to understand (nonmathematical background) and simple to teach (teaching time). As a consequence, there can be no complex modelling. Examples are height determinations of trees, simple financial mathematics (especially interest calculations), volume of a conically shaped wine glass, volume of a cigar, the Pythagorean knotted rope, pieces of cake (fractions) and some applications in physics. These examples correspond very well with Teacher A's *subjective definition of application* derived from the interviews: "Everything (!), that students know from their everyday life (or can at least imagine) and that can be associated with mathematics, is (!) an interesting application."

This brings us to the *right packaging*: Teacher A mentions a task from a schoolbook: In a picture, an expander is hanging from the ceiling of the room. By Hooke's law with a proportional function the elongation of the expander, respectively, the weight-force, can be computed by elongation = const. × weight of person. After a short explanation of the expander and after introducing "Silke and Dirk," this is a quite normal, fairly boring word problem. So what's the point to Teacher A that's worth mentioning this task in the interview? Teacher A transfers the situation in the picture into his classroom. He brings an expander with him. He lets the student try out the expander and finally he screws the expander to the ceiling of the classroom and comes to the same questions – but with the real expander. Surely not a real-world problem, but his students are interested by the packaging. As mentioned before, time for applications is limited by the main goals – so for this teacher applications are merely a *tool* for motivation.

To explain (or at least illustrate) the origin of the three different types of "teacher's subjective substructures according to applications," we are closing with a brief look at the position of applications in the hierarchy of goals of two other teachers and summarize these three types in the following Fig. 8.5.

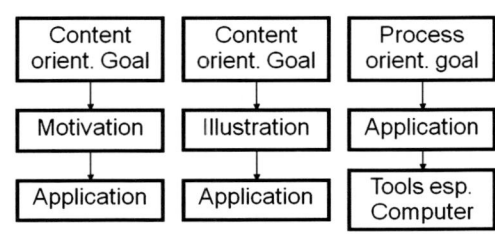

Fig. 8.5 Position of applications in the hierarchy of goals

Teacher B sees *applications* as a central goal of teaching corresponding with the goal (and tool) of *teaching applied mathematics*. The goals have higher aims as *empowered citizen*, *problem solving* (inside and outside mathematics), *a positive attitude towards mathematics*, and *learning to ask questions*. This teacher also has high *professional pretension* and he gets the needed teaching time by using the handheld *computer* as a tool in the hands of the students. *Motivation* is not a goal, but a consequence of the other goals.

In contrast, Teacher C is a *structuralist* sensu (Eichler 2007), happy with context-free mathematics. Therefore, you have to search for applications, which are merely a *tool* for *illustrating mathematical content* and quite isolated from the rest of the goal hierarchy.

The first column of Fig. 8.5 leads to attaching word problems to a traditional form of mathematical teaching without modelling. The second propagates context-free problems – and sporadic use of applications – corresponding with very high, nearly unrealisable expectations in terms of a realistic context for the applications. The third approach allows more complex applications, sometimes corresponding with high expectations of the mathematics involved in the applications. For type one and three, there will be no search for tools to overcome the barriers that hinder these teachers to have applications in their teaching.

5 Conclusion

> Quite clearly, there is a fundamental need to understand everything that underlies the way in which mathematics teachers approach their subject before suggestions and recommendations concerning good classroom practice can be made. (Eichler 2007, p. 208).

There is an unexpected *high consistency* between the picture of mathematics, the global goals concerning teaching and the selection, and reasons for or against applications. It was not expected to be so clearly defined. This is also important because one aspect of the research on teachers' beliefs is the conclusion that they have a high impact on students' beliefs (Chapman 2001). And, the students of today are the teachers of tomorrow. *Motivation* is *the* dominant argument for applications. The assumption, mathematics is per se formative toward active, creative, and flexible individuals, makes applications as an independent and clear goal abundant. And, therefore *potential applicability* of mathematics is sufficient for these teachers – in correspondence with their picture of mathematics. *Realistic modelling* and further educational demands do not play any (important) role in the classroom and there is often a mixture of *applications* and *applied mathematics*.

Interesting is the role of the *second teaching subject* of the German teachers. We expected higher competence in applications especially with physics and other natural science teachers, but we also found limiting factors as the immense time exposure for the second subject (especially physics) when taught as an experimental subject and different approaches that teachers have to applications in their different subjects – "applications in physics: of course" but "applications in mathematics: no need for them."

Tertiary Education has effects on teaching of applications: For instance, application examples often come from the teachers' own university education. Also the *German traineeship for teachers* has effects, because frequently the trainees are encouraged by the instructors to teach applications. However, if the closer contact with applications does not start until this traineeship, this "retrofitting" of competencies in applications is considered by the teachers as amateurish (dilettantish) and after their traineeship they will not teach applications any further. So a positive attitude to applications and knowledge about applications and modelling should be set up in school or in university study at the latest. Starting late will be too late!

References

BIQUA (2007). Die Bildungsqualität von Schule. Abschlussbericht. http://www.ipn.uni-kiel.de/projekte/biqua/index.html. Accessed 10 September 2010.

Calderhead, J. (1996). Teachers: Beliefs and knowledge. In D. C. Berliner (Ed.), *Handbook of education* (pp. 709–725). New York: Macmillan.

Chapman, O. (2001). Understanding high school mathematics teacher growth. In M. Heuvel-Panhuizen (Ed.), *Proceeding of the 25th PME Conference* (Vol. 2, pp. 233–240). Utrecht: PME.

Corbin, J., & Strauss, A. L. (2008). *Basics of qualitative research: Techniques and procedures for developing grounded theory*. Thousand Oaks: Sage.

Eichler, A. (2007). Individual curricula: Teachers' beliefs concerning stochastics instruction. *International Electronic Journal of Mathematics Education, 2*(3), 208–225.

Fernandes, D. (1995). Analysing four preservice teachers' knowledge and thoughts through their biographical histories. In L. Meira & D. Carraher (Eds.), *Proceedings of the 19th PME Conference* (Vol. 2, pp. 162–169). Recife: PME.

Förster, F. (2008). Subjektive Strukturen von Mathematiklehrerinnen und -lehrern zu Anwendungen und Realitätsbezügen im Mathematikunterricht. In H. Jungwirth & G. Krummheuer (Eds.), *Der Blick nach innen: Aspekte der alltäglichen Lebenswelt Mathematikunterricht – Band 1/2*. Münster: Waxmann.

Glaser, B., & Strauss, A. (1967). *The discovery of grounded theory*. Chicago: Aldine.

Grigutsch, S., Raatz, U., & Törner, G. (1998). Einstellungen gegenüber Mathematik bei Mathematiklehrern. *Journal für Mathematik-Didaktik, 19*(1), 3–45.

Groeben, N., & Scheele, B. (1977). *Argumente für eine Psychologie des reflexiven Subjekts*. Darmstadt: Steinkopff.

Groeben, N., & Scheele, B. (2000). Dialogue-hermeneutic method and the "research program subjective theories". *Forum Qualitative Social Research 1*(2). http://www.qualitative-research.net/index.php/fqs/article/view/1079/2354. Accessed 10 September 2010.

Groeben, N., Wahl, D., Schlee, J., & Scheele, B. (1988). *Forschungsprogramm Subjektive Theorien – Eine Einführung in die Psychologie des reflexiven Subjekts*. Tübingen: Francke.

Hiebert, J., et al. (2003). *Teaching mathematics in seven countries*. Washington: NCES.

Hofer, M. (1986). *Sozialpsychologie erzieherischen Handelns*. Göttingen: Hogrefe.

Hugener, I. (2008). *Inszenierungsmuster im Unterricht und Lernqualität*. Münster: Waxmann.

Humenberger, H. (1997). Anwendungsorientierung im Mathematikunterricht – erste Resultate eines Forschungsprojekts. *Journal für Mathematik-Didaktik, 18*(1), 3–50.

Jungwirth, H., & Krummheuer, G. (2006/2008). *Der Blick nach innen: Aspekte der alltäglichen Lebenswelt Mathematikunterricht – Band 1/2*. Münster: Waxmann.

Kaiser, G. (1995). Realitätsbezüge im Mathematikunterricht – Ein Überblick über die aktuelle und historische Diskussion. In G. Graumann et al. (Eds.), *ISTRON* (Vol. 2, pp. 66–84). Bad Salzdetfurth: Franzbecker.

Kaiser, G. (1999). *Unterrichtswirklichkeit in England und Deutschland*. Weinheim: Beltz, Deutscher Studien Verlag.
Kempinsky, H. (1920, 1928). *Sachaufgaben für das erste Schuljahr*. Leipzig: Dürr'sche Buchhandlung.
KMK. (2004). *Bildungsstandards im Fach Mathematik für den Mittleren Schulabschluss*. München: Wolters Kluwer.
König, E. (1975). *Theorie der Erziehungswissenschaft Band 2 – Normen und ihre Rechtfertigung*. München: Fink.
Leder, G. C., Pehkonen, E., & Törner, G. (Eds.). (2002). *Beliefs: A hidden variable in mathematics education?* Dodrecht: Kluwer.
Maaß, K. (2004). *Mathematisches Modellieren im Unterricht – Ergebnisse einer empirischen Studie*. Hildesheim: Verlag Franzbecker.
Mayring, P. (1995). *Qualitative Inhaltsanalyse: Grundlagen und Techniken*. Weinheim: Deutscher Studien Verlag.
Neubrand, M. (Ed.). (2004). *Mathematische Kompetenzen von Schülerinnen und Schülern in Deutschland. Vertiefende Analysen im Rahmen von PISA 2000*. Wiesbaden: Verlag für Sozialwissenschaften.
Niedersächsisches Kultusministerium (Ed.). (2006). Kerncurriculum für das Gymnasium Schuljahrgänge 5–10 Mathematik Niedersachen. Hannover: Unidruck. http://db2.nibis.de/1db/cuvo/datei/kc_gym_mathe_nib.pdf. Accessed 10 September 2010.
Niss, M. (2000). Applications of mathematics "2000". Moments of mathematics education in the twentieth century pp. 271–284. http://www.mathunion.org/fileadmin/ICMI/files/ Digital_Library/Other_ICMI_Conferences_Proceedings/Proc_EM_ICMI_Symp.pdf. Accessed 7 August 2009.
Scheele, B., & Groeben, N. (1988). *Leitfaden zur Ziel-Mittel-Analyse (ZMA)*. Tübingen: Francke.
Stake, R. E. (2000). Case studies. In N. K. Denzin & Y. S. Lincoln (Eds.), *Handbook of qualitative research* (pp. 435–508). Thousand Oaks: Sage.
Stigler, J. W., & Hiebert, J. (1999). *The teaching gap*. New York: The Free Press.
Thompson, A. G. (1984). The relationship of teachers' conceptions of mathematics and mathematics teaching to instructional practice. *Educational Studies in Mathematics, 15*(2), 105–127.
Thompson, A. G. (1992). Teachers' beliefs and conceptions: A synthesis of the research. In D. A. Grouws (Ed.), *Handbook of research on mathematics teaching and learning*. New York: Macmillan.
Tietze, U.-P. (1990). Der Mathematiklehrer an der gymnasialen Oberstufe – Zur Erfassung berufsbezogener Kognitionen. *Journal für Mathematik-Didaktik, 11*(3), 177–243.
Tietze, U.-P. (1992). *Materialien zu: Berufsbezogene Kognitionen, Einstellungen und Subjektive Theorien von Mathematiklehrern an der gymnasialen Oberstufe, Band 1–3*. Göttingen: Universität Göttingen.
Tietze, U.-P., Klika, M., Wolpers, H., & Förster, F. (1997). *Mathematikunterricht in der Sekundarstufe II*. Braunschweig/Wiesbaden: Vieweg.
Wilson, T. P. (1973). Theorien der Interaktion und Modelle soziologischer Erklärung. In Arbeitsgruppe Bielefelder Soziologen (Ed.), *Alltagswissen, Interaktion und gesellschaftliche Wirklichkeit, Bd. 1* (pp. 54–79). Reinbek: Rowohlt.
Wilson, M. C., & Cooney, T. (2002). Mathematics teacher change and development. In G. C. Leder, E. Pehkonen, & G. Törner (Eds.), *Beliefs: A hidden variable in mathematics education?* (pp. 127–147). Dordrecht: Kluwer.
Zimmermann, B. (2002). Vorstellungen über Mathematik und Mathematikunterricht von Lehrerinnen und Lehrern verschiedener Schularten. *MU, 48*(4/5), 7–25.

Chapter 9
Secondary Teachers' Beliefs on Modelling in Geometry and Stochastics

Boris Girnat and Andreas Eichler

Abstract This chapter presents two combined qualitative studies on secondary teachers' beliefs on modelling in geometry and stochastics. The teachers' views on modelling, which are described in detail, differ considerably in both parts of mathematics from a pragmatic approach to modelling. In case of elementary geometry, a conflict with a traditional view on geometry is detected and elucidated. In case of stochastics, the need for data and real situations are revealed as controversial. The chapter ends with the invitation to analyse the parts of factual school mathematics including teachers' beliefs more specifically, that is, to compare applied-oriented aims with other didactical requests, and to design tasks which are supposed to be a response to the teachers' hesitations on modelling analysed before.

1 Teachers' Beliefs and Individual Curricula

"That what teachers believe is a significant determiner of what gets taught, how it gets taught, and what gets learned in the classroom" (Wilson and Cooney 2002). Based on this rationale, teachers' beliefs have become a vivid research focus of mathematics education (Philipp 2007). In this chapter, we will present the core results of two combined studies concerning secondary teachers' beliefs on applications in geometry and stochastics, respectively. The studies rest upon small samples (less than 18) and follow a qualitative methodology based on in-depth interpretations of semi-structured interviews. All the teachers consulted are employed at German higher-level secondary schools (Gymnasien).

The studies share the same theoretical framework and research question, namely, the reconstruction of teachers' individual curricula on teaching geometry and stochastics. Individual curricula are supposed to possess similar constituents and the same purpose as written curricula to guide the instructional practice to specific goals of

---

B. Girnat (✉) and A. Eichler
University of Education Freiburg, Freiburg, Germany
e-mail: boris.girnat@ph-freiburg.de; andreas.eichler@ph-freiburg.de

education. Intended curricula are the "blue prints" of individual curricula, that is, the teachers' instructional intentions whether they can be implemented in classroom practice exactly or not (Eichler 2007). Hence, our questions were directed to teaching goals, teaching methods, and the students' learning. The studies suggest that applied-oriented goals are seen as subordinate ones among others and that there are significant differences in elementary geometry, analytical geometry, and stochastics. These differences, but also some unsuspected similarities between analytical geometry and stochastics, lead to the decision to present these studies combined and to deliberate on the special position of elementary geometry.

2 Theoretical Background, Data, and Evaluation

The design of the interviews and the interpretation of the data are based on the *research programme of subjective theories* (Groeben et al. 1988), which is intended to reconstruct the background theories that professionals use to manage their job-related behaviour. According to this background theory, the interviews are designed as semi-structured ones. They start with open questions on the professionals' intentions and knowledge and lead to confronting questions, derived from literature, subsequently. All the interviews were held and transcribed by the authors. The participants were not chosen by specific criteria, but volunteered for the interviews in response to an impersonal invitation. In our cases, we began with open questions on the teachers' goal of education and confronted them with divergent opinions (cf. 3). To summarise, the interviews involve questions about (1) goals of the mathematics curriculum, (2) goals of the geometry and stochastics curriculum, (3) content of the geometry or stochastics curriculum, and (4) students' learning and teaching methods. The evaluation process is guided by a so-called *dialogue-hermeneutic methodology* (Groeben and Scheele 2001), which contains two steps: Firstly, the interpreter explicates the central subjective notions by "defining" paraphrases and links between them similar to a concept map and reconstructs the argumentative structure of each interview by a hierarchical diagram, containing top-level goals of education on the highest level and derivative goals, contents, and methods on lower levels (Eichler 2007, cf. Fig. 3 for an example). These diagrams are intended to express the implicit means-ends relations the teachers take for granted when structuring their classroom practice. Secondly, the paraphrases, concept maps, and diagrams are discussed with the teachers; and the teachers have the chance to approve, to dismiss or to change the researcher's suggestion. This is the dialogical part of the methodology, which is intended to enforce the reliability.

3 Applications and Model Building

We now describe the topics used for confronting questions on the applied-oriented aspects during the interviews. We assume that the main questions of teaching applied-oriented mathematics are as follows: What is the *relationship*

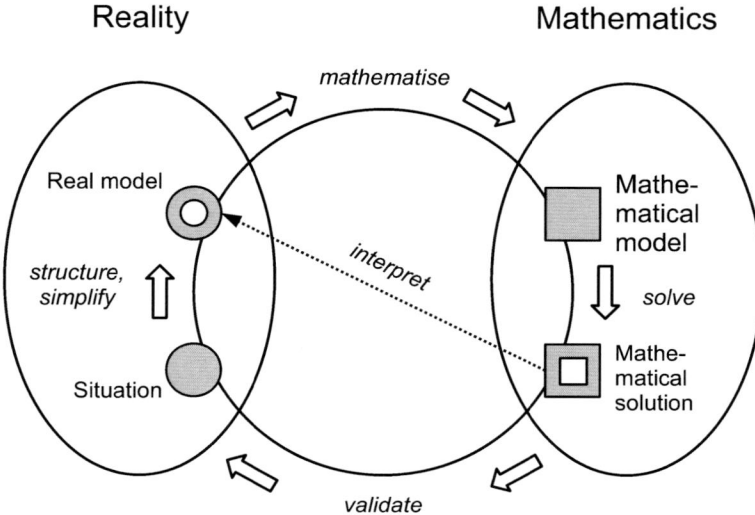

Fig. 9.1 Modelling cycle used in our studies

between *general* mathematical concepts or theories and *specific empirical knowledge* on singular situations (Kaiser 1995)? Is mathematics seen from a *static* or *dynamic* point of view (Hersh 1986)? In how far do the teachers' ideas match the concept of modelling, which is often seen as "one of the main components of the theory for teaching and learning mathematical modelling" (Kaiser et al. 2006, p. 82)?

The concept of modelling is typically explained by one of the common *modelling cycles* (Kaiser 1995). To leave the teachers room for personal perspectives, we used a simple version of these cycles (Fig. 9.1). In addition to *conceptual* topics, we were also interested in *normative* aspects. These opinions were analysed against the background of Kaiser-Meßmer's classification (Kaiser-Meßmer 1986): The extremities are seen in the *pragmatic* and the *scientific-humanistic* approach. Whereas the latter emphasises mathematical concepts, theories, and taxonomies, using real-world situations as subordinate tools to develop mathematical concepts based on manifold realistic associations, the pragmatic view stresses empirical knowledge and a reflection on the relationship between mathematics and reality on a meta-level: (1) Utilitarian aims: The real-world situation and the gain of empirical knowledge are taken seriously. (2) Methodological aims: It is a goal to achieve general competencies and meta-knowledge about applying mathematics. (3) Scientific aims: Applied mathematics is to be perceived as model building, which includes reflections on modelling and an introduction of its basic concepts in classroom practice.

We were interested in localising the teachers' standpoints in this area of tension and in finding reasons *why* a teacher prefers one or the other position by posing questions derived from the topics above during the interview. The questions are not quoted here literally, since they vary from interview to interview in some minor details according to the open structure of the method, which provides questions merely as adapted responses to the teachers' former statements. In terms of intended curricula, this task consists of in reconstructing the location of applying mathematics within the teachers' intended curricula and in revealing connections and conflicts with other goals.

4 Geometry

The study on geometry consists of nine interviews. We refer to the corresponding teachers by the letters A to I. The findings on elementary geometry, taught from grade 7 to grade 10, differs from the ones on analytical geometry, taught from grade 11 to grade 13. Hence, we present them separately.

4.1 Elementary Geometry

Seven of nine teachers express a seemingly paradoxical opinion: They regard geometry as an applied part of mathematics par excellence, but not as very suitable for model building, though being open-minded about modelling in other parts of mathematics.

Mr. A: I think, the better applications can be found in algebra or stochastics, per cent calculations, linear optimisation. It is important to get a deeper insight into reality by modelling. In geometry, there are such things as dividing a pizza by a compass. I saw a trainee teacher do so. That's ridiculous.

Mr. B: Geometry as a tool to get access to the real world is not in the first place, and it is rightly not in the first place. An application is useful to introduce a new subject, to legitimise it, and to test the competencies of this field by realistic tasks in the end. But in between, a lot has to be done without any reference to the real world, detached from these accessory parts which are not important to the mathematical model.

Mr. F: Applications are motivating, but it is important to me that my pupils also switch to an abstract level, practise pure geometry. In order to do so, concrete figures, measuring and so on are rather obstacles than aids.

Mr. C: If someone asserted in [the] case of the Pythagorean Theorem "Proved by measuring, the theorem holds", then something of value would disappear, something which is genuinely mathematical. ... If geometry just consisted of measuring, calculations, drawing, constructing, and land surveying, then I would regard it as poor.

Mrs. G: Besides proof abilities, problem solving is in fact the most important thing I want to convey in my lessons on geometry.

Table 9.1 Differences between modelling and proving or problem solving

	Model building	Proving/problem-solving task
Object of interest	Singular situation	General theorem or configuration
Access to objects	By measurement/experience	By construction descriptions
Building a real model	By simplification	Not allowed
Mathematical treatment	Inventing a mathematical model	Using known operators
Validation	Empirically	By deductive arguments

Summarising these quotations, our teachers do not see mathematics education from a *comprehensive applied-oriented approach*, but as *split* into the common disciplines of school mathematics. Insofar, applied-oriented goals are not top-level aims, but have to find their places within the *local curricula of the particular disciplines*. The range of goals is occupied by several categories, mainly abilities in proving, defining, problem solving, and constructing. Applying geometry is only a further goal among others; and deduction and problem solving are seen as the main objectives of geometry than getting "access to the real world". Insofar, certain unease about teaching geometry applied-oriented arises from the various goals of education to handle in conjunction. Additionally some conflicts go deeper and are bounded to a *classical Euclidean view* on geometry (Girnat 2009a): Even though geometry is applied, the justification of every assertion has to be done purely deductively on known axioms and theorems, whereas referring to experience is regarded as a sign of a deficient understanding. Hence, some essential parts of a modelling cycle are in contrary to the settings of a proving or problem-solving task (cf. Holland 2007, pp. 170–195) (Table 9.1).

It may be comprehensible to avoid geometrical applications "in between" to prevent students getting confused by different standards and challenges of modelling, proving, and problem solving. This challenge is suspected to be unique to geometry, since it seems to be the only part of school mathematics which allows regarding its objects "naturally" from two different perspectives (Girnat 2009a): from a theoretical Euclidean point of view and from a more empirical perspective of modelling. Since teachers have to fulfil both of them, the academic debate is requested to state an answer on how to deal with these disparities.

As a further finding, it is interesting to see what most of our teachers perceive as "good" geometrical applications. Typically, the examples possess a two-step structure: In the first step, geometry is used to calculate some boundary conditions, for example, some lengths, areas, or volumes. Afterwards, these values are committed to a second step, which normally includes a non-geometrical question, for example, some price, weight, or velocity calculations. Especially Mr. A mentioned that the interesting insights primarily arise in the second step. Even in case of optimisations (when a geometrical value is adjusted afterwards), the issues and structure of model building typically arise only in the second step, whereas in the first one, the geometrical background is taken for granted. Here, a static

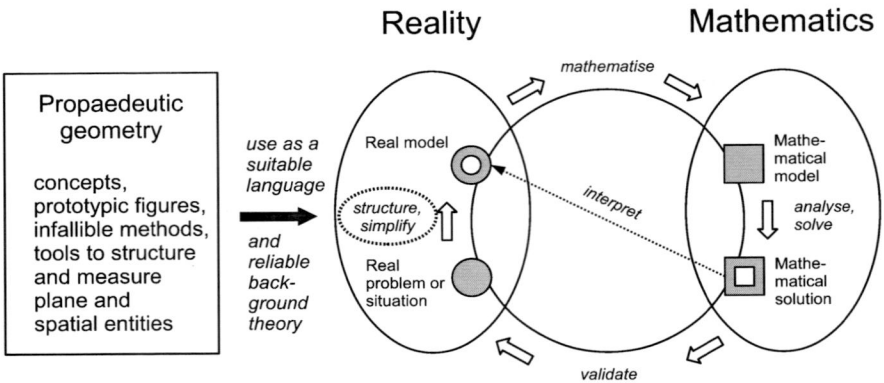

Fig. 9.2 Geometry as propaedeutic to model building

view of mathematics is predominant, forming geometry as being "propaedeutic" to model building. Insofar, this view of modelling can be illustrated as follows (Fig. 9.2).

This observation is interesting for two reasons: It could give some advice to manage the disparities between modelling and a Euclidean view on geometry: In joining both steps, a static and dynamic view and a pragmatic and scientific-humanistic approach can be combined in the same task. The reason why we call this use of geometry propaedeutic and why this function cannot be integrated into the modelling cycle under "mathematise" is as follows: Geometrical concepts and theorems are already used to structure and to simplify the real situation, that is, to build the *real* model. Hence, they are prior to any kind of mathematisation in the sense of the modelling cycle. This observation seems to be unique to geometry, since geometrical terms are part of the vocabulary we naturally use to describe the objects surrounding us and, therefore, geometry has a different, and quasi-unavoidable reference to reality, more than other parts of mathematics (Girnat 2009b). Thus, it is questionable if it makes sense to distinguish between a real model and a mathematical model or even to use the word "model" at all as far as geometry alone is concerned.

4.2 Analytical Geometry

Goals of education are manifold in elementary geometry, and, hence, our teachers' opinions cannot to be described by a single model. On the contrary, in the case of analytical geometry (AG), it is possible to present a *single* curriculum which eight of nine teachers possess. The basic structure can be described by a hierarchical diagram of educational goals (Fig. 9.3).

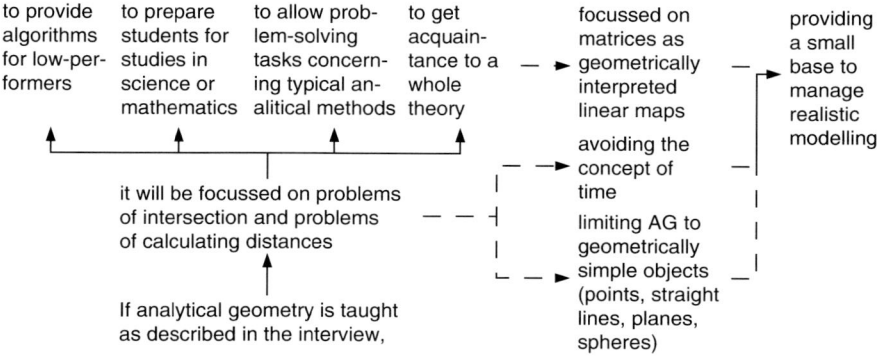

Fig. 9.3 Main aspects of the predominant intended curriculum in analytical geometry

The uniformity may be enhanced by the fact that the school leaving certificate is more standardised than the examinations in the lower secondary school. Nevertheless, the limitation to issues of intersection and distances and the focus on problem solving is openly approved:

Mr. B: Analytical geometry, that is just the metric Euclidean geometry: relative positions, calculating angles, distances, intersections.

Mr. C: Calculating distances without knowing the methods completely and, finally, inventing a formula to calculate distances, these are the things the focus has to be on.

In Fig. 9.3, the disadvantageous consequences of such a curriculum on applied mathematics are marked by dotted lines, forming the "trilemma" of application in analytical geometry:

1. If AG is taught as described in Fig. 9.3, it will be difficult to find realistic applications.
2. If AG is enriched by the concept of time and some basic physical theory instead, most of our teachers will regard AG as unfair to pupils who have not chosen physics and to teachers who do not teach science as a second subject.
3. If AG is enriched by parts of linear algebra which are not interpreted geometrically, but as tools of social and biological science instead, most of our teachers will regard AG as mathematically too simple or will fear an inappropriate restriction of the "real core" of AG or will accuse these applications as being not realistic, since "not everything in the world is linear" (Mr. A).

In contrast to elementary geometry, the main obstacles for applications do not bear on two opposing approaches to geometrical objects, but on the focus on problem solving and on preparing pupils for academic studies.

5 Stochastics

The settings of the second study are equal to the first one, but rest upon 17 interviews. The teachers are denoted by a to q. In contrast to geometry, the teachers' intended curricula are more diversified, and all the possible combinations of Kaiser's classification are instantiated, leading to the following prototypes (Eichler 2007): Shown in Table 9.2.

Both types of the humanistic-scientific approach provide statements which are familiar from the geometrical part of this chapter: Traditionalists approve insights into formalism and mathematical theories; structuralists tend to stress problem solving. The only difference is the fact that these opinions are not supported by a traditional educational theory, as it exists in the case of a Euclidean view on geometry and provides some kind of legitimisation to these points of view. Much more interesting are the more pragmatic types, which are in principle open minded to an applied-oriented approach, including "real" model building. But two of Mrs. f's tasks and Mr. d's comments are indicators for a different interpretation:

Mrs. f: Task 1: "In a German city, 30% of the population are infected with HIV, …"; task 2: "The probability of a hamburger having two slices of tomatoes is 10%. In case you buy three hamburgers, …"

Mr. d: And that's what I am trying to illustrate here as well, that you get models of approach this way, but of course become better afterwards … that there are quite often problems you can solve with maths, … that students are enabled to categorise mathematical models better.

Similar to this illustration, even the pragmatic teachers of our study differ from some essential properties of the applied-oriented approach mentioned above: At first, some of the teachers, like Mrs. f, take empirical knowledge on a specific situation not very seriously and replace real data by partly ludicrous dummy data, starting the modelling cycle at a simulated, not realistic "real model" and taking the "recognition of the need for data" as an essential topic of the current debate on stochastics ad absurdum (Wild and Pfannkuch 1999).

Table 9.2 Prototypes of teachers' intended curricula

	Static	Dynamic
Humanistic-scientific	*Traditionalists*: establishing a theoretical base, including algorithmic skills and insights into the abstract structure of mathematics, not involving applications.	*Structuralists*: encouraging students' understanding of the abstract system of mathematics in a process of abstraction, starting from applications.
Pragmatic	*Application preparers*: making students grasp the interplay between theory and applications (first theory, then applications).	*Every-day-life-preparers*: developing statistical methods in a process, making students cope with real stochastic problems and criticise them.

Furthermore, as seen in Mr d's quotation, the methodological aims are partly turned into their opposite: There is no process which consists of analysing a situation and *inventing* a fitting model. Instead, there are some *pre-established* models, and the students' task is to recognise properties of the situation in order to *choose* an "adequate" model and, then, to work the chosen one over for deepening the *mathematical* understanding of this model. Why this model may be empirically adequate is not discussed. For these reasons, "absurd" data are sufficient as "model indicators"; but exactly the aspect of *building* a model and evaluating its *empirical* relevance is suspended, which leads to omitting its most important meta-scientific feature: Building a model is typically not determined into one direction; but just the insight that there are many possibilities to treat a situation mathematically and that there is no "unique solution" is avoided by several teachers. As a result, the intended or unconscious scientific aim of such an approach is not perceiving applied mathematics as model building, but as choosing fitting operators (in disguise of known models) in the sense of problem solving. Only a few teachers mentioned some data-related aspects, like Mr. d, connected to real situations in a process where mathematics is seen as a tool to describe the world. But even in these cases in which developing mathematical methods is not the primary goal in itself, but as a tool to enable students to cope with real problems, the process of building a model is not detectable. Overall, this is an interesting consequence, also perceivable in geometry: Scientific aims of applying mathematics are typically not pursued on an *abstract level* (like the process of model building as a *general* approach to applied mathematics), but on more concrete ones which are *bounded to specific disciplines*: In stochastics, that means the selection of the fitting model (like the correct urn problem or the adequate average); in geometry, there are problems of measurement, choosing an adequate formula, or dividing an object into known figures.

6 Conclusions

Our studies underline the importance on empirical investigations of teachers' beliefs: Although our teachers try to match the same written curriculum, the outcome differs considerably. The focus on intended curricula has served as a useful tool to reveal the reasons why the written curriculum is interpreted differently. These findings are not only a preliminary work to design representative studies on larger samples, but highlight some crucial topics worth discussion: In case of elementary geometry, model building is in conflict with aspects of traditional approaches to geometry and with educational goals of proving and problem-solving tasks. These oppositional requests have to be clarified in the academic debate and to be balanced for a realisable combination in practice (cf. 2.1). In case of stochastics and analytical geometry, the main question is: Do we have convincing rationales for emphasising modelling in a strict sense instead of only using mathematical applications to motivate and illustrate mathematical content? Our studies namely suggest that teachers mostly plan their lessons in view of the mathematical content and think in separated mathematical subdisciplines, leading to a preference for content specific, purely mathematical problem solving or even schematic tasks (cf. 2.2 and 3). This invokes two challenges: Firstly, it seems advisable to consider the various parts of school mathematics more differentiated and to integrate and balance didactical requests which are not focussed on applications. Secondly, it poses the question if there are really convincing examples which are both realistic applications and fruitful occasions to establish a broad theoretical background of mathematics for every discipline (i.e., algebra, geometry, stochastics, and analysis), for every grade, and for proving and problem-solving tasks or if the teachers' hesitation indicates a lack of mathematically rich applied-oriented tasks.

References

Eichler, A. (2007). Individual curricula – Teachers' beliefs concerning stochastics instruction. *International Electronical Journal of Mathematics Education* 2(3). Online http://www.iejme.com/.
Girnat, B. (2009a). Ontological belief and their impact on teaching elementary geometry. In M. Tzekaki, M. Kaldrimidou, & H. Sakonidis (Eds.), *Proceedings of the 33rd IGPME Conference* (Vol. 3, pp. 89–96). Greece: Thessaloniki.
Girnat, B. (2009b). The necessity of two different types of applications in elementary geometry. In *Proceedings of the 6th CERME Conference,* Lyon Online http://ermeweb.free.fr/.
Groeben, N., Wahl, J., Schlee, D., & Scheele, B. (1988). *Das Forschungsprogramm Subjektive Theorien (The research programme of subjective theories)*. Tübingen: Francke Verlag.
Groeben, N. & Scheele, B. (2001). Dialogue-hermeneutic method and the "research program subjective theories". *Forum: Qualitative Social Research,* 2(1), Art. 10, http://nbn-resolving.de/urn:nbn:de:0114-fqs0002105.
Hersh, R. (1986). Some proposals for revising the philosophy of mathematics. In T. Tymoczko (Ed.), *New directions in the philosophy of mathematics* (pp. 9–28). Boston: Birkhauser.
Holland, G. (2007). *Geometrie in der Sekundarstufe (Geometry in lower secondary schools)*. Hildesheim und Berlin: Verlag Franzbecker.
Kaiser, G. (1995). Realitätsbezüge im Mathematikunterricht – Ein Überblick über die aktuelle und historische Diskussion (Reality relationship in mathematics education – A survey on the contemporary and historical debate). In G. Graumann, T. Jahnke, G. Kaiser, & J. Meyer (Eds.), *Materialien für einen realitätsbezogenen Unterricht (ISTRON)* (pp. 66–84). Hildesheim/Berlin: Franzbecker.
Kaiser, G., Blomhøj, M., & Sriraman, B. (2006). Towards a didactical theory for mathematical modelling. *Zentralblatt für Didaktik der Mathematik,* 38(2), 82–85.
Kaiser-Meßmer, G. (1986). *Anwendungen im Mathematikunterricht (Applications in mathematics classrooms)*. Bad Salzdetfurth: Franzbecker.
Philipp, R. A. (2007). Mathematics teachers' beliefs and affect. In F. K. Lester (Ed.), *Second handbook of research on mathematics teaching and learning* (pp. 257–315). Charlotte: Information Age Publishing.
Wild, C., & Pfannkuch, M. (1999). Statistical thinking in empirical enquiry. *International Statistical Review,* 67(3), 223–265.
Wilson, R., & Cooney, T. (2002). Mathematics teacher change and development. The role of beliefs. In G. Leder, E. Pehkonen, & G. Törner (Eds.), *Beliefs: A hidden variable in mathematics education?* (pp. 127–148). Dodrecht: Kluwer.

Chapter 10
Examining Mathematising Activities in Modelling Tasks with a Hidden Mathematical Character

Roxana Grigoraş, Fco. Javier Garcia, and Stefan Halverscheid

Abstract Modelling tasks without numbers break with the usual mathematical contract on modelling tasks. At the same time, the approach provokes modelling actions by making it less evident to employ standard procedures in mathematics. This partial rupture of the didactical contract is analysed with the help of the Anthropological Theory of Didactics. Having established the a priori analysis for certain tasks without numbers, the theory of epistemic actions is used to describe at a micro-level modelling actions which appear in this setting.

1 Introduction

Mathematising activities take place while modelling contextual situations by mathematical means. These mathematising activities can emerge without the engaged persons, in our case, the students, being aware of the processes taking place during their work, in order to get to a solution of the initial problem. It is therefore even more desirable to obtain a better understanding and analysis of students' mathematical behaviour in solving modelling tasks. Modelling engages many processes, and mathematising is the crucial one, if we want students to become independent modellers. In general, the didactical contract that rules modelling in school is directing students to a specific mathematical topic, usually previously introduced or dealt with in the classroom. Therefore, the process of mathematising is often eased by

R. Grigoraş (✉)
University of Bremen, Bremen, Germany
e-mail: roxana@math.uni-bremen.de

F.J. Garcia
University of Jaén, Jaén, Spain
e-mail: fjgarcia@ujaen.es

S. Halverscheid
University of Göttingen, Göttingen, Germany
e-mail: sth@uni-math.gwdg.de

the fact that students are implicitly pointed to go in a particular direction. But this facility can have as effect a superficial treating of the problem, or even hinder developing modelling competencies.

The present study positions itself in the modelling interest area and more exactly in the meta-perspective segment, that is, where the cognitive processes of students while modelling, are addressed. Other authors opted for similar cognitive approaches, for example Galbraith and Stillman (2006), where attention is drawn on the kinds of mental activity that the individuals have engaged in during transitions between real and mathematical world in the modelling process. Matos and Carreira's (1995) research stresses learners' cognitive processes. A particular modelling cycle perspective is offered by Crouch and Haines (2004). It regards the basic stages in modelling, namely transition from real world to mathematical world, formulating and working with a mathematical model, then moving back from the mathematical model to the real world. Borromeo Ferri's (2007) view lies probably closest to our micro-level perspective of looking at the process; nevertheless a categorisation in terms of thinking styles or modelling routes is, so far, beyond the scope of our study. In this paper, whilst drawing on these previous results, we will go further, through a combination of different theoretical frameworks, in order to find answers to questions like the following: How does the formulation of a modelling task without numbers influence the degree of mathematics that students will use to handle the problem? If comparing results coming from slightly different task formulations, which phase(s) of mathematisation can be identified as making the distinction in students' approaches? Are there other reasons which make students treat this particular kind of modelling tasks in a specific way? How do they decide using mathematics?

2 Framework

Two main different frameworks have been used for our case study. In this paper, modelling is mainly conceived as a cyclic process where continuous transitions between mathematics and the rest of the world occur (for a detailed description, see Niss et al. 2007). For a deeper empirical analysis, in this work we will consider, on the one hand, a structural approach, which was developed to identify epistemic actions within the modelling cycle (Halverscheid 2008). This will provide us with a tool for visualising students' utterances. On the other hand, the Anthropological Theory of Didactics (ATD) (Chevallard 1999) will help us grasp more about the performed experiments, from an institutional perspective, and provides some tools for explaining how the mathematising process emerges.

2.1 ATD and Modelling

The most specific particularity of the ATD is that mathematics is considered as a human activity. As any other human activity (Chevallard 1999), it is modelled in terms of praxeologies. A mathematical praxeology is basically made up of two parts, the praxis (practical, the 'know how' block) and the logos (theoretical, the 'know why' block), each of the parts consisting of two components. The practical block includes the type of tasks and techniques for solving them, while the theoretical block contains the technology, which justifies the technique, and the theory, which testifies the technology itself (Chevallard 2007).

In the ATD framework (Bosch et al. 2006a), doing mathematics and learning mathematics are considered in an integrated way. The *process of study* tries to capture the nature of mathematical praxeologies both (A) as a process and (B) as the result of this process. In every process of study, some specific *moments* can be potentially identified, called the *didactic moments*, which shape the dynamic nature of any mathematical activity (see Fig. 10.1).

There are therefore two aspects with respect to the didactic praxeologies (Bosch et al. 2006b): first one (A) is the process of study, or the mathematical construction, and the second one (B) is the result of this construction, which in turn is the mathematical praxeology itself. Didactical praxeologies aim at creating the conditions for the mathematical praxeologies to emerge and evolve. Figure 10.1 visualises the relation between mathematical and didactic praxeologies.

Artaud (2007) uses the expression 'mixed praxeologies' when there is a praxeology that involves extra-mathematical elements. Given that the modelling process, by its nature, is made up of intra- and extra-mathematical elements, we could say that each time modelling occurs, and therefore mathematising processes with it, mixed praxeologies are activated.

> Praxeology always arises as an answer A to a question Q; when … question Q is a non-mathematical one, the answer A, if including mathematics, is a mixed mathematical praxeology, that is a praxeology in which mathematics is mingled with the 'real world area' that can be other academic science (Artaud 2007, p. 373).

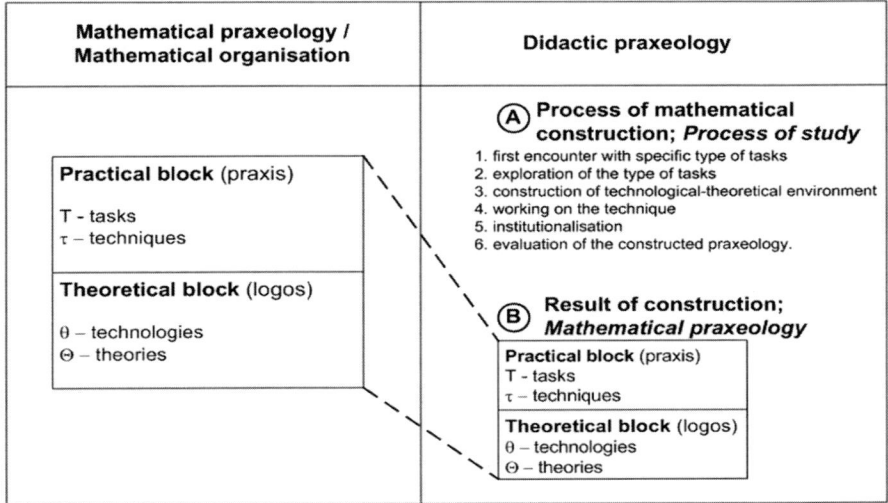

Fig. 10.1 Mathematical and didactic praxeology in a nutshell

Modelling is then the process making possible to come from the initial question Q to a mathematical question Q_M, to which an answer A_M (which could be well-known) will be produced. Then a mathematical praxeology will be brought out or might be generated by using didactic praxeology.

2.2 Structural Analysis

The structural analysis involves actions that are observable empirically. These observed actions are structured according to the role the experiments play in these. The structural analysis divides the actions in 'the rest of the world' and the 'mathematical world' according to the basic modelling framework (see, e.g. Blum and Niss 1991). Data will be encoded in a diagram which structures the processes according to the mathematical modelling framework. The diagrams are organised along the timescale (see Figs. 10.5 and 10.6).

If an epistemic action concerns the nature of a real setting and not primarily its model in the mathematical world, it is considered as belonging to the 'rest of the world'. For instance, this is the case if the results of an experiment are described or if data in the experiment is discussed. Then, the action is depicted as a straight vertical line on the left-hand side. Analogously, actions which work out mathematical problems within a mathematical model are placed as straight dashed vertical lines at the right-hand side of the diagram.

In addition, there are actions which link the mathematical model and the real world. These actions are placed in the centre of the diagram. If a mathematical model is developed from a real model, this is described by a line from the upper left to the lower right. If mathematical results are interpreted in a real model or in a real situation, respectively, this is expressed by a line from the upper right to the lower left (Fig. 10.2).

It is obvious that mixed forms of this will appear. We do not introduce an intermediate category for those actions which cannot be definitely classified to either of the two worlds, that is the dichotomy of the modelling framework is maintained. Either an action is classified to 'the rest of the world' or to 'the mathematical world'. In graphical illustrations of the modelling cycle, the 'rest of the world' is by tradition sketched on the left-hand side and 'the mathematical world' on the right-hand side. The lines move back and forth from the left-hand side to the right-hand side. Actions on the same topic are represented by a continuous path.

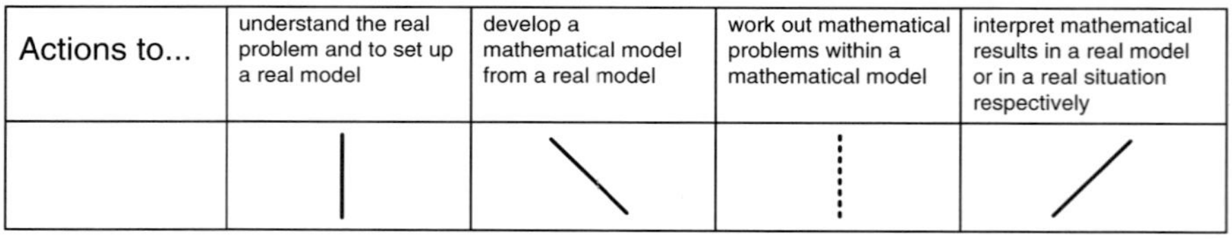

Actions to...	understand the real problem and to set up a real model	develop a mathematical model from a real model	work out mathematical problems within a mathematical model	interpret mathematical results in a real model or in a real situation respectively
	|	\	┊	/

Fig. 10.2 Symbols for different types of actions

3 Empirical Design

The present study is part of a wider research, where various tasks with a hidden mathematical character have been previously experimented. Because the space is limited, we will be focusing on the 'Mars task'. The research includes videotaped sessions, post-interviews and sometimes questionnaires. Three variants of the following assignment were the basic material for both bachelor students (pre-service mathematics teachers) and 8th graders who worked on it.

> An astronaut is sent to Mars. His landing place is where the square can be seen. His mission is to drive on the surface of Mars and to investigate the craters. From the Earth, you can pilot his mobile-station. In which order would you let the astronaut study the craters? Find out which aspects are to be considered. Keep in mind that the astronaut should come back to the landing-place (square as it can be seen in Fig. 10.3), from where he is supposed to fly back to the Earth.

A second variant of this task was different through the fact that it lacked the requirement 'Find out which aspects are to be considered', whereas the third variant was exactly as the second one, but it was printed out on millimetre paper, before being handed to students.

Teams of two to three volunteer students were videotaped and were given about 50 min at their disposal for working on different variants of the tasks, every team worked on a single variant, not being aware of other existing versions. Through the team work, the so-called peer mediation (Chevallard 2007) was intended to be stimulated, this going hand in hand with the 'diffusion of knowledge' which didactics is supposed to be, as ATD sustains. The 8th grade students had no special experience with modelling tasks, though they were occasionally asked to solve

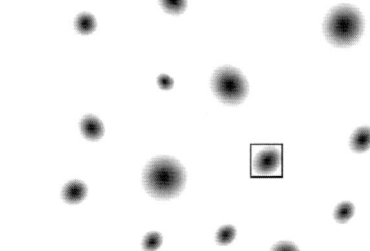

Fig. 10.3 Craters land on Mars surface

such type of problems before, whereas the bachelor students turned out to have had met and quite intensively worked on modelling tasks previously. Teachers had to be there for eventual clarifications of task formulation matters, but not giving hints as to how to solve it. Each student was in his (her) usual environment, the 8th graders in the school classroom and the mathematics students in the university laboratory.

4 A Priori Analysis

The given task offers a problem embedded in a situation that is open enough to let students develop different mathematical techniques. The generative question is: Given a set of craters in Mars, which is the best (optimal) way to visit those, considering that a starting and ending point are given? Various ways of structuring the Mars situation are conceivable for giving rise to diverse mathematical praxeologies, depending on how the word 'best' is interpreted. In an a priori analysis, two different activities can be considered. We will refer to them as a topological activity and a metric activity.

Figuring that 'best' means the possibility of going through every crater without visiting one twice or more times, a topological activity can be considered as an enumeration-like activity that would activate graph theory techniques. The picture of Mars' craters can lead to different graphs, each crater being a vertex and an edge being a path between two craters (similar to the famous Königsberg problem). In this context, 'optimal' can be viewed as finding a path which does not repeat a crater (except the starting and ending point, which is the same), visiting all of them. 'Optimal' could also be interpreted by the students as finding the path for which the number of visited craters is minimal. In this way of modelling the situation, distance is not taken into account, which makes the model interesting, but somehow weak. Exploring two distinct solutions for the task, one could be assessed better than the other, if fewer craters are repeated. But if both repeat the same number of craters (and obviously, do not omit any crater, and start and finish where the square is located), there is no mean to compare them. From the logos, graph theory and topological properties are involved.

Combined with the topological activity, a metric activity will examine not only the possibility of visiting all the craters, without repeating, but also the measure of the distance between craters. That makes the task even more complex: topological techniques have to be combined with metric ones. 'Optimal' means now not only to cover all the craters, but also with a minimal distance. In terms of the modelling process, the impossibility of the first praxeology to compare various solutions leads to the necessity of new tools (praxeologies of increasing complexity). Now there are new ways to determine whether one solution is better than another, including those cases that could not be contrasted in the topological activity. From the logos, also measure theory elements are involved.

5 Findings

Several bachelor student teams, as well as 8th graders were examined, but only a few of their results turned out worthwhile to be studied. On a macro-level and in a comparative view, it cannot be said that a considerable background knowledge progress was observed, at least not concerning the mathematical methods and algorithms used. The young students proceeded basically in choosing various paths, which were afterwards compared with respect to the minimal distance contained in it. The bachelor students have approached the given task, and hence have generated the corresponding mathematical problem, in a rather more mature manner than the young students, but did not make use of clearly stated mathematical concepts. Particularly interesting in their solution was the noticeable awareness that they were working on a mathematical modelling task, and knew exactly what modelling implies: simplifying, making assumptions, excluding, validating against reality.

Assumptions turned out to be modelling specific elements which clearly differentiated the students' solving approach: younger students have made very few and just tacit assumptions, whereas the bachelor students turned out to have made substantially more assumptions, most of them being explicitly stated. This helped them a lot while modelling, gave a mature style to their work behaviour and qualified them, in turn, as experienced modellers.

When being investigated with ATD tools, the mathematical praxeologies of the analysed transcripts of two teams look like in Fig. 10.4 (one pre-service mathematics teacher – left column, the other one of the 8th grade students – right column). As can be seen in their table form representation, their results differ in the techniques and technologies they have made use of. No real difference in complexity of the mathematical levels is showing between the teams, although the younger students followed a somehow simplistic and straightforward approach. The pre-service mathematics teachers seemed to be formed to go for elaborated mathematical solutions, but superior mathematics did not provide them the actual tools to solve the problem, in this case. Many Bachelor students mentioned the Dijkstra algorithm as being able to help for finding the optimal path, but none of the teams succeeded in practically applying the algorithm, so it remained at the stage of suggestion, or, in the best case, a good idea.

In the Figs. 10.5 and 10.6 it can be seen how the separation between the real world and the mathematical world (see details in Grigoraş and Halverscheid 2008) is depicted, with the aid of the epistemic actions within the modelling process. The topological and metric activities from our a priori analysis have occurred, as expected, in a combined manner. Students were arguing about how to follow the suggested path, first, the outer circle (the imaginary circle the craters points are circumscribed to), then the inner circle. Then they compared this idea with an eventual zigzag pattern, which seemed to have been unfavourable, concerning the optimality criteria.

Passages from students' transcripts, as in Fig. 10.5, where the entire path is 'real world → passing to the mathematical world → working in the mathematical world → then back to the real world', are occasional occurrences. It was observed that bachelor

Practical block (praxis)	It integrates types of (problems and) tasks T and techniques τ to solve them.	**Task (min1):** Finding the optimal (the term will be defined by the solvers and treated according to the particular meaning given) way between a number of 'points' (places of slightly different shape, geometrical-like)	
		Techniques	
		Choosing basically two paths, then by means of comparison, selecting the 'best' one (path 1): 1. the given 'points' imagining as lying on an inner and an outer circle, the path being the route covering the outer circle, followed by the inner one (min 02:23) 2. imagining the 'points' as lying on parallel lines, which can be covered one by one, from left to the right (min 02:53)	Choosing basically two paths in form of a circle (min 02:22), then by means of comparison, selecting the 'best' one ('from inside to outside, in circle' – min 04:06), meaning 'the fastest' (min 03:38), 'the most direct' (min 03:44), 'the shortest' (min 07:58)
Theoretical block (*logos*)	It integrates the **technologies** θ and **theories** Θ used to describe, explain and justify the practical block.	**Technologies**	
		Contain properties of a circle, respectively calculating the circumference of the circle (path 1, min 10:42), as well as measuring the length of the parallel lines (path 2, min 15:17)	Measuring distances between neighbouring craters with the ruler (min 05:32), then summing up, and deciding for the minimal total distance (min 14:56)
		Theories: Measuring	

Fig. 10.4 Mathematical praxeologies constructed from bachelor, respectively, 8th grade students' transcripts

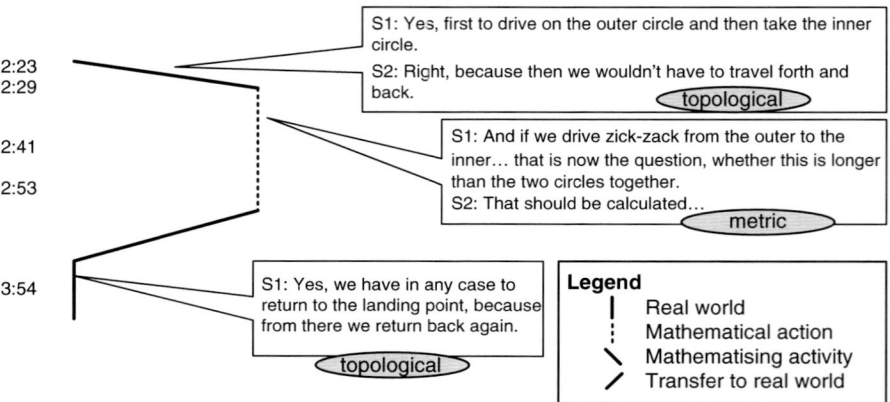

Fig. 10.5 A priori ATD analysis reflected in the epistemic actions by bachelor students

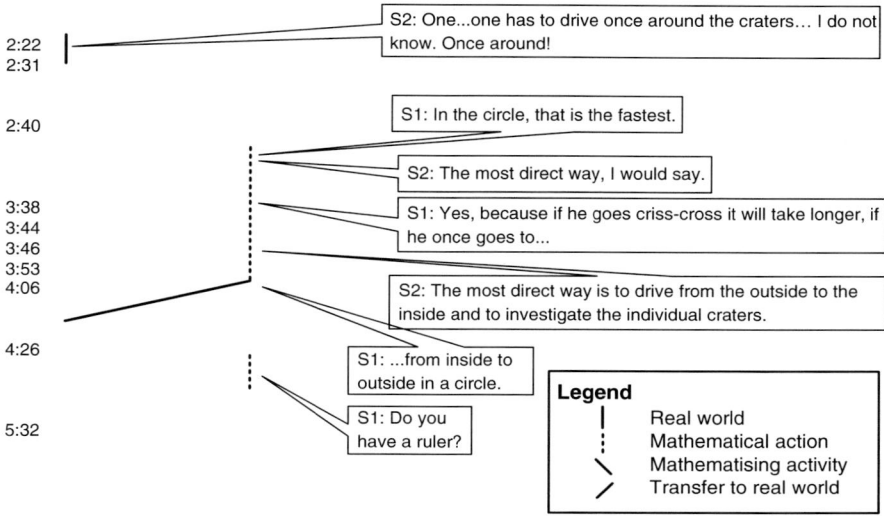

Fig. 10.6 Epistemic actions by the 8th grade students

students delivered considerably more such complete and continuous paths than the younger students, where the transfer from one world to the other is sometimes done instantaneously. In younger students' transcripts, fewer paths between the worlds have been found (Fig. 10.6), but single instances either in the real or in the mathematical world. It is questionable whether the modelling experience played a definite role in such path discontinuities, or the task itself does particularly stimulate a thorough sequel from one modelling environment to the other, including the tracks in between.

6 Summary and Conclusions

It was expected that the three variations of the same task will have as effect different levels of mathematising in students' solutions. Unlike this prediction, different teams of students working on different variants of the task delivered solutions where no indication about this distinction was found. Nevertheless, other interesting aspects came out, like for example, background and modelling experience led to a significantly deeper approach and more sophisticated solution, where the used terminology (e.g. 'modelling means simplifying and excluding', 'we would simply neglect the minimal deviation', 'we went so long through the modelling cycle, so that it cannot be done anymore'), as well as the solving strategy denoted the team of bachelor students as experts in modelling, even though their solution was not a perfect one. The present study showed that, it could be the case, as our bachelor students sample turned out to show, that some solvers know very well that they are modelling, and therefore proceed with the whole 'modelling arsenal' they possess.

ATD helped in organising the solving strategy in such terms, so that institutional items can be understood and checked for, as can be seen in Fig. 10.4. Using the epistemic actions within modelling processes, a conceptual and visualisation tool was developed, yielding a scheme as seen in the Figs. 10.5 and 10.6, where students' utterances hold as mathematising indicators.

The structural analysis scheme is used to gather and structure empirical data within the modelling framework. As an educational target, conforming with the view of learning as 'learning from the situation', it is hoped that our specific kind of tasks encapsulates potential knowledge (Chevallard 2007) and carries the students through those situations where they can learn. Since modelling situations are very complex and since it is not yet evident what modelling knowledge is, the structural analysis was used to identify relevant actions in the framework. Deeper analyses of the modelling situations are necessary to understand the interplay of tasks, techniques, technology and theories in modelling situations.

In the ATD, mathematising is not identified as a separate process, but it is inserted inside the praxeology. Mathematising is not considered as a general process (involving some cognitive differentiated schemas), but as a part of the process that the praxeology is describing in itself. It is different to try to mathematise a situation than when trying to work inside the model. From the ATD perspective, what is different here is the kind of problem the solver faces and the techniques used for solving. That is, what is different is the kind of praxeology, but not the theoretical construct.

For a better integration of the two theoretical frameworks, on one side, the existing modelling cycles together with transitions between worlds, and on the other side, the ATD with its mathematical and didactical praxeologies, we currently see as meaningful a correspondence between the steps in the modelling cycle and the six stages of the didactical praxeologies that could be built up. Such a mapping could serve as an analysis tool of what children at a certain age know, and more importantly do not know for solving certain tasks and problems coming from the real world and not having obvious mathematical character. On this basis, the process of didactical transposition (institution → teacher → student) could be improved, by considering the weaknesses and enriching the teaching model, more precisely, the mathematical praxeology to be used in order to meet the students' mathematical knowledge level.

As for a teaching practice suggestion, modelling tasks where no numbers are given and apparently nothing to be calculated is asked for, appear to break up the classical mathematical culture of the classroom. The emphasis falls on elaborating strategies, learning to think, and in fact the biggest challenge is how the real situation comes out to be translated in a mathematical problem, and not the mathematical solving itself. Therefore, it would probably be good to contemplate, so that this kind of task could become employed as a teaching method, too. Students would benefit for sure, at least from two viewpoints, one being the elaborate and systematic plan of action which they are demanded to develop, as well as the immediate gain offered by mathematics, even in situations which do not seem to ask for applying it.

References

Artaud, M. (2007). Some conditions for modelling to exist in mathematics classrooms. In W. Blum, P. L. Galbraith, H.-W. Henn, & M. Niss (Eds.), *Modelling and applications in mathematics education. The 14th ICMI study* (pp. 371–378). New York: Springer.

Blum, W., & Niss, M. (1991). Applied mathematical problem solving, modelling, applications, and links to other subjects – State, trends and issues in mathematics instruction. *Educational Studies in Mathematics, 22*(1), 37–68.

Borromeo Ferri, R. (2007). Personal experiences and extra-mathematical knowledge as an influence factor on modelling routes of pupils. In D. Pitta-Pantazi & G. Philippou (Eds.), *CERME 5 – Proceedings of the Fifth Congress of the European Society for Research in Mathematics Education* (pp. 2080–2089). Larnaca: University of Cyprus.

Bosch, M., García, F. J., Gascón, J., & Ruiz Higueras, L. (2006a). Reformulating "mathematical modelling" in the framework of the anthropological theory of didactics. In J. Novotná, H. Moraová, M. Krátká, & N. Stehlíková (Eds.), *Proceedings of the 30th Conference of the International Group for the Psychology of Mathematics Education* (Vol. 2, pp. 209–216). Prague: Charles University.

Bosch, M., Chevallard, Y., & Gascón, J. (2006b). Science or magic? The use of models and theories in didactics of mathematics. In M. Bosch (Ed.), *Proceedings of the Fourth Congress of the European Society for Research in Mathematics Education* (pp. 1254–1263). Barcelona: Fundemi IQS – Universitat.

Chevallard, Y. (1999). L'analyse des pratiques enseignantes en théorie anthropologique du didactique. *Recherches en Didactique des Mathématiques, 19*(2), 221–226.

Chevallard, Y. (2007). Readjusting didactics to a changing epistemology. *European Educational Research Journal, 6*(2), 131–134.

Crouch, R., & Haines, C. (2004). Mathematical modelling: Transitions between the real world and the mathematical model. *International Journal of Mathematical Education in Science and Technology, 35*(2), 197–206.

Galbraith, P., & Stillman, G. (2006). A framework for identifying student blockages during transitions in the modelling process. *Zentralblatt für Didaktik der Mathematik (ZDM), 38*(2), 143–162.

Grigoraş, R., & Halverscheid, S. (2008). Modelling the travelling salesman problem: Relations between the world of mathematics and the rest of the world. In O. Figueras, J. L. Cortina, S. Alatorre, T. Rojano, & A. Sepúlveda (Eds.), *Proceedings of the Joint Meeting of PME 32 and PME-NA XXX* (Vol. 3, pp. 105–112). Morelia: Cinvestav-UMSNH.

Halverscheid, S. (2008). Building a local conceptual framework for epistemic actions in a modelling environment with experiments. *ZDM – The International Journal on Mathematics Education, 40*(2), 225–234.

Matos, J., & Carreira, S. (1995). Cognitive processes and representations involved in applied problem solving. In C. Sloyer, W. Blum, & I. Huntley (Eds.), *Advances and perspectives in the teaching of mathematical modelling and applications (ICTMA-6)* (pp. 71–80). Chichester: Ellis Horwood.

Niss, M., Blum, W., & Galbraith, P. (2007). Part 1. Introduction. In W. Blum, P. L. Galbraith, H.-W. Henn, & M. Niss (Eds.), *Modelling and applications in mathematics education. The 14th ICMI study* (pp. 3–32). New York: Springer.

Chapter 11
The Sun Hour Project

Thomas Lingefjärd and Stefanie Meier

Abstract A mutual teaching project was set up between a gymnasium in Sweden and a gymnasium in Germany within the aims of the Comenius Network project "Developing Quality in Mathematics Education II". The main objective was that the two teachers wanted to let their students work on the same modelling assignment to observe the similarities and the differences between how the teachers and the students handled this project. We will deal with the question: How does the communication between students, teacher, and computer influence the modelling process?

1 Introduction

In the Sun hour project, the students were expected to find or develop a suitable mathematical model for the phenomenon of possible daily sun hours at different geographical positions around the world.

The modelling assignment was given as an open problem formulation:

> Form 4 groups and work on a suitable mathematical model to describe the phenomenon sunrise/sunset and the change of the daylight time for your town and others. The model should be able to account for differences in location. Describe situations in which this model could be useful and how it could be used.

The Earth's orbit around the sun is an ellipse with the sun at its focus. At certain times of the year, the Earth is nearer to the sun and moves faster than when it is

---

T. Lingefjärd (✉)
University of Gothenburg, Gothenburg, Sweden
e-mail: Thomas.Lingefjard@gu.se

S. Meier
Technische Universität Dortmund, Dortmund, Germany
e-mail: Stefanie.meier@math.tu-dortmund.de

further away, so that the sun appears to move faster across the sky. Furthermore, the axis of the Earth is tilted 23.5° to the direction of motion around the orbit and this also affects the time it takes for the sun to appear and move around the Earth.

Kirsch (1994) develops in a first step a formula for calculating the sun hours at given latitude by using measured data. The result is: $L_0(n) = 12.23 + 4.37 * \cos(n)$, where 12.23 is the average of the longest and shortest, and 4.37 the difference between the longest and the average sun hours per day. To come to a better justified formula, he suggests visualizing what is happening here with a globe and a lamp as the sun. Instead of using the real Earth's orbit in the shape of an ellipse, he uses a circle. Further on, he simplifies the number of days per month to 30 days. The sun hours for 1 day at specific latitude can be calculated with the relation: length of the illuminated part to circumference of the latitude times 24 h.

With that and the model in Fig. 11.1, he develops the formula: $L_{max} = \frac{2}{15} * \arccos(-\tan(\varepsilon)\tan(\beta))$ for the maximum of sun hours at given latitude. Connecting this to the first formula he developed a first sufficient, but not exact heuristic approach, which is given by: $L_1(n) = 12 + (L_{max} - 12) + \cos(n)$.

This is a first conclusion about the sun hours per day in relation to date and latitude. The graphs in Fig. 11.2 show the graphs for Dortmund and Gothenburg calculated with L_1 and also included: data from the Internet. The obvious differences give many reasons for discussing the mathematical model with students. For a detailed analysis see Kirsch (1994).

Another approach is to take a picture of the Sun every week at a fixed time for a whole year. What kind of curve would the Sun trace out in the course of the year? If you think carefully you would probably, after some considerations, decide that the Sun would move up and down along a straight line parallel to the Earth's axis. If you took the picture at noon, you would expect a symmetric curve up and down

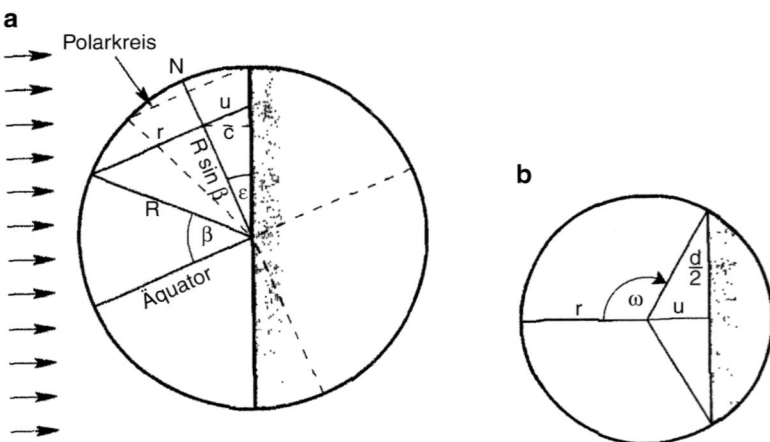

Fig. 11.1 Model of the Earth (Kirsch 1994, p. 6)

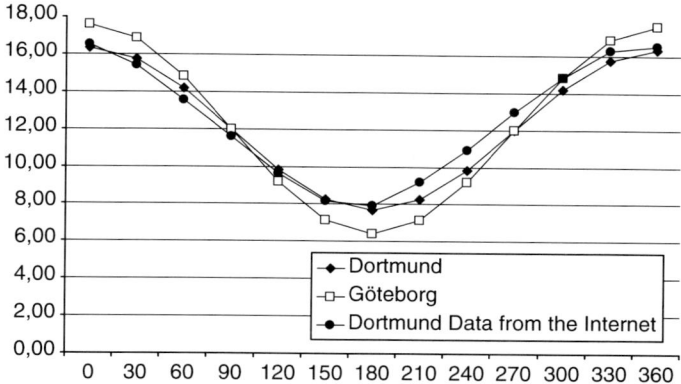

Fig. 11.2 Sun hours for Dortmund and Gothenburg

along the meridian. The meridian is the great semicircle between the north and south points on the horizon that passes through the zenith, the point straight above you. However, we instead get a strange sort of curve called the "analemma," which represents the equation of time. The equation of time is the difference between true local time and local mean time, both taken at a given place at the same time.

This implies a rather complex model both in terms of epistemological competence and in cognitive competence for determination of the daylight time in relation to different values of longitude and/or latitude of the globe.

The final presentations of the students were videotaped by the authors of this chapter. The videotaped presentations play a vital role in our research-based conclusions further on in this chapter.

2 Classroom Circumstances

2.1 Classroom Experiences in Sweden

The Swedish students were enrolled in the upper secondary school science program and were the age of 17–18 years. The assignment was given as part of the Mathematics course D. The above instructions were handed out together with a short introduction to the project. During the videotaped lesson, the students had 90 min to work on their models and afterwards the results were presented. The Swedish students were divided into groups of 3–4 students during a time frame from 09.00 to 14.00. Most groups managed to find appropriate data for the times of sunrise and sunset for Gothenburg city. The Swedish meteorological institute, SMHI, seemed to possess the most comprehensive database. A few groups had also collected data for other towns as well. None of the groups had started to think about possible functions for modelling the data. (See Andersson et al. 2009).

Most groups had focused on trying to model the daylight hour variation through a trigonometric model. Once all the data has been entered into Graphmatica, this is done quite easily. The challenge here lies more in adapting the model to different latitudes. This was only done completely by one group. This group was also able to identify possible limitations of their model, such as that it is only valid between the polar circles for instance. One group got hooked up in the formulation about atmospheric refraction in the instructions. Most of the groups managed to develop a mathematical model that would model the variations in daylight hours, even if not all groups were able to find a more general model that could be used for different latitudes. See Andersson et al. (2009) for example graphs of students' work.

2.2 Classroom Experiences in Germany

The mathematics class consisted of 13 students (age 17–18 years). The project day began at 7.50 and lasted until 13.10. The last 90 min were used to present results of the mathematical modelling work the students had done earlier.

The students worked together in 5 groups with 2 or 3 students in every group. The teacher did not intervene in the work of the students so their interests and first results varied a lot. One group was interested in a functional correlation. They investigated what "the change of daylight" means in a first approach. They generated different questions like: To what extent does the daylight time change at different dates and at a constant point of time to different positions in longitude or latitude? (See Andersson et al. 2009).

Another group decided to investigate the declination of the sun related to the rotation of the earth around the sun (which means over the year at the same daytime) and also the rotation around itself over a day. They designed an illustrative and possible geometrical model for the change of the declination of the sun when viewed by an observer at a fixed point on the Earth's surface during one day.

A third group also searched the Internet and focused on the equation of time. From their search results, two students analyzed the equation to understand the effect of a linear combination of two trigonometric functions. With the use of Excel and scroll bars, they examined the effect of different parameters of the equation.

In both classrooms, it was obvious that the students started with collecting data. These data were then represented by graphs with the help of a computer algebra system. This approach has been suggested by Kirsch (1994): "A first access can be a list e.g. in a calendar of the times of sunrise and dawn" (Kirsch 1994, p. 1). Kirsch's approach is then to represent this data with trigonometric functions. This is what the Swedish students did as well. But entering data into the computer and receive graphs in return, does not at all reduce the complexity of the problem as we will see in the dialogues from the two classrooms. See Fig. 11.3 for a sense of this complexity.

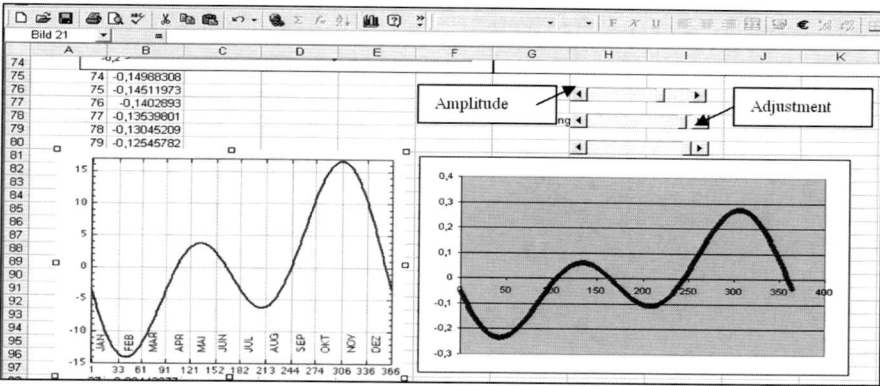

Fig. 11.3 The complexity of different representations of the sun hour phenomenon

3 Researchers' View of the Project

3.1 *Theoretical Frameworks*

In a sophisticated mathematical modelling task such as the sun hour project, there is a third silent partner in the didactical contract (Brousseau 1997) beside the teacher and the students – the computer. Another way of understanding the role and function of the didactic contract is to view it as a part of what Goffman (1986) calls the framing of a social situation. In Goffman's analytical perspective, the concept frame implies that there is a definition of a situation, which the participants share and most often take for granted. A frame can be seen as the participants' shared response to the question "what is going on here" (Goffman 1986, p. 8). We humans constantly produce and construe events, actions, and utterances in line with the framing we perceive as relevant.

> Given their understanding of what it is that is going on, individuals fit their actions to this understanding and ordinarily find that the ongoing world support[s] this fitting. These organizational premises – sustained both in the mind and in activity – I call the frame of the activity. (Goffman 1986, p. 247).

The focus on framing thus put an emphasis on the actors' perspective on the situation they are engaged in. The fact that students are engaged in mathematics learning in a classroom context is hence not extraneous to their activity; this is what structures what they are attempting to accomplish. Their meaning-making practices, including how they read the task, what is relevant to attend to, and what is an expected outcome, are all embedded in, intertwined with, and structured by, their involvement in this particular setting.

We will also reflect on what role the computer has in the framing of this situation, according to our notion of the computer as a silent partner in the modelling process.

3.2 Analyzing the Work in the Swedish Classroom

The observed group included both girls and boys who all were fluent in Swedish. For ethical reasons, we do not indicate the gender of any of the students we refer to. The group members are labeled Student 1, Student 2, etc. In the translation of what the students said from Swedish into English, the essence of what the students said has been considered. In many cases, the translation has been verbatim and with an effort to use translated words that seemed suitable for the context at hand. Figures in brackets are numbered items from the transcript, illustrating that this chapter only accounts for some of the discussion that took place. Obviously, it is a major challenge to avoid using one's a priori assumptions to select and analyze some parts of the transcripts, thereby also neglecting other parts. It is also important to realize that the observer's presence most likely affects the framing or social structure of the situation. To make the analysis of the transcripts more objective, it has been done in mutual exchange of both authors.

Student 1 and Student 2 start powering up their notebooks while the other three organize their various notes. After a while, the group has two computers running and a lot of notes around them as the following discussion starts.

Student 1: Hey, have you done your part of the Graphmatica stuff you were supposed to? And have you brought the Earth globe you talked about? We need to get organized now, right? [1]

Student 4: The Graphmatica presentation is almost finished; I just need some minutes to get the new data set into the curve fitting. [2]

We interpret this conversation as framing by Student 1 and Student 4, bordering the problem and the presentation task within a specific social contextual framing that is well known to all group members: They have to get organized and prepare their presentation.

Student 3 gives Student 4 a USB stick and the data set is transferred into the computer. Student 4 starts to work on the computer; Student 5 presents an Earth globe.

Student 3: He (the teacher) said something about the curve fitting procedure last time in class, didn't he? He said something about quality control. Do you remember? [3]

Student 3 is framing the structure of their activity within a mathematical framing, a framing of the mathematical theory of curve fitting, mathematical modelling, and regression coefficient. He/she wants to give a starting point for the mathematical discussion.

Student 3 quickly enters some figures into the calculator and responds:

Student 3: I have another number, called r, here in the calculator. Is that the same? [7]

Student 1: I do not understand what this χ^2 means. We have a value of $\chi^2 = 1200$ or something. Is that good? [8]

Student 3: Well, I think that he (the teacher) told us last week that any mathematical model can fit better or worse and therefore you need to have a quality control and that is what we have in this χ^2. But what is r? Is it the same? [11]

Student 3 looks around, eyeing the classmates in the group in order to see whether his or her framing of the problem makes sense and if they understand his/her question. The group grunts in a positive but not explicitly clear way, although our mutual interpretation is that they are accepting Student 3's framing. Some of the students are also stumbling between frames in utterances [14]–[17].

Student 1: Right. So we have a quality of 1200, is that what you are a saying? [14]
Student 3: Well, yes, and maybe we need to check that number somehow. Wasn't there a table or something? [15]
Student 5: Yes, you are right! Good work! We can check it through a table, but we do not need to do that now. Let's see. We just tell the others that our model is good, since we have χ^2 equal to 1200. Great! Are we done now? [16]
Student 3: Here, check my calculator. This *r* value is pretty low? See here. [17]

In [14]–[17], Students 1, 3, and 5 are working inside the mathematical framing while Students 2 and 4 are listening and making notes on paper. The students are stumbling within the mathematical framing, more or less lost in a totally wrong assumption about the meaning of χ^2. From the inside, it sounds as if all the students are lost. Student 5 gives a solution for the problem which depends on just ignoring it. Student 3 still tries to understand what χ^2 and *r* mean, but obviously has no idea how to solve that problem. So, one of the students walks over to the teachers and asks him something. The teacher responds and the student comes back to the group and says:

Hey, go out to the Internet and Google on "the meaning of χ^2". It seems as if we are deadly wrong.

There is a minute of active reading until the group concludes that they were wrong.

Student 5: But then it is catastrophic! We have 1200 and should have 12! Help! What should we do? We need to do something! [22]

Student 5 looks around, but all the students in the group are silent and say nothing. Even the dominant Student 1 is silent and seems lost.

Goffman (1974/1986) suggests that when we face situations where the framing is problematic, we end up in uncertainty about how to act in that situation; we do not understand the situation. As a result, the group becomes silent and especially Students 1 and 3 seem rather lost and quite uncertain about how to frame this feedback. There is a silence and a feeling of tragedy in the group. Suddenly, Student 1 takes command.

Student 1: This is what we must do. You and you (Student 1 points at the two students with their own notebook) present the Graphmatica result with the projector and make sure that the χ^2 value is hidden on the monitor by changing colors or something! Alright? [25]

Student 1: continues: You and you (Student 1 points at the two other students) must hold up the Earth globe and the large wall map now and then. I will talk about the difference in terms of longitude, difference in sun hours between Gothenburg, Paris and Addis Abbaba. And no one answer any questions about χ^2 or regression coefficients. Got it? [26]

By this shifting from an uncertain mathematical framing into a more secure social framing, Student 1 gets the group back into the play and they actually deliver a very good presentation in which they succeed to talk not about the mathematical modelling experience but a lot about the number of sun hours in Addis Abbaba. And also the teacher did not ask the students if they solved their problem with the χ^2 value.

3.3 *Analyzing the Work in the German Classroom*

The following transcript has been taken from the video of the preparation part for the presentation in the German classroom. The above-mentioned points for videotaping, the transcription and the translation into English apply here, too.

The students had already worked some time on the task, when one group has a question and called the teacher for help. The group consists of three students, but only two are talking with the teacher.

Student 1: We have the problem that we do not know what these variables mean (points to the following function: $y^n \cdot \sin(x \cdot day + z)$). [1]
Teacher: Hmm. [2]
Student 1: This is an assumption, but we do not know, if that is right. We use y^n for Height over equator and longitude is z. And we do not know what "x" means. We thought it might be the latitude, because we have here the longitude. So what we would like to know first is the meaning of the variables and the numbers here. [3]
Teacher: Yes. [4]

Student 1 is, in Goffman's words, framing the problem the group has with the task in a mathematical framing. Student 2 is completing this framing.

Student 2: We have these single sine functions... And those have different variables (points on the monitor). [5]
Teacher: Yes [6]
Student 2: But a reason for that cannot be found on the Internet. [7]

Now their problem is completed. They had found an equation on the Internet, and realized that it is composed of two sin functions and try now to understand what the variables in these functions mean.

In the next part of the conversation, the teacher is framing the problem and clarifying that the equation has something to do with time.

In the following discussion, the students figured out, with the help of the teacher, that the variables are correcting factors and that x is "the" variable for the function. After that Student 1 is once again framing the problem:

Student 1: But what are the factors correcting? [14]

After a short discussion Student 2 shifted the discussion to the content of the lesson before the holidays:

Student 2: You (the teacher) draw something on a sheet before the holidays. (He points on that drawing). [15].

Teacher: Yes, that is the idea behind that. [16]

Student 2 is then shifting to something he learned in physics:

Student 2: Probably y^n is the highest amplitude. [17]
Teacher: Yes [18]

The teacher explains why his assumption is correct in the following. Afterwards Student 2 is again framing the problem.

Student 2: That is the graph of two sinus curves (points at the monitor). It shows the deviation in minutes. [19]
Teacher: Of what? [20]
Student 2: Of some kind of time [21]
Teacher: So that is what you have to figure out next. You can ask Student 4 (at another table) for help. He knows what the deviation is and you know much about the sine curves. So you can try to bring that together. [22]

So at the end, the students and the teacher are framing a new problem which the students work on. In their presentation, they only showed what they found out about superimposing sine curves, because the other student already talked about the deviation. But the problem, what the single factors in the equations actually mean, has been left aside.

4 Conclusions

To reproduce the meaning of a classroom discussion is naturally a delicate mission. Even with videotaped classroom experiences, we are responsible for the interpretation we have done. Our interpretation was done in the spirit of Goffman's theory of framing. We considered this framework as a suitable framework for analyzing class room discussions like the ones we have done. Nevertheless, it is important to remember that our interpretations are all there are.

In Sweden, the students obtained their main model by the computer which shows them an equation which they cannot handle. So the computer was not only a tool to work with, but also something which responded to the students' actions and thereby gave the students a problem to solve. Their way of solving was first thinking about it, and then asking the teacher who told them to ask Google. With the explanation from the Internet, they realized that the result of their calculation was wrong; but they did not try to find the mistake in their calculation, instead they decided not to talk about that in their presentation. No one, not even the teacher, missed this point in their presentation. So what they did in Goffman's words was that they framed a problem and solved the problem in the, for them, easiest way: in the social framing. They left the mathematical framing aside. They shifted the frame from mathematics to social life.

In the German class, the information from the Internet gave the students the problem. They searched for answers, but only found more questions. Together with their teacher they framed the problem by talking about it, making it more concrete and also managed to clear frames in between. It is the same situation as in the Swedish classroom: the tool influences and shapes the thinking of the user, in this example, framing a problem. However, in their presentation, they did not talk about that problem. Another student explained the idea behind that, but the question about what the variables stand for, was not answered. So a part of the mathematical framing was also left aside here. Due to the fact that the teacher framed the problem together with the students, he did not ask for the meaning of the variables.

We can summarize according to this case study of two classrooms that it is essential for the observed students and teachers to be aware of how they actually frame problems when working with a complex modelling task. People in general, and maybe perhaps specifically students, are always rational and try to find a way out from a problematic situation, or from a problematic framing. The teachers need to make sure that they clear the framing the way the teacher intended to. Further on the computer as a silent partner in the group work is not always the one solving problems, but also raising problems. So the shaping of mental schemes by a computer does not always produce solutions, but also problems that sometimes cannot be solved by the user.

References

Andersson, M., Lingefjärd, T., Meier, S., & Müller, J. (2009). The sun hour project. In W. Henn & S. Meier (Eds.), *Growing mathematics*. Dortmund: Print on demand.

Brousseau, G. (1997). *Theory of didactical situations in mathematics: Didactique des mathématiques 1970–1990*. Dordrecht: Kluwer Academic.

Goffman, E. (1986). *Frame analysis: An essay on the organization of experience*. Boston: Northeastern University Press. (Original work published 1974.)

Kirsch, A. (1994). Das Problem der täglichen Sonnenscheindauer als Thema für den Mathematikunterricht. *Didaktik der Mathematik, 22*(1), 1–19.

Chapter 12
Mathematical Knowledge Application and Student Difficulties in a Design-Based Interdisciplinary Project

Kit Ee Dawn Ng

Abstract This chapter presents types of mathematical knowledge application during a design-based interdisciplinary project as displayed through the work of student-group cases from grades 7 and 8 (aged 13–14) in two educational streams across three Singapore government secondary schools. It was found that the students did not apply all the expected mathematical knowledge and skills afforded by the project. They also lacked awareness of the purpose of scale and displayed limited activation of real-world knowledge for mathematical decision making. Findings presented have implications on the facilitation of quality mathematical application in contextualised tasks.

1 Background of Research

Interdisciplinary project work (PW) was introduced as an educational initiative in Singapore primary, secondary, and pre-university institutions in 2000. In PW, explicit links between different subject knowledge are made so that students can learn to "appreciate the inter-connectedness of disciplines and see the relevance of classroom learning to their current or future interests" (Chan 2001, p. 1). This aspect of PW is also a key focus of the most recent Singapore mathematics syllabus (Curriculum Planning and Development Division [CPDD] 2006), which highlights "connections" (i.e. within-subject, between-subjects, and real-world links) (p. 5) as an important process during mathematical learning.

A major focus of PW in Singapore is *knowledge application*. Anchoring subjects for any PW are explicitly stated such that students "extend their knowledge from their immediate environment to perform, apply, and transfer new tasks to a variety

K.E.D. Ng (✉)
Mathematics and Mathematics Education Academic Group, National Institute of Education,
Nanyang Technological University, NIE-03-10, 1 Nanyang Walk, Singapore 637616, Singapore
e-mail: dawn.ng@nie.edu.sg

of circumstances" (Quek et al. 2006, p. 14). It is assumed that most of the required content knowledge and skills are taught in traditional subject-specific lessons and that students apply relevant learnt knowledge and skills to further their understanding during PW. Hence, PW is essentially an interdisciplinary applications task situated in a real-world context.

Although PW is generally assessed informally in most primary and secondary schools, it became an entry requirement into university in 2005 (Ministry of Education 2001). PW lessons in secondary schools are conducted within curriculum time and run concurrently with traditional subject-based lessons in at least one of the semesters for selected year levels (i.e. students aged 13–15) not involved in national examinations. Students, in groups of 3–4, usually engage in a PW task for 3–4 months, holding weekly meetings in class with teacher facilitation.

2 Rationale for Research

Contextualised mathematical tasks allow students to experience problem solving within real-world constraints in meaningful ways (Stillman 2000). However, it is challenging to ensure *quality mathematical outcomes* during contextualised tasks. Firstly, students can choose not to engage with the mathematical aspects of the task as intended (Gravemeijer 1994). Secondly, students may not find their subject-specific knowledge useful and therefore use them sparingly in the task, against expectations (Venville et al. 2004). Thirdly, students may ignore the contexts provided (Verschaffel et al. 2000) and choose not to activate their "real-world knowledge and realistic considerations" (van den Heuvel-Panhuizen 1999, p. 137) for decision making, verification, and sense making during mathematical knowledge application. Lastly, not all students (i.e. low and high mathematical achievers) engage with the context presented by the task in the same way and the forms of task engagements impact on the types of mathematical outcomes (Kramarski et al. 2001). The challenges outlined can be further complicated by the presence of unpredictable group dynamics during PW. An individual's mathematical engagement, reasoning, task interpretation (especially perceptions of inter-subject connections and real-world links with the task), and knowledge application approaches may well be influenced by group members.

Research studies on PW involving the nature of mathematics application are limited to date. Only two were found. Tan's (2002) study did not provide detailed analysis on the nature and process of mathematical knowledge application. On the other hand, Chan (2008) found that mathematical modelling tasks appeared to promote mathematical processes and transfer of domain-specific mathematical knowledge in grade 6 students (aged 12). Hence, one of the aims of a larger study undertaken was to answer the following research question: *What are the types of mathematical knowledge application and student difficulties faced during participation in a design-based interdisciplinary project?*

3 Research Design

3.1 Research Task

A design-based PW involving mathematics, science, and geography was implemented in 16 classes of students ($N=617$) from grades 7 and 8 (aged 13–14) in two educational streams (high and average) across three Singapore government secondary schools. It followed the theme of *environmental conservation* and was completed through 15 weekly meeting sessions. The goal of this PW was for students to work in groups of four to design an environmentally friendly (EF) building at a location of their choice within Singapore (see Ng 2006). Student groups were given mini tasks (e.g. library research) during the meeting sessions to help them work towards the goal using supporting materials developed by the researcher in accordance with the guidelines set by the Singapore Ministry of Education. Among the mini tasks were three mathematical tasks with written components, which are the focuses of this chapter: (a) *decision making* about the various aspects of the building (i.e. size, dimensions, location, purpose, EF features, design), (b) *cost of furnishing* and fitting out a selected area in the building (i.e. budgeting including flooring, painting, choice of electrical appliances, and furniture), and (c) hand-drawn *scale drawings* of the actual building with the number and types of drawings decided by students. Each student group was also expected to construct a physical scale model of their building from recycled materials based on their drawings. With reference to the mathematics syllabus and the schools' teaching plans, it was assumed that the students involved in the research had prior knowledge and skills on area and perimeter measurements, basic arithmetic calculations for budgeting, and making scale drawings on isometric or graph paper. In addition, the students also attended design and technology classes where they experienced making wood work pieces from 2D plans.

3.2 Setting and Sample

As in many other Singapore secondary schools, student participation during the research project was facilitated using the project-based approach (see Quek et al. 2006). No special instructions for teaching intervention were given. The participating classes had at least one teacher with mathematics, science, geography, and design and technology specialisation facilitating each session at one time. The teachers continued with their usual facilitation methods as in other interdisciplinary tasks. Except for the stated tasks (a)–(c) above, the teachers could also reorganise the sequence of the proposed materials prepared by the researcher and re-craft the resources provided.

Out of 16 participating classes in the three schools, the researcher tracked the progress of 10 case-study groups ($n=38$, two students excluded due to technical difficulties) throughout the project in their weekly discussion sessions during curriculum time. There were five groups in each stream with only one group from a particular class.

Students formed their own groups with some help from their teachers within their classes. Each group consisted of students with mixed ability in mathematics.

3.3 Data Collection Methods

A multi-site multi-case-based approach (Yin 2003) was adopted in data collection and analysis. Data consisted of documentary (i.e. copies of students' work, field notes, memos, and email correspondence with participating teachers), audio-visual (i.e. video-generated data), and verbal evidence (i.e. interview-generated data).

Each group was videotaped during their in-class discussions on (a)–(c) above. Individual group members then participated in video-stimulated recall interviews within 1 week of their discussions. Work from the groups (i.e. notes, research materials, resources, drafts, drawings, and task sheets), their group project files, and final products of the project were collected for analysis along with their teachers' comments and grades. The researcher recorded lesson observation notes including the instructions and the nature of teacher scaffolding. All project queries from students made to the researcher were redirected to the facilitating teachers.

3.4 Analysis Procedures

Both documentary and audio-visual evidence were taken to be the main source of data for analysis. Open and axial coding based on Strauss and Corbin's (1998) reformulated grounded theory was used. Students' written work in (a)–(c) were first classified according to the types of mathematical knowledge and skills applied (e.g. area measurement, use of addition algorithm, proportional reasoning). These were triangulated with video excerpts and transcripts of students' actual use of mathematics during the tasks. In addition, various types of students' affective reactions to the tasks (e.g. mathematical decision making assigned to perceive more competent group members, negative attitudes towards the mathematical demands of the task) were also elicited from video documentations as these not only influenced the types of mathematics used, but also they gave an indication of the nature of mathematical difficulties the student groups faced.

4 Findings

4.1 Coverage of Mathematical Knowledge and Skills

Five out of ten groups used mathematics during the *decision making task*. Table 12.1 shows the types of mathematical concepts and skills afforded by the task, the purposes for applying them, and those actually applied by the five groups. Each of

Table 12.1 Coverage of expected types of mathematics during decision making task

Mathematics afforded by task	Purposes	Group 1	4	7	5[a]	9[a]
Estimation	Dimensions of actual building	•	•	•		•
	Dimensions of model			•	•	
	Budget for making scale model[b]	•				
Proportional reasoning	Determining scale for actual building	•	•	•	•	

[a]Groups from high educational stream
[b]Budget for making scale model was not an expected outcome of the task

the five groups did not apply all the expected mathematical concepts and skills for the task. For example, Groups 5 and 9 decided on the dimensions and scales for their buildings during the scale drawing task instead. Of note was that the two high-stream groups (Groups 5 and 9) only achieved minimal coverage of the expected mathematical concepts and skills in the task, despite having comparatively longer periods of discussion times than others. Moreover, fostered by the teacher (Ms Amy), the task was misinterpreted by Group 1 who wrongly assumed that they were to budget for the cost of making a model of their building:

Chi: What is "budget"?
Ms Amy: The budget is determined by the students, they should choose to build the model out of recycled materials to save money.

During the *cost of furnishing task*, six of the groups made the expected arithmetic calculations (e.g. addition, multiplication, and area) of total furnishing costs involving mainly floor tiling, furniture items, and appliances. The other three groups simply listed the cost of furnishing items without further computation. In addition, each group was also encouraged to make drawings (with dimensions included) of the selected area for furnishing in the task (e.g. Fig. 12.1). Nonetheless, whether the diagrams were completed or not did not impede progress in deciding furnishing items and calculating the budget for some groups. The need for a diagram of how the selected area should be furnished depended on two factors: whether group members shared real-life experiences involving furnishing and the decision making patterns of the groups. Members from Group 1, for example, had shared understandings or "common mental pictures" of how their selected area (i.e. school hall) should be furnished based on their own life experiences, removing the need for a drawing. In contrast, only one member each from Groups 4, 5, and 7 worked on the calculations as part of their assigned duties. These members either made quick, non-evaluative decisions about the types of furnishings and estimated costs based on their own life experiences or were members of the group (self-assigned or otherwise) who monopolised decision making for the task.

Four of the groups (three from high-stream classes) produced scale drawings (Fig. 12.2) whereas four other groups (all from average-stream classes) did measured drawings (i.e. some apparent use of measurement but lack of scale interpretation) for the *scale drawing task*. This suggests that the average-stream groups might have

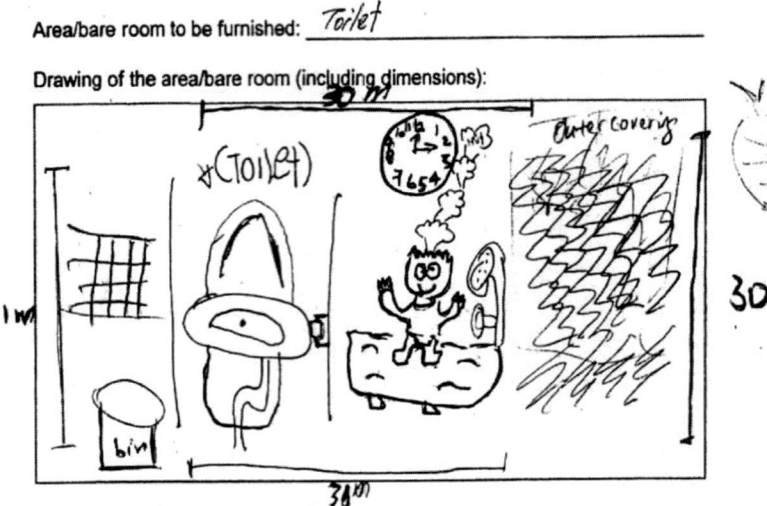

Fig. 12.1 Sketch by Group 8

interpreted scale drawings to mean measured drawings, apparently due to a lack of reference to prior knowledge. Additional investigations revealed that through the various drawings produced (i.e. sketches, measured, and scale drawings), the student groups here used three out of four types of projections of their buildings as found by Athanasopoulos et al. (1993) in another similar design activity. Firstly, evidence of orthogonal projection (mainly top and front views of buildings as well as floor plans of building interiors) was found. Secondly, some forms of rotational projections of the buildings were attempted (Fig. 12.3). Lastly, representations of the façade of the buildings without any appearance of depth were produced.

4.2 Mathematical Difficulties

Three mathematical difficulties faced by students during the project are highlighted here. Firstly, some students displayed *a lack of awareness of the purpose of scale*. However, only four groups (three of which were from high-stream classes) completed proper scale drawings for the project. Others made sketches or measured drawings, showing no indications of their awareness of the purpose of using scales. Student-interview data revealed that some students were not aware that two scales had to be selected to represent the dimensions of the physical scale model and the real-life building in the drawings. For instance, a length of 100 m can be represented by 1 cm and 10 cm on the drawing and model, respectively. This means two scales (i.e. 1:10 000 from drawing to building and 1:1000 from model to building) could be used in the same drawing. Only two groups (Groups 2 and 7) managed this. Unfortunately, their scales for the model did not realistically represent the

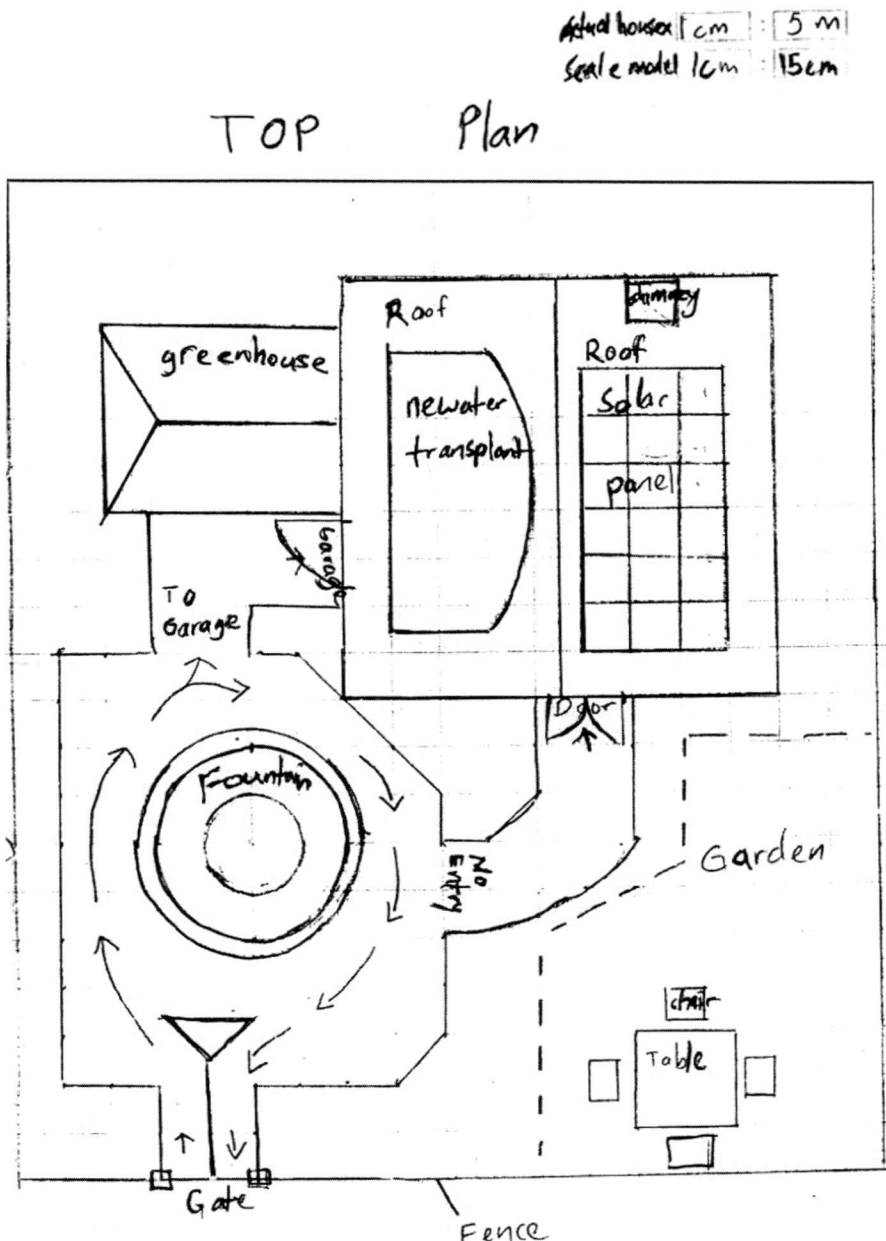

Fig. 12.2 Projection of building from the top illustrated by Group 7

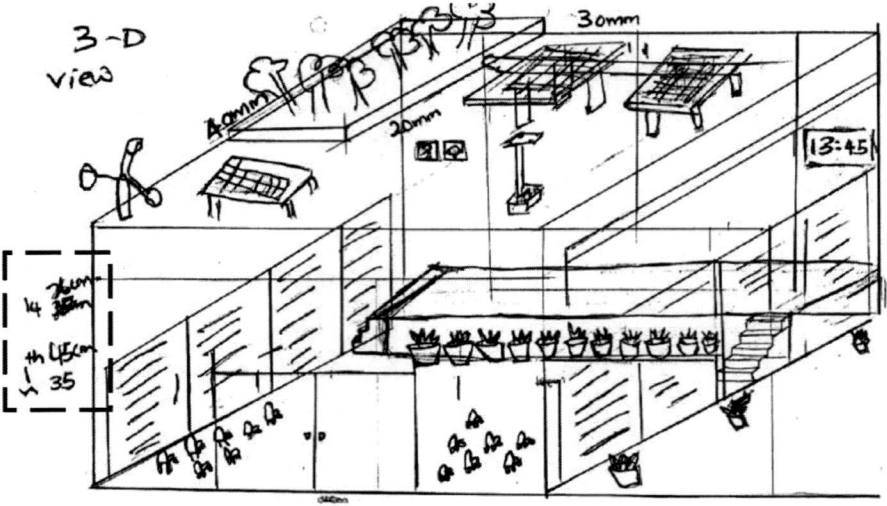

Fig. 12.3 A 3D sketch by Group 1

dimensions of their buildings. Group 2 used the scales 1:400 and 1:200 to represent the real-life building and the model, respectively, in their drawings. This worked out to be 1 cm on the model representing 2 cm on the building. This was neither realistic nor reasonable in a real-world context. Hence, it was questionable whether students from the sample were aware of the relationship between the two scales.

Secondly, some students used *different scales for the same building*. For example, Group 3 found themselves having to adjust their drawings at a later stage to accommodate real-life measurements and proportional reasoning of the different parts of the building in order to form a coherent image.

Thirdly, there was *limited activation of real-world knowledge* by the groups. Despite the presentation of the project within a real-world scenario, only Groups 2, 3, and 9 from the high-stream applied real-world knowledge during the three tasks of the project. For this study, activation of real-world knowledge refers to the incorporation of EF features in building design and furnishings as well as displays of real-world considerations in mathematical decision making. An example of activation came from Group 2, who debated intensively about the possible dimensions of the actual building citing examples from real-life experiences bearing in mind space constraints in their chosen location for the building.

5 Discussion and Conclusion

As found in other studies (e.g. Kramarski et al. 2001), not all students engage with the context of the research project in the same way. Examining the extent of coverage of expected mathematics during engagement with the tasks and student difficulties with the tasks not only revealed whether the tasks were interpreted as predicted but also raised possible misinterpretations of the project requirements. Such information is crucial for subsequent design and facilitation of open-ended interdisciplinary contextualised tasks with purposive weak scaffolding structures.

The findings presented above have significant implications for future work with applications tasks such as interdisciplinary project work in the Singapore context. For one, students did not apply all the expected mathematical knowledge and skills afforded by the tasks even upon the assumption that they had the relevant prerequisites. Gravemeijer (1994) and Venville et al. (2004) also reported that student samples grappled with whether and how to engage mathematically with such tasks. Interestingly, some students from this study went beyond expectations, albeit due to task misinterpretations. To a large extent, coverage of mathematics by each group during the tasks depends on task sensitivity, task engagement, task scaffolding, and a shared repertoire of mathematical concepts and skills, among other factors. It was assumed that students working in groups would complement each other on these aspects and promote higher quality mathematical outcomes. However, this did not and would not happen automatically without frequent and appropriate facilitation by their teachers. Yet, too regimented scaffolding might not do justice to the purposive open-ended nature of such tasks for creative problem solving and interconnected meaningful learning. The question is thus how do teachers achieve a "balance" in scaffolding during such tasks in order to retain the mathematical rigour of the tasks in the eyes of the teachers?

Secondly, several groups clearly had difficulties with scale drawings, particularly with the purpose and flexibility of scale usage in order to represent real-life objects. This brings to mind that some students may also have difficulties with other related concepts such as estimation, proportional reasoning, and spatial visualisation, as also reported in Kordaki and Potari (1998). Further investigations are needed to verify this.

Thirdly, limited activation of real-world knowledge in mathematical application and decision making by the majority of the groups in this study echoed the findings of Verschaffel et al. (2000) and van den Heuvel-Panhuizen (1999). This suggests that students may learn mathematics in isolation of its use in the real world and hence form certain beliefs about the nature of mathematics. Although contextualised tasks were postulated to provide a more meaningful learning experience for students, it seems the challenge still exists for teachers and curriculum planners to bridge the gap between school mathematics and the mathematics used in real life.

Lastly, group dynamics came into play in determining the nature of mathematical application during the project. Indeed, effective groups spent time fruitfully on attaining project requirements and advanced steadily towards project goals. These groups were also observed to be collaborative learners, where members added value to discussions, worked together for more accurate and appropriate mathematical knowledge application, and hence engaged more mathematically with the tasks. Nonetheless, it was discovered that even effective groups had difficulty sustaining their interest in the project and maintaining quality mathematical outcomes towards the end of the project. Two reasons could account for this. The long implementation period of the project could have delayed gratification and a sense of achievement from the project. In addition, some students were asked to complete smaller tasks along the way in a piecemeal manner in order to progress to the next stage as part of teacher facilitation. At times, such measures may hinder progress and prevent students from perceiving the project in a holistic way.

References

Athanasopoulos, P., Patronis, T., Potari, D., Spanos, D., & Spiliotopoulou, V. (1993). Constructing a village: A cross-curricular activity. In T. Breiteig, I. Huntley, & G. Kaiser-Messmer (Eds.), *Teaching and learning mathematics in context* (pp. 62–80). Chichester: Ellis Horwood.

Chan, J. K. (2001, June). *A curriculum for the knowledge age: The Singapore approach*. Paper presented at the eighth annual Curriculum Corporation Conference, Sydney, Australia.

Chan, C. M. E. (2008). Using model-eliciting activities for primary mathematics classrooms. *The Mathematics Educator, 11*(1/2), 47–66.

Curriculum Planning and Development Division [CPDD]. (2006). *Mathematics Syllabus*. Singapore: Ministry of Education, CPDD.

Gravemeijer, K. P. E. (1994). *Developing realistic mathematics education*. Utrecht: Freudenthal Institute.

Kordaki, M., & Potari, D. (1998). Children's approaches to area measurement through different contexts. *Journal of Mathematical Behavior, 17*(3), 303–316.

Kramarski, B., Mevarech, Z. R., & Liberman, A. (2001). The effects of multilevel – Versus unilevel-metacognitive training on mathematical reasoning. *Journal of Educational Research, 94*(5), 292–300.

Ministry of Education (MOE). (2001, February 14). *Overview*. Retrieved February 21, 2007, from http://www.moe.gov.sg/corporate/overview_01.htm.

Ng, K. E. D. (2006). Interdisciplinary task: Designing an environmentally friendly building. In J. Ocean, C. Walta, M. Breed, J. Virgona, & J. Horwood (Eds.), *Mathematics, the way forward: MAV Conference 2006* (pp. 250–259). Melbourne: The Mathematical Association of Victoria.

Quek, C. L., Divaharan, S., Liu, W. C., Peer, J., Williams, M. D., Wong, A. F. L., et al. (2006). *Engaging in project work*. Singapore: McGraw Hill.

Stillman, G. (2000). Impact of prior knowledge of task context on approaches to applications tasks. *Journal of Mathematical Behavior, 19*, 333–361.

Strauss, A., & Corbin, J. (1998). *Basics of qualitative research: Techniques and procedures for developing grounded theory*. Thousand Oaks: Sage.

Tan, T. L. S. (2002). *Using project work as a motivating factor in lower secondary mathematics*. Unpublished masters thesis, Nanyang Technological University, National Institute of Education, Singapore.

van den Heuvel-Panhuizen, M. (1999). Context problems and assessment: Ideas from the Netherlands. In I. Thompson (Ed.), *Issues in teaching numeracy in primary schools* (pp. 130–142). Buckingham: Open University Press.

Venville, G., Rennie, L., & Wallace, J. (2004). Decision making and sources of knowledge: How students tackle integrated tasks in science, technology and mathematics. *Research in Science Education, 34*(2), 115–135.

Verschaffel, L., Greer, B., & De Corte, E. (2000). *Making sense of word problems*. Lisse: Swets & Zeitlinger.

Yin, R. K. (2003). *Case study research: Design and methods* (3rd ed.). Thousand Oaks: Sage.

Chapter 13
Evaluation of Teaching Activities with Multi-Variable Functions in Context

Yoshiki Nisawa and Seiji Moriya

Abstract In mathematics education in Japan, learning by practical problem solving is not emphasized because high school students and teachers focus more on the university entrance exam, leading to a lack of understanding of the essential meaning of mathematics and of relations between mathematics and daily life.

These problems affect the teaching of various mathematical concepts. For instance, it cannot be said that the current mathematical education content in Japan sufficiently teaches functions. We propose teaching materials on functions that relate to and explain familiar events and function phenomena and in doing so deepen the students' understanding of functions.

1 The Background of the Research

Japanese secondary mathematics education for the Junior High School (ages 12–15 years) and for the High School (ages 15–18 years) includes two targets: (1) Mathematics as science is studied. (2) The use of mathematics (use in science and use in daily life) is learnt. However, high school students and teachers tend to focus on university entrance examinations and the skills necessary for their entrance exam problems rather than address these targets. In practice, the main study method in mathematics is to solve a given problems repeatedly. High school students can solve the problems, but they do not understand the meaning of many theorems, formulae, and concepts of functions because of this method. As a result, many university students have difficulty with understanding mathematics at this level or they simply lose interest in mathematics. These concerns are not being addressed although they

Y. Nisawa (✉)
Kyoto Prefectural Rakuhoku Senior High School, 59 Shimogamo-Umenoki-cho, Sakyo-ku, Kyoto 6060851, Japan
e-mail: yoshiki@cc.kyoto-su.ac.jp

S. Moriya
Tamagawa University, 6-1-1 Tamagawagakuen, Machida, Tokyo 1948610, Japan

are frequently pointed out. However, these concerns can be tackled directly at the high school level and in the early years of university.

An example of university students not understanding mathematics at this level because of their previous education and study methods is when they are taught functions. When learning about functions, students learn about single-variable functions in junior high school and high school and they meet multi-variable functions in university. An awareness survey of the concept of functions was given to more than 30 second year students in a science course. Looking at the results, there were no students who were able to answer the question "What is a function?" They only recognized functions that had been shown through expressions.

We believed this was a result of the students studying with a bias toward examination mathematics in high school. We paid attention to the concept of functions to address this problem. Educational content that allowed students to understand the meaning and use of functions was researched and practised. This content was chiefly intended for seniors of high school and first and second year university students.

We wanted to make students understand that the concept of a function is not limited to expressions, already learned theorems can be thought of as multi-variable functions, and that when functions appear in various familiar situations and are analyzed, it is useful. We thought that mathematical modelling was best for these goals. Two-variable functions were primarily adopted as teaching material, because we believed multi-variable functions, not single-variable functions, were best. The reason for believing this is in the following two points. (1) Phenomena in our lives with two factors or more are more common than ones with one factor. (2) We can consider two-variable functions as single-variable functions when we think of one variable as constant, and then we can use previously learned knowledge of functions.

Research on function education examining both single-variable functions and multi-variable functions was carried out by Yokochi (1983). This research, at the junior high school level, did not sufficiently address the larger problems surrounding function education and mathematical education generally in Japan. There is little research on function education at the high school level; therefore, teaching material that can solve the above problems of mathematics education and the problem of function education in the high school must be sought.

Multi-variable functions are not in the high school curriculum. We want to introduce mathematical modelling containing multi-variable functions at the high school level to solve the problem of university students not understanding the concept of functions due to their previous mathematical education and teaching and learning methods. We tested the proposed course content on university students learning multi-variable functions for the first time to see if the content can develop understanding of functions.

2 The Educational Course Content

The multi-variable function course content tested on university students is reported in this chapter. These were students of Kyoto Sangyo University who aim to become mathematics teachers in junior high schools or high schools. This course was taught in four classes between April and May 2009. There were about 30 students who took these classes. The content is appropriate for teaching high school students.

The educational course comprised:

1. Two-variable functions – Analytic geometry approach
2. Mathematical modelling – The Vehicle Stopping Distance Example
3. Mathematical modelling – The Sound Propagation Example

The content was intended to make the students understand that functions are not limited to expressions, that previously learned theorems can be considered multi-variable functions, and that functions appearing in daily phenomena are useful.

2.1 Two-Variable Functions – Analytic Geometry Approach

We wanted the students to notice that many previously studied formulae are multi-variable functions, so the example of a rectangle was introduced.

A rectangle's area is obtained from the product of its vertical length and horizontal length.

Example: Find the maximum value of the function $z = xy$ when $x+y = 1$

The given problem of calculating the maximum of a quadratic function is studied in high schools. The example is usually solved by calculating the maximum of a single-variable function. However, this example is raised as a two-variable function.

In order to analyze the character of the two-variable function $z = xy$, a table relating the group (x,y) to z was created (Fig. 13.1).

The following main characteristics were discovered.

1. The graph is a continuous curved surface.
2. The intersection of $z = xy$ and the plane $y = 2$ is the straight line $z = 2x$. It helps to consider one variable as a constant when considering two-variable functions.
3. The intersection of $z = xy$ and the plane $y = x$ is a parabola. The example used shows that the intersection of $z = xy$ and the plane $x + y = 1$ is the parabola $z = -x^2 + x$.

2.2 Mathematical Modelling

Many phenomena from our daily lives can be considered as examples of multi-variable functions. To help students to understand this, models of the stopping distance of a car and the propagation of sound were introduced.

2.2.1 The Vehicle Stopping Distance Example

We considered modelling the *stopping distance* of a car. The effects of air resistance and ABS (Antilock Brake System) were not taken into consideration.

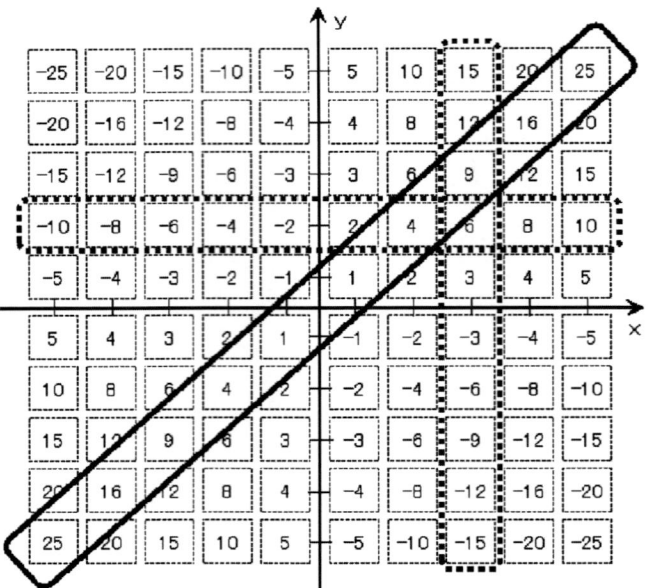

Fig. 13.1 A table relating the group (x, y) to z

The stopping distance is the sum of the driver's *thinking distance* and the car's *breaking distance*.

Set v m s^{-1} as the car's velocity; μ, the coefficient of friction between the tyre and the road surface; and g, the acceleration due to gravity taken as 9.8 m s^{-2} and estimate the average thinking time at 0.7–1.0 s. When the thinking time is 1 s, the thinking distance is v m. The braking distance, d, can be found straightforwardly assuming the final speed is zero (of course!) so that $d = \dfrac{v^2}{2\mu g}$ and the total stopping distance, z, is given by $z = \dfrac{v^2}{2\mu g} + v$.

This formula gives the following characteristics for the stopping distance of a car. (1) The mass of the car is not related to the braking distance. (2) If a car has a low speed, the stopping distance is short. (3) If a car has a high coefficient of friction between its tyres and the road surface, the stopping distance is short.

Using the usual coordinate system, the stopping distance may be expressed as $z = \dfrac{X^2}{2gy} + x$. z is a two-variable function with respect to x and y, where $x \geq 0, y \geq 0$.

The following characteristics of the function were found: (1) Function z is a quadratic function with respect to x and it is a fractional function with respect to y. (2) Braking distance is proportional to the square of the car's velocity. (3) Braking distance is inversely proportional to the coefficient of friction between the tyres and the road surface.

After analyzing the function, the stopping distance of a car, when the coefficient of friction between its tyres and the road surface changes, was found using Excel. Students set a realistic situation for the car by having it run at 40 km h^{-1} on a tarmac road on a bright day when they found the stopping distance. Through this experiment using Excel, students realized that it was a realistic problem and that such functions appear in everyday phenomena. Furthermore, to visually confirm the relation between x and z and the relation between y and z, a 3-D graph was drawn with the aid of a computer and presented to the students (Fig. 13.2).

Finally, a curved surface model was presented to students, because it visually confirmed that a two-variable function can be considered a single-variable function when one variable is assumed to be a constant, building on previously learned knowledge of functions (Fig. 13.3).

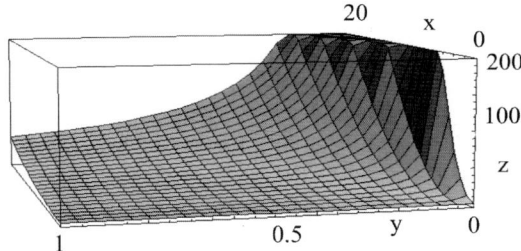

Fig. 13.2 Graph of stopping distance

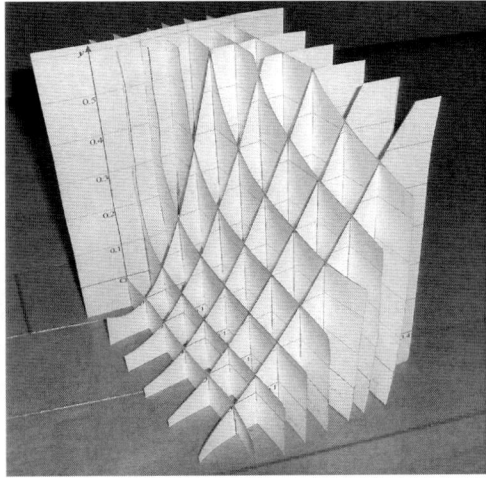

Fig. 13.3 The physical model of the curved surface

2.2.2 The Sound Propagation Example

Next, the students were taught about mathematical modelling of sound propagation. The minimum unit which constitutes a tone is called a *pure sound*. In this example, pure sound was considered.

First, the students were shown that a single vibration can be expressed with trigonometric functions of the form $y = \sin \theta$. Then, the example in which sound is transmitted from a speaker was shown, and it was explained that sound is a vibration of air. $\theta = 2\pi ft$ was used with f being the number of times a wave vibrates in 1 s (frequency, measured in Hertz (Hz)) and t being the time to vibrate. Sound can then be expressed as $y = A \sin 2\pi ft$ where A is an amplitude.

In order to clarify the relation between sound and function, an experiment was conducted which made a real sound using a personal computer. The following characteristics of sound were found from this experiment. (1) If the frequency of a sound is large, it will produce a high sound. Conversely, it will produce a low sound if its frequency is small. (2) If the amplitude of a sound is large, it will produce a loud sound. If the amplitude of a sound is small, it will produce a quiet sound.

Sound can be expressed as the function $z = x \sin aty$, where $a = 2\pi$, x is amplitude, and y is frequency. The following characteristics of the function were found. (1) Sound is a 3-variable function dependent on time, t; frequency, y; and amplitude. (2) Sound is a trigonometric function with respect to time t; a primary function with respect to amplitude, x; and a trigonometric function with respect to frequency, y. (3) When one of the time, the frequency, and the amplitude is considered as a constant, it is a two-variable function.

Sound can actually be built up by combining various pure sound waves. Students made an expression that added two or more pure tones, and experimented on the resulting sound using a personal computer. Moreover, they experimented on the sound of *resonance* by overlapping two sound waves of slightly different frequency with a personal computer.

In order to deepen their understanding of the function of sound, students created a formula for sound using a multi-variable function. Finally, students made a curved surface model of function $z = x \sin y$ and $x = \sin xy$. The surface was constructed by combining pieces of paper one by one to get a curved surface section. Through this construction activity, students better understood the characteristics of the function by creating them in this way.

3 Results from an Evaluation Questionnaire

An evaluation was carried out using a questionnaire which was completed by students after the course to assess their understanding of the course material.

3.1 Content of Investigation

The questionnaire explored the following themes:

- Two-variable functions: Did students become better able to consider the behavior of the function and in particular the degree of the function?
- The example of the multi-variable function: Can students consider the theorems and the formulae that have been studied up to now as multi-variable functions?
- The Vehicle Stopping Distance Example: Can students model the stopping distance of the car through an expression and can they analyze that expression?
- The Sound Propagation Example: Can students model the phenomenon of sound through an expression and can they analyze that expression? Moreover, when students compare the results of their analysis with the phenomena, can they verify those results?
- The Sound Propagation Example: Can students build an expression for the propagation of the sound based on the results of their analysis?
- Constructing a curved surface model: After this activity, do students understand two-variable functions better?

The questionnaire contained the following questions (in addition to others not listed):

[A] Answer the next question about the function $z = \pi x^2 y$. Which is the variable that affects the value of z greatest, x or y?

[B] Write an example of a multi-variable function from the theorems and the formulae that have been studied up to now.

[C] (1) Do you understand that stopping distance is a three-variable function depending on the car's speed, the coefficient of friction between the tyres and the road surface, and thinking time?
    (2) What relations do the braking distance of a car and the following have?
        (a) Speed
        (b) The coefficient of friction between the tyres and the road surface.

[D] (1) Have you understood that the propagation of sound is caused by vibrations in air?
    (2) Have you understood that sound (pure sound) can be expressed with trigonometric functions?
    (3) Sound (pure sound) can be expressed with the function $y = A \sin 2\pi f t$.
        (a) When the value of the frequency f is changed, how does the sound change?
        (b) If the value of the amplitude A is changed, how does the sound change?
    (4) When time is made into a constant, what are the remaining variables in the function describing sound?

[E] Please construct an expression that shows the sound from the function.

[F] Please make a curved surface model of a two-variable function.

3.2 Results from the Evaluation Questionnaire and Discussion

[A] The correct answer rate was 90% or more. Students considered the problem using the knowledge of single-variable functions that had already been learned by thinking of one of the two variables as a constant.

[B] In an investigation before the class, few students could answer a question similar to this. Multi-variable functions given by students as answers included: Conic volume, Sine Theorem, Combined gas law, etc.

Students were now able to give various theorems and formulae as examples of multi-variable functions. It assumed that each student's idea of what functions are changed.

[C] As for the result of (1), the influence of each student's physics ability was large. As for the result of (2), when the function expression was completed, even students not good at physics understood the character of the function expression (Table 13.1).

[D] The ratio of students who understood (1) and (2) (Table 13.2) was better than the ratio in the previous question, [C]. The propagation of sound seemed to have been familiar and it was a more comprehensible example for the students than the stopping distance of the car. With regards to (3) (a) and (b), and (4) of this question, the students' rate of correct answers was 90% or more.

It can be inferred that most students understood the relation between the behavior of the function and the phenomena.

[E] The following examples are functions which the university students created as multi-variable functions.

$$y = \sin 2\pi n^{\frac{1}{3}}(t^n + 4000t), \quad y = \frac{1}{t}\sin 2\pi \, 4000\frac{1}{t}, \quad y = e^t \log t \sin 2\pi \, 2000 t^2$$

The sounds of the functions that the students constructed were emitted using a PC. In one example, the sound from one of the functions that the

Table 13.1 The results of Q. [C] (%)

Level of understanding	(1)	[C]		(2)(a)	(2)(b)
Understood very well	46.4	Correct answer		92.9	89.3
Understood	42.9	Mistake		7.1	10.7
Not understood well	10.7				
Not understood at all	0.0				

Table 13.2 The results of Q. [D] (%)

Level of understanding	(1)	(2)	[D]		(3)(a)	(3)(b)	(4)
Understood very well	60.7	64.3	Correct answer		92.9	92.9	96.2
Understood	32.1	35.7	Mistake		7.1	7.1	3.8
Not understood well	7.1	0.0					
Not understood at all	0.0	0.0					

students constructed could not be heard. In that case, the students mutually discussed and improved the function expression. Such a class is extremely rare in Japan. We recognized the importance of a class in which students can engage in discussion in this way.

The importance of the process of analyzing the character of the function expression that models the phenomenon, and applying the expression to the phenomenon again is demonstrated by the following students' impressions: "By making sound with a PC, we felt the expression of propagation of sound more realistic." "We made some numerical expressions of sound, and by those sounds made with a personal computer, we could understand the amplitude and the frequency very well."

[F] When we presented a 3D graph on a PC, many students said, "The graph was able to be understood" and similar expressions. Students, however, understood the concepts more by constructing a curved surface model than by just looking at presentations of 3D graphs on a PC. A curved surface model was created by combining pieces of paper one by one to get a curved surface section. In this way, they could understand the structure and they understood the importance of actually making a curved surface model (Fig. 13.4). This is clear from the students' remarks. "I have understood more by building a model of a curved surface than having seen a graph with a PC. A 3-D figure is easily made by putting the sections of the graph in order." "I forgot the time, concentrated, and worked on making my model. By making the model, I understood the graph of my 2-variable function correctly." Moreover, when students learn various other mathematical concepts, such as partial differentiation and double integration, the construction process of a curved surface model is useful.

The results of the evaluation questionnaire indicate that there was a change in the students' ideas about functions. The change appears in the students' impressions of the class as follows: "When I was a high school student, I merely remembered the calculation of the function. In this class I acquired various concepts and how to use functions. " "It has been understood that there are functions in our surroundings

Fig. 13.4 Models of the curved surfaces which students constructed

and not only in mathematics." More than half of the students who undertook this class made remarks such as this. They were able to clearly consider and use functions outside of expressions. They were able to actually understand the functions and not merely find answers. Also, their view of the usefulness of functions increased because they came to understand that functions are a significant part of everyday life. Mathematical modelling is an important method because it builds and uses concepts linked to functions.

4 Conclusion

From the results of the questionnaire, students came to consider theorems and formulae which they had already learned as multi-variable functions. As a result, they understood functions are not only shown by expressions and the phenomena of our surroundings exist in theorems and formulae. Especially, students understood through mathematical modelling that the events in their surroundings are not decided from one factor but from two or more factors, and, therefore, can be shown through multi-variable functions. As a result, a change was seen in the students' attitude to functions.

By changing the current education content and study methods, students were able to develop their understanding of functions and a solution to a problem in Japanese mathematical education was investigated. We believe these changes should be utilized at the high school and early university level to maximize their effectiveness. We confirmed the effectiveness of mathematical modelling.

We are convinced that we encouraged and prepared students to voluntarily think about and solve various mathematical problems by themselves by teaching unusual and innovative classes with a functions theme and exploring ideas and problems related to functions.

Reference

Yokochi, K. (1983). *Su Daisu Kaiseki No Taikeika To Jissen [Systematization and practice of a number, algebra, and analysis]*. Tokyo: Gyousei.

Chapter 14
Mathematical Modelling in Secondary Education: A Case Study

José Ortiz and Aldora Dos Santos

Abstract We are interested in how students solve problems related to their physical and social world. The participants are five first year high-school students (11–13 years old) from an urban area in Venezuela. The study considers the processes and the representations used in the problem solving. This case study research uses a qualitative approach, results from which reveal that the problem solving schemes demonstrated in this study by students new to modelling are mostly linked to the accepted descriptions of mathematical modelling, particularly in the case of identifying the problem situation and interpreting the solution in the real world context. However, the students do progress, relying on the structuring of numerical answers with measure units and verbal representations. There is absence of graphical representations.

1 Introduction

Mathematical modelling offers an organized and dynamic alternative method by which the gap between mathematics and the real world may be reduced. With this in mind and to explore the extent to which this is so, this chapter has been focused on the analysis of the mathematical modelling and the representations utilized by first year high-school students in solving problems of the physical, natural, and social world. It is assumed that the process of mathematical modelling is more effective within the context of the students' own environment (Biembengut 2007); in that way, school practice is enriched and students learning of mathematical knowledge is greater (Bonotto 2007). At the same time, the appropriate use of

---

J. Ortiz (✉)
University of Carabobo, Valencia, Venezuela
e-mail: ortizjo@cantv.net

A.D. Santos
Monseñor Francisco Miguel Seijas Bolivarian High-School, Tinaquillo, Venezuela

different representational systems helps good modelers in their understanding of the problem situation and the realization of different stages in the modelling process (Garcia and Ortiz 2007).

We consider that the school curriculum should encourage students' understanding and comprehension of the world. Therefore, the inclusion of mathematical modelling in high-school education furthers this aim: enabling students to set out and solve real problems with objective criteria and consonant with their social and cultural environment and current scientific advances. From this perspective, the didactic utility of strategies which include the context of the student in teaching and learning is recognized (Kaiser and Schwarz 2006; Ortiz et al. 2007).

However, in spite of the fact that modelling clearly enhances teaching and learning leading to higher levels of attainment, in the Venezuelan curriculum, the utilization of mathematical modelling is not included, even though it recommends that pupils and teachers should work with problems of everyday life (Ortiz and Sánchez 2002). On the other hand, we know that mathematics teachers in service do not have a developed knowledge and experience of modelling and this, in itself, affects the utilization of modelling as a learning strategy in the classroom. It is therefore important to answer the question: What is the current reality in the mathematics classroom? Consequently, a group of students was identified to be given tasks of solving of certain problem situations; their answers were to be analyzed and possible links with mathematical modelling established. In light of the above discussion, we ask the following questions: How do students represent mathematical problems related to their physical and social environment? What modelling schemes are used by students when solving real life problems? What answers are given by the students to problems posed within a context of mathematical modelling?

2 Methodology

This investigation was carried out as a case study following Yin (2003). In this study, carried out in the 2005–2006 academic school year, the voluntary participants were five students, from Francisco Miguel Seijas Bolivarian High-School, Tinaquillo, State of Cojedes. Five problems were given to the students. No induction or previous workshop was implemented to help the students in the resolution of the problem situations. The only requirement that the students should have was to be attending the first year of high school (11–13 years of age). The research instruments comprised a problems questionnaire and an interview for which a standard script was used. The data analysis was concerned with the cognitive answers and contributions that the students made when faced with the problem situation. The analysis took account of the representations assigned to them and the possible modelling schemes used. To assist in the analysis, the modelling cycle used by Blum and Leiss (2007) was adopted. This cycle which structures modelling through seven stages: (1) building, (2) simplifying/structuring, (3) mathematising, (4) working mathematically, (5) interpreting, (6) validating, and (7) exposing.

These steps are given following the trajectory: real situation and problem → model of the situation → real model and problem → mathematical model and problem → mathematical results → real results → situation model.

For the selection and formulation of the problems, included in the questionnaire, the following were considered: (1) opening approach, (2) mathematical content for students of first year (whole numbers, divisibility criteria, minimum common multiple, maximum common divisor, percentage, capacity, and volume), (3) contextualization with the reality of the participating students in the study, and (4) applicability of the mathematical modelling and the representation systems. The selected problems were exposed, discussed, and validated in expert group meetings.

The analysis was focused on the answers, solutions and representations, opinions and affirmations of the students. In this way, it was possible to prove what mathematical modelling schemes were used, if the model followed was known or conceived by some authors in particular, or if on the contrary, the model being presented is the product of their proper thoughts and deliberations, structured and conceived through their own life experiences.

Each student was free to develop his own work style at the moment of dealing with the problem. None of the students who participated in the study was a student of the researchers; that is, they did not have the impression of being evaluated by their teachers at the moment of solving the problems. At the beginning of the activity, they were informed that they did not have limited time to solve the five problems proposed, and that they had the authorization of the institution to develop that activity.

The problems set out to the students were the following:

P1. In the Nuestra Señora del Socorro Church, there are three bells. One bell sounds with a 10-s interval between two ringings, another one with a 20, and another one with a 24-s interval. Now, if they give the first ring simultaneously, after 120 s, how many ringings has each bell rung?

P2. A house drinking water tank in the Tamanaco neighborhood, Tinaquillo, has a capacity of 2,000 l of water necessary to supply the Sanchez family. The approach is the following: (a) One of the inlet regulator valves which supplies 10 l of water a minute has been opened for 1 h. (b) Another inlet regulator valve which supplies 20 l of water a minute has been opened for 5 min. How many litres of water are still needed to fill the tank?

P3. In the La Plaza bookstore, Tinaquillo a decision was made so that on the last day of each month, a discount was offered when buying any educational text, where for each Bs.100 spent by the buyer, he or she will save Bs.20. On the day appointed for this offer, Miguel Reyes went to the bookstore and at the end of his purchase, he noticed that by buying the book, he had saved Bs.1,800. What was the price of the book? While in La Esperanza bookstore for each Bs.100 that the buyer spent, he or she would save Bs.30. How much would Miguel Reyes save if his book would have been bought in La Esperanza bookstore? (Bs. means Bolivars).

P4. An Oriental wholesaler, in his supplies store, offers the Tinaquillo merchants, a curious coffee mixture made up in the following way. He mixed 100 kg of coffee at the price of Bs.8,000 per kg with 60 kg of coffee at the price of Bs.4,400 per kg, with 40 kg of coffee at the price of Bs.10,800 per kg. How much will 20 kg of this mixture of coffee cost?

P5. Twelve dentists were selected at random from the Tinaquillo area. Each dentist treats at least two students from a class of 29 students at Monseñor Francisco Miguel Seijas Bolivarian High-School. How many students can each dentist treat?

For the interviews, two students were selected from the five participants. The criteria used for their selection was the type of answers given to the formulated problems. Through the interview it was possible to explore deeper the way that students faced the problems; that is, to investigate procedures followed by the students in their search for solutions to the given questions in situation. Students were told that the interview was to take place in the library of the educational center and that it would be recorded. Interviews were carried out individually for each student selected (S1 and S2).

3 Results and Discussion

In solving the problems, some students were cautious, while others acted in a fast and determined manner at the moment of answering. It was observed that students (S1, S2, S3, S4, S5) solved the problems (P1, P2, P3, P4, P5) in different ways; that is, in some cases (S1, S2), they set out arguments and procedures to explain their answers. While, in other cases (S3, S4, S5), no well-sustained answers were observed, nor were the students sufficiently explicit in their justification of the results.

In the first problem (P1), the students presented numerical answers with verbal explanations. For example, S1 divided the problem situation into three "phases": (a) He considers that "the first bell rings with ten (10) seconds between two ringings, now it will ring 24 ringings in 120 seconds". In order to solve P1, he puts into columns (numerical representation) his interpretation of the phenomenon, starting from number 10 to 120 (Table 14.1). It is observed that S1 does not count the bell ringings adequately; as at 20 s he considers that 4 ringings have taken place; at 30, 6 ringings have occurred; and so successively until reaching 120 s with 24 ringings, which does not correspond with the assumed conditions. This erroneous reasoning could have been picked up by the student S1 if he had resorted to

Table 14.1 First phase of S1 for solving P1

10:2	50:10	90:18
20:4	60:12	100:20
30:6	70:14	110:22
40:8	80:16	120:24

other representations or if he had had the opportunity to discuss the results with other students and to compare them with the real situation of P1.

In phases (b) and (c), he makes the same reasoning error as in phase (a). In (b), S1 states that the "second bell rings with an interval of 20 seconds between two ringings, now it will ring twenty-two ringings in 120 seconds". Likewise, he writes in column style, starting the counting from number 20, adding 10 each time through to 120. In (c), S1 states that "the third bell rings with a 24 (twenty-four) second interval between two ringings, now it will ring 50 times in 120 seconds". On the other hand, S2 and S3 set out and solve utilizing the division of 120 by 10, 20 and 24. They come to the conclusion that the 3 bells give 12, 6 and 5 ringings respectively. In other words, these students show competence to utilize their mathematical knowledge in order to solve the problem (Maaß 2006). With regards to students S4 and S5, it seems that they did not understand the situation and, consequently, gave erroneous answers, not consistent with the conditions of the problem and without specifying the steps taken.

In the approach to the P2 situation, the students assumed that the tank was initially empty when the inlet regulator valves were opened, corresponding with the comprehension of the phenomenon and the simplification of the situation in order to structure the real model (Blum and Leiss 2007). Then, S1, S2, and S3 solved and found that the inlet regulator valves provide 200 l of water, and so in this manner, 1,300 l of water is needed to fill the tank.

In situation P3, student S1 arrives at a numerical representation (percentages table) in order to understand what was happening with the discounts given for each bookstore. Finally, he finds that Manuel Reyes could have saved Bs.2,700 if he had bought the book in La Esperanza bookstore. Student S2 uses arithmetical knowledge and also solved the problems. Students S3 and S4 had the intention of dealing with P3, but they did not finalize the results. Student S5 did not appear to be able to solve the situation. The work, without the results of students S3, S4, and S5, could have been caused by conceptual deficiencies in knowledge of percentages. In the same way, this could be a consequence of lack of knowledge and expertise in solving of modelling problems during the process of teaching and learning mathematics.

In problem P4, it seems that students tackled it with low comprehension of the given situation. For example, S1 considers the cost of 20 kg of each type of coffee with the mistaken intention of obtaining the cost of the 20 kg of the mixture. On the other hand, a low command of the arithmetical operations was observed, specifically with regards to the multiplication by the units followed by zeros, as there were errors in the calculations carried out. This problem, P4, became the most difficult one for the five students, as it had different amounts of coffee as well as different prices per kg. This perception was confirmed in the interviews.

With regards to P5, students S1, S2, and S3 assumed that all students were treated at the same time, as they only asserted that "seven dentists treated two students each, and five dentists treated three students each." Besides, in their results, they did not show other working arrangements for the dentists. Student S4 assumed that not necessarily all the students are treated at once; in that regard, he simply answered that "each dentist treats two students." In the case of student S5, there was no evidence that the problem was understood. He stated that "each dentist could treat 14 students" which is not consistent with the given conditions in the problem. In general, in the solution of P5, the students did not show that they understood and could work with the real situation described (Kaiser 2007); despite this, there were systematic intuitive attempts to solve P5.

The answers given by the students revealed a lack of experience in problem solving, confirmed in their interviews when they stated that: "In the school we seldom solve problems like these (S1, S2)".

Table 14.2 displays a synthesis of the answers given by the students for each of the problem situations which was given to them. It can be seen that the students, even though they do not have the benefit of experience or know the structures of mathematical modelling, do have an intuitive idea that helps them to solve the proposed situations. They do in fact have a tendency, even partial or ingenious, toward descriptions of modelling behaviour equivalent to the classic ones, such as those described by Blum and Leiss (2007). Note that S1 and S2 are the students who understand, simplify, mathematise, carry out mathematical procedures, interpret solutions, but do not validate nor explain the results obtained (Table 14.2). Student S3 only shows signs of comprehension and simplification. Students S4 and S5 set out to understand the situations, but do not make progress in the modelling process. In regard to the representations used, they only made use of the numerical tables. This reveals that the students require a style of teaching which will give priority to the multiple representation systems in the solving of problems. In general, the answers given by the students to each one of the problems of the questionnaire reinforce the researchers' view that the incorporation of the modelling in the training of teachers, and, even more, in the everyday school practice in the mathematics classroom is a priority (Lingefjärd 2006; Maaß 2006).

4 Conclusions and Recommendations

In relation to the representations, it was found that the students used numerical representations, with or without measure units. Likewise, they stated their answers in a verbal manner without using graphic representations. This could be an indication that, in the classroom, the teachers do not resort to the several representation systems to which the student could appeal at the moment of solving the problem.

On the other hand, the schemes of mathematical modelling, which emerged from the responses of the students, can be partially framed in the proposal of Blum and Leiss (2007); that is, the students looked for the comprehension, simplification, mathematization, mathematical solutions and interpretation, even though they did not validate nor put into effect the results obtained in each problem. For the case of understanding the problem, the students showed competence to discern between relevant and non-relevant data (Maaß 2006). Likewise, they showed evidence of coming to numerical representations and to the use of mathematical knowledge to

Table 14.2 Summary of the answers given by the students

	P1	P2	P3	P4	P5
S1	Identifies the situation Builds models Chooses mathematical contents Does not do validation	Builds the model and solves	Realizes identification of the problem Uses mathematical methods Interprets conclusions	After solving, makes a validation of the model	Adopts a convenient model and makes numeric representations
S2	Builds a model and solves Was the problem he liked most	Builds the model and solves	Builds the model and solves Interprets the solution Realizes validation	Identifies the problem situation Builds the model Chooses mathematical contents Interprets the obtained conclusions	Builds the model and solves Interprets the solution Validates the model Explains his arguments
S3	Builds a model and solves Does not interpret solutions	Solves and does not interpret solutions	Identifies the problem situation but does not advance	Does not overcome the doubt in the interpretation of the variable "product price"	Does not explain much, but he does it with logic
S4	He cannot follow the utilized scheme Answers without procedures nor written analyses	He cannot follow the utilized scheme	Does not identify the problem situation	Does not overcome the doubt in the interpretation of the variable "product price"	He cannot follow the utilized scheme Gives wrong answers and without arguments
S5	Builds a model Presents errors in the mathematical resolution	Cannot follow the utilized scheme	Does not answer Lack of analysis is perceived	Does not overcome the doubt in the interpretation of the variable "product price"	Cannot follow the utilized scheme. Gives wrong answers and without arguments

solve problems, even though in this last point, lack of skill was observed to realize arithmetical calculation. However, it must not be overlooked that in their academic learning, the students have not been taught how to deal with this type of problem. This might mean that the participants possess an intuitive sense of modelling, which could be improved by a teacher who has an adequate preparation and who utilizes modelling strategies in the mathematics classroom. Another important aspect to be considered is that the participants always assumed that there would be one answer only, and no more than one. This can be a consequence of "old schemes" utilized in classrooms, where many teachers assert as an "error" the fact that a problem provides more than one answer.

In regard to the limitations of the students when faced with problems embedded in a particular context, they could be linked to traditional teaching methods and schemes still prevail in the high-school mathematics classroom. In this regard, Ikeda (2007) shows that the training of the teachers is one of the recognized aspects to be attended to in the future so that the adequate utilization of mathematical modelling occurs in the mathematics classrooms. This could help students to become competent in dealing with and solving problem situations. So, in the Venezuelan case, a process of dissemination mathematics modelling among the teachers and in parallel their incorporation in the training programs of the mathematics teachers is desirable.

Finally, the adequate incorporation of mathematical modelling in the classroom will motivate and open further possibilities for the learning and teaching of mathematics at high-school levels. The curriculum needs to be organized so that physical, natural, and social contexts are considered in proposing real and meaningful situations to the students. This could contribute to students critically understanding and appreciating their environment with the aim of finding possible improvements to, and preservation of, the environment, as well as understanding the importance of mathematics in this regard.

References

Biembengut, M. S. (2007). Modelling and applications in primary education. In W. Blum, P. Galbraith, H. Henn, & M. Niss (Eds.), *Modelling and applications in mathematics education (The 14th ICMI study)*. New York: Springer.

Blum, W., & Leiss, D. (2007). How do students and teachers deal with modelling problems? In C. Haines, P. Galbraith, W. Blum, & S. Khan (Eds.), *Mathematical modelling (ICTMA 12): Education, engineering and economics* (pp. 222–231). Chichester: Horwood Publishing.

Bonotto, C. (2007). How to replace word problems with activities of realistic mathematical modelling. In W. Blum, P. Galbraith, H. Henn, & M. Niss (Eds.), *Modelling and applications in mathematics education (The 14th ICMI study)*. New York: Springer.

Garcia, R., & Ortiz, J. (2007). Representaciones y Modelización Matemática en la Resolución de Problemas. In E. Castro & J. L. Lupiañez (Eds.), *Investigación en Educación Matemática: Pensamiento numérico (Libro homenaje a Jorge Cázares Solórzano)* (pp. 283–302). Granada: University of Granada.

Ikeda, T. (2007). Possibilities for, and obstacles to teaching applications and modelling in the lower secondary levels. In W. Blum, P. Galbraith, H. Henn, & M. Niss (Eds.), *Modelling and applications in mathematics education (The 14th ICMI study)*. New York: Springer.

Kaiser, G. (2007). Modelling and modeling competencies in school. In C. Haines, P. Galbraith, W. Blum, & S. Khan (Eds.), *Mathematical modelling (ICTMA 12): Education, engineering and economics* (pp. 110–119). Chichester: Horwood Publishing.
Kaiser, G., & Schwarz, B. (2006). Mathematical modelling as bridge between school and university. *Zentralblatt für Didaktik der Mathematik, 38*(2), 196–208.
Lingefjärd, T. (2006). Faces of mathematical modelling. *Zentralblatt für Didaktik der Mathematik, 38*(2), 96–112.
Maaß, K. (2006). What are modelling competencies? *Zentralblatt für Didaktik der Mathematik, 38*(2), 113–142.
Ortiz, J., & Sánchez, L. (2002). La Educación Básica en Venezuela. Consideraciones sobre el Currículo de Matemática en la Tercera Etapa (13–15 años). In A. Maz, M. Torralbo, & C. Abraira (Eds.), *Currículo y Matemáticas en la Enseñanza Secundaria en Iberoamérica*. Córdoba: University of Córdoba.
Ortiz, J., Rico, L., & Castro. (2007). Mathematical modelling: A teachers' training study. In C. Haines, P. Galbraith, W. Blum, & S. Khan (Eds.), *Mathematical modelling (ICTMA 12): Education, engineering and economics* (pp. 241–249). Chichester: Horwood Publishing.
Yin, R. (2003). *Case study research. Design and methods (Applied social research methods series)* (Vol. 5, 3rd ed.,). Beverly Hills: Sage.

Chapter 15
Students Overcoming Blockages While Building a Mathematical Model: Exploring a Framework

Sanne Schaap, Pauline Vos, and Martin Goedhart

Abstract In the Netherlands, modelling is a compulsory topic for all pre-university science-stream students. Nevertheless, these students have difficulties in building a mathematical model. Our research aims at identifying the occurrence and removal of blockages when students create mathematical models. By means of a pilot study, we looked for an appropriate framework to identify students' obstacles and opportunities during this process. The results show that the initially chosen framework, which describes modelling as a cyclic process, needs addition from frameworks referring to problem solving, metacognition and beliefs.

1 Introduction

Modelling was made a compulsory topic in the mathematics curriculum of the pre-university science-stream in the Netherlands, because many students, in particular science-stream students, will encounter mathematical models in further studies and professions. However, building mathematical models is a difficult curriculum topic for students: Mathematical models (formulas) are often given ready-made in written tests (Vos 2007), and strategies for building mathematical models are not taught in mathematics classes. As a consequence, students do not learn to create their own models.

S. Schaap (✉)
University of Groningen, Groningen, The Netherlands
and
University of Amsterdam, Amsterdam, The Netherlands
e-mail: s.schaap@uva.nl

P. Vos
University of Amsterdam, Amsterdam, The Netherlands

M. Goedhart
University of Groningen, Groningen, The Netherlands

To gain insight into the strategies students use in building models, we want to identify the opportunities and blockages pre-university science-stream students encounter. In this chapter, we report a pilot study that aimed at finding an appropriate framework to identify these opportunities and blockages.

2 Theoretical Framework

This section will present the framework that was initially selected to identify opportunities and blockages when students create mathematical models. We conceive the activity 'creating a mathematical model' as a sub-activity of the activity 'modelling'. The presentation of the initial framework is therefore preceded by explaining what we mean by these activities.

2.1 *Modelling*

Blum and Leiß (2005) describe modelling as a chain of activities, starting from a 'real situation'. When the problem in this situation is understood (*understanding*), the 'situation model' is structured and simplified (*structuring* and *simplifying*) into a 'real model' and thereupon converted into a 'mathematical model' (*mathematising*). 'Mathematical results' are then generated and embedded in the context of the initial problem. If these 'real results' do not answer the initial problem, the modelling cycle is 'run through' another time. Besides Blum and Leiß, other researchers have described the modelling process with a cyclic model (Blomhøj and Jensen 2007; Galbraith and Stillman 2006; Maaß 2006; Vos and Roorda 2007), and termed this as a *modelling cycle*. All these researchers include in their modelling cycles the activity of translating reality to mathematics. In our research we, too, conceive modelling as a complex process, in which the translation process of reality to mathematics plays an important role. Building a mathematical model therefore is a complex set of activities within the entire modelling process.

Kaiser and Sriraman (2006) have overviewed international perspectives on modelling in mathematics education. In their terminology, our research links to directions in which mathematical knowledge is used for modelling. Our research does not link to directions in which modelling is a vehicle to learn mathematical concepts. Bliss (1994) used the terms *explorative modelling* and *expressive modelling*. With the first type, students discover and test given mathematical models, and with the second type, students build a mathematical model themselves by pragmatically using mathematics to solve a problem. Our research focuses on expressive modelling.

2.2 Mathematising

Our research focuses on the activities that are needed to build a mathematical model, starting from a situation. We will denote these activities by the term *mathematising*, following the definition of Vos and Roorda (2007). We choose to mark the term mathematisation with a clear starting point (a situation) and a clear endpoint (the mathematical model), so that these activities are recognisable and operational for research purposes. We have a broader definition than that of Blum and Leiß. Our reasons for this choice are: (1) The terms 'situation model' and 'real model' are not always recognisable stages in the modelling process and may be problem-dependent. (2) We take activities such as understanding, structuring, simplifying and formalising as interrelated. (3) Students do not sequentially follow the steps in the modelling cycle, but have their own 'modelling routes' (Borromeo Ferri 2007).

In literature, we found that mathematising can consist of the following activities: *analysing the situation, understanding, interpreting the context, defining the problem, structuring, simplifying, assuming* and *formalising*. These activities are not necessarily performed in this very order by the student.

2.3 Opportunities and Blockages

First, this sub-section will discuss a framework on opportunities and blockages that consider modelling as a cyclic process and that focus on activities that are part of this cycle. Second, it will discuss frameworks that focus on modelling competencies or skills which lead to successful modelling. Third, it will make a choice of initial framework for our pilot study.

Galbraith and Stillman (2006) and Stillman et al. (2010) developed a framework to identify the occurrence or removal of blockages during the modelling process. They depart from a modelling cycle, and build their framework by investigating for several modelling tasks what obstacles students experience during the modelling process, and how students remove these obstacles. They focus on blockages during transitions between stages in the cycle, and connect *metacognitive activities* with reflection on these transitions (reflective metacognitive activities). They mention reflection as a potential way to overcome blockages of low intensity. The framework does not focus on those *opportunities* that stimulate the modelling process without necessarily removing blockages.

Maaß (2006) also starts from the modelling cycle, but focuses on competencies as opportunities in the modelling process, instead of focussing on blockages. Besides competencies needed to execute activities from the modelling cycle, she describes other necessary modelling competencies: the competency to use knowledge about the modelling process; competencies to '*keep an overview over their proceedings and aim at a goal*' (p. 137); competencies to work with a sense of direction to a solution; competencies to see that mathematics offers opportunities to solve real world problems.

To adequately describe modelling competencies and in order to let students develop modelling competencies, Blomhøj and Jensen (2007) distinguish three dimensions to analyse modelling competencies: (1) the elements of the modelling process used and the extent of reflection by students, (2) the technical level of activities, and (3) the types of situations that activate modelling competencies.

Galbraith and Stillman (2006) and Stillman et al. (2010) use their modelling cycle as a starting point to identify and document the occurrence or removal of blockages, whereas the research by Blomhøj and Jensen (2007) and Maaß (2006) primarily focuses on competencies. Galbraith and Stillman (2006) notified that their framework still needed further validation. Because our research also focuses on the occurrence and removal of obstacles, we will consider this as an invitation to use, specify and to expand their approach, but restricted to the subset of activities that are needed for the mathematisation process (understanding, interpreting the context, structuring, simplifying, formalising, etc.). In comparison to their work, our research will also include the search for opportunities without blockages.

3 Method

To explore the framework suggested in the previous section, we organised a pilot study. Six pre-university science-stream students in grade 11 of the Marecollege in Leiden (the Netherlands) participated in the study. The data were gathered through *task-based interviews* of 60–90 min. In order to stimulate the students to enter in a dialogue, the students were interviewed in pairs. The three pairs were Jonathan and Tim, Darlene and Dianne, and Michelle and Irene (not their real names).

We used five tasks, which were divided among the pairs, such that each pair worked on four of them, and such that each task was attempted at least once. We selected tasks on their mathematisation demand (either implicitly or explicitly asked for) and on time demand (to be completed within a short time span). The tasks were selected from different sources: the 'Swimming Pool Task' from a course on problem solving, the 'Horizon Task' from a modelling course for teacher training and the remaining tasks were taken from a national examination, a textbook and a journal for mathematics teachers. Due to the page limitations, we will primarily focus on the 'Swimming pool task' and the 'Horizon task' (see Fig. 15.1); the results on the other tasks will be summarised without further discussion.

The video recordings were transcribed verbatim. Transcript-fragments were then characterised in terms of student activities. If an activity was identified as an opportunity or a blockage, then we determined if the activity could be located in Blum and Leiß's modelling cycle. We also used students' notes and field notes of the interviewer to support our findings. The analysis of the three interviews thus aims to verify to what extent the activities from the modelling cycle are apt to identify the occurrence and removal of blockages during the mathematisation process.

15 Students Overcoming Blockages While Building a Mathematical Model   141

Swimming Pool	Horizon
In a garden of 10 m x 15 m we dig a swimming pool (length 6 m and width 3 m). The pile of soil hereby created is spread evenly over the garden. This has as effect that when you dig 1 m and spread the soil, the swimming pool will be deeper than 1 m. How deep do we have to dig to obtain a swimming pool with depth 2 m?	You are standing on the beach and are looking at the horizon. Visibility is excellent. How far away is the horizon? Take 6370 km as radius of the earth.

Fig. 15.1 The 'Swimming pool task' and the 'Horizon task'

4 Results

This section reports on observed opportunities and blockages when students worked on the two tasks mentioned above. Two pairs worked on the Swimming Pool Task, and all three pairs worked on the Horizon Task.

4.1 The Swimming Pool Task

Dianne and Darlene used the symbol x to describe the height of the layer of soil spread over the garden. They subsequently derived a correct equation to calculate the volume of soil A that has to be excavated ($6 \cdot 3 \cdot (2 - A/132) = A$). They were not hindered by any blockages and attained a mathematical model. This differed from the process of the second pair, who started to calculate from a concrete example:

Tim:	So then… let's first calculate how d… deep the pile of soil becomes that gets on top, when you dig one metre …
Tim:	So that will then be six times three equals… eighteen. So then you have…
Both:	Eighteen cubic metres of soil.
Jonathan:	And you spread it out.
Tim:	You spread it out.
Jonathan:	But we first have to calculate the surface of this bit.
Tim:	So it is 150 minus eighteen metres.
Jonathan:	Is 132.
Jonathan:	So when you dig one metre, then you dig eighteen cubic metres of soil. We are going to spread this out over 132… ehrm is one metre deep… so when you dig one metre… and then you remove the soil and then you spread it out so then it isn't really one metre deep… so it becomes ehrm more.

Tim: So let me see. So you have 132 m squared times x equals eighteen cubic metre.
Jonathan: So x equals eighteen divided by 132.
Jonathan: Then it becomes one metre and fourteen centimetres.

In this example, they moved through the modelling cycle using a concrete example (calculating the depth of a pool for 1 m digging), knowing that they will not immediately attain the required answer. We termed this as *scouting the problem by working with concrete examples* and identified it as an opportunity. Subsequently, they also calculated the resulting depth of the swimming pool after digging 2 m deep:

Jonathan: … and then times two (here, Jonathan multiplies the number that belongs to one metre digging by two). Yes that is indeed linear.

From the calculations with two concrete examples, Tim and Jonathan conjectured that the excavation depth is directly proportional to the depth of the pool. We have identified this opportunity as *searching for a model using an inductive method*.

Thereafter, Jonathan and Tim used variables and Tim instructed Jonathan to write down his intermediate results ('also put that under it'). Jonathan followed Tim's directions and thereafter saw in his writing (see Fig. 15.2) a way to determine the desired depth:

Jonathan: But look… here it says y plus x and ehrm x is… yes but look here you have it. y plus x equals two. x equals zero point fourteen y. So y plus zero point fourteen y equals two. And then we have one point one fourteen y equals two. And y then is two divided by one point fourteen.
Tim: Jonathan, you are a genius. (laughs)

Writing down systematically intermediate results enabled Jonathan to wind up the formalising process.

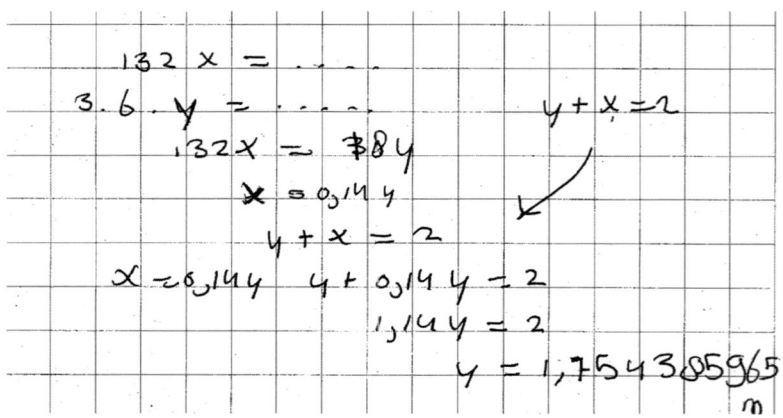

Fig. 15.2 Excerpt from Tim and Jonathan's computation for the swimming pool task

4.2 The Horizon Task

The following discussion took place after Tim and Jonathan made a sketch of the situation:

Tim: But I think, Jonathan, that we have to make a tangent with the circle, such...
Jonathan: Yes.
Tim: that we... and then at this point this is where you see the horizon.

Next, Jonathan drew the situation again – this time neatly with set square and compass – and took his time. In this way, Jonathan found that the angle between the radius of the earth and the line of sight was 90°. After this, they successfully solved the task. We regard *drawing the problem situation* as an opportunity because a sketch gives students the possibility to visualise the situation.

The second pair, Darlene and Dianne, discussed the term horizon. Dianne described it as 'the spot where it disappears like this', while gesturing with her hands. We consider this as an opportunity for *exploring the problem*. Subsequently, Darlene made an *erroneous assumption* that the horizon is one-eighth of the perimeter of the earth, which hindered their progress. Dianne noticed that 'when you are taller the horizon will be further away', but they were not able to translate this idea into a mathematical expression. We identified *not recognising a relevant variable*, and *not being able to convert the specifications of a relevant variable into a relation* as blockages. Later on, they talked about the formulation of the problem situation, where they were looking at the text for something to hold on to:

Darlene: What shall we do with the radius of the earth? Or is that useless information?
Dianne: Well I actually don't think so because it is also the only information so
Darlene: Yes (laughs)
Dianne: (laughs) So when it is useless then we have nothing left.

This was identified as a blockage in *picking up the problem situation.*

Irene, member of the third pair, concluded that they did not need to consider the spherical shape of the earth. She said: 'What is not mentioned in the problem text does not have to be used in calculations'. This assumption hindered her in solving the problem, because she could not find relevant directions in the problem text. Irene was looking for cues in the problem text to help her proceed through the task; this behaviour is termed as *cue-based* (Boaler 2002).

4.3 Additional Results

Analysing the five tasks, we additionally identified the following blockages: *lacking algebraic skills, overlooking essential elements in the problem text, impeding formulation of the problem text, blockage in communication.* Furthermore, we observed the following opportunities: *re-reading the problem text, subconsciously simplifying the situation, slowing down the process by taking one's time, verifying the model by estimation* and *assuming that a solution is valid*.

Michelle was very skeptical about one of the tasks. She explained 'I cannot stand this and because of this I would like to finish it as soon as possible' and during the solving process, she uttered 'I am always performing very poor on this and I always find it irritating' after which she pushed all papers over to Irene. We classified her attitude as a hindrance due to a *lack of self-confidence*, or due to considering the task as mathematically too demanding; but we also think that a *non-appealing task* can be hindering the mathematisation process.

5 Conclusion and Discussion

In our pilot study, we investigated which components a framework should contain for identifying opportunities and blockages of students who try to formulate a mathematical model. Galbraith and Stillman (2006) used a modelling cycle as a framework for their analysis on the occurrence and removal of student blockages during transitions in the modelling process. We decided to also start from a modelling cycle, that is to say, Blum and Leiß's modelling cycle.

We were able to identify several blockages in terms of the modelling cycle. In the phase of understanding, for example, we identified a blockage in *picking up the problem statement*, caused by an *impeding formulation of the problem text*, *overlooking essential parts in the problem text*, or *expecting hints, guidelines and necessary data in the problem text*. In the phase of structuring, we observed *making erroneous assumptions* and *not recognising a relevant variable*. In the phase of formalising, we perceived *not being able to convert specifications into a relation between variables* and *lacking algebraic skills* as blockages.

Besides blockages, we were also able to identify opportunities in terms of the cycle. In the phase of understanding: *picking up the problem text* by describing the problem situation, and *exploring the problem*. In the phase of structuring and simplifying: *drawing the problem situation*, and *subconsciously simplifying the situation*. In the phase of formalising: *scouting the problem using concrete examples*, *searching for a model by using an inductive method*, and *verifying the model by estimation*.

The Blum and Leiß modelling cycle describes activities that contribute to successful modelling. Successfully performing an activity is therefore an opportunity and unsuccessfully performing an activity is therefore a blockage; however, besides successfully performing an activity, students can also seek for strategies to overcome blockages (e.g. *scouting the problem using concrete examples*, and *verifying the model by estimation*). These opportunities are described in the literature as *heuristics* or *problem solving strategies* (Polya 1988; Schoenfeld 1992).

Given that modelling cycles focus on activities, types of models, transitions or stages, but not on general problem solving strategies, we conclude that we can identify a number of opportunities with a framework of problem solving strategies (heuristics).

Note that unsuccessful activities are an opportunity to develop more deeply metaknowledge about modelling and mathematisation in particular. Thus, unsuccessful activity has a double nature – a blockage and an opportunity.

The identified blockage due to *cue-based* behaviour can be located in the modelling cycle. However, this blockage does not relate to *activities* during the modelling process but relates to *knowledge* that students have about task expectations and how to handle tasks (the tasks in the pilot differed from standard tasks in Dutch textbooks). According to Maaß (1996), having knowledge about activities is a *metacognitive modelling competency*. *Slowing down the process by taking your time, re-reading the problem text* and *writing down systematically intermediate results* are examples of opportunities that cannot be identified by the modelling cycle. These opportunities overarch the cycle, because they focus on process regulation, and they differ from reflective metacognitive activities that focus locally on transitions in the cycle. In the literature, these opportunities are described with the term *metacognitive competencies* (Maaß 2006; Meijer et al. 2006; Schoenfeld 1992). Therefore, it is meaningful to extend our framework also with *metacognitive competencies* that overarch the modelling cycle.

One student was also hindered by her earlier experience with and her attitude towards mathematics. This blockage occurred during the entire modelling process, and therefore is not identifiable by means of the modelling cycle. We further noticed that a task can be hindering, when a student finds a task non-appealing. This blockage is linked to what in the literature is known as beliefs (Schoenfeld 1992). Because of this observation, we think it is meaningful to extend our framework with *beliefs*.

In conclusion, we state that the description of activities in the modelling cycle has proven to be useful: When activities from the cycle are completed, they are an opportunity, and when these activities were not performed successfully, they formed a hindrance to the process. However, we also identified occurrence and removal of blockages outside this framework; here we identified these in terms of *problem solving strategies*, *metacognitive competencies* and *beliefs*. In order to optimally identify the occurrence and the removal of blockages during the mathematisation process, we therefore need to extend the initial framework. We will use this extended framework in a follow-up study to describe the occurrence and removal of blockages during the mathematisation process, and regard problem solving strategies, metacognitive competencies and beliefs as important aspects for analysing students' behaviour.

We would like to stress that the newly proposed framework might still be incomplete, because we have interviewed only a few students, using only five tasks. It can also be argued that the selected tasks and method (task-based interview) are not appropriate to identify certain types of obstacles and opportunities.

References

Bliss, J. (1994). From mental models to modelling. In H. Mellar, J. Bliss, R. Boohan, J. Ogborn, & C. Tompsett (Eds.), *Learning with artificial worlds: Computer based modelling in the curriculum*. London: The Falmer Press.

Blomhøj, M., & Jensen, T. H. (2007). What's all the fuss about competencies? In W. Blum, P. Galbraith, H. Henn, & M. Niss (Eds.), *Applications and modelling in mathematics education* (pp. 45–56). New York: Springer.

Blum, W., & Leiß, D. (2005). "Filling up" – the problem of independence-preserving teacher interventions in lessons with demanding modelling tasks. In M. Bosch (Ed.), *Proceedings of the 4th European Congress of Mathematics Education* (pp. 1623–1633). Gerona: FUNDEMI IQS – Universitat Ramon Llull.

Boaler, J. (2002). *Experiencing school mathematics: Traditional and reform approaches to teaching and their impact on student learning*. Mahwah: Lawrence Erlbaum Associates.

Borromeo Ferri, R. (2007). Modelling problems from a cognitive perspective. In C. Haines, P. Galbraith, W. Blum, & S. Khan (Eds.), *Mathematical modelling (ICTMA 12): Education, engineering and economics* (pp. 260–270). Chichester: Horwood Publishing Limited.

Galbraith, P., & Stillman, G. (2006). A framework for identifying student blockages during transitions in the modelling process. *Zentralblatt für Didaktik der Mathematik, 38*(2), 143–162.

Kaiser, G., & Sriraman, B. (2006). A global survey of international perspectives on modelling in mathematic education. *Zentralblatt für Didaktik der Mathematik, 38*(3), 302–310.

Maaß, K. (2006). What are modelling competencies? *Zentralblatt für Didaktik der Mathematik, 38*(2), 113–142.

Meijer, J., Veenman, M. V. J., & Hout-Wolters, B. H. A. M. (2006). Metacognitive activities in text-studying and problem-solving: Development of a taxonomy. *Educational Research & Evaluation, 12*(3), 209–237.

Polya, G. (1988). *How to solve it* (2nd ed.). Princeton: Princeton University Press.

Schoenfeld, A. H. (1992). Learning to think mathematically: Problem solving, metacognition, and sense-making in mathematics. In D. Grouws (Ed.), *Handbook for research on mathematics teaching and learning* (pp. 334–370). New York: Macmillan.

Stillman, G., Brown, J., & Galbraith, P. (2010). Identifying challenges within transition phases of mathematical modelling activities at year 9. In R. Lesh, P. Galbraith, C. Haines, & A. Hurford (Eds.), *Modeling students' mathematical modeling competencies* (pp. 385–398). New York: Springer.

Vos, P. (2007). Assessment of applied mathematics and modelling: Using a laboratory-like environment. In W. Blum, P. Galbraith, H. Henn, & M. Niss (Eds.), *Applications and modelling in mathematics education* (pp. 441–448). New York: Springer.

Vos, P., & Roorda, G. (2007). Interpreting velocity and stopping distance; complementarity, context and mathematics. In G. Kaiser, F. Garcia, & B. Sriraman (Eds.), *Proceedings of the working group on mathematical modelling and applications at the 5th conference on European Research in Mathematics Education (CERME-5)*. Nicosia, Cyprus: University of Cyprus.

Chapter 16
What Did Taiwan Mathematics Teachers Think of Model-Eliciting Activities and Modelling Teaching?

Shih-Yi Yu and Ching-Kuch Chang

Abstract This chapter reports perceptions and obstacles of 16 secondary mathematics teachers after experiencing three model-eliciting activies (MEAs) and designing one MEA in a 9-week course linked to a master's degree program in education for in-service teachers. Data collections included the learning sheets that showed teachers' strategies of the three MEAs and the results of the MEA they designed, observation journals, reflection journals, questionnaires, interview reports, and video tapes of the classes. The results showed that teachers regarded modelling as a problem-solving process, and agreed with the advantages of implementing the MEAs in mathematics classrooms; they also mentioned obstacles of implementing MEA and designing MEA.

1 Background

Researchers have emphasized more and more on the issues of enhancing students' mathematical competency recently in mathematics education (Lesh and Zawojewski 2007; Niss 2003). Developing students' modelling ability is one effective teaching strategy (Lesh and Doerr 2003; Niss 2003).

Recently, the issues of the model and modelling perspective gradually receive more and more attention in Taiwan. Some empirical research has focused on the investigations of modelling contests. Some talked about modelling teaching that related to specific mathematical contents such as linear function, parabolic equations. Their results showed students' positive learning motivations or learning effects increased after modelling teaching. Other empirical research of teachers'

S.-Y. Yu (✉) and C.-K. Chang
Graduate Institute of Science Education, National Changhua University of Education, Changhua, Taiwan
e-mails: sheree318@yahoo.com.tw; ckuchang@gmail.com

education has focused on the problem-solving strategies of pre-service teachers and the results show the modelling activities and teaching need to correspond to students' experience. Also, technology is worth considering. Other research has investigated the latent mechanism underlying the case teacher's reflection in the modelling context whilst collaborating with the researcher and the results showed the effects of teachers' professional development in practice. These empirical studies reveal the approval of modelling teaching in Taiwan.

On the other hand, most mathematics teachers have taught in lecture style and transference of mathematical knowledge to students in the school context. Students just need to listen to what teachers say and there is a lack of thinking by themselves. Over a long period of time, students have become used to memorising the formula and solving routine mathematical problems. Under these circumstances, students hardly develop multiple mathematical competencies in their school mathematics classes.

Recently, we tried to phase in the MEAs to our mathematics classrooms in order to amend the situation and promote students' thinking, explaining and interpreting opportunities. The crucial reason we used modelling activities was that in such activities, students have to describe, manipulate, predict, and verify (Lesh and Doerr 2003). We hoped that students could enhance the descriptions and interpretations of what they saw and observed and the ability of problem solving through MEAs. Although empirical studies related to modelling teaching showed positive results of enhancing students' mathematical learning, this still was an uncommon teaching strategy in the Taiwanese context. So, the first problem we needed to face and overcome was to let these teachers learn how to implement MEAs in their mathematics classes. We designed a course to foster these teachers to become involved in MEAs and modelling teaching. The purpose of the study was to know what the Taiwan mathematics teachers thought of MEAs and modelling teaching.

2 Theoretical Frame

To answer the question, the theoretical approach focused on the discussion of MEA and the six principles of designing MEAs. Lesh and Doerr (2003) refer to "Case Studies for Kids" as many cases of MEAs. Each case consists of four main parts: newspaper articles, readiness questions, problem statements, and the process of sharing solutions. The purpose of the newspaper articles and readiness questions was to introduce the students to the context of the problem. Students can become more familiar with the situations of the case via reading the article and readiness questions just like a warm-up period. The problem statements should be the central part of the teaching and teachers present these to the students according to the grade level and previous experiences they have. Whether the students could identify the client they were working for and the product they should create must be verified. Then comes the process of sharing solutions and it is the stage of presentations of solutions when the teacher tries to encourage students to not only listen to the other groups' presentations but also to try to understand the other groups' solutions and consider how well these solutions meet the needs of the client.

On the other hand, Lesh and Doerr (2003) mention six principles to evaluate the quality of a modelling activity and these were also crucial points that we considered. The construction principle ensured that the solutions to the activity required the construction of an explicit description, explanation, procedures, or justified prediction for a given mathematically significant situation. The reality principle, also called the meaningfulness principle, required the activity to be designed so that students can interpret it meaningfully from their different levels of mathematical ability and general knowledge, and also pose a problem that could happen in real life. The self-assessment principle ensures that the activity contains criteria that students can identify and use to test and revise their solutions and also include information that students can assess the usefulness of their alternative solutions. The documentation principle ensures the activity requires students to create some form of documentation that can reveal explicitly how they are thinking about the situation. Share-ability and re-usability principles require students to produce more generalized solutions that others can also use or solutions that can be reused in other similar situations. Effective prototype principle ensures the solution of the activity is as simple as possible yet mathematical and significant and provides useful prototypes for interpreting other similar situations.

3 Methodological Approach

Methodologically, the research is qualitatively oriented and the applied empirical methods concerning choice of samples, data collection, data analysis, and data interpretation are based on the theoretical attempts of Grounded Theory (Strauss and Corbin 1998).

3.1 Samples

A total of 16 secondary mathematics teachers participated in the study. The information on these teachers' background is given in Table 16.1.

3.2 Research Process

The process of this study is linked to a master's degree program in education for in-service teachers with 2 h per week for 9 weeks and includes two stages. First,

Table 16.1 Background information of samples

Categories	Background information (N: category)		
Sex	8: male	8: female	
School	8: junior high school	5: senior high school	3: vocational school
Teaching years	6: <5 years	5: 6–10 years	5: >11 years
Age	6: 26–30 years old	7: 31–35 years old	5: 36–40 years old

Fig. 16.1 Teachers engaged in the "Parking Lot" problem

these teachers were divided into four groups with three to five teachers in a group as the role of students engaged in three MEAs, such as "Who saved the oriental cherry trees?", "Parking Lot" and "Volleyball problems." They cooperatively discussed solving one MEA every 2 weeks (see Fig. 16.1). They also wrote reflection journals to compare these MEAs and show their understanding of modelling pedagogy. Secondly, each group was asked to design one MEA and used the Six Principles of designing a MEA (Lesh and Doerr 2003) to evaluate these MEAs by themselves. The evaluative process also showed their perception of mathematics and understanding of MEAs.

3.3 Data Collection

The sources of data collections included the learning sheets that showed teachers' strategies for the three MEAs and the result of the MEA they designed, researchers' observation journals, reflection journals, open-ended questionnaires, interview reports, and video tapes and audio tapes of the classes.

Learning sheets. These teachers solved three MEAs in groups and wrote down their solving strategies, and letters to the client of these MEAs. These sheets showed teachers' thinking during modelling activities.

MEA. Every group of teachers designed one MEA, according to their experience of solving MEA and their understanding of MEA. They also introduced their designed MEA to other teachers and evaluated these MEAs with the six principles. These MEAs showed these teachers' perceptions of MEA.

Observation journals. Researchers kept observation journals every week. We wrote our reflections about every class in our program and kept notes about these teachers' questions and changes.

Reflection journals. Teachers' reflection journals were collected every week. The participants were encouraged to reflect on these MEAs and the experiences of solving MEAs. They were also asked to compare the similarities and differences between these three MEAs.

Open-ended questionnaire. The open-ended questionnaire was administrated on the first day and the last day of the program, in order to give an opportunity for teachers to reflect on their beliefs about mathematics, teaching, and learning.

Interview reports. Four teachers, especially those who were interested in modelling teaching and activities, joined the interview and shared their ideas about modelling teaching and MEAs with us. Much interesting data were provided for the research through these informal talks.

Video tapes and audio tapes. Every class for 9 weeks was videotaped. When these teachers discussed in groups, we also audiotaped the process of their discussion to keep the details of their group discussion.

3.4 Data Analysis

These qualitative data such as teachers' reflection journals, interview reports, and video tapes of the classes were read, coded, and categorised repeatedly by the two authors. In doing analysis data, we started to make sense of the data by "making interpretations." This process continued using "open coding" to discover categories. In this process, we used the "make comparisons" procedure to conceptualize our data by taking apart each observation, every oral or written comment, and we gave each emerging category a "name".

The teachers' thinking about modelling teaching and MEAs were the main focus of the study. The richest part of the data came from the reflection journals of these teachers which were written during the program. First, we quoted the teachers' writings that mentioned MEAs and modelling teaching and checked the frequencies of different quotations. Then, we categorized these quotations into the same categories. These categories conceptualized these teachers' thoughts about MEAs and modelling teaching. Second, we also compared the results of the MEA they designed with Six Principles of designing a MEA (Lesh and Doerr 2003) and attempted to reveal these teachers' thinking about MEAs and modelling teaching. The following was the result of the analysis of data regarding the focal point. The quotations and paraphrases included in the following paragraphs are representatives of the range of the teachers' thinking.

In the process of analysis, many themes emerged from the data. The following themes represent the nature of the findings of the research. Finally, we analyzed and interpreted these data into three themes: positive thinking about MEAs and modelling teaching, negative thinking about modelling teaching, and weaknesses of designing MEAs.

3.4.1 Positive Thinking about MEAs and Modelling Teaching

According to these teachers' reflection journals, interview reports, and video tapes of the classes, we grouped the advantages into four aspects, which are shown as follows:

Close-in Real Life Situation

A total of 8 of 14 teachers expressed this point of view. They regarded MEAs and modelling teaching relates to a real life situation intently. For example, T22 said "when she experienced MEA, she felt that mathematics can be constructed from the learning activity of real life experience". T3 mentioned that "It was full of math in the modelling process and we used mathematical language to deal with the problem which was in connection with real life." T7 and T24 both voiced the idea that MEAs are closer to real life situations than the textbooks.

Enhancement of Mathematical Competencies

As, MEAs are all open-ended problems and are accompanied by a lot of information, so teachers approved for enhancing students' competencies relating to learning mathematics. T10 said that "developing the modelling ability can promote students' problem solving ability." T7 referred to "creative thinking ability, conjectural ability, induction and categorization, built the model." T21 thought that "we ask students to think an integral problem with the concepts which they learned before, and they needed to use mathematical competencies of logical thinking, data gathering and data analyzing…. MEAs are divergent problems and the abilities that students developed were comprehensive. It made students to learn, search for information and analyzed data actively." T30 pointed out that "the focus of mathematical modelling was different than traditional problem solving and changed into, transformed, and explained the situation, recognized potential problems, built the model, re-interpreted the premise, hypothesis and biases of mathematical solution."

Advantages of Modelling Teaching

On the other hand, they also spoke of the advantages of implementing MEAs in school mathematics classes. T12 noted that "In the process, students needed to talk to each other and utilised peers' thought to inspire themselves to think the problem." T3 emphasized that "students can learn how to communicate with others, establish good relationship with peers and understand the importance of respect." T28 pointed out that "the mathematical content was not too difficult for students and it didn't make students feel scary." T30 liked the way of group discussion and he thought it was helpful to students to think through problems and also convey their opinions.

MEAs as Supplementary Materials

As for the possibilities for implementing MEAs in school mathematics classes, teachers in the interview thought that MEAs could work well as supplementary materials, and modelling activities and teaching could be regarded as the corporation or training for supplementary curriculum. We found that the teachers who taught the mathematical corporation accepted MEAs and they agreed with MEAs as supplementary materials more easily.

3.4.2 Negative Thinking about Modelling Teaching

These teachers mentioned many obstacles to implementing MEAs in school mathematics classes. This included the weak connection to the current school curriculum, the influence of the entrance examinations, and mathematical content being too easy.

Out of School Curriculum

In Taiwan, the main curriculum standards come from the government. Although there are different versions of textbooks, most content in textbooks is in traditional mode, such as examples for teachers and exercises for students. Also, teachers in Taiwan are used to teaching with textbooks and relying on the content of textbooks. The most common obstacle these teachers mentioned was the weak connection to the current school curriculum. T5 mentioned that "so far, I doubted that whether MEAs can [be] put into the current mathematics classes and maybe this will be one of my goals in the future." T21 said that "MEAs [had] almost no direct connections with current mathematical textbooks of junior high school." T24 pointed out that "It seemed not helpful for students to learn school mathematics." T3 wondered "how to transform the materials in school math into appropriate MEAs?" T10 thought that "not all units in school math were suitable for transforming into MEAs and it was not necessary to use modelling teaching." So, these teachers were really concerned with the connection between MEAs and the current school mathematics curriculum, and they would accept modelling teaching into their classes only when they were sure that the connection was close.

Out of Entrance Examinations

In Taiwan, students need to pass the entrance examinations to enter senior high schools and universities. These teachers always emphasise students' grades in these examinations as the purpose of their teaching. So, the entrance examinations of senior high schools and colleges are also the main factor why teachers resist modelling teaching coming into their classes. T22 said "my school is a typical private senior high school that emphasized the rate of entering colleges, so students' grades were the most important thing." T23 mentioned that "how to connect [the prevailing] education system (exam system) will be the first barrier in reality!" T30 referred to "the first consideration of students and teachers was to get higher scores in the entrance exams to colleges."

Other Obstacles of Modelling Teaching

Other obstacles such as "I cannot convey the mathematical concepts which students wanted most. (T3)" "We spent too much time letting students solve and discuss the MEA, so that we cannot achieve the rate of progress of school math. (T24)" "Students and I were not familiar with modelling teaching and MEAs, so we may have the attitude of rejecting this teaching mode.... The group discussion makes chaos in the classroom, and students can't keep their concentration. (T30)" "I thought that was a challenge for me to end the open MEA. I don't know what to do and it seemed not very interesting. (T23)". Therefore, it will be a tough challenge to tune these teachers' minds to accept modelling teaching in this kind of background and trend. After arranging these data and themes, we found that teachers in senior high schools or vocational schools displayed more obstacles to modelling teaching than did teachers in junior high school according to the teachers' reflection journals.

3.4.3 Weaknesses of Designing MEAs

In the second part of this course, these teachers designed one MEA in every group. Four groups produced four MEAs, and it deserved to mention that we just introduced to these teachers the six principles of designing MEAs. Also, they did not read the literature about the model and modelling perspective. The understanding of MEAs they showed was simply according to the experience which they engaged in during the three MEAs.

Here, we describe four MEAs that they designed; they are shown in Table 16.2.

In general, these teachers feel that MEAs need to be realistic situations and relating to students' life experiences. Also, they keep the problem as open as possible in order to get a model of the solution. Furthermore, the authors tried to use the six principles (Lesh and Doerr 2003) of designing MEAs to check the MEAs produced by these teachers. Table 16.3 showed the authors' interpretation based on the corresponding six principles. They also revealed obstacles to designing MEAs. The principles which were achieved easily are the Reality Principle and the Model. Construction Principle, but the other four were hardly present. It meant that the ways of promoting teachers' ability for designing MEAs will still be an issue to be addressed in the future.

4 Conclusions and Implications

After the 9-week course, these teachers revealed many thoughts about MEAs and modelling teaching. We summarized the teachers' thinking into three aspects shown in Table 16.4. We found that these teachers agreed that MEAs were useful

Table 16.2 MEAs designed by each group

Group	Title	MEA
1	The trip to Taichung	Four junior students (A, B, C, D) plan to travel from Changhua to Taichung together. They bring 4000 NT with them and need to buy specific gifts, such as famous cookies, and they also have time limitation to finish their task. The problem is to arrange a timetable of their trip and have to consider the money at the same time.
2	The reconstruction of the campus wall	Ask for students' help to use tiles to reconstruct the wall of the campus. The length of the wall is 12 m and the width is 3 m. The shapes of the tiles are rectangles and squares. The colors of the tiles are white and black, and the areas of the black tiles need to be 1/3 to 1/4 of the whole wall. The arrangements of the tiles are all designed by students themselves.
3	The procedure design of sports meeting	Ask students to design the procedures of the whole day sports meeting. The areas include a playground with 200-m athletic track, four basketball courts. There are 39 classes of the senior high school and 1 teacher group joining in the sport meetings. They have to consider the time and sequences of three games, seven races, the opening and closing ceremonies. Students need to plan well in order to get the results of all contests on that day.
4	Sampling	Let students decide the number of students and the distribution of schools for sampling to understand the percentage of students who got involved in drugs. They also want students to consider the budgets of their sampling.

Table 16.3 MEAs and the corresponding six principles

Group	1	2	3	4
Title	The trip of Taichung	The reconstruction of the campus wall	The procedure design of sports meeting	Sampling
Construction	"The design of the route of the trip" was open but not precise enough to ask for a general model.	There were too many limitations of students' solution.	"The design of procedure of the sports meeting" and "the rules of scoring" were too fuzzy.	The problem "How to sample" was too fuzzy.
Reality	Corresponds with students' life experience.	Corresponds with students' life experience.	Corresponds with students' life experience, but the description was too simplified.	Not close to students' life experience.

to enhance students' problem-solving ability and they had positive attitudes toward MEAs and modelling teaching. But they thought that there were still many obstacles to implementing MEAs in their mathematics classes. On the basis of the MEAs

Table 16.4 Teachers' thinking about MEAs and modelling teaching

Positive aspects	Negative aspects	Designing MEA
Close-in real life situation	Out of school curriculum	Fit in construction principle
Enhance students' mathematical competencies	Out of entrance exam	
Advantages of modelling teaching	Other obstacles in modelling teaching	Fit in reality principle
MEAs as supplement materials		

that they designed in the end, it was shown that they were still lacking the ability to design MEAs to fit with the six principles.

In terms of this study, we noticed that a literature review of MEAs and modelling may be important for these teachers to understand how to implement MEAs in their classrooms and to design MEAs. Because of the lack of theoretical background, they just pay attention to the surface characteristic of MEAs. Besides, we found that strengthening the connections between MEAs and the school mathematics curriculum and improving the modelling teaching so as to relate to teaching practice closely are two important factors which influence these teachers' ideas regarding MEAs and modelling.

References

Lesh, R., & Doerr, H. M. (2003). *Beyond constructivism: Models and modeling perspectives on mathematics problem solving, learning, and teaching*. Mahwah: Lawrence Erlbaum.

Lesh, R., & Zawojewski, J. (2007). Problem solving and modeling. In F. K. Lester (Ed.), *Second handbook of research on mathematics teaching and learning* (pp. 763–804). Charlotte: Information Age.

Niss, M. (2003). Mathematical competencies and the learning of mathematics: The Danish KOM project. In A. Gagarsis & S. Papastavridis (Eds.), *3rd Mediterranean conference on mathematical education* (pp. 115–124). Athens: Hellenic Mathematical Society and Cyprus Mathematical Society.

Part II
Looking Deeper into Modelling Processes: Studies with a Cognitive Perspective

Chapter 17
Looking Deeper into Modelling Processes: Studies with a Cognitive Perspective – Overview

Susana Carreira

As pointed out by Galbraith and Stillman (2001), the issue of the role played by extra-mathematical knowledge in modelling and applications activities is part of an agenda shared by researchers and educators who are interested in the specific nature and characteristics of modelling activities within educational settings. An important aspect of such mathematical activities is the actual structure and elements of the context situation that are embedded (either explicitly or implicitly) in the task formulation, thus opening significant room for assumptions, interpretations and considerations concerning the so-called extra-mathematical world. Nurturing a long-standing debate on the nature of mathematical models and on the types of knowledge and experiences activated by mathematical modelling, the six articles composing this chapter highlight in very different and yet strikingly coherent ways the paramount role of knowledge, besides mathematics, in illuminating students' processes, decisions and understandings while working on modelling tasks.

Brown and Edwards show how secondary school students develop mathematical modelling with rich tasks, where genuine links to a real-world context are expected and encouraged. In addition, mathematical modelling and mathematical understanding are conceived as running side-by-side. This entails a view of the modelling tasks and processes which are at odds with the often endorsed conception that sees mathematics as a system into which reality must be translated and symbolised. By underlining the importance and intervention of students' prior knowledge about the anchoring context of the task, their work offers a much broader image of the modelling process. Engagement with the task context, as long as students perceive it as realistic and significant to formulate a model and to derive one or more possible solutions, proves to be an essential condition for meaningful interpretative mathematical modelling. Drawing on Stillman's (2000) categorisation of different types of prior knowledge – academic, encyclopaedic and episodic – Brown and Edwards' results show how students' prior knowledge (especially academic and encyclopaedic) was activated and operated to enhance understanding,

---

S. Carreira (✉)
Universidade do Algarve, Faro & UIDEF, Lisboa, Portugal
e-mail: scarrei@ualg.pt

to facilitate checking, to inform decisions and to validate results. Therefore, mathematics and the prior knowledge of the context were integrated, enabling deeper understanding. In a way, it seems possible to compare the previous paper with the one proposed by Carreira, Amado and Lecoq, where they investigate how knowledge is elicited through mathematical modelling of daily life situations within adult education mathematics classes. In stressing the socio-cultural nature of the learning that takes place in adult education, adults' life experiences are viewed as major resources in their formative learning process. Adults mathematics education is expected to promote individuals' professional, social and daily life experiences. The production of bridges thus becomes an essential part of the mathematical activity in modelling real situations. From this standpoint, looking at the relationship between schooling and other practices entails focusing on the production of meaning as part of the process of boundary-crossing between a lived-in world and a figured world (the conversion of a particular real context into a school situation) (Boaler 2000). A group of adult students solving a task on making Margaritas according to four different recipes, where the ratios of the ingredients varied, revealed how a student's prior experience of having tasted Margaritas previously became a source of reliability for the others in the group. In general, to the adult students, the mathematical language and the technical numeracy constituted a thick and impermeable boundary between the mathematical world and the lived-in world. Matters other than a mathematical analysis of the problem came into play and both the domain of inquiry and the mathematisation involved in the modelling cycle were clearly influenced by specific context-knowledge and particular forms of agency within the perceived reality.

Mousoulides and English stress the importance of implementing interdisciplinary problem-solving activities in mathematics and science school curricula as a way to empower students to tackle real-world problems emerging in our present society and in the foreseeable future. Their work reports on a study with 12-year-old students working on Engineering Model Eliciting Activities characterised as authentic, meaningful, realistic and client-driven problems. In their paper, particular attention is devoted to the analysis of four models produced by different groups of students working on a task about the exhaustion of natural gas reserves. As the authors concluded, students had to elicit their own mathematical and engineering ideas to work on the problem; but more importantly, the students' models may be seen as actual embodiments of all the factors, ideas, relationships and interpretations that they considered important in the creation of their models. Again, we should recognise that looking at mathematical models includes gaining a perception of the human factor in the modelling process (Araújo 2008; Skovsmose 1994).

In a similar vein, Carreira and Baioa describe the modelling routes of 9th grade students in tasks that comprise manipulating and experimenting with real objects. They suggest such tasks generate particular kinds of modelling approaches, namely in accounting for the role of experimentation in formulating and testing hypotheses for a solution to the problem. The results obtained from a group of students dealing with the problem of folding cardboard cake boxes showed that while experimentation may appear to keep students at a level of horizontal mathematisation, the final stages of the modelling process actually indicate the potential of doing experiments to promote a deeper conceptual understanding of the problem and of the mathematical model. Their conclusions indicate that hands-on activities contributed to making students familiar with the conditions, relationships and factors involved in the real situation and helped them correct and tune the intermediate mathematical models underlying their successive attempts to create a model of the folded cake boxes.

Also focusing on students' modelling routes within realistic mathematics projects, Sol, Giménez and Rosich analyse the written reports produced by classes of students ranging from 12 to 26 years old. The study highlights the theoretical phases and actions of the modelling cycle, as attributed to the expert modeller, and compares it with the actions performed by students when working on open-ended real problems. The results pointed to the importance of a particular form of mathematical knowledge, namely functional reasoning, responsible for advancing the initial steps of modelling. Moreover, one fundamental idea regarding students' knowledge and experience concerns the absence of an awareness of the overall modelling process that was consistent through age groups. The lack of experience in realistic projects and open-ended problems in school mathematics seems to be accountable for the students' limited view of the modelling process and for their transformation in a chain of smaller problems.

Finally Stillman, in linking metacognition with the transitions between the modelling stages, offers a detailed investigation on what the modeller does. Drawing on the results of a study with secondary school students, she uses the modelling cycle as the underpinning frame of analysis and fills it with the cognitive and metacognitive activity that is going on mentally as the modellers work. Stillman points out that not all metacognitive acts are productive in the sense of giving a better orientation to the modelling process. However, productive metacognitive acts were identified at three levels: (1) recognition that particular strategies are relevant, (2) choice of the strategy and (3) successful implementation. Not surprisingly, the first level is heavily connected with the individual's assessment of personal resources and notably it relates to the personal knowledge and competence with respect to the real situation. This raises the issue of scaffolding both for the teacher's role and for the modelling task design regarding the desired balance between students' immediate possible ideas and their freedom to make decisions about the pathways to choose in the modelling process. As Stillman concludes from the data on a group of two students creating a model of the best goal opening in a soccer field, it is possible to help students to make an optimal use of their metacognitive knowledge and strategies within the modelling context. In particular, attention has to be paid to the difference between lack of reflection or incorrect and incomplete knowledge. As in the revealing transcription of the dialogue between the students and the teacher, one central question to be asked in face of a mathematically extracted result or conclusion should be the following: "Does that really make sense?".

In reacting to Stillman's paper, Borromeo Ferri takes in several developments on metacognition to substantiate the importance of carefully examining the notion of 'your knowledge' when referring to young children who do not normally engage in a description of their thinking processes. Hence, starting with modelling tasks in primary school should be taken as an important goal both as a way to introduce young pupils to the process of solving realistic problems and as a form of initiating them into the kind of thinking that is essential to modelling. Indeed the key issue seems to be not only in 'your knowledge' but beyond that in 'how you use your knowledge'. On the other hand, the 'metacognitive modelling cycle', in Borromeo Ferri's terms, seems to be as a vital and strategic instrument to describe modelling from the meta-level point of view. However, the profusion of variations on the modelling cycle offered in the literature points to the flexible and somewhat adaptive nature of the cycle itself. In fact, in one way or another, the various papers in this set embrace the modelling cycle as a theoretical and analytical instrument. Extending the ideas proposed by Stillman, the presented reaction suggests linking knowledge, control and beliefs in the modelling process, emphasising the role of the teacher and adopting the modelling cycle to work on describing thinking and sharing it with others.

To summarise the overall tone of this collection of papers, a special focus is placed on knowledge, meta-knowledge and experience in the process of devising mathematical models from rich, inspiring and authentic tasks in mathematics classes. From an epistemological point of view, the chapters highlight the human nature of modelling and the conceptual and interpretative nature of mathematical models. Many other researchers have stressed this essential aspect of modelling and have drawn consequences to the educational setting. Let me start by annotating Freudenthal's (1991) notion of an intermediate model between reality and mathematics ensuing from the idealisation and interpretation of reality. Additionally, I would refer to the perspective of metaphorical thinking as a lens to understand how concepts from mathematics are interconnected with concepts from other conceptual systems (Carreira 2001). All the above converge to the view that prior knowledge, experience and sense making of the task context are cornerstones in the modelling process and give mathematical modelling its subject-centred quality. Thus, the interplay between the real world and mathematics in school authentic tasks is obviously a decisive matter to be discussed and researched. But certainly this has to be done against the background of the culture of mathematics classrooms. As Schwarzkopf (2007) argues, students often follow the logic of the classroom culture instead of the logic of the problem solving; actually, the opposite is also true when blockages are severe enough to hinder the fluid 'translations' from one language to another language or otherwise from one system of meanings to another one.

References

Araújo, J. L. (2008). Formatting real data in mathematical modelling projects. Paper presented at the *Topic study group 21 at the 11th International Congress on Mathematics Education (ICME 11)*, Monterrey, Mexico. Retrieved from http://tsg.icme11.org/document/get/449.
Boaler, J. (2000). Introduction: Intricacies of knowledge, practice and theory. In J. Boaler (Ed.), *Multiple perspectives on mathematics teaching and learning* (pp. 1–17). London: Ablex Publishing.

Carreira, S. (2001). The mountain is the utility – On the metaphorical nature of mathematical models. In J. F. Matos, W. Blum, K. Houston, & S. P. Carreira (Eds.), *Modelling and mathematics education – ICTMA 9: Applications in science and technology* (pp. 15–29). Chichester: Horwood Publishing.

Freudenthal, H. (1991). *Revisiting mathematics education: China lectures*. Dordrecht: Kluwer.

Galbraith, P., & Stillman, G. (2001). Assumptions and context: Pursuing their role in modelling activity. In J. F. Matos, W. Blum, K. Houston, & S. P. Carreira (Eds.), *Modelling and mathematics education – ICTMA 9: Applications in science and technology* (pp. 300–310). Chichester: Horwood Publishing.

Schwarzkopf, R. (2007). Elementary modelling in mathematics lessons: The interplay between "real-world" knowledge and "mathematical structures". In W. Blum, P. Galbraith, H.-W. Henn, & M. Niss (Eds.), *Modelling and applications in mathematics education: The 14th ICMI study* (pp. 209–216). New York: Springer.

Skovsmose, O. (1994). *Towards a philosophy of critical mathematics education*. Dordrecht: Kluwer.

Stillman, G. (2000). Impact of prior knowledge of task context on approaches to applications tasks. *Journal of Mathematical Behavior, 19*, 333–361.

Chapter 18
Applying Metacognitive Knowledge and Strategies in Applications and Modelling Tasks at Secondary School

Gloria Stillman

Abstract The importance of reflective metacognitive activity during mathematical modelling activity has been recognised by scholars and researchers over the years. The metacognitive activity (or lack of it) of secondary students associated with transitions between stages in the modelling process – especially in relation to the identification and release of blockages to progress – is considered. Productive metacognitive acts are identified as occurring at three levels. Routine metacognition together with metacognitive responses to Goos' red flag situations are elaborated together with the notion of meta-metacognition being engaged in by teachers trying to foster students' development of independent modelling competencies especially their metacognitive competencies.

1 Introduction

Studies in metacognition in Australia on applications and mathematical modelling tasks mainly resulted from research on problem solving in the late 1980s and early 1990s following on from Schoenfeld's work on mathematical problem solving, in particular, decision making during task solution and links to metacognition (1985, 1987, 1992). Another influence was the work of Garofalo and Lester (1985) and the development of their cognitive/metacognitive framework and its use in mathematical problem solving research. Definitions of *metacognition* and operational constructs followed those of Flavell, in particular his characterisation of metacognition as:

> one's knowledge concerning one's own cognitive processes and products or anything related to them ... metacognition refers, among other things, to the active monitoring and consequent

G. Stillman (✉)
School of Education (Victoria), Australian Catholic University (Ballarat),
PO Box 650, Ballarat VIC 3350, Australia
e-mail: gloria.stillman@acu.edu.au

regulation and orchestration of these processes in relation to the cognitive objects on which they bear, usually in the service of some concrete goal or objective. (Flavell 1976, p. 232)

Flavell's (1979) model of metacognition and cognitive monitoring has underpinned much research on metacognition since he first articulated it. According to his model, a person's ability to control 'a wide variety of cognitive enterprises occurs through the actions and interactions among four classes of phenomena: (a) metacognitive knowledge, (b) metacognitive experiences, (c) goals (or tasks), and (d) actions (or strategies)' (p. 906).

Metacognitive knowledge incorporates three interacting categories of knowledge, namely, person, task and strategy knowledge. Examples include personal knowledge about oneself as a modeller (e.g. awareness of difficulty in easily formulating reasonable estimates), knowledge about task variables (i.e. awareness of task characteristics affecting the task solution) and pertinent knowledge about cognitive or metacognitive strategies such as awareness of their effectiveness when used in the past (see Stillman 2004; Stillman and Galbraith 1998, for further examples in the context of mathematical applications).

Metacognitive experiences are any conscious cognitive or affective experiences which control or regulate cognitive activity. Efklides (2002) characterises metacognitive experiences as involving feelings and judgements or estimates, such as a feeling of difficulty, as a person responds to processing a task prospectively, throughout the task, and retrospectively (Efklides et al. 2006). Metacognitive experiences 'inform the person of task processing demands…based on task perception' and previous experience with similar tasks as well as informing 'future decisions regarding involvement with similar tasks' (Efklides et al. 2006, p. 16). However, metacognitive experiences might not be reliable as 'it is…the student's perceived difficulty level that guides actions, whether it is correct or not' (Stillman 2002, p. 291).

Metacognitive goals are the objectives of any metacognitive activity. An example comes from the situation where a student is engaged in mathematical modelling and has begun to explore the real world situation and is wondering (metacognitive experience) if he or she has adequate domain knowledge of the situation. The student then has the metacognitive goal of assessing his/her current knowledge of the problem domain. To reach this goal, a metacognitive strategy such as self-questioning might be employed. *Metacognitive strategies* are strategies used to regulate and monitor cognitive processes and thus achieve metacognitive goals.

When it comes to solving mathematical applications and engaging in mathematical modelling, effective use of metacognition attains crucial importance (Maaß 2007; Stillman 2004) as coordination and integration of information and representations and allocation of attention resources are vital to efficient functioning of working memory during solutions (Carlson et al. 1989, 1990). Executive control decisions directed at controlling or monitoring cognitive activities during the application of mathematical knowledge or the modelling process may initiate metacognitive strategies and monitoring of cognitive progress (Kluwe 1987) informed by metacognitive knowledge.

2 Modelling and Metacognition

2.1 *The Modelling Process and Reflection*

The following theoretical framework for studying modelling used in the author's more recent work is oriented towards the modelling individual to give not only a better understanding of what the modeller does when attempting modelling problems, but also a better basis for teachers' decision making and interventions. Figure 18.1, an adaptation of the cycle used in Stillman et al. (2007), encompasses both the task orientation of many diagrammatic representations of the modelling cycle and the need to capture what is going on mentally as modellers work on modelling tasks. The respective entries A–G represent stages in the modelling process, where the thicker arrows signify transitions between the stages, and the total solution process is described by following these arrows clockwise around the diagram from the top left. It culminates either in the report of a successful modelling outcome, or a further cycle of modelling if evaluation indicates that the solution is unsatisfactory in some way. The kinds of mental activity that individuals engage in as modellers attempt to make the transition from one modelling stage to the next are given by the broad descriptors of cognitive activity 1–7 in Fig. 18.1. The double headed light arrows emphasise that the modelling process is far from linear, or unidirectional as has been confirmed empirically recently by Borromeo Ferri (2007) in her reconstructions of students' modelling routes during task solution in a variety of modelling tasks and earlier by Oke and Bajpai (1986) using relationship level graphs. The light arrows also indicate the presence of reflective metacognitive activity as proposed by several researchers (e.g. Maaß 2007; Stillman and Galbraith 1998). Such reflective activity can look both forwards and backwards with respect to stages in the modelling process, hence the double directional arrows. However, the

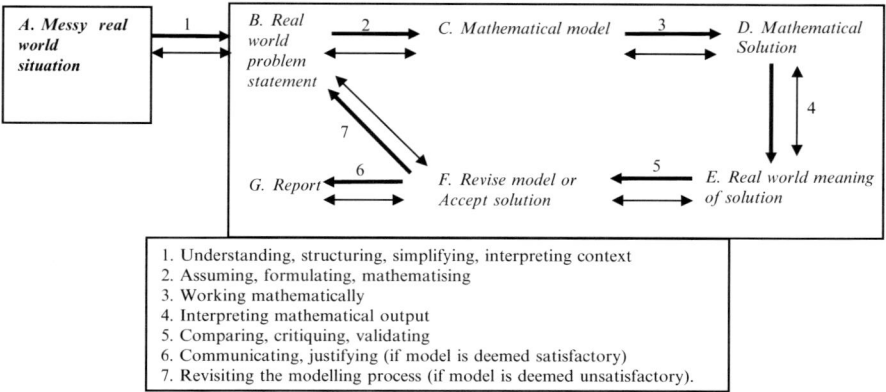

Fig. 18.1 Modelling cycle adapted from (Stillman et al. 2007)

cycle in the diagram is a simplification as modellers can deviate from the cycle as well as backtrack and to capture this diagrammatically leads to a far more complex diagram (see Stillman 1998, for such examples).

In modelling, reflection can make sense only when related to mathematical content and the processing decisions by means of which the content is evoked and implemented. Both monitoring and planning are seen as both subject to, and products of, reflection. When considering reflection during the actual activity of modelling, it is not viewed as dealing with hindsight because in mathematical modelling, the present constantly changes with the phase of the modelling cycle, so terms such as past and present and future are temporary and very fluid.

Interest in reflection relates exclusively to its role in facilitating metacognitive activity within the modelling process. In the *modelling as content* approach (Stillman and Galbraith, this volume) advocated here, the immediate goal is to obtain a solution to a real world problem, but simultaneously and cumulatively, there is a desire to develop consistent and robust meta-knowledge about modelling and applying mathematics (Blum and Kaiser 1984; Maaß 2007). The educational goal is for students to become better modellers not just solvers of separate problems.

As an illustration, when a Year 9 student who had participated in several extended modelling tasks well spaced over a 10-month period (see Galbraith et al. 2007, for details) was asked what was the teacher's purpose in using such tasks, his response showed that he had developed meta-knowledge about the modelling process by engaging in it several times.

Cai: Probably to give us an understanding of how applied mathematics works. You find data, you investigate the data and you investigate any trends in the data and if there are any trends; and try and formulate an equation or statement about the data which can be used to predict, data that will come, or data that can be used, to predict other data outside the given data you found. Applied mathematics is needed, is necessary to complete many tasks in the world. And sometimes you don't see that some, even little, even a task such as soccer would involve mathematics but in fact it does.

Clearly, this student had developed a well-grounded meta-knowledge about the modelling processes he had been involved in. He was a student who did particularly well in mathematics and was more used to doing repeated exercises before he became involved in the programme, and he was able to appreciate both abstract and real world tasks.

Cai: Real world tasks provide me with an understanding of mathematics in the world and how they use mathematics. Abstract tasks are totally abstract and sort of stretch your mind a bit further and just sort of have some fun on the paper.

It is expected that the quality of such meta-knowledge developed through modelling programmes is related to students' competencies in modelling with good modellers like Cai displaying high meta-knowledge about the modelling process as was found by Maaß (2007).

In the programme adopted in my research, metacognition is located heavily in the transitions as metacognitive activity is anticipated as being necessary to monitor and contribute to the cognitive activities shown in Fig. 18.1. How teaching addresses the fostering of associated metacognitive competencies is crucial to achieving the goal of fostering students who can consistently model.

2.2 Meta-Metacognition and Modelling

During mathematical modelling activities in class, the teacher must monitor the progress of individuals or groups to intervene strategically only when necessary if the ultimate goal is to facilitate independent modelling. Thus, the teacher has to appraise the enactment of metacognitive activities by students, for example, whether students are undertaking sufficiently perceptive and rigorous reflection in considering the approach to, or the quality of, a solution. The teacher reflects on the students' metacognitive activity both within the specific situation and with respect to its role in the modelling process. The teacher is thus engaging in a *meta-metacognitive process*. At the macro-level, how a teacher generally undertakes this meta-metacognition with respect to student activities and reacts are crucial to whether modelling is nurtured or stifled in the classroom overall whatever the teacher's intentions with respect to mathematical modelling. At the micro-level, the capacity of students to develop skills in making transitions between phases in the modelling cycle and to release blockages in the solution process depends on how they are facilitated in learning and applying the modelling process and metacognitive strategies central to it. This is dependent on the perceptiveness and skill with which teachers assess, mediate and provide for the metacognitive activity of students during mathematical modelling activities.

2.3 Productive Metacognitive Acts and Modelling

Not all metacognitive acts are productive. Productive metacognitive acts occur at three levels during (1) recognition that particular strategies are relevant, (2) choice of strategy for implementation and (3) successful implementation. The first level involves assessment of personal resources (e.g. knowledge and competence in relation to the task). The second involves assessment of viability of alternatives. The third is impacted by various sub-competencies of the modeller related to identifying and correcting intermediate errors, and procedural efficiency in obtaining a successful solution.

2.3.1 Routine Metacognitive Activity

Depending on the metacognitive stores of students and their perceptiveness in using these (i.e. knowing what, how, when and to use these metacognitive knowledge and strategies), this can be manifested as routine metacognitive activity during problem solving.

To illustrate this, the transcript of a videotape of two students (Jim and Ahmed) in a Year 9 class is examined. The boys were working as a pair at their desks beside a second group of three boys (the first of whom is Ozzie) in the same row of desks on the modelling task in Fig. 18.2. They had worked out that in order to find the shot angle, BSC in Fig. 18.3, they should use trigonometry. However, Jim, despite being able to recall the correct technique of inverse tan, was unable to use this new information as his self-assessment of his current knowledge revealed he had forgotten how to use it.

Ozzie: No the angle formula. It's like reverse, you do that minus one sign. You know how to figure out the angle?
Jim: Sine over?
Ozzie: No the angle not the [stops]. It's like reverse.
Ahmed: Sin?
Jim: I remember it's reverse. You do that. Yeah, you do the one on top [pointing to tan^{-1} on his graphing calculator]. *[New information]*
Jim: I have forgotten how to do it. Trigonometry is the easiest one. *[Assessment of resources (knowledge)]*

Once they overcame this temporary blockage by consulting their notes from the previous semester's work, Jim's partner, Ahmed, suggested that an alternative strategy might be to use Pythagoras' Theorem when Jim raised doubts about the validity of an interim result. Thus, as they had a second possible strategy to use, the boys had to make an assessment of the viability of these alternatives.

Shot on Goal[1]: You have become a strategy advisor to the new football recruits. Their field of dreams will be the FOOTBALL FIELD. Your task is to educate them about the positions on the field that maximise their chance of scoring. This means—when they are taking the ball down the field, running parallel to the SIDE LINE, where is the position that allows them to have the maximum amount of the goal exposed for their shot on the goal?
Initially you will assume the player is running on the wing (that is, close to the side line) and is not running in the GOAL-to-GOAL corridor (that is, running from one goal mouth to the other). Find the position for the maximum goal opening if the run line is a given distance from the near post.

Fig. 18.2 The *Shot on Goal* Task [SOG] (Image source: http://images.sportsnetwork.com/soccer/wc/2006/stadium/gelsenkirchen.jpg)

---

[1] This task was based on a task originally designed by Ian Edwards, Luther College.

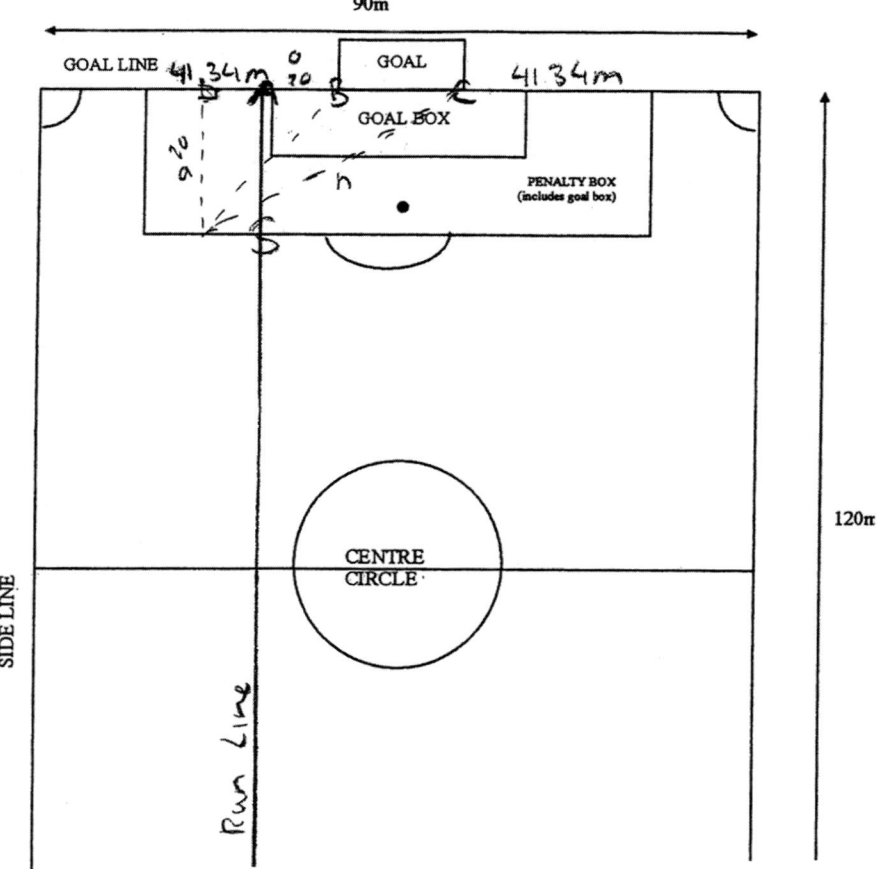

Fig. 18.3 Jim's representation of the situation being modelled

Jim: 20.75, I don't think that's right. *(Assessment of Interim Result)*
Ahmed: [laughs] Maybe it's not Tan [pause].
Jim: That's right, that's a 90 degree angle there. We have to figure out the length of there using SOCATOA first. So the hypotenuse is [pause as he thinks this through]. Tan… *(Assessment of appropriateness of strategy)*
Ahmed: Can't we use Pythagoras to figure out the hypotenuse?
Jim: Do we *need* the hypotenuse? Because if you've got the length of that right? It has to be the right length or it's not going to work. *(Dismisses alternative strategy)*
Ahmed: Oh yeah.

Successful implementation was impacted by the boys' competencies in identifying and correcting intermediate errors. Their procedural efficiency was assisted by the use of the calculator.

Ahmed: So this always has to say 10 [distance from near post to run line.]
Jim: So that should be 10?
Ahmed: Yeah. That's what I did.
Jim: Did you? I thought I did...
Ahmed: You did yours as 10 as well.
Jim: Did I?
Ahmed: Yeah. You just wrote 15 there.
Jim: We have to...[looking and pointing at where he has done a by-hand calculation.]
Ahmed: Jim, did you have breakfast?
Jim: Yes, I did, Ahmed. [Jim works on his graphing calculator checking his calculation.] Maybe I did do 10. I just wrote 15. [Confirms error.]
Ahmed: Yes. I did every calculation after you, Jim, so I know.

2.3.2 Responses to Red Flag Situations

Goos (1998) associated metacognitive success with productive responses to 'red flag situations'. Red flags are metacognitive triggers 'when students become aware of specific difficulties' (p. 226). Red flag situations occur when there is lack of progress, errors occur and are detected, and anomalous results arise as shown in these three examples respectively where they are accompanied routinely by metacognitive monitoring.

Jim: I think no one has got past this question. *[Assessment of progress]*
Jim: [to Ahmed] That was the old stuff, that's wrong. *[Red flag – detection of error]*
Ozzie: You do tan that.
Jim: You can't because that's my answer. It doesn't make sense. I tried it. It ends up being negative 40. *[Red flag —anomalous result]*

In modelling, situations such as these that should result in a red flag being raised could result in blockages, so the nature of the modeller's response to the situation is crucial. Responses by modellers to such situations where there is the potential for a red flag situation to arise could be: (1) routine metacognition, (2) metacognitive blindness, (3) metacognitive vandalism, (4) metacognitive mirages or (5) metacognitive misdirection. Instances of routine metacognition were illustrated in Sect. 2.3.1.

Metacognitive blindness occurs when a red flag situation is not recognised, so no appropriate action is taken. The modellers fail to notice something is wrong, persisting with a wrong strategy and/or overlooking calculation errors. For example, at one point in their modelling, Ahmed and Jim were trying to find the marked angle in Fig. 18.4 using inverse tan.

Ahmed: The opposite?
Jim: That angle [referring to the marked angle] so it is 20 over 10. *[Error – mixing opposite and adjacent sides of triangle.]*

Fig. 18.4 Jim's diagram with angle marked

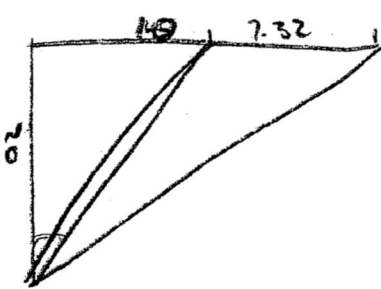

Ahmed: Yeah.
Jim: [works out $\tan^{-1}(20/10)$] So it is 63. [long pause]
Ahmed: To find this angle would you still use 20 over, that would still be right, wouldn't it? [*Checking procedure*]
Jim: No that's to figure out that…
Ahmed: That's to figure out that 'cause that's 20 [pause] Like that one's 20?
Jim: Yeah the other one would be wrong.
Ahmed: The top one's wrong, yeah [meaning the side length marked 10 on their diagram]. [pause as he works $\tan^{-1}(20/10)$ on graphing calculator] So it's 63.43.

The boys engaged in routine monitoring by checking the procedure but failed to recognise that an error was made to the input.

Metacognitive vandalism occurs when the response to a perceived red flag involves taking drastic and often destructive actions that may not only fail to address the issue, but also alter the task itself.

Metacognitive mirage describes a situation when unnecessary actions are taken that derail a solution, because the modellers perceive a difficulty that does not exist. Summer and Sui, for example, were working on the same task together but had been given distances by their teacher from the near goal post to the run line of 11 and 12 m, respectively. They had calculated shot angles for spots 20, 15, 10 and 5 m from the goal line when Summer asked if she could compare her answers to Sui's. Summer's angles were 13.68°, 14.44°, 39.76° and 9.17° whereas Sui's were 13.05°, 13.51°, 12.44° and 8.11°.

Summer: Can I check my answers? Awh, yours is different. They are nearly the same [meaning their angle values for 20 m, 15 m and 5 m].
Sui: Is there a problem?
Summer: These are kind of similar. Yours should be more than mine shouldn't it? I definitely got that one wrong [the angle for 10 m]. [Pause] No they are both less [referring to the component angles used in the calculation]. This one is wrong [indicating her angle of 86.88° from the runline to the far goal post as it is much larger than Sui's 62.63°]. [*Metacognitive red flag recognised.*]

Sui: No, I think some of mine are wrong because the first one with 20 was less than that [for 15 m]. [*metacognitive mirage*]

Summer: Yeah, see I got this one was more than that [indicating her shot angle for 15 m was also larger than for 20 m]. [*metacognitive mirage dismissed*]

Metacognitive blindness, vandalism and mirage were all described by Goos (1998). To cover all responses observed in classrooms, however, there needs to be another category of responses, metacognitive misdirection. *Metacognitive misdirection* describes the common situation of a potentially relevant but inappropriate response to a perceived red flag that represents inadequacy, rather than vandalism.

From a teaching perspective, if teachers are going to be able to engage in meta-metacognition effectively, it is important that they be aware of the nature of both routine and non-productive metacognitive acts such as those described that have been observed to occur during modelling especially when trying to encourage development of students' reflective metacognitive activity.

3 Development of Modellers' Reflective Metacognitive Activity Through Meta-Metacognition

At this point, attention to reflection relates exclusively to its role in facilitating metacognitive activity within the modelling process. Teachers use a variety of strategies to attempt to initiate reflective activities in students. The process foreshadowed earlier of the teacher engaging in reflecting on the students' metacognitive activity with respect to the current task and their development of metacognitive competencies with respect to the modelling process comes to the fore. This meta-metacognition can be related to (a) task design and (b) teacher decisions about scaffolding of implementation, pre-planned intervention activities and actions on the fly. How desirable or appropriate these are from the perspective of developing modelling competencies and meta-knowledge of students (as in Sect. 2.1) depends on the teacher's pedagogical content knowledge/mathematical content knowledge about modelling at this level of schooling and their ability to respond quickly to sudden realisations of possibilities and opportunities.

In task booklets, for example, when beginning a programme of modelling, reflective questions can be inserted at strategic points to provoke students to pause and reflect on their immediate progress in relation to the total modelling purpose as shown in Fig. 18.5. Some young students, such as 13 year-old Cai, do realise the task writer's intention for such activities and appreciate their self-regulatory activity being scaffolded, but they also want the freedom that comes as the scaffolding is faded across a series of tasks so that they can make their own decisions about their modelling pathways. When Cai was asked if he liked doing such reflective questions he replied:

Cai: To an extent, yes, but if there are too many questions you are sort of led along a certain path and you cannot, say, try and formulate your own theories, put

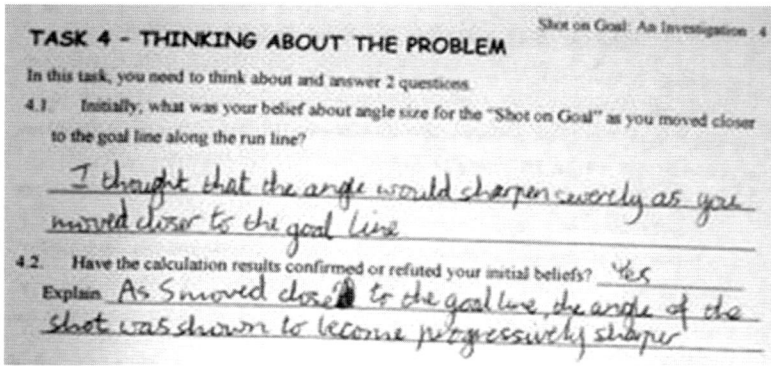

Fig. 18.5 Reflective activity

it forward and see if they are correct or not but interpretive questions to a good extent are very useful I think.... and this task [SOG] had less pointers, had less pointers to what you should do next. It sort of gave you more thinking space.

Others, however, were oblivious to these purposes; so, this becomes another consideration for the teacher – how to convey the pedagogical intentions to young students of modelling and its relation to the other mathematical activities in which they engage in the classroom.

If teachers are teaching reflective learning, when they decide to intervene with students who are experiencing a persistent blockage to their progress, they try to alter first the student's current mental model through reflection and then the actions of the student. Meta-metacognition plays a significant role in this process as is shown in the following example.

Three students, Stella, Mia and Gabi, who were working as a group on the Shot on Goal task encountered a persistent blockage although for most of the time they were in a state of metacognitive blindness with respect to this. They used an incorrect specification for their model (see Fig. 18.6) but metacognitive red flags when they were presented with conflicting results only served to increase their belief in the validity of their model. They were so convinced that their particular model was correct they did everything possible other than review their model.

This trio made genuine attempts at all reflection questions; however, the end result was that they forced their data into fitting with their formulation, rather than revising their model. In response to the question: Initially, what was your belief about angle size for the Shot On Goal as you moved closer to the goal line along the run line? Stella wrote, 'Initially, I believed the angle would get smaller because it was a tighter angle to score a goal so I thought the angle would be smaller.' In response to: Have the calculation results confirmed or refuted your initial beliefs? Stella wrote, 'No!' Meaning they were not confirmed as indicated by video evidence. However, she explained this contradiction away by writing 'As you run

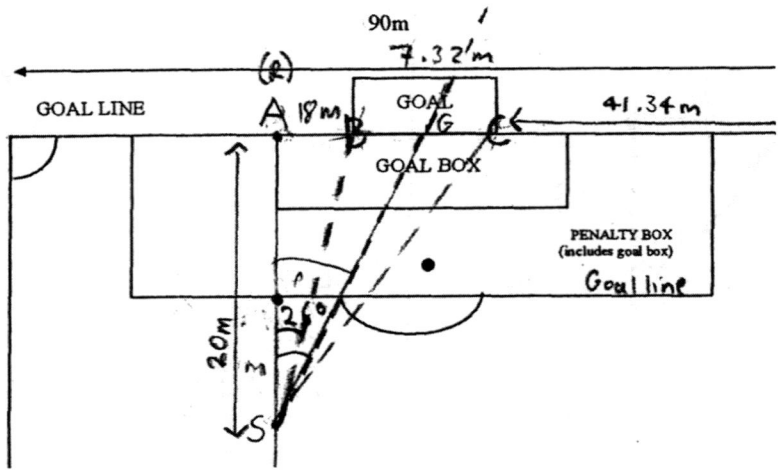

Fig. 18.6 Mia's diagram for finding the shot angle

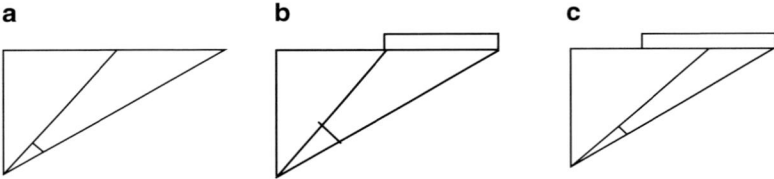

Fig. 18.7 Representing the shot angle

along the run line towards the goal the angle got bigger because you'd have to turn more in order to face the goal.' The angle she was talking about was angle ASG on Mia's diagram when in fact she was supposed to be finding angle BSC.

It was not until after they had been working on the task in mathematics lessons for 2 days that any chink in the defence of their model began to appear as they were puzzled by an apparent need in the task sheet for a multi-step process to find the shot angle whereas they had found their angle in one step. They called the teacher over, and in the discussion that ensued, it became apparent that they had not done things in the way the teacher had. Both the group and the teacher saw the angle marked in Fig. 18.7a as representing the angle they were finding, but the teacher's interpretation was that the angle was to the goal posts covering the entire face of the goal as in Fig. 18.7b whereas the girls saw it as covering only half of the goal face as in Fig. 18.7c. As Mia pointed out: 'But we didn't do it like that, we didn't do it for the whole goal.' The girls saw this as an alternative representation of the problem – not an incorrect specification of the angle. The bell rang to end the lesson at this point; so, the teacher had overnight to consider what he might do. This

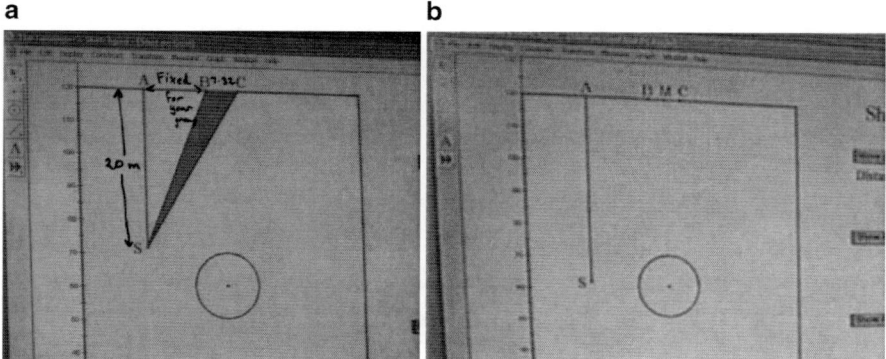

Fig. 18.8 Dynamic geometry representations of the situation being modelled

clearly gave him time to reflect on the nature of their mental representation and to weigh up how best to facilitate their reflection on their own model to realise its shortcomings for themselves.

Throughout the modelling sessions there had been a dynamic geometry display (Fig. 18.8a) projected onto the whiteboard at the front of the room which the teacher had used at the beginning of the mathematics sessions each day to bring the class together and look at possibilities together in relation to their representation of the situation. The girls were observed making use of this display on several occasions in their discussions of the task as they were seated at the front of the room. The teacher came prepared with a dynamic geometry diagram (Fig. 18.8b) to represent their model which had the subtle, but insightful, change of a midpoint, M, marked between the points representing the bases of the goal posts. The girls had also had time to reflect and were unconvinced that they needed to change their model at all with Gabi, in particular, suggesting that it would not have to be changed. Mia also declared that she did not see there was anything wrong with their model.

The teacher used the software dynamically to show them how unsatisfactory their model was by allowing them to realise that the maximum for their angle would occur on the goal line. As the girls clung to their mental model until the final moment when they finally had to admit it did not make sense, the teacher had to allow them to make this connection as they reflected on what they were seeing on the display, how this connected to their model and how this connected to the real situation of shooting a goal on a soccer field as a player moved towards the goal line on such a run line.

Teacher: So you are calculating that angle to the mid-point, aren't you? Is that what you're doing? Let's move along a bit. [On the computer the teacher moves the point, A, representing where the shot will be taken closer to the goal line along the run line SA.]

Stella: Isn't that what we are meant to do?

Teacher: Is it getting bigger?
Stella: Yeah.
Teacher: Okay. So when is your angle going to be at it's maximum?
Stella: When it is like right up close. When it is like right up close to A.
Teacher: Yeah, so when you measure it. [Starts measuring the angle using GSP.]
Stella: What are we doing?
Teacher: So at the moment it is 41 degrees.
Stella & Mia: Yeah.
Teacher: So, let's move a little bit closer.
Stella: [excitedly] Oh my god, I was right! [*Believing their model was correct.*]
Teacher: Now it is up to 69, it's quite a big one isn't it? All right. So move a bit closer. It is up to 85.
Mia: It is still harder.
Teacher: Where's the best place to take the shot?
Stella: Not there. [*Realising there is an inconsistency with the real situation.*]
Teacher: A, isn't it? What would the angle be when you're at A?
Mia: 180 degrees [sic].
Teacher: Well, what's it getting close to at the moment?
Gabi: 90.
Teacher: How about a little bit further?
Gabi: All right. [*Conceding their model must be wrong.*]
Stella: I knew that.
Teacher: [Runs S through the goal line off the field.] Too far, all right, back in. But yes, so according to your model, the best place to take the shot would be from A wouldn't it? Does that really make sense?
Stella: No.
Mia: No it doesn't. [*Agreeing their model leads to an incongruity.*]

As has been pointed out previously in (Stillman et al. 2010, p. 395), this is

> an example of resistance of schemas to the need for accommodation in Piagetian terms. These students persisted in attempting to assimilate, rather than accommodate, (Piaget 1950) new contradictory information into their chosen [model], and resisted the necessity to consider the [model] itself as deficient – the students engaged in cognitive dissonance (Atherton 2003; Festinger 1957) which prevented them from activating procedures to unblock their progress.

From a teaching perspective, the teacher had to distinguish between blockages induced by lack of reflection, or incorrect or incomplete knowledge, and those involving the need to revise schemas. It is contended that he was able to do this by engaging in the process of meta-metacognition where he considered both the students' mental models and their metacognitive activity as observed by himself during their modelling activity to this point. Meta-metacognition allowed him to recognise the type of intervention needed to overcome this blockage that resulted from a resistance to accommodate new contradictory information, not a lack of reflection on the part of the students. Understanding and identifying blockages as well as the metacognitive activity in which their students engage are crucially important for teachers as they facilitate students' development as independent mathematical modellers.

4 Conclusions

The focus of this chapter has not been on the nature of specific metacognitive strategies and metacognitive knowledge displayed or used by secondary students when engaging with mathematical applications or during mathematical modelling activities. Rather, interest has been in illustrating the meta-knowledge about modelling and application of mathematics to real world tasks and the types of productive and nonproductive metacognitive responses that students might exhibit in such situations for the purpose of informing teaching decision making through reflective activity in the form of meta-metacognition. What is desired is orchestration by teachers of the optimal use by their student modellers of metacognitive knowledge and strategies so as to develop students' competencies in their productive use of these within the modelling context to obtain not only a satisfactory outcome to the current modelling activity but also to further their long-term reflective activity for modelling purposes.

Acknowledgement Examples used in this chapter are from research that was funded by the Australian Research Council linkage project, RITEMATHS (LP0453701), industry partner secondary schools and Texas Instruments.

References

Blum, W., & Kaiser, G. (1984). Analysis of applications and of conceptions for an application oriented mathematics instruction. In J. S. Berry, D. N. Burghes, I. D. Huntley, D. J. G. James, & A.O. Moscardini (Eds.), *Teaching and applying mathematical modelling* (pp. 201–214). Chichester: Ellis Horwood.

Borromeo Ferri, R. (2007). Modelling problems from a cognitive perspective. In C. Haines, P. Galbraith, W. Blum, & S. Khan (Eds.), *Mathematical modelling (ICTMA12): Education, engineering and economics* (pp. 260–270). Chichester: Horwood.

Carlson, R. A., Sullivan, M. A., & Schneider, W. (1989). Practice and working memory effects in building procedural skill. *Journal of Experimental Psychology: Learning, Memory, and Cognition, 15*, 517–526.

Carlson, R.A., Khoo, B. H., Yaure, R. G., & Schneider, W. (1990). Acquisition of a problem-solving skill: Levels of organisation and use of working memory. *Journal of Experimental Psychology: General, 119*(2), 193–214.

Efklides, A. (2002). The systemic nature of metacognitive experiences: Feelings, judgements, and their interactions. In M. Izaute, P. Chambres, & P.-J. Marescaux (Eds.), *Metacognition: Process, function, and use* (pp. 19–34). Dordrecht: Kluwer.

Efklides, A., Kiorpelidou, K., & Kiosseoglou, G. (2006). Worked-out examples in mathematic: Effects on performance and metacognitive experiences. In A. Desoete & M. Veenman (Eds.), *Metacognition in mathematics education* (pp. 11–33). New York: Nova Science.

Flavell, J. H. (1976). Metacognitive aspects of problem solving. In L. B. Resnick (Ed.), *The nature of intelligence* (pp. 231–235). Hillsdale: Erlbaum.

Flavell, J. H. (1979). Metacognition and cognitive monitoring: A new area of cognitive-developmental inquiry. *The American Psychologist, 34*, 906–911.

Galbraith, P., Stillman, G., Brown, J., & Edwards, I. (2007). Facilitating middle secondary modelling competencies. In C. Haines, P. Galbraith, W. Blum, & S. Khan (Eds.), *Mathematical modelling (ICTMA12): Education, engineering and economics* (pp. 130–140). Chichester: Horwood.

Garofalo, J., & Lester, F. K. (1985). Metacognition, cognitive monitoring, and mathematical performance. *Journal for Research in Mathematics Education, 16*(3), 163–176.

Goos, M. (1998). 'I don't know if I'm doing it right or I'm doing it wrong!' Unresolved uncertainty in the collaborative learning of mathematics. In C. Kanes, M. Goos, & E. Warren (Eds.), *Teaching mathematics in new times. (Proceedings of the twenty-first annual conference of the Mathematics Education Research Group of Australasia)* (Vol. 1, pp. 225–232). Gold Coast: MERGA.

Kluwe, R. H. (1987). Executive decisions and regulation of problem solving behaviour. In F. E. Weinert & R. H. Kluwe (Eds.), *Metacognition, motivation and understanding*. Hillsdale: Erlbaum.

Maaß, K. (2007). Modelling in class: What do we want the students to learn? In C. Haines, P. Galbraith, W. Blum, & S. Khan (Eds.), *Mathematical modelling (ICTMA12): Education, engineering and economics* (pp. 63–78). Chichester: Horwood.

Oke, K. H., & Bajpai, A. C. (1986). Formulation – Solution processes in mathematical modelling. In J. S. Berry, D. N. Burghes, I. D. Huntley, D. J. G. James, & A. O. Moscardini (Eds.), *Mathematical modeling methodology, models and micros* (pp. 61–79). Chichester: Ellis Horwood & Wiley.

Schoenfeld, A. H. (1985). *Mathematical problem solving*. Orlando: Academic.

Schoenfeld, A. H. (1987). *Cognitive science and mathematics education*. Hillsdale: Lawrence Erlbaum.

Schoenfeld, A. H. (1992). Learning to think mathematically: Problem solving, metacognition, and sense making in mathematics. In D. A. Grouws (Ed.), *Handbook of research on mathematics teaching and learning* (pp. 334–370). New York: Macmillan.

Stillman, G. A. (1998). The emperor's new clothes? Teaching and assessment of mathematical applications at the senior secondary level. In P. Galbraith, W. Blum, G. Booker, & I. D. Huntley (Eds.), *Mathematical modelling: Teaching and assessment in a technology-rich world* (pp. 243–253). Chichester, UK: Horwood.

Stillman, G. A. (2002). Assessing higher order mathematical thinking through applications. Unpublished Doctor of Philosophy thesis, Brisbane, Australia: University of Queensland.

Stillman, G. A. (2004). Strategies employed by upper secondary students for overcoming or exploiting conditions affecting accessibility of applications tasks. *Mathematics Education Research Journal, 16*(1), 41–70.

Stillman, G. A., & Galbraith, P. L. (1998). Applying mathematics with real world connections: Metacognitive characteristics of secondary students. *Educational Studies in Mathematics, 36*(2), 157–195.

Stillman, G., Galbraith, P., Brown, J., & Edwards, I. (2007). A framework for success in implementing mathematical modelling in the secondary classroom. In J. Watson & K. Beswick (Eds.), *Mathematics: Essential research, essential practice. (Proceedings of the 30th Annual Conference of the Mathematics Research Group of Australasia (MERGA))* (Vol. 2, pp. 688–707). Adelaide: MERGA.

Stillman, G., Brown, J., & Galbraith, P. (2010). Identifying challenges within transition phases of mathematical modeling activities at year 9. In R. Lesh, P. L. Galbraith, C. R. Haines, & A. Hurford (Eds.), *Modelling students' mathematical modeling competencies ICTMA13* (pp. 385–398). New York: Springer.

Chapter 19
Effective Mathematical Modelling without Blockages – A Commentary

Rita Borromeo Ferri

1 Meta-Cognition – A Fuzzy Word?

In this chapter, I comment on the paper by Stillman on "Applying metacognitive knowledge and strategies in applications and modelling tasks in secondary school." In her paper, Stillman highlighted very important aspects concerning learning, teaching, and understanding modelling in the classroom. It is again impressive to see which important role meta-cognitive-activity plays while modelling and also to note that not all meta-cognitive acts are productive for getting a solution. It is quite common that the field of research in meta-cognition is very wide. A lot of research can be found concerning meta-cognition and problem solving. There also exists research on mathematical modelling and meta-cognition, as it was pointed out in the paper and the described project is a wonderful example for this. In my opinion, it is important that we have to learn a lot more on how several aspects of meta-cognition interplay with the learning and understanding of mathematical modelling. So, in my view, one central result of the study of Stillman et al. is that (productive) meta-cognitive activities can be seen as key for effective modelling behavior without "blockages," respectively for providing "blockages." Because I see this also as one important goal for teaching and learning mathematical modelling, it will be the starting point of my commentary as well. In the following lines, I discuss the paper by Stillman on the basis of well-known research areas of meta-cognition.

In the eighties of the last century, Schoenfeld discussed in his well-known article "What's all the fuss about metacognition" the contributions by Henry Pollak and others after the Second Conference on Problem Solving, challenging the fuzzy word "meta-cognition". Schoenfeld rephrased this critique as follows:

> Meta-cognition is a buzzword for your researchers…The word has been used in almost every talk…But the plain fact is that it's jargon doesn't communicate anything to us

R. Borromeo Ferri (✉)
Department of Mathematics, University of Kassel, Germany
e-mail: borromeo@mathematik.uni-kassel.de

non-researchers…If meta-cognition is so important you have a responsibility to explain to us …all in clear language that we can understand. (Schoenfeld 1987, p. 189)

Schoenfeld defined meta-cognition and explained why it is important and he also made clear that research on meta-cognition focuses on three related but *distinct* categories of intellectual behavior, namely:

Your knowledge about your own thought processes. How accurate are you in describing your own thinking?
Control, or self regulation. How well do you keep track of what you're doing when (for example) your're solving problems, and how well (if at all) do you use the input from those observations to guide your problem solving actions?
Beliefs and intuition. What ideas do you bring to your work in mathematics, and how does that shape the way that you do mathematics? (Schoenfeld 1987, p. 190)

In the current discussion on meta-cognition in mathematical education (Veenmann (2006) or Mevarech et al. (2006)), meta-cognition can be differentiated into two central components: *meta-cognitive knowledge and meta-cognitive processes or skills*. But you can integrate these components also in the three research areas Schoenfeld differentiated. I will use these three components for my line of argumentation – to be more concrete: I like to sum up some theoretical and empirical ideas on how these components and their connections can promote effective modelling without "blockages."

2 Effective Modelling Without Blockages: But How?

I like to start with the first component: "Your Knowledge." There is a large body of research on this aspect. But to sum it up with Schoenfelds words:

The research indicates that children are not very good at describing their own mental abilities, but that they get better (though nowhere near perfect) as they get older. (Schoenfeld 1987, p. 190)

In the paper of Stillman, it became obvious that the presence of meta-cognitive activity has been confirmed by young modelers, young means modelers of grade 7–9. Stillman made clear that teachers use a variety of strategies to attempt to initiate such reflective activities in students. My question is: Why don't we request primary teachers to use these strategies as soon as possible? Starting with modelling in early age groups could help the youngest to learn and to describe their own thinking processes with the help of the teacher so that they can recognize that these reflections can be helpful for solving modelling tasks. This could be a wonderful beginning for early age groups.

Modelling activity in primary school is an exception in Germany for example, but as we know from research of Lesh (Lesh and Doerr 2003), English (2006) and others, it is important to start with modelling in primary school. These young pupils do not only learn to solve such realistic tasks; but also they get an idea, or is it better to say *a feeling*, of the way of thinking which is essential for modelling.

That is a good basis for learning to describe our own thinking processes on the one hand but on the other hand for the teacher of course hard work, because he or she has to explain how to describe own thinking processes in general and concerning the modelling process. So my opinion is that the teacher is engaged in a *meta-meta-cognitive process* as well because the pupils and teachers have to talk about meta-cognition itself and how it can be adopted specially on the modelling process.

I turn to the aspect o control, or self-regulation. The former aspect "Your knowledge" can be seen as an important basis, but "It is not only what you know, but how you use is that matters" in Schoenfeld's (1987, p. 190) words so to speak. So, pupils have to think about management issues, for example, the time management while working on complex modelling tasks and so on. In most cases, modelling tasks are solved in groups of pupils, which makes sense, because even in reality, complex problems are not solved alone. I think the way groups of pupils work together on modelling tasks is important, for this can promote or hinder applying meta-cognitive activities like the meta-cognitive skills, which are *task analysis, planning, monitoring, checking, and reflections* (see e.g., Veenman 2006).

Stillman pointed out which meta-cognitive acts could be productive, so the recognition and the choice of a strategy are relevant. That is what Schoenfeld means with "how you use is that matters." But there is another important aspect Stillman mentioned for discussion: "A focus on the modelling cycle that explicitly includes recognition of a modeller's non-linear process, the presence of metacognitive activity at each part of the process has identified a potential means to overcome blockages of low intensity – genuine reflection."

My question here is: How does this, I call it "meta-cognitive cycle," look like for practical issues?

The first thing which comes to my mind is taking a modelling cycle as a strategic instrument and a necessary "material for modelling," because solving processes for modelling tasks are described on a meta-level simultaneously. But which kind of modelling cycle can be appropriate especially for including meta-cognitive activity at each part of the process? You can find a lot of different cycles in the discussion on modelling for every purpose (Borromeo Ferri 2006). Kaiser (2005) and also Maaß (2007) pointed out that a cycle with four phases can be seen as a good instrument for building meta-cognitive modelling competency. Blum (2007) created within the DISUM-project, a so-called solution plan for modelling tasks in which the general skills fit in perfectly. Of course, this is a helpful instrument, but it does not guarantee that meta-cognitive acts will be productive or that relevant strategies will be recognized and implemented successfully at the end. I think the connection between the knowledge about the cycle from a normative and descriptive perspective and the related "sub-competencies" which I would call the heart or content of the cycle is crucial. Furthermore, if "sub-competencies" are considered as strategies besides "meta-cognitive strategies," this should be a good basis for providing blockages of high intensity. There is so much more to say about that aspect, but I will change to the last one: beliefs and intuition. In my opinion, this aspect will enrich some thoughts Stillman pointed out, in particular the following one, because beliefs and intuitions have also central impacts on the development of blockages and concerning the level:

"Analysis has led us to infer the cause of the more robust blockages that are different in type and cognitive demand and what type of intervention is needed to overcome them."

Maaß (2007) and other researchers showed on an empirical basis that implementing mathematical modelling tasks in mathematics lessons can change mathematical beliefs of pupils over time. In general discussions there are different views, if the person variables, except for the universals of cognition, are better seen as motivational constructs rather than as meta-cognitive ones. Other researchers have integrated the motivational variables in their meta-cognitive models.

> The basic argument is that strategy-based actions directly influence self-concept, attitudes about learning and attributional beliefs about personal control. (Borkowski et al. 1990, p. 54)

In turn, these personal motivational states determine the course of new strategy acquisition and, more importantly, the likelihood of strategy transfer and the quality of self-understanding about the nature and the function of mental processes. Implications, especially practical ones, are not really easy to characterize, because the belief system of an individual is very complex as it is known. The part of the teacher is a crucial one, as Stillman made clear. Nevertheless, a practical idea for effective modelling is exploring with the pupils questions like "What does mathematics mean to me?" or supporting reflections like "What is the relationship between myself and mathematics"? or "How do I like to learn and understand mathematics in the best way?" This is again a discussion on the meta-level. I think that this debate could be very helpful to get a clearer view for pupils: why there could be blockages while solving modelling tasks. That is a good point for connecting all the three aspects.

3 Connecting and Acting

Beliefs and intuitions can be very powerful for effective modelling as I mentioned earlier, because belief systems are one's mathematical worldview and the perspective with which one approaches mathematics and mathematical tasks.

> Beliefs establish the context within which resources, heuristics, and control operate. (Schoenfeld 1985, p. 45)

Knowledge about meta-cognitive skills and relevant strategies or sub-competencies is on the one hand important, but on the other hand one's own beliefs about mathematics can determine how one chooses to approach problems, which of these techniques will be used or avoided, and how long and how hard one will work on it. The connection between "Your Knowledge" and "Control" was already mentioned at the beginning, so the knowledge about my own thoughts is a basis for meta-cognition, but the decision, and that is what I want to stress, is important as well.

Let me briefly sum up my theoretical and practical ideas for supporting an effective modelling behavior of our students, so that "blockages" can be avoided: All three aspects, "Your Knowledge," "Control," and "Beliefs" must be connected and supported through the help of the teacher. If you want your students to do a cooperative learning task, for example, you have to learn with them the method of doing that. That is the same with meta-cognition: Pupils have to learn how to describe their thinking processes and how to share them with others in the group. Especially in mathematical modelling, the modelling cycle is a wonderful instrument for that. Pupils do not think about their relationship between themselves and mathematics normally. So, teachers have to give them suggestions to think about that.

Applying meta-cognitive activities while modelling must be a central part of learning and teaching modelling starting in early ages.

References

Blum, W. (2007). Modellierungsaufgaben im Mathematikunterricht – Herausforderung für Schüler und Lehrer. In A. Büchter et al. (Eds.), *Realitätsnaher Mathematikunterricht – vom Fach aus und für die Praxis* (pp. 8–23). Hildesheim: Franzbecker.

Borkowski, J. G., Carr, M., Rellinger, E., & Pressley, M. (1990). Self-regulated cognition: Interdepence of metacognition, attributions, and self-esteem. In B. F. Jones & L. Idol (Eds.), *Dimensions of thinking and cognition instruction* (pp. 53–92). Hillsdale: Lawrence Erlbaum Associates.

Borromeo Ferri, R. (2006). Theoretical and empirical differentiations of phases in the modelling process. *Zentralblatt für Didaktik der Mathematik, 38*(2), 86–95.

English, L. D. (2006). Mathematical modelling in the primary school: Children's construction of a consumer guide. *Educational Studies in Mathematics, 63*, 303–323.

Kaiser, G. (2005). Mathematical modelling in school – Examples and experiences. In H.-W. Henn & G. Kaiser (Eds.), *Mathematikunterricht im Spannungsfeld von Evolution und Evaluation. Festband für Werner Blum* (pp. 99–108). Hildesheim: Franzbecker.

Lesh, R., & Doerr, H. (Eds.). (2003). *Beyond constructivsm – Models and modelling perspectives on mathematics problem solving, learning and teaching*. Mahwah: Lawrence Erlbaum.

Maaß, K. (2007). Modelling in class: What do we want the students to learn? In C. Haines et al. (Eds.), *Mathematical modelling: Education, engineering and economics* (pp. 63–78). Chichester: Horwood.

Mevarech, Z., Tabuk, A., & Sinai, O. (2006). Meta-cognitive instruction in mathematics classrooms: Effects on the solution of different kinds of problems. In A. Desoete & M. Veenman (Eds.), *Metacognition in mathematics education* (pp. 73–82). New York: Nova Science Publishers, Inc.

Schoenfeld, A. H. (1985). *Mathematical problem solving*. Orlando: Academic.

Schoenfeld, A. H. (1987). What's all the fuss about metacognition? In A. H. Schoenfeld (Ed.), *Cognitive science and mathematics education* (pp. 189–215). London: Lawrence.

Veenman, M. (2006). The role of intellectual and metacognitive skills in math problem solving. In A. Desoete & M. Veenman (Eds.), *Metacognition in mathematics education* (pp. 35–50). New York: Nova Science Publishers, Inc.

Chapter 20
Modelling Tasks: Insight into Mathematical Understanding

Jill P. Brown and Ian Edwards

Abstract It is claimed, students' communication of their solutions to modelling tasks gives insight into the depth of their mathematical understandings and how they use prior knowledge of the context of a task in their solution. In the example given, both students, Tabitha and Tanya, take an integrating approach to dealing with mathematics and reality in such tasks and their manner of dealing with the context of real world tasks remained stable from Year 9 to Year 11. In addition, communicative artefacts required by the tasks help reveal the students' deepening understanding of mathematics.

School modelling tasks by their nature involve "some genuine link(s)" with a real world context (Galbraith 2007, p. 55), this in turn requires some level of complexity of mathematical thinking – higher order thinking – when attempting to solve such tasks. This then suggests students' communication of their solutions to modelling tasks may allow insight into their mathematical understanding and that nontrivial solution attempts to such tasks should be indicative of deep understanding which might have developed during engagement with the task, or the complexity of the task, by its very nature, allows this deep understanding to be demonstrated. Of importance are the relationships existing between mathematical modelling [MM] and mathematical understanding. The study reported here explores this relationship in addressing: What aspects of MM support demonstration or deepening of understanding of mathematics?

Three approaches to analysing such tasks will be employed in investigating this assertion. The first is the use of prior knowledge of task context based on the work of Stillman (2000). The second is the application of Busse's (2005) typology

---

J.P. Brown (✉)
School of Education (Victoria), Australian Catholic University, 115 Victoria Parade, Locked Bag 4115, FITZROY 3065, Australia
e-mail: jill.brown@acu.edu.au

I. Edwards
Luther College, Victoria, Australia
e-mail: ie@luther.vic.edu.au

of ideal types when dealing with such contexts. Both of these will be linked to notions of higher order thinking. Finally, a framework for higher order thinking developed by the first author for analysing student understanding of functions in a *technology-rich teaching and learning environment* (TRTLE) (Brown 2007) is used as an integrating lens.

1 Prior Knowledge of Task Context

In school modelling and applications tasks, the explicit sources of information are the problem statement and any included visual representations (graphs, tables, diagrams, etc.). These sources provide data to the task solver, which may be relevant, however, "not all data in the problem presentation will have the same strength in cueing facts, concepts, processes, prior experiences, semantic knowledge (Tulving 1985) or metacognitive knowledge, and strategies (Stillman and Galbraith 1998) from long-term memory" (Stillman 2000, p. 335).

The source of prior knowledge of task context was used by Stillman (2000, p. 333) to classify such knowledge as (a) academic, that is, "vicarious experiences in other academic subject areas," (b) encyclopaedic – "general encyclopaedic knowledge of the world," or (c) episodic – "truly experiential knowledge developed from personal experiences outside school or in practical school subjects". Stillman (2000) details a range of purposes reported by students for using prior knowledge. These include: *enhancing understanding*, by confirming other forms of prior knowledge, developing a mental picture, and visualizing; *enabling the student to relate to the context* (making the student comfortable with the task, confirming feasibility of the task); *enhancing decision making during execution, selecting a mathematical model* including choosing between two mathematical options; and *facilitating the checking of progress*, by keeping the student on track or judging the reasonableness of interim or final results.

In a study investigating senior secondary mathematics students' approaches to, and performance on, applications tasks, Stillman (1998) did not find a clear link between high engagement with the task context and the level of success on a task. However, "moderate to high engagement with the task was not often associated with poor performance" (p. 51). In addition, "poor performance was more likely to be associated with no to low engagement" (p. 51). In most cases prior knowledge enhanced students' understanding of, and engagement with, the task. In some cases "prior knowledge was used to check progress or the reasonableness of interim or final results". There were *"few cases* where prior knowledge was actually used to *enhance decision making, facilitating students' selection* of an appropriate mathematical model or *choice between* two mathematical options" (Stillman 2000, p. 335).

Stillman (1998) reports two conditions upper secondary students in her study believed helped them engage with the context of a task. These were having *an objective to work towards*, and a *sense of realism* in the task. An example of the first occurred where students engaged in a *Road Accident Problem* "were required to check if a driver was telling the truth about the speed of a car involved in an accident" (p. 63). By a sense of realism, Stillman was referring to students reporting they believed the task was realistic.

2 Ways of Dealing with the Context

A second relevant study was undertaken by Busse (2005) with eight upper secondary students in Germany. Busse reported that "students deal very individually with the context" (p. 354) of a task. His study of 16–17-year-old students (2 female and 2 male pairs from 4 schools) addressed the question of "how an *individual* deals with the context and how the context in a given task-text is *internalised*" (p. 354). He identified four ideal types, allowing him to contrast the actual behaviour of individuals at a given point in time against the ideal type. Busse derived his ideal types by considering how the task solver dealt with "*reality* on the one hand and *mathematics* on the other hand" (p. 355) and the way they were related to each other. Busse's four ideal types are: reality bound, mathematics bound, integrating, and ambivalent.

A reality-bound approach would not entail the application of mathematical methods, rather extramathematical concepts and methods would be used. A mathematics-bound approach takes the context of a realistic task as decoration; therefore, the task must be solved by mathematics exclusively. An integrating approach is indicated by the use of both mathematical methods and personal knowledge of the task context beyond that stated in the mathematical solution of the task. An ambivalent approach means neither the mathematical nor extramathematical methods are given precedence in the solution and both mathematics and reality aspects of the task are perceived but not synthesised.

For brevity, only one type, integrating, will be discussed in depth as it was found to be sufficient for our analytical purpose. However, from a modelling perspective, this describes a critically important way of thinking as both mathematics and reality are considered in task solving. It is in the integrating ideal type that the task solver uses "contextual ideas in productive combination with mathematical methods" (Busse 2005, p. 356). Here Busse draws on the work of Stillman (1998, 2000) and the use of prior knowledge of the task context to support problem solution. Task solvers draw on the context of the task both in mathematising the problem situation and in validating the solution. Busse points out it is not clear whether displaying a typical way of dealing with task context is a constant characteristic of a modeller or if it is more related to the context in which the modelling or application task is set, suggesting that "the types are not necessarily invariably linked to persons" (p. 358). Hence, his categories describe way(s) of dealing with the context, rather than task solvers. However, Busse (2005) notes gender differences in his study, two of four females, but none of the four males were assigned to the ambivalent category. In contrast, three of the four males (and none of the females) were assigned to the integrating type based on their approaches to a *Home for aged people task*. This is in agreement with the findings of Kaiser-Messmer (1993) that gender differences exist in student preferences for particular contexts.

3 Higher Order Thinking

Higher order thinking in its simplest terms can be used to describe thinking beyond the retrieval of information, that is, any transformation of information (Baker 1990, p. 7). However, Resnick (1987) would strongly caution against a misinterpretation of the notion of higher order thinking as suggesting that there is some lower order thinking that must precede this. She argues that higher order thinking should be occurring at all levels of development. The term has also been used to describe communication skills, reasoning skills (Romberg et al. 1990), and metacognitive skills (Baker). Clearly, deep understanding involves thinking processes related to subject knowledge and this knowledge includes interactions of strategic, procedural, and content (or declarative) knowledge (Baker). Romberg et al. in elaborating on higher order thinking in mathematics proposed it involves some, but not necessarily all (at a particular point in time), of the following: a solution path that is not initially obvious, its complex nature, a tendency for multiple rather than unique solutions, judgment and interpretation, application of multiple and possibly conflicting criteria, uncertainty (not everything required is given), the imposition of meaning, effort, and finally, an essential element is the self-regulation of thinking (pp. 22–23). Self-regulation is part of the metacognitive competencies necessary for successful modelling (Maaß 2006).

In considering *higher order thinking* when students are engaged in tasks involving functions in a *technology-rich teaching and learning environment* (TRTLE), the first author as part of a larger study[1] (e.g., Brown 2007) developed the following framework – to be used in searching for evidence of higher order thinking. In light of the above, this is taken to mean instances where there is evidence that a student appropriately makes choices about the solution path (e.g., decisions about processes, representations, technology use and type); makes links across representations; expects to verify a conjectured solution; appreciates the value of, or need for, verification; is aware of the value of verification occurring in a yet unused representation or in multiple representations; and/or differentiates between global verification and local checking. Clearly, these are crucial for student success in modelling and applications tasks in secondary school mathematics.

---

[1] Jill Brown was a doctoral student of University of Melbourne on Australian Research Council funded Linkage Project – LP0453701 when these data were collected.

4 Methods

The main study, from which the data are drawn, follows a qualitative inquiry where a case-based approach (Stake 2005) was adopted, with the units of analysis being individual students. This is an intrinsic case study where "case study serves to help us understand the phenomena or relationships within it". The case or cases studied play "a supportive role [as they are] facilitat[ing] our understanding of something else" (p. 445), here, the depth of understanding shown by students as they engage with the context of modelling tasks and the higher order thinking involved as their solution progresses.

Two students, Tabitha and Tanya, were selected from the larger study. They were selected for pragmatic reasons – both were in the same Year 9 class, taught by the second author, both high performing in mathematics, both female, and both were in the larger research study which occurred when they were in Year 11.

The students' responses to two tasks are the basis for the data collection. The first task is an open modelling task designed by the classroom teacher. However, as noted by the teacher, constraints of external assessment by examination loom large in upper secondary; hence the second task designed by the first author, although attempted 2 years after the first task, involved more structured modelling. Data for this analysis includes student scripts, and for the second task also includes recordings of graphing calculator screens used by students during modelling, post-task interviews, and for Tabitha audio and video recording during solution of the modelling task.

4.1 The Tasks

The first task, *Tommy Tinn's Trout Farm* [TT], sees environmental issues and concerns impinging on economic models, whereas in the second task, *Save the Platypus* [PP], environmental issues and concerns impact on population models. Both tasks are set in the local geographical environment of the school which is near the base of the Dandenong Ranges close to the imaginary setting of the trout farm, although there are trout farms in the area, and close to the Yarra River, the home of platypus.

Whereas in the *Road Accident Problem*, used by Stillman (1998), the situation saw students being given a benchmark against which to test results, both tasks in this study provided an objective, albeit of a different nature. In TT, students had to write a letter to the fishery manager recommending one of four proposed farming strategies, whereas in PP, they were required to examine the results of a mathematical analysis of the data and make recommendations about the continuation of an intervention project, in the form of a powerpoint presentation. Additional details are shown in Fig. 20.1.

The teacher's rationale for developing and implementing the TT task was: "if mathematics teachers *fill the allotted time with drilling* their students in *routine operations*, they *kill their interest*, hamper their intellectual development, *and*

> **Tommy Tinn's Trout Farm** (Year 9 task)
> A lake in a Lilydale National Park was stocked with approximately 10000 trout. In similar lakes, when left to natural factors trout numbers increase on average by 20% per year. For the fishery to remain viable there needs to be at least 750 fish in the lake. Overpopulation can cause a dramatic fish kill, where up to 95% of the fish stock may die. The carrying capacity of the lake is assumed to be approximately 5 times the current capacity of 10000 fish. Consider the 4 strategies and make a recommendation.
> *1:* Do nothing to disrupt with the normal population control of the fish in the lake. Fishing in the lake shall be on a catch and release basis.
> *2:* Fishing permitted. Total catch allowed for all fishing licenses is 1800 fish per year.
> *3:* One approved contractor licensed to remove up to 2500 fish each season.
> *4:* Permit the fish population to reach 25000 fish, then issue and monitor amateur fishing licenses that would maintain fish stock at this level.
>
> **Platypus in Danger** (Year 11 task)
> The platypus is an endangered species that may become extinct unless action is taken to save it. An annual survey held in a nearby national park showed an alarming decrease in the number of platypus over the years 1993-1998. Two sets of data representing a platypus population before and after an intervention project, were presented. Find a model to represent platypus numbers over time for both data sets. Questions then considered included: Did the intervention improve the situation, what was the predicted population a decade later, and When would the population return to the initial value?

Fig. 20.1 Further details of the *Tommy Tinn Trout Farm* and the *Save the Platypus Task*

misuse their great opportunity." However, if they challenge the curiosity of their students by setting them problems proportionate to their knowledge, and help them to solve their problems with stimulating questions, they may give them a taste for, and some means of independent thinking. To facilitate this, the task required students to suggest appropriate management procedures for a potential new fishery. The teacher believes "mathematics does not readily produce artefacts which can display the originality and versatility of students' creative energies. For this reason, the opportunities for students to see the results of others should be seized." The TT report is proposed as an exemplar of this approach. He argues, "prior to focusing on this approach, the final report (received from his students) would be minimal. Little evaluation of results was attempted. Subsequently, the complexity of the mathematics and incorporation of the mathematics to validate opinion have increased. The need to have a single correct answer is gone."

5 Findings

5.1 Evidence of Prior Knowledge of Task Context

In identifying the use of prior knowledge by the students in the Year 9 task, it was found that whilst both students activated prior knowledge differences in the source

20 Modelling Tasks: Insight into Mathematical Understanding

Table 20.1 Identification of student use of prior knowledge in the tasks

	Type and use of prior knowledge in TT and PP task	
Student	Academic	Encyclopaedic
Tanya	–	Enhancing decision making [TT]
		Enables student to relate to context [both]
		Facilitates the checking of progress [PP]
Tabitha	Enhances understanding [TT]	Enhancing decision making [both]
	Enables student to relate to context [TT]	Facilitates the checking of progress [PP]
	Selects mathematical model [PP]	Enhances understanding [PP]
	Enhancing decision making [PP]	

were identified (see Table 20.1). Both used encyclopedic prior knowledge, but for different purposes. Both used it to enhance decision making during execution, for example, when Tanya made reasonable estimates and Tabitha introduced alternative ideas as she argued it is more humane to control the fish population by fishing than to allow overcrowding and subsequent mass fish deaths. In addition, Tanya used this type of prior knowledge to enable her to relate to the task context. For Tabitha, encyclopedic prior knowledge was also used to enhance her understanding and decision making. Only Tabitha showed academic prior knowledge, activating it to enhance understanding and to allow her to relate to the task context. No episodic prior knowledge was evident for either student.

In identifying the use of prior knowledge by the students in the Year 11 task, again differences in the source were noticed (see Table 20.1). Both used encyclopedic knowledge, again only Tabitha used academic prior knowledge and neither used episodic knowledge. Both used encyclopedic knowledge for facilitating checking of their progress. In addition, Tanya used it to relate to the task context, and Tabitha to facilitate checking of her progress and to enhance her understanding. In addition, Tanya used academic prior knowledge, for selecting her mathematical model (in conjunction with mathematical knowledge), and to enhance decision making during execution.

5.2 Busse's Ideal Types

Following Busse's schema, both students' actions were classified as dealing with both tasks in an *integrating way*. In each task, both students applied mathematical methods to solve the task in conjunction with perceiving the task to be realistic. Both brought prior knowledge of the task context to support their task solution. In TT, Tanya considered the use of the Trout Farm as a nature park to increase income particularly in the years when the proposed strategies resulted in little income. She proposed charging for fish by weight, then rejected this strategy, "even though some trout will be heavier and worth more than $8, on the average, this is the right price, and will cause fewer hassles with price." In PP, she was so concerned with saving the platypus from extinction that she deliberately constructed a mathematical model that predicted their survival. Tabitha, on the other hand, discussed the need to consider the costs involved in maintaining the lake and feeding the trout in TT. She proposed a bag limit for fishers, and considered a strategy resulting in mass fish deaths as inhumane. In PP, she realised a population is technically extinct before it reaches zero, and used this in her mathematical analysis. Both used a range of mathematical methods to their solutions, thus "using contextual ideas in productive combination with mathematical methods" (Busse 2005, p. 356).

Certainly their way of dealing with tasks was influenced by teachers and teaching, however, when offered the opportunity an integrated approach was taken and enjoyed – much to Tanya's surprise – as in (her Year 11 mathematics) class, she tended to focus on actions resulting in positive teacher feedback rather than her engagement with a task per se. When asked in the post-task interview what she thought of the second task, her response indicated that not only did she appreciate undertaking the task, she spontaneously made connections with the mathematics she engaged in happily in Year 9. *I thought it was all right, ... No, it was good. It was like what I did when I was doing Year 9.* When Tabitha was asked what she thought of tasks like Platypus compared to other mathematics tasks she had done in mathematics, she replied:

> *I actually found it enjoyable to do this kind of thing. It is challenging and it puts to work the ability to decide where [pause] like you have got so many mathematical tools at your disposal and to be able to find out how you can apply them and how to know when to use them and that kind of thing.*

She also elaborated on her thinking:

> *As I was going through all of this [indicating her solution], ... I started to think, 'Surely something positive should be occurring here' and then I started to doubt my exponential function, yeah. So then I went through the whole presentation with the exponential models because that was what I had gotten but then on reflection I decided that perhaps that wasn't the best models to be using.*

5.3 Modelling and Higher Order Thinking

The focus of the analysis presented next is on the use of context and the demonstration of higher order thinking by the students during modelling tasks. The intention of the teacher, in his Year 9 task selection, was for students to have a series of rich enjoyable tasks where they were working mathematically from their knowledge base to a new level of expertise (Edwards 2005). He aimed to set tasks that addressed multiple concepts that provided opportunities for concepts to be strongly connected to students' understandings to form deeper understanding of the connections within mathematics and with the real world. He always asked himself: *Are there sufficient opportunities where students can propose alternate pathways*

20 Modelling Tasks: Insight into Mathematical Understanding

Table 20.2 Evidence of higher order thinking

Higher order thinking	Tanya	Tabitha
(a) Makes choices about her/his solution path (processes, representations, technology use and type);	Yes	Yes
(b) Makes links across representations;	No	Yes
(c) Expects to verify a conjectured solution;	Yes	Yes
(d) Appreciates the value of or need for verification;	No	Yes
(e) Is aware of the value of verification occurring in a yet unused representation or in multiple representations;	No	Yes
(f) Differentiates between global verification and local checking	No	Yes

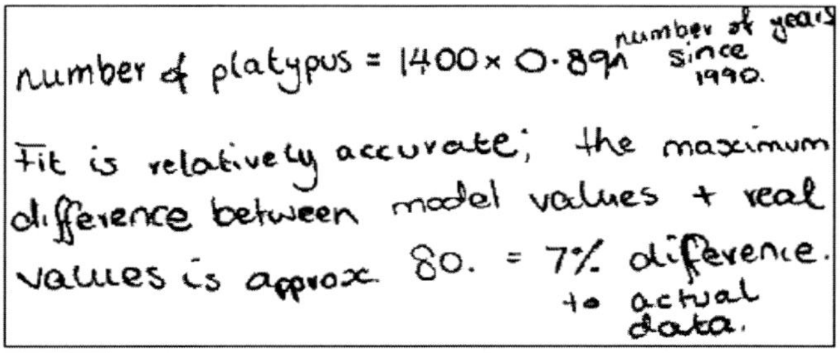

> number of platypus = 1400×0.89^n number of years since 1990.
>
> Fit is relatively accurate; the maximum difference between model values + real values is approx 80. = 7% difference to actual data.

Fig. 20.2 Evidence of higher order thinking

for the investigation? This suggests that one teacher aim was to have his students engage in tasks that allowed opportunities for higher order thinking. However, as the data for the first task was limited to the student script, it was not possible to analyse the students' approaches to this task using the *Higher Order Thinking Framework*. This was however possible for the second task as shown in Table 20.2, where clear differences between the students are evident.

Tabitha displayed all categories in the *Higher Order Thinking Framework*. However, for Tanya, evidence was found for only two of the six categories, namely, making choices about her solution path, and that she expected to verify a conjectured solution. Figure 20.2 shows an example, based on the interview and task script, where Tabitha demonstrates awareness of the value of verification occurring in multiple representations as she checks her intermediate solution, a model for the data, both graphically and numerically. In the interview she explained:

> I did a STATPLOT of the data and checked the function [graph] against it .... Yes, this is by eye and then calculating values and finding out [pause] what kind of agreement between the values there was. ... I had the actual values of data and then I used the CALC function.... For that point on the graph.... I could work out percentage differences. Yeah, I checked that it fitted the points and it did fit the points very well.

6 Discussion and Conclusion

Both tasks in this study had an objective to work towards, built into the task design, and both were found by students to be realistic – mirroring the conditions identified by Stillman (1998) to help students engage with task context. Both students drew on their prior knowledge for a range of purposes. Similarly to Stillman (2000), prior knowledge of task context was used to enhance understanding of, and engagement with, the task. Tanya drew only on encyclopaedic knowledge whereas Tabitha also drew on academic prior knowledge. There was evidence that both students used prior knowledge to enhance decision making and to relate to the context in both tasks, and to facilitate the checking of progress in the PP task. Tabitha also activated prior knowledge to enhance understanding in both tasks and to facilitate selection of her model [PP task]. Thus, both drew on prior knowledge for a range of purposes that facilitated their modelling.

Building on this engagement with the task context saw both students responding to both task contexts in an integrating fashion. Given the gender of these students, this is in contrast to Busse's findings. In Busse's study (2005), no female students were identified as dealing with the task context in an *integrating way* but both were in this study. Furthermore, this appears to be a stable approach for these two students from Year 9 to Year 11, although it is acknowledged only two tasks have been considered. Clearly modelling tasks such as these allow students to deepen their understanding of mathematics. Both students in this study integrated mathematics and the real world in their approach to the task. That both took an integrating approach is further evidence of a deeper understanding of modelling. Communication artefacts allowed the demonstration of this deepening understanding – the letter allowed the teacher and researcher greater insight, whereas the powerpoint presentation was more time efficient. Thus, both were useful aspects of the modelling task making deepening understanding transparent.

One might suspect that the *Tommy Tinn Trout Farm Task* where students are presented with the basis of a model and assume the role of having to report the most appropriate strategy to the manager almost imposes engagement with task context through task design. However, Gruenwald et al. (2007) have demonstrated with engineering students that such design features, of themselves, do not ensure engagement.

The differences in type and use of prior knowledge and the integrating approach to the task context were echoed and expanded in the analysis based on the higher order thinking framework. For Tabitha, evidence of each thinking category was evident. In contrast, Tanya made choices about her solution path, but there was no evidence indicating she made connections across representations, even for verification purposes, nor did she appear to differentiate between global verification and local checking.

Modelling tasks, particularly those with the required communication artefacts described in this chapter, provide opportunities for the development of mathematical understanding – as mathematical and additional prior knowledge of the context are integrated. In addition, they provide opportunities for revealing this deep understanding. The setting of modelling tasks in a *technology-rich teaching and learning environment* facilitated these opportunities – for further examination of this point, see Brown (2007). However, not all modellers take up these opportunities to the same extent.

References

Baker, E. (1990). Developing comprehensive assessments of higher order thinking. In G. Kulm (Ed.), *Assessing higher order thinking in mathematics* (pp. 7–20). Washington: American Association for the Advancement of Science.

Brown, J. (2007). Early notions of functions in a technology-rich teaching and learning environment (TRTLE). In J. Watson & K. Beswick (Eds.), *Proceedings of MERGA30* (Vol. 1, pp. 153–162). Adelaide: MERGA.

Busse, A. (2005). Individual ways of dealing with the context of realistic tasks – First steps towards a typology. *Zentralblatt für Didaktik der Mathematik, 37*(5), 354–360.

Edwards, I. (2005). New wine in old skins. In W. Moroney & C. Stocks (Eds.), *Quality mathematics in the middle years* (pp. 73–81). Adelaide: AAMT.

Galbraith, P. (2007). Dreaming a 'possible dream': More windmills to conquer. In C. Haines, P. Galbraith, W. Blum, & S. Khan (Eds.), *Mathematical modelling: Education, engineering and economics* (pp. 44–62). Chichester: Horwood.

Gruenwald, N., Sauerbier, G., Zverkova, T., & Klymchuk, S. (2007). Models of ecology in teaching engineering mathematics. In C. Haines, P. Galbraith, W. Blum, & S. Khan (Eds.), *Mathematical modelling: Education, engineering and economics* (pp. 314–322). Chichester: Horwood.

Kaiser-Messmer, G. (1993). Results of an empirical study into gender differences in attitudes towards mathematics. *Educational Studies in Mathematics, 25*, 209–233.

Maaβ, K. (2006). What are modelling competencies? *Zentralblatt für Didaktik der Mathematik, 38*(2), 113–142.

Resnick, L. B. (1987). *Education and learning to think*. Washington: National Academy Press.

Romberg, T., Zarinna, A., & Collis, K. (1990). A new world view of mathematics. In G. Kulm (Ed.), *Assessing higher order thinking in mathematics* (pp. 21–38). Washington: American Association for the Advancement of Science.

Stake, R. (2005). Qualitative case studies. In N. Denzin & Y. Lincoln (Eds.), *The Sage handbook of qualitative research* (3rd ed., pp. 443–466). Thousand Oaks: Sage.

Stillman, G. (1998). Engagement with task context of application tasks: Student performance and teacher beliefs. *Nordic Studies in Mathematics Education, 6*(3–4), 51–70.

Stillman, G. (2000). Impact of prior knowledge of task context on approaches to applications tasks. *Journal of Mathematical Behavior, 19*(1), 333–361.

Chapter 21
Mathematical Modelling of Daily Life in Adult Education: Focusing on the Notion of Knowledge

Susana Carreira, Nélia Amado, and Filipa Lecoq

Abstract In our research, we aim to look at the notion of knowledge as it is elicited through mathematical modelling of daily life situations, within the context of adult education. In the school scenario of adult education, notions from situated cognition will be brought into play to examine the meaning of mathematisation and of mathematical modelling competence. The empirical data refer to a 2-month period of work on the theme of cookery, one that was chosen by the students. Data were collected in a school environment within the subject "Mathematics for Life", a course in Adult Education, for certification of compulsory general education (i.e., 9th grade in regular school).

1 Introduction

The concept of mathematical literacy has been evolving over time. The PISA 2003 (OECD 2004) defines mathematical literacy as

> an individual's capacity to identify and understand the role that mathematics plays in the world, to make wellfounded judgements and to use and engage with mathematics in ways that meet the needs of that individual's life as a constructive, concerned and reflective citizen. (p. 37)

Matos (2002) stresses that literacy "focuses on knowledge used by adults in daily life matters, while basic school knowledge is essentially about newly acquired ideas, inserted in a school context, despite being expected to become applied in students' future life" (p. 3). Zevenbergen (2002) talks about three levels of numeracy: technical

---

S. Carreira (✉) and N. Amado
Universidade do Algarve, Faro & UIDEF, Lisboa, Portugal
e-mail: scarrei@ualg.pt; namado@ualg.pt

F. Lecoq
Escola Básica Prof. Paula Nogueira
e-mail: filipalecoq@gmail.com

numeracy, where basis skills can be seen as central (mastery of calculation processes, measuring); practical knowing (being able to apply the technical knowledge in a context); and critical knowledge (using numeracy in the development of social and ideological critique). Accordingly, there are many forms of numeracy needed today, requiring significant shifts from traditional numeracy mainly aimed at basic skills.

The Portuguese Agency for Adult Education and Qualification states that nowadays to be mathematically competent means to know how to solve problems and to deal with real situations. One of the tenets of Adult Education courses is to value adults' life experiences, viewed as major resources for their formative learning process. So, curricular guidelines recommend considering such contributions in organizing learning through genuine problem solving. While acknowledging the importance of adults' knowledge and experience, we intend to address specific questions concerning the nature of knowledge. This is a matter whose complexity has been rightly postulated: "I regard 'adults and mathematics' as a complex subject for mathematics education, whether the focus be in teaching, learning, or knowledge (Wedege 1999, p. 206)".

In our work, we have selected an overarching problem: "What elements of the modelling process are mostly affected by the scope of adults' mathematical literacy and reversely by their social, professional or personal ways of dealing with reality?"

2 Theoretical Framework

2.1 A Discussion on the Real World

Developing mathematics education for adults, namely with a target on mathematical literacy, and endorsing the perspective of modelling daily life situations to generate mathematical competencies, involves dealing with the issues of *context* and *practice*.

Mathematics classrooms in adult education are expected to promote adults' professional, social and daily life experiences. The aim is to call upon the students' *lived-in world*, because it means a useful path for students' mathematical practice. However, another version of the real world, not less important than the former and certainly valuable too, appears in our classrooms, when frequently, we figure it, that is, we try to convert the lived-in world into a particular schooled environment – the *figured world* of the classroom, as described by Boaler (2000). This idea echoes an important clarification offered by Wedege (1999) that distinguishes the different pedagogical meanings assigned to context: the task-context and the situation-context. The task-context for a particular problem may be the same, yet the problem can be tackled differently depending on the situations where the individuals are participating in.

Our theoretical framework intends to connect a socio-cultural view on knowledge (based on situated cognition) and a modelling perspective where the modelling cycle is central and includes a fundamental element broadly named "reality". One of our purposes is to call upon the theory on modelling and applications, since there is not "a total paradigm or a 'grand narrative' concerning adults and mathematics" (Wedege 1999, p. 208). With real life eliciting problems, the production of bridges between different ways of perceiving the world and acting in it becomes essential. The classroom's figured worlds are key elements in the process of knowing and thus have to be minded. Recognising barriers and specificities of practices across contexts and social organizations becomes a precondition. Yet the possibility of integrating different forms of knowledge and experiences has also to be seen as feasible.

> Those who support abstract procedure repetition as the most efficient way to learn mathematics (Becker and Jacob 2000) overlook the fact that students are not only learning an efficient set of procedures, but an esoteric set of practices that are not well represented outside of mathematics classrooms. (Boaler 2000, p. 4)

> Whilst the neo-Vygotskian work recognizes context specificity, it, perhaps allows insufficient room for leakage between contexts. (Dowling 1991, p. 116)

The above contrasting claims place a strong challenge in looking for a way out of the tension between specificity and permeability of contexts and practices. Therefore, examining how knowledge may be conceptualised is one of our concerns.

2.2 *Knowledge, Practice and Context*

Situated perspectives recognize learning as a social phenomenon constituted in the experienced, lived-in world, through *legitimate peripheral participation* in ongoing social practices. Lave (1988) highlights the existing discontinuities between mathematical practices in and out of school. Such discontinuities are a clear sign that mathematics learning belongs to different social practices when in school or out of school. Whilst we may find similarities in the problems of both practices, we also notice that school methods often become inaccessible. In fact, Lave mentions the fact that adults, in their daily lives, usually do not resort to algorithms learned at school to solve problems. In contrast with schooling, other practices are experienced in a concrete way, where adults can control their activities, interact with the environment, and enjoy the freedom of choosing the solving processes. Lave and Wenger (1991) also explain the relation between participation and learning: Learning takes place in communities of practice which are the most adequate places to obtain knowledge; and practice is the specific knowledge that is developed and shared in the community. As pointed out by Wedege (1999), from this standpoint, meaning is not created by the individual but has a relational character that refers to the concrete connectedness of the activity.

As the well-known study Adult Math Project (Lave 1988) revealed, by observing adults who failed in formal school settings while acting in a competent way in everyday life settings, the meaning varies as the situation-context varies. This immediately raises the question of transfer; that is, the application of knowledge from school contexts to work or everyday life and vice versa (Evans 2000).

Concerning the question of knowledge transfer, three different positions can be outlined (Carreira et al. 2002; Evans 2000; Muller and Taylor 1995):

1. Boundaries between practices are impermeable; transfer is something at odds with situated learning; knowledge is located in the social practice, not in the individual.
2. Despite the frontiers between practices, knowledge is always the same and naturally flows from each practice to the other; knowledge is located in the individual who simply has acquired it or not (if someone holds a particular knowledge, then she will apply it in any context).
3. The transfer of knowledge is possible but problematic; boundaries exist yet they can be crossed; transfer depends on the individual but not exclusively (boundary crossing is a process of establishing chains of signification).

Following Evans (2000), chains of signification derive from people's interpretations and the ways they make sense of the problem situation. Eventually, depending on the task-contexts, chains are interrupted and part of the teacher's role includes to restore and remake them (e.g., in promoting dialogue within the class and among students). A related account on repairing the breakdowns between discourses is offered by Williams and Wake (2004), who speak about bridging the gaps between mathematical practices and discourses through the introduction of signs (like metaphors) to "afford 'new' links between signs which result in new chains and interpretants, and hence meaning and understanding" (p. 414). Such diverse discourses are difficult to isolate in ordinary life activities where clear-cut problems are not present.

2.3 A Connection with Modelling and Competence

Mathematical competence can be defined as "someone's insightful readiness to act in response to *a certain kind of mathematical challenge* of a given situation" (Blomhøj and Jensen 2007, p. 47). In particular, mathematical modelling competence would consist of one particular mathematical competence and could be described as "someone's insightful readiness to carry through all parts of a mathematical modelling process in a certain context" (Blomhøj and Jensen 2007, p. 48).

Drawing on Blomhøj and Jensen (2007), we devised a picture of the modelling cycle (Fig. 21.1) that we find suitable for our analysis. Some mathematical modelling features like mathematization (transforming something that is not mathematical into another which is), critical interpretation (decoding given information), manipulation (ability and skill to handle mathematical and non-mathematical entities), and communication (continuous interplay between participants) will be examined for the intervention of knowledge.

In parallel, we will assume the constructed nature of competence as explored by Gresalfi et al. (2009). They problematise the general view on competence that attributes a number of skills, abilities or dispositions to individuals apart from the specific contexts in which they participate. The redefined concept of competence consists of

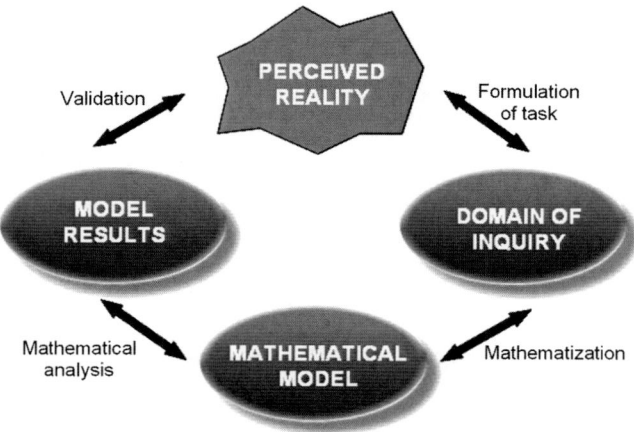

Fig. 21.1 Modelling cycle

something that gets constructed in particular classrooms. It is locally and individually defined through participation with the teacher and the other students. Therefore, "the teacher is not the only participant who is able to shape the construction of competence in a classroom; students also play a role in the negotiation" (Gresalfi et al. 2009, p. 51). In their view, the idea of distribution of agency is vital:

> An individual's agency refers to the way in which he or she acts, or refrains from acting, and the way in which her or his action contributes to the joint action of the group in which he or she is participating. (p. 53)

Therefore, when returning to the modelling cycle, as presented in Fig. 21.1, a number of questions emerge. How is the competence to decode the task defined? Who is accountable for making the transition from the *figured* world to the *lived-in world*? How is the distribution of agency in giving meaning to mathematical and non-mathematical entities? What is the knowledge that is introduced in the mathematical analysis of the problem? How does individuals' knowledge play a role in validation? What kinds of agency can students exercise: offer an idea, critique an idea, engage in argumentation, contribute to sense making, find solutions, and settle for the results?

3 Methodology and Data Analysis

As in the work of Gresalfi et al. (2009), we offer a micro-analysis of adult students engaged in a modelling task and look at how the modelling competence depends on the task, and on the agency and accountability with which students are positioned.

Three classes of Adult Education courses (only one third of the adults, which represents about 17 students with a regular assiduity), organized in five groups of 3–4 students were involved in this case study. Students' ages as well as social and

The task

Margarita is the most common tequila-based cocktail, made with tequila mixed with orange liquor and lime or lemon juice, often served salt on the glass rim. From a search on the Internet the given recipes were found.

Questions

RECIPE A
2:1:1 = (tequila : orange liquor: lemon juice)
RECIPE B
3:2:1 = (tequila : orange liquor: lemon juice)
RECIPE C
3:1:1 = (tequila : orange liquor: lemon juice)
RECIPE D
1:1:1 = (tequila : orange liquor: lemon juice)
RECIPE IBA *(International Bartenders Association) - Standard*
7:4:3 = (tequila : orange liquor: lemon juice)

1. Which of the given recipes will give us a stronger tequila flavour in the *Margarita*? Explain your answer.
2. Which of the given recipes will give us a lighter tequila flavour in the *Margarita*? Explain your answer.
3. Explain how you would prepare a glass of *Margarita* at home according to recipe B. To answer this question you may use a glass, a measuring glass and the 3 bottles containing: plain water (tequila), blue coloured water (orange liquor) and yellow coloured water (lemon juice).
4. At the opening of a Mexican restaurant, 120 glasses of *Margarita* will be necessary. How many glasses of tequila, orange liquor and lemon juice will therefore be needed if considering each recipe? Explain your answer.

Fig. 21.2 Margarita task

professional backgrounds were wide-ranging. The empirical data were collected in a 2-month period of the school-year (when working on the theme of "Cookery"). During this period, several tasks were proposed: each of them taking in average two 90 minutes lessons to be solved. Our present analysis focuses in just one of the tasks – "Making Margaritas" (Fig. 21.2) – carried out by group 2.

This research adopts a qualitative approach, in the form of case study. Data were gathered through participant observation and individual questionnaires to all participants (before and following the task developed). With these questionnaires, we acquired information on professional and life experiences, and previous knowledge relevant to the proposed problems. Audio records of groups and whole class dialogues, photographs, documents from the sessions, and field notes were also collected.

The purpose is to look at the notion of knowledge as it is elicited through mathematical modelling of daily life situations in adult education. The empirical data are analysed and discussed in light of the concepts described in the theoretical framework. Students' activity is segmented into several pieces that are thoroughly examined in search of the presence of different types of knowledge and of the constructed competence.

4 Results

4.1 Students' First Chains of Meanings

Students' initial idea was seeing all recipes as identical, since they all had the same three ingredients. Understanding the different proportions in each recipe was a critical obstacle. The representation of the given ratios was not familiar to the students. So, they started to give meaning to the numbers by making a simple association. Then, the values given could *mean* quantities, but the absence of volume units was problematic.

4.2 Responses to Question 1 – Number Magnitude

The answer to question 1 was based on the magnitude of the number rather than on proportion. Student A6, for example, looked at the number 7 in the ratio 7:4:3 and realised it was the larger number that was assigned to tequila in all the recipes (see Fig. 21.3). No concept of ratio or part/whole relation is used. However, she *knows* that tequila is the major drink in a Margarita. This was the only student who had *tried* it before.

4.3 A "Different" Answer in the Group

Another adult student (A4) justified her answer with her *knowledge* of "concentration". She took the notion of mixture and she started to compare 1 portion of orange liquor to 3 of tequila (recipe C) with 4 portions of orange liquor to 7 of tequila – recipe IBA (see Fig. 21.4). She tried to find out which one was the most concentrated and thought about it for a while but soon she abandoned her thinking in face of the more *knowledgeable* mate – the one who had already *tasted Margaritas*.

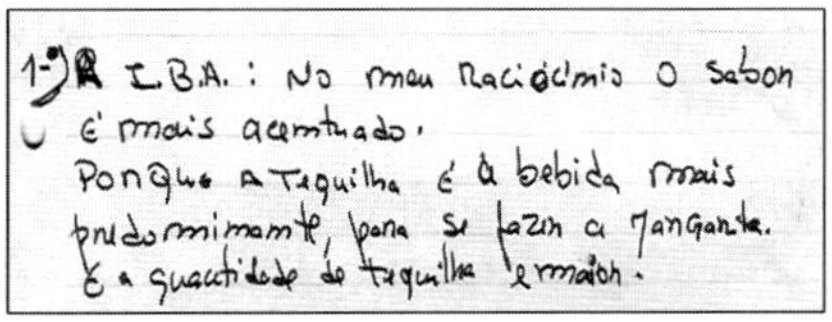

| [handwritten Portuguese response] | The IBA recipe (stronger tequila flavour):
In my thinking, the flavour is stronger.
Because tequila is the most predominant drink in making a Margarita.
Here, the quantity of tequila is bigger. |

Fig. 21.3 Answer to question 1 – A6

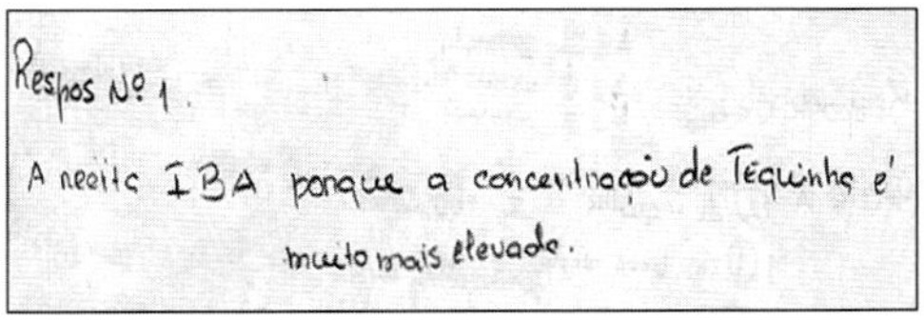

| [handwritten Portuguese response] | Answer n.° 1

The recipe IBA because the concentration of tequila is much higher. |

Fig. 21.4 Answer to question 1 – A4

4.4 Responses to Question 2 – The Best Flavour

Sometime in her life, this adult (A6) had actually tried Margaritas; therefore, she justified her judgement about the lighter tequila flavour based on her *life-experience*. She also realised that the reciped (1:1:1) was the only one where tequila was not predominant, and therefore, she spoke of a *balance* (Fig. 21.5).

4.5 Explaining to the Group – The Medium Flavour

She continued to analyse the recipes in terms of her *life-experience* focusing on issues that she *knew* of, like acidity, to decide about the less intense flavour of tequila (Fig. 21.6).

4.6 Responses to Question 3 – The Size of the Glass

The concrete materials available (Fig. 21.7) turned out to be helpful to overcome difficulties (millilitre, centilitre, c.c. were considered). The group decided that one portion would be 5 ml. They did not care about the size of the Margarita glass. In fact, this unity was suggested by a mother in the group *based* on the amount of medicine that she *used to* give to her children.

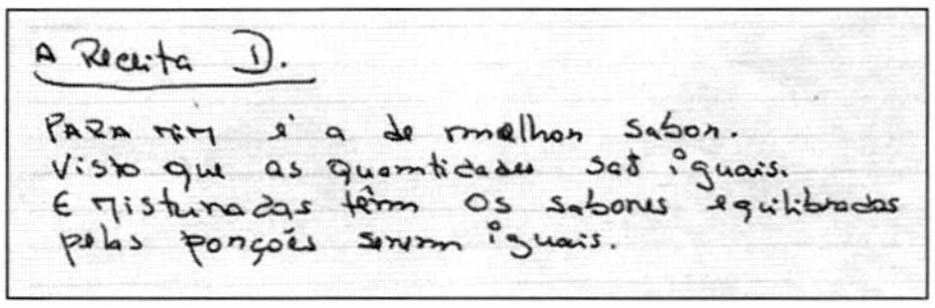
A Receita D.
Para mim é a de melhor sabor.
Visto que as quantidades são iguais.
E misturadas têm os sabores equilibrados
pelas porções serem iguais.

The recipe D (lighter tequila flavour):
To me it is the one with the best flavour because all the quantities are equal.
When mixed the flavours are balanced since the portions are the same.

Fig. 21.5 Answer to question 2 – A6

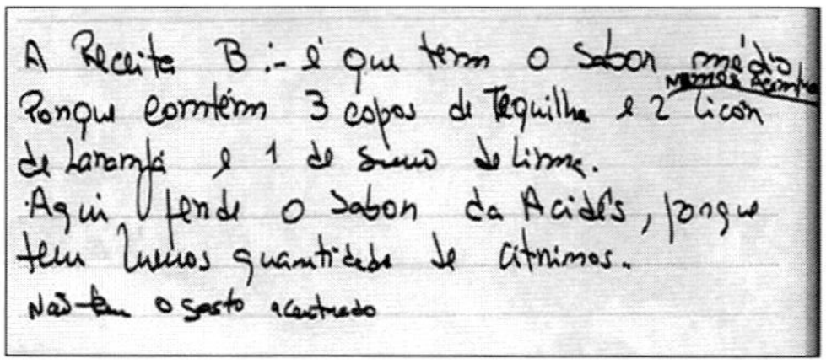
A Receita B: é que tem o sabor médio menos acentuado
Porque contém 3 copos de Tequilha e 2 licor
de laranja e 1 de sumo de limão.
Aqui perde o sabor da acidês, porque
tem menos quantidade de citrinos.
Não tem o gosto acentuado

The recipe B:
Is the one that has a medium flavour (less intense) because it includes 3 glasses of tequila and 2 of orange liquor and 1 of lemon juice.
It looses the acid flavour because it has less quantity of citrons.
It hasn't a strong flavour.

Fig. 21.6 Additional justification – A6

21 Mathematical Modelling of Daily Life in Adult Education

Fig. 21.7 Groups with materials

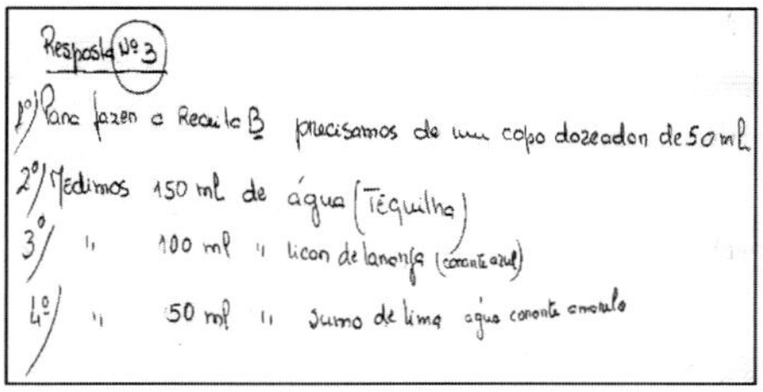

Answer n.° 3
1°) to make the recipe B we need one measuring glass of 50 ml
2°) we measure 150 ml of water (tequila)
3°) " 100 ml of orange liquor (blue coloured)
4°) " 50 ml of lemon juice yellow coloured water

Fig. 21.8 Answer to question 3

Later on, students felt the need to *know* the size of a real glass of Margarita; each group had a different sized and shaped glass.

A4: These glasses should take about half a litre...
A4: Are these cocktails served in big or small doses?
A4: Millilitres are too short for this glass, aren't they?

After some discussion, they decided to change the amount of one portion from 5 to 50 ml (total volume of the measuring glass). With some luck, they succeeded in their trial, filling up the glass without any waste (Fig. 21.8).

4.7 Responses to Question 4 – Part/Whole Relation Versus a Total of 120

Question 4 motivated a long discussion in the class, as the dialogue shows:

T: Tell me about the relation between the orange liquor and the lemon juice, in the recipe A.
A4: There is no difference. They are equal.
A6: It's the same.

T: Ah... it's the same. If I tell you that this recipe A takes 5 glasses of lemon juice, then how many glasses of orange liquor will it take?
A6: Five.
T: And how many glasses of tequila?
A4 & A6: Ten.
T: It is the double. So, I have 5 glasses of lemon juice, another 5 of orange liquor and 10 of tequila. I shake it all together and I get a mixture that would give us... How many glasses of Margarita?
A5: 120 glasses.
T: No, I mean in this particular example.
A6: 20 glasses.
T: But is that what I want? 20 glasses of Margarita?
A4: No. You want 120 glasses.
A6: In that case, it's 6 times more, isn't it? 20 times 6 are 120.

After the discussion, students could only solve the question for recipes A (60:30:30) and D (40:40:40). They *knew* that the total number of glasses had to be 120. In recipe B, students associated 1 part of lemon juice with 30 glasses, 2 parts of orange liquor with 60 glasses, and finally, just by adding, they assigned the additional 30 glasses to tequila in order to get a total of 120. Additive reasoning overrode proportion, thus exhibiting a difficulty that is well documented in numerous studies on proportional reasoning.

5 Discussion and Final Comments

We argue that mathematical tasks, even those that make some kind of reference to real situations – like the recipes for Margaritas – are part of a figured world. The questions posed have an underlying mathematical model, which we will call a ratio model. Regardless of the fact that students have several liquids, glasses and measuring glasses to perform experiments, ratio and proportionality are a central element of the task. Behind the production of Margaritas and the inquiry about the stronger or the lighter flavour of tequila, there is a mathematical discourse involved. The proportion model says very little and rather symbolically about making Margaritas. Therefore, the language of proportions means a high and thick boundary between the figured world and the lived-in world of making Margaritas.

Students' first encounter with the task seems to show how boundaries can be problematic. The situation presented mentions a Margarita, a drink made of several ingredients according to different mathematically coded recipes. When the mathematical point of view is too inaccessible, experience from the lived-in world comes into play. Being one in the group who has tried a Margarita allocates agency and involves revealing knowledge: the balanced mixture, the acidity of the lemon, the best flavour. Someone who had used measuring cups for taking a medicine represents another agency: it offers a possible idea for a volume unit. Knowledge crosses into the figured world, even if colliding with the intended mathematics in the domain of inquiry.

"Making Margaritas" was devised to promote students' experience with ideas, objects and tools where mathematics was blended. However, a removal from the lived-in world was always induced. The final question in the task can actually be seen as a typical school mathematics problem. Hence, the use and relevance of *life-knowledge* depends on what people see as reality and has clear implications in the modelling competence shown. Transfer of knowledge also relates to the meaning given to the "model". In that sense, an analysis of the model may become other than mathematical.

References

Becker, J., & Jacob, B. (2000). The politics of California school mathematics: The anti-reform of 1997–1999. *Phi Delta Kappan, 81*(7), 529–537.

Blomhøj, M., & Jensen, T. H. (2007). What's all the fuss about competencies? In W. Blum, P. L. Galbraith, H. Henn, & M. Niss (Eds.), *Modelling and applications in mathematics education – The 14th ICMI study* (pp. 45–56). New York: Springer.

Boaler, J. (2000). Introduction: Intricacies of knowledge, practice, and theory. In J. Boaler (Ed.), *Multiple perspectives on mathematics teaching and learning* (pp. 1–17). London: Ablex Publishing.

Carreira, S., Evans, J., Lerman, S., & Morgan, C. (2002). Mathematical thinking: Studying the notion of 'transfer'. In A. D. Cockburn & E. Nardi (Eds.), *Proceedings of the 26th Conference of the international group for the psychology of mathematics education* (pp. 185–192). Norwich: University of East Anglia.

Dowling, P. (1991). The contextualizing of mathematics: Towards a theoretical map. In M. Harris (Ed.), *Schools, mathematics and work* (pp. 93–120). London: The Falmer Press.

Evans, J. (2000). *Adult's mathematical thinking and emotions: A study of numerate practices*. London: Routledge Falmer.

Gresalfi, M., Martin, T., Hand, V., & Greeno, J. (2009). Constructing competence: An analysis of student participation in the activity systems of mathematics classrooms. *Educational Studies in Mathematics, 70*, 49–70.

Lave, J. (1988). *Cognition in practice: Mind, mathematics and culture in everyday life*. Cambridge: Cambridge University Press.

Lave, J., & Wenger, E. (1991). *Situated learning: Legitimate peripheral participation*. Cambridge: Cambridge University Press.

Matos, J. F. (2002). Educação matemática e cidadania. *Quadrante, 11*(1), 1–6.

Muller, J., & Taylor, N. (1995). Schooling and everyday life: Knowledges sacred and profane. *Social Epistemology, 9*(3), 5–275.

OECD. (2004). Learning for tomorrow's world – First results from PISA 2003. Retrieved from http://www.pisa.oecd.org/dataoecd/58/41/33917867.pdf, July 2009.

Wedege, T. (1999). To know or not to know – Mathematics, that is a question of context. *Educational Studies in Mathematics, 39*, 205–227.

Williams, J. S., & Wake, G. D. (2004). Metaphors and cultural models afford communication repairs of breakdowns between mathematical discourses. In M. J. Hoines & A. B. Fuglestad (Eds.), *Proceedings of the 28th conference of the international group for the psychology of mathematics education* (Vol. 4, pp. 409–416). Bergen: Bergen University College.

Zevenbergen, R. (2002). Citizenship and numeracy: Implications for youth, employment and life beyond school yard. *Quadrante, 11*(1), 29–39.

Chapter 22
Students' Modelling Routes in the Context of Object Manipulation and Experimentation in Mathematics

Susana Carreira and Ana Margarida Baioa

Abstract The present study is a classroom-based research where students develop mathematical modelling tasks that involve manipulating and experimenting with real objects. The research was developed in two 9th grade classes of students aged 14/15 years old. These students never had this kind of modelling activities in their mathematics classes before. Our purpose is to discuss the modelling routes produced by middle school students in an experimental mathematics environment – both from the point of view of realistic mathematics education and of the model-eliciting perspective.

1 Introduction

The modelling cycle, as described in mainstream approaches of Applications and Modelling, consists of a sequence of stages: identification of the real problem, formulation of the mathematical model, production of the mathematical solution or solutions from the mathematical model, interpretation of the solutions, evaluation of the solutions in terms of the real setting, and, if necessary, the revision of the model and a new cycle performed. Finally, a report with the results and analysis of the problem is produced (Blum and Niss 1991; Niss et al. 2007). In this chapter, we intend to see how hands-on experience in situations that involve using and manipulating objects to solve real problems has a role in students' modelling thinking and in their modelling routes. In particular, the real object and furthermore manipulation and experimentation are seen as a powerful tool to "find" an answer

S. Carreira (✉)
Universidade do Algarve, Faro & UIDEF, Lisboa, Portugal
e-mail: scarrei@ualg.pt

A.M. Baioa
Escola Básica D. Manuel I, Tavira, Portugal
e-mail: ambaioa@gmail.com

to the problem. In a sense, we are exploring the possibility of seeing experimental mathematics as a particular kind of modelling of real situations in school mathematics. This perspective is being explored by other researchers on mathematical modelling and applications to support mathematics education (see Alsina 2002, 2007; Bonotto and Basso 2001; Halverscheid 2008). Yet, as remarked by Alsina (2002), hands-on materials have been generally neglected on the basis of their irrelevance to levels of teaching other than the very elementary ones. Refusing this argument, the author claims:

> If we want to show applications and modelling procedures, we can find in our home-made materials great opportunities to bring 'real' objects into the class and to provoke an experimental research approach by modelling by means of specific materials. (p. 246)

2 Connecting Modelling to Experimentation in Mathematics Classroom

2.1 *From the Point of View of Mathematics Education*

The introduction of mathematical modelling in mathematics teaching and learning has been advocated on the basis of different arguments. In particular, from the five arguments to include mathematical modelling in curricula presented by Blum and Niss (1991), we find the following three especially important in our work:

- *The formative argument* (emphasised in the Portuguese curricula) – modelling is a means to develop students' general skills and attitudes, namely, problem-solving ability, inquiring attitudes, creativity, mathematical reasoning and communication.
- *The picture of mathematics argument* – modelling helps to provide students with a richer and wider picture of mathematics, in all its facets, as a science and as a field of activity in society.
- *The learning of mathematics argument* – modelling assists students' learning of mathematical concepts and procedures; in particular, it strengthens mathematics understanding when it is applied to new problem situations.

The use of real objects and the process of experimentation in mathematics learning meet the above arguments, namely, the development of inquiring attitudes, the image of mathematics as useful and relevant to interpret daily instruments as well as the concreteness of mathematical ideas in real world environments. Additionally, we see modelling activities as considerably "rich" mathematical contexts in the sense that they typically include three important components from a mathematical thinking point of view: *problem solving, investigating/exploring, validating and extending solutions.*

Modelling activities involve a goal (finding an answer), focusing on some part of the world, finding patterns, devising an adjusted and good model, testing and validating the solutions and also analysing extensions of the model. Having this in mind, we claim that the manipulation of real objects is a way to engage in a mathematical activity closer to the experimental sciences methods, with the modelling purpose being to produce and/or explain particular features of real and common objects, by means of hands-on and conceptual work.

2.2 From the Point of View of Learning by Doing

Real objects, real places and real challenges may play an important role in mathematical modelling when moving towards manipulation and experience and reducing the "talk & chalk" (Alsina 2007, p. 35). Interesting modelling activities can focus on objects, instruments, and everyday situations. But objects motivate a concrete visual approach to mathematics giving students the opportunity to explore the potentials and limitations of tangible things when trying out their properties and characteristics (Alsina 2007).

In this respect, Bonotto (2007) argues that mathematical facts embedded in cultural artefacts and in everyday life are relevant to students as they offer references to concrete situations. The dual nature of artefacts – belonging to the world of everyday life and to the world of symbols – gives children the opportunity to recognize situations as mathematical or more precisely as mathematizable situations. In a similar tone, Vos and Kuiper (2002) note that manipulatives are useful to organise hands-on activities that link mathematics to other areas because mental acts (manipulation of objects in the mind) develop from material acts (manipulation of tangible objects). An *experimental investigative approach* to real situations by modelling, as proposed by Alsina (2002), is coherent with the perspective on *experimental modelling environments* described by Halverscheid (2008). The latter realises the need to build a local conceptual framework for the construction of mathematical knowledge in learning environments with experiments, in a study of pre-service teachers' activity with the motion of a ball on a circular billiard table. The meaning and role of experiments is clearly identified:

> Experiments, which the students themselves carry out, are considered when the task is to explain the experiments by setting up a suitable mathematical model. (Halverscheid 2008, p. 225).

> Experiments related to mathematics find their natural place in the framework of modelling because they represent 'the rest of the world' for which mathematical models are built. (Halverscheid 2008, p. 226).

Learning by doing (Dewey 1938/1997) emerges as a natural learning perspective when we look at modelling as a kind of work closer to the methods of experimental sciences.

> Although experiments as such may be considered typical for science rather than mathematics, many mathematical activities representations and models are strongly connected with experiments. (Halverscheid 2008, p. 225).

According to Dewey, experience as a basis for education must be one that positively influences future experiences in productive and creative ways. And experiences in education take up something from past learning and change subsequent learning in some way. So, the educator needs to see in what direction an experience is heading (Dewey 1938/1997). Experience is the result of a transaction between us and what constitutes our environment at the time. Environment is therefore part of the situation.

In summary, the possibility of seeing experiments with real objects as a particular kind of modelling stems from the following facts: (1) Students have the opportunity of learning by doing (while performing actual manipulation and experimentation and engaging in conjecturing and validation). (2) Working on physical concrete materials is a way of inquiry into the mathematical properties of objects. (3) To investigate through experimentation reflects on mental actions and on past and subsequent learning of mathematical ideas and becomes a way to develop understanding of mathematical models.

In our theoretical approach, we then embrace the possibility of connecting modelling to experimentation and consider the implications of such an approach in different theoretical perspectives on applications and modelling in mathematics teaching, namely, *realistic mathematics education* and *model-eliciting perspective*.

3 Theoretical Perspectives on Applications and Modelling

Realistic Mathematics Education (RME) is a teaching and learning theory in mathematics rooted in Freudenthal's interpretation of mathematics as a human activity through mathematizing. The process of mathematization includes both horizontal and vertical developments. Freudenthal (1991) states that horizontal mathematization involves the passage from the world of life into the world of symbols and vertical mathematization means moving within the world of symbols.

In school mathematics, one fundamental approach of RME is to look for real contexts that students can use as starting points for progressive mathematization, in going from informal mathematical knowledge by using a *model of* to formal mathematical knowledge by using a *model for* (Gravemeijer 1994) (Fig. 22.1).

Models and modelling (M&M) perspectives emphasize the fact that "thinking mathematically" is about interpreting situations mathematically. In modelling activities, students use their initial ideas to make sense of the situation; they model the situation and develop the underlying mathematical concepts, promoting a conceptual change in their understanding of mathematical ideas and of the specific situations (Lesh and Doerr 2003).

A model-eliciting activity leads students to express their thinking and refining of it several times. Their mathematical models are the result of a recursive process where students articulate ideas, test, revise and extend their interpretations. The model development happens along with conceptual development (Fig. 22.2).

The two perspectives take into account different stages of models, either under the name of *model of* and *model for* or under the designation of *models* and *constructs*. In both cases, we acknowledge the fact that models evolve in their degree of generality as they move from contextual problems towards more formal

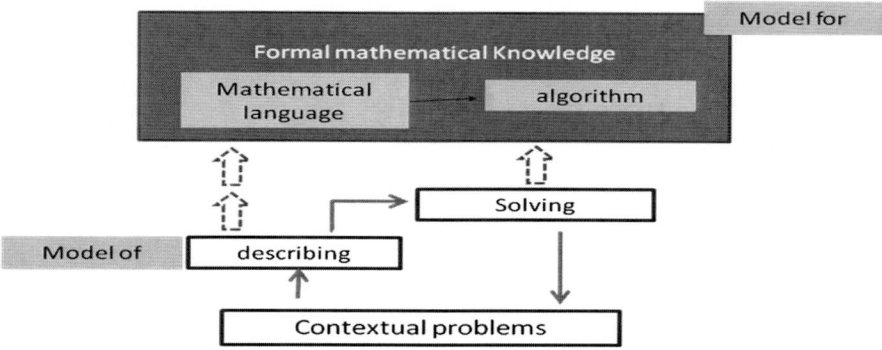

Fig. 22.1 Modelling from RME perspective (Adapted from Gravemeijer 1994)

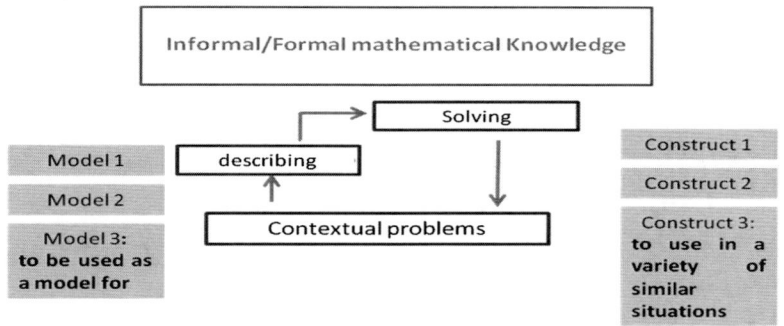

Fig. 22.2 Modelling from M&M perspective

mathematical knowledge impelled by the need to solve particular problems or questions and to reflect on the solutions back in to the contextual situation. One of our concerns is to see how real objects and experimentation fit into both theories and how these relate to the nature of the models that students produce.

4 The Research Empirical Work

Our classroom-based research is a teaching experiment in a regular curriculum environment with two 9th grade classes (14–15 years old). The students involved in the research had never worked on modelling tasks before. Lessons of 90 min on modelling activities were conducted from January to June 2009, once a week in each class.

The main question of our work is: How does hands-on experience in situations that involve using and manipulating objects play a role in shaping students' modelling routes?

In refining the focus of our leading question, we have formulated two other sub-questions: (1) What can we say about experience with concrete objects from the RME perspective? (2) What can we say about experience with concrete objects from the M&M perspective?

Procedures for collecting qualitative data were participant observation, video and audio recording of students' work, and written reports of each activity. The teacher was also the researcher who undertook participant observation in the class and compiled field notes.

The data presented refer to the second modelling task in a total of five and the task was presented in late January. "The cake box" task includes three parts (see below). In the first part, the real situation is presented. The second part consists of experimental work with physical objects and the third one is the setting where a mathematical model "comes out".

5 Description and Data Analysis

Students went through several modelling cycles in an attempt to find a solution to the problem by performing successive experiments. The following description concerns one group of four students (I, F, M, R) who exhibited both a strong concentration on experimenting and a demand to go beyond concreteness to formal mathematics.

The group started to measure the sheet, made drawings and wrote down the dimensions of the sheets in their notebooks; they folded and made three boxes from three different sized sheets. Next they tried to pack some real cookies (circular shaped and assembled in sets by a plastic wrap) in the boxes. They realized that one of the boxes would be perfect to pack two sets of cookies and another one would be appropriate for only half a set of cookies.

Their first hypothesis was: One of the boxes works well for two sets of cookies, so half the sheet would solve the packing of one set of cookies. They drew a scheme of their hypothesis for creating another box (Fig. 22.3), made it and tested it. It did not fit their aim.

Their second hypothesis was: Two small boxes united would do, so doubling the sheet could be the solution. They put together two small sheets and taped them in one single sheet. They made the new box and tested it (Fig. 22.4). Again, the cookies did not fit. All the work developed to this point was mainly experimental and consisted in constructing boxes with different sized sheets, according to hypothesis where relations for the volume were immediately translated into similar relations for the area of the sheet (doubling or halving).

Eventually students decided to record the dimensions of the boxes, unfolded the boxes and tried to relate the dimensions of the box with the dimensions of the sheet and the resulting creases. The boxes were measured again more than once. An algebraic relation started to be considered by one of the students (I) and she tried to approach the problem with the formal relation found. The final answer only appeared later on, as students continued to work on the task at home. The written

"The Cake Box"

In bakeries today there are standard cake boxes. In the past, the bakers made these boxes themselves knowing the size of the cardboard sheet to make a box for a certain number of cakes. Let's try making some cake boxes.

From experience...

1. Measure the dimensions of the given sheets of paper (length and width). (Don't forget to register in your notebook).
2. Make the cake box following the instructions below.
3. Measure the three dimensions of the boxes you made (length, width and height).

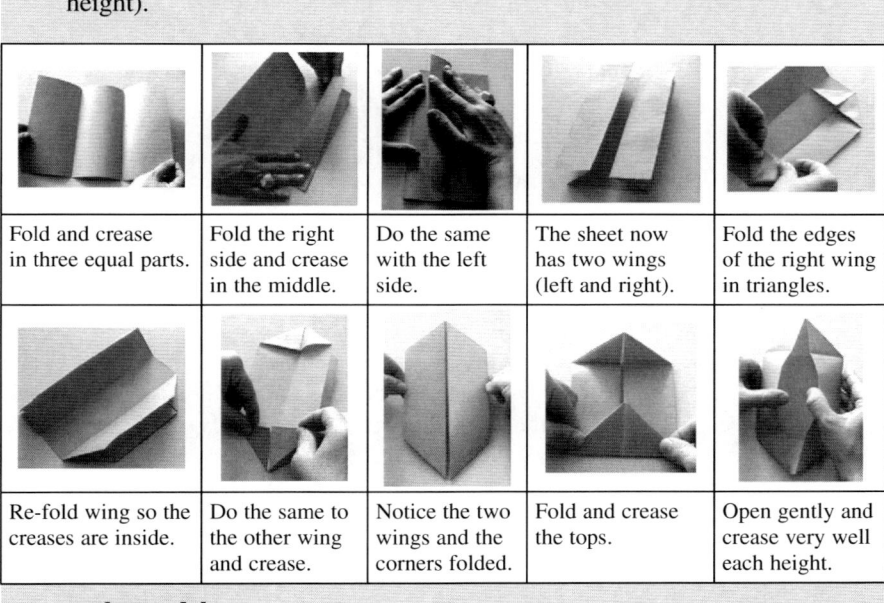

Fold and crease in three equal parts.	Fold the right side and crease in the middle.	Do the same with the left side.	The sheet now has two wings (left and right).	Fold the edges of the right wing in triangles.
Re-fold wing so the creases are inside.	Do the same to the other wing and crease.	Notice the two wings and the corners folded.	Fold and crease the tops.	Open gently and crease very well each height.

... to the model.

4. Find the relations between the dimensions of the initial sheets and the dimensions of the resultant boxes, looking at the creases.
5. Find a relation between the dimensions of sheet and box (look for a mathematical expression).
6. Now you want to tightly pack in a box the cookies you have on your table. What must be the dimensions of this cookie box and what should be the dimensions of the sheet?
7. Suppose you also want to pack a birthday cake in a box. This cake has 26 cm diameter and 10 cm height. What must the dimensions of the sheet be?
8. Elaborate a report focusing on the next topics: (1) An explanation of the experimental situation; (2) Hypotheses formulated; (3) The exploratory work done; (4) The results; (5) Task evaluation; (6) Difficulties encountered.

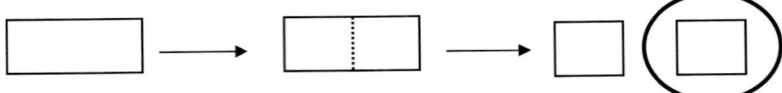

Fig. 22.3 Plan of the first hypothesis

Fig. 22.4 Plan of the second hypothesis

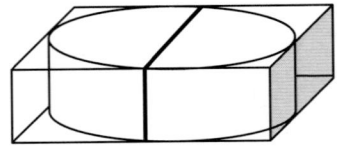

report shows how they found a model of the correspondence between the dimensions of the box and the dimensions of the sheet, as follows:

$$\boldsymbol{boxlength} = sheetwidth - \frac{2}{6} sheetlength$$

$$\boldsymbol{boxwidth} = \frac{1}{3} sheetlength \quad \boldsymbol{boxheight} = \frac{1}{6} sheetlength$$

In the following lesson, students addressed question 7, related to packing a birthday cake, and they decided to use an A4 sheet to start with. The box was made and measured, and the dimensions recorded. The lengths of the sheet were again related with the dimensions of the box. The creases were analyzed and one of the students suggested using mathematical relations as she finished checking sums on her calculator. Another student however continued with successive experiments with the real object, folding and unfolding the box, each time increasing the size of the sheet by taping more strips of paper. After some discussion, they decided to stick together four A4 sheets, having realized that two would not be large enough, according to the diameter of the cake and the relations already found.

A big box was made and measured, but the result failed to match the dimensions required for the birthday cake. Student I then started to work individually, using the formal model, while the other students (F, M, R) added another piece of paper to the big sheet. A bigger box was made by students F, M, and R. Before they finished, student I came up with a numerical solution and the others finally stopped the experiments. They all turned to their notebooks and continued with the algebraic exploration of the problem. Drawings and mathematical expressions were presented in the group's report (Fig. 22.5a–d).

The empirical data show intense experimental work with the real object from most students in the group. A lengthy time was spent analysing and understanding the real situation. The identification of variables and the relations between them came about quite slowly. Much time was devoted to trial and error with the physical manipulation, and folding and unfolding of boxes. Several tangible models of cake

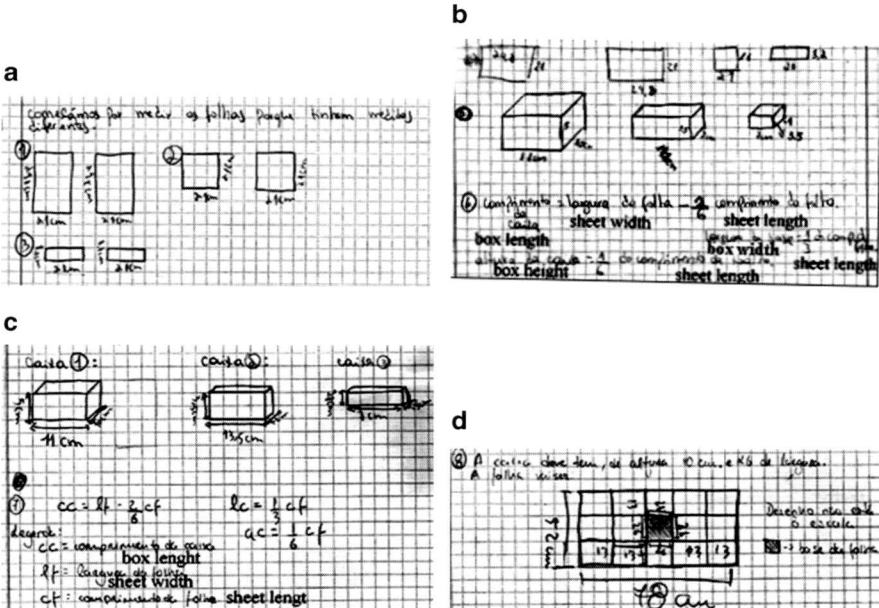

Fig. 22.5 (a) Initial sheets with dimensions recorded, (b) sheets and boxes with dimensions and relations between variables in words, (c) relations in mathematical language and (d) the solution for the big cake

boxes were produced but often missed the desired target. During such experimental processes, a "model of" was being repeatedly tested. This horizontal mathematization was carried out iteratively until a "model for" started to be investigated. Then mathematical relations emerged rapidly and a mathematical model was set up to obtain the size of the sheet for any possible cake box.

6 Synthesis of Findings

Both the theory undertaken and our preliminary results indicate that hands-on activity gives students a chance to develop a more solid understanding and familiarity with the situation (including with the mathematics involved). Experimentation with concrete objects supports students' search for a solution to the problem, as far as consecutive trials and testing are possible before stronger mathematization. Although experimentation may seem to keep some students at the level of horizontal mathematization, the final stages of the modelling process reveal that it is actually promoting a deeper conceptual development.

Looking at the data from the two perspectives, realistic mathematics education and the models and modelling perspective, we find them to be compatible although highlighting different features of the modelling activity. From RME, we looked at the movement from horizontal to vertical mathematization, concluding that manipulation can result in a longer process of horizontal mathematization. From the standpoint of Models and Modelling, we observed a series of micro-cycles depicted in the successive experiments performed, which represented important conceptual fine-tuning to arrive at a formal mathematical model.

References

Alsina, C. (2002). Too much is not enough: Teaching maths through useful applications with local and global perspectives. *Educational Studies in Mathematics, 50*, 239–250.

Alsina, C. (2007). Less chalk, less words, less symbols… more objects, more context, more actions. In W. Blum, P. Galbraith, H.-W. Henn, & M. Niss (Eds.), *Modelling and applications in mathematics education. The 14th ICMI study* (pp. 35–44). New York: Springer.

Blum, W., & Niss, M. (1991). Applied mathematical problem solving, modelling, applications and links to other subjects – State, trends and issues in mathematics instruction. *Educational Studies in Mathematics, 22*(1), 37–68.

Bonotto, C. (2007). How to replace word problems with activities of realistic mathematical modelling. In W. Blum, P. Galbraith, H.-W. Henn, & M. Niss (Eds.), *Modelling and applications in mathematics education. The 14th ICMI study* (pp. 185–192). New York: Springer.

Bonotto, C., & Basso, M. (2001). Is it possible to change the classroom activities in which we delegate the process of connecting mathematics with reality? *International Journal of Mathematical Education in Science and Technology, 32*(3), 385–399.

Dewey, J. (1938/1997). *Experience and education.* New York: Touchstone.

Freudenthal, H. (1991). *Revisiting mathematics education – China lectures.* Utrecht: Kluwer.

Gravemeijer, K. (1994). *Developing realistic mathematics education.* Utrecht: CD-β Press.

Halverscheid, S. (2008). Building a local conceptual framework for epistemic actions in a modelling environment with experiments. *ZDM – The International Journal on Mathematics Education, 40*, 225–234.

Lesh, R., & Doerr, H. (Eds.). (2003). *Beyond constructivism: Models and modelling perspectives on mathematics problem solving, learning, and teaching.* Mahwah: Lawrence Erlbaum.

Niss, M., Blum, W., & Galbraith, P. (2007). Introduction. In W. Blum, P. Galbraith, H.-W. Henn, & M. Niss (Eds.), *Modelling and applications in mathematics education. The 14th ICMI Study* (pp. 3–32). New York: Springer.

Vos, P., & Kuiper, W. (2002). Exploring the potentials of hands-on investigative tasks for curriculum evaluations. In A. D. Cockburn & E. Nardi (Eds.), *Proceeding of the 26th Annual Conference of the International Group for the Psychology of Mathematics Education* (Vol. 4, pp. 329–336). Norwich: University of East Anglia.

Chapter 23
Engineering Model Eliciting Activities for Elementary School Students

Nicholas G. Mousoulides and Lyn D. English

Abstract This chapter argues for a future-oriented, inclusion of Engineering Model Eliciting Activities (EngMEAs) in elementary mathematics curricula. In EngMEAs, students work with meaningful engineering problems that capitalise on and extend their existing mathematics and science learning, to develop, revise and document powerful models, while working in groups. The models developed by six groups of 12-year students in solving the Natural Gas activity are presented. Results showed that student models adequately solved the problem, although student models did not take into account all the data provided. Student solutions varied to the extent students employed the engineering context in their models and to their understanding of the mathematical concepts involved in the problem. Finally, recommendations for implementing EngMEAs and for further research are discussed.

1 Introduction

The world's demand for skills in science, technology, engineering, and mathematics is increasing rapidly, yet supply is declining across several nations in EU and USA (National Academy of Sciences 2007; OECD 2006). Further, recent research findings in a number of countries revealed that school students' mathematical and problem-solving skills are rather poor and stressed the importance of implementing interdisciplinary problem-solving activities in mathematics and science school curricula (Kaiser and Sriraman 2006; Mousoulides and English 2008; Zawojewski et al. 2008). We need young scholars to be involved in the next generation of

N.G. Mousoulides (✉)
University of Cyprus, Nicosia, Cyprus
e-mail: n.mousoulides@ucy.ac.cy

L.D. English
Queensland University of Technology, Brisbane, Australia
e-mail: l.english@qut.edu.au

innovative ideas that support our society's needs. Interdisciplinary problem solving that involves core ideas from engineering, mathematics, and science can empower students to tackle the many real-world problems society faces now and in the future.

Following recommendations for ensuring school students' early exposure to interdisciplinary problem solving and a sense of the role of mathematics in solving real-world problems (OECD 2006), we have introduced a sequence of Engineering Model Eliciting Activities (EngMEAs) for elementary school students in a 3-year longitudinal study. In the activity presented in this chapter, students worked on an engineering modelling activity related to natural gas consumption and reserves. Student models and solutions building on their mathematical and engineering ideas and processes in solving the *Natural Gas Modelling Activity* are presented in this study.

2 Theoretical Framework

Despite more than five decades of research, it seems that students' problem-solving abilities still require substantial improvement (Lesh and Zawojewski 2007). Much-needed, recent calls for new perspectives regarding the nature of problem solving and its role in the mathematics curriculum have appeared (e.g., Lesh and Zawojewski 2007; Mousoulides et al. 2008). One such perspective involves interdisciplinary problem solving. It is being increasingly recognized that future-oriented problem-solving experiences in mathematics and science require interdisciplinary contexts (Zawojewski et al. 2008). These findings present interesting challenges for mathematics (and science) educators. Among the core questions that arise, questions like how we might assist students in better understanding how their mathematics and science learning in school relates to the solving of real problems outside the classroom and how we might broaden students' problem-solving experiences to promote creative and flexible use of mathematical ideas in interdisciplinary contexts can be addressed through the discipline of engineering.

More than ever before, we need to increase the profile and relevance of mathematics and science education in solving problems of the real world, and we need to begin this in the primary and middle schools (National Academy of Engineering, and Institute of Medicine 2007). Engineering provides an exceptional context in which to showcase the relevance of students' learning in mathematics and science to dealing with authentic problems meaningful to them in their everyday lives (Petroski 2003). By incorporating engineering-based problems within both the primary and middle school mathematics curriculum, we can: (a) engage students in creative and innovative real-world problem solving involving engineering principles, design processes, and mathematical modelling that build on the students' existing mathematics and science learning; (b) show students' how their learning in mathematics and science applies to the solution of real-world problems; and (c) promote group work where students learn to communicate and work collaboratively in solving complex problems (English and Mousoulides 2009).

One manner of addressing engineering-based problems is through the use of Engineering Model Eliciting Activities (EngMEAs) – realistic, client-driven problems based on the theoretical framework of models and modelling (Lesh and Doerr 2003). An EngMEA is a complex problem set in a realistic context with a client, characteristics that place EngMEAs in the authentic assessment category. Solutions to EngMEAs are generalizable models which reveal the thought processes of the students. The models created include procedures for doing things and, more importantly, metaphors for seeing or interpreting things. The activities are such that student teams of three to four express their mathematical model, test it using sample data under the possible engineering constraints, and revise their procedure to meet the needs of their client (Lesh and Doerr 2003; Mousoulides et al. 2008). In sum, from the EngMEAs perspective, engineering-based problems are realistically complex situations where the problem solver engages in mathematical and engineering thinking beyond the usual mathematical classes experience and where the products to be generated often include complex artifacts or conceptual tools. The problems present a future-oriented approach to learning, where students are given opportunities to elicit their own mathematical and scientific ideas as they interpret the problem and work towards its solution (Lesh and Zawojewski 2007; Zawojewski et al. 2008).

3 The Present Study

3.1 Participants and Procedures

The participants were a class of twenty 12-year-old high achiever elementary school students, who participated in a 3-year longitudinal study of children's mathematical modelling and engineering thinking. The students were from a public K-6 elementary school in the urban area of the capital city of Nicosia.

The data reported here are from the second year of the respective study and are drawn from one of the problem activities the students completed during the second semester of the second year. The engineering modelling problem, namely the *Natural Gas* activity, focuses on the natural gas resources and consumption. The activity presented data related to the worldwide reserves of natural gas in 1993, and the annual average consumption for the next 15 years. Specifically, the activity provided students with the following information:

> In 1993 the worldwide reserves of natural gas were estimated to be 141.8 billion cubic metres. Since then 2.5 billion cubic metres have been used every year on average. The Ministry of Communications and Works is thinking on placing a large investment on building natural gas and oil refinery stations. Calculate when the reserves of natural gas will be exhausted, so as to advise them whether they should proceed with the investment or not.

The engineering problem presented in the activity required students to use different assumptions and develop model(s) for calculating when the reserves of natural gas will be exhausted. In implementing the activity, we were primarily interested in: (a) how the students interpreted the problem, (b) the ways in which the students worked with the provided data and the extra data they retrieved from the Web, and (c) the nature of the models the students generated in solving the problem.

The problem entails: (a) a warm-up task comprising a mathematically rich newspaper article designed to familiarize the children with the context of the modelling activity. This article was published in a local newspaper few months previously, presenting some facts and figures about the explorations in the sea between Cyprus and Egypt for natural gas and oil. (b) "Readiness" questions to be answered about the article, and (c) the problem to be solved, including the figures and text mentioned above.

Since this problem was part of a sequence of engineering modelling activities, students were familiar with working in groups, developing models for solving quite complex problems, and presenting and documenting their results. The problem was implemented by the first author and one postgraduate student. Working in groups of three to four, the students spent four 40-min sessions on the activity. During the first session, the children worked on the newspaper article and the readiness questions. In the next three sessions, the children developed their models, wrote their letters that explained their models, and presented their work to the class for questioning and constructive feedback. During their explorations, students could freely search the Web for finding useful data in further unfolding the complex engineering problem. Finally, a class discussion followed that focused on the key engineering and mathematical ideas and relationships the children had generated.

3.2 Data Sources and Analysis

The data sources for this study were collected through (a) videotapes of students' responses during whole class discussions, (b) audiotapes of students' work in their groups, (c) students' final models, students' worksheets and final reports detailing the processes used in developing models, and (d) researchers' field notes. The analysis of the data, using interpretative techniques (Miles and Huberman 1994), was completed in the following steps. First, the transcripts were reviewed by both researchers several times to identify and trace developments in the model creations of the students with respect to: (a) the ways in which the students interpreted and understood the problem, (b) their initial approaches to dealing with the data sets, and (c) the ways in which they selected data sets, and applied mathematical operations. Second, the transcripts were reviewed to identify how students interacted in their groups, and how discussions within the groups resulted in their final models. In the next section, we summarize the model creations of the student groups in solving the *Natural Gas* activity.

4 Results

Students found the problem interesting and challenging and developed a number of different models for solving the problem. Quite surprisingly, a number of students experienced difficulties in fully understanding and using the "concept of average" in developing their models. Four out of the six groups developed models appropriate for solving to some extent the provided problem. Further, many students easily calculated the remaining reserves of natural gas in 2008, by multiplying 2.5 billion cubic meters by 15 years (1993–2008). Some students, however, failed to understand that consumption in recent years was not 2.5. A number of students successfully used the data provided in the activity and other resources from the Web to make assumptions about the future reserves and consumption of natural gas, and how the use of renewable energy sources might have an impact on natural gas consumption. Two groups of students developed more coherent models, taking into consideration current reserves, how the consumption of natural gas will be increased (using data from the Web) and how the use of renewable energy sources will affect the consumption of natural gas. The four appropriate models presented by the four groups of students are summarized below.

4.1 Model A

The first group commenced the problem in a rather simplified way. They discussed the provided data, but partially failed to fully understand the concept of average. As a consequence, although they correctly calculated the remaining reserves, they did not discuss how annual consumption changed during the last 15 years. In discussing the problem, they reported in their worksheets that natural gas consumption will be increased. They further supported that hypothesis by documenting that due to a number of environmental and economical issues oil consumption is decreased and as a consequence, natural gas consumption and renewable energy sources use be increased. However, students failed to clearly document how these hypotheses affected the average consumption of natural gas. Their approach can be partially explained by the intense discussion on TV and newspapers in Cyprus on the natural gas reserves Cyprus might have in its coastal zone.

Students reported in their worksheets that the new annual average consumption for the following years will be 3.0 billion cubic meters. Similar to other groups, students in this group did not provide any support for this conclusion. Even when prompted by the researchers, students in this group failed to explain or try to predict how consumption will change, for instance, in 5, 10 or 12 years. Further, when asked by the researchers, students reported that the new average was reasonable, and the increase was not that big. In terms of the mathematical developments, groups calculated the remaining reserves in 2009 and then divided the remaining reserves by the new average consumption, by providing simple linear functions. Their final model was: $(141.8 - (2.5 * 15)/3)$.

Students' work in this group resulted in finding when the natural gas reserves will be exhausted. As a consequence, students reported in their final letter that it might not be a good investment to finance a natural gas plant, since the reserves will be exhausted in less than 35 years. They concluded by underlining that it would be better to invest in renewable energy use, because "solar power is free and will never be exhausted".

4.2 Model B

Similar to Model A, this group developed a quite similar model. However, a number of differences can be tracked between this model and Model A. Since students did not attempted to retrieve any data from the Web, they ended by proposing two different and even contradictory hypotheses, in an attempt to solve the engineering problem. Specifically, based on their first hypothesis, they documented that natural gas consumption will be increased. They reported that natural gas consumption might increase, since there is a global shift from oil to natural gas use. Thinking more locally, they also reported that the new reserves of natural gas that have been found in the coastal zone of Cyprus would also have an impact on natural gas use. They further exemplified their thought by stressing the importance of getting cheap energy; new natural gas reserves will have an impact on oil use and will provide Cyprus with cheap natural gas.

Since students in this group did not succeed in retrieving more data from the Web, they set a second hypothesis, namely, that natural gas consumption will be decreased. When prompted by the researchers, students reported that there is a global shift to renewable energy resources, like solar and wind power. They also underlined the importance of taking measures for saving the environment and that people are getting more and more aware of environmental issues. They concluded that the above situation will have a direct impact on decreasing nonrenewable energy consumption.

Based on the two hypotheses, mentioned above, students developed three different solutions. Their first model was similar to the one presented earlier (Model A). However, students in this group used 2.8 billion cubic metres as the new annual average. This model resulted in finding that the natural gas reserves will be exhausted in 38 years. Based on their second hypothesis, students again resulted in estimating that the new annual average would be 2.2 billion cubic metres and consequently that natural gas reserves would be exhausted in 47.5 years. In referring to both solutions, students reported in their worksheets that 47.5 or even 38 years were good enough for placing the investment. The latter was another difference between this group's work and the first group's work, who decided that it would be better not to invest in such a project. Finally, following a student's suggestion from this group, students ended with a third approach and a model. They reported that: "since there are factors that might increase and other factors that might decrease natural gas consumption, we could assume that annual average will remain 2.5 billion cubic metres". Consequently, they calculated that reserves will be exhausted in 57 years.

Quite confusing, 2 out of the 4 students in this group reported in their worksheets that "since 2.5 annual average was the same for the last 15 years, then it is reasonable that average will be the same for the following years". The latter was disappointing, considering that all students appeared quite confident and successfully used the average in making calculations. However, it appeared that these students did not have a conceptual understanding of the concept of average and how this concept could be appropriately used in the context of a real-world problem. Further, although students in this group explicitly discussed the existence of new reserves, they did not adopt and use this discussion in their models. Finally, similar to Model A group's work, students in this group did not manage to retrieve or use data from the Web.

4.3 Model C

The group who developed this model commenced the problem by listing all possible factors that might have had an impact on natural gas consumption. Students in this group reported the following possible factors: use of renewable energy sources, natural gas price, natural gas reserves and availability, and people's awareness of environmental issues. It should be noticed here that some of these factors were retrieved from the Web. Similar to the previous group, students in this group also concluded with two different hypotheses: the first hypothesis was based on the assumption that natural gas consumption will be increased and the second hypothesis on the assumption that consumption will be decreased. An important difference that can be tracked from the previous group's work was students' documentation that reserves will be increased (new reserves appear every year), no matter how the consumption of natural gas will change.

They decided, after a long debate within the group, that natural gas consumption will be increased. They reported in their worksheets that on one hand there is a shift to renewable energy sources, but this shift is still not so important; they concluded so the shift from oil to natural gas will increase gas consumption much more. In developing their final model, they calculated the remaining reserves in 2009 (without incorporating in their model any new reserves found between 1993 and 2009) by performing the following calculations: $141.8 - (2.5 * 15)$. Their next step was to divide the remaining reserves plus new reserves by 3.2. Their final model was: $(R_{2009} + \text{New } R)/3.2 =$ They explained that the new increased annual consumption will be 3.2 billion cubic metres, since the 2008 annual consumption was 2.8 billion cubic metres.

In documenting their results and preparing their final report, students underlined that they could not be sure when (and if) natural gas reserves will be exhausted, since this is directly related to the existence of new reserves and the new annual average consumption. Quite surprisingly, students did not incorporate in their model the shift to renewable energy sources, although they explicitly discussed it. The latter was among the additional information included in Model D, which is presented below.

4.4 Model D

Model D students started the problem by listing all possible factors that might have an impact on natural gas consumption. In preparing their list, they also (similar to Model C) used Web resources. Using the data provided and also data they retrieved from the Web, they concluded that the natural gas consumption will be increased. Differently from Model C group's work and using data they retrieved from the Web, students in this group found the natural gas reserves during the last 30 years and documented that during the last 20 years, the natural gas reserves remained the same.

The above findings and students' explicit discussion on the shift from oil to natural gas and renewable energy resulted in documenting that new reserves will cover new increased needs. As a consequence, students developed a more qualitative model, reporting that natural gas reserves will never be exhausted and therefore, government should invest in building a natural gas plant in Cyprus.

5 Conclusions

In this chapter, we have argued that the inclusion of engineering model eliciting activities in elementary school mathematics curricula can engage students in creative and innovative real-world problem solving and can increase their awareness of the different aspects of mathematical problem solving in engineering. The problem we have implemented has been developed from a models and modelling perspective, which takes students beyond their usual mathematical problem-solving experiences to encounter situations that require substantial interpretation of the problem goal and associated more complex data. Students have to elicit their own mathematical and engineering ideas and mathematical operations as they work the problem; this usually involves a cyclic process of interpreting the problem information, selecting relevant quantities, identifying operations that may lead to new quantities, and creating meaningful representations (Zawojewski et al. 2008). Because students' final products embody the factors, relationships, and operations that they considered important in creating their model, powerful insights can be gained into the growth of their mathematical and engineering thinking.

The students who participated in the present study developed a number of different models that adequately solved the problem, although not all models took into account all possible data and relations between mathematics and the engineering world. Further, it was explicit that a number of students failed to fully understand the concept of average and effectively use it into their models. Besides the two groups that failed to understand the problem situation and to use the concept of average, at least five more students showed that they did not have a conceptual understanding of average. Although students were only 12 years old, we give consideration here to the various inconsistencies in students' work and we underline the importance of teaching mathematical concepts and processes like average through complex problem solving (Mousoulides and English 2008).

Engineering-based modelling problems can present students with situations that reflect real-life scenarios that build on students' existing knowledge and experiences from mathematics and engineering. Furthermore, such problems should address topics that require students to develop a model that integrates the underlying structural characteristics of the engineering problem being addressed. It is important that students' model constructions be documented so that their thinking and reasoning can be externalised in a variety of ways including tables, lists, graphs, and diagrams. Furthermore, the models students construct need to involve more than a brief answer: descriptions and explanations of the steps taken in constructing their models should be included. Finally, it is imperative that the models students create should be applicable to other related engineering problems. Although the problem presented here did not afford the students the opportunity to generalise their models to a related problem situation, it nevertheless enabled them to revise their initial models to accommodate new information they could retrieve from the Web.

Substantially more research is clearly needed in the design and implementation of engineering model eliciting activities for elementary school students and the learning generated. We need to know, for example, (a) the developments in elementary school students' learning in solving a range of engineering-based problems; (b) the ways in which the nature of engineering and engineering practice can best be made visible to these students; and (c) the types of engineering contexts that are meaningful, engaging, and inspiring for these learners.

References

English, L., & Mousoulides, N. (2009). Integrating engineering education in the middle school mathematics curriculum. In B. Sriraman, V. Freiman, & N. Lirette-Pitre (Eds.), *Interdisciplinarity, creativity, and learning* (pp. 165–175). The Montana Mathematics Enthusiast Monograph Series, Monograph 7. Charlotte: Information Age Publishing Inc.

Kaiser, G., & Sriraman, B. (2006). A global survey of international perspectives on modelling in mathematics education. *Zentralblatt für Didaktik der Mathematik, 38*(3), 302–310.

Lesh, R., & Doerr, H. M. (2003). *Beyond constructivism: A models and modeling perspective on mathematics problem solving, learning and teaching*. Mahwah: Lawrence Erlbaum Associates.

Lesh, R. A., & Zawojewski, J. S. (2007). Problem solving and modeling. In F. Lester (Ed.), *Second handbook of research on mathematics teaching and learning* (pp. 763–804). Greenwich: Information Age Publishing.

Miles, M., & Huberman, A. (1994). *Qualitative data analysis* (2nd ed.). London: Sage.

Mousoulides, N., & English, L. D. (2008). Modeling with data in cypriot and Australian classrooms. In O. Figueras, J. L. Cortina, S. Alatorre, T. Rojano, & A. Sepulveda (Eds.), *Proceedings of the 32nd international conference of the international group for the psychology of mathematics education,* (Vol. 3, pp. 423–430). Morelia: University of Morelia.

Mousoulides, N., Sriraman, B., & Lesh, R. (2008). The philosophy and practicality of modeling involving complex systems. *The Philosophy of Mathematics Education Journal, 23,* 134–157.

National Academy of Sciences, National Academy of Engineering, and Institute of Medicine. (2007). *Rising above the gathering storm: Energizing and employing America for a brighter economic future.* Washington: The National Academies Press.

Organization for Economic Cooperation and Development (OECD). (2006). Education at a glance: OECD indicators 2006. Retrieved 2nd of June from http://www.oecd.org/document/52/0,3343,en_2649_39263238_37328564_1_1_1_1,00.html.

Petroski, H. (2003). Early education. *American Scientist, 91,* 206–209.

Zawojewski, J. S., Hjalmarson, J. S., Bowman, K., & Lesh, R. (2008). A modeling perspective on learning and teaching in engineering education. In J. Zawojewski, H. Diefes-Dux, & K. Bowman (Eds.), *Models and modeling in engineering education: Designing experiences for all students.* Rotterdam: Sense Publications.

Chapter 24
Project Modelling Routes in 12–16-Year-Old Pupils*

Manuel Sol, Joaquin Giménez, and Núria Rosich

Abstract The modelling behaviour of 12–16-year-old pupils was studied on the basis of written reports about realistic mathematics projects. These were analysed by using a hypothetical project modelling route involving 16 actions. Application of this tool was useful in understanding the difficulties pupils have in carrying out the initial steps and the validation process.

1 Context and Aims

For over 15 years, we have been interested in studying how pupils develop modelling competencies through working on realistic mathematics projects (Vilatzara Grup 2001), and have sought to do so by analysing their productions. As teachers and researchers, our aim is to understand the emergence of early modelling behaviour and its application in heterogeneous and multicultural classrooms of the state education system. The research presented here was conducted within the framework of a curricular reform in Catalonia (Spain), one which explicitly set out the need for mathematics education at the secondary level (12–16 years) to help pupils to model real-life situations that were linked to other knowledge areas. This type of activity was not traditionally carried out in Spanish schools at this level, since teachers considered that mathematical modelling was a task for older pupils.

---

*Part of the research is granted by the I+D+I project EDU2008 5050, by Spanish Ministry of Education.

M. Sol (✉)
Vilatzara Secondary School, Vilassar de Mar, Barcelona, Spain
e-mail: msol@xtec.cat

J. Giménez and N. Rosich
University of Barcelona, Barcelona, Spain
e-mails: quimgimenez@ub.edu; nuriarosich@ub.edu

We regard a realistic mathematics project as a *lengthy activity that is rich in mathematisation and which, starting from a real perspective, fosters the processes of modelling and autonomy in pupils* (Sol and Giménez 2004). The fundamental aim of this type of activity is that students develop the mathematical modelling competencies (Blomhøj and Jensen 2003) required to tackle realistic problems (whether personal, school-related, local, regional or of general interest). This development takes place through guided reconstruction (Treffers 1987) lasting 1 month, throughout which time pupils are free to make their own decisions. It is important to note the key role played by this autonomy with respect to the initial decisions about the questions/problems to be tackled, which are never formulated in advance by the teacher. In this way, it is possible to simulate the expert modelling process and observe a range of behaviours in pupils (Giménez and Sol 2005; Sol et al. 2007) by analysing the texts they produce.

Many studies have focused on a priori ideas of experts doing modelling. Therefore, it is not known whether the regular cycles of modelling are really used by junior secondary pupils when they start doing activities such as realistic mathematical projects, or what may specifically appear when we observe their written reports. In a previous study, we analysed these activities in a group of pupils who were weak at mathematics and found that one of the main difficulties appears at the start of the modelling process (Sol et al. 2009). This finding is consistent with several previous reports (Haines and Crouch 2010; Haines and Houston 2003; Maaß 2007). It is also known that not all pupils show the same behaviour when doing project work (e.g. Borromeo Ferri 2007; Galbraith and Stillman 2001). Given that the type of activity proposed here is different to that used by these authors, it seemed important to conduct a more detailed empirical analysis with a larger group of pupils. Therefore, the present study sought to answer the following question: *What kinds of modelling behaviours are revealed by 12–16-year-old pupils in their written reports about realistic mathematical projects?*

Although this analysis has similarities with the cognitive perspective on modelling routes described by Borromeo Ferri (2006), the interpretations made here regarding pupils' understanding of the modelling process are based solely on their written material.

2 Theoretical Background

In discussing mathematical modelling and associated activities, authors and researchers often represent the modelling process as a cycle of activity so as to understand pupils' behaviours (Haines and Crouch 2010). We started our research using the Mason cycle (Mason and Johnston-Wilder 2004, p. 190). In line with the proposal of Haines and Crouch (2010), each of the phases or stages in this cycle is associated with a set of actions. Some of these actions are defined according to what is proposed in other known modelling cycles (Blum and Leiß 2007; Borromeo Ferri 2007; Voskoglou 2007).

Blum & Leiß phases	Voskoglou cycle	Mason cycle	Observable hypothetical actions
1, 2	Analyse	Specify	1. Understand and recognise a mathematically manageable problem. 2. Simplify and structure. Recognise restrictions and specifications. Make decisions about a statement.
3	Mathematise	Build a model	3. Identify objects and relevant relationships. 4. Choose relevant variables, distinguishing from others. 5. State assumptions. Recognise the mathematical background that is needed. 6. Explain relationships between real objects and mathematical knowledge. 7. Check the coherence in the set of assumptions and mathematical relationships according to the real situation.
4		Formulate mathema tically	8. State the relationship among variables using mathematical language. 9. Formulate hypotheses mathematically. 10. Formulate problems and/or sub-problems in a mathematical way.
	Solve & Interpret	Find mathematic al solutions	11. Problem-solving processes involved in finding the solution.
5		Interpret	12. Find and interpret solutions mathematically in the model used.
6	Validate	Compare with the original	13. Recognise the meaning and extent of the solutions and conclusions in the real situation. Pupils can also state the model. 14. Validate the model itself. Change the model if necessary. 15. Promote reflection about results.
7	---	Write a report	16. Communicate the process and results when the model is valid.

Fig. 24.1 Hypothetical modelling actions related to several modelling cycles

The first set of actions (see Fig. 24.1) corresponds to the two phases of the cycle proposed by Blum and Leiß, in that pupils must show they are capable of formulating a statement/question associated with a real situation, having already understood the complexity of that situation. Actions 3–7 set out different aspects of the mathematisation process, which appear in almost all modelling cycles. These actions had been observed empirically when we compared the behaviour of 12–13-year-old pupils who found mathematics difficult with that of pupils without such difficulties (Sol 2009). Actions 8, 9 and 10 are directly associated with the notion of working mathematically in the cycle of Blum and Leiß (2007). Actions 11 and 12 correspond to phase 3 of Mason's cycle. Actions 13, 14 and 15 refer to the validation process and were also observed empirically in the abovementioned study (Sol 2009). Finally, action 16 clearly indicates what is proposed by all modelling cycles.

In order to analyse pupils' behaviour, Borromeo Ferri (2006, p. 91) developed the construct of *individual modelling route*, which refers to pupils' passage through the different phases of the modelling cycle, which is revealed in both their verbal expressions and other external representations. In our study, the term *project modelling route* refers to a representation (corresponding to a discursive development) that expresses a set of actions performed by a group of pupils when producing the written report about a realistic mathematical project. This route can be represented either symbolically or graphically. The notion is therefore a technical construct, rather than a strictly cognitive one, and it enables pupils' written discourse to be analysed in order to understand their modelling behaviour.

3 Study Methodology

The modelling behaviour of 12–16-year-old pupils engaged in project work was analysed by observing their written material (as proposed by Haines and Houston 2001). The methodological approach taken is that of a qualitative, ethnographic case study, which provides an overall understanding of pupils' behaviour (Marshall and Rossman 1998). Indeed, we believe that the texts analysed enable a holistic view to be taken of the modelling process, in accordance with the nomenclature of Blomhøj and Jensen (2003). The research, conducted in a state school in a town near Barcelona, focused on almost all the project work carried out by 12–16-year-old pupils with the same teacher during the period 2005–2008. Pupils who had previously carried out activities of this kind were excluded from the study. The final research population is described in Table 24.1.

Each project required pupils to work in groups of two to four (which they formed voluntarily) over a period of around 4 weeks. The following description of a project carried out by 14–15-year-old pupils illustrates how mathematical projects are developed.

(a) Pupils choose a context that interests them and are then asked to set some objectives/challenges for their work. The group begins outside the classroom, but this is then followed by a brainstorming session involving all the class pupils, the result of which is the selection of a series of questions by each group, this being the first decisions they make. These questions do not necessarily have to be mathematical. In fact, in this example, they were: *What is a watchtower? Who or what is being watched? What is gained by having a vantage point on the hilltop rather than on the coastline?*

Once pupils have found the information they need they can then begin to discuss the relevance and scope of these questions in the problem context. In this example, the pupils learned/recognised that they lived in a coastal town of around 20,000 inhabitants (25 km to the north of Barcelona) which, during the sixteenth century, was vulnerable to attack by pirates who came to plunder and kidnap. In order to defend themselves, the townspeople built a number of watchtowers, one of which (the Nadal tower) remains standing to this day. Around 4 km inland, there is a hilltop castle (Burriac castle, also dating back to the sixteenth century) and another watchtower that was also built for the same purpose. It is in this context that the questions posed by pupils take on meaning and interest. Initially, we allow non-mathematical questions to be posed and limit ourselves to helping them to see the possibilities of the topic at hand.

Table 24.1 Research population

	Secondary school				
	1st sec school (12–13)	2nd year (13–14)	3rd year (14–15)	4th year (15–16)	Total
Núm Projectes	7	12	9	4	32
Núm Alumnes	15	26	20	11	72

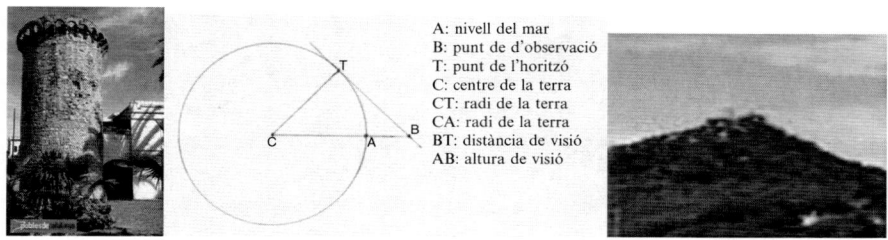

Fig. 24.2 Nadal tower, Burriac castle and the scheme proposed by the pupils

(b) In the next stage, we begin to guide a process of construction/deconstruction (Gellert and Jablonka 2006) of the real model. To this end, pupils need to understand the possibility of describing the situation in mathematical terms and simplifying it. They must select those objects which are relevant to the proposed objective and discover the relationships between them. In the example, the relevant variables were the distance to the horizon and the elevation from which the observation was made.

(c) The next step is for pupils to express a question or problem in mathematical terms. They must decide what data they need and think about how to obtain them; for example, Will they have to be measured directly, obtained from an information source or calculated? They have to recognise the real objects that are involved, choose the variables, describe the relationship between real objects and mathematical objects, explain the dependency relationship between variables, and explain their assumptions. In the example, the pupils had to consult maps to know the elevation of Burriac castle, and they represented the mathematical situation as shown in Fig. 24.2.

(d) Next, pupils are helped to make explicit the process of mathematisation and to show how they interpret the mathematical models used. In the example, the pupils went to the Nadal tower in order to calculate its height by applying Thales' theorem. In Fig. 24.2, they applied Pythagoras' theorem to calculate the distance to the horizon, BT. Using this scheme, they made the calculation from three different positions: at sea level, from the top of the Nadal tower and from the top of the Burriac castle. The results were presented in a table and compared. Having calculated the height of the Nadal tower as 12.5 m and the distance to the horizon as 12.6 km, they then had to work out the tower height required for each kilometre that one moves away from the coastline. This was done by considering a relationship of similarity and, by using various data, they came up with the equation $x = h*y/12{,}600$, where y is the distance from the foot of the Nadal tower to the point on the horizon that is seen from the top of the tower, h is the height of the Nadal tower, x is the height to be calculated and 12,600 is the distance to the horizon from the Nadal tower.

(e) In the next stage, they are asked to explain how they approached and arrived at their answer to the question posed at the start. In the example, they explained that they needed to know the speed at which the pirate ships could travel. After consulting a number of books, and deciding to ignore the wind variable, they decided that the ships travelled at a speed of 18 knots. They then converted this into km/h and, applying a rule of three, calculated how long the town's inhabitants would have to react once a ship was spotted on the horizon.

(f) In a meeting with the group prior to the presentation of their project report, they are told that it is good to justify the mathematical solutions they have found in the real context from which they started. In the example, they explained why it made sense to build the watchtowers on the hilltop, but they did not see the need to make clear that they realised this when they discovered the relationship between elevation and range.

(g) Finally, the pupils communicate their results. Each group submits a written report and makes an oral presentation to the whole class.

4 From Analysis to Results

The modelling routes were identified on the basis of the pupils' written work. Specifically, the statements and actions contained in this mathematical text and which formed part of the modelling process were identified and coded, the final decision being reached by consensus in the research team (Blommaert and Bulcaen 2000).

Thus, for example, in a project about the school common room, one pupil asked: *How does the number of people who fit [a] round a table of the same size vary according to whether the table is round or rectangular?* Our interpretation here is that the pupil is posing a mathematical question, corresponding to action 10 in Table 24.1. At another point, a pupil wrote: *We decided to use round tables because that way we would make use of the space that is lost at the corners of a rectangular table.* Here we consider that the pupil is illustrating the purpose of a problem statement, corresponding to action 2. The text analysis continues in this way.

Pupils do not always present the calculations and relationships between variables in a general way, as one would expect from an understanding of the model as such. We also note that the same written report may include both symbolic and graphical elements as evidence of various actions. For example, in the example of the school common room, our interpretation is that they are stating a relationship between the variables involved, making use of mathematical language and, therefore, performing action 8 of Table 24.1. In addition, the graphical information they provide explains relationships between real objects (the table and the space required around it for people to be seated) and mathematical objects such as radius and diameter (of the table and the seating space), and we interpret this as evidence of action 6. Finally, in the last lines of the report, they show their interpretation of the solution, stating that their chosen design allows four rows of tables, before going on to say that only an extra 15 cm would be required to include a fifth row. They then state that "taking a further 1.5 cm from each side of the table would not have a noticeable effect and this would allow a fifth row of tables to be included," thus

Table 24.2 Vector representation of some modelling routes

Title	Year	Project modelling route	Neglected actions
Smoking	1st	1,2,2,10,11,11,11,13,12,1,13,1,10,11,13	3,4,5,6,7,8,9,14,15,16
School canteen	2nd	10,1,2,3,12,9,3,4,12,4,4,8,6,11,4,8,6,11	5,7,14,15
Nadal tower	3rd	2,5,10,5,11,12,10,11,8,12,13,2,3,10,11, 8,1 2,13,2,9,11,12,13,1,10,4,6,11,12,13	1,3,7,14,15
Tins and cans	4th	1,2,4,6,11,12,11,13,8,13,3,4,11,12,11, 10,13,8,13,13	3,5,7,9,14,15

improving upon their initial solution. An explanation such as this suggests that the pupils have recognised the meaning and extent of the solutions in the real situation. In our view, this is consistent with the presence of action 13 in Table 24.1, which one would not usually expect to observe in 12–13-year-old pupils.

Continuing with this process, we now represent the sequence of actions observed in each written report by means of a matrix or graph, which enables the modelling routes to be compared visually. In the above example of the school common room, we represented the modelling route followed by pupils in terms of vectors, writing the number that corresponds to each action observed: 10, 1, 2, 3, 12, 9, 3, 4, 12, 4, 4, 8, 6, 11, 4, 8, 6, 11 and 13. By repeating this process with all the analysed projects, it is possible to construct tables such as Table 24.2. Thus, one can see the actions that are followed or neglected by pupils.

The analysis shows that the model does not have linear continuity. Greater importance is placed on description, and it can be seen that pupils do not know how to represent the overall problem; furthermore, they find it difficult to simplify the text mathematically, as an expert would. Therefore, we believe that they interpret the project as a *set of problems* that they wish to explain one after another. Many pupils also find it difficult to pass through certain actions. As mentioned above, these can also be represented with a vector, as shown in Table 24.2.

5 Results

Having analysed all 32 projects, the data in Table 24.3 show the percentage of reports whose content provides evidence that the pupils have carried out each action in the hypothetical modelling route.

Most routes start at action 1 or 2 and continue at 10 or 11, which seems to reflect the classical structure of the problem-solving process. In contrast, actions 5, 7, 9, 14 and 15 hardly appear in the project modelling routes. We interpret this to mean that the pupils see the project as a chain of small problems rather than as one big problem. As such, they are not aware of the modelling process as a global mathematical process but a large set of problem-solving actions. The set of actions 10, 11 and 13, which recalls the classical interpretation of a problem-solving classroom activity, is frequently repeated. Action 6 poses some difficulties: The relationship

Table 24.3 Percentage of reports that pass through each action of the modelling route

Actions	1	2	3	4	5	6	7	8
12–13 Years	71.43	42.86	71.43	14.29	0	0	0	0
13–14	66.67	91.67	75	91.67	0	25	8.33	41.67
14–15	55.56	100	44.44	66.67	8.33	22.22	0	22.22
15–16	100	75	75	100	0	50	0	100
Actions	9	10	11	12	13	14	15	16
12–13	0	85.71	100	28.57	100	0	0	0
13–14	8.33	75	100	100	100	0	8.33	0
14–15	11.11	66.67	100	100	100	0	0	0
15–16	25	100	100	100	100	0	0	0

between real objects and mathematical knowledge is easy to observe in the case of a real object, for instance, when we think about a cake we can speak about shape and weight; however, distance is not a tangible object and it is therefore more difficult for pupils to relate it to a mathematical object.

The 12–14-year-old pupils go through fewer actions than do the 14–16-year-olds, and they also find it more difficult to recognise variables and their relationships (actions 3 and 4). In addition, they need help to carry out actions 1 and 2. The 15–16-year-old pupils go through actions 4, 6 and 8, whereas 12–13-year-olds do not use these actions. The use of action 4 appears explicitly in the projects carried out by 15–16-year-old pupils. As can be seen in Table 24.3, some actions were not passed through by the pupils. It seems that abstraction is a barrier to the recognition of certain hypothetical actions.

6 Conclusions

We believe that the results provide a complement to previous findings, not least in that they are derived from realistic mathematics projects. The following conclusions can be drawn.

Firstly, we found that the hypothetical project modelling route involving 16 actions enabled us, when used as a methodological tool, to identify the modelling process that pupils follow. It can also help to find ways of improving the interaction between pupils and teacher, and may provide clues to new forms of classroom assessment. Although the teacher guided the process in the initial stages, pupils found it difficult to perform certain actions that one would ideally observe in the modelling cycle. For example, 12–16-year-old pupils do not have an overall view of the problem situation and the use of variables (Blum and Leiβ 2007). It seems that this is partly due to the fact that the primary curriculum lacks a functional view involving pupil initiative, the usual focus being on solving problems set by the teacher.

The 12–16-year-old pupils found several actions difficult to perform: communicating the social role of their mathematical problem; controlling mathematical relationships (they are unable to state a real formulation of the problem [as proposed by Maaß 2007]); and validating a model, possibly due to a lack of awareness of having done it (Maaß 2007). These results suggest that it is important to work with functional reasoning in order to improve the initial modelling steps. Mathematical knowledge, which is a necessary feature of a realistic mathematics project such as that described here, should be analysed carefully and adjusted according to the competency level of pupils.

Differences were observed between 12–14- and 14–16-year-old pupils. However, no general patterns were found regarding which modelling characteristics relate more to younger pupils overall. A similar result was reported by Haines and Crouch (2010).

Pupils showed no awareness of the overall modelling process. They interpret projects as a set of small problems, perhaps because primary school does not prepare them to work on open-ended problems. Moreover, they particularly failed to recognise the validation process. In contrast, they did evaluate the results of partial problems which they had worked on previously in the classroom.

The present research suggests the need for further studies in order to clarify the role of teachers as regards improving the mathematical communication of pupils, helping them to become more aware of the processes involved in the modelling cycle.

References

Blomhøj, M., & Jensen, T. H. (2003). Developing mathematical modelling competence: Conceptual clarification and educational planning. *Teaching Mathematics and Its Applications, 22*, 123–139.

Blommaert, J., & Bulcaen, C. (2000). Critical discourse analysis. *Annual Review of Anthropology, 29*, 447–466.

Blum, W., & Leiß, D. (2007). How do students and teachers deal with modelling problems? In C. Haines et al. (Eds.), *Mathematical modelling. Education, engineering and economics* (pp. 222–231). Chichester: Horwood.

Borromeo Ferri, R. (2006). Theoretical and empirical differentiations of phases in the modelling process. *Zentralblatt für Didaktik der Mathematik, 38*(2), 86–95.

Borromeo Ferri, R. (2007). Modelling problems from a cognitive perspective. In C. Haines, P. Galbraith, W. Blum, & S. Khan (Eds.), *Mathematical modelling (ICTMA12): Education, engineering and economics* (pp. 260–270). Chichester: Horwood.

Crouch, R. M., & Haines, C. R. (2003). Do you know which students are good mathemathical modellers? Some research developments. Technical Report No. 83, Department of Physics, Astronomy and Mathematics, University of Hertfordshire Hatfield, p. 25.

Galbraith, P. L., & Stillman, G. (2001). Assumptions and context: Pursuing their role in modelling activity. In J. F. Matos, W. Blum, K. Houston, & S. P. Carreira (Eds.), *Modelling and mathematics education: ICTMA9 applications in science and technology* (pp. 300–310). Chichester: Horwood.

Gellert, U., & Jablonka, E. (Eds.). (2006). *Mathematisation and demathematisation*. London: Sense.

Giménez, J., & Sol, M. (2005). Students' difficulties when starting with mathematical projects. In L. Santos, A. P. Canavarro, & J. Brocardo (Eds.), *Paths and crossroads* (pp. 217–230). Lisboa: APM.

Haines, C., & Crouch, R. (2010). Remarks on a modelling cycle and interpretation of behaviours. In R. Lesh, P. L. Galbraith, C. R. Haines, & A. Hurford (Eds.), *Modeling students' mathematical modeling competencies* (pp. 145–154). New York: Springer.

Haines, C., & Houston, K. (2001). Assessing students project work. In D. Holton (Ed.), *The teaching and learning of mathematics at university level: An ICMI study* (pp. 431–442). Dordrecht: Kluwer.

Maaß, K. (2007). Modelling taks for low achieving students. First results of an empirical study. In D. Pitta-Pantazi & G. Philippou (Eds.), *CERME 5 – Proceedings of the Fifth Congress of the European Society for Research in Mathematics Education* (pp. 2120–2129). Larnaca: University of Cyprus.

Marshall, C., & Rossman, G. B. (1998). *Designing qualitative research*. Thousand Oaks: Sage.

Mason, J., & Johnston-Wilder, S. (2004). *Fundamental constructs in mathematics education*. London: Routledge Falmer & Open University.

Sol, M. (2009). Anàlisis de les competències i habilitats en el treball de projectes matemàtics amb alumnes de 12-16 anys en una aula heterogènia. Doctoral thesis. UB, Barcelona.

Sol, M., & Giménez, J. (2004). Proyectos matemáticos realistas y resolución de problemas. In J. Gimenez, J. P. Ponte, & L. Santos (Eds.), *La actividad matemática en el aula* (pp. 35–47). Barcelona: Graó.

Sol, M., Giménez, J., & Rosich, N. (2007). Competencias y proyectos matemáticos realistas en la ESO. *UNO Revista de Didáctica de las Matemáticas, 46*, 43–67.

Sol, M., Giménez, J., & Rosich, N. (2009). Analyzing modelling as a process: A teaching experiment with immigrant students. In H. Labelle, D. Gauthier (Eds.), *Proceedings CIEAEM 61*, Montréal, Canada, July 26–31, 2009. *Quaderni di Ricerca in Didattica*, Supplemento n. 2. Palermo, Italy: G.R.I.M., Department of Mathematics, University of Palermo.

Treffers, A. (1987). *Three dimensions: A model of goal and theory descriptions in mathematics instruction – The Wiskobas project*. Dordrecht: Kluwer.

Vilatzara, G. (2001). Experiencias sobre proyectos e investigaciones matemáticas en secundaria. *Números. Revista de Didáctica de las Matemáticas, 46*, 29–47.

Voskoglou, M. (2007). A stochastic model for the modelling process. In C. Haines, P. Galbraith, W. Blum, & S. Khan (Eds.), *Mathematical modelling (ICTMA12): Education, engineering and economics* (pp. 149–157). Chichester: Horwood.

Part III
Mathematical Modelling in Teacher Education

Chapter 25
Mathematical Modelling in Teacher Education – Overview

Jill P. Brown

1 Introduction

Within ICTMA and beyond, it is essential that authors clearly present their understanding of the term mathematical modelling. There are multiple interpretations and associated understandings, which contribute to fruitful discourse, but at times, it is left to the reader to identify the views of the author. The same is true of the term teacher education, but here the differences relate to practices rather than interpretation. Firstly, the meaning of mathematical modelling is examined. With respect to purposes for teaching, Julie and Mudaly (2007, p. 504) identified *modelling as vehicle* and *modelling as content*. For the former, real world problems are used for both motivation and the development of specific mathematical content. For the latter, the focus shifts to 'developing the capacity of students to address problems located in the external world, and to evaluate the quality of their solutions' (Galbraith 2007, p. 181). These approaches are not dichotomous, whilst the emphasis may be different the goals may intersect as, in order to solve a genuine problem, the need for new mathematical content may emerge (Stillman et al. 2008, p. 145). In addition, what distinguishes a modelling task from an application task? Stillman positions applications tasks 'between structured word problems and open modelling problems' (2000, p. 334) such that the problem must be embedded in a real world context. Moreover, 'in an applications task, the primary sources of information that are external to the task solver are the problem statement and any accompanying visual representations' (p. 335).

Finally, the term 'teacher education' needs to be considered. There is no generic teacher education program. Authors need to report sufficient details for the reader to understand the mechanisms of their systems. In some countries, pre-service teachers undertake teaching experiences in schools as part of their teaching preparation,

---

J.P. Brown (✉)
School of Education (Victoria), Australian Catholic University, 115 Victoria Parade, Locked Bag 4115, FITZROY 3065, Australia
e-mail: jill.brown@acu.edu.au

concurrent with their studies of education and mathematics (if any). We would expect a greater understanding of classroom practices from these students when compared to others who had no experience in schools in their undergraduate teacher education as is the case in other settings. In-service teacher education takes on many forms as well. It is important to be aware of the level of schooling for which they are preparing or currently teaching.

2 Research Using Different Perspectives of Modelling

Kaiser and Sriraman (2006) note that within modelling in mathematics education 'the apparent uniform terminology and its usage masks a great variety of approaches' (p. 302) and so propose a classification to distinguish the 'various perspectives according to their central aims' (p. 303). The perspectives include

> *Realistic or applied modelling* (pragmatic-utilitarian goals, i.e. solving real world problems, understanding of the real world, promotion of modelling competencies);
>
> *Educational modelling*: differentiated in (a) didactical modelling and (b) conceptual modelling (Pedagogical and subject-related goals:(a) Structuring of learning processes and its promotion (b) Concept introduction and development) (p. 304).
>
> *Model-eliciting approach* (Psychological goals, i.e. apply model elicited through solving the original problem to a new problem) is added in a revised classification (Kaiser et al. 2007, p. 8).

These perspectives are used to consider the chapters presented in this section.

Kuntze investigated 79 in-service and 230 pre-service German teachers' views about mathematical modelling tasks in the mathematics classroom. This study involved tasks of two types, described as those with substantial/higher modelling requirements (which require at least one translation step between the situational context and a mathematical model, and different solutions are possible) and those with low modelling requirements (where the mathematical model is provided to the solver, translation processes are less important, and only one solution is possible). The sample tasks presented appear to be applications rather than modelling tasks in Stillman's (2000) classification. The low level sample task presents an octagon for which the area needs to be calculated and compared. The importance of authors presenting their definition of modelling is not unproblematic. The focus of this study relates to modelling competencies and the importance of task selection is related to this. Thus, this study is classified as a realistic or applied modelling perspective. Kuntze found the majority of pre-service teachers expressed a preference for low level tasks and classified tasks with higher modelling requirements as having lower learning potential than tasks with lower modelling requirements. Kuntze found the lack of compatibility between mathematical modelling and the stated need in mathematics for an exact answer to be the cause rather than the level of complexity involved in higher level tasks. In contrast, in-service teachers rated higher level tasks more positively. He suggests in-service teachers may be more aware of the learning opportunities inherent in the higher level tasks; however, he notes with concern the uncertainty of these teachers 'concerning the modelling cycle'. Modelling appears to be a vehicle for teaching mathematics in this study – all tasks used are in a single content area, namely, measuring areas.

The study of Biembengut, in Brazil, had a dual focus. As the participants (pre-service and in-service teachers) engaged in tasks involving classical mathematical models (the restricted growth of an organism and cooling of a liquid), the aim was realistic or applied modelling. However, when the teachers later engaged in modelling tasks (creating a parking lot and packaging for kitchen oil) and then in pairs, chose a situation of interest, posed a question that was amenable to solution using mathematics, and engaged in the modelling cycle to investigate their problem, the aim was clearly educational modelling. To engage one's students in modelling, a teacher must themselves be able to model. Clearly, the intent here is modelling as content and the latter tasks are modelling tasks. Biembengut found that the role of the teacher, as perceived by the teachers themselves, is to be one who tells their students what and how to do mathematics. Moreover, this mathematics has been prepared by someone else (perhaps presented in a textbook or similar) as the role of the teacher does not generally include the notion of designing instructional materials. Poor salaries and resulting high workloads, in the South American context, further exacerbated this even in an environment where students are performing at an increasingly worse level. Modelling is seen only as a vehicle for teaching mathematics by these teachers.

Stillman and Brown also focus on teacher beliefs; however, in this case, all teachers in the study were pre-service secondary teachers in Australia. The approach here is educational modelling as the researchers seek to identify the affinity of pre-service teachers to using modelling tasks and if the length of the teacher preparation (a 1-year program undertaken after an undergraduate degree, or a 4-year program where teaching preparation was undertaken concurrently with the study of mathematics) was a factor. Nearly all pre-service teachers believed modelling tasks are a part of mathematics. There were, however, interesting differences associated with program length and the authors speculate on possible reasons for these.

The perspective of Doerr and Lesh would be classified by Kaiser et al. (2007) under the model-eliciting approach. Doerr and Lesh are quick to point out that there is more to modelling, however, than just model-eliciting activities. They describe modelling as also including 'model exploration activities and model adaptation activities'. They argue that model-eliciting activities were developed so that those engaged 'are likely to make significant adaptations to their initial interpretations or conceptualisations of the situations' in a short period of time. Moreover, as thought revealing instruments, these allow insight into students' thinking and increase opportunities for teacher effectiveness. The approach taken by Doerr and Lesh seems to have elements of both the modelling as vehicle and modelling as content perspectives. On the applications – modelling task continuum (Stillman 2000) – Doerr and Lesh describe their Ferris Wheel Task as a 'model application task'. Stillman no doubt would agree. Doerr and Lesh's descriptions of the teacher actions in the pre-calculus classroom included: reluctance to intervene when progress was occurring; interactions that did occur focused on the meaning of quantities calculated; and differing assumptions were seen as opportunities for future class discussion. Furthermore, the task provided opportunities for productive work, multiple representations were involved, and students were able to self-check solutions for reasonableness.

3 Reflections

Although somewhat disappointing to see so few studies focussed on teacher education, the studies themselves are not a disappointment. Although, naturally sharing some common themes, they are as diverse as they are interesting. The theoretical classifications used in this chapter have proved fruitful in allowing a comparison of key similarities and differences. All classifications are tentative and were undertaken on the basis of information presented in the chapters. However, not unexpectedly, mathematical modelling in education and in teacher education is complex, and the studies were not all able to be neatly classified. It is this complexity that makes mathematical modelling, its enactment in educational settings, and research related to these forever interesting.

References

Galbraith, P. (2007). Authenticity and goals: Overview. In W. Blum, P. Galbraith, H.-W. Henn, & M. Niss (Eds.), *Modelling and applications in mathematics education: The 14th ICMI study* (pp. 181–184). New York: Springer.

Julie, C., & Mudaly, V. (2007). Mathematical modelling of social issues in school mathematics in South Africa. In W. Blum, P. Galbraith, H.-W. Henn, & M. Niss (Eds.), *Modelling and applications in mathematics education: The 14th ICMI study* (pp. 503–510). New York: Springer.

Kaiser, G., & Sriraman, B. (2006). A global survey of international perspectives on modelling in mathematics education. *Zentralblatt für Didaktik der Mathematik, 63*(9), 302–310.

Kaiser, G., Sriraman, B., Blomhøj, M., & Garcia, J. (2007). Report from the CERME5 working group modelling and applications – Differentiating perspectives and delineating commonalties. *ICTMA Newsletter, 1*(1), 6–10.

Stillman, G. A. (2000). Impact of prior knowledge of task context on approaches to applications tasks. *The Journal of Mathematical Behavior, 19*(3), 333–361.

Stillman, G. A., Brown, J. P., & Galbraith, P. L. (2008). Research into the teaching and learning of applications and modelling in Australasia. In H. Forgasz et al. (Eds.), *Research in mathematics education in Australasia 2004–2007* (pp. 141–164). Rotterdam: Sense.

Chapter 26
Models and Modelling Perspectives on Teaching and Learning Mathematics in the Twenty-First Century

Helen M. Doerr and Richard Lesh

Abstract Research based on models and modelling perspectives (MMP) has shown that, in order for mathematical concepts and abilities to be useful beyond school, new levels and types of understandings are needed beyond those that have been emphasized in even the most innovative and future-oriented statements of curriculum standards. Similarly, teacher-level knowledge and abilities consists of a great deal more than the kind of beliefs, dispositions, and pedagogical content knowledge that have been emphasized in most past research on teacher development.

1 Introduction

When investigating the development of both students' and teachers' thinking, one of the most important distinguishing characteristics of research based on models and modelling perspectives (MMP) is its emphasis on the fact that in virtually every field where learning science researchers have investigated what it means to develop competence, it has become clear that highly competent individuals not only *do* things differently but they also *see* (or *interpret*, or *conceptualise*) things differently. In particular, in mathematics education, we assume that this claim also applies to many levels and types of subjects who range from individual students, to groups of students, to teachers as well as from educational researchers, to curriculum developers, to policy makers. For example, some of the most important interpretation systems that mathematics teachers need to develop involve making sense of students' ways of thinking about mathematical learning and problem-solving

---

H.M. Doerr (✉)
Mathematics Department, Syracuse University, Syracuse, New York, USA
e-mail: hmdoerr@syr.edu

R. Lesh
School of Education, Indiana University, Bloomington, Indiana, USA
e-mail: ralesh@indiana.edu

activities. Therefore, because of the interacting nature of the conceptual systems that are developed by students, teachers, and other educational decision-makers, MMP research often involves *multi-tier studies* in which: (a) Students go through sequences of cycles in which they iteratively express, test, and revise their interpretations or conceptualisations of mathematical learning or problem-solving situations. That is, students are engaged in model development activities. (b) Teachers go through sequences of cycles in which they iteratively express, test, and revise their interpretations or conceptualisations of students' model development activities. (c) Researchers go through sequences of cycles in which they iteratively express, test, and revise their interpretations or conceptualizations of students' and teachers' model development activities. MMP research is mainly about developing models of teachers' and students' models and modelling abilities. We expect similar principles to apply at all three levels. Since we have found that many traditional research methodologies are based on assumptions which are not appropriate for the dynamically adapting and interdependent "subjects" that we investigate, this chapter also describes some of the ways that MMP research designs provide alternatives to a variety of traditional research methodologies (Kelly and Lesh 2000; Kelly et al. 2008; Lesh 2002).

2 MMP Research Investigates What It Means to "Understand" Important Concepts and Abilities

How can teachers be expected to teach important concepts or abilities effectively if it is not clear what it means for students to "understand"? Yet, documents that define curriculum standards routinely encourage teachers to teach vague constructs such as problem solving, connections, dispositions, and metacognitive ideas and procedures – even though there is little clarity about what it means for a given student to "understand" such constructs, and even less clarity about how the development of such constructs can be documented or assessed. For example, in a recent analysis and critique of the past 50 years of research on problem solving in mathematics education, Lesh and Zawojewski (2007) noted that virtually every past comprehensive review of the literature has concluded that attempts simply have not worked which have tried to improve students' problem-solving abilities by teaching some manageably short list of heuristics, strategies, beliefs, metacognitive processes, dispositions, or habits of mind. Short lists of rules tend to lack prescriptive power, even though they often seem to have face validity for giving after-the-fact descriptions of past problem-solving behaviours. Longer lists of more detailed prescriptive processes tend to be unproductive because learning them clearly involves not only knowing how to do them in important settings, but also knowing *when, where, why*, with *whom*, and for *what* purpose to use them. Learning such constructs tends to be highly context specific; transfer of learning has been unimpressive; and, long-term changes in behaviour have not been documented. Even in cases where some evidence exists to suggest that students might have improved their problem-solving abilities within some small domain of problem-solving tasks, such achievements seldom occur unless world class teachers teach semester-sized treatments, where the complexity of the experiences ensures that the causes for improvements will be impossible to isolate. Perhaps students' problem-solving abilities improved because they learned some mathematics rather than because they learned some problem-solving processes. Yet, traditional research on problem solving usually begins with the assumption that the researchers already possessed clear and complete understandings about what it means for students to "understand" the most important concepts and abilities that they believe to be important. Showing that "*it*" *can be taught* (where "it" is the researchers' preconceived notions about what it means to understand "problem solving") seems to be the issue that needs to be resolved. When treatments fail, the assumption generally is made that something was wrong with the treatment itself, rather than with flaws or fuzziness in preconceived notions about the nature of the concepts or abilities students were expected to learn.

MMP research emphasizes that, before educators rush ahead to teach things, more clarity is needed about: (a) what it means to "understand" the things we want students to learn and (b) how the development of such understandings can be measured and assessed. For example, in mathematics education, most traditional research has defined problem solving to be about situations that involve *getting from givens to goals when the path is not obvious.* In MMP research about what it means to "understand" various aspects of problem solving, results suggest that a more appropriate conception of problem solving should be about *goal-directed activities in which significant adaptation needs to be made to the problem solvers' initial interpretation of the situation.* The distinction may sound subtle, but it is large.

According to MMP perspectives, interpretations of givens and goals are expected to change during solution processes. Progress is not expected to occur along a single path. Most relevant concepts are expected to be as intermediate stages of development – not fully developed, and not totally undeveloped. Problem solvers are expected to be interpretation developers (i.e., model developers) at least as much as they are information processors. Strategies and heuristics that are most productive are expected to be those that help problem solvers advance beyond early interpretations of problem situations, not those that provide clues about next steps when paths are blocked, or when productive concepts or processes are not readily apparent.

Is the essence of problem solving about learning concepts and processes separately and then putting them together to solve problems? Is there any reason to expect that the heuristics, strategies, and processes that are useful for describing experts' past problem-solving experiences should also be useful for prescribing "next steps" for novices? Are useful heuristics, strategies, processes, beliefs, or dispositions reducible to lists of condition-action rules or declarative statements (i.e., facts, rules, habits)? Do useful heuristics, strategies, processes, beliefs, and dispositions function explicitly and analytically, or do they often function tacitly and non-analytically? In MMP research, when students engage in meaningful problem-solving activities, results have shown that most relevant concepts and processes are at intermediate stages of development. When problem solvers develop a language for describing their own problem-solving activities, what they do when they are most successful is similar to what athletes, performing artists, and others do when they engage in complex activities and then watch videotapes and reflect about past activities. Their goal is not to reduce future activities to strings of rule-governed behaviors. Instead, their goal is mainly to improve what they "see" when they are actively engaged in problem-solving and decision-making activities.

In more mature domains of research, scientists often devote large portions of their time and energies toward developing tools for their own use. Many of their efforts aim at developing tools for measuring important constructs. Similarly, in MMP research, *model-eliciting activities* (MEAs) were developed so that: (a) relevant subjects (students, teachers, researchers) are likely to make significant adaptations to their initial interpretations or conceptualizations of the situations and (b) these adaptations are likely to occur during 60–90 min time periods which are sufficiently brief so that the processes that lead to change can be observed directly (rather than observing only a few intermediate states of development and trying to infer from these what processes might have led from one state to another). In MMP research, MEAs are *thought-revealing activities* that function similarly to Petri dishes in biochemistry laboratories.

3 Do Model-Eliciting Activities (MEAs) Work?

Even though research on MEAs is an important part of MMP research, and even though detailed design principles have been clearly specified for developing MEAs (Lesh and Doerr 2003), many of the most common misconceptions about the underlying theory (MMP) have resulted from the following two false assumptions about the nature of MEAs:

- MEAs often are mistakenly thought of as functioning in isolation rather than being small parts within more comprehensive *model development sequences* that include a variety of other types of activities such as *model exploration activities* (MXAs) and *model adaptation activities* (MAAs) as has been described in detail in a number of past publications (e.g., Doerr and English 2003; Lesh and Doerr 2003). All emphasize that concept development does not end when acceptable solutions have been produced for MEAs. For example, when students finish MEAs: (a) The models produced usually integrate ideas and procedures from a variety of textbook topic areas and/or disciplines. To further advance understandings, these interpretation systems usually need to be unpacked. (b) The models that are developed usually have meanings that are highly situated. That is, they are shaped by constraints and affordances in the given contexts, and they draw heavily on students' personal experiences related to the contexts. To further advance understandings, there usually are important ways that these interpretation systems need to be decontextualized. (c) The models usually are expressed using a variety of representational media each of which focus on somewhat different aspects of the situation. To further empower thinking, more elegant and efficient representations usually need to be introduced which students should not be expected to invent for themselves. (d) Problem solvers often think *with* interpretation systems that they have not yet explicitly thought *about* as objects of thought. Further development often needs to involve more formal and abstract explorations in which interactions with teachers tend to be important.

- MEAs often are mistakenly thought of as instructional treatments whose worth depends mainly on evidence showing that "they work" to produce significant learning gains. However, MEAs were developed first and foremost to be used as research tools to investigate the nature of students' and teachers' thinking. Asking MMP researchers whether MEAs "work" when they are used as teaching and learning activities is like asking Piaget whether his famous conservation tasks and clinical interviews work if they are used as the basis for teaching and learning activities. If either Piaget's tasks or MEAs are used in teaching and learning situations, their ability to "work" depends greatly on the fact that teachers tend to be more effective when: (a) they develop sound conceptions of what it means to "understand" the concepts they are trying to help students learn and (b) they know as much as possible about the strengths and weaknesses of their students' thinking. However, because MEAs are designed to be both *model-eliciting* and *thought-revealing*, they promote learning in direct ways. This is, MEAs were designed so that: (a) significant conceptual adaptations (i.e., conceptual change) are likely to occur during brief periods of time and (b) important aspects of the thinking are directly observable. Consequently, many research studies involving MEAs have shown that students' thinking often progresses through several Piaget-like stages during single 60–90-min problem-solving activities (e.g., Lesh and Zawojewski 2007). This is why MEAs sometimes have been referred to as *local conceptual development activities* (Lesh and Harel 2003). Because of their *thought-revealing* nature, they often make it possible to document and assess levels and types of understanding that have seldom been assessed in other ways (Lesh and Doerr 2003, Lesh and Lamon 1992). MMP studies have shown that, without guidance or scaffolding from teachers, average ability children often are able to develop (or make significant adaptations to) powerful elementary-but-deep concepts and abilities which have been thought to be beyond their grasp. Furthermore, because MEAs emphasize a broader range of deeper understandings than those assessed on most tests, a broader range of students often emerge as having extraordinary abilities. MEAs provide powerful tools to promote diversity and democratic access to powerful ideas (Lesh and Lamon 1992).

As an example of a study showing that MEAs work, one recent study involved graduate students in an intense 5-week summer school course designed to cover the equivalent of one full 15-week semester on statistics for research in education (Lesh et al. 2010). Forty-eight students were randomly selected into a 24-student control group and a comparable 24-student treatment group. The treatment group devoted more than half of its class time to MEAs. Results demonstrated that the treatment group outperformed the comparable-but-traditionally-taught class by fully six standard deviations – a level of significance that is far beyond the 0.0001 level! Furthermore, a second phase of this study was conducted 1 year after the end of the course when, eight students were found in each of the two sections of the course who had taken no more statistics courses. When these 16 students were interviewed concerning what they remembered from their courses, the results showed that: (a) nearly all of the students in the MEA group were able to give impressively detailed descriptions of what they had learned from MEAs, whereas (b) among the students in the control group, recollections about the topics they had studied were exceedingly unimpressive.

Larger scale success stories have been reported in current and past ICTMA meetings and have been conducted in places ranging from Texas, to Cyprus, to Israel, to Australia, to Mexico (Lesh et al. 2010), and to a number of the USA's leading graduate schools of engineering (Zawojewski et al. 2009). Even though such demonstrations are gratifying, they are of little practical use unless we understand *when*, *where*, *why*, with *whom*, and for *what* purposes MEAs work. The probability of successful learning experiences is strongly influenced by all of these factors. For example, a recent study compared the results produced in two comparable statistics classes for graduate students in education. One class, the MEA-to-MXA group, emphasized "mathematizing reality" by using learning activities in which a MEA was followed by teacher-directed model exploration activities (MXAs). The other class, the MXA-to-MEA group, emphasized "realizing mathematics" by using learning activities in which a MXA was followed by a MEA. The exact same curriculum pieces were used in both classes. But, in the MEA-to-MXA class, MEAs were used before MXAs to challenge students to go through a series of iterative cycles in which they expressed, tested, and revised their current ways of thinking in order to develop interpretation systems that were powerful for the situation at hand, re-useable in other situations, and sharable with other people. In contrast, in the MXA-to-MEA class, the MEAs were used after MXAs and served as applications of concepts that were taught by the teacher during the MXAs. Results of the study showed that students in the MEA-to-MXA class not only significantly outperformed their counterparts in the MXA-to-MEA class, but also they demonstrated a mastery of many higher-order understandings about issues which seldom arose in the MXA-to-MEA group (Lesh et al. 2007). Thus, the impact of MEAs on learning depends not only on *whether* they are used, but also on *when* and *why*. The study also reaffirmed that, at the end of MEAs, a great deal more development still needs to occur. Nonetheless, because MEAs are designed to focus on the big ideas in any given course in which they are used and because MEAs are designed to be simulations of real life situations beyond school, the concepts which are embodied in the models that students produce often exhibit usual levels of durability and transferability.

Other studies have shown that many of the most important impacts of MEAs on learning result from second-order effects. For example, because MEAs are designed to be *thought-revealing activities*, and because insightfulness about students' thinking is one of the primary characteristics of effective teachers, MEAs contribute to teacher development which, in turn, encourages student development. When MEAs are used at the beginning of curriculum units, the kind of information that teachers see is similar to what they might have seen if they had been able to conduct clinical interviews with each of their students before starting the unit.

Beyond the preceding evidence documenting the successful use of MEAs when they are used for teaching and learning (rather than for research), another answer can be given for those who want to know do MEAs "work"? MEAs clearly satisfy even the most strongly specified definitions of problem-based learning (PBL). It is widely believed that a great deal of evidence has accumulated to show that PBL "works." To the extent that readers believe this literature, it surely is a straightforward inference that MEAs work. Unfortunately, we ourselves are highly skeptical of this literature, mainly because such studies vary widely concerning what is considered to be a "problem," how the "problem" is used to promote learning, and what kind of measures are used to assess success. Some PBL "problems" are more like "applications" which occur after traditional teaching activities. Others problems are more like highly guided project-sized collections of teacher-directed activities (similar to the manuals that accompany modern software packages such as *Photoshop* or *Dreamweaver*). In fact, in a recent review of the literature on PBL (Lesh and Caylor 2007), relatively few problems were found which engaged students in MEA-like experiences where they expressed, tested, and revised their own ways of thinking, rather than being guided along narrow paths toward idealized versions of teachers' and textbooks' ways of thinking. While some PBL lessons clearly were intended to use problem solving as a way for students to develop important mathematical constructs, many other PBL lessons simply used stories about real life decision-making situations as contexts in which quite traditional teacher-centered teaching was used. Still other PBL lessons were aimed at teaching problem solving itself, in ways that appeared to be no different than those that have failed in the previously described mathematics education research literature on problem solving. Nonetheless, even with all of the preceding ambiguity about what PBL really means and when, where, why, for whom, and in what ways it is intended to work, claims that it "works" seem to be generally accepted.

4 In What Ways Do MEAs (and MMP in General) Provide Alternatives to Traditional Research?

In a number of past publications, MMP researchers have described how MEAs can be used to provide alternatives or supplements to Piaget-style clinical interviews, teaching experiments, and video analyses. One significant difference between Piaget's research and MMP research is that Piaget deemphasized the importance of variations in students' thinking when they move from one task to another characterized by the same underlying structure. MMP research recognizes that (a) small changes in tasks often significantly change their level of difficulty, (b) students' thinking often varies significantly across tasks that Piagetians would think of as having the same structure, and (c) within MEAs, students' thinking often advances through several Piagetian stages during a single 60–90-min problem-solving episode. While Piaget referred to such variations as relatively uninteresting *decalages*, other researchers such as Vygotsky refer to them as *zones of proximal development*. In MMP research, we use MEAs to investigate students' thinking as it develops along a variety of dimensions such as concrete-abstract, preoperational-operational, intuition-formalization, situated-decontextualized, global-analytic, or specific-general.

4.1 MEAs Provide Alternatives to Naturalistic Observations

In MMP research, researchers are considered to be model developers who have characteristics similar to those that we attribute to students and teachers. If we recognize that students' initial interpretations of situations tend to be remarkably barren and distorted compared with later interpretations, then similar characteristics should be expected for researchers as well. For example, in early MMP research investigating the nature of mathematical thinking in real life situations beyond school, we became dissatisfied with our own naturalistic observations because the results seemed to be far too dependent on our own preconceived notions about *what kinds* of situations to observe (e.g., grocery shoppers, street vendors, and other ordinary folks, or business managers, aeronautical engineers, or medical decision-makers), *when* to observe (e.g., while the preceding people are planning, or monitoring, or explaining their work), or *what* to observe (e.g., calculations, deductions, or explanations). Hence, we developed a research methodology which recognized that decisions about who, when, where, and what to observe significantly influence results. We shifted toward MEAs which were designed to be *simulations of "real life" situations* where mathematical thinking is useful beyond schools (Lesh and Landau 1983). To design these activities, we enlisted help from people who were knowledgeable about new ways that mathematical thinking is needed beyond school in fields such as medicine, business management, or engineering. But, we did not simply accept without question the opinions that these experts expressed. Instead, we enlisted them to participate with us as "evolving experts" whose opinions were expected to change as they were expressed in forms that went through several iterative cycles of testing and revision.

We engaged a team of "evolving experts" to develop MEAs which they believed involved the kind of mathematical thinking to be increasingly important in the twenty-first century. As these teams worked together over a semester to develop a series of MEAs for middle school students, the thinking of the experts tended to evolve significantly. While their first-draft activities often focused on traditional basic skills and word problems, later drafts emphasized that, in the twenty-first century, people increasingly live in worlds where the conceptual systems that are developed to make sense of experiences also mold and shape those experiences through the development of tools and artifacts. Regardless of whether we investigate ordinary citizens or specialists in future-oriented sciences with hyphenated names that integrate a variety of traditional sciences, the abilities that are emerging as important include: (a) designing and describing complex systems, (b) working in teams of diverse specialists, (c) adapting to rapidly evolving tools for conceptualization and communication, and (d) functioning effectively within multi-stage projects where planning, monitoring, and assessing are nontrivial tasks. Furthermore, many such systems cannot be modeled using single one-way input–output functions that are differentiable and solvable. Instead, they often involve multiple interacting agents, feedback loops, second-order effects, and other characteristics associated with dynamic and continually adapting complex systems – where the kinds of issues that arise involve modularization, systematization, maximization, minimization, stabilization, and other issues which once required calculus to deal with effectively but which are now handled computationally and graphically (Lesh et al. 2007a).

4.2 MEAs Provide Alternatives to Expert-Novice Studies

In MMP research, we often investigate problem solvers who are in fact three-person teams of individuals. We often use results from this research to inform research on problem solvers who are individuals. We sometimes compare problem solvers who have access to powerful tools and resources with problem solvers who do not have access to such capability amplifiers. One reason why such comparisons have been productive is because, when we look beyond school in the twenty-first century, it is obvious that learning organizations and other learners-who-are-groups are important. But, a second reason is because many abilities that characterize productive problem-solving teams also characterize productive problem-solving individuals. Note that we are not assuming that there are not significant differences between individuals and groups or between problem solvers with tools and resources and problem solvers without tools and resources.

In MMP research on problem solving, when the subject is a team of problem solvers, it is generally obvious that, as a team, the problem solvers' early interpretations of problems typically consist of a hodge-podge of relatively fuzzy, undifferentiated, and unintegrated ways of thinking. Progress for *problem-solvers-who-are-groups* typically involves gradually clarifying, sorting out, and establishing connections among a variety of ways of thinking. Furthermore, the same tends to be true for problem-solvers-who-are-isolated-individuals. This is one reason why engineers and other design scientists often emphasize that realistic solutions to complex problems usually require the problem solver to integrate ideas and procedures drawn from more than a single textbook topic area or theory. Furthermore, real life problems tend to involve constraints. In fact, in design sciences such as engineering, experts often joke that theirs is a science of situations where you never have enough money, enough time, or enough of other resources and where multiple stakeholders emphasize partly conflicting criteria for success (e.g., low costs but high quality, or low risks but high gains). In problem solving which sorts out and integrates ideas and procedures from multiple disciplines, a "ballistic model" of learning and problem solving (where a single point moves along a single path) generally fails to capture the facts that (a) concepts tend to evolve along a variety of dimensions and (b) at every stage in the solution process, interpretations associated with a variety of textbook topic areas tend to be engaged. This means that, for most purposes, the kind of single-track cyclic diagram of modelling (where modelling cycles involve description, manipulation, prediction, and verification) represents a misleading conception of the way models develop, especially during early stages of development.

MEA-based evolving expert studies also suggest that the use of metacognitive processes and understandings depend on how situations are interpreted or conceptualized. In other words, they are associated with models; and, when students engage a given model (or interpretation) of a problem-solving situation, they do not just engage a logical/mathematical system, they also engage a variety of feelings, values, heuristics, strategies, and metacognitive processes and understandings. Instead of imagining that these metacognitive processes and understandings are learned in the abstract and then connected to specific concepts and contexts, MMP suggests that: (a) feelings, values, and metacognitive processes are learned as parts of the models that students develop, (b) the meanings of statements of beliefs (or values, or feelings) depend on systems of beliefs, and (c) the language that problem solvers develop to describe past problem-solving behaviors functions mainly to help elaborate and refine their interpretations and conceptualizations, rather than providing explicit rules specifying what to do next. In much the same way that athletes, performing artists, teachers, and others who are engaged in complex decision-making situations find it useful to watch videotapes of past performances, and to develop a metacognitive language to describe what they have done, this does not imply that they expect this language to function as a list of explicit rules which should be executed in upcoming performances. In fact, successful athletes and performing artists usually understand quite well that, during ongoing performances, it tends to be disruptive to pay attention to such explicit rules. In fact, during ongoing complex decision-making situations, both interpretation systems and the rules associated with them often function implicitly rather than explicitly; and, follow-up reflection activities function to develop implicitly and intuitively functioning interpretation systems, rather than functioning as devices to trigger explicitly functioning rules of behavior.

5 Teaching from a Models and Modelling Perspective

Our starting point for conceptualizing the knowledge that is needed to teach mathematics from a modelling perspective is that the distinguishing characteristics of excellent teaching are reflected in the richness of ways in which a teacher *interprets* her practice not only in the *actions* that she takes. This practice includes choosing appropriate modelling tasks for students; selecting activities and curricular materials that will foster the development of students' models over the course of several lessons; and devising strategies for engaging students in assessing and revising their models. Teachers have models for teaching! These models are what teachers use to see students' ways of thinking, to respond to students' ideas, to differentiate among the nuances of contexts in their practice, and to see generalized strategies that cut across contexts. These models are what a teacher uses to make sense of her practice and, as we shall describe below, includes both the ways in which the teacher *sees* her practice and what she *does* in practice. We wish to argue that a modelling perspective on teachers' knowledge suggests some ways in which the knowledge needed for teaching modelling goes beyond traditional and reform-based methods for teaching mathematics and provides some new ways for thinking about how teachers develop that knowledge. We will suggest some principles for the design of professional experiences that will elicit and foster the development of teachers' models.

5.1 The Nature of Teachers' Knowledge

What do teachers need to know in order to teach from a modelling perspective? Much current research in the USA, driven in large part by the NCTM's call for Standards-based teaching, is placing heavy emphasis on the kinds of mathematics that teachers need to know, especially at the elementary level and often referred to as the mathematical knowledge for teaching (Davis and Simmt 2006; Hill et al. 2008). From a modelling perspective, we would emphasize three interrelated characteristics of teachers' knowledge: (a) an understanding of mathematical content; (b) an understanding of the multiplicity of ways that students' thinking might develop; and (c) a knowledge of pedagogical strategies that can be drawn on in varying contexts to support the development of students' mathematical thinking. But these characteristics in and of themselves do not fully capture a more fundamental distinction in the nature of teachers' knowledge between the *actions* that teachers take in their practices and the ways in which they *see* and *interpret* their practices. We wish to argue that a distinguishing characteristic of excellent teaching is reflected in the richness of ways in which the teacher *interprets* her practice, not only in the *actions* that she takes. In 1976, Skemp made a key distinction about students' mathematical understanding, namely, that it is not only instrumental (the how) but also relational (the when and why). Other researchers (Baroody et al. 2007; Hiebert and Lefevre 1986; Star 2005) have described how both procedural knowledge and conceptual knowledge are of critical importance in students' learning of mathematics. There is much empirical evidence that students' procedural and conceptual knowledge are interdependent and develop iteratively, with an interplay from one to the other. Here, we wish to apply this same distinction to teachers' knowledge. That is to say, teachers' knowledge is both procedural and conceptual.

Much of the knowledge needed for teaching can be captured by the skills, processes, and guidelines that are used in action in the classroom. Teachers acquire and develop specific pedagogical procedures (or strategies) to deal with the very specific issues that arise for them in practice. Such procedures are specific responses to problem situations involving particular students, materials, and mathematical goals. But at the same time, some of these procedures (or strategies) can become more conceptually grounded as rationales for the strategies become articulated, as teachers modify and adapt particular strategies across a range of contexts and problem situations, and as strategies become shared among colleagues. Pedagogical strategies, refined through experimentation in practice and by interactions among teachers as the strategies are shared across contexts and problem situations, shift from being merely local procedures to becoming conceptually ground principles for seeing and responding to the tasks of teaching.

To illustrate the interplay between the procedural and conceptual aspects of teachers' knowledge, consider the following example from a modelling task in secondary mathematics. We will use this task to also illustrate two critically important aspects of teachers' knowledge from a modelling perspective: (1) Teachers need to know how to engage students with tasks that will evoke the development of significant mathematical models. (2) Teachers need to recognize and respond to the development of students' models (i.e., student learning) as it occurs.

5.1.1 Engaging Students with Modelling Tasks

Consider the following example from a pre-calculus classroom where students were investigating periodic motion, including the relationship between periodic position graphs and periodic velocity graphs. This particular task occurred about two-thirds of the way through a sequence of model development activities designed to support students' understanding of periodic motion. This particular task focused on the decomposition of the circular motion of a Ferris Wheel into its vertical and horizontal components. To engage students with this question, the teacher began the lesson with a physical model of a Ferris Wheel that she used to project the vertical motion onto to wall with a flashlight. The speed of the wheel and the radius of the wheel could be varied. While demonstrating the projected motion, the teacher asked for students' observations about the motion and, in particular, their sense of how the speed of the projected vertical motion changed as the wheel rotated. Following the demonstration, the teacher provided physical "toy" models that the students had at their desks. These toy models were a reference point for many (but not all) of the groups in their work in analyzing the vertical component of the circular motion of the wheel. As the students worked, they referred back and forth (in a nonlinear way) among the (a) physical model, which they had marked with a starting point, (b) their own sketches and diagrams of the wheel, and (c) their algebraic equations and calculations.

As a particular pedagogical strategy (or procedure), the teacher's use of a physical model to demonstrate the relationship between circular motion and the vertical component of that motion would seem like merely a rule for engaging the students with the task at hand. The teacher's rationale for this decision becomes more evident as we observed how she engaged the students with the physical device. The importance of physical models is as a support for students' sense-making activities of situations that are realistic for them (though not necessarily physically present). The teacher asked for students' observations about the motion as she varied both the speed of the device and the radius. She focused their attention on the crucial aspect of the phenomena, namely, how the speed of the projected vertical motion changed as the wheel rotated. This would build on their previous modelling tasks and apply their developing ideas in a next context. Finally, the "toy" Ferris Wheels, along with the demonstration wheel, provided representations that the students would translate into their own diagrams and sketches and could reason about in ways that made sense to them.

5.1.2 Recognizing and Responding to Students' Thinking

Teachers' knowledge also includes an understanding of the multiplicity of ways students' models might develop and the ability to respond to students' thinking as it occurs. That is, teachers need to be able to *recognize* and *respond to* model development (i.e., student learning) as it occurs. Following the demonstration of the Ferris Wheel and the students' observations about its motion, the students were given a "model application" task (see Fig. 26.1). This task followed a model-eliciting activity and model exploration tasks that had occurred earlier. The students' overall model development continues to occur in cycles that are more or less visible to the teacher.

One group of three students created a sketch (Fig. 26.2) and went through several cycles of modelling as they reasoned about the problem:

- *Cycle One*: The students assume that the starting position was at 3 o'clock. Thus, when $t=3$ s, the seat is at an angle of 45° above the hub, since $\frac{3}{24} = \frac{\theta}{360}$. This is correct proportional reasoning.
- *Cycle Two*: The students decide that the seat has to be 10 ft above the hub because the diameter is 80 ft and $\frac{45}{360} = \frac{x}{80}$. This is incorrect proportional reasoning, and perhaps an overgeneralization of the previous correct reasoning.

Suppose a Ferris Wheel with an 80 foot diameter makes one revolution every 24 seconds in a counterclockwise direction. The Ferris Wheel is built so that the lowest seat on the wheel is 10 feet off the ground. This particular Ferris Wheel has a boarding platform which is located at a height that is exactly level with the center (or hub) of the Ferris Wheel. You take your seat level with the hub as the ride begins. What is your height above the hub after 3 seconds?

Fig. 26.1 Model application task

Fig. 26.2 Student sketch of the Ferris Wheel

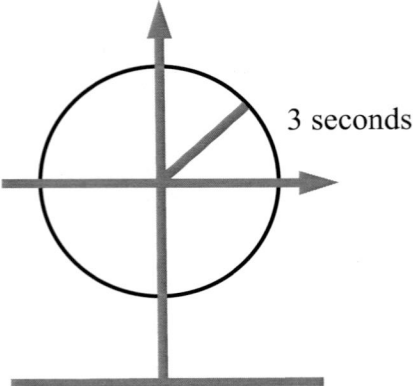

- *Cycle Three*: The students decide that the seat has to be 20 ft above the hub because the entire 80 ft of diameter is traversed as the wheel turns through half the circle or 180°. A turn of 360° would mean a vertical motion that was both up and down. Hence, the height should be $\frac{45}{180} = \frac{x}{80}$ which means $x=20$ ft. The students confirm that this seems reasonable on their physical model.
- *Cycle Four*: The teacher talks with the students about the units associated with the quantities that they are considering: feet, degrees, and seconds. Shortly after that conversation, the students shift to a new view, namely, $\sin(45) = \frac{x}{40}$ which means $x = 20\sqrt{2}$ or 28.28 ft.

We do not want to focus on the students' reasoning, but rather we want to use the distinction between procedural knowledge and conceptual knowledge to examine the teacher's knowledge. Several observations are to be made, based on the teacher's actions during the episode and on her comments on the lesson following the episode.

1. The teacher (as is her practice) seldom intervenes as long as "students are making progress." Through the first three cycles, the students worked quite independently, sorting out their own ideas and interpretations of the problem situation.
2. When the teacher did interact with the students, her interactions were focused on the *meaning* of the quantities that they had calculated. After the lesson, the teacher observed that she thought that "it was the focus on units" that helped the students to move to thinking about using a trigonometric ratio.
3. Another group of students had assumed that the starting position was at 9 o'clock and had calculated a position of $y = -20\sqrt{2}$ or -28.28 ft. The teacher did not encourage the students to shift their reasoning to the 3 o'clock position. After the lesson, she explained that "this will make for good discussion" when the students later share their solutions.

These observations provide some insight into both the procedural and conceptual knowledge of the teacher. The notion that one should "let" students engage in working through their own ideas – expressing them and revising them – is a common one among the teachers we have worked with. Teachers often articulate this as "just let them work," "let them struggle," or "do not capitulate and give them the answers." Such a guideline can be a useful heuristic for making decisions as students are working on modelling tasks. But underlying this pedagogical strategy is a conceptually based rationale for when and why this "procedure" is useful. For example, in this case, the teacher chose a problem that the students were able to productively work on. As described earlier, this problem was introduced via a physical model that students were able to make sense of (the Reality Principle from MEAs). The representational space for working on this task was relatively rich: an actual physical model, diagrams and sketches made by students, and algebraic equations and computations (the Documentation Principle from MEAs). This provided several mediums for the students to express their ideas. Finally, the task was one where students were able to self-evaluate whether their answers were reasonable or made sense (the Self-Evaluation Principle from MEAs). This (along with the various representations) enabled the students themselves to revise and refine their ideas. The teacher's understanding of these characteristics of a good task, along with her own understanding of how students might engage with the task, is an aspect of the teacher's conceptual understanding that is a part of the interpretive scheme that enables the teacher to use this heuristic in ways that support student engagement with the task and their learning.

Similarly, we consider the reasoning that underlies the guideline for "intervening" by asking students about the meaning of their work. This question will often prompt student's to give explanations that will move their thinking forward – by clarifying their representations, by identifying a mismatch between representations, by pointing to an erroneous assumption about the problem, or by suggesting an alternative perspective. From the teacher's point of view, this question requires much more than a procedural response. The teacher needs the ability to follow students' reasoning when often that reasoning is only partially developed. Listening to students' emerging thinking is not an easy task for the teacher (Davis 1997; Doerr 2006; Wallach and Even 2005). As students develop their ideas, the paths that students take are less like particular learning trajectories and more like "meandering" or "roving" over the problem space of a system (Davis and Simmt 2003). The practice of asking the students about the meaning of their work signals an important departure from a pedagogy of problem solving oriented along known solution paths. Instead of evaluating students' work and guiding their movement along known paths, the teacher's role is recast as of one of engaging students in the self-evaluation of their work and encouraging them to revise their thinking in ways that make sense to them (Doerr 2006).

When students work through the kinds of model development sequences described earlier, most often the process produces a diversity of ideas. From the teachers' perspective, encouraging a diversity of ideas will "make for good discussion." Discussing students' solutions to modelling tasks is a useful heuristic (or procedure). The underlying rationales for such discussion become visible as teachers use these discussions as opportunities for students to sort, select, and compare ideas. For example, in the case of the Ferris Wheel modelling task, the variation in assumptions about the starting position affords an opportunity to discuss horizontal shifts and position as a signed quantity. At other times, the discussion can lead students to have to reconcile apparently different symbolic forms describing the same phenomena. Students' partial solutions can lead to fruitful discussion about "dead ends" or possible next steps. Teachers need the ability to recognize productive paths versus less useful paths, to engage the students in sorting out more useful ideas from those that are less useful, and to support students in making connections between ideas. If we understand learning as a complex interactive system, then the presence of a diversity of approaches is necessary, since one cannot generally specify in advance what sorts of variation will be necessary for appropriate action. The conceptual knowledge of the teacher, in this case, is about how she sees and interprets the variation of student ideas and then uses this variation to drive learning. Because the variation cannot be fully anticipated in advance, the conceptual demands on the teacher are substantial.

Teachers' models are significantly more complex and broader in scope than students' models. As we have illustrated above, teachers' models certainly include rules or heuristics that provide guidance for acting in the classroom. But teachers' models are also conceptual principles grounded in rationales and interpretations of classroom events. Teachers' models include descriptions of students' ideas, ways of engaging students in describing their own ideas, abilities to plan and sequence tasks, knowing useful symbolizations in various contexts, highlighting underlying mathematical structure, supporting students in making connections to other ideas, fostering productive use of diverse ideas, and ways of engaging students in self-assessment.

5.2 The Development of Teachers' Knowledge

How do teachers acquire the expertise that is needed for teaching? In other words, how do teachers' models develop? This question focuses our attention not on the static picture of what teachers know, but on the dynamics of how teachers' models or systems of interpretation develop over time. Here we find it useful to think of teachers as "evolving experts" who work within a complex system. Is teaching a complex system or is it merely complicated? The following characteristics of teaching and of classrooms give credence to an argument for considering teaching as a complex system: (a) There are multiple interacting agents. (b) Cause and effect are often distant from each other in time and often inadequate for explaining learning phenomena. (c) Teaching is goal directed, but often goals are changing or teachers must address conflicting goals. (d) Teachers must make decisions based on partial and incomplete information. (e) Teachers must make decisions in constrained environments and consequences may be unclear. Expertise in a complex domain requires the flexible use of cognitive structures to accommodate partial information, changing or unclear goals, multiple perspectives, and uncertain consequences (Feltovich et al. 1997). This perspective on teachers' knowledge suggests that expertise in teaching does not conform to a single, uniform image of a "good" teacher. Rather, expertise in teaching is highly variable, both across and within individuals and across multiple settings. Hence, teaching needs to be viewed as evolving expertise that grows and develops along multiple dimensions in varying contexts for particular purposes. The central problem, then, is to understand the kinds of experiences that are likely to promote this growth and development in ways that are continually better.

As teachers draw on their knowledge in the acts of teaching, this is a potential site for teachers' continued learning (or model development). Eraut (1994) argues that the predominant context for teacher learning is the classroom and that classroom instruction is characterized by "hot action" – it is "a place where teachers must decide courses of action quickly with minimal time to reflect on past knowledge or memory and in a profession/craft filled with ambiguity and uncertainty.

Teachers operating in hot action contexts rely proportionately more on personal knowledge – that is, knowledge gained through experience, often in isolation, and routinized into tacit behaviors" (p. 242). Needing to act in contexts that are ambiguous and uncertain (that is, in a complex system) leads to the development of routinized, tacit behaviors (what we have been calling procedures). One way to promote the development of teachers' models is to make these routinized behaviors "visible" to both teachers and researchers and subject to the kind of scrutiny that will lead to adaptation and transferability. In other words, as strategies become shared, revised, and reused, pedagogical "procedures" move from being merely strategies that can be used in a particular setting to pedagogical "concepts" that provide teachers with increasingly sophisticated ways of interpreting and responding to classroom events. We will illustrate this with a brief example involving the collaborative design of sequences of modelling tasks and then suggest some principles for the design of experiences for teachers that promote the development of their models for teaching mathematics.

5.2.1 The Collaborative Design of Sequences of Modelling Tasks

The collaborative design and revision of lessons has long been the foundation of Japanese lesson study and is currently being adapted and studied in American classrooms (Lewis et al. 2006) as a means of improving instructional practice. Certainly, the design of the lesson is central to a teacher's practice and, as such, is a potential site for teachers' learning. Key to lesson study is the implicit assumption that engaging teachers in the collaborative design of a lesson with a clear goal or "end in view" and in cycles of teaching and revising the lesson will help teachers to grow professionally (Zawojewski et al. 2008). Such cycles are likely to lead to changes both in the goals of the lesson and in the teachers' ways of thinking about the lesson. In other words, this approach to teacher development mirrors a complex system that changes over time as artifacts (in this case a lesson) are being created. We illustrate this dynamic with a brief example from two pre-service teachers who were engaged in collaboratively designing a sequence of four modelling tasks for secondary school students.

The pre-service teachers went through multiple cycles of design in developing their sequences of modelling tasks. They began by identifying the goals of each sequence. Each of the pre-service teachers then worked on two of the sequences, with multiple iterations as they shared these sequences with each other and offered critique and suggestions. They created brief "teacher notes" for each sequence and provided a written summary that described their reasoning for the decisions they made in designing the tasks. Following this, there was a brief discussion about the process of developing the sequence of tasks. The "teacher notes" captured the pedagogical strategies (or procedures) that the pre-service teachers envisioned for the tasks; the cycles of revision and their written rationales captured much of the underlying principles (or conceptual basis) of their developing thinking about tasks for learners.

This collaborative process illuminates three points about the development of the pre-service teachers' thinking:

(a) The design of the tasks made visible the pre-service teachers' thinking about the goals for students' learning. One pre-service teacher described the difficulties in seeing the tasks from a teaching and a learning perspective: "It's hard trying to think from multiple perspectives. Because you're thinking about what you want as a teacher, but then you're [thinking] about...what the main goal is and how you're going to get there. And then you're thinking about it from the student's perspective. Is this going to make sense? Are they going to get it? Is it going to be worthwhile? Is it worthwhile for them to get there?" The pre-service teachers also struggled with identifying the knowledge that students would bring to the tasks: "We didn't exactly know where the students would be at each level. So in some ways, we just made assumptions about where we thought they should be." To address this difficulty, the pre-service teachers designed "pre-labs" that were intended to motivate the tasks, but also to "bring out some different ideas" and to "have an actual discussion."

(b) The revisions of the tasks led to new goals for the pre-service teachers that would help them to continue their professional learning (or model development). After several cycles of revision, the pre-service teachers realized that they would need to teach the tasks before they could become any better. This became a new goal and a new insight for the pre-service teachers: They now saw that the tasks themselves were structured so that they could learn from the teaching of the tasks. As one pre-service teacher observed "I think it's an on-going process to revise these." The other observed that by examining students' written work, she will be able to "see they didn't get much out of this question. Or they didn't get anything. Or wow look at these answers! They're really rich. So these [questions] are really good. I want to keep these. Maybe I should reword these and completely toss these." The learning from the tasks now included both the students' learning and the pre-service teachers' learning. One of the pre-service teachers described this dual aspect of the learning: "I can get even more out of this [set of tasks] when I get to [teach] them. Will they work the way that I want them to work? Or the way that I'm thinking that they're going to work? And are the students really going to get something out of writing? So I think that there's more that is still going to come from what am I going to get out of these. Or what am I going to learn from them. I think that these are a really good [way] to learn from the students. So a lot [more] can come out of it." The pre-service teachers now see that the tasks have been designed in a way that will enable them to continue to revise the tasks and enable them to learn from teaching the tasks.

(c) The pre-service teachers' interactions with each other revealed their thinking in ways that led to the revision and redesign of the modelling tasks for students. When describing their interactions, these pre-service teachers highlighted how they asked critical questions of each other, rather than "just scanning the thing." As one of the pre-service teachers commented, they pressed each other to think through how students might respond and the intentionality behind their questions: "How to really work through what's going to work and what's not going to work. [We asked each other] how do you think students would respond? ... What are you trying to get at here?...Is that what we're really trying to get at here? Just going through everything with a critical eye." Along with this critical eye, the pre-service teachers pointed to the value of designing for generalizations. By designing for generalizations, the pre-service teachers felt that they were better able to understand the connections between ideas. One of the pre-service teachers said that this process "helps you connect different units. This is a big idea here, but how does it link to the previous big idea? How does it link to the next big idea? And the big ideas of the course?" In this way, the pre-service teachers were attending to the development of students' thinking across tasks and instructional units. These pre-service teachers are beginning to see teaching as emphasizing students' abilities to make adaptations to existing concepts in ways that will support the development of those concepts across instructional tasks.

5.3 *Principles for the Design of Model Development Experiences for Teachers*

A key assumption from a modelling perspective on the development of teachers' knowledge is that teachers must be provided with opportunities to develop their own models (or systems of interpretation). These opportunities need to be organized around experiences that engage teachers in expressing their current ways of thinking, such as making explicit the implicit routines of practice or providing rationales for particular instructional strategies. Once current ways of thinking are revealed, teachers need to engage in multiple cycles of testing and revising those ways of thinking in particular contexts for specific goals and sharing their ideas with colleagues for replication and reuse in multiple contexts. By designing model development experiences for teachers that are embedded in the complex settings of schools, researchers can engage with the complexity and the dynamics of the system in which teachers (and students) learn. The following three principles are intended as guidelines for the design of model development experiences for teachers.

1. Diversity is a source of possible interpretations and alternative responses to classroom events. A modelling perspective forefronts the variability in the abilities, the pedagogical strategies, and interpretations that teachers bring to their practices. Not all teachers begin at the same "starting point" nor do they all follow pre-determined paths for learning to interpret and analyze the complex and ill-structured domain of practice. Teachers vary in their skills, knowledge, and growth in ways that are contextualized and highly variable. By bringing together multiple perspectives on particular teaching situations, teachers can notice more about the details of the situation, can use considerably different schemes to think about the important patterns and regularities, and can act differently in the situation. Encountering variation can shift the teachers' focus from the details to the big picture, from isolated elements in a situation to interacting relationships, or from particular events to more generalized relationships. Model development experiences for teachers should provide opportunities for teachers to express alternative perspectives on particular teaching situations. This juxtaposition of multiple perspectives is a source of new ideas and can lead to the revision of teachers' systems of interpretation.
2. Redundancy is the complement of diversity with its emphasis on differences. Among any group of learners, there must also be some similarities or commonalities in background, vocabulary, experiences, and purpose. These commonalities support the communication of ideas. Strategies and concepts about teaching and learning need to be shared among multiple teachers and reused in other contexts that require reinterpretation and further analysis. Particularly powerful strategies and concepts for teachers are those that come from other teachers and can be used in one's own practice. Model development experiences for teachers must provide opportunities for teachers to share and reuse ideas in multiple contexts. This leads to more flexible ways of reasoning about practice and to potentially more generalized ways of interpreting their practices.
3. In any complex system, agents or neighbours interact with one another. Here, we draw on Davis and Simmt who point out that in the context of teaching and learning mathematics, "these neighbors that must 'bump' against one another are ideas, hunches, queries, and other manners of representations." (2003, p. 156). Model development experiences for teaching must make ideas visible so that they can bump against one another. This interaction of ideas can lead to testing and revising teachers' systems of interpretation.

6 Concluding Remarks

We began by arguing that a critical feature of teachers' knowledge is how they see and interpret the teaching and learning of students. Teachers' models are characterized by a complexity that exceeds that which we usually associate with students' models. Teachers' models include both pedagogical strategies that teachers use to respond to students' thinking and conceptual systems that teachers use to design instructional sequences, to interpret students' solutions, to draw on useful symbolizations in various contexts, to highlight mathematical structure, and to engage students in representing and evaluating their own ideas. When students are engaged in modelling activities, teachers are likely to encounter substantial diversity in student thinking. This places new demands on teachers for listening to students, responding with useful representations, hearing unexpected approaches, and making connections to other mathematical ideas. At the same time, such classroom settings potentially provide an opportunity for teachers' models to develop. We have suggested several principles for the design of experiences for teachers that can foster this development by focusing attention on teachers' systems for interpreting their practices. Much practical work remains to be done that meets the substantial challenges of revealing teachers' thinking so that it can be tested, revised, and shared in ways that will contribute to a growing body of knowledge about teaching and learning in the twenty-first century.

References

Baroody, A. J., Feil, Y., & Johnson, A. R. (2007). An alternative reconceptualization of procedural and conceptual knowledge. *Journal for Research in Mathematics Education, 38*(2), 115–131.
Davis, B. (1997). Listening for differences: An evolving conception of mathematics teaching. *Journal for Research in Mathematics Education, 28*(3), 355–376.
Davis, B., & Simmt, E. (2003). Understanding learning systems: Mathematics education and complexity science. *Journal for Research in Mathematics Education, 34*(2), 137–167.
Davis, B., & Simmt, E. (2006). Mathematics-for-teaching: An ongoing investigation of the mathematics that teachers (need to) know. *Educational Studies in Mathematics, 61*, 93–319.
Doerr, H. M. (2006). Examining the tasks of teaching when using students' mathematical thinking. *Educational Studies in Mathematics, 62*(1), 3–24.
Doerr, H. M., & English, L. D. (2003). A modeling perspective on students' mathematical reasoning about data. *Journal for Research in Mathematics Education, 34*(2), 110–136.
Eraut, M. (1994). *Developing professional knowledge and competence*. London: Falmer.
Feltovich, P. J., Spiro, R. J., & Coulson, R. L. (1997). Issues of expert flexibility in contexts characterized by complexity and change. In P. J. Feltovich, K. M. Ford, & R. R. Hoffman (Eds.), *Expertise in context: Human and machine* (pp. 125–146). Cambridge: AAAI/MIT Press.
Hiebert, J., & Lefevre, P. (1986). Procedural and conceptual knowledge. In J. Hiebert (Ed.), *Conceptual and procedural knowledge: The case of mathematics* (pp. 1–27). Hillsdale: Erlbaum.

Hill, H. C., Ball, D. L., & Schilling, S. G. (2008). Unpacking pedagogical content knowledge: Conceptualizing and measuring teachers' topic-specific knowledge of students. *Journal for Research in Mathematics Education, 39*(4), 372–400.

Kelly, A., & Lesh, R. (Eds.). (2000). *Handbook of research design in mathematics and science education.* Mahwah: Lawrence Erlbaum Associates.

Kelly, A. E., Lesh, R. A., & Baek, J. Y. (Eds.). (2008). *Handbook of design research methods in education: Innovations in science, technology, engineering and mathematics learning and teaching.* New York: Routledge.

Lesh, R. (2002). Research design in mathematics education: Focusing on design experiments. In L. English (Ed.), *Handbook of international research in mathematics education* (pp. 27–49). Mahwah: Lawrence Erlbaum Associates.

Lesh, R., & Caylor, E. (2007). Modeling as application versus modeling as a way to create mathematics. *International Journal of Computers for Mathematical Learning, 12,* 173–194.

Lesh, R. A., & Doerr, H. M. (Eds.). (2003). *Beyond constructivism: Models and modeling perspectives on mathematics problem solving, learning, and teaching.* Mahwah: Lawrence Erlbaum Associates.

Lesh, R., & Harel, G. (2003). Problem solving, modeling and local conceptual development. *Mathematical Thinking and Learning, 5*(2–3), 157–189.

Lesh, R., & Lamon, S. (Eds.). (1992). *Assessment of authentic performance in school mathematics.* Washington: American Association for the Advancement of Science.

Lesh, R., & Landau, M. (1983). *Acquisitions of mathematics concepts and processes.* New York: Academic.

Lesh, R., & Zawojewski, J. S. (2007). Problem solving and modeling. In F. K. Lester (Ed.), *Second handbook of research on mathematics teaching and learning* (pp. 763–804). Charlotte: Information Age Publishing.

Lesh, R., Hamilton, E., & Kaput, J. (Eds.). (2007a). *Foundations for the future in mathematics education.* Mahwah: Lawrence Erlbaum Associates.

Lesh, R., Yoon, C., & Zawojewski, J. (2007b). John Dewey revisited – Making mathematics practical versus making practice mathematical. In R. Lesh, R. Hamilton, & J. Kaput (Eds.), *Foundations for the future in mathematics education* (pp. 313–349). Mahwah: Lawrence Erlbaum Associates.

Lesh, R., Galbraith, P., Haines, C., & Hurford, A. (Eds.). (2010). *Modeling students mathematical modeling competencies.* New York: Springer.

Lewis, C., Perry, R., & Murata, A. (2006). How should research contribute to instructional improvement? The case of lesson study. *Educational Researcher, 35*(3), 3–14.

Skemp, R. R. (1976). Relational understanding and instrumental understanding. *Mathematics Teaching, 77,* 20–26.

Star, J. R. (2005). Reconceptualizing procedural knowledge. *Journal for Research in Mathematics Education, 36,* 401–411.

Wallach, T., & Even, R. (2005). Hearing students: The complexity of understanding what they are saying, showing, and doing. *Journal of Mathematics Teacher Education, 8*(5), 393–417.

Zawojewski, J., Chamberlin, M., Hjalmarson, M., & Lewis, C. (2008). Developing design studies in mathematics education professional development: Studying teachers' interpretive systems. In A. E. Kelly, R. A. Lesh, & J. Y. Baek (Eds.), *Handbook of design research methods in education: Innovations in science, technology, engineering, and mathematics learning and teaching* (pp. 216–245). New York: Routledge.

Zawojewski, J. S., Diefes-Dux, H., & Bowman, K. (Eds.). (2009). *Models and modeling in engineering education: Designing experiences for all students.* Rotterdam: Sense.

Chapter 27
Mathematical Modelling in a Distance Course for Teachers

Maria Salett Biembengut and Thaís Mariane Biembengut Faria

Abstract In this chapter, we present the principal results of a research project in which empirical data was obtained from a distance course of mathematical modelling (MM) to teachers and students of mathematics teacher-training. The objective was to understand the limitations and possibilities that a distance course of MM offers. Based on this understanding, the goal was to reorient the project's proposals and actions with the expectations to make MM effective for Mathematics Education at any scholarly level. In preparation, we needed instructional materials and all the educational elements available on the Web site in order to use it to teach. We obtained data from interviews, observations, and participants' questions and difficulties. Indicators of the participants' difficulties were categorized as concerning teacher's education or the necessity of teacher's education.

1 Introduction

The Brazilian government has set up National Curricular Lines for courses to prepare elementary education teachers so that courses that give students better knowledge about the socio-cultural context have an appropriate orientation within the curriculum. The National Curricular Lines seek to promote the development and understanding of mathematics and the integration of mathematics into other areas of knowledge, such as mathematical modelling.

According to the government data, there are 413 preparatory courses for training mathematics teachers in Brazil; of these, the authors have identified (to March 2010) about 30% that include modelling in the curriculum. In courses in which Mathematical Modelling (MM) is included, the teacher responsible for instruction in this area should know how to model mathematically in topics from a wide area

---

M.S. Biembengut (✉) and T.M.B. Faria
Department of Mathematics, Universidade Regional de Blumenau, Blumenau, Brazil
e-mail: salett@furb.br; thaismarianeb@gmail.com

of knowledge, and also he or she should know how to adapt such mathematical models to other school phases. Furthermore, these courses prepare mathematics teachers from the same region to serve a significant number of children and young people from different socio-cultural backgrounds who need to receive sufficient general instruction relevant to their own situations.

Thus, there is a real need for teachers to learn modelling. With the possibilities offered by technologies that facilitate distance courses, it is now possible to address this demand. For Mathematics teachers to be able to use MM in their practice in the classroom, with the purpose that their students learn the art of modelling situations from some area of knowledge, similarly, a distance course needs to have processes and methods to achieve this purpose. That is, it has to have a distance system of teaching modelling that allows the reorientation of the education and the understanding of the participants using the actual structural conditions available (physical space, human, and technological resources) (Biembengut 2007).

It is not sufficient to promote courses of continuing professional education without knowing the potential for the teachers to improve or change their pedagogical practice. This research aims at understanding the limitations and possibilities offered by a continuous professional education distance course in Mathematical Modelling for teachers. Based on this understanding, the goal is to reorient the proposals and actions with the expectation of making MM effective for mathematics education at all educational levels.

The methodology of this research took into account the literature on teacher education and the empirical understanding of a group of 29 mathematics teachers. We sought to identify, describe, and analyze the difficulties and advances during the MM course. These goals were accomplished in two phases: (a) *organization of the course* and (b) *analytical procedures*.

(a) *Organization of the course*: This phase involved: production of the didactic material, elaboration of tools to collect data, classification of data, organization of the virtual structure, and planning. It was necessary to form a pedagogical and a technical team. The aims of this course were to teach the participants how to use Modelling both in research and in mathematics education to achieve such a goal, we explained and discussed material of didactic support that is organised in four phases: (1) to formulate and solve two situational-problems for which data had to be collected, (2) to verify, experimentally, two classical models of differential equations (restricted growing and liquid cooling), (3) to model a topic of interest to the teachers, and (4) to adapt the modelling performed, for the teaching of mathematics appropriate to a particular school year. The 40-hour course was divided into eight simultaneous meetings over 60 days (September–November, 2008).

(b) *Analytical procedures*: The research data were collected from the observation and description of actions and events during the course, from the work done by the participants and from the interviews given by them about the validity of the course and about the possibilities and difficulties experienced in making modelling a pedagogical practice. The events during the course raised two issues:

interest in the proposal and the need to learn modelling to use in their practice. To better understand these events, we have searched the literature on interest and need for conceptions and definitions. From these theoretical concepts, it was possible to understand the underlying structures that are expressed, often hidden from the educational realities of the participants, who voluntarily participated in this research.

2 MM Course: Main and Common Events

The course was developed in four phases, in increasing order of difficulty, with the intention of guiding the participants how to carry out research and, once that was accomplished, the aim was to make them know how to implement this process in classroom practice. As the issues, doubts, and advances from the participants were similar, we opted to describe these in conjunction with the main and most common events, making general comparisons, but not ignoring the qualitative aspects of reality and the differences in learning and in motivation of each participant. What follows is a summary of the main participant events during the course, the two main difficulties, and the possibilities.

Introductory class: Through videoconferencing, everybody was present and in doing so their motivation and interest was confirmed. Only 5 of the 29 participants had prior knowledge about the material and the many and various instructions regarding the topic and had downloaded or printed the material in preparation for the course. According to Dewey (1922), interest is dynamic, compelling the participant to action, and has the goal itself in some object or finality and, still, it provides the means to internal achievement or gives an important feeling. Herbart (1806) said that interest permeates the being as a spectator of the facts and interacting with them; it is attached to the image, to the relationships. We can say that interest for most of the participants was attached to "being a spectator" of the facts and of the proposals to be presented.

First Phase: Mathematics and language: In this phase, the two activities aimed at inspiring the participants regarding the researcher spirit: the will to collect data, to decide on the mathematical language to be used in the formulation and solution, and to analyze the validity of the solution found. None of the tasks demanded mathematics aside from that taught in the Elementary School. From the first moment, many of the participants did not understand the task; they expected data, despite the given set of suggestions and orientations about how to deal with the data. Their lack of understanding indicated how deeply these participants were rooted in the way of teaching they had lived through in their school lives, accomplishing the task according to "what the teachers ask for" and not because it is desired or necessary to know.

In this phase, the participants were faced with two open tasks, requiring them to search for data and then formulate, solve, and decide the best way of creating a parking lot and packaging for kitchen oil. They stated that they had difficulties not only reading and attending to the prescribed task descriptions but also in remembering from their previous education the mathematical knowledge demanded for the solution, despite the fact that the data to be collected were basic content, an integral part of everyday teaching. This difficulty, faced by most of the participants, influenced their motivation and their interest in the course: 12 of the 29 participants gave up, citing lack of time to accomplish the tasks as an excuse. Learning depends on a set of reasons for each individual person, such as perception and interpretation according to the context, to the interest, and especially to the need. The interest can be created through the process of teaching, but, it is the need of having specific knowledge that truly engages the person in learning.

Second Phase: Classical mathematical models: In this phase, a brief explanation was made about two classical models of differential equations: one, about the restricted growing of a live organism; and another, about the cooling of a liquid or Newton's Law of Cooling. These models are included in the Integral Differential Calculus (IDC) course for mathematics teachers. In general, in this subject, the teacher first presents the classical models and then suggests the students solve problems applying the data to the models (mathematics formula). The application is done, many times, in a mechanical way, and the students do not evaluate the validity of the result. The experiences should allow them to establish the constants of the respective classical models and check the validity of the experiments made. As in the two activities of the first phase, the participants were given a set of task descriptions indicating how to proceed. The experimental data was obtained without difficulty by all the participants of both groups. The difficulty occurred in the formulation: knowing how to obtain the constants from the data.

Even though the participants were teachers with a good understanding of Calculus (IDC), among other content, they did not know how to use this "knowledge" and did not have the skill needed to make use of the learned mathematics in their own school situation. This demonstrates that many education courses for teachers still do not provide consistent and broad preparation for future teachers that allows them to use effectively at least part of the content developed in such courses. Of our group of participant teachers, three more gave up, citing lack of time to accomplish the tasks, which left the course with only half of the participants originally enrolled.

Third Phase: Mathematical modelling: In this phase, we asked the participants to work in pairs and choose a theme of their own interest and empathy and construct a mathematical model. This request generated many questions, among them: What would be a good choice as a theme of interest? This question showed that they were neither acquainted with, nor did they understand the preliminary text about modelling that was in the instructional material. This condition – that they must be able to do modelling in order to teach mathematics through modelling – contributed to five more participants giving up without trying to accomplish the task. According to Claparède (1958), to make a person act, it is necessary that he or she is in proper condition for the rising of a need, and that it raises his or her interest in satisfying that need. Thus, it is assumed that the need to "learn to teach" was remote according to the value that was attributed by these five participants and that the "need" was not recognized by them.

Fourth Phase: Mathematics integrating modelling: In this phase, the participants were asked to adapt their work in MM for teaching; that is, they had to prepare instructional materials using the MM process for teaching. For these participants to have their own instructional material, more or less according to the reality of their students or future students, we asked them to adapt their modelling work to some appropriate group and school year and to some programmed mathematical content. Most students/participants did not see the need to know how to create their own instructional material, saying that these materials already existed in schools: The topic is in the instructional book from the beginning to the end, and it is just necessary to present it. We realized again that school experience had overwhelmed them, making it difficult to change the teachers' ways. After having prepared the materials, many teachers still looked for a good instructional textbook; in general, they searched out the same type of textbook that was used in their own education, reproducing in their own teaching the same procedures and content that they experienced as students (Bonotto 2007). Despite this tendency toward teaching as they had been taught, the nine participants finished the task.

At the end of the course, we sent a questionnaire to all participants originally enrolled. They were asked about the course organization, their performance, and the validity of the course. Most of those who dropped out answered, and we received responses from 26 of the 29 enrolled. In relation to the answers about their performance and the validity of the course in continuing professional education, two distinct aspects were evident: one concerned the teacher's education and, the other concerned the necessity for teacher education.

2.1 Modelling in the Classroom: Possibilities and Challenges

The student's comments on the incidents during the course and their inability to complete the MM course, the tasks accomplished by them, and their statements about the validity and viability of the course all indicate a potential for mathematical learning through modelling, but, there are challenges to using MM in pedagogical practice. We found ourselves in agreement with their statements: The process of modelling contributed to the stimulation of the perception and understanding of the concepts. These participants tried to design the task so that they understood what was possible and imagined an environment around them. The skills demanded from the participants were in effect tools in this process.

Although MM is an important method of teaching, some aspects should be catiously emphasized, bearing in mind the limitations of the educational structures within which the teacher works and the students learn. We are currently having an educational structure with curricula across many disciplines; each discipline is under the responsibility of a teacher who is also under the restrictions of schedules and periods to accomplish each phase, all of which contribute to the main difficulty of turning MM into a method of teaching and learning in the classroom.

The events that were indicators of the participants' difficulties (teachers and future teachers), were put into two categories: participants' education and necessity for education. The necessity can be intrinsically related to a personal interest, or extrinsically related to surviving in one's difficult environment. In either case, time, availability, and planning are involved.

Participants' education: One of the main problems of the Brazilian school system structure, from the Elementary to the Superior, is that few students learn how to carry out research and they are not held to be responsible for their own learning, except in isolated cases. The structure makes the teacher alone responsible for the student's learning. In this scenario, many students assume their physical space, transfer the content somewhere, and answer the questions or exercises that the teacher presents only if they are asked to take a test for a specific evaluation and are given a grade. When a student has a question or a doubt, he or she expresses it during a class and, in general, it is answered at the same moment. Rarely, is he or she encouraged to search for an answer independently, in a kind of "doing to know" and "knowing how to do it", as suggested by Maturana and Varela (2001).

Most of the participants in this course, even being teachers, assumed the same position as the students: waiting for the teacher in this case to tell them what, how, and which results he or she "would like" to receive. As this was their first experience in a distance course, related to the time experienced in the actual structure, most expected to be given the teacher's direction in each phase, even for something simple. In the process involved in the distance course, auto-didacticism is fundamental. The times and spaces are diverse. If, from one point of view, it is easier for the person who is taking the course in this mode to do the activities according to his or her availability, from another point of view, his or her questions or doubts are answered according to the availability of the instructional team.

All the participants reported: They worked for at least two periods of time, and had little time to study and to do the activities involved in the course. The restricted time to accomplish the tasks and the explanations, plus the questions and doubts raised, contributed to raising their interest in many cases in continuing with the course. In this sense, interest is a kind of feeling that prevents the action. It comes from the different circumstances that one person faces and that incites him or her to give an answer; interest develops through observation and is associated with conceptualization: with its contrasts and interconnections. "The interest only transcends the simple perception, by the fact that in it the thing observed conquers preferably the spirit and it imposes certain causality among other representations" (Herbart 1806, p. 73). Thus, the participants of this course, when seeing themselves hindered by the time and by their own school experience, were motivated to revise their needs for learning, based on the same interest that took them to participate in this experience in the first place.

Necessity for education: The act of learning depends on the interest and the necessity of the person involved. Most of all, it demands from the person: diligence, discipline, and perseverance. According to Habermas (1987), knowledge is found at the peak of development through the life of a person and is part of his or her process of being human. The activities undertaken come from his or her interests and needs, "The interest implies a need, or then the interest generates a need" (Habermas 1987, p. 220). According to Claparède (1958), every human being tends to keep stable until something disturbs his or her interior balance and promotes necessary acts leading to his or her own reconstruction. It is about "a continuous readjustment of a balance perpetually broken"; a search to reach "an objective and not to vanish the needs that show up" (Claparède 1958, p. 40). In the functional perspective of larger relevance, defended by Claparède, it is the need that makes human beings move; it is this that is the interior stimulus for doing the activities.

There are two kinds of needs: primary – vital and secondary – motivational, happening according to the primary needs derived from them. That is, intrinsic need is instigated by physical survival, and extrinsic need comes from some requirement and demands that it is accomplished because of survival (physical, professional, familial, social). If the interest is involved with a feeling, or a wish, it comes from one need, and when the interest is free from feelings, it provokes a need (Habermas 1987). The need depends on the action or the experience that is produced on the outside in a kind of reflection. In general, it depends on the goals, the interests, and the experiences of each person. The interaction of each person with the environment in which each is surrounded promotes the need that provokes the action with the intention of completing it or gaining balance; from this action comes learning and, as a consequence, it helps the formation.

The personal life, particularly in this type of virtual time, is found to be multifaceted in occupations of multiple interests and needs. The participants, when faced with the proposal of the modelling course and the demands required in order to accomplish it, such as having enough experience or understanding to be able to describe and refine this description, were induced to think again about other needs of the multiple tasks they were involved with: to produce a kind of reflection. On the one hand, their difficulties in being auto-didactic, for instance, in reading and interpreting different contexts from the questions of the textbooks, defy their available time for the accomplishment of the tasks and the course. On the other hand, the diverse requirements or factors coming from the actual education system (Biembengut 2009) might have influenced them.

This reflection was fundamental for almost 70% of the enrolled students who gave up. The justifications for giving up were lack of time to attend to multiple tasks and educational politics that presented certain contradictions between the proposals and the actions. From one side, the official Brazilian documents emphasized the importance of turning mathematics into something meaningful for the students, in promoting learning, skills, and critical senses; they prescribe a pedagogical orientation that respects the socio-cultural differences among the students; on the other side, there remains inconsistency on what this is about: educational structure and professional standing.

Educational structure: The educational structure at all levels (from Basic Education to Superior) operates with curriculum prescribed in many subjects; there is insufficient time to work deeply with the subjects and having each one of these subjects under the responsibility of one teacher makes it difficult for meaningful changes to happen in the students' teaching and learning styles. It is unsurprising, therefore, that students without interest and without realizing any need in getting this academic knowledge achieve increasingly poor results in the examinations and do not fare well in the job market. The teacher in this context keeps using his or her techniques and strategies. It is fact that a majority of teachers search for efficient ways for their students to learn. The didactic resources vary according to the topic with which they are intended to work, applying a method that is considered adequate to promote learning. In the meantime, they face a lack of motivation and intellectual curiosity from many students in knowing and understanding the curricular themes.

Professional standing: The interaction of each person with the environment in which he or she is surrounded promotes needs that impel him or her to act in the sense of accomplishing these needs or regaining a balance; from this action comes learning and, as a consequence, it helps in his or her formation. The need depends on the objects, interests, and experiences of each person. But the external variables exercise meaningful influence on people's actions. These variables have incited Brazilian teachers to an awareness of two kinds of needs: a need for continuing their own education for better work conditions and a need for keeping themselves in the school institution.

In the last decades, public education teachers' salaries in many Brazilian states reducing. To attain a reasonable standard of living, teachers take on the biggest number of classes and/or hours that are possible in public and private schools. As a consequence, teachers do not have time to study, nor even to prepare the classes; this causes them to reproduce the same teaching that they experienced in their own education, transmitting directly what is in the textbooks, without promoting any interest from the students or addressing their learning needs. Facing that, as it was said by Granger (1969, p. 38), "the taste of knowledge and the desire to reason are turned off, then face to the wish of an effective freedom, after this time of strong pressure, and proclaims the absurd or vain of the human world."

Given these conditions, the educational proposals with results that tend to improve mathematics education are tenuous, in particular. If guided by these stagnating signs, it is possible to move on to other educational areas: from the needs to the interests of survival and from the interests to the needs of survival, in a continuous circle, without bringing improvements in the academic formation of the people. In the attempt to give words to the problems coming from these contradictions in the public educational politics, it is prudent to have an analysis of the objective manifestations from the students' and teachers' reality. It is not less meaningful that "to take the consciousness of your reach, of the doctrinal autonomy, and the extension of your success to a popular auditorium, it is precisely situated" nowadays (Granger 1969, p. 38).

3 Conclusion

Research indicates how fertile the use of mathematical modelling in teaching is for students' education. Once this method is allowed to raise their interests in learning, improving their understanding about the world around them and inspiring consciousness about the environment, they then have the possibility of contributing to a better way of life. The research continues to show how the school community works as an educational institution, the role that it plays in a person's education, the group's, and the society's. It calls to attention some facets in educational politics that promulgate actions expressed in research for the improvement in the education, but that are not complete due to power of educational sanctions. This research also highlights facets that could be identified during the period of the distance modelling course.

A majority of the teachers did not have an adequate background that allows alternative practices in the classroom. When they started noticing the unsatisfactory results of teaching, they become stuck in the actual system even though they are aware of relevant pedagogical proposals, citing the time experienced over teaching across several disciplines causing a poor understanding of the relationships among the mathematical topics and other areas of knowledge together with the influence of multi-occupations (professional, familial, and personal). Thus, there remain few professional teachers who dare stop acting the same way as their predecessors did, assuring the continuing historical reproduction of the same teaching procedures.

Factors that contribute to a mathematics teacher's inability to alter his or her practice despite the difficulties presented by many of the students are: the multi-occupations that most of these teachers are involved in due to interests and diverse needs; the disinterest of many students at all levels to learn; and, the command structure in educational politics. A change of any magnitude would demand a commitment from the authorities to restructure the schools at all levels, giving teachers the need to alter their practices and prompting the students to have a renewed interest in learning.

If the researcher looks back at a period long ago, he or she will be able to trace the changes that took place but exacted almost no meaningful improvement in the process of teaching and learning within the educational system in Brazil. In today's case, what is expected to contribute to the educational process is that each teacher or researcher would be able to correct, continuously, the education process, and without amendment, to transform the scientific knowledge to attend to the new requirements that come out and are unexpected at each moment and still, teachers must also have options to extend their studies to other fields of interests or the like.

The accumulated ideas in this research allowed us to improve the comprehension of each variable analyzed, permitting the results of our experiences to create a new sense of awareness of the importance of changing our educational system in Brazil; the research establishes another reference-point for these needed changes to be realized in a new way. In the momentary difficulty of changing the actual educational structure, we call for researcher teachers of mathematics education to continue searching for new ways, processes, and necessary methods to acquire necessary knowledge for the maintenance of life and, moreover, to instigate the interest and the need of teachers and students in this cause.

References

Biembengut, M. S. (2007). *Mapeamento da modelagem matemática no Ensino Brasileiro*. Blumenau: Relatório de Pesquisa. Conselho Nacional de Desenvolvimento Científico – CNPq.
Biembengut, M. S. (2009). *Processos e métodos de Ensino e aprendizagem matemática na formação continuada dos professores*. Blumenau: Relatório de Pesquisa – Conselho Nacional de Desenvolvimento Científico e Tecnológico – CNPq.
Bonotto, C., et al. (2007). How to replace the word problem with activities of realistic mathematical modelling. In W. Blum, P. Galbraith, H.-W. Henn & M. Niss (Eds.), *Modelling and applications in mathematics education* (pp. 185–192). New York: Springer.
Claparède, E. (1958). *A educação funcional*. Tradução e notas de PENHA, J. B. D. 5ª Ed. São Paulo: Cia Editora Nacional.
Dewey, J. (1922). *Human nature and conduct*. New York: Henry Holt and Co.
George, F. (1973). *Modelos de pensamento*. Trad. Mario Guerreiro. Petrópolis: Vozes.
Granger, G. (1969). *A Razão* (2ªth ed.). São Paulo: Difusão Européia do Livro.
Habermas, J. (1987). *Conhecimento e interesse: com um novo posfácio*. Rio de Janeiro: Guanabara.
Herbart, F. (1806). *Pedagogía general derivada del fin de la educación*. Madrid: Humanitas.
Maturana, H. R., & Varela, F. G. (2001). *A Árvore do Conhecimento*, tradução de Humberto Mariotti e Lia Diskin. São Paulo: Palas Athena.

Chapter 28
In-Service and Prospective Teachers' Views About Modelling Tasks in the Mathematics Classroom – Results of a Quantitative Empirical Study

Sebastian Kuntze

Abstract Views of mathematics teachers about tasks with modelling relevance are likely to influence the ways teachers create learning opportunities in the classroom. As quantitative empirical evidence about such task-related views is scarce, this chapter reports on corresponding findings. In particular, views of prospective and in-service teachers are compared and possibilities of improving professional knowledge are identified.

1 Introduction

Professional knowledge and beliefs of mathematics teachers concerning the role of modelling for the mathematics classroom may have impacts on the way teachers conceive learning opportunities for their students. In particular, views about tasks requiring modelling steps might be relevant for instruction-related decisions of teachers.

However, there is still a need for quantitative empirical research about such views and corresponding professional knowledge. Consequently, this study concentrates on this area. A total of 230 prospective and 79 in-service secondary teachers were asked about their instruction-related views, including views about modelling and about their views on characteristics of different tasks with higher or lower modelling requirements.

The results indicate that the prospective teachers preferred tasks with rather low modelling relevance to those requiring more intensive modelling activities. These views might be linked to the fear of the prospective teachers that the goal of mathematical exactness may be in conflict with modelling tasks. However, the in-service teachers' views differed significantly: The tasks requiring more intensive modelling activities were rated more positively. These results indicate that the in-service

---

S. Kuntze (✉)
Ludwigsburg University of Education, Ludwigsburg, Germany
e-mail: kuntze@ph-ludwigsburg.de

teachers might be more aware of the learning opportunities linked to modelling tasks on the basis of their more extensive classroom experience. Yet, on the level of more global views on modelling, the in-service teachers felt, for example, on average rather unsure about their knowledge concerning the modelling cycle. These findings can help to identify possibilities of fostering the teachers' professional knowledge linked to modelling in the mathematics classroom.

In the following, the chapter will give an overview on its theoretical background, present research questions, information about design of the study, and results, which will be discussed in a concluding section.

2 Theoretical Background

Fostering modelling competency is an important goal for mathematics instruction (e.g., Blum et al. 2007; KMK 2004; OECD 2003), and modelling competencies are even seen as essential for mathematical literacy (e.g., OECD 2003). When fostering modelling competencies in the classroom, tasks play a crucial role. Indeed, learning opportunities linked to modelling largely depend on tasks and ways of dealing with them in the classroom (Blum et al. 2007; Maaß 2006; Reiss et al. 2008). From the perspective of research on instruction, tasks can be seen as an indicator for aspects of instructional quality (Bromme 1992). Corresponding studies (e.g., Neubrand 2002) have identified a potential of improving instruction with respect to modelling. For example, Neubrand (2002) found that in German classrooms, tasks that require multi-step solutions, as is frequently the case for modelling tasks, are rare. Accordingly, research about reasons for the teachers' preference for tasks with lower modelling relevance could not only improve our understanding of these observations, but also help to define goals of teacher training. For this reason, views of mathematics teachers about modelling and modelling tasks are in the focus of this study.

However, the definitions of "modelling task" are partly divergent (e.g., Blomhøj and Jensen 2003; v. Hofe 2008; Maaß 2006). Consequently, a pragmatic distinction is used:

- *Tasks with substantial/higher modelling requirements*: tasks that require at least one translation step between a given situational context and a mathematical model, and that allow different solutions.
- In *tasks with low modelling requirements*, the mathematical model is already given and translation processes are less important. For these tasks, only one correct solution is possible.

This study concentrates on views of mathematics teachers about such tasks with higher or lower modelling requirements. These views are considered as being part of the professional knowledge of mathematics teachers. For components of professional knowledge, we refer to a three-dimensional model (Kuntze and Zöttl 2008). This model includes the distinction of areas of professional knowledge according to Shulman (1986; for the possibility of further refinement into domains see

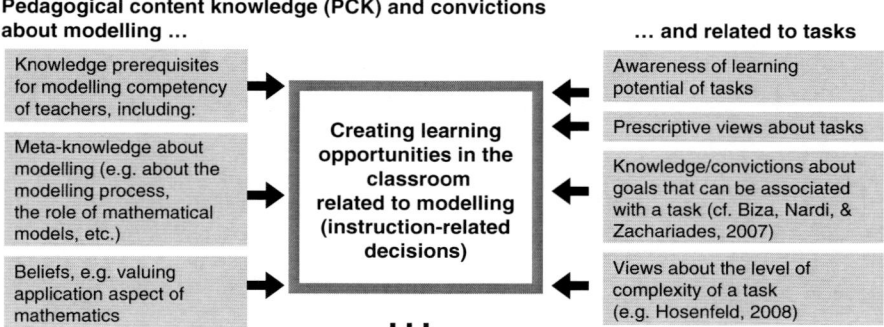

Fig. 28.1 Components of PCK as influencing factors on creating learning opportunities

Ball et al. 2008), the spectrum between declarative/procedural knowledge and prescriptive convictions/beliefs, as well as a distinction of levels of globality or content-relatedness of components of professional knowledge (cf. Kuntze and Reiss 2005).

Task-related views can be classified as rather content-specific components of pedagogical content beliefs, whereas meta-knowledge about modelling is on a much more global level. Figure 28.1 gives an overview on important components of professional knowledge relevant for creating learning opportunities in the classroom related to modelling (cf. also results of the study of Schwarz et al. 2008). Compared to the model referred to above (Kuntze and Zöttl 2008), Fig. 28.1 gives a detailed perspective which selects crucial aspects from the more complex interplay of components of professional knowledge. As instructional practice may have an influence on professional knowledge related to tasks, both views of in-service and pre-service teachers are interesting. The views of in-service teachers are likely to be based on more experience.

3 Research Questions

Against the background of the considerations made above, the study focuses on the following research questions:

- Which views about tasks with lower or higher modelling requirements do prospective and in-service mathematics teachers hold? How are these views structured and are there differences between in-service and prospective teachers?
- What meta-knowledge on modelling do in-service teachers report to have?

4 Design and Sample

The study is based on two samples. The first sample consists of 230 prospective teachers with context data given in Table 28.1.

Table 28.1 Data on the sample of prospective teachers

Prospective teacher career	Academic track	Technical track	General school	School for students with special needs	No data about school type
Number of prospective teachers	55	61	62	43	9
Female	32	48	46	31	5
Male	23	13	16	12	4
Mean number of semesters (SD)	4.98 (2.01)	1.62 (1.11)	2.51 (1.98)	1.95 (1.72)	2.89 (3.33)
Mean age (SD)	22.1 (3.3)	22.5 (4.7)	22.9 (3.5)	21.6 (1.9)	27.3 (7.8)

The second sample encompassed 79 German in-service secondary mathematics teachers working at academic-track schools (35 female, 43 male, one without data; 25 teachers aged up to 35 years, 21 teachers aged 36–45 years, 23 teachers aged 46–55 years, nine teachers aged more than 55 years, one teacher without data). The recruitment of the in-service teachers was done via the school administration in the framework of an empirical study about student achievement. For comparisons with the academic-track in-service teachers, we will refer only to the 55 academic-track prospective teachers.

These prospective and in-service teachers were asked to fill in a questionnaire (multiple-choice, four-point Likert scale) containing scales referring to task-related and more global views. The sub-questionnaire about tasks with lower or higher modelling requirements referred to six tasks in the content area of measuring areas: Four of these tasks were marked by higher modelling requirements, whereas two tasks were characterized by lower modelling requirements. For these tasks, identical scales of four items per task focused on the positive judgment on the learning potential of the task, respectively. Sample items for these scales are "I see the role that mathematics plays for the solution of this task as meaningful for building up mathematical competency" and "students can learn a lot when working on this task." Beyond these scales, one-item-indicators concentrated on the perceived level of complexity of the task and on the compatibility of the task with the goal of exactness in mathematics instruction, respectively (as additional aspects in Fig. 28.1).

Figure 28.2 shows two of the tasks the teachers were asked to comment on, representing one each of the considered types of task, respectively. Task 1 in Fig. 28.2 was designed to have a rather "application" character with relatively lower modelling requirements, as the mathematical model is mainly given, whereas Task 2 has relatively higher modelling requirements (Kuntze and Zöttl 2008).

5 Results

The results presented in this section focus firstly on the prospective teachers' task-related views and their structure and secondly on the findings regarding the in-service teachers. The task-related scales were reliable, as shown in Table 28.2.

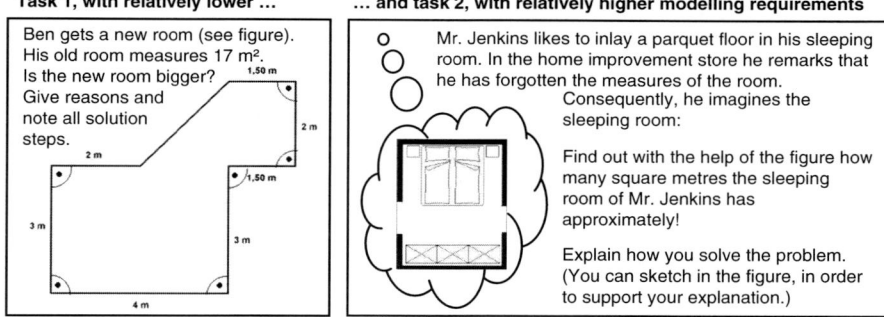

Fig. 28.2 Tasks with lower and higher modelling requirements (Kuntze and Zöttl 2008)

Table 28.2 Scales and reliability values

Scale: Positive judgment on the learning potential of…		Number of items	Prospective teachers α (Cronbach)	In-service teachers α (Cronbach)
…Task 1	Tasks with lower modelling requirements	4	0.69	0.55
…Task 5		4	0.82	0.84
…Task 2	Tasks with higher modelling requirements	4	0.86	0.82
…Task 3		4	0.88	0.77
…Task 4		4	0.87	0.73
…Task 6		4	0.88	0.81

Only the reliability value of the scale of task 1 in the case of the in-service teachers was rather low, but still tolerable, given the low number of items.

In order to find out about the structure of the task-related views, correlations between the scales in Table 28.2 were calculated (prospective teachers). The scales of the tasks 1 and 5 correlated with $r=0.54$ (tasks with lower modelling requirements). Similarly, the scales for the tasks 2, 3, 4, and 6 (higher modelling requirements) were interdependent (correlation coefficients between 0.41 and 0.60). Correlations between scales of different task type were lower (correlation coefficients between n. s. and 0.33). As evaluations of tasks had higher correlations within task types than across task types, the design is supported empirically.

For providing an insight into different views within the sample of prospective teachers, Fig. 28.3 presents results of a cluster analysis based on the scales presented in Table 28.2. Especially the prospective teachers in cluster 2 rated tasks with lower modelling relevance much more positively. In this cluster, negative judgments of the learning potential of tasks with higher modelling requirements were frequent (s. also Kuntze and Zöttl 2008). In order to identify possible reasons for the discrepancy between the types of tasks, the prospective teachers' answers for the indicators of the level of complexity of the tasks were considered. However, no consistent pattern

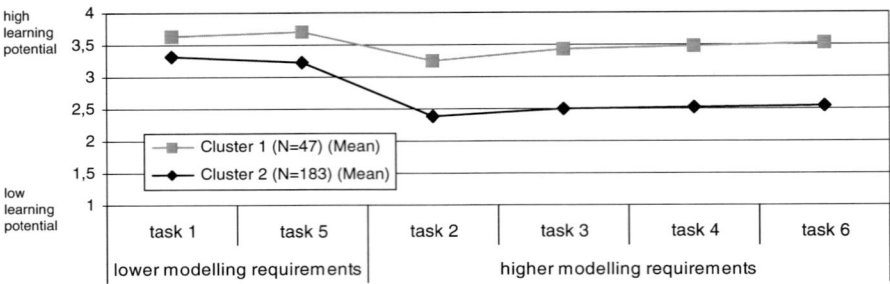

Fig. 28.3 Learning potential of tasks for clusters (Ward method) of prospective teachers

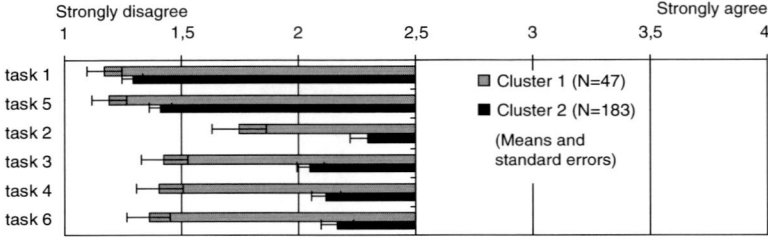

Fig. 28.4 Incompatibility with goal of exactness for clusters of prospective teachers

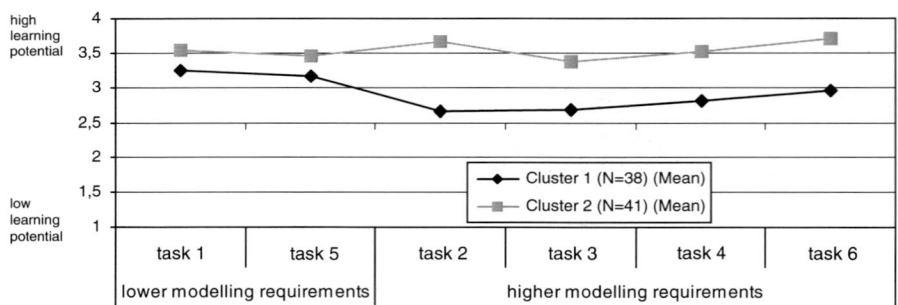

Fig. 28.5 Learning potential of tasks for clusters (Ward method) of in-service teachers

could be observed: In particular, cluster 2 did not rate the tasks with higher modelling requirements to be generally of a higher level of complexity.

Yet, the data in Fig. 28.4 shows that the prospective teachers of cluster 2 were more in fear of an incompatibility of the tasks with higher modelling requirement with the goal of exactness in mathematics education than their counterparts in cluster 1.

These structures in the task-related views can also be seen for the sample of the in-service teachers (cf. Figs. 28.5 and 28.6). However, the in-service teachers rated

Fig. 28.6 Incompatibility with goal of exactness for clusters of in-service teachers

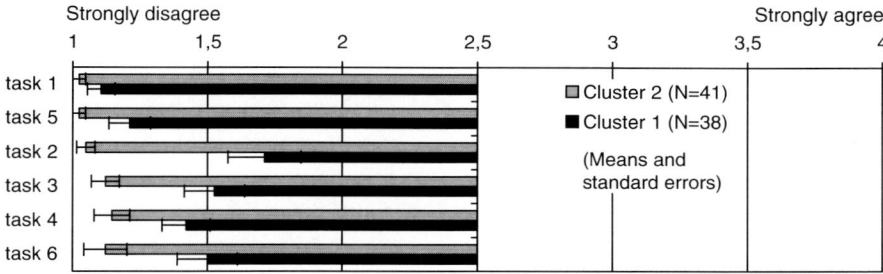

Fig. 28.7 Learning potential of tasks: Comparisons of prospective and in-service teachers

the tasks with higher modelling requirements more positively (see also Fig. 28.7, mostly strong effects) and they showed less fear of incompatibility of these tasks with the goal of exactness (Fig. 28.6). Again, there was no consistent pattern for the perception of the level of complexity of the tasks.

The second research question focuses on the more global scales, which included the scale about reported meta-knowledge about modelling. The scales were reliable (cf. Table 28.3). The mean values of these scales are displayed in Fig. 28.8: The in-service teachers do not report a good meta-knowledge about mathematical modelling. The reported influence of standards (in Germany, the national standards include modelling as one of six competency areas) was rather low, too. In comparison, the importance attributed to algorithmic goals was significantly higher.

6 Discussion and Conclusions

From the methodological point of view, a first important result was that the scales included in the questionnaire instrument were reliable for task-specific and more global views related to modelling. Hence, the instrument can be used in further studies.

Table 28.3 Scales on more global views, including reported meta-knowledge about modelling

Scale	Sample item	Number of items	α (Cronbach)
Reported meta-knowledge about modelling	I could explain what is happening in the different phases of mathematical modelling	5	0.91
Reported influence of standards	When choosing or designing tasks for my classroom, I bear in mind fostering the competencies described in the KMK standards	6	0.88
Importance of algorithmic goals	Above all, a solid mathematics knowledge means that the students are able to process certain solution procedures	4	0.82
Importance of learning by rote	In mathematics, it is necessary for the students to learn some content by rote	4	0.85

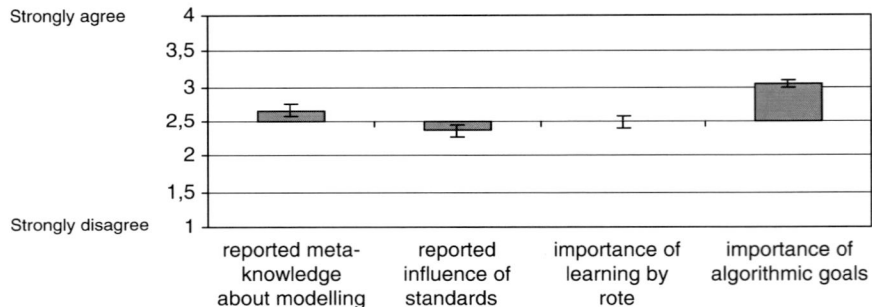

Fig. 28.8 More global scales included in the study (in-service teachers, M and standard errors)

The results show possibilities of professional development necessary for enhancing learning opportunities related to modelling in the classroom. As the majority of the prospective teachers asserted tasks with higher modelling requirements consistently to have a lower learning potential than tasks with lower modelling relevance, the need to include work on tasks in mathematics teacher training is highlighted. The results might reflect a low awareness of possibilities of fostering modelling competencies in the content area of the tasks. The fear of incompatibility of modelling tasks with the goal of mathematical exactness might be an obstacle for seeing productive learning opportunities in the tasks with higher modelling relevance. This fear played a more important role than the fear of a high level of complexity of modelling tasks. For prospective teachers in cluster 2 (see Fig. 28.3), this observation could be explained by a lack of professional knowledge about the role of mathematics in the modelling process, as modelling, as required in the tasks, can even offer learning opportunities for building up knowledge about the exact nature of mathematics.

The in-service teachers showed a tendency of similar structures of task-specific views, but they saw a higher learning potential for tasks with higher modelling requirements. The T-test data in Fig. 28.7 shows effect sizes indicating strong effects. Consistently, the in-service teachers showed less fear of incompatibility with the goal of mathematical exactness. An explanation for this might be an increased tolerance with respect to alternative tasks in general, or more real-life application-friendly task-related convictions in the content domain of measuring areas. Perhaps the instructional experience of the in-service teachers with teaching this content facilitated these views, whereas the prospective teachers referred to their experience at school from a student's perspective.

An alternative explanation could be that the in-service teachers had different expectations about the intentions behind the items of the questionnaire, so that they showed more openness toward tasks with higher modelling relevance. Yet, this second interpretation also implies that the in-service teachers saw the characteristics of the tasks with higher modelling requirements differently, which reflects corresponding professional knowledge.

However, the in-service teachers did not report good meta-knowledge coverage of mathematical modelling. In line with this, the reported influence of national standards on instructional practice was not high. Hence, even the in-service teachers are not prepared in an optimal way for creating rich learning opportunities linked to modelling in their classrooms. The results suggest that some teachers will hardly be able to provide their students with meta-knowledge support when dealing with modelling tasks in the classroom.

Moreover, the study raises points of interest for further research. For example, the question, which other task-specific views of teachers can be implemented in studies and which interdependencies are there between different aspects, could deepen our understanding of task-related professional knowledge. Questions linked to this are: How much task-specific views of mathematics teachers depend on particular tasks, which characteristics of a task have an influence on judgments of tasks, and how task-specific views develop. In addition, impacts of task-specific views of mathematics teachers on instructional practice and on competency growth of students related to modelling should be investigated in studies that include observations of classrooms and achievement data of the students. Finally, the question of how culture-dependent task-specific convictions of mathematics teachers related to modelling are merits attention in corresponding comparative research.

References

Ball, D., Thames, L., & Phelps, G. (2008). Content knowledge for teaching: What makes it special? *Journal of Teacher Education, 59*(5), 389–407.
Biza, I., Nardi, E., & Zachariades, T. (2007). Using tasks to explore teacher knowledge in situation-specific contexts. *Journal of Mathematics Teacher Education, 10*, 301–309.
Blomhøj, M., & Jensen, T. H. (2003). Developing mathematical modeling competence: Conceptual clarification and educational planning. *Teaching Mathematics and Its Applications, 22*(3), 123–139.

Blum, W., Galbraith, P. L., Henn, H. W., & Niss, M. (Eds.). (2007). *Modeling and applications in mathematics education. The 14th ICMI study*. New York: Springer.
Bromme, R. (1992). *Der Lehrer als Experte. Zur Psychologie des professionellen Wissens*. [The teacher as an expert. On the psychology of professional knowledge]. Bern: Hans Huber.
Hosenfeld, I. (2008). Diagnostische Kompetenzen von Mathematiklehrkräften und Leistung. Presentation on 26.08.2009, *71st AEPF Conference*, Kiel.
Kultusministerkonferenz (KMK). (2004). *Bildungsstandards im Fach Mathematik für den mittleren Schulabschluss*. München: Wolters Kluwer.
Kuntze, S. & Reiss, K. (2005). Situation-specific and generalized components of professional knowledge of mathematics teachers. In H. L. Chick & J. L. Vincent (Eds.), *Proceedings of the 29th conference of the international group for the psychology of mathematics education (PME)* (Vol. 3, pp. 225–232). Melbourne: University.
Kuntze, S., & Zöttl, L. (2008). Überzeugungen von Lehramtsstudierenden zum Lernpotential von Aufgaben mit Modellierungsgehalt. *Mathematica Didactica, 31*, 46–71.
Maaß, K. (2006). What are modeling competencies? *Zentralblatt für Didaktik der Mathematik, 38*(2), 115–118.
Neubrand, J. (2002). *Eine Klassifikation mathematischer Aufgaben zur Analyse von Unterrichtssituationen*. Hildesheim: Franzbecker.
OECD. (2003). *The PISA 2003 assessment framework – Mathematics, reading, science and problem solving knowledge and skills*. Retrieved from http://www.pisa.oecd.org/dataoecd/46/14/33694881.pdf, 15 January 2010.
Reiss, K., Kuntze, S., Pekrun, R., & Ufer, S. (2008). Die Kompetenz "Modellieren" in Verbindung mit unterschiedlichen Leitideen – von Zielen der Bildungsstandards zu Fragen der Konzeption von Kompetenzmodellen. In E. Vasarhélyi (Ed.), *Beiträge zum Mathematikunterricht 2008* (pp. 185–188). Münster: WTM-Verlag.
Schwarz, B., Kaiser, G., & Buchholtz, N. (2008). Vertiefende qualitative Analysen zur professionellen Kompetenz angehender Mathematiklehrkräfte am Beispiel von Modellierung und Realitätsbezügen. In S. Blömeke, G. Kaiser, & R. Lehmann (Eds.), *Professionelle Kompetenz angehender Lehrerinnen und Lehrer* (pp. 391–424). Münster: Waxmann.
Shulman, L. (1986). Those who understand: Knowledge growth in teaching. *Educational Researcher, 15*(2), 4–14.
v. Hofe, R. (2008). *Zur Entwicklung mathematischer Grundbildung in der Sekundarstufe I – Ergebnisse aus der Längsschnittstudie PALMA*. [Presentation on 5 June 2008, University of Munich].

Chapter 29
Pre-service Secondary Mathematics Teachers' Affinity with Using Modelling Tasks in Teaching Years 8–10

Gloria Stillman and Jill P. Brown

Abstract First results with respect to teaching modelling are presented from Australian data collected as part of an international study of pre-service mathematics teachers, Competencies of Future Mathematics Teachers. Data were collected from 73 volunteer pre-service secondary mathematics teachers from six cohorts at five university sites in three Australian states. Questionnaire responses targeting affinity of pre-service teachers by using modelling tasks in Years 8–10 are analysed from the perspective of possible differences associated with the length of teacher preparation program being undertaken.

1 Introduction

In recent years, the preparation of mathematics teachers for secondary school has come under scrutiny and comparative international studies of professional competence of pre-service teachers have been conducted in several countries (e.g., *Mathematics Teaching for the 21st century [MT21]*, see Schmidt et al. 2007, and the IEA *Teacher Education and Development Study: Learning to Teach Mathematics [TEDS-M]*). Blömeke et al. (2008) view professional competence of pre-service teachers as "a complex hypothetical construct that underlies teacher performance" (p. 723) consisting of several knowledge and belief components as well as personal characteristics. Personal characteristics are not likely to be changed in short teacher preparation courses (e.g., those lasting 12–18 months) but act as mediators of what pre-service teachers attend to, take on

G. Stillman (✉)
School of Education (Victoria), Australian Catholic University, PO Box 650, Ballarat, VIC 3350, Australia
e-mail: gloria.stillman@acu.edu.au

J.P. Brown
School of Education (Victoria), Australian Catholic University, 115 Victoria Parade Melbourne, VIC 3065, Australia
e-mail: jill.brown@acu.edu.au

board, and ultimately take up in their practice. Professional knowledge and beliefs are far more likely to undergo change during pre-service teacher education and thus are the focus of our research, in this instance, relating to the teaching of mathematical modelling in Years 8–10, lower secondary schooling in Australia.

2 Background

In the research study, *Competencies of Future Mathematics Teachers* [*CFMT*], the connection between the various components of professional knowledge of pre-service mathematics teachers with respect to mathematical modelling and real-world contexts at the lower secondary level has been investigated at sites in Germany, Australia, Hong Kong, the Chinese mainland and Taiwan. The overall aim of this study is to evaluate the professional competencies of pre-service secondary mathematics teachers at universities in several countries by taking a qualitatively oriented approach and developing detailed in-depth studies of the professional knowledge of pre-service teachers. With respect to competencies to teach mathematical modelling, Kaiser et al. (2007, 2010) report preliminary results from German pre-service teachers. This chapter reports the first findings from the Australian university sites.

3 Theoretical Framework

CFMT has been developed within the framework of *MT21* and *TEDS-M*. Following Blömeke et al. (2008), professional competence is seen as involving beliefs about teaching and the nature of mathematics as a discipline and mathematics in schooling, professional knowledge and personality characteristics. Based on the work of Shulman (1986, 1987) and others, professional knowledge consists of *mathematical content knowledge (MCK)*, *pedagogical content knowledge (PCK)* and *general pedagogical knowledge* (*PK*). This categorisation of teachers' professional knowledge is, however, a convenience for discussing various aspects of teacher professional competencies rather than easily distinguishable components of knowledge as in practice these are often combined and difficult to separate. These terms have several interpretations in the literature. Bullough (2001) and Krauss et al. (2008) overview the on-going debate. The interpretations used here follow those of Schmidt et al. (2007) in *MT21*. MCK for pre-service teachers refers to knowledge of mathematical facts, concepts, and processes for the various topics at the level of schooling being targeted. PCK in mathematics includes knowledge of appropriate content and form of mathematics for the particular schooling level, how this might be presented and represented, how to analyse student responses, student difficulties, and the place of current content and skill development in the overall development of mathematics through schooling. PK refers to generic knowledge in teaching and learning such as knowledge of classroom organisation, assessment, and dealing with diversity of learners. The main categories of professional competence for teaching mathematical modelling to be reported in this chapter are: (a) beliefs about the nature of mathematics and affinity with modelling in teaching (Beliefs) and (b) didactical reflections about modelling (i.e., an amalgam of PCK and MCK).

4 The Study

The major research questions for the larger study that are of interest here are:

1. What are the professional competencies of future teachers with respect to the teaching of modelling and real-world applications for the lower secondary level (Years 8–10)?
2. How distinctive are the different routes of mathematics teacher education?

Investigation of the first question will focus on affinity with using modelling tasks and the latter question will be explored only from the perspective of length of teacher preparation program. The aim is to profile the pre-service teachers with respect to the variables of interest as one pooled group and then to investigate any associations between program length and particular differences in profiles when the group is divided along these lines.

Data were collected from 73 volunteer pre-service teachers at five university sites in three east coast Australian states: three in Victoria, and one each in New South Wales and Queensland. The universities chosen were a mixture of older well-established and more recently established institutions in inner city and metropolitan areas. The pre-service teachers were enrolled in a variety of teacher preparation programs of different lengths which could be broadly described as being:

(a) One-year programs ($n=46$) of general teacher education and mathematics education subjects (e.g., Postgraduate Diplomas of Education or Postgraduate Diplomas of Teaching) following a qualifying degree with a mathematics component (e.g., Bachelor of Science or Bachelor of Engineering).
(b) Four-year programs ($n=27$) with students doing double degrees in which mathematics is learnt concurrently with general teacher education and mathematics education (e.g., Bachelor of Arts/Bachelor of Teaching or Bachelor of Science/Bachelor of Education).

Regardless of whether the students were enrolled in either program type, the mathematics was taught within university mathematics departments or faculties and the mathematics education within university education faculties. All pre-service teachers concurrently were involved in a teacher practice program in secondary schools (Years 7/8–12) where they observed classes and planned and taught lessons. On successful graduation, they would be qualified to teach mathematics at secondary school in their state and possibly other states.

Towards the end of their programs, the pre-service teachers completed a written questionnaire, consisting of open items bridging professional knowledge domains and beliefs, which was designed by Kaiser and Schwarz (Schwarz et al. 2008). In particular, one item consisting of several sub-items was based on a modelling example about the takings of an ice cream shop and the suggested task solution of a Year 8 student, Leo. This was the item that is the basis for the responses analysed in this chapter.

> There are four ice-cream shops in Leo's city, Springfield. Leo is standing in front of his favourite ice-cream shop "Sorrento" as he does often in summertime. One scoop of ice-cream costs 60c. He asks himself how much money the owner of the ice-cream shop gets by selling ice-cream on one hot summer Sunday.
>
> To solve the problem Leo does the following: On the next day he asks his best three friends how many scoops of ice-cream they bought last Sunday and gets the following answers: Marcus: 3 scoops, Peter: 5 scoops, Tom: 4 scoops.
>
> Leo calculates on average $(3+4+5) \div 3 = 4$ scoops per day. He multiplies the result by the number of citizens in Springfield (30,000) and divides the result by 4 because there are four ice-cream shops in Springfield. So 30,000 scoops of ice-cream are sold in the ice-cream shop Sorrento per day. Income: $30,000*\$0.60 = \$18,000$. What do you think about this?

5 Findings

5.1 Diagnostic Competencies with Respect to Modelling

In order to diagnose student difficulties and choose necessary interventions, teachers need to be able to analyse students' attempts at modelling appropriately. To test this competency, after attempting the Ice Cream Task themselves, pre-service teachers were asked to analyse, from a teaching perspective, dialogue from parts of interviews with four Year 8 students who had been asked how they would respond to the Ice cream Task, to see whether the suggested modelling approaches were appropriate. The pre-service teacher's response to each student was classified as appropriate [+] or not [0]. To be classified as appropriate, the pre-service teacher's response was expected to involve correct analysis of the modelling approach used and comments on the adequacy of the response naming one strong or one weak point. Pre-service teacher AU3_08S1,[1] for example, noted that one student broadened the assumptions of the real model through the representativeness of the sample used. Further, a weakness of this student's response was noted in that other unrealistic assumptions in the approach given in the task statement were not critiqued. Naming either of these points was all that was necessary for the + coding to be given. In responding to another students' response, AU4_08S2's analysis correctly identified that the student named two aspects to improve the real model, namely, competition between shops related to location and expected consumption of ice cream. The aspect of asking the shop owners, which is not part of a modelling approach, was named also, but this did not prevent the positive coding being given. Both these pre-service teachers were from 4-year programs.

---

[1] Each response has an identifier for a university site, year of data collection, and student.

Inappropriate responses (a) failed to identify a strong or a weak point for a particular student's response (e.g., AU5_08S24), (b) gave an incorrect analysis (e.g., AU6_08S9), or (c) simply described the student's solution (e.g., AU5_08S7) as shown here:

> *The task will be complicated, will not be able to come to any conclusion. [0, AU5_08S4, 1 year program]*
>
> *Interview 1 The student is unaware of sample sizes, how to choose one and has agreed with Leo's approach of sampling only 3 of his friends. He has approximated that half of the population would not eat ice-cream at all. This approach has limitations and is not right, and I would encourage him to think over it. [0, AU6_08S9, 1 year program]*
>
> *Asking the right questions & the variations that apply to each scenario. Looking at recording the correct information + the solving. [0, AU5_08S7, 1 year program]*

In addition, there were inappropriate responses, such as the following by AU5_08S1, where only general comments were made and no specific student's response was identified. In this case, the pre-service teacher's response would be coded as 0 for all four analyses.

> *Some student[s] talk about the fact that not all the population would buy ice cream. I would praise them and then encourage them to think deeper, about the initial cost of the ice cream the ice cream man pays, i.e. his income would be the profit. [0 for all analyses, AU5_08S1, 4 year program]*

The pre-service teacher's overall response was then classified as displaying very low, low, medium, high, and very high competencies with respect to the analysis of the four student solutions from a modelling perspective. Where no appropriate response to any of the four student dialogues presented was identified, the overall response was coded as showing very low diagnostic competencies. One appropriate response was coded as low, two as medium, three as high and four as very high.

As shown in Fig. 29.1a, almost two thirds [62%] of the pre-service teachers from the pooled sites demonstrated medium or above levels of knowledge from their adequate reflection on the students' solutions, with approximately half [51%]

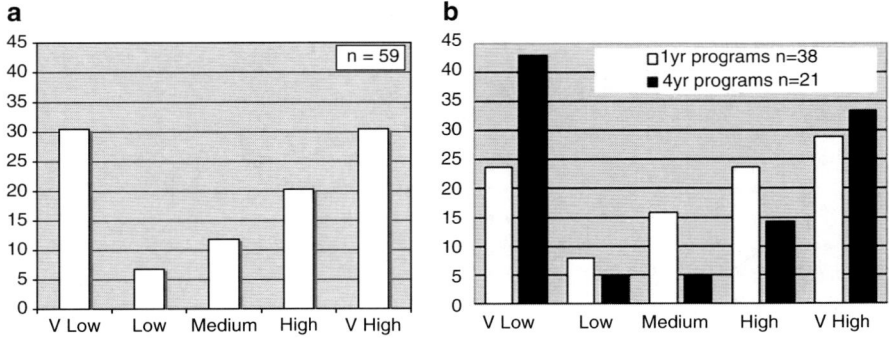

Fig. 29.1 Diagnostic competencies about modelling across sites (**a**) and by program length (**b**)

demonstrating this at a high or very high level. Almost one-third of the students demonstrated very high competencies through their adequate evaluation of all the students' solutions. When the responses were grouped according to teacher preparation program length (Fig. 29.1b), however, the distributions of responses differed in that almost half [43%] of the pre-service students in the 4-year programs were unable to provide adequate evaluation for any of the students' solutions compared to just less than 24% in the 1-year programs. For both program types, approximately 30% of all students were able to adequately evaluate all four student responses.

5.2 Competencies in Didactical Reflections About Modelling – Appropriateness of Task

Pre-service teachers' responses to the questions: *Is such a task appropriate for secondary school at this level? If yes, why? If no, why not?* were classified as displaying low, medium and high competencies in didactical reflections about modelling. The responses of those categorised as displaying low competencies showed no appreciation of modelling at this level of schooling stating that such a task was not appropriate (e.g., AU5_08S4, 1-year program). Those classified as indicative of medium competencies indicated that such a modelling task was appropriate at this level of schooling but either did not include a reason or, if one was included, it was not a didactical aim of modelling (e.g., AU3_08S12, 4-year program). The responses of those classified as displaying high competencies showed an appreciation of modelling and named at least one didactical aim of modelling as a reason for this (e.g., preparing students for life, AU3_08S10, 4-year program).

As shown in Fig. 29.2a, the majority (85%) of the pre-service teachers agreed that such tasks were suitable for Year 8 students with just less than half (40%) also demonstrating that they had competency in didactical reflections about aims of

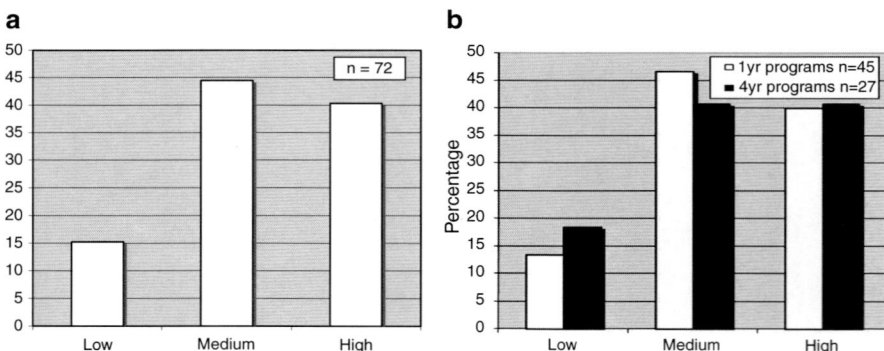

Fig. 29.2 Competencies in didatical reflections about appropriateness of a modelling task across sites (**a**) and by program length (**b**)

modelling. When the responses were grouped according to teacher preparation program length, it is seen that a slightly greater proportion of responses of pre-service teachers from 4-year programs did not agree that the task was appropriate for this level of schooling (Fig. 29.2b).

5.3 Affinity with Modelling in Teaching in Years 8–10

When asked if they would use this kind of task in their mathematics lessons at this level explaining their position and specifying reasons for their responses, only 11 out of 71 (15.5%) who gave useable responses to the question indicated they would not. These responses (e.g., AU5_08S9, 4-year program) were classified as showing low affinity with modelling in teaching in Years 8–10. The vast majority of student teachers, 60 (84.5%), at these sites agreed they would use this kind of task in their mathematics classes at this level. Those responses that showed agreement but did not include a reason or the reason given was not a reflection about any educational aim of mathematical modelling (e.g., only a reference to the motivational content of the task, AU5_08S19, 1-year program, or mathematical competencies, AU3_08S2, 4-year program) were categorised as showing medium affinity. In addition, responses which placed conditions on using modelling tasks but still showed these were appreciated, but no educational aim of mathematical modelling was named (e.g., AU5_08S17, 4-year program), were also classified as showing medium affinity with modelling in teaching at this level of schooling. Furthermore, 28 pre-service teachers (39.4% of 71) were able to adequately explain their position and specify at least one reason for their answer which was a meaningful didactical reflection about an educational aim for mathematical modelling (e.g., communication about mathematics, AU1_08S7, 1-year program, or promotion of understanding of mathematical content, AU3_08S15, 4-year program). These responses were classified as showing high affinity with mathematical modelling in teaching in Years 8–10.

The distribution of these responses from the pooled sites is shown in Fig. 29.3a. When this variable was examined by program length (Fig. 29.3b), there was very little difference across the distributions.

5.4 Affinity with Modelling Related to Beliefs About the Nature of Mathematics

Pre-service teachers were asked whether modelling tasks were part of mathematics because they represent experimental, applied mathematics or should mathematics be a deductive, abstract science. They were then asked to explain their position. Three positions were possible: (a) agreement with modelling being part of mathematics on this basis [A codes], (b) modelling tasks were seen as part of mathematics because mathematics has both an experimental, applied nature and an abstract,

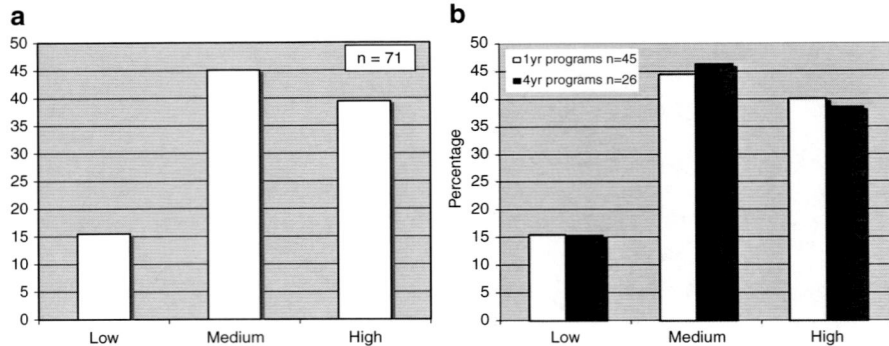

Fig. 29.3 Affinity with mathematical modelling across sites (**a**) and by program length (**b**)

deductive nature [B codes]. (c) modelling tasks have no place in mathematics as it is a pure abstract science [C codes]. Responses were coded to reflect these three positions and also with respect to the reasons given (e.g., A1 mathematically oriented agreement, A2 didactically oriented agreement from a student perspective, A3 didactically oriented agreement from a teacher perspective) or lack of such reasons [Y codes, e.g., Y1 position (a)]. A cross section of examples follows.

> *Clearly this is a part of mathematics. Mathematics is simply the analysis of observable elements of life that can be solved, aided or simplified through numerical-or-otherwise problem solving approaches. [A1, AU3_08S16, 1 year program]*
>
> *Yes, and it also shows a students' understanding and interpretation of the results. [A3, AU3_08S8, 1 year program]*
>
> *Modelling tasks are definitely part of mathematics, and particularly at the high school level [maths] should rarely be deductive, abstract science. [Y1, AU3_08S6, 4 year program]*
>
> *Modelling tasks should be a part of mathematics in the classroom but not the only thing. It should be a part...because they will develop skills that can be used in the workforce. Also, we need to show students that Maths is beautiful to simply think about and it should therefore be taught as an abstract ART (not science) as well. [B2, AU1_08S7, 1 year program]*
>
> *I feel 'mathematics' is deductive and abstract and that modelling really describes arithmetic and focuses on calculations. 'Mathematics' for me, is more closely linked with philosophy. [C1, AU1_08S4, 1 year program]*

Figure 29.4a shows 68% of the pre-service teachers believed modelling tasks are part of mathematics because it is experimental and applied. A further 29% believed this to be the case but also indicated mathematics has a deductive abstract aspect as well. Only 3% of the pre-service teachers suggested modelling tasks were not part of mathematics instead viewing mathematics as being only a deductive abstract science. When responses were examined according to program length, there were some differences. The few students who did not consider modelling tasks to be part of mathematics all came from the 1-year programs. The proportion of students seeing modelling tasks as belonging to mathematics

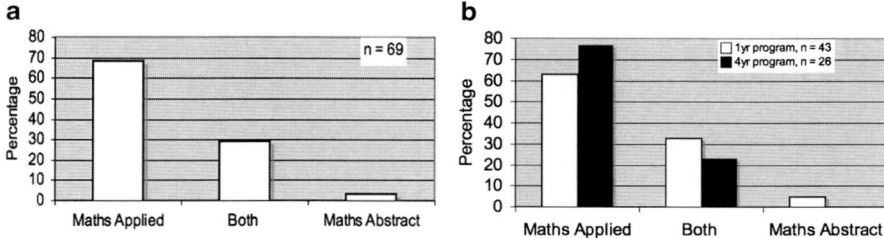

Fig. 29.4 Affinity with mathematical modelling related to beliefs across sites (**a**) and by program length (**b**)

because of the dual aspects of position (b) was higher in 1-year programs. There were also differences with respect to how students explained their position. Didactical reasons were most common in both groups (61.5% 4 years; 56% 1 year). Mathematically oriented reasons were a higher proportion in the 1-year group (19% 4 years; 35% 1 year) but giving no reason was higher in the 4-year group (23% 4 years; 11.5% 1 year).

6 Discussion and Conclusion

The general belief is that longer pre-service preparation programs such as a 4-year double degree are of more benefit and that pre-service teachers exiting from short programs such as 1-year postgraduate diplomas are underprepared for teaching. In some areas such as discerning whether the modelling task was appropriate for the target level of schooling and whether they were prepared to use such a task in classrooms at this level, there was little difference in the responses of these groups with a high level of agreement reflecting the emphasis on these types of tasks in Australian curricula at this level of schooling. Pre-service teachers who were educated in Australia would perhaps have experienced such tasks when they were in schooling as well as in their school experience.

With respect to diagnostic competencies and the ability to analyse student responses for appropriateness of modelling approach, the much higher proportion of 4-year program students who could not do this was surprising as they have more experience in schools over an extended period and thus would be expected to acquire more PCK in this regard than their 1-year counterparts. However, all students in 4-year programs prepare for two teaching areas, one of which is mathematics and they can do this with a minor focus in mathematics. Some thus might place less emphasis on their preparation to teach mathematics as their other teaching area is of more personal interest. The 1-year programs in contrast included some students who were preparing as only mathematics teachers. In general, but not always, these students had a strong mathematics background and would be expected to place a strong emphasis on being as prepared as possible to teach mathematics.

The differences with respect to beliefs about the nature of mathematics were also surprising especially the higher proportion of pre-service teachers in 1-year programs acknowledging the dual nature of mathematics. Perhaps this is related to the higher proportion of pre-service teachers in the 1-year groups basing this view on mathematically oriented arguments being commensurate with a deeper orientation toward mathematics teaching. The number of 4-year program students who gave no reasons for their position is disappointing as pre-service programs emphasize the importance of reflection on practice. These are, however, first results and further analysis needs to be completed to answer the questions these preliminary results raise.

Acknowledgement This research was funded by a University of Melbourne Joint Research Grant (International). The coding assistance of Björn Schwarz, Nils Buchholtz, Björn Wissmach, Ling Schuller, and Tak Wai Ip is acknowledged. Thanks also to those who collected data at various sites.

References

Blömeke, S., Felbrich, A., Müller, C., Kaiser, G., & Lehmann, R. (2008). Effectiveness of teacher education: State of research, measurement issues and consequences for future studies. *ZDM – The International Journal of Mathematics Education, 40*(5), 719–734.

Bullough, R. V. (2001). Pedagogical content knowledge circa 1907 and 1987: A study in the history of an idea. *Teaching and Teacher Education, 17*(6), 655–666.

Kaiser, G., Schwarz, B., & Krackowitz, S. (2007). The role of beliefs on future teacher's professional knowledge. In B. Sriraman (Ed.), *Beliefs and mathematics: The Montana mathematics enthusiast monograph 3* (pp. 99–116). Charlotte: IAP.

Kaiser, G., Schwarz, B., & Tiedemann, S. (2010). Future teachers' professional knowledge on modelling. In R. Lesh, P. Galbraith, C. Haines, & A. Hurford (Eds.), *Modelling students' mathematical modelling competencies* (pp. 433–444). New York: Springer.

Krauss, S., Baumert, J., & Blum, W. (2008). Secondary mathematics teachers' pedagogical content knowledge and content: Validation of the COACTIV constructs. *ZDM – The International Journal of Mathematics Education, 40*(5), 873–892.

Schmidt, W. H., et al. (2007). *The preparation gap: Teacher education for middle school mathematics in six countries (MT21 report)*. East Lansing: Centre for Research in Mathematics and Science Education, Michigan State University.

Schwarz, B., Kaiser, G., & Buchholtz, N. (2008). Vertiefende qualitative Analysen zur professionellen Kompetenz angehender Mathematiklehrkräfte am Beispiel von Modellierung und Realitätsbezügen. In S. Blömeke, G. Kaiser, & R. Lehmann (Eds.), *Professionelle Kompetenz angehender Lehrerinnen und Lehrer* (pp. 391–424). Münster: Waxmann Verlag.

Shulman, L. S. (1986). Those who understand: Knowledge growth in teaching. *Educational Researcher, 15*(2), 4–14.

Shulman, L. S. (1987). Knowledge and teaching: Foundations of the new reform. *Harvard Educational Review, 57*, 1–22.

Part IV
Using Technologies: New Possibilities of Teaching and Learning Modelling

Chapter 30
Using Technologies: New Possibilities of Teaching and Learning Modelling – Overview

Gilbert Greefrath

1 Modelling Using Digital Tools

The solution of modelling problems using digital tools requires that two important translation processes take place. Firstly, the real situation of the problem has to be understood and translated into mathematical language. This translation concerns, depending on the modelling cycle used, different steps, for example, understanding the task, simplifying and mathematising.

The digital tool, for example, a computer algebra system calculator, though, cannot be used before the mathematical expressions have been translated into the language used by the computer. So a special computer model has to be built. The computer results then have to be translated into mathematical expressions again. Finally, the problem can be solved by relating the mathematical results to the given real situation. Using digital tools broadens the possibilities to solve certain mathematical models, which would not be used and solved if digital tools were not available. But the use of digital tools like in Fig. 30.1 gives a restricted view of using digital tools in applications and modelling. It is also possible to use the digital tools in many phases of the modelling process.

One type of these applications of digital tools in the modelling cycle is experimenting. For example, one can transform with the help of dynamic geometry software or a spreadsheet a real situation into a geometrical or numerical model. A very similar activity to experimenting is simulating real situations with digital tools. Experiments with a mathematical model are conducted, if the real situation is too complex. A common use of digital tools, particularly computer algebra systems, is the computation of numeric or algebraic results, which can not be reached by students without these tools or not in appropriate time.

G. Greefrath (✉)
University of Münster, Münster, Germany
e-mail: greefrath@uni-muenster.de

To the sector of computations with digital tools belongs also finding algebraic representations from given data. This so-called "algebracising" (Brown 2007) is characterised by the fact that real data are entered into the computer and the computer supplies an algebraic representation. In addition, digital tools can achieve the task of visualising. For example, given data can be represented with the help of a computer algebra system or a statistic tool in a coordinate system. This is then, for example, the starting point for the development of mathematical models.

In addition, the results of the computations can be visualised likewise. Digital tools can support control processes, for example, when operating with discrete functional models. The mathematical model can thus be numerically controlled. It is however just a graphic control with the help of the graph and the real data or – in other cases – also an algebraic control is conceivable. If one does not use handheld devices, but instead computers with Internet connection in mathematics education, then these can be used also for investigating information, for example, in connection with a real problem. In this way, real problems can be first understood and simplified. The different functions of the digital tools in mathematics lessons are important for modelling problems in different phases in the modelling cycle. So, control processes are usually settled in the last steps of the modelling cycle. Some possibilities for the employment of digital tools during a modelling process are represented in the following modelling cycle by Blum and Leiß (2006) (see Fig. 30.2). Therefore, the use of digital tools does not only create an important appendix to the modelling cycle (see Fig. 30.1), but also influences each part of the cycle (see Fig. 30.2). So the technology is relating to the real world and mathematical world of the modelling cycle.

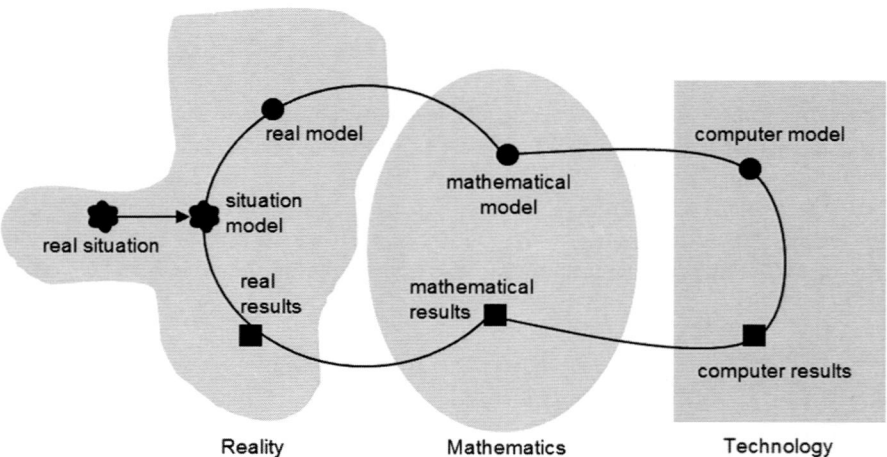

Fig. 30.1 Modelling cycle (Blum and Leiß 2006) with added computer model

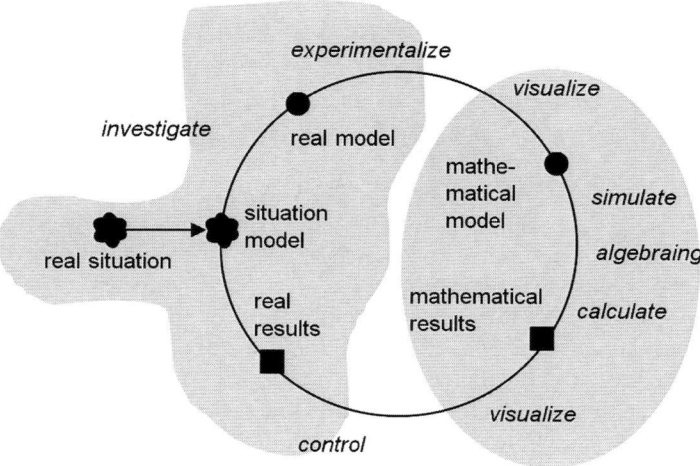

Fig. 30.2 Modelling cycle (Blum and Leiß 2006) with added influence of digital tools

2 Empirical Studies and Experiences on Modelling Using Digital Tools

One important idea of using digital tools in mathematics and especially in modelling lessons is the fact that the integrated numeric, graphic and symbolic tools of modern calculators and computers provide new ways of learning and understanding mathematics. Students following a programme, which emphasises multiple representations of algebraic ideas, are in fact better able to deal with mathematical tasks requiring representational fluency (Huntley et al. 2000).

But the digital tools are not only tools to support modelling activities of students. In many projects, teachers try to implement digital tools like notebooks with computer algebra software to bring more applications and modelling into the every day teaching experience (Henn 1998).

Geiger in this volume deals with experiences of secondary teachers with different adoption of computer algebra systems for supporting mathematical modelling. His study was completed over a 12-month period and involved six teachers, all from different schools in Australia. In this country, both CAS technology and mathematical modelling are strongly encouraged in recently developed mathematics syllabus documents. In the focus of the chapter are two teachers with differences in experience and expertise. The first teacher had taught mathematics for approximately 15 years although he had limited experience, prior to the project, of teaching mathematical applications and modelling or the use of technology when teaching mathematics. The other teacher is a very experienced user of CAS technology but had also embraced the teaching of mathematical modelling and applications to this extent. Geiger points out the different beliefs of these teachers and the influence on

CAS-enhanced modelling introduced in their classrooms. Geiger describes a very interesting adapted model of digital tools–enhanced mathematical modelling. He also sees the role of technology in every step of the modelling cycle (s. Fig. 30.2).

Neves, Silva and Teodoro in this volume present a new way to develop computational modelling learning activities in the context of physics. It is shown how Modellus can be used to develop modelling activities and the activities are reflected in student answers of a questionnaire. They describe examples in introductory mechanics which were implemented in the general physics course taken by first-year biomedical engineering students in Portugal. It is shown that they have a real-time visible correspondence between the animations with interactive objects and the object's mathematical properties defined in the model, and also the possibility of manipulating simultaneously several different representations like graphs and tables. The successful class implementation of the computational modelling activities was reflected in the student answers to the Likert scale questionnaire.

Greefrath, Siller and Weitendorf in this volume discuss the role of digital tools in the modelling cycle and present some interesting examples for modelling with digital tools. It is shown that digital tools can be helpful at any step of the modelling cycle. An example shows that CAS could be necessary to interpret and validate the solution. They also consider the special problem of good examination tasks with modelling problems and use of digital tools. Some criteria for good examination tasks with modelling problems and the use of technology have been found.

Some studies suggest that calculator use during instruction should be long term (i.e. 9 or more weeks) and calculators should be available during evaluations of middle and high school students' problem-solving skills (Ellington 2003). So we need more long-term studies concerning modelling with digital tools to get a better understanding of the interrelation of these two important aspects of learning mathematics.

References

Blum, W., & Leiß, D. (2006). How do students and teachers deal with modelling problems? In C. Haines, P. Galbraith, W. Blum, & S. Khan (Eds.), *Mathematical modelling (ICTMA12): Education, engineering and economics* (pp. 222–231). Chichester: Horwood.

Brown, J. (2007). Early notions of functions in a technology-rich teaching and learning environment (TRTLE). In J. Watson & K. Beswick (Eds.), *Mathematics: Essential research, essential practice. Proceedings of the 30th annual conference of the Mathematics Education Research Group of Australasia*, Hobart (Vol. 1, pp. 153–162). Adelaide: MERGA.

Ellington, A. J. (2003). A meta-analysis of the effects of calculators on students' achievement and attitude levels in precollege mathematics classes. *Journal for Research in Mathematics Education, 34*(5), 433–463.

Henn, H. W. (1998). The impact of computer algebra systems on modelling activities. In P. Galbraith, W. Blum, G. Booker, & I. Huntley (Eds.), *Mathematical modelling: Teaching and assessing in a technology rich world* (pp. 115–123). Chichester: Horwood.

Huntley, M. A., Rasmussen, C. L., Villarubi, R. S., Santong, J., & Fey, J. T. (2000). Effects of standards-based mathematics education: A study of the core-plus mathematics project algebra and function strand. *Journal for Research in Mathematics Education, 31*, 328–361.

Chapter 31
Factors Affecting Teachers' Adoption of Innovative Practices with Technology and Mathematical Modelling

Vince Geiger

Abstract This chapter contrasts the experiences of two secondary teachers from different education jurisdictions in relation to the adoption of computer algebra systems (CAS) as a supporting technology for teaching mathematical modelling. The study reveals that the differing dispositions and beliefs of these teachers were highly influential in the degree to which CAS-enhanced mathematical modelling was introduced into their classrooms. Thus, the role of technology, specifically CAS, within theoretical models of the process of mathematical modelling, can be viewed as variable and situational rather than fixed.

1 Introduction

Mathematical modelling and/or applications of mathematics appear in the curriculum documents of most Australian states. While there is significant research related to solving contextualized problems through the use of the multiple representational facilities offered by digital technologies, and substantive argument to support the use of CAS to enhance the process of mathematical modelling, literature that deals with teachers' adoption of technology to enhance mathematical modelling is only just emerging. This chapter investigates the affordances and constraints that influence the adoption of computer algebra systems (CAS) as a means of enhancing students' learning experiences with mathematical modelling by two teachers from different Australian educational jurisdictions.

V. Geiger (✉)
Australian Catholic University, McAuley Campus, Brisbane, Australia
e-mail: vincent.geiger@acu.edu.au

2 The Use of CAS in Mathematical Modelling

Over the past two decades, researchers have argued that digital technologies have the potential to enhance the teaching and learning of mathematics (e.g., Zbiek et al. 2007). Although research into the transformative power of digital technologies in relation to mathematical instruction has proliferated (Hoyles and Noss 2003), studies into how the availability of technology impacts on the teaching and learning of mathematical modelling are less prevalent. For example, the Study Volume produced from the 14th Study of the International Commission for Mathematical Instruction entitled *Modelling and Applications in Mathematics Education* (Blum et al. 2007) contains only one chapter out of 58 that focuses on technology use in mathematical modelling.

While abstraction is one of the cornerstones of the discipline of mathematics, the capacity to make use of mathematics in real world contexts is recognized internationally as an equally valuable capability (e.g., National Council of Teachers of Mathematics 2000). At the same time, the incorporation of mathematical modelling and applications of mathematics into mathematics curricula varies internationally and this is generally restricted to upper secondary school when implemented (Stillman 2007).

Encouragingly, a number of researchers have found that the multiple representational facilities offered by digital technology can enhance the capacity of students to solve contextualized mathematics problems (see for example, Huntley et al. 2000), although Kiernan and Yerushalmy (2004) caution that such benefits may not be realized without supportive changes to curriculum and to modes of instruction. Thus, research into the potential benefits offered by technologies which are relatively new to school contexts, such as CAS-enabled technologies, must also consider the affordances and constraints encountered within authentic classroom settings when such technologies are introduced. While it has been argued that CAS has the potential to provide access to more sophisticated life-related problems (Thomas 2001) through the highly integrated nature of representational facilities and the enhanced computational power offered by these devices, the potentials of modelling and CAS have generally been considered separately in mathematics education research (Thomas et al. 2004). The increasing introduction of CAS-enabled technologies into mainstream mathematics classrooms means there is a need to understand the implications of this technology for all aspects of classroom practice including that of mathematical applications and modelling.

While CAS appears to offer potential advantages to the teaching and learning of mathematical modelling, how this is realized in the classroom is dependent on the disposition of teachers toward both technology and mathematical modelling. While the influence of teachers' dispositions and beliefs toward the uptake of technology into school mathematics classrooms is well documented, teachers' attitudes to the incorporation of mathematical modelling into instruction is also a factor that affects the degree of implementation of modelling activity into school classrooms (Stillman and Galbraith 2009).

3 Models of the Use of Technology in Mathematical Modelling

Mathematical modelling is often presented as a cyclic process that starts with a problem set in a life-related context which is abstracted into a mathematical representation of the contextualized situation and solved through the application of mathematical routines and processes. The solution is then brought into relief against the original problem to consider its fit with the original context. If the fit is not considered sufficient, adjustments are made to the model and the process repeated until a satisfactory fit is achieved. The role of technology in this process has been described by Galbraith et al. (2003). Here, we argued that mathematical routines and processes, students, and technology are engaged in partnership during the *Solve* phase of a problem, which follows from the abstraction of a problem from its contextualized state into a mathematical model. This view identifies the conceptualization of a mathematical model as an exclusively human activity while the act of finding a solution to the abstracted model can be enhanced via the incorporation of technology. Thus, technology is seen as a tool used to interact with mathematical ideas only after a mathematical model is developed, rather than as a tool for the exploration and development of a model or its validation as a reliable representation of a life-related situation.

Our recent research, and that of others, indicates that this is a limited view of the role of technology in mathematical modelling. Confrey and Maloney (2007), for example, argue that the process of modelling is founded on two activities: inquiry and reasoning. Confrey and Maloney (2007) claim that it is through the coordination of these artifacts and the processes of inquiry, reasoning, and experiment that an indeterminate situation is transformed into a determinate situation. While technology in this model can incorporate and generate representations which assist in the transformation of an indeterminate to a determinate situation, it also plays a central role in coordinating the inquiry, reasoning, and systematizing that lead to a determinate situation.

In a recent study into the role of technology in mathematical modelling, Geiger et al. (2008) found that CAS technology could be used by teachers as a provocative agent for stimulating secondary school students' exploration of mathematical concepts within life-related problems. While not a primary focus of this study, it was noted that CAS was used by students in a broader range of modes than simply as a tool to effect a solution to an already mathematized problem situation.

4 Context of the Study

The study was completed over a 12-month period and involved six teachers, all from different schools, comprising three teachers from Brisbane in Queensland, and three from Canberra in the Australian Capital Territory (ACT). In Australia, education is administered by jurisdictions based on state boundaries. States are responsible for the independent development of curriculum documents and resources as well as statewide assessment regimes. As a result, curriculum and assessment varies between states. The curriculum contexts of both Queensland and ACT are outlined below.

4.1 Curriculum Contexts

In the Australian Capital Territory, the study of mathematics through applications and modelling and the use of technology are strongly encouraged in recently developed mathematics syllabus documents. There is no limit on the type of technology that can be used. Consequently, teachers in Canberra have the freedom to teach mathematical modelling and make use of any available technology within mathematics, but the depth to which these aspects of mathematics teaching are implemented is the prerogative of individual schools.

In Brisbane, Queensland, the teaching and assessment of mathematical modelling and applications is a mandatory objective of all state mathematics syllabuses. The use of technology is required to a minimum level: specified as a graphing calculator. No upper limit, in relation to the type of technology that can be used, is identified. Thus, for teachers in Brisbane, it is mandatory to teach mathematical modelling and to make use of technology in mathematics classes. While the degree to which technology is used is the prerogative of an individual school, teachers are free to choose any available technology to study mathematics.

In summary, both CAS technology and mathematical modelling were important elements within the syllabuses of both educational jurisdictions.

4.2 Teachers' Backgrounds

Teachers were invited into the program on the basis of recommendations from school systems or professional teachers' associations or through other professional networks of the researchers. Participation in the project was entirely voluntary. There were differences in the experience and expertise of the two teachers who are the focus of this paper.

Teacher 1 is from Canberra and had taught mathematics for approximately 15 years although he had limited experience, prior to the project, of teaching mathematical applications and modelling or the use of technology when teaching mathematics. This was because he had only recently taught the subject in which these aspects of mathematics instruction had been introduced into ACT's mathematics syllabuses. His students had some experience with using spreadsheets and with the use of graphing calculators for graphing functions but had only encountered CAS-enabled technologies through the project.

Teacher 2 is from Brisbane and is a very experienced user of CAS technology (specifically TI-Nspire). He had also embraced the teaching of mathematical modelling and applications to the extent that he had developed a statewide reputation for expertise in this area. In his school, students are required to use graphing calculator technology from early secondary school (Year 9). While the students in this teacher's class had made use of graphing calculators for 2 years prior to the study, they had only been introduced to CAS-enabled technology from the beginning of the year of the study.

5 Method and Approach to Data Collection

A case study approach (Stake 2005) was used to document the actions and interactions of the two teachers who are the focus of this report. Sampling was purposive as cases were chosen for the capacity to illuminate and enhance understanding rather than for representativeness (Stake 2005). In particular, the cases reported in this chapter were selected because of the differences between the views and actions of the teachers in relation to the implementation of CAS technology in their classrooms as a means of enhancing the study of mathematical modelling.

The project commenced with each participant teacher attending an event, one in Brisbane and one in Canberra, where they were provided with instruction on the use of a hand-held CAS-enabled technology, TI-Nspire, and also examined pre-prepared resources designed for the use of CAS within mathematical modelling tasks. Teachers were provided with class sets of TI-Nspire and asked to look for opportunities to make use of CAS when teaching modelling and applications.

It was anticipated that each teacher's classroom would be observed and videotaped three times across the duration of the project and audiotaped semistructured interviews, with the teacher and also with three to four students nominated by the teacher, conducted after each video session. In addition, a teleconference was planned at the end of the project where all teacher participants would meet to discuss personal observations drawn from their participation in the study. While all Brisbane teachers participated in the project fully, the teachers in Canberra appeared to struggle to prepare lessons that they felt comfortable being observed in their classrooms. The Canberra-based researcher was only able to arrange one audio interview with one of the teacher participants from Canberra within the initial phase of the project. Even after the project team provided additional assistance to the teachers in Canberra, they found it difficult to design lessons they believed were worth observing. As a result, the only data collected in Canberra was an initial audio interview and one final interview between one of the teachers and a researcher.

Data reported upon here are selected from teachers' semi-structured interviews and the final videoconference. The video session and audio interviews were transcribed and examined for evidence of different teachers' views on the following three issues: the value of using technology to enhance mathematics learning; the importance, or otherwise, of mathematical modelling within a students' mathematics education; and the potential of CAS technology to enhance mathematical modelling. Individual teachers' responses were studied across data collection instances for consistency of a teacher's position on these three issues. The excerpts presented below are representative of opinions consistently expressed in the two selected cases.

6 A Tale of Two Cities and Two Teachers

The following excerpts are taken from the second (and final) interview of Teacher 1 in Canberra and from comments made by Teacher 2 during the teleconference session at the conclusion of the Brisbane phase of the project. These comments reveal both teachers' dispositions toward teaching mathematical modelling and the use of CAS.

6.1 Teacher 1

Teacher 1: I think they're a tool that's great but I'm just thinking – targeting at my group they're not up to using CAS at the moment and yet there are other groups that possibly are in the same class that I'd like to see gain a bit more of the basics before they start using the black box. Just thinking of their background – and this goes back to what maths teaching is going on anyway – they're just missing some basics and fundamentals to worry about this end of things. It's probably not necessary at the moment for some groups. It's great in some instances but this is me not having taught much *Methods* before either but in the main I couldn't really see myself using it a great deal. I'll push myself to use it in certain applications then I'm sure I'd be able to answer that question a lot better.

Researcher: And those lesson ideas that were provided; were any of those of any potential relevance? Would you use them for assignments or class lessons or things like that?

Teacher 1: Yeah, I would, I would. I've read through them but I can't think of them per se. I've seen a couple of them before and yeah I would like to give that a go but.... The right group? The topics in the class at the moment haven't really lent themselves to that...

In this excerpt, the teacher reveals he has concerns about introducing CAS and mathematical modelling because he believes that students should learn the basics of mathematics before they engage in the use of technology or explore mathematical modelling activities. He perceives there is a danger that CAS might be used as a *Blackbox* where students perform mathematical procedures they do not fully understand. Teacher 1 also expresses concern that his own inexperience in the subject matter has limited his capacity to think of ways of using technology and finding appropriate applications of mathematics. He also indicates that "the right group" is required to work with technology and mathematical modelling which implies he does not believe these aspects of learning mathematics are appropriate to all students. Finally, Teacher 1 believes that only certain topics in mathematics lend themselves to the use of technology and mathematical modelling activity. It seems that these beliefs restricted this teacher's capacity to implement a teaching program which incorporates the use of CAS and mathematical modelling and applications.

6.2 Teacher 2

Researcher: So what difference does it make?
Teacher 2: It gives the kids the access to the problems. If we didn't have the CAS calculators we couldn't do half the stuff that we do.
Researcher: Because it allows them to work with algebra? Or because…?
Teacher 2: It is the integration of the whole box and dice. From my perspective it is the integration of the whole lot together. We have a set of data and we try and build a model from that. And we do a scatter plot of that and we just make some sorts of decisions about the model. We go away and we build a model and we want to do and we make some sorts of predictions about that.
Researcher: So you can differentiate and integrate functions that the kids wouldn't be able to attempt without it?
Teacher 2: Yeah that is right. It allows access for the lower achieving kids too. Your lower achievers may be struggling with differentiation or integration at that particular point in time…but they can still have access to the problem. I would expect my high achieving kids to do it both ways. They can do it pen on paper and they can do it using the CAS calculators. But then my lower achieving kids can still engage in the problem and still make some meaningful contributions. If they don't get caught up in all that manipulation they can still be thoughtful about it.

This teacher indicates he believes CAS should be available to all students at all levels because it is an enabling technology at a number of levels. In the excerpt above, he indicates that CAS can provide a scaffold over gaps in basic mathematical knowledge and understanding that allows a greater number of students access to interesting and authentic problems that require complex mathematics. Further, CAS allows students to make connections between different aspects of mathematics which can then be brought to bear simultaneously on life-related problems. In Teacher 2's classroom, CAS is used in nearly all phases of the modelling cycle to: explore a life-related context by representing data in a graphical format; develop initial models based on the data; refine the model; and use the model to make predictions. These views are highly supportive of the use of CAS in mathematical modelling activities as CAS is seen to offer potential benefits to all students when attempting to explore life-related contexts through mathematical modelling. Observation of Teacher 2's classroom demonstrated his commitment to the use of CAS in mathematical modelling activities as he made use of a range of authentic life-related problems, including some very sophisticated examples, in his teaching on a regular basis.

7 Discussion and Conclusions

Despite working in educational jurisdictions that supported the teaching of mathematical modelling and placed no limit on the type of technology that could be employed in teaching and learning, the views of these two teachers and the degree to which they engaged in the project were very different.

Teacher 1's views appear to have prevented him from engaging in the project in any genuine sense. His views on what it means to learn mathematics seem to make him suspicious of the use of technology to support student learning as he indicates students should learn the "basics" first. He appears suspicious that CAS is likely to support a *Blackbox* approach to solving problems in mathematics and there is no indication that he would encourage students to use CAS to explore the possibilities for finding a solution within a contextualized mathematical situation. This position seems to be consistent with a view that the use of technology is appropriate after the "mathematics" of developing a model has first taken place which is similar to the model of technological use in mathematical modelling that is portrayed in Fig. 31.1. In addition, Teacher 1 believes that the use of technology and the study of mathematical modelling are only appropriate to groups of students who have mastered the "basics." This implies that some students should never have exposure to either aspect of the syllabus that is current in the Australian Capital Territory.

By contrast, Teacher 2 believes that technology, specifically CAS, can provide students with the opportunity to investigate authentic and complex applications of mathematics. He also sees CAS as a tool that enables students to continue to engage in sophisticated contextualised problems even though they may have gaps in their existing mathematical knowledge. This teacher indicates that CAS can be used to

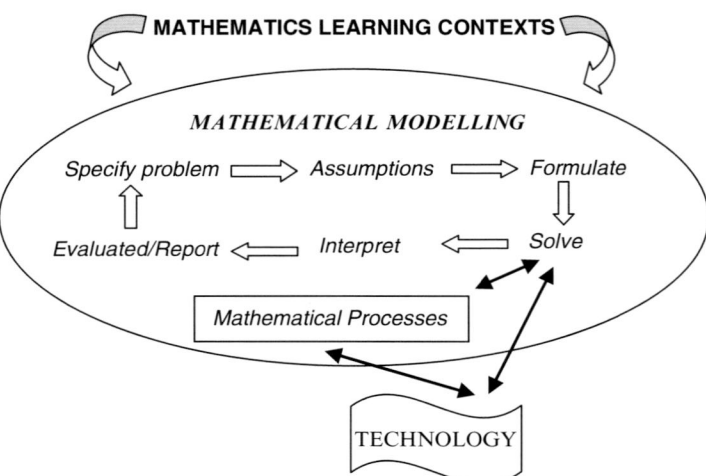

Fig. 31.1 Some technological and mathematical interrelationships from Galbraith et al. (2003, p. 114)

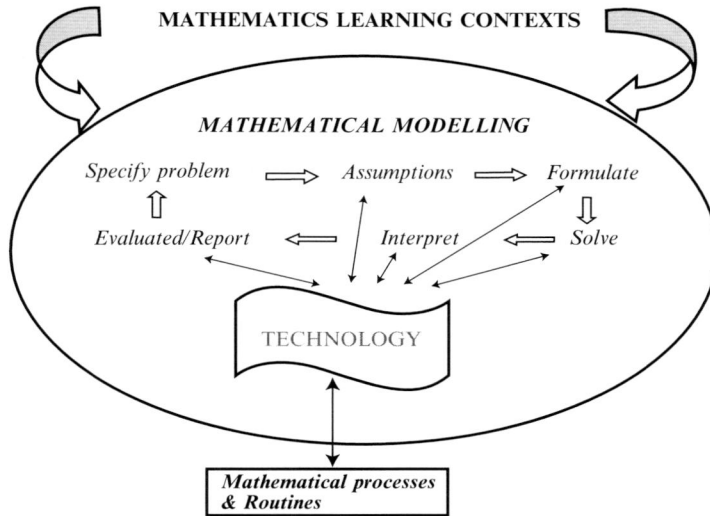

Fig. 31.2 Adapted model of technology-enhanced mathematical modelling

assist students to explore data associated with contextualized situations in order to formulate and then validate a model. He also appears to have an expectation that technology has a role if it is necessary to reformulate a model because the validation phase indicates a model is in need of improvement. In this view, technology has a role to play in mediating mathematical processes at nearly all points within the modelling cycle as illustrated in Fig. 31.2 which sits in contrast to the original model proposed by Galbraith et al. (2003). This representation of the role of technology in the modelling cycle is consistent with that of Confrey and Maloney (2007) who view technology as vehicle for the coordination of the inquiry and reasoning necessary to engage with a problem set in a real-life context.

The different perspectives, described above, indicate that the role of technology, and specifically CAS, within the modelling cycle is situational, that is, dependent on teachers' personal views on the use of technology and of mathematical modelling and their perception of the circumstances in which they work. The implementation of CAS-active mathematical activities and of the teaching of mathematical modelling seems less influenced by curriculum documents which offer strong encouragement for the use of both aspects of mathematical teaching and learning.

While there is no attempt to generalize findings from the views of the two teachers described in this chapter, their positions on the use of CAS and mathematical modelling provide insight into the affordances and constraints that may be encountered while attempting to implement initiatives that aim to enhance the teaching of mathematical modelling through the use of CAS-enabled technologies. In doing so, this chapter highlights the essential role teachers play in relation to pedagogical change and, it would seem, especially in relation to technological development and when traditional views of what it means to learn mathematics are challenged.

References

Blum, W., Galbraith, P., Henn, H., & Niss, M. (Eds.). (2007). *Modelling and applications in mathematics education: The 14th ICMI study*. New York: Springer.

Confrey, J., & Maloney, A. (2007). A theory of mathematical modelling in technological settings. In W. Blum, P. Galbraith, H. Henn, & M. Niss (Eds.), *Modelling and applications in mathematics education: The 14th ICMI study* (pp. 57–68). New York: Springer.

Galbraith, P., Renshaw, P., Goos, M., & Geiger, V. (2003). Technology-enriched classrooms: Some implications for teaching applications and modelling. In Y. Qi-Xiao, W. Blum, S. K. Houston, & J. Qi-Yuan (Eds.), *Mathematical modelling in education and culture* (pp. 111–125). Chichester: Horwood.

Geiger, V., Faragher, R., Redmond, T., & Lowe, J. (2008). CAS enabled devices as provocative agents in the process of mathematical modelling. In M. Goos, R. Brown, & K. Maker (Eds.), *Navigating currents and charting directions (Proceedings of the 31st Annual Conference of the Mathematics Education Research Group of Australasia, Brisbane, Qld)* (pp. 246–253). Brisbane: MERGA.

Hoyles, C., & Noss, R. (2003). What can digital technologies take from and bring to research in mathematics education? In A. J. Bishop (Ed.), *Second international handbook of mathematics education* (pp. 323–349). Dordrecht: Kluwer.

Huntley, M. A., Rasmussen, C. L., Villarubi, R. S., Santong, J., & Fey, J. T. (2000). Effects of standards-based mathematics education: A study of the Core-Plus Mathematics Project Algebra and Function Strand. *Journal for Research in Mathematics Education, 31*, 328–361.

Kiernan, C., & Yerushalmy, M. (2004). Research on the role of technological environments in algebra learning and teaching. In H. Chick, K. Stacey, & J. Vincent (Eds.), *The future of teaching and learning of algebra (Proceedings of the 12th ICMI Study Conference)* (pp. 99–152). Melbourne: The University of Melbourne.

National Council of Teachers of Mathematics. (2000). *Principles and standards for school mathematics*. Reston: NCTM.

Stake, R. (2005). Qualitative case studies. In N. Denzin & Y. Lincoln (Eds.), *The Sage handbook of qualitative research* (3rd ed.). Thousand Oaks: Sage Publications, Inc.

Stillman, G. (2007). Implementing case study: Sustaining curriculum change. In W. Blum, P. Galbraith, H. Henn, & M. Niss (Eds.), *Modelling and applications in mathematics education: The 14th ICMI study* (pp. 497–502). New York: Springer.

Stillman, G., & Galbraith, P. (2009). Softly, softly: Curriculum change in applications and modeling in the senior secondary curriculum in Queensland. In R. Hunter, B. Bicknell, & T. Burgess (Eds.), *Crossing divides (Proceedings of the 32nd Annual Conference of the Mathematics Education Research Group of Australasia, Wellington, NZ)* (pp. 201–208). Wellington: MERGA.

Thomas, M. O. J. (2001). Building a conceptual algebra curriculum: The role of technological tools. In H. Chick, K. Stacey, J. Vincent & J. Vincent (Eds.), *The future of teaching and learning of algebra (Proceedings of the 12th ICMI Study Conference)* (pp. 582–589). Melbourne: The University of Melbourne.

Thomas, M. O. J., Monaghan, J., & Pierce, R. T. (2004). Computer algebra systems and algebra: Curriculum, assessment, teaching and learning. In H. Chick, K. Stacey, & M. Kendal (Eds.), *The future of teaching and learning of algebra (Proceedings of the 12th ICMI Study Conference)* (pp. 151–186). Dordrecht: Kluwer.

Zbiek, R., Heid, M., Blume, G., & Dick, T. (2007). Research on technology in mathematics education: A perspective on constructs. In F. K. Lester (Ed.), *Second handbook of research on mathematics teaching and learning: A project of the National Council of Teachers of Mathematics* (pp. 3–38). Charlotte: Information Age Pub.

Chapter 32
Modelling Considering the Influence of Technology

Gilbert Greefrath, Hans-Stefan Siller, and Jens Weitendorf

Abstract In this chapter, we discuss the specifics about modelling with technology. First, some general points are mentioned for a better understanding of the role of technology in the modelling cycle. In the second part, we describe a detailed example for using technology for a modelling problem. Then, we exemplarily show some technical possibilities of software tools for teaching mathematical modelling to reflect the role of technology in the modelling cycle. In the fourth part, we consider the special problem of good examination tasks with modelling problems and use of technology. In all those examples, which are described and discussed in detail, the connection of modelling in terms of using technology becomes obvious.

1 Introduction

The development of mathematics has been influenced by the development of technology from the beginning. In the age of information technology and 'New Media', electronic additives are broadening the horizon in both the education and teaching of mathematics.

Electronic devices are supporting cognition; they are part of the cognition and changing cognition in mathematics education. The support is given by the possibility of transferring complex operations to technology and the use of different computer

---

G. Greefrath (✉)
University of Münster, Münster, Germany
e-mail: greefrath@uni-muenster.de

H.-S. Siller
University of Salzburg, Salzburg, Austria
e-mail: hans-stefan.siller@sbg.ac.at

J. Weitendorf
Gymnasium Harksheide, Norderstedt, Germany
e-mail: JWeitendorf@t-online.de

models. The efficient use of methods involving intensive computation and/or models is guaranteed.

By using technology in mathematics education, an enormous shift from the accomplishment to the planning of problem-solving can be done. Therefore, a useful shift of emphasis from mathematical operations to the use of mathematical knowledge and reflections can be realised. If you use technology, for example, CAS (Computer-Algebra-Systems), Graphical Calculators, Spreadsheets, DGS (Dynamical Geometry Software), or Modelling Software, some inner-mathematical reflections have to be done, because the solution created by technology has to be kept in mind.

Looking back in history, the use of technology in mathematics education started in the 1970s (Siller 2008). At that time, numerical calculators were state of the art. Today the development in technology has been very successful, and from a large range of software, the adequate technology for the respective problem must be selected. Numerical calculators still exist in many classrooms, but other electronic devices such as CAS-calculators (e.g., CASIO ClassPad) are more powerful instruments for education. Such additives should not only be used as 'number crunchers', they can be used, for example to experiment, to check solutions and as a method for communication in mathematics education. In addition, abstract mathematical objects can be visualised very easily using these instruments by using numerical, graphical or symbolic manipulation.

As the role of technology can strongly influence mathematics education, in particular mathematical modelling, it is necessary to include the role of technology-additives in the modelling cycle. An approach has been conceptualised by Siller and Greefrath (2010). In this first approach, we focus on the translations from mathematics to technology and back, which forms an obstacle for students in the modelling process (Fig. 32.1).

Based on this graphical illustration, we have to discuss the use of technology in terms of modelling in a more detailed way. By the help of some examples, which can be found in tests, textbooks or in education suggestions, problems, possibilities and ideas shall be shown.

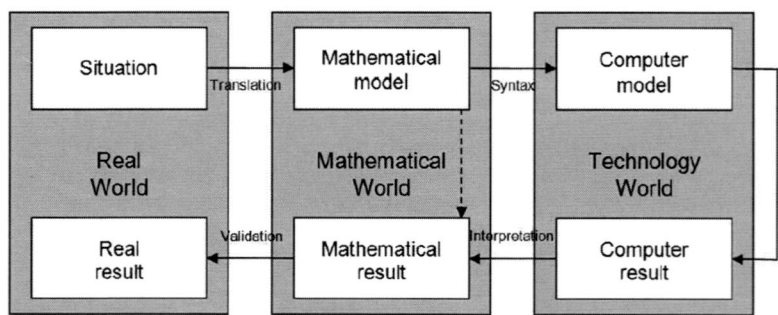

Fig. 32.1 Modelling cycle concerning technology

2 Alcohol in Blood – A Prospective Example for Modelling with Technology

The starting point of a lot of modelling examples is a real-life situation, at best procedures, scenes or incidents in real life, shown in Siller and Maaß (2009) or the ISTRON-series (http://istron.ph-freiburg.de – last access: 24.06.2010). But it is possible to find 'simple' examples in school-books, too. The only condition a teacher has to provide is that students are allowed to discuss such examples in an extraordinary way so that they are thinking about several opportunities that could happen. By looking through Austrian school-books for Mathematics, an interesting example on the topic of alcohol can be found. Students are confronted with it, nearly their whole life long because in the press, it is possible to find different announcements on this topic very often. Links to teaching mathematics can easily be found, and a starting point for cross-curricular teaching could be met. In Malle et al. (2006, p. 113), such an example is stated (Fig. 32.2):

> In the adjoining graph you can see the chronological sequence of the concentration of alcohol in blood (in ‰) after the consumption of a particular abundance of beer with 5% or alternatively 3%. Describe the progress in your own words. Can you find similarities; in what way are they different?

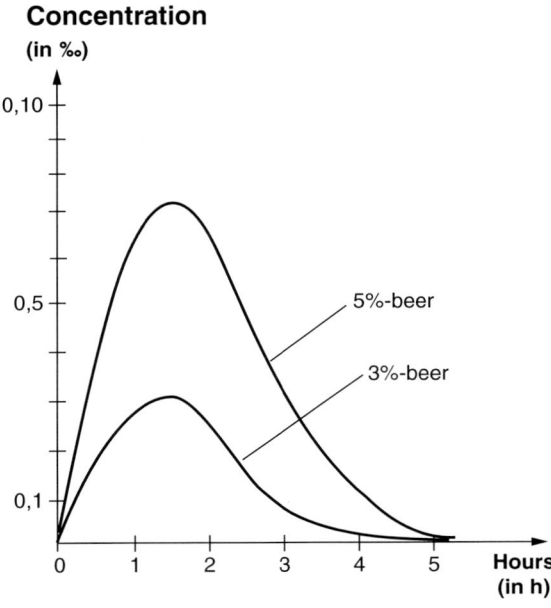

Fig. 32.2 Graph in the example

2.1 Modelling a Theoretical Concentration

By discussing this example with students, several different questions may arise. They could be taken as a starting point for modelling activities in class. Some questions which could be interesting for students are:

- We have learned that the reduction of alcohol in blood takes place at a constant rate: In every hour, the same amount of alcohol is reduced! Is that right?
- In the case of the same amount of alcohol in blood, the reduction of the 5%-beer is faster than with the 3%-beer. Why?
- In the graph it is shown: if you consume alcohol once, it will never disappear from the blood. Is that right?

This leads to the question: How does the degradation and absorption of alcohol take place in real life? By searching for an answer, a lot of different and inconsistent results can be found – especially by using the world-wide-web. A definite answer cannot be given. Some more questions, which are of mathematical relevance, may help to find a possible answer and are the starting point for this modelling approach:

- Somebody is drinking ½l of beer. How much alcohol in blood remains in this person after 2 h?
- Somebody has caused an accident. Two hours later the driver has to deliver a blood-sample. The alcohol test produced a blood alcohol level of 0.7. What was the alcohol-concentration in the blood when the accident happened?

A first and very rough calculation can be done by the Widmark-formula (Widmark 1932) (Fig. 32.3). The concentration of alcohol only depends on the

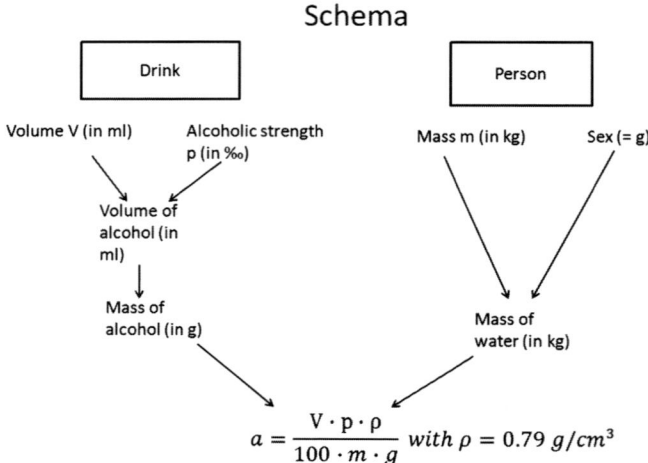

Fig. 32.3 Scheme of the Widmark formula

person (drinking it) and the alcoholic drink itself. The parameters which are included are the amount of alcoholic drinks consumed and the alcoholic strength, as well as the mass of the person drinking alcohol and the sex. The age or the height is not considered, as in the Watson-formula (Watson et al. 1980). By following the scheme first considerations are possible. Think about a man ($g=0.7$) with a mass of 63 kg drinking half a litre of beer ($p=4.5\%$). Taking this formula, you will find that this person has a theoretical amount of $a=0.4$ alcohol level in blood. Now it is possible to think about the reduction of alcohol. Some assumptions that are made are

- The constant reduction of alcohol per hour (is an empirical value of 0.1 alcohol level/h $\leq d \leq$ 0.2 alcohol level/h depending on sex, constitution etc.)
- $b(t)$ as a linear model for the concentration of alcohol after t hours ($b(t) = a - d \cdot t \Rightarrow b(t) = 0.4 - 0.12t$).

However, by using these assumptions, it can be seen that these ideas may be simple but are not very realistic. In reality, the process is more complex because consumed alcohol will not fade into the blood suddenly; it is mostly absorbed by the gastrointestinal tract and the absorption and reduction of alcohol are two overlapping procedures. So we have to think about using another approach.

2.2 Absorption and Reduction Shown as a Mathematical Process

Considering how to show these two procedures leads to two different models. The first one describes the situation as a possible starting point to the problem; the second one leads to some interesting solutions.

2.2.1 First Model – Linear Approach

The time until the whole alcohol is faded into the blood is a well-known empirical value – it lasts about 60 min. Knowing this value, it is possible to make an assumption, which should help to find an appropriate linear model. It should be that in the same unit of time, the same amount of alcohol is absorbed. Therefore, the following linear model is constructed:

For the reduction of alcohol, the well-known linear function $b(t) = 0.4 - 0.12 \cdot t$ can be found. The absorption of alcohol can be shown through $a*(t) = 0.4 \cdot t$. But alcohol is already reduced after 1 h, so we get the following (linear) function for the real absorption $a(t) = 0.28 \cdot t$. By looking at the graphs of these functions (Fig. 32.4), it is possible to see that although the progress is similar to the one shown in the schoolbook example, the curve does not fit at all. So another model has to be found.

The technology in this part of the example is only used for validating the results. The mathematical objects, that is the linear functions, are displayed and by comparing these graphs with the graph of the example, everybody is able to recognise that this interpretation is not possible. The technological help allows students to recognise efficiently that the chosen model is not appropriate at all.

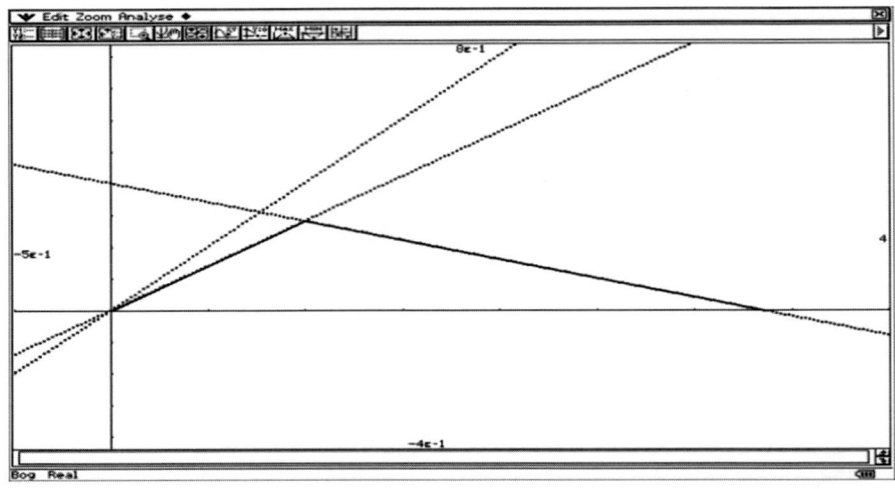

Fig. 32.4 Linear processes

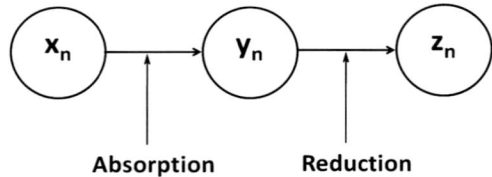

Fig. 32.5 Absorption and reduction

2.2.2 Second Model – Semi-linear Approach

The absorption of alcohol in blood is realised by a process of diffusion. That means that a fixed portion r, which can be found in the gastrointestinal tract, is absorbed by the body. Not the whole amount of alcohol is absorbed in 1 h, but nearly all of it, let us say about 95% (because $0.95 \approx 1 - 0.95^{60}$, that is about 5%/min). Therefore, the factor r is equal to 0.05. Now a discrete model with the parameters x_n (amount of alcohol in the gastrointestinal tract after n minutes), y_n (amount of alcohol in blood after n minutes) and z_n (reduced alcohol after n minutes) with the initial values $x_0 = a$ and $y_0 = z_0 = 0$ can be constructed. By thinking about the process, the absorption and reduction could be shown as in Fig. 32.5. The following equations show the results:

$$x_{n+1} = x_n - 0.05 x_n = 0.95 x_n$$

$$y_{n+1} = y_n + 0.05 x_n - 0.002$$

32 Modelling Considering the Influence of Technology

	A	B	C	D	E
3			Absorption-Ra...	r=	0.05
4			Reduction-Rate	d=	2E-3
5					
6					
7					
8			Supply	Bowel	Change-... Blood
9					
10	0		0	0	0
11	1		0	0	0
12	2		0	0	0
13	3		0	0	0
14	4		0	0	0
15	5		0	0	0
16	6	0.4	0	0	0
17	7		0.4	0.02	-2E-3
18	8		0.38	0.019	0.016
19	9		0.361	0.01805	0.033
20	10		0.34295	0.017148	0.0491
21	11		0.3258025	0.016290	0.0642
22	12		0.309512375	0.015476	0.0785
23	13		0.2940367563	0.014702	0.0920
24	14		0.2793349184	0.013967	0.1047
25	15		0.2653681725	0.013268	0.1166
26	16		0.2520997639	0.012605	0.1279
27	17		0.2394947757	0.011975	0.1385
28	18		0.2275200369	0.011376	0.1485
29	19		0.2161440351	0.010807	0.1579
30	20		0.2053368333	0.010267	0.1667

Fig. 32.6 Spreadsheet solution

$$z_{n+1} = z_n + 0.002$$

It is valid that $x_{n+1} + y_{n+1} + z_{n+1} = x_n + y_n + z_n = \cdots = a \ \forall n \in N$. By recognising x_n as a geometrical sequence and z_n as an arithmetical sequence, the possible solution $y_n = a \cdot (1 - 0.95^n) - 0.002n$ can be found. By using technology, this model can be simulated in many different ways. One is shown in Fig. 32.6 with a graphical representation in Fig. 32.7.

By thinking about such natural processes, a lot of other models (e.g. continuous model or a model by differential equations) can be found, too. Of course, validation is necessary. It can be done easily with Wilkinson et al. (1977, Fig. 3, p. 221).

The role of technology in this part is multilayered. It is used for validating, interpreting as well as experimenting. The constructed model is implemented as it is demanded by the tool used. The example shows that different computer applications are helpful to construct and work with varying mathematical models. Someone will be aware that the use of spreadsheets, CAS or other tools supports different mathematical models. So the use of technology in modelling needs not only a flexible use of models but also flexibility in using computer applications.

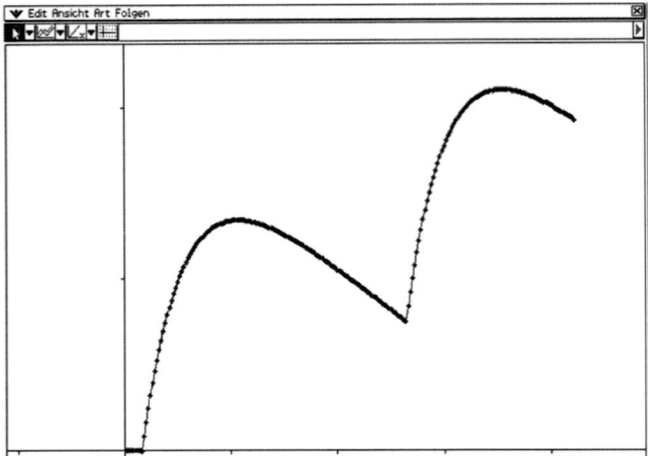

Fig. 32.7 Graphical output of the solution

2.3 The Role of Technology in the Modelling Cycle

Technological tools are often helpful or sometimes even necessary in the modelling process. Some problems are solved faster (as it can be seen in the example before) or even trivialised; others can only be solved by using technology. In connection with this, it is important to discuss the influence of technology on the modelling cycle. One aspect, the role of technology in the step from the mathematical model to the mathematical result, was discussed in Sect. 1. The question is whether technology only helps to deal with complex formulas or whether technology can even be helpful in the process of understanding a problem. Therefore, the relationship between reality and the model of reality and the step to the mathematical model can be influenced by the type of technology that is used. For a scientific approach, you have to choose between CAS, DGS, spreadsheets and other software, for example for dynamic systems.

We would like to show examples of different kinds of technology which can be used and show that technology can be helpful for each part of the modelling cycle. Especially when using the Casio ClassPad 330, you have the opportunity to use different kinds of technology which are interconnected. For example, it is possible to use the spreadsheet with the power of CAS. The next two examples show that technology is helpful in developing an idea of how to start and how to validate results.

2.4 An Example Where Technology Is Helpful to Get an Idea

> Build a new waste pipe to connect the two villages A and B with the sewer shown in Fig. 32.8, where it is only possible to connect them at one point of the sewer and it is not possible to connect three pipes at one point.

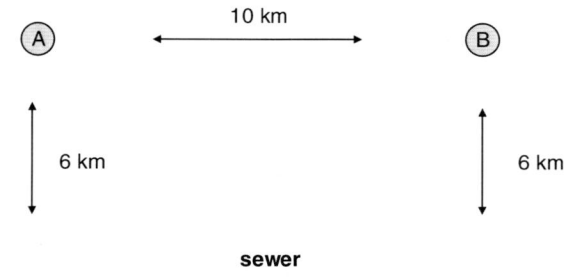

Fig. 32.8 Waste pipe problem

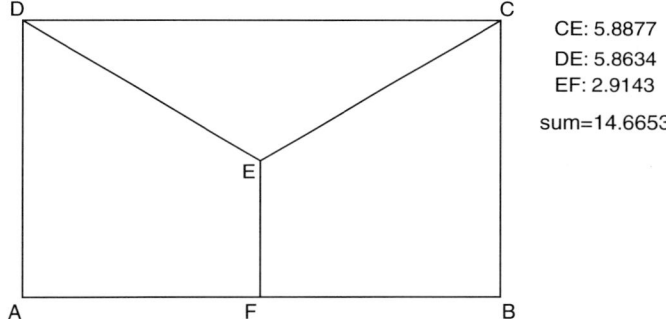

CE: 5.8877
DE: 5.8634
EF: 2.9143

sum=14.6653

Fig. 32.9 Construction with ClassPad

In Fig. 32.8, there is no question given. So we first have to find one: At which point should the two waste pipes which come from points A and B be connected so that the length of all three pipes is as short as possible?

You have to create a real and a mathematical model in connection with the modelling cycle. The use of DGS is helpful when you do this (see Fig. 32.9).

When you work with a ClassPad, it is necessary to find out that point F (see Fig. 32.9) is the midpoint of AB. If you put the origin of a coordinate system as A, the coordinates of E are (5/y). Now the problem has become a problem in one dimension and it can be solved easily by CAS.

The function $d(x,y)$ is a two-dimensional function for the length of the pipe. It can be shown that the partial derivative of x is zero for $x=5$. If you replace x with 5, you have the function $d(5,y)$, a one-dimensional function which is much easier to handle. If you differentiate the function $d(5,y)$ by y and if you solve the equation $\frac{\partial d(5,y)}{\partial y} = 0$ you get the solution for the optimal point E(5, 3.113248654) The equation is solved numerically, so the solution is not exact. It is clear by the problem itself that there has to be a minimum. Therefore, we do not have to prove this mathematically. If you use a good quality CAS such as Maple, Mathematica or MuPad, you can solve the problem directly.

A system like this has the power to solve a non-linear system of equations numerically. So this example shows that one kind of technology is suited to developing an idea of how to solve the problem and the other to solving it easily.

Technology offers the opportunity to write programmes. Therefore, you can write programmes for simulations which are helpful for developing possible solutions dependant on different initial conditions. The problem 'Finding the best husband' or rather 'Finding the best secretary' is an example of a problem for which a simulation is helpful.

2.5 The Fuel Tank – An Example for Using Technology to Validate

What kind of shape fits the following data?							
Level of the dipstick in cm	20	40	60	80	100	120	140	159
Volume of the tank in Litres	355	983	1,747	2,574	3,398	4,158	4,776	5,105

There are two suitable ways to deal with this problem. You can think about suitable shapes of existing tanks in reality or you can solve the problem by using the fundamental theorem of calculus. First, it is helpful to have a look at the graph of the given table. When you discuss the data, a cubic regression seems to be suitable (Fig. 32.10).

When you look at the graph, the shape of the tank should be symmetric. Shapes like a sphere (A), a lying cylinder (B), a cylinder with parts of a sphere at top and base (lying) (C) and a cylinder with two hemispheres (lying) (D) are possible. When you compare the functions $V(h)$ of the different shapes with the given data, you see that the model C is the best one (see Greefrath 2007a). For this task, the use of a spreadsheet with the power of a CAS is very helpful. If you know the function, which describes the edge of a rotation of a symmetric shape, you get the volume by $V(h) = \pi \cdot \int_0^h f(x)^2 \, dx$. This formula can be used in reverse too, which means you use the fundamental

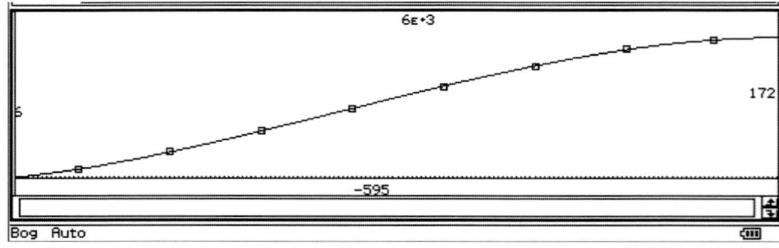

Fig. 32.10 Cubic regression

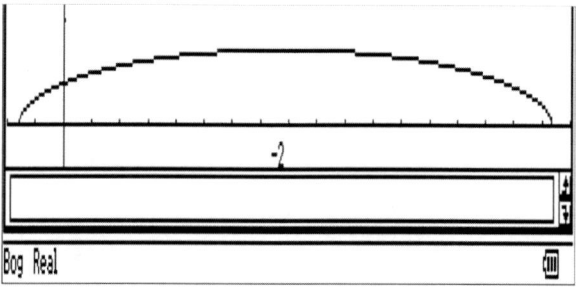

Fig. 32.11 Graph of the edge

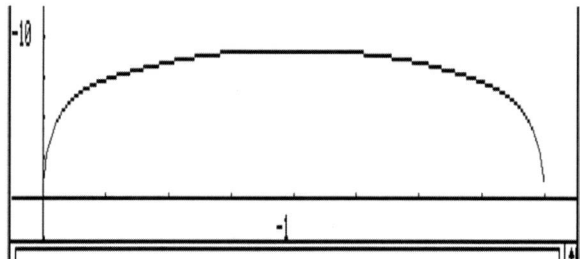

Fig. 32.12 Graph of the edge

theorem in reverse, that means: $f(h) = \sqrt{\dfrac{V'(h)}{\pi}}$. As we have seen before, you can do a cubic regression to get the function V(h).

The graph (Fig. 32.11) shows that the solution does not match reality. It should be zero for $x=0$ and $x=159$. So we have to make a fresh attempt at the problem. We have to insist that the function V(h) has to fit the given data and $V(0)=V'(0)=0$. It will be shown by the solution that you do not have to ask for $V(159)=V'(159)=0$. This is met by the symmetry of the data.

We set $V(x) = x^2\left(ax^7 + bx^6 + cx^5 + dx^4 + ex^3 + fx^2 + gx + h\right)$ and the rest of the work can be done by the ClassPad. Having a look at the graphs (Fig. 32.12), you can see that it matches to reality.

3 Examination Tasks – With Modelling Problems and Use of Technology?

Modelling problems can be well-established in many classrooms, when we have adequate examination tasks to test not only the modelling competencies but also the use of technology. When creating examination questions, many aspects should be considered. In particular, examination questions containing modelling problems require special attention. As a basic principle, applications in education should only be used in authentic real situations or if they bring advantages in understanding the problem. Examination tasks containing a whole modelling process are in most cases not possible due to the complexity factor. We want to consider the challenges of using technology in such examinations. Looking at the German situation, we illustrate two aspects of creating examination tasks concerning modelling problems and the use of technology. The first aspect concerns problems resulting from the real *situation* described in the task; the second aspect concerns the use of digital media, that is the *computer model* (see Fig. 32.1).

In Germany, there are specific practical regulations for using computers in upper secondary school examinations. Two calculator-specific versions of the mathematics examination have been in existence since 2007, for example, in the federal state of North Rhine-Westphalia (NRW). One group is for standard or graphing calculators and another group for CAS-calculators. The mathematical content is (nearly) the same for both groups, but some interesting aspects are different. Figure 32.13 shows a typical example for a task with and without CAS (Greefrath 2007b).

There are certainly not too many differences between these tasks, but a trend in a positive direction can be seen. Tasks with CAS are more open-structured. In the example, there is no given equation of the function and no coordinate system.

Without CAS

A canoe club would like to acquire a property for a new club house with a landing place to the Wupper river. The past owner [...] offers that property... at a price of 12 € per m². [...]

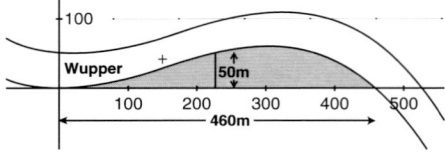

a) Explain, that the function $f(x) = ax^2(x - 460)$ describes the waterside in the given coordinate system. Calculate the variable a.

(result: $a = -\dfrac{1}{243340}$)

b) Calculate the price for the parcel of land.

With CAS

A canoe club would like to acquire a property for a new club house with a landing place to the Wupper river. The past owner [...] offers that property... at a price of 12 € per m². [...]

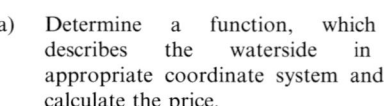

a) Determine a function, which describes the waterside in appropriate coordinate system and calculate the price.

Fig. 32.13 Test exercise with and without CAS

Fig. 32.14 Numerical solution

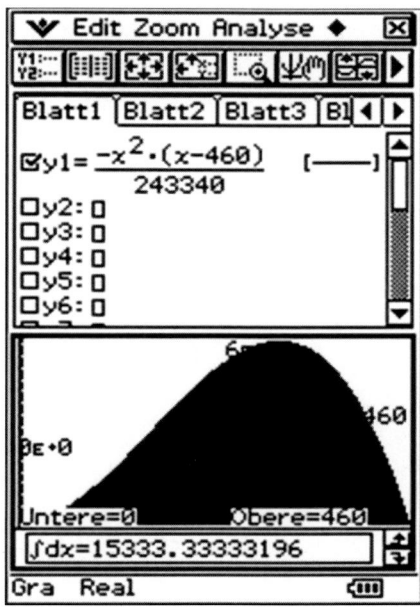

Hence, the task is more open and the students have to take more steps in the modelling cycle.

Another aspect in using technology in examinations is the difference in the computer models used. In the given task, we can determine the function with a linear equation system or a statistic regression. The solution could be calculated by an integral numerically (Fig. 32.14) or algebraically (Fig. 32.15). This is a good starting point for an interesting discussion in a mathematics lesson, but the evaluation in central examinations is not that easy.

An analysis of current examinations in the NRW part of Germany shows that there are typical varieties of tasks depending on the mathematical domain. In Stochastics, the tasks for examinations are application-oriented. The stochastic models used, for example, the binomial distribution, are well known. So the applications are not really modelling problems, but standard tasks. The students in both groups have nearly the same or exactly the same tasks, so the use of CAS does not change tests in Stochastics.

Exercises in Analytic Geometry are usually unrealistic. A typical question, for example, is in the context of an excessively simplified tower, if the temperature sensor at a special point T (4 | 10 | 2,6) is in the shadow of the tower. No one will think that this is a meaningful use of mathematics. There is no improvement possible by the use of CAS. Examination tasks should show authentic use of mathematics. So, in this area, the improvement can be an inner-mathematical task with or without

Fig. 32.15 Algebraical solution

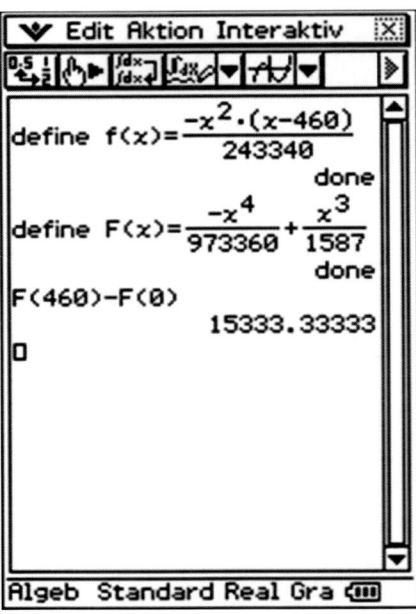

CAS. As a consequence, we have no modelling problems in this part of an examination, just an authentic use of mathematics.

The most common area for modelling problems in examinations is Analysis. There, tasks should include real-world problems. The use of CAS shows the right trend, but actual tasks contain too many standard parts. For example, in 2008, there was a problem about a model for concentration of a medicine in blood. The main part of the task was to calculate interesting points of the graph (e.g. maximum and inflection point). Considering all Analysis exams for CAS in NRW, the proportion of parts that really need a CAS and not only a graphing calculator is less than 5%. High potential for improvements here can be found easily. A realistic and interesting modelling example with technology is shown in Sect. 2.

Some criteria for good examination tasks with modelling problems and the use of technology have been found. The first is the authentic use of mathematics. If necessary – like in Analytic Geometry – it is better to have authentic use of mathematics rather than simple word problems with unrealistic contexts. The second is a good choice of interesting and relevant real-world problems to have an essential part of the modelling cycle in examinations. The third is a real use of CAS with different computer models (see the modelling cycle concerning technology above). But not all parts of examinations need to be solved with a computer. For the examination of mathematical competencies, which were acquired in education with digital tools, it does not necessarily require the digital tools in the test situation (Greefrath et al. 2008).

4 Conclusion

The discussed examples show that technology can be helpful at any step of the modelling cycle. Thinking of the pipe and the shapes of the tank, we see that technology helps to reduce reality to a model of itself. The example of the pipe shows that DGS helps to translate the model of reality to the mathematical model. Although in our opinion, you cannot separate these two from each other. CAS and DGS are designed to solve mathematical problems. Therefore, the assistance of CAS and DGS for the step towards a solution does not have to be discussed. The last example shows that CAS could be necessary to interpret and validate the solution. Therefore, the use of technology not only creates an important appendix to the modelling cycle (see Fig. 32.1), but also influences each part of the cycle. So the technology world is relating to the real world and mathematical world as shown in the alcohol example. This multiple influence of technology in solving modelling problems can also be found and integrated in examination tasks. Hence, the important role of technology can be seen by all the examples exemplarily. The role of technology for modelling activities is as important in tests as in process-related educational situations.

References

Greefrath, G. (2007a). Mathematisch Modellieren lernen – ein Beispiel aus der Integral rechnung. In *Schriftenreihe der ISTRON-Gruppe* (Vol. 11, pp. 113–122). Hildesheim: Verlag Franzbecker.

Greefrath, G. (2007b). Computeralgebrasysteme und Prüfungen, Beiträge zum Mathematikunterricht 2007 (pp. 55–58). Hildesheim: Verlag Franzbecker.

Greefrath, G., Leuders, T., & Pallack, A. (2008). Gute Abituraufgaben – (ob) mit oder ohne Neue Medien. *Der Mathematische und Naturwissenschaftliche Unterricht, 61 Bd. 2,* 79–83.

Malle, G., Ramharter, E., Ulovec, A., & Kandl, S. (2006). *Mathematik verstehen 5.* Wien: Öbv & hpt.

Siller, H.-St. (2008). Informatics – A subject developing out of mathematics – A review from 1970 to 2007. In C. Tzanakis (Ed.), *HPM Conference Proceedings*, (pp. 1–12), Mexico City.

Siller, H.-St., & Greefrath, G. (2010). Mathematical modelling in class regarding to technology. In V. Durand-Guerrier, S. Soury-Lavergne, & F. Arzarello (Eds.), *Proceedings of the Sixth Congress of the European Society for Research in Mathematics Education*. January 28th–February 1st 2009, France: Lyon.

Siller, H.-St., & Maaß, J. (2009). Fußball EM mit Sportwetten. In *Schriftenreihe der ISTRON-Gruppe* (Vol. 14, pp. 95–112). Hildesheim: Verlag Franzbecker.

Watson, P. E., Watson, R., & Batt, R. D. (1980). Total body water volumes for adult males and females estimated from simple antropometric measurements. *The American Journal of Clinical Nutrition, 33,* 27–39.

Widmark, E. M. P. (1932). *Die theoretischen Grundlagen und die praktische Verwendbarkeit der gerichtlich-medizinischen Alkoholbestimmung.* Berlin/Wien: Verlag Urban und Schwarzenberg.

Wilkinson, P. K., Sedman, A. J., Sakmar, E., Kay, D. R., & Wagner, J. G. (1977). Pharmacokinetics of ethanol after oral administration in the fasting state. *Journal of Pharmacokinetics and Biopharmaceutics, 5*(3), 207–224.

Chapter 33
Improving Learning in Science and Mathematics with Exploratory and Interactive Computational Modelling

Rui Gomes Neves, Jorge Carvalho Silva, and Vítor Duarte Teodoro

Abstract Scientific research involves mathematical modelling in the context of an interactive balance between theory, experiment and computation. However, computational methods and tools are still far from being appropriately integrated in the high school and university curricula in science and mathematics. In this chapter, we present a new way to develop computational modelling learning activities in science and mathematics which may be fruitfully adopted by high school and university curricula. These activities may also be a valuable instrument for the professional development of teachers. Focusing on mathematical modelling in the context of physics, we describe a selection of exploratory and interactive computational modelling activities in introductory mechanics and discuss their impact on student learning of key physical and mathematical concepts in mechanics.

1 Introduction

Science is an evolving structure of knowledge based on hypotheses and models which lead to theories whose explanations and predictions about the universe must be consistent with the results of systematic and reliable experiments (see, e.g. Chalmers 1999; Feynman 1967). The process of creating scientific knowledge is an interactive blend of individual and group reflections which involve modelling

---

R.G. Neves (✉) and V.D. Teodoro
Unidade de Investigação Educação e Desenvolvimento (UIED) e Departamento de Ciências Sociais Aplicadas (DCSA), Faculdade de Ciências e Tecnologia (FCT), Universidade Nova de Lisboa (UNL), Monte da Caparica, 2829-516 Caparica, Portugal
e-mail: rgn@fct.unl.pt; vdt@fct.unl.pt

J.C. Silva
Departamento de Física, Faculdade de Ciências e Tecnologia (FCT), Centro de Física e Investigação Tecnológica (CEFITEC), Universidade Nova de Lisboa (UNL), Monte da Caparica, 2829-516 Caparica, Portugal
e-mail: jcs@fct.unl.pt

processes that balance theory, experiment and computation (Blum et al. 2007; Schwartz 2007; Slooten et al. 2006). This cognitive frame of action has a strong mathematical character, since scientific reasoning embeds mathematical reasoning as scientific concepts and laws are represented by mathematical entities and relations. In this process, computational modelling plays a key role in the expansion of the science and mathematics cognitive horizon through enhanced calculation, exploration and visualisation capabilities.

Although clearly linked to real world phenomena, science and mathematics are thus based on abstract and subtle conceptual and methodological frameworks which change along far from straightforward evolution timelines. These cognitive features make science and mathematics difficult subjects to learn, to develop and to teach. In an approach to science and mathematics education meant to be effective and in phase with the rapid scientific and technological development, an early integration of computational modelling in learning environments which reflect the exploratory and interactive nature of modern scientific research is of crucial importance (Ogborn 1994). However, computational knowledge and technologies, as well as exploratory and interactive learning environments, are still far from being appropriately integrated into high school and university curricula in science and mathematics. As a consequence, these curricula are generally outdated and most tend to transmit to students a sense of detachment from the real world. These are contributing factors to the development of negative views about science and mathematics education, leading to an increase in student failure.

Physics is a good illustrative example. Consider the general physics courses taken by first year university students. These are courses which usually cover a large number of difficult physics topics following a traditional lecture plus laboratory instruction approach. Due to a lack of understanding of fundamental concepts in physics and mathematics, the number of students that fail in examination tests is usually very high. Moreover, many students that eventually succeed also reveal several weaknesses in their understanding of elementary physics and mathematics (Halloun and Hestenes 1985; Hestenes 1987; Hestenes et al. 1992; McDermott 1991; McDermott and Redish 1999).

Although it is clear that there are many reasons behind this problem, it is also clear the solution has to involve changes in the physics education model. Indeed, many research studies have shown that the process of learning can be effectively enhanced when students are involved in the learning activities as scientists are involved in research (Beichner et al. 1999; Handelsman et al. 2005; Keiner and Burns 2010; Mazur 1997; McDermott 1997; McDermott and Redish 1999; Redish 2004). In addition, several attempts have been made to introduce computational modelling in research-inspired learning environments. The starting emphasis was on professional programming languages such as Fortran (Bork 1967) and Pascal (Redish and Wilson 1993). Although more recently this approach has evolved to Python (Chabay and Sherwood 2008), it still requires students to develop a working knowledge of programming, a generally time-consuming and dispersive task which can hinder the process of learning physics. The same happens when using scientific computation software such as Mathematica and Matlab. To avoid overloading students with programming notions or syntax, and focus the learning process on the relevant physics and mathematics, several computer modelling systems were created, for example, Dynamical Modelling System (Ogborn 1985), Stella (High Performance Systems 1997), Easy Java Simulations (Christian and Esquembre 2007) and Modellus (Teodoro 2002).

In this chapter, we discuss how Modellus (see http://modellus.fct.unl.pt) can be used to develop exploratory and interactive computational modelling activities which can be adopted by high school and university curricula in science and mathematics as well as be a valuable instrument for the professional development of teachers. Focusing on mathematical modelling in the context of physics, we describe activities in introductory mechanics which were implemented in a new course component of the general physics course taken by first year biomedical engineering students at the Faculty of Sciences and Technology of the New Lisbon University (FCT/UNL). For mathematics education, these activities are relevant as concrete applications of mathematical modelling (Carson 1999; Garcia et al. 2006; National Research Council 1989).

2 Course Organisation, Methodology and Student Evaluation Procedures

Let us start by describing the implementation context for the computational modelling activities. The organisation, methodology and evaluation strategies used in general physics can serve as a model to be adapted to other areas of science and to mathematics.

The 2009 general physics course for biomedical engineering involved 115 students, 59 of them taking the course for the first time. The structure and programme themes were those of the 2008 edition (Neves et al. 2009). In the computational modelling classes, students were organised in groups of two or three, one group for each available computer. In each class, the groups worked on a set of five computational modelling activities conceived to be interactive and exploratory learning experiences about challenging but easily observed physical phenomena. An example is the motion of a swimmer in a river with a current (Neves et al. 2009). The teams were motivated to solve the problems on their own using the physical, mathematical and computational modelling guidelines provided by the class documentation. To ensure adequate working rhythm with appropriate conceptual, analytical and computational understanding, the students were continuously helped during the exploration of the activities.

All activities were created as computational modelling experiments with Modellus. Each class activity was presented in a PDF document, with text and embedded video support to help students both in class or at home in a collaborative online context based on the Moodle online learning platform. To design the activities, emphasis was placed on cognitive conflicts in the understanding of physical concepts, the manipulation of multiple representations of mathematical models and the interplay between the analytical and numerical approaches applied to solve problems in physics and mathematics. In this course, the majority of the supporting text and videos presented complete step-by-step instructions to build the Modellus mathematical models, animations, graphs and tables. After constructing the models, students explored the multiple representations available to answer several questions about the proposed general physics problems. Some activities involved modelling problems where students saw only videos of the Modellus animations or graphs. After this they constructed the mathematical models to reproduce the animations or graphs, and answer proposed questions. Modellus was particularly effective in these classes because of the following main advantages: (1) an easy and intuitive creation of mathematical models using standard mathematical notation, (2) the possibility to create animations with interactive objects that have mathematical properties expressed in the model and (3) the simultaneous exploration of images, tables, graphs and object animations.

The student evaluation procedures in the computational modelling classes involved group evaluation and individual evaluation. For each class, all groups had to build five Modellus models and complete a Moodle online test answering the questions of the corresponding activity PDF document. The individual evaluation consisted of the solution of two homework activities and a final test, both with new problems based on those covered in class but with only partial text and video instructions on how to build the models and solve the problems. Students also took pre-instruction and post-instruction Force Concept Inventory (FCI) tests (Hestenes et al. 1992) which did not count for their final classification. At the end of the semester, students answered a Likert scale questionnaire to access their degree of receptivity to this new computational modelling component of the general physics course.

3 Computational Modelling Activities with Modellus

Let us now discuss, as illustrative examples, two of the computational modelling activities about circular motion and oscillations, the theme opening the second part of the course. Again, these are thought not only from the point of physics but also from the point of view of mathematics in order to *help students make connections between different subjects.*

A particle in circular motion (representing, for instance, a runner going around a circular track) describes a circle of radius R, a mathematical curve defined by $x^2+y^2=R^2$ in a Cartesian reference frame Oxy whose origin is at the centre of the circle. In this frame, x and y are the Cartesian coordinates of the position vector $\vec{r}$. This vector has magnitude R and specifies where the particle is on the curve. As the particle moves around the circle, the magnitude R is kept constant but the direction of $\vec{r}$ changes with time. This direction is given by the angle θ that $\vec{r}$ makes with the Ox axis. The variables R and θ define the polar coordinates of $\vec{r}$. The coordinates x and y are also time dependent and are related to R and θ by trigonometric functions: $x = R\cos(\theta)$ and $y = \sin(\theta)$.

To explore circular motion, students started with uniform circular motion. When the circular motion is uniform, the particle traces one circle in every constant time interval T. This time interval is the period of the motion and its inverse $f = 1/T$ is the frequency of the motion. The angle θ is then a linear parametric function of the time t, $\theta = \omega\, t + \theta_0$ where $\omega = 2\pi/T$ is the motion angular frequency, measured in radians per second, and θ_0 is the initial direction of $\vec{r}$. The velocity $\vec{v}$ is tangent to the circular trajectory, always orthogonal to $\vec{r}$, and has constant magnitude $v = \omega R$. The acceleration $\vec{\alpha}$ has magnitude $\alpha = \omega^2 R$ and a centripetal direction, that is, opposite to $\vec{r}$. The uniform circular motion is the composition of two simple harmonic oscillations: one along the Ox axis and the other along the Oy axis. These oscillations are characterised by the same amplitude $A = R$ and the same frequency $f = 1/T$. The initial phase of the Ox oscillation is θ_0, and between them, there is a time-independent $\pi/2$ phase difference.

To model this type of motion, students had to recall what they learnt in the first part of the course during the computational modelling activities about vectors, parametric equations of motion, velocity and acceleration (Neves et al. 2009). Building on this prior knowledge, students were able to construct a model associating the Cartesian coordinates of $\vec{r}$ to the corresponding polar trigonometric functions with the angle θ given by the linear parametric equation $\theta = \omega\, t + \theta_0$. They were also able to define the coordinates of $\vec{v}$ and $\vec{\alpha}$ (see Fig. 33.1). This mathematical model was complemented with graphs and tables of the different coordinate variables as functions of time, and by an animation allowing direct manipulation of the independent parameters of the model, R, T, θ_0, as well as real time visual display of the trajectory of the moving particle, $\vec{r}$, $\vec{v}$, and $\vec{\alpha}$. The harmonic oscillatory motions along the coordinate axis were also represented (see Fig. 33.1). With this model, students were able to explore, visualise and reify the initially abstract physical and mathematical concepts associated with uniform circular motion. For example, by combining the information from the several different simultaneous representations, they analysed the motion of a particle tracing a circle of radius $R = 150$ m every 2.5 min, and were able to compare $\vec{v}$ and $\vec{\alpha}$ as functions of time and to calculate these vectors at time $t = 7$ min.

During these activities, students showed difficulties in distinguishing between a vector, like $\vec{v}$ or $\vec{\alpha}$, and its magnitude. They were also puzzled when asked to solve the same problem considering the angles measured in degrees instead of radians. Indeed, at first, students were frequently unable to create $\vec{v}$ and $\vec{\alpha}$ with the correct magnitude and direction. Similarly, they did not place the angle conversion factor in the correct place everywhere in the mathematical model. For example, in their corrrect attempt, they incorrectly multiplied the speed by $180/\pi$. To be able to correct the models and at the same time visualise the effect of the change in the animation and other model representations was for the students an essential advantage of the modelling process with Modellus in helping them to solve these learning difficulties.

Using and extending this trigonometric model, students were then able to construct a model in Modellus to estimate the solution to the following astronomical problem: What is the time interval between two successive oppositions of the Earth and Mars? To help students, we suggested the assumption of considering the motions

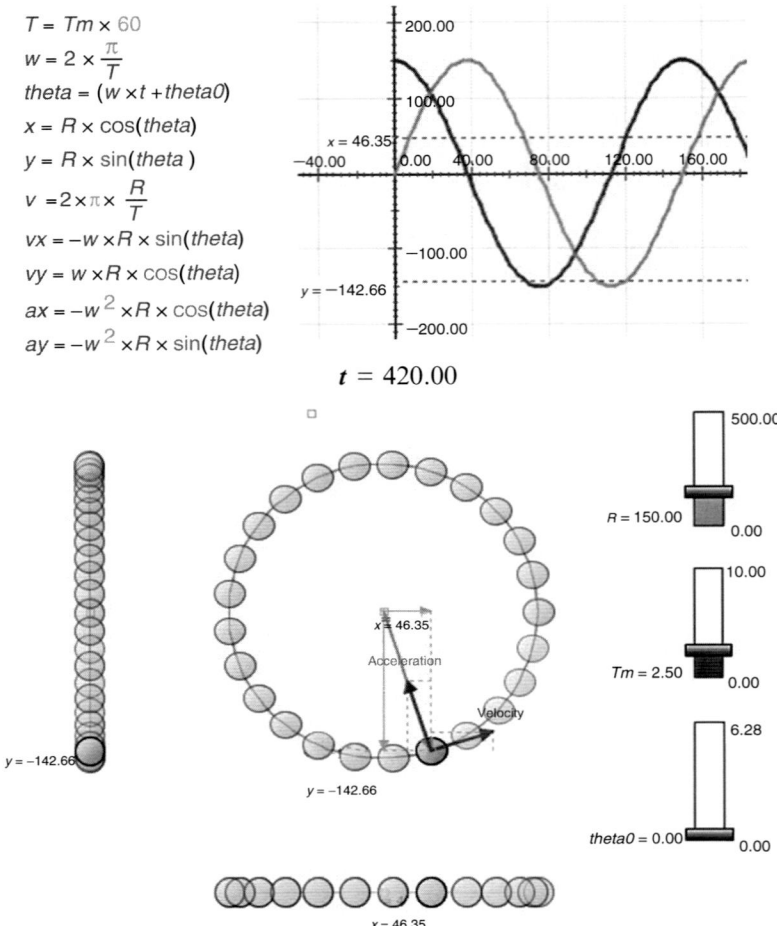

Fig. 33.1 Uniform circular motion: equations as seen in the Modellus Mathematical Model window, examples of coordinate-time graphs and the Modellus animation

of the Earth and Mars around the Sun to be uniform circular motions. We also taught them to use the average Earth–Sun distance (known as the astronomical unit and denoted by AU) as the distance scale for the problem. In this scale, the average Earth–Sun distance is simply 1 AU and the average Mars–Sun distance is 1.53 AU. Taking into account that the approximate motion periods of the Earth and Mars are, respectively, 1 year and 1.89 years and using the year as the unit of time, students were able to develop a mathematical model and an animation representing the motions of the Earth and Mars around the Sun. In the process, they were able to determine the angular velocities of both planets and the time interval between two successive oppositions. Using the conversion factors $1\ AU = 1.50 \times 10^8$ km and

1 year = 3.15×10^7 s, they were also able to find in km/s the orbital velocities of the Earth and Mars at the time of the model first occurring opposition. To achieve the precision required by the Moodle online test, students used a position vector or velocity coincidence method. The adjustment of the numerical step was an important numerical technique students learned to apply to obtain animations with realistic trajectories and correct answers to the questions of this astronomical challenge.

4 Conclusions

In this chapter, we have shown how Modellus can be used to develop exploratory and interactive computational modelling activities for science and mathematics education. We have described examples in introductory mechanics which were implemented in the general physics course taken by first year biomedical engineering students at FCT/UNL. We have shown that during class, the computational modelling activities with Modellus were successful in identifying and resolving several student difficulties in key physical and mathematical concepts of the course. Of crucial importance in this process, was the possibility to have a real time visible correspondence between the animations with interactive objects and the object's mathematical properties defined in the model, and also the possibility of manipulating simultaneously several different representations such as graphs and tables. Thus with Modellus, students can be exploring authors of models and animations, and not just simple browsers of computer simulations.

The successful class implementation of the computational modelling activities was reflected in the student answers to a Likert scale questionnaire (see Fig. 33.2), with results improving slightly on those of the 2008 edition (Neves et al. 2008). Globally, students reacted positively to the activities, considering them to be helpful in the learning process of mathematical and physical models. For them, Modellus was easy enough to learn and user-friendly. In this course, students showed a clear preference to work in teams in an interactive and exploratory learning environment. The computational modelling activities with Modellus presented in PDF documents with embedded video guidance were also considered to be interesting and well designed. A natural sense of caution in relation to novelty and to evaluation procedures was nevertheless detected. Students also felt that the content load was heavy and that the available time spent on the computational modelling activities was insufficient.

In spite of global success during the class implementation phase, the FCI test results led to an average FCI gain of 22%, an indication that the general physics course with the computational modelling component is just performing as a traditional instruction course (Hake 1998). Although this performance score refers to the general physics course as a whole, the results of the questionnaire and students' opinions about the computational modelling component also indicate that some aspects of the implementation approach should be changed. In this context, possible ways forward are: (1) Increase the relative importance and value of the computational modelling component. (2) Reduce the heavy content load (as perceived by

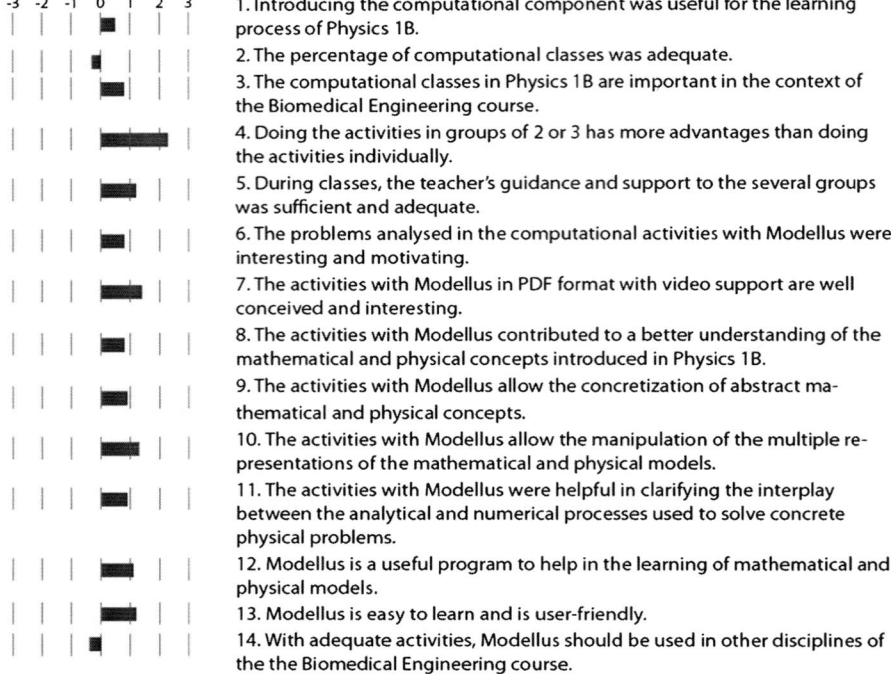

Fig. 33.2 Questionnaire and results, shown graphically by the average bar over all answers

students). (3) Increase time spent on the modelling tasks. (4) Choose problems more closely related with the specific subject of the student's course major. (5) Introduce less guided, more discovery-oriented instruction guidelines as well as computational modelling problem finding.

Acknowledgements Work partially supported by Unidade de Investigação Educação e Desenvolvimento (UIED) and Fundação para a Ciência e a Tecnologia (FCT), Programa Compromisso com a Ciência, Ciência 2007.

References

Beichner, R., Bernold, L., Burniston, E., Dail, P., Felder, R., Gastineau, J., Gjertsen, M., & Risley, J. (1999). Case study of the physics component of an integrated curriculum. *Physics Education Research, American Journal of Physics Supplement, 67*, 16–24.
Blum, W., Galbraith, P., Henn, H.-W., & Niss, M. (Eds.). (2007). *Modelling and applications in mathematics education.* New York: Springer.
Bork, A. (1967). *Fortran for physics.* Reading: Addison-Wesley.
Carson, S. (Ed.). (1999). *Shaping the future: Physics in mathematical mood.* Bristol: Institute of Physics.

Chabay, R., & Sherwood, B. (2008). Computational physics in the introductory calculus-based course. *American Journal of Physics, 76*, 307–313.
Chalmers, A. (1999). *What is this thing called science?* London: Open University Press.
Christian, W., & Esquembre, F. (2007). Modeling physics with easy Java simulations. *Physics Teacher, 45*(8), 475–480.
Feynman, R. (1967). *The character of physical law*. New York: MIT Press.
Garcia, F., Gascón, J., Higueras, L., & Bosch, M. (2006). Mathematical modelling as a tool for the connection of school mathematics. *Zentralblatt für Didaktik der Mathematik, 38*, 226–246.
Hake, R. (1998). Interactive-engagement versus traditional methods: A six-thousand-student survey of mechanics test data for introductory physics courses. *American Journal of Physics, 66*, 64–74.
Halloun, I., & Hestenes, D. (1985). The initial knowledge state of college physics students. *American Journal of Physics, 53*, 1043–1048.
Handelsman, J., Ebert-May, D., Beichner, R., Bruns, P., Chang, A., DeHaan, R., Gentile, J., Lauffer, S., Stewart, J., Tilghmen, S., & Wood, W. (2005). Scientific teaching. *Science, 304*, 521–522.
Hestenes, D. (1987). Toward a modelling theory of physics instruction. *American Journal of Physics, 55*, 440–454.
Hestenes, D., Wells, M., & Swackhamer, G. (1992). Force concept inventory. *Physics Teacher, 30*, 141–158.
High Performance Systems. (1997). *Stella (version 5) [Computer software]*. Hannover: High Performance Systems.
Keiner, L., & Burns, T. (2010). Interactive engagement: How much is enough? *Physics Teacher, 48*, 108–111.
Mazur, E. (1997). *Peer instruction: A user's manual*. Upper Saddle River: Prentice-Hall.
McDermott, L. (1991). Millikan lecture 1990: What we teach and what is learned–closing the gap. *American Journal of Physics, 59*, 301–315.
McDermott, L. (1997). *Physics by inquiry*. New York: Wiley.
McDermott, L., & Redish, E. (1999). Resource letter: PER-1: Physics education research. *American Journal of Physics, 67*, 755–767.
National Research Council. (1989). *Everybody counts*. Washington: National Academies Press.
Neves, R., Silva, J., & Teodoro, V. (2008). Improving the general physics university course with computational modelling. Poster presented at the *2008 Gordon Research Conference, Physics Research and Education: Computation and Computer-Based Instruction*, Bryant University, Smithfield.
Neves, R., Silva, J., & Teodoro, V. (2009). Computational modelling with Modellus: An enhancement vector for the general university physics course. In A. Bilsel & M. Garip (Eds.), *Frontiers in science education research* (pp. 461–470). Famagusta: Eastern Mediterranean University Press.
Ogborn, J. (1985). *Dynamic modelling system*. Harlow: Longman.
Ogborn, J. (1994). Modelling clay for computers. In B. Jennison & J. Ogborn (Eds.), *Wonder and delight, essays in science education in honour of the life and work of Eric Rogers 1902–1990* (pp. 103–114). Bristol: Institute of Physics Publishing.
Redish, E. (2004). A theoretical framework for physics education research: Modeling student thinking. In E. Redish & M. Vicentini (Eds.), *Proceedings of the International School of Physics: Enrico Fermi Course CLVI* (pp. 1–63). Bologna: Italian Physical Society.
Redish, E., & Wilson, J. (1993). Student programming in the introductory physics course: M.U.P.P.E.T. *American Journal of Physics, 61*, 222–232.
Schwartz, J. (2007). Models, simulations, and exploratory environments: A tentative taxonomy. In R. Lesh, E. Hamilton, & J. Kaput (Eds.), *Foundations for the future in mathematics education* (pp. 161–172). Mahwah: Lawrence Erlbaum Associates.
Slooten, O., van den Berg, E., & Ellermeijer, T. (Eds.). (2006). *Proceedings of the International Group on Research on Physics Education (GIREP) 2006 Conference: Modelling in Physics and Physics Education*. Amsterdam: European Physical Society.
Teodoro, V. (2002). *Modellus: Learning physics with mathematical modelling*. PhD thesis. Faculdade de Ciências e Tecnologia, Universidade Nova de Lisboa, Lisboa, Portugal.

Part V
Modelling Competency: Teaching, Learning and Assessing Competencies

Chapter 34
Modelling Competency: Teaching, Learning and Assessing Competencies – Overview

Morten Blomhøj

This chapter frames eight papers which are all addressing questions and issues related to the teaching, learning and assessing of mathematical modelling competency. As a main commonality in the papers, the authors take point of departure in the existence of (such a thing as) mathematical modelling competency (as a concept). A competency in general is understood as a person's mental capacity to cope with a certain type of challenge in a knowledgeable and reflective way. According to this, a person possesses mathematical modelling competency if he or she is capable of carrying through a mathematical modelling process in order to solve a problem or to understand a situation within a certain domain. Thus, modelling competency is attached to a modelling process, and therefore, it is also no surprise that all the papers refer – some more explicitly than others – to a modelling cycle – and in many cases, the modelling cycle from Blum and Leiß (2007) is used as a reference. However, as argued by Haines in his plenary address, this modelling cycle does not necessarily provide a good representation of real modelling processes carried through by students in learning situations nor should it be perceived as ideal for mathematical modelling processes. The different variations of the modelling cycle are analytical tools for analysing the sub-processes and the related sub-competencies which – in principle – are involved in mathematical modelling competency. It may also serve as a tool for analysing and describing differences among the modelling processes performed by individual students or groups of students as done by Borromeo Ferri (2006). However, it is quite clear from the papers in this chapter and from research on the teaching and learning of mathematical modelling in general that there is a need for further development of the conceptualisation of mathematical modelling competency so as to understand and assess progression in the learning of modelling competency. In such endeavour, one will unavoidably be confronted with the dilemma between maintaining the holistic understanding of mathematical modelling competency, which is in the cor e of the concept, and the quest for reducing modelling competency to a number of

---

M. Blomhøj (✉)
IMFUFA, NSM, Roskilde University, Roskilde, Denmark
e-mail: blomhoej@ruc.dk

sub-competencies which can be detected and assessed independently. A few of the papers recognise this challenge and refer to the idea of Niss and Jensen (2006) to develop the holistic approach to modelling competence by suggesting a 3-dimensional model which captures the progression in the development of the concept. The three dimensions are: degree of coverage with respect to the elements in the mathematical modelling process, radius of action in terms of domains and situations in which modelling competency is activated, and technical level in relation to the mathematical concept and methods involved in the modelling process. However, further research is needed in order to reveal if, and how, this model can be used to characterise and assess progress in the development of students' modelling competency in the practices of mathematics teaching.

1 Presentation of the Papers

The plenary address attached to this chapter was given by one of the grand old men in the educational field of mathematical modelling, namely, Christopher Haines from London, UK. It is reported in the paper 'Drivers for Mathematical Modelling: Pragmatism in Practice' with a commentary by Katja Maaß who acted as discussant at the plenary session. In the paper, Haines argues through a row of interesting examples for the importance of a close connection to modelling and applications of models in real life for all educational aspects. With respect to researching, teaching and learning mathematical modelling, the most important factor to take into account, according to Haines, is the connection to authentic problems and to real life practices of modelling. As presented in the paper, the idea is that teachers should take a pragmatic position towards the integration of mathematical modelling in daily teaching, which is often framed by a tight curriculum, and look for good examples of real life problems or situations for modelling. (And) educational research should provide teachers with good and well-analysed examples of real life applications of mathematical modelling. Students' motivation for working with genuine and challenging real life problems should be used as a drive for mathematical modelling in teaching. In my reading of the paper, Haines is arguing strongly but implicitly for seeing the development of good modellers as an educational end in its own right for mathematics teaching. However, as pointed to by Maaß in her commentary, if mathematical modelling is the educational end we need to be able to support mathematics teachers through their teacher education and subsequent professional development programmes in order to achieve this end. How can teachers really support their students in developing modelling competency, how can we assess the students' progression in modelling competency, and what should be accepted as a standard for the common level of modelling competency at the different levels in the educational system? In order to answer these questions, we need research close to the day-to-day teaching *practices*, including context and framework of school life. In this way, the approach of pragmatism to mathematical modelling can/could be extended to the practice of teaching mathematical modelling.

In the paper 'Documenting the development of modelling competencies of grade 7 mathematics students' by Biccard and Wessels, you find a report of a detailed empirical study on 12 students working in groups of four with three consecutive modelling eliciting tasks over a period of 12 weeks with a weekly 1-h meeting with the researcher. Two of the three groups are characterised as weak in mathematics, while the third group is characterised as strong. Based on research literature, the authors refine the list of the sub-competencies involved in mathematical modelling by including: understanding, simplifying, mathematising, working mathematically, interpreting, validating, presenting, arguing, sensing of direction, using informal knowledge, planning and monitoring and students' beliefs. The developments of these sub-competencies in the three groups are detected. In general, the analysis shows a progression of competencies over the test period. In particular, there was a clear development in the students' beliefs about mathematics and its relevance for describing and handling daily life situations. The validation and interpretation competencies were weak throughout the period in all three groups. The findings are related to specific features of the tasks and the setting. It is argued that in order to develop all the many aspects of modelling competency, the students should be exposed to a broad variation of modelling tasks and experience working with peers in different settings.

The paper 'Students' reflections in mathematical modelling projects' by Blomhøj and Kjeldsen is a theoretical paper introducing internal and external reflections as two different types of reflections that need to be challenged and developed in different ways in relation to students' modelling work. The two types of reflections are defined by their relation to the modelling process. Internal reflections have the modelling process with its sub-processes as its object, while external reflections have the process of application of a model as its object. The two types of reflections are illustrated and discussed through the analysis of two students' modelling projects from the introduction study programme in natural science at Roskilde University. In general, external reflections are related to (a) the reformulation of the problem caused by the application of a model, (b) changes in the discourse about the problem towards a model discourse about possible adjustments of the model, (c) the limitation of the possible actions taken into consideration to those that can be evaluated in the model and (d) the delimitation of the group of people that can take part in the discussion. The students' reflections in two projects illustrated all these four types of reflections.

The paper 'From data to functions: Connecting modelling competencies and statistical literacy' by Engel and Kuntze begins with the claim that gathering, handling and interpretation of data are given too little attention in modelling activities in teaching. In fact, it is a missing element in many descriptions of the modelling process. It is argued that data can be – and in real life often are – a starting point for modelling and that serious validation of model results in most cases has to involve comparison with data. From analysing descriptions of modelling competency and proficiency and the notion of statistical literacy, the authors argue that statistical literacy should be seen as and treated didactically as closely intertwined with mathematical modelling. Behind this argument lies the strong ontological position concerning the nature of numerical data. Data are always a sum of signal and noise or of structure and random variation or model fit and residual. Hence, handling data always involves modelling. On the ground of such theoretical reflections, the authors investigate, through an empirical study, whether data-based modelling activities support the development of students' statistical literacy. The results from a pre- and post-test setup involving 179 second year student teachers are reported and indicate that data-based modelling activities have a positive effect on the students' statistical literacy.

The paper 'First results from a study investigating Swedish upper secondary students' mathematical modelling competencies' by Frejd and Bergman Ärlebäck reports on an empirical investigation of modelling competencies among nearly 400 students. The authors use a test instrument which is a slight modification of the one constructed by Haines et al. (2001). The instrument consists of 22 items covering the eight aspects of the mathematical modelling process identified by Blum and Leiß (2007). In addition to each item, the students are asked to give their opinion on: the relevance of the problem for a mathematics class; whether it is of interest for them; and if it is connected to reality. Several non-parametric statistical tests have been applied and the analyses have yielded some significant results. The sub-competencies related to the initial parts of the modelling process – understanding and simplifying the problem – and sub-competence related to the choice of a mathematical model are the ones where the students exhibit the highest degree of difficulties, while sub-competencies related to the formulation of a mathematical problem and to the identification of key variables and parameters in a model are the sub-competencies where the students display the highest level of proficiency. Even though models and modelling are included in Swedish upper secondary curriculum, only 22.5% of the students stated that they had heard about these issues in their previous mathematics teaching, and those who had previous experiences with modelling did not perform any better than those who had never worked with modelling before. Moreover, the students in general did not find the modelling tasks relevant for their mathematics classes nor did they find them of personal interest. A possible explanation for these findings is that the students' modelling competency is really not of any significance for the students' success in the system.

In the paper 'Why cats happen to fall from the sky or on good and bad models' Hans-Wolfgang Henn discusses how *central examinations, the use of computers, and the motivation of the teachers* can either support the development of the students' modelling competency or form obstacles for such development. The discussion is illustrated with concrete examples of modelling tasks from central examinations, mathematics textbooks and modelling projects for teachers' professional development. It is argued that many examination tasks involving modelling activities or applications of models are meaningless in the sense that mathematical modelling in these tasks does not really help solve or understand a real life problem. The situation contexts in which the problems are given are not being treated seriously – in fact, the students' are better off disregarding the situation context. Computer-supported modelling activities often lead to senseless curve fitting without any discussion of the explanatory or pragmatic value of the mathematical model in the given context.

Professional development activities for teachers often involve authentic cases of mathematical modelling from the public media or from industries. However, as shown in the paper with an example of a mathematical model regulating the minimal free space for calves held in herds, in such cases it is also relevant to discuss with the students if a model is good or bad and if the use of a mathematical model is relevant at all. Therefore, such critical reflections need to be included in the professional development activities for teachers. In general, the paper argues that teaching models and modelling at schools should always include a serious and broad discussion with the students about the quality and relevance of models. Is it a good or a bad model? This is always a relevant question in teaching when the development of mathematical modelling competency is in focus.

The paper 'Assessing modelling competencies using a multidimensional IRT-approach' by Zöttl, Ufer and Reiss addresses the question of how to assess the students' modelling competency. The overall interest is to investigate the relationship between the students' performance on test items which challenge the students to work with an entire modelling process and their performance on items which challenge particular sub-competencies related to specific sub-processes in the modelling cycle. For this purpose, the authors apply the methodology of an item–response–theory approach and, in particular, the Rasch model approach. The authors have developed a test instrument focusing on circumference and area of simple geometrical figures consisting of a total of 36 test items and administered to 1,657 lower secondary students in pre-, post and follow-up tests. The items cover four types of challenge of which the first three are connected to sub-processes in the modelling cycle, while the fourth type challenges the students to perform an entire modelling process. Following the Rasch model approach, each individual student's modelling competency is measured on a unidimensional scale and on a sub-dimensional scale taking the sub-competencies into account. Comparing these two measurements, it is concluded that the test with sub-dimensional scaling seems to be a promising approach to testing students modelling competency.

References

Blum, W., & Leiß, D. (2007). How do students and teachers deal with modelling problems? In C. Haines, P. Galbraith, W. Blum, & S. Khan (Eds.), *Mathematical modelling: Education, engineering and economics* (pp. 222–231). Chichester: Horwood.

Borromeo Ferri, R. (2006). Theoretical and empirical differentiations of phases in the modelling process. *Zentralblatt für Didaktik der Mathematik, 38*(2), 86–95.

Haines, C., Crouch, R., & Davis, J. (2001). Understanding students' modelling skills. In J. F. Matos, W. Blum, K. S. Houston, & S. P. Carreira (Eds.), *Modelling and mathematics education: Applications in science and technology* (pp. 366–380). Chichester: Horwood.

Niss, M., & Jensen, T. H. (Eds.). (2006). Competencies and mathematical learning – Ideas and inspiration for the development of mathematics teaching and learning in Denmark. In the series of IMFUFA texts, Roskilde University, Denmark. To be obtained from imfufa@ruc.dk.

Chapter 35
Drivers for Mathematical Modelling: Pragmatism in Practice

Christopher Haines

Abstract In mathematical modelling, the way that certain topics are introduced depends upon many complex interacting factors, but for the teacher, learner or researcher being in touch with the real world is a key factor. Behaviours of students when faced with real-world problems are commonly represented in terms of activity within a modelling cycle but not all behaviours fit such a model; students exhibit non-linear behaviours and even within such cycles they can, and do, follow individual modelling routes. In this context, with competing and varied drivers for mathematical modelling and recognising issues of assessment, this chapter addresses the following questions: How well do students link mathematical knowledge to the task? How far away is the real world? Is mathematical modelling itself a driver for mathematical modelling?

1 Being in Touch with the Real World

Even with decades of experience in teaching mathematics and mathematical modelling, the complexities of mathematical modelling itself still led Werner Blum (Ch. 3), an acknowledged expert in education, to ask: *Can modelling be taught and learnt*? His questioning of how modelling skills of pupils in schools might be better developed touched upon the need for the teacher to adopt a role as an exemplar modeller. Within higher education, Houston and Neill (2003) had previously found a disappointingly low increase in expertise amongst undergraduate students following mathematical modelling courses over 2 or 3 years. In post-education in business, industry and in government, mathematical modelling features large; specific mathematics and mathematical methods and strategies are used to solve real

---

C. Haines (✉)
School of Engineering and Mathematical Sciences, City University, Northampton Square, London EC1V 0HB, UK
e-mail: c.r.haines@city.ac.uk

problems such as the distribution of synthetic fibres on a conveyor, tackling big problems in applied mathematics (Hunt 2007) or dealing with forecasts in banking (Barker 2007). This illustrates the diversity of applications and levels at which mathematical modelling is encountered, and so even after being involved in the teaching of engineers and mathematicians for more than 40 years in schools, in colleges and in universities, it is still difficult to say what is 'the best way' to introduce certain topics and the more so where mathematical modelling is concerned. 'The best way' must depend upon a multitude of complex interacting factors, but one thing is certain, that as a teacher, learner or researcher being *in touch with the real world* is tremendously important.

Much has been written about *what mathematical modelling is*, and *how it should or could be embedded within the curriculum*, whether in primary schools, secondary education or in colleges and universities. The reality is that many teachers and lecturers have to deliver a particular curriculum and they are bound by constraints placed upon them, but this should not stop them doing what they can, adopting pragmatic views and, as Lamon et al. (2003) put it: *regarding mathematical modelling as a way of life*. It is important to consider what are regarded as drivers for mathematical modelling and how teachers and lecturers in different contexts may or may not be pragmatic in the practice of mathematical modelling. Where promoting active learning through modelling is concerned, it is easy to see that whatever model of modelling is adopted, being in touch with the real world is crucial. The following well-known examples are being used by practitioners to good effect and have strong links with reality. This does not mean that the pupil or student sees that link or that the teacher understands how it is viewed by the pupil or student (Haines and Crouch 2005).

The London Ring Main (Fig. 35.1) is an accessible resource for modelling, which is easy to embed in curriculum work on volumes and capacity. The Sugar Loaf Cableway in Rio de Janeiro, reported through a newspaper article, has excited pupils engaged in modelling tasks developed by teachers in schools (Blum and Leiβ 2007). The Red Sand lighthouse, off the coast at Bremen, Germany, provides interesting exercises possibly involving the curvature of the earth and accessible to lower secondary school pupils (Borromeo Ferri 2007). Now that the era of cheap air travel has arrived, Kaiser (2007) has adapted modelling tasks that investigate how such flights are priced on the web.

In these examples, being *in touch with the real world* relies on the teacher and pupils or students being motivated and engaged in tasks which materialise through a large-scale engineering project, a newspaper article about a cableway engineer, the utility of a purpose built tower and market-driven aspects of using the web respectively. When attempting tasks in modelling and applications, pupils and students do not always see things as the teacher does or as the medium describing the background to the task intended. The real world for some pupils can remain firmly entrenched in the classroom and simplifications introduced in a modelling task sometimes elicit unexpected and/or inappropriate responses not dissimilar to so-called howlers seen in student responses in examinations. The examiner asking the examinee to 'explain the shape of the curve' in a graph question can hardly be

35 Drivers for Mathematical Modelling

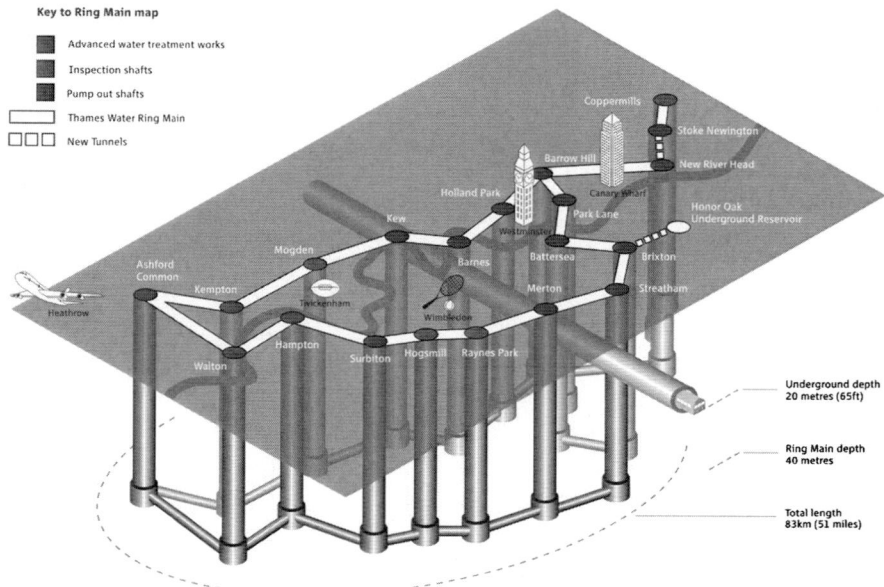

Fig. 35.1 The London Ring Main (Thames Water 2009)

surprised when the response begins: *Its curvy with a higher bit at the end and a rather aesthetically pleasing slope downwards towards a pretty flat strai(gh)t bit....* Then in a mechanics question what exactly did the examiner expect on asking: 'Does the object continue to move after it comes to rest?' Would a simple *No!* response suffice? These are simple illustrations that serve to highlight that greater difficulties occur in mathematical modelling tasks where complexity and open-endedness of the task description often leads to a rich diversity of responses. That richness should not be confused with an absence of good working knowledge of the subject matter and background application material which sometimes itself leads to lateral thinking. For example, consider a sub-task where a student is given two or three Smarties and then asked to estimate how many Smarties there are in a tube of Smarties. The student, unable to establish a simple mathematical model relating to the task, might reach a solution merely by telephoning the manufacturing company. Lateral thinking might or might not be a good thing in mathematical modelling!

Pollak (1983) commented that: *Society provides the time to teach mathematics in our schools every year. Why? Not because mathematics is beautiful – which it is – or because it provides great training for the mind, but because it is so useful.* This observation is as relevant now, a quarter of a century later, as it was then and so we find applications and modelling in the mathematics curriculum at all levels. This is reflected in the above examples, the way that they are used and, new skills and understandings brought by undergraduates to their engineering and mathematics courses. Understanding the processes used by students when faced with real-world problems for which practical outcomes might be achieved by constructing a mathematical model has been the subject of a great deal of research. It is common to represent such behaviours in terms of activity within a modelling cycle but not all fit such a model; students exhibit non-linear behaviours and even within such representational cycles they can, and do, follow individual modelling routes (Borromeo Ferri 2007; Doerr 2007; Galbraith and Stillman 2001). Against this background, with competing and varied drivers for mathematical modelling locally and recognising issues of assessment, what happens in practice? In order to answer this, consider the following questions: *How well do students link mathematical knowledge to the task at hand? How far away is the real world? Is mathematical modelling a driver for mathematical modelling?*

2 How Well Do Students Link Mathematical Knowledge to the Task at Hand?

Teachers and lecturers can think of modelling tasks in their own situations, in primary, secondary or in tertiary education, of the mathematics that could be necessary for the task and the pupils' or students' capability in that regard. Some time ago, Burley and Trowbridge (1984) put forward a thesis that modelling in the students' current year should be done with the previous year's mathematics. In expressing this view, they recognised that modelling itself in their university context needs both maturity and technical skill. One is driven to ask: *How might this apply to modelling tasks in the secondary school? Does it have relevance in primary education? How is maturity in approaching a particular problem defined? What technical skills are required?* Some recent research on modelling drug administration regimes with students in Romania (András and Szilágyi 2010) tends to support this view: Maturity in modelling a solid prior grounding in mathematical skills is required. In András and Szilágyi (2010), students were required to learn about solving systems of coupled differential equations on the same time scale as modelling the regimes, there were very mixed results in terms of understanding and application of mathematics.

Even final year undergraduate students struggle to link key mathematical results, learned earlier in their mathematics education, to building and refining a mathematical model. For example, in the context of a mathematical models and modelling course, students consider, develop and extend models concerned with kidney dialysis (a two-compartment model), rocket satellite systems (conservation of momentum model) and the aggregation of slime mould (input-output model). Each of these requires a firm understanding of the limit definition of a derivative, usually learned in the upper secondary school and reinforced in the first year at university. Similar difficulties are faced with the limit definition of the exponential function, the mean value theorem for integrals, linear dependence and notions of eigenvalues and eigenvectors, Laplace transforms—all introduced early in university courses.

Even more surprising is the lack of confidence shown by some students in analysing quadratic equations arising in particular models where real roots are required. Students have difficulty recalling such definitions and using them; the mathematical modelling seems to obscure the link to previously learned mathematics. The above observations are consistent with the research of Anderson et al. (1998) who tested 155 final year undergraduates in 15 higher education institutions. It is evident that foundations of mathematics in school, in the first year at university and subsequently are often very flimsy and that the retention of foundation material in mathematics is weak—all of which inhibits the practice and development of mathematical modelling.

Several important questions arise (Haines and Crouch 2009), even with flexible and pragmatic approaches to mathematical modelling locally. In giving a particular modelling task to students, teachers and lecturers need to understand the attainment levels of the students and to have an idea of the mathematics necessary for different outcomes from the task. They need to ask the question: *To what extent is a particular level of mathematics competence assumed?* Faced with having to deal with specific mathematics within the modelling task, the student's behaviour will be affected by his or her level of competence and so the teacher or lecturer needs to consider: *How does the latent mathematical content of a modelling situation alter learning precedences for the pupil? If, when faced with a model for which the pupil's mathematics is inadequate, does learning the mathematics take precedence?* It is clear that more research needs to be done on mathematics as a precursor to mathematical modelling, but, *what happens now in practice?* Given the Burley and Trowbridge (1984) premise and the recent inconclusive experiences of András and Szilágyi (2010), for example, *Can students learn new mathematics simultaneously with modelling experiences? Is learning content a driver for mathematical modelling?*

3 How Far Away Is the Real World?

It is unsurprising that current mathematics teaching emphasises applications of mathematics and the ability of pupils and students to address real-world problems. Mathematical modelling, encompassing projects, investigations, open-ended problem solving and other tasks, engages pupils and students directly with the links between the real world and mathematics itself, and the transitions between them, making the usefulness of mathematics obvious to them. That is the theory, but now try to look at the reality, try to look at the real world as far as the pupils and students are concerned. Teachers and lecturers try to create a real world for the students by providing a proxy for it through a range of mechanisms some of which are illustrated in the examples of kidney dialysis, rocket satellite systems, aggregation of slime mould amoebae and road traffic flows which follow. The first example is discussed in more detail; other examples concentrate on the proxy for the real world.

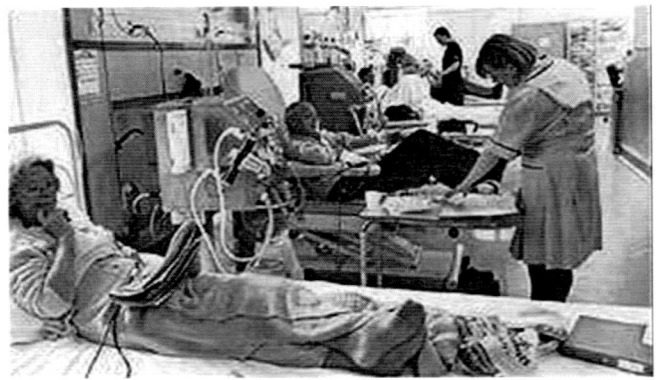

Fig. 35.2 Patients undergoing dialysis through a kidney machine (BBC News 2000)

3.1 Kidney Dialysis

Using the modelling cycle described by Blum and Leiβ (2007), it is easy to identify stages involved in the modelling process. Patients hooked up to a kidney machine (Fig. 35.2) provide a startling indication of the *real situation*, and this can be supplemented by text and other material on kidney dialysis.

A schematic representation of this process is shown in Fig. 35.3, the *real situation model* of Blum and Leiβ (2007). Figures 35.2 and 35.3 help to create a 'student' real world, for the majority of them would have little or no knowledge of kidney dialysis and neither are they likely to have experienced it. Moving from the real world into a mathematical world, Fig. 35.4 is integral to the *mathematisation* and construction of the *mathematical model*.

A further stage in the modelling process is taking a section of the machine of length δx, and completing the mathematisation through a simple input-output *mathematical model* (Fig. 35.5). Figures 35.2–35.5 show clearly that it is not a simple matter for the student to move from the real world to the mathematical world, and neither is it easy for students to relate behaviours of their mathematical model to the real world. This is consistent with Crouch and Haines (2004), who demonstrated that transitions between the real world and the mathematical world, in either direction, cause great difficulty for students.

In this modelling situation, the proxy for reality described above does not work for all students, because in discussing the model, dealing with the mathematics and obtaining results from it, errors and misunderstandings occur. The mathematical boundary conditions chosen by the student, such as: $u(0)=u_0$ and $u(L)=0$, ignoring the steady state nature of the model and the role of the dialysate, or $u(0)=0$, $v(L)=u_0$, assuming that the blood has no waste products in it on entry to the machine, show that the student does not relate the mathematical model to the real situation. Further misunderstandings occur over the modelling of the passage of waste material from the blood to the dialysate across the membrane (Fig. 35.4) represented by the line DC (Fig. 35.5).

35 Drivers for Mathematical Modelling

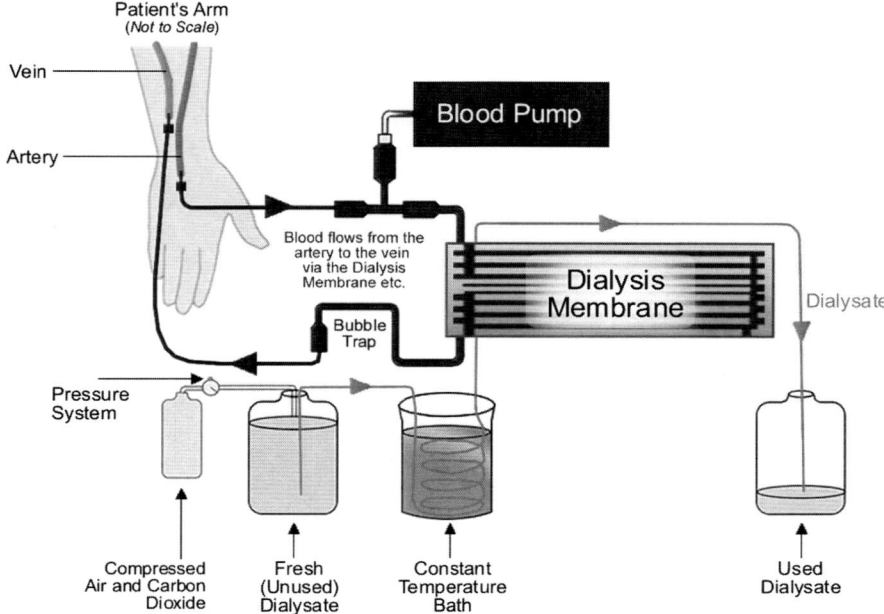

Fig. 35.3 A schematic representation of kidney dialysis (Ivy Rose Holistic 2009)

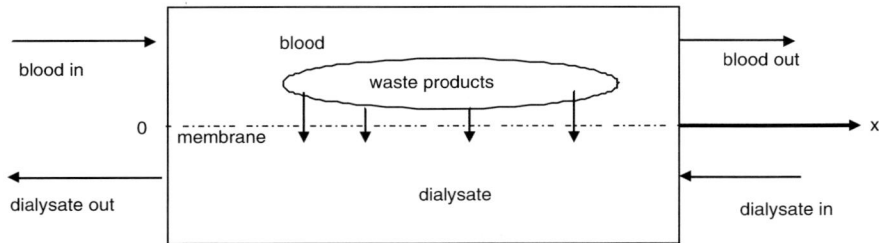

Fig. 35.4 Developing the kidney machine model as a two-compartment model

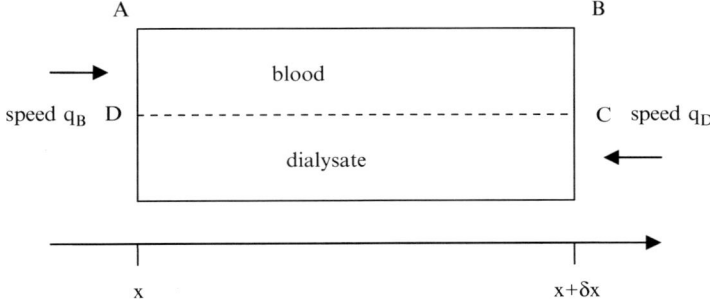

Fig. 35.5 A section, length δx, through a kidney machine

3.2 Rocket Satellite Systems

Another example concerns rocket satellite systems, in which an initial model, grounded in circular motion, discusses the speeds of satellites in various orbits of the earth, the moon or other planets. A second model discusses size and structure, in terms of stages, of rockets necessary to place such satellites in orbit using straightforward ideas of conservation of momentum. The reality here is provided by news journalism in various media; the constantly changing stories in the news are usually inherently interesting.

During a modelling course in 2009, *The Times* (2009a) reported on the failed deployment of the Orbiting Carbon Observatory. The story itself and the graphics (Fig. 35.6) provided by the newspaper helped the students to understand the initial model for speeds of satellites in orbits, the implications for the required speed of the delivery rocket and how the different stages of the rocket are deployed. The currency of the event and the media reports influenced *reality* of the situation for the students.

Prior to that on 13 February 2009, *The Times* (Fig. 35.7) reported that: '...*a US Iridium 33 satellite collided with a Russian Kosmos satellite over Siberia. The*

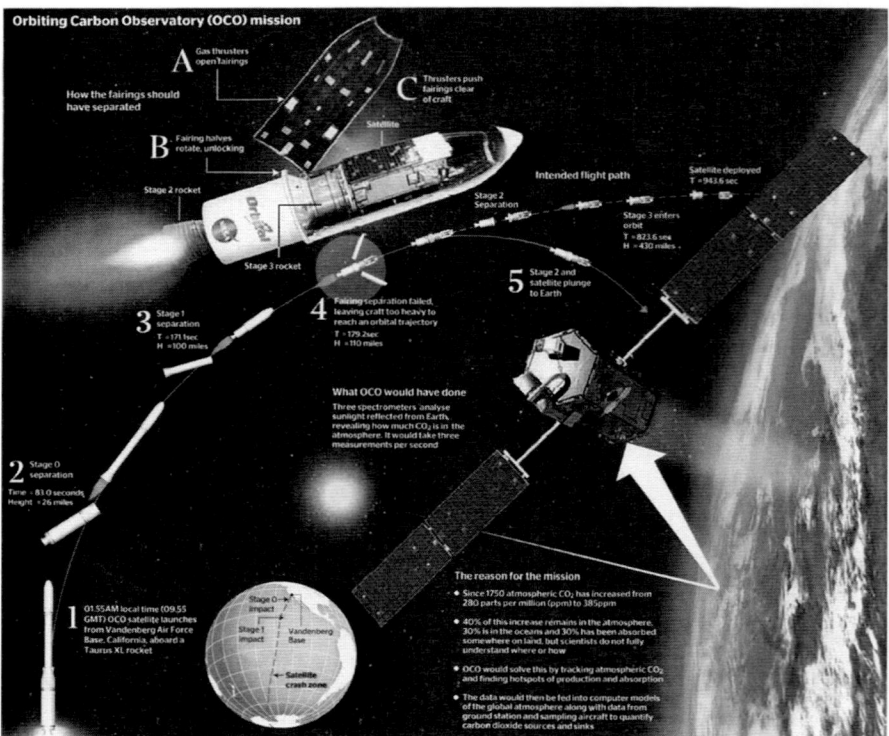

Fig. 35.6 The failed deployment of the orbiting carbon observatory (*The Times* 2009a)

35 Drivers for Mathematical Modelling

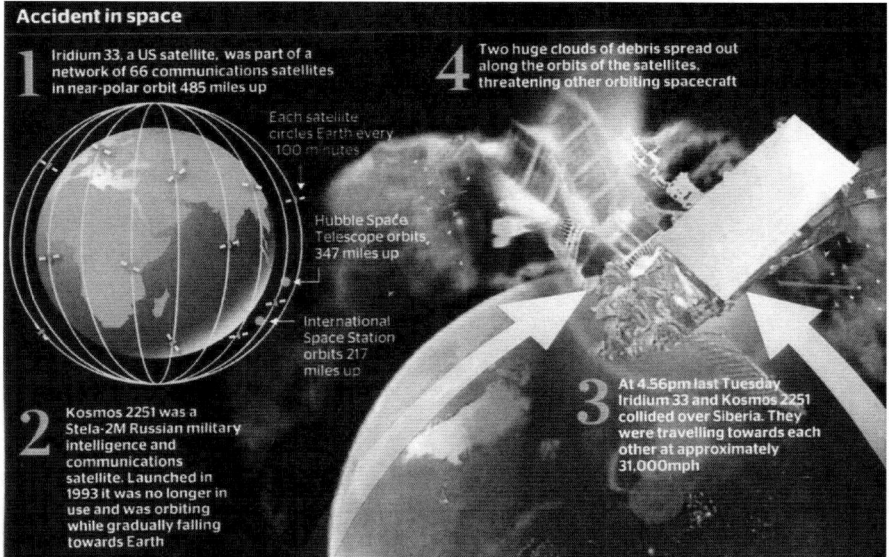

Fig. 35.7 Satellite collision (*The Times* 2009b)

Iridium satellite was in a near polar orbit 775 km above the Earth each orbit taking about 100 minutes. The two satellites were travelling towards each other at about 14 km/sec'. In this situation, the proxy for reality again does not work for all students. Comprehending the report and dealing with the mathematics resulted in simple errors and misunderstandings.

One student suggested that: *'The report should say that the satellite was travelling at about 7.5 km/sec and not 14 km/sec'* failing to understand the concept of collision speed. A second student also made this error: *'The speed of the satellite is 7.479 km/sec so the time to orbit 100 minutes is ok. The collision speed is not very accurate'*. In the context of the satellite collision, the students did not interpret collision speed correctly. Would they have made this mistake had the collision been between two cars racing towards each other?

Satellites and rocket systems remain a rich source of material for modelling, certainly at upper secondary and tertiary level. Even now, there has been a resurgence of interest in the moon landing programmes of the 1960s and early 1970s. It would be interesting to hear of space applications at primary and lower secondary levels.

3.3 Aggregation of Slime Mould Amoebae

Although rockets and satellites provide a large-scale modelling experience for which media reports may be a proxy for *reality*, and for which the modelling may be readily understood, it is not so easy with applications in microbiology, such as the aggregation of slime mould amoebae. Here the real world is interpreted through reports of laboratory experiments and the background is hard to understand for students without grounding in biology.

The aggregation process for the amoebae, moving from a free-living state to aggregation, is the small section (labelled b) of the complete life cycle illustrated in Fig. 35.8, whilst a laboratory image of the aggregation is shown in Fig. 35.9. Individual amoebae aggregate like this to form a 'motile' slug (National Science Foundation 2003).The modelling route is usually through a much simplified one-dimensional model in which the aggregation itself is modelled sinusoidally and is viewed as an instability emerging from a steady state. At first sight, the real world is some distance away from the student, but in practice, students readily identify with the proxy provided. Here student problems usually relate to basic theorems of analysis and the behaviour of mathematical functions described in Sect. 2, rather than indentifying with the real world provided by the microbiology background.

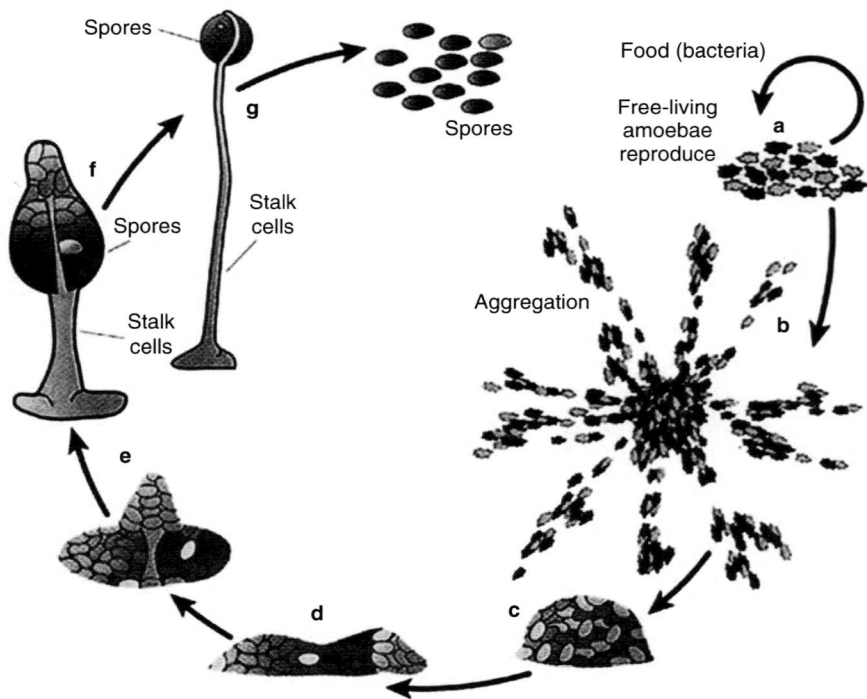

Fig. 35.8 Life cycle of slime mould amoebae (National Science Foundation 2003)

Fig. 35.9 Amoebae aggregating (National Science Foundation 2003)

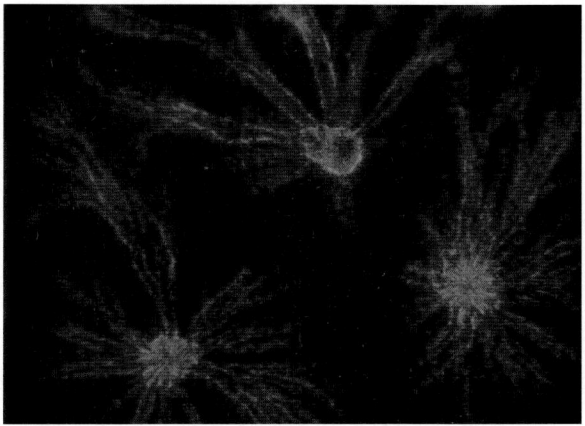

Fig. 35.10 Crossing a road *without delay*

3.4 Road Traffic Flows

Road traffic flows and pedestrians crossing roads are everyday experiences that will be familiar to students. Here, the real world is close to personal experience and there is no need to have a proxy for reality. These may be probability models for crossing a road along which traffic flows. In some of these models, the pedestrian looks for a gap in the flow of traffic that is large enough for the crossing to be made *without delay* (Fig. 35.10).

Ideas in these simple models can be used as a basis for vehicles crossing a priority intersection or GIVEWAY junction such as is shown in Fig. 35.11. In these cases, in addition to finding the probability of crossing the intersection without delay, the modeller can find the *mean delay* experienced by a vehicle waiting to cross to the other side. Notwithstanding the engagement of students with everyday experience and its motivation for the modelling taking place, students find it

Fig. 35.11 A GIVEWAY junction

difficult to distinguish between the probability of pedestrians crossing a road, or a vehicle crossing a GIVEWAY junction, *without delay* and the *mean delay* for a vehicle waiting to cross a GIVEWAY junction.

3.5 Local Models

The road traffic example of Sect. 3.4 is a good one in that it brings students' everyday experience into the modelling. Sunshine and shadows provide a rich field for investigation by students with problems that can be embedded in the immediate environment of the school or university and which benefit from the personal experiences of the student. Galbraith et al. (1998) reported on a workshop in which the movement of the sun was modelled in a task that looked at whether canopies provided better protection from the rain or from the sun. Haines (2009) has generalised the workshop so that teachers can focus on the passage of the sun in their own locality and the shadows cast by the sun in this situation. Data is easily downloaded from the GAISMA web site from which the sunpath for Hamburg on 16 June 2009 illustrated in Fig. 35.12 has been obtained. Fortunately, the download from the web is in colour and is rather easier to read. Suffice to say that Hamburg is at the centre of the diagram and the transept of the sun follows the line from about N45E round to N45W. Other downloads give, for example, sunrise and sunset times.

Models that are situated in the local environment and with everyday experience will often prove successful. In Hamburg, The Old Elbe Tunnel (Fig. 35.13) has the potential for another local source of inspiration for teachers bringing modelling tasks to their pupils. Whether they concentrate on basic engineering applications, traffic flow through the tunnel or logistics of the Hamburg Tunnel Marathon (2009),

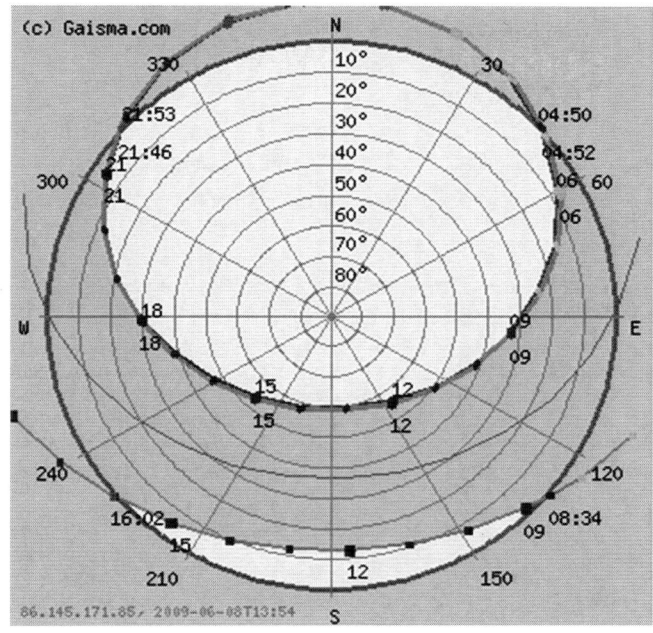

Fig. 35.12 Sun path diagram for Hamburg, Germany on 16 June 2009 (GAISMA 2009)

Fig. 35.13 The Old Elbe Tunnel, Hamburg

here it is certain that motivation and engagement of student will be strong. There is no need for a proxy real-world situation.

In all these examples, it is easy to see that the real world has different meanings and different manifestations for different problems. Students in tertiary education might at one time be further away from the practical reality (c.f. rocket satellite systems) and at the same time closer to it through experimental projects in the laboratory on sub-models of that practical reality. At one extreme, the real world is everyday personal experience of a structure or an event; the closeness of the real world is self-evident. However, major questions remain: *What constitutes the real world for students? How close is the real world for students and can we define and measure descriptors of that closeness? Do attitude, motivation and engagement occur in such descriptors? Is there confusion between a proxy for a real-world situation and the real world itself? Are we confusing the real world and the classroom context? Are the problems that we ask students to consider really real-world problems?*

4 Is Mathematical Modelling a Driver for Mathematical Modelling?

Recall some assessment objectives in mathematics (Table 35.1) developed by lecturers in a consortium of 15 UK universities engaged in an HEFCE project (Haines and Dunthorne 1996). Most of these objectives can be achieved *within* a mathematical modelling framework, but a question that does arise is whether mathematical modelling itself is necessary for the achievement of such objectives. Of course, it is necessary in some way because it forms the basis for at least two of the objectives, but does mathematical modelling form an essential part of the curriculum in all cases?

If a purpose of mathematical modelling is mathematical modelling itself, that is, pupils and students are being encouraged to become good mathematical modellers,

Table 35.1 Assessment objectives in mathematics (Haines and Dunthorne 1996, p. 5.3)

- Recall select and use mathematical facts, concepts, techniques
- Construct mathematical arguments
- Formulate mathematical models
- Evaluate mathematical models
- Develop skills of criticism
- Organise mathematical information
- Interpret mathematical information
- Communicate mathematical ideas
- Develop oral and written communication skills
- Read and comprehend mathematics
- Develop logical thinking
- Provide students with vocational education
- Encourage independence of thought and initiative
- Develop group-working skills

then care must be taken about the assessment of mathematical modelling. The process itself must be assessable rather than simply concentrating on the outcomes of one or two individual or group projects (say). This is not easy since in most situations and certainly, within mathematics courses at university, students' experiences of mathematical modelling may be restricted to one, two or three cases, case studies or projects.

What should be involved in the assessment of modelling? Niss and Jensen (2006) take a holistic approach to modelling competence, suggesting a 3-dimensional model for which the dimensions are: degree of coverage, radius of action and technical level. Degree of coverage relates to understanding stages of modelling and how they are linked. Haines and Crouch (2001, 2005) have demonstrated that assessment of this dimension could be done using multiple-choice questions that focus on each stage of a modelling cycle. It is possible that a rating scale of this dimension applicable to postgraduate, undergraduate, pre-university and school levels could be constructed (Haines and Crouch 2007). Radius of action refers to the experience that the learner has in the variety and complexity of models. Such experience in schools will necessarily be more limited and narrower than that of students in undergraduate courses. But whilst that experience is necessarily restricted in comparison to an expert, it might be extensive in respect of other children of the same age. Whilst there is a difficulty with the 'Radius of action' dimension, Haines and Crouch (2007) recognise that in a structure of a mathematical modelling expertise continuum, behaviours in school, for example, might be qualitatively the same as in pre-university, undergraduate or postgraduate sectors, but differ substantially in terms of acquired expertise. This area is extremely important because we know that a weak knowledge base in students and a lack of experience in abstraction cause difficulties in the transition from the real world to the mathematical world (Crouch and Haines 2004). The knowledge base and experience can be improved by repeated graded exposure to models and modelling. We are not convinced that current curriculum practice does give graded continuous exposure in this way. The dimension 'Technical level of mathematics' is usually regarded as adequately assessed outside modelling activities. However, the technical level of mathematics influences the modelling that is accessible to the learner.

If it is accepted that becoming a good mathematical modeller is a driving force for mathematical modelling itself then it follows that, under the Niss Jensen model, each of these dimensions must be assessed combining them in some way to give an overall assessment of mathematical modelling competence. It is arguable that this approach is in fact a practical one within the classroom, but Zöttl, Ufer and Reiss (Ch. 42) have demonstrated a research tool that seeks to assess mathematical modelling as a whole. They use multidimensional Rasch modelling with dichotomous responses that does not include partial credit and have achieved some interesting results in a restricted field. An alternate view is that mathematical modelling is a driver for the acquisition and deployment of mathematical knowledge in which case the assessment of mathematical modelling as whole could be a secondary issue.

Acknowledgements I thank Professor Gabriele Kaiser for her support in inviting me to give this plenary lecture at ICTMA 14; it is an honour that is much appreciated; Dr A.M. Fairhurst of IvyRose Ltd for providing Fig. 35.3 and Rosalind Crouch (formerly University of Hertfordshire, Hatfield, UK) with whom many ideas in this chapter have been discussed.

References

András, S., & Szilágyi, J. (2010). Modelling drug administration regimes for asthma: A Romanian experience. *Teaching Mathematics and Its Applications, 29*(1), 1–13.

Anderson, J., Austin, K., Barnard, A., & Jagger, J. (1998). Do third-year mathematics undergraduates know what they are supposed to know? *International Journal of Mathematics Education in Science and Technology, 29*(3), 401–420.

Barker, K. (2007). Economic modelling: Theory, reality, uncertainty and decision-making. In C. Haines, P. Galbraith, W. Blum, & S. Khan (Eds.), *Mathematical modelling ICTMA12: Education, engineering and economics* (pp. 25–42). Chichester: Horwood.

BBC News (2000). Kidney dialysis patients. http://news.bbc.co.uk/1/hi/health/991860.stm. Accessed 7 Oct 2009.

Blum, W., & Leiβ, D. (2007). How do students and teachers deal with modelling problems? In C. Haines, P. Galbraith, W. Blum, & S. Khan (Eds.), *Mathematical modelling ICTMA12: Education, engineering and economics* (pp. 222–231). Chichester: Horwood.

Borromeo Ferri, R. (2007). Modelling problems from a cognitive perspective. In C. Haines, P. Galbraith, W. Blum, & S. Khan (Eds.), *Mathematical modelling ICTMA12: Education, engineering and economics* (pp. 260–270). Chichester: Horwood.

Burley, D. M., & Trowbridge, E. A. (1984). Experiences of mathematical modelling at Sheffield University. In J. S. Berry, D. N. Burghes, D. J. G. Jones, & A. O. Moscardini (Eds.), *Teaching and applying mathematical modelling* (pp. 131–142). Chichester: Ellis Horwood Ltd.

Crouch, R. M., & Haines, C. R. (2004). Mathematical modelling: Transitions between the real world and the mathematical model. *International Journal of Mathematics Education in Science and Technology, 35*(2), 197–206.

Doerr, H. M. (2007). What knowledge do teachers need for teaching mathematics through applications and modelling? In W. Blum, P. L. Galbraith, H.-W. Henn, & M. Niss (Eds.), *Modelling and applications in mathematics education* (pp. 69–78). New York: Springer.

GAISMA (2009). Sunrise, sunset, sun path information for locations worldwide. http://www.gaisma.com. Accessed 16 June 2009.

Galbraith, P. L., & Stillman, G. (2001). Assumptions and context: Pursuing their role in modelling activity. In J. F. Matos, W. Blum, K. Houston, & S. P. Carreira (Eds.), *Modelling and mathematics education: ICTMA9 applications in science and technology* (pp. 300–310). Chichester: Horwood.

Galbraith, P., Haines, C., & Izard, J. (1998). How do students' attitudes to mathematics influence the modelling activity? In P. Galbraith, W. Blum, G. Booker, & I. D. Huntley (Eds.), *Mathematical modelling: Teaching and assessment in a technology-rich world* (pp. 265–278). Chichester: Horwood.

Haines, C. R. (2009). *Sunshine in DQME?* Paper prepared for the Comenius project: Developing quality in mathematics education II. University of Dortmund, Dortmund, Germany.

Haines, C. R., & Crouch, R. M. (2001). Recognising constructs within mathematical modelling. *Teaching Mathematics and Its Applications, 20*(3), 129–138.

Haines, C., & Crouch, R. (2005). Applying mathematics: Making multiple-choice questions work. *Teaching Mathematics and Its Applications, 24*, 107–113.

Haines, C. R., & Crouch, R. M. (2007). Mathematical modelling and applications: Ability and competence frameworks. In W. Blum, P. L. Galbraith, H.-W. Henn, & M. Niss (Eds.), *Modelling and applications in mathematics education* (pp. 417–424). New York: Springer.

Haines, C. R., & Crouch, R. M. (2009). Remarks on a modelling cycle and interpreting behaviours. In R. A. Lesh, P. L. Galbraith, & C. R. Haines (Eds.), *Modelling students' mathematical modelling competencies*. New York: Springer.

Haines, C. R., & Dunthorne, S. (Eds.). (1996). *Mathematics learning and assessment: Sharing innovative practices*. London: Arnold.

Hamburg Tunnel Marathon. (2009). Diagrams and details of the marathon can be found at http://www.100marathon-club.de.

Houston, K., & Neill, N. (2003). Investigating students' modelling skills. In Q.-X. Ye, W. Blum, K. Houston, & Q.-Y. Jiang (Eds.), *Mathematical modelling in education and culture: ICTMA 10* (pp. 54–66). Chichester: Horwood.

Hunt, J. (2007). Communicating big themes in applied mathematics. In C. Haines, P. Galbraith, W. Blum, & S. Khan (Eds.), *Mathematical modelling ICTMA12: Education, engineering and economics* (pp. 2–24). Chichester: Horwood.

Ivy Rose Holistic. (2009). Kidney dialysis diagram. http://www.ivy-rose.co.uk.

Kaiser, G. (2007). Modelling and modelling competencies in school. In C. Haines, P. Galbraith, W. Blum, & S. Khan (Eds.), *Mathematical modelling ICTMA12: Education, engineering and economics* (pp. 110–119). Chichester: Horwood.

Lamon, S. J., Parker, W. A., & Houston, S. K. (2003). Preface. In S. J. Lamon, W. A. Parker, & S. K. Houston (Eds.), *Mathematical modelling: A way of life* (pp. ix–x). Chichester: Horwood.

National Science Foundation. (2003). NSF PR 03-106-September 24 2003. Photo of aggregation by Kevin Foster (Strassmann/Queller), Rice University, USA, Illustration of life cycle by Mary Wu and Rich Kessin, Columbia University, USA. http://www.nsf.gov/od/lpa/news/03/pr03106_images.htm. Accessed 7 Oct 2009.

Niss, M., & Jensen, T. H. (Eds.). (2006). Competencies and mathematical learning – Ideas and inspiration for the development of mathematics teaching and learning in Denmark. Under preparation for publication in the series Tekster fra IMFUFA, Roskilde University, Denmark. To be obtained from imfufa@ruc.dk.

Pollak, H. O. (1983). Applications and teaching of mathematics. In J. S. Berry, D. N. Burghes, D. J. G. Jones, & A. O. Moscardini (Eds.), *Teaching and applying mathematical modelling* (pp. xv–xvi). Chichester: Ellis Horwood Ltd.

Thames Water (2009). London Ring Main Map. http://www.thameswater.co.uk/cps/rde/xbcr/corp/ringmain-high-extensions-map.jpg. Accessed 30 Sept 2009.

The Times (2009a). Orbiting Carbon Observatory mission. *The Times*, 25 Feb 2009.

The Times (2009b). Accident in space. *The Times*, 13 Feb 2009.

Chapter 36
Identifying Drivers for Mathematical Modelling – A Commentary

Katja Maaß

In the following, I will comment on the chapter by Haines on "Drivers for Mathematical Modelling: Pragmatism in Practice" in this volume. The examples described by Haines exemplify in an expressive way that modelling can be quite challenging and many reality-related problems can be only dealt with at tertiary level. Subsequently, it may be questioned if modelling is possible in lower secondary school and even in primary school. When it comes to lower secondary level, this has been widely accepted (see e.g. Blum and Leiß 2007). Additionally, modelling is also possible at primary level, when appropriate tasks for children are chosen (see, e.g. Biembengut 2007).

In his paper, Haines is guided by the questions, to what extent modelling can serve as a motivation for learning mathematics and what has to be taken into account when modelling is to serve as a motivation for learning mathematics. He reflects on these aspects by raising three key-questions: (1) How well do students link mathematical knowledge to the task at hand? (2) How far away is the real world? (3) Is mathematical modelling a driver for mathematical modelling? In the following, I will look at the questions from the perspective of lower secondary and primary education followed by reflections on the title.

1 How Well Do Students Link Mathematical Knowledge to the Task at Hand?

Haines points out that students often cannot use the mathematics they learnt directly before dealing with the modelling task, but that the mathematical knowledge needs to be solid. When students cannot apply the mathematical content they need, it has

---

K. Maaß (✉)
University of Education Freiburg, Freiburg, Germany
e-mail: katja.maass@ph-freiburg.de

to be relearned and in this case, the priority of the students' actions shifts away from the modelling task and to the mathematics. He also raises the question to what extent a particular level of mathematics is required to solve a certain task.

Let us now look at students at primary and lower secondary level. Two examples will illustrate some issues that might occur. The first example is from primary school and here students aged six (first class) attempt a modelling task. They have no experience in modelling (Maaß 2009).

Teacher:	Alina wants to invite 12 children to her birthday party. Do you think they can all sit at this table?
Jona:	Yes, they can.
Sabine:	No!
Teacher:	What shall we do now?
Sabine:	No!
Teacher:	Why?
Patricia:	6 per bench (measuring with her fingers)
Patricia:	30 all together, because of the 5 benches

At first, these children just guess and do not use any mathematics. After a while, however, one girl sees a relation between this task and mathematics and uses the mathematics she knows to solve the task.

The second example comes from lower secondary school. Here the students dealt with the question "How big is the surface of a Porsche 911?" One girl intended to use a cube as a model – which is correct – but then uses a wrong formula to calculate the surface of a cube ($O = a + 2b + 2c$) (Fig. 36.1).

The two examples show that at primary or lower secondary level, students may, in some cases, not use mathematics at all or use mathematics incorrectly. When we look at the question to what extent a particular level of mathematics is required to

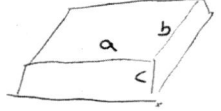

Fig. 36.1 One girl intended to use a cube as a model – which is correct – but then uses a wrong formula to calculate the surface of a cube ($O = a + 2b + 2c$)

solve a certain task, we need to say that for the school levels discussed here, this is a relevant question directed at the teacher.

On a more *general level,* the question of how well students link knowledge to the task at hand depends on several factors, for example students' mathematical competencies (Blomhøj and Jensen 2007), the cognitive demand of the task (Jordan et al. 2008), students' basic mental ideas – so-called Grundvorstellungen – which means that mental objects are necessary for transitions between reality and mathematics because they mediate between the two worlds (Kleine et al. 2005), students' beliefs about mathematics (Maaß 2004), the didactical contract (Verschaffel et al. 2000) and students' experience in modelling (Maaß 2004).

2   How Far Away Is the Real World?

Haines raised in his paper the questions to what respect, the real world actually is a real world to the students, how can we measure whether a context is reality or not for students and which constitutes the real world for the students. Again, we will look at these questions from the primary and lower secondary level. A well-known task for primary school is the so-called captain's task: *There are 26 sheep and 10 goats on a ship. How old is the captain?* The large majority of primary school pupils, who were given the task, gave a numerical answer by adding the two numbers given. This however is not due to an inability of the students but the didactical contract in the classes, as this procedure normally works for traditional word problems (Verschaffel et al. 2000). Also at lower secondary level, some students distinguish clearly between mathematics and their general knowledge about the context. The students were given the following task: *In 1993, the worldwide reserves of natural gas were estimated to be 141.8 billion cubic metres. Since then 2.5 billion cubic metres have been used every year on average. Calculate when the reserves of natural gas will be exhausted. Use different assumptions and models. Explain all your steps.* A student, Albert, provided only one and the most obvious model: He divided the 141.8 billion by 2.5 billion and got a result. Then he drew a line and talked about his experiences with renewable energies. His text showed that he knew a lot about this topic, but he was not able to link his knowledge to a mathematical task. Altogether, these two examples show that the context does not form a real world for the students because it is given to students within the context of mathematics education. So, when reflecting on what constitutes the real world for students, we also have to take into account the context of mathematics education.

On a more *general level,* the question of how well students link knowledge to the task at hand depends on several factors, for example the students' experience in modelling and their beliefs (Maaß 2004), the complexity of the situation (Jordan et al. 2008), students' thinking styles (Borromeo Ferri 2007), the didactical contract and the context in which the students learn (Lave and Wenger 1991).

3 Is Mathematical Modelling a Driver for Mathematical Modelling?

Haines raised this question without answering it, but emphasised that if this is the case, there are implications for practice. We need structured approaches on how to test modelling (Haines and Crouch 2005) and we need models for modelling competencies (Blomhøj and Jensen 2007). These requirements are also requirements for primary and secondary level.

To sum up: Given these issues, raised when reflecting on the questions of Haines, Is modelling a driver for learning mathematics at primary and secondary level? the difficulties named above are mainly difficulties when students start to model. Links to the real world can make mathematics more concrete for many students, especially for low achieving students. They learn mathematics more meaningfully (Maaß 2004). Subsequently, modelling can be a driver for learning mathematics, but students need time to adapt to modelling tasks. In order to deal with students' different ways of working, a variety of contexts and different types of tasks should be used (Blomhøj and Jensen 2003; Kaiser 1995). Once students are adapted to modelling, it can also be a driver for itself.

4 What the Title Made Me Think of…

At the end of this chapter, I would like to reflect on the title of Haines' paper which I found very inspiring *"Drivers for mathematical modelling: Pragmatism in practice"*. So far, we have discussed modelling as a driver for mathematics and modelling as a driver for modelling itself. Thinking of drivers, the link to the aims of the integration of modelling is close (Kaiser 1995). As aims we find among others the application of mathematics in life, understanding the world around us, developing problem-solving competencies, getting insight into the usefulness of mathematics (learning mathematics meaningfully), understanding and memorising mathematics and developing positive attitudes toward mathematics. The important question is: Are these aims also drivers for the implementation into day-to-day practice? *What are drivers for modelling?*

When looking at this question from primary and lower secondary school, we have to differentiate between two levels: drivers for students and drivers for teachers to implement modelling. For *students,* drivers for modelling may be learning mathematics meaningfully as well as external drivers such as parents, teachers or grades. However, what really happens in class is determined by the teacher. So what works as drivers for *teachers*?

Apparently, the aims are not drivers in a way we would wish them to be, as modelling is not widely implemented in daily teaching practice (Blum, this volume). Actually, there is a theory-practice-gap (Bruder 2009) as ideas which are developed within research are not implemented in practice in a way that they should be. The reasons for this seem to belong to a complex system of components, among which we find, for example, centralised assessment, the curriculum and teachers' beliefs about mathematics (Kaiser 2006). Altogether, change is hard to be achieved (Tirosh and Graeber 2003; Wilson and Cooney 2002) and facilitators or drivers to do this may be helpful.

Drivers for teachers, which may help to implement modelling and which can be provided by research, can be materials (tasks and guidelines for implementation, based in theory and addressing teachers' needs) and courses of professional development. Factors which we should analyse more carefully are the contexts and frameworks (see below).

In relation to materials, we need modelling tasks for all school levels (including primary school) which cover the main areas of mathematical school content. We also need different types of modelling tasks including an overview about different features of tasks (Blomhøj and Jensen 2003; Kaiser 1995) as students have different styles of learning (Prenzel et al. 2004).

There is not much empirical evidence on how to really teach modelling (Leiß 2007). Here we need more research. However, a variety of studies show that working in small groups, discussion in groups and students working independently can support the development of modelling competencies (see e.g. Ikeda and Stephens 2001). Based on these results, teaching concepts for all levels of education need to be designed and published in a way and in media that teachers read. Teachers also need to get materials and information on how modelling can actually be assessed, how a written class test can be designed, what other forms of assessment exist and how to assess students' solutions.

In relation to professional development, we also need materials for this based in theory and taking into account teachers' competences, beliefs and needs (Tirosh and Graeber 2003, Wilson and Cooney 2002) for example as have been developed within the European project LEMA (Maaß and Gurlitt, this volume). Last but not least, we need effective methods for large-scale training (Adler and Jaworksi 2009), which also considers context factors such as curriculum, external assessment, politics, parents, school principals, etc.

To sum up, in order to set up or identify *drivers for modelling*, we need more designed materials based in theory and tested on a large scale, we need design research (Burkhardt 2006). Further, we need more research close to day-to-day teaching *practice* (Bruder 2009), including context and framework of school life. When carrying out such research, we need to be *pragmatic*, because when working with real classes instead of laboratories, it may turn out to be complicated to have a perfect research design.

References

Adler, J., & Jaworski, B. (2009). Public writing in the field of mathematics teacher education. In R. Even & D. Loewenberg Ball (Eds.), *The professional education and development of teachers of mathematics – The 15th ICMI study* (pp. 249–254). New York: Springer.

Biembengut, M. S. (2007). Modelling and applications in primary education. In W. Blum, P. Galbraith, H.-W. Henn, & M. Niss (Eds.), *Modelling and applications in mathematics education* (pp. 451–456). New York: Springer.

Blomhøj, M., & Jensen, T. (2003). Developing mathematical modelling competence: Conceptual clarification and educational planning. *Teaching Mathematics and Its Applications, 22*(3), 123–139.

Blomhøj, M., & Jensen, T. (2007). What's all the fuss about competencies? In W. Blum, P. Galbraith, H.-W. Henn, & M. Niss (Eds.), *Modelling and applications in mathematics education* (pp. 45–56). New York: Springer.

Blum, W., & Leiβ, D. (2007). How do students and teachers deal with modelling problems? In C. Haines, P. Galbraith, W. Blum, & S. Khan (Eds.), *Mathematical modelling ICTMA12: Education, engineering and economics* (pp. 222–231). Chichester: Horwood.

Borromeo Ferri, R. (2007). Modelling problems from a cognitive perspective. In C. Haines, P. Galbraith, W. Blum, & S. Khan (Eds.), *Mathematical modelling ICTMA12: Education, engineering and economics* (pp. 260–270). Chichester: Horwood.

Bruder, R. (2009). Langfristige fachdidaktische Forschungsprojekte zur mathematischen Unterrichtsentwicklung in der Sekundarstufe I. Paper presented at the Tagung der Gesellschaft für Didaktik der Mathematik. In M. Neubrand (Ed.), *Beiträge zum Mathematikunterricht* (pp. 23–30). Münster: WTM-Verlag.

Burkhardt, H. (2006). From design research to large-scale impact: Engineering research in education. Paper presented at the International Society of Design and Development in Education. In J. Van den Akker, K. Gravemeijer, S. McKenney, & N. Nieveen (Eds.), *Educational design research* (pp. 185–228). London: Routledge.

Haines, C., & Crouch, R. (2005). Applying mathematics: Making multiple-choice questions work. *Teaching Mathematics and Its Applications, 24*, 107–113.

Ikeda, T., & Stephens, M. (2001). The effects of students' discussion in mathematical modelling. In J. F. Matos, W. Blum, K. Houston, & S. P. Carreira (Eds.), *Modelling and mathematics education: ICTMA 9: Applications in science and technology* (pp. 381–390). Chichester: Horwood.

Jordan, A., Krauss, S., Löwen, K., Blum, W., Neubrand, M., Brunner, M., et al. (2008). Aufgaben im COACTIV-Projekt: Zeugnisse des kognitiven Aktivierungspotentials im deutschen Mathematikunterricht. *Journal für Mathematik-Didaktik, 29*(2), 83–107.

Kaiser, G. (1995). Realitätsbezüge im Mathematikunterricht – Ein Überblick über die aktuelle und historische Diskussion. In G. Graumann, T. Jahnke, G. Kaiser, & J. Meyer (Eds.), *Materialien für einen realitätsbezogenen Mathematikunterricht* (Vol. 2, pp. 66–84). Hildesheim: Franzbecker.

Kaiser, G. (2006). The mathematical beliefs of teachers about application and modelling – Results of an empirical study. In J. Novotaná, H. Moraová, M. Krátká, & N. Stehliková (Eds.), Paper presented at the *30th Conference of the International Group for the Psychology of Mathematics Education* (pp. 393–400). Prague: PME.

Kleine, M., Jordan, A., & Harvey, E. (2005). With a focus on 'Grundvorstellungen' as a theoretical and empirical criterion. *Zentralblatt für Didaktik der Mathematik, 37*(3), 234–239.

Lave, J., & Wenger, E. (1991). *Situated learning. Legitimate peripheral participation*. Cambridge: Cambridge University Press.

Leiß, D. (2007). *"Hilf mir es selbst zu tun" – Lehrerinterventionen beim mathematischen Modellieren*. Hildesheim: Franzbecker.

Maaß, K. (2004). *Mathematisches Modellieren im Unterricht – Ergebnisse einer empirischen Studie*. Hildesheim: Franzbecker.

Maaß, K. (2009). *Mathematikunterricht weiterentwickeln*. Berlin: Cornelsen Scriptor.

Prenzel, M., Baumert, J., Blum, W., Lehmann, R., Leutner, D., Neubrand, M., et al. (2004). *PISA 2003 – Der Bildungstand der Jugendlichen in Deutschland – Ergebnisse des zweiten internationalen Vergleichs*. Münster: Waxmann.

Tirosh, D., & Graeber, A. O. (2003). Challenging and changing mathematics teaching practises. In A. Bishop, M. A. Clements, C. Keitel, J. Kilpatrick, & F. Leung (Eds.), *Second international handbook of mathematics education* (pp. 643–688). Dordrecht/Boston/London: Kluwer.

Verschaffel, L., Greer, B., & De Corte, E. (2000). *Making sense of word problems.* Lisse: Swets & Zeitlinger.

Wilson, M., & Cooney, T. J. (2002). Mathematics teacher change and development. The role of beliefs. In G. C. Leder, E. Pehkonen, & G. Törner (Eds.), *Beliefs: A hidden variable in mathematics education?* (pp. 127–148). Dordrecht: Kluwer.

Chapter 37
Documenting the Development of Modelling Competencies of Grade 7 Mathematics Students

Piera Biccard and Dirk C.J. Wessels

Abstract Modelling is a mathematical competence and a means of learning significant mathematics. Preliminary findings on the nature of competence and modelling competencies, as well as the development of mathematical modelling competencies of Grade 7 students are presented in this chapter. Twelve students solved three model-eliciting tasks in groups over a period of 12 weeks. The development of competencies is considered across the group as a whole. Analysis of the data shows that modelling competencies are activated and do develop when students take part in well-orchestrated modelling activities.

1 Introduction

Mathematical modelling entails the resolution of a specific type of task, based on a certain context in reality, set in an environment which makes students' conceptual and procedural knowledge visible. These real, complex tasks have come to be known as model-eliciting tasks (English and Lesh 2003; Lesh et al. 2000, p. 608, p. 298). The process of resolving these tasks is formulated as a cycle. This cycle has been formulated in many ways. The cycle of Blum and Leiss (Borromeo Ferri 2006, p. 87) is considered for this study and each node of the cycle was used as a starting point for competency identification in depicting modelling competencies.

This cycle (as many of the others) assists in recognising the students' path through a modelling problem. Borromeo Ferri (2006, p. 91) reminds us that these phases are normative and seen as an ideal way of modelling; modelling does not

---

P. Biccard (✉) and D.C.J. Wessels
Department of Curriculum Studies, Stellenbosch University, Stellenbosch, South Africa
e-mail: pbiccard@yahoo.com; dirkwessels@sun.ac.za

develop in the perfect circular order presented. Very often students move back and forward in their need to understand, structure and to mathematise the problem. The ease with which students are able to move through the cycle is dependent on a number of factors such as their level of experience in modelling problems, context of the problem and group dynamics.

2 A Perspective for Modelling

Kaiser and Sriraman (2006, p. 302) reveal that there is no common understanding of modelling within an international discussion on modelling, but that certain perspectives can be found. It is important to embed this study within or amid theoretical and conceptual frameworks that currently exist and to formulate a justification around these. The theoretical and conceptual framework adopted essentially reveals what it means to the authors to 'learn mathematics by doing mathematics'. Since this study displays (in Kaiser and Sriraman's terms) pedagogical aims, psychological aims and subject-related aims, it can be considered as an example of an 'integrative perspective' (Kaiser and Sriraman 2006, p. 302). This study integrates and advances 'contextual', 'educational' and 'cognitive' perspectives (Kaiser and Sriraman 2006, p. 304) to mathematical modelling. A contextual perspective emphasises model-eliciting problems starting from meaningful situations while an educational perspective focuses on the integration of mathematical modelling in mathematics teaching (Blomhøj nd.). We merge cognitive and educational perspectives in our conceptualisation of modelling competencies. A cognitive perspective forms the main focus of the empirical component of the study. This study focuses on group routes and competencies by using visible external representations (Borromeo Ferri 2006, p. 92) of group modelling sessions while we also focus on the meaningful integration of modelling in mathematics education as a significant means of learning mathematics.

3 What Is Competence and What Are Modelling Competencies?

Establishing just what modelling competencies are at micro-level forms part of the essence of this study. The latter question stated in Galbraith (2007, p. 84) that modelling ability is better estimated as a pattern of performance across several problems than on a single common problem is accepted. This view is also held by Goldin (1987, p. 138) in his work on problem-solving competence based on cognitive representation. He describes competence as the capability to perform successfully over a class of tasks. This definition is directly relevant to this study, where competence is taken to develop over a series of tasks. Goldin further adds that competence does not necessarily imply that the student can perform successfully every time a task is encountered. Competence is better seen as the conceptual net that encompasses the problem situation and not as isolated areas of ability. Henning and Keune (2007, p. 225) use Weinert's definition of competence as being the sum of available abilities and skills – the willingness of a student to solve problems and act responsibly concerning the solution. Maaß (2006, p. 139) laid out separate areas of modelling competencies which we used as a broad framework of modelling competence. These broad areas provided structure to our characterisation of modelling competencies. It was decided to delineate three distinct areas of competence: cognitive competencies, affective competencies and meta-cognitive competencies. Cognitive competencies encompass the entire modelling cycle. They pertain to the conscious activities students are involved in while modelling. Affective competencies relate to student beliefs about mathematics, the nature of problems and the value of mathematics in solving real problems while meta-cognitive competencies relate to those factors that support cognition.

The seven phases of the modelling cycle (see Fig. 37.1) led to the clarification of what comprised cognitive competencies (understanding, simplifying, mathematising, working mathematically, interpreting, validating and presenting). Arguing was identified by Maaß (2006) as another area of competence necessary for modelling and was therefore included as a cognitive competency. The pilot study led to the distinction of important meta-cognitive competencies for modelling being: a 'sense of direction' (Treilibs et al. 1980); using informal knowledge (Mousoulides et al. 2007) and planning and monitoring as an indication of general task organisational abilities. We selected 'beliefs' as an important affective competency based on a central question in Maaß's (2006) study. A brief generic elaboration on each of the identified competencies follows; for each modelling problem, these competencies have to be specified and elaborated.

Understanding: Means to know the nature of something. It will involve assuming information that is implied. Understanding can only be determined in conjunction with context and experience.

Simplifying: Means seeing the essential features of the problem. This would also mean using a significant sample of the data and the reasons for this selection.

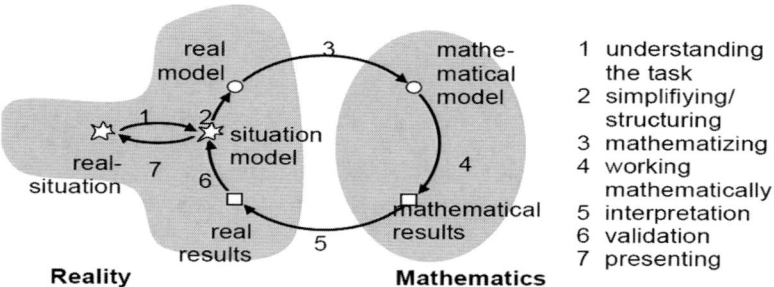

Fig. 37.1 The modelling cycle (Blum and Leiss in Borromeo Ferri 2006, p. 87)

Mathematising: Is translating from the real world to the mathematical world. Detecting features in the real world that corresponds to mathematical concepts.

Working mathematically: Involves the ease with which the chosen mathematics is applied and used. One must also pay attention to the 'type' of mathematics selected to solve the problem.

Interpreting: Borromeo Ferri (2006) sees interpreting as mathematical results that must be reinterpreted in the real situation. The mathematical results that the groups have worked with now have to be re-evaluated in terms of the real problem.

Validating: Validating refers to students securing that their model is consistent and that it satisfies the conditions of the real situation context.

Presenting: Involves communication – where students must clearly describe their thinking (Mousoulides et al. 2007). Students will need to refer to the 'trail of documentation' (Lesh and Doerr 2003, p. 31) they created and include their fellow students and the teacher into the dynamic domain of their group interaction.

Arguing: Arguing as a process and not a finished product is the focus of this competence. It is reasoning that is communicated to convince or explain.

Sense of direction: This was described by Treilibs et al. (1980) as the ability of a group to anticipate the underlying structure of their solution. This would include clarity by the group of what to do next and how this is related to what they wanted to achieve at the end of the task.

Using informal knowledge: Mousoulides et al. (2007) found a significant factor in students modelling to be the use of their informal knowledge. This is student knowledge used that is not specifically from a mathematical realm.

Planning and monitoring: Planning and monitoring refers to groups organising and overseeing their solution route. These competencies were viewed as one in terms of how the group managed the problem.

Beliefs: Blomhøj and Jensen (2007) describe a key to modelling as learning to cope with feelings of 'perplexity due to too many roads to take and no compass given'. This relates to student affective competencies: their beliefs about mathematics, the nature of problems, how problems are solved and the value of mathematics in solving real problems.

4 Methodology

Twelve students were selected for the study based on their previous year's mathematics results and met with the researcher weekly for sessions of 1 h over 4 months. The groups consisted of two groups of students who achieved lower results in traditional instruction and one group of students who achieved higher results in traditional instruction. The weak students comprise what Kuhn and Udell termed a 'difficult to work with population' (Kuhn and Udell 2003, p. 1246). We accept their reasoning in that the competencies that did develop were not the ones emphasised in their schoolwork or were developing anyway as may be the case with academically stronger students. Each group comprised two girls and two boys.

Data were coded using the above competency identification. A table was drawn up that incorporated the competencies and levels. A score for each group was determined after the transcriptions of the audio recording were made. Competencies were identified from student text and coded in the written transcriptions. A competency index was allocated per weekly session according to how much progress the group had made during that session. The indexing was done before the next week's session to ensure a valid assessment. Although a competency was observed a number of times throughout any session (e.g. understanding), the overall effect of the group's understanding was indexed and not individual utterances. A zero (0) was indexed if the competency was present (or observed) but in a very fragile or barren state. So a group's attempts to understand but not making real progress in understanding was allocated a zero. A one (1) was indexed if the competency was observed and was contributing to group progress in some way. So if a group did understand the problem and were able to move to a new or different level of understanding they were allocated a one. A two (2) was indexed if the competency observed was contributing to the group's progress in a significant way. So a group being able to simplify or mathematise or bring another competency into play from their understanding meant a two was allocated. A three (3) was indexed if the competency observed was enabling the group to make considerable progress in developing a generalisable model. So students had to bring other situations into their discussions for a three to be allocated. Once a score was determined, the resulting graphs were drawn up. The score should be seen not as an absolute measurement but rather as an index of their competency. The word index is used as a manifestation or indicator and not as a fixed measurement. This enabled a visual clue to the development of group modelling competencies (Biccard and Wessels 2009). In this chapter, we report on only a few of the competencies.

5   Results

Students initially displayed weak competencies in all areas of modelling. However, these developed slowly and gradually. Competencies do not display themselves in a linear way, nor do they develop in a linear way. Competencies are interrelated and interdependent. Just as groups displayed individual modelling routes (Borromeo Ferri 2006, p. 91), so competency development for each group follows an individual path. It can however be said that the groups' competencies do develop over a period of time. It would also seem as if cognitive competencies develop sooner than meta-cognitive competencies while a marked shift in student beliefs was evident. Table 37.1 illustrates the development of student beliefs over the period that they were exposed to modelling tasks.

For some competencies, an improvement is noted after the first task and plateaux thereafter. It would be of value to determine group competencies over a longer period of time to ascertain when the next improvement occurs. The following table (Table 37.2) shows the indexed competencies for the first four competencies in the

Table 37.1 Changing beliefs

How is mathematics used in our daily lives? 2009-02-03	How is mathematics used in our daily lives? What did you learn about mathematics during the programme? 2009-06-02
• Measuring, counting and costing • Distances travelled in a day and paying bills • It is most important for multiplying, adding, subtracting and dividing • Accounting, buying, selling and many more • I use it when baking, cooking, buying things • We use it when we go shopping	• Counting money, working out problems. I learned that mathematics is not only individual numbers but also it is a sociable subject • Money, bills, taxes, profits and income • I learnt that we could find the real answers in real life by adding, subtracting, scaling down/up, etc. • I learnt that there are many different ways to use mathematics • I learnt that maths can even be used when sewing a quilt together

normative modelling cycle. Students revisited and revised these aspects continually during the solution process.

Task 1 ran from week 1 to week 3 while week 4 was the presentation session. Task 2 ran from week 5 to week 7 while week 8 was the presentation session. Task 3 ran from week 9 to week 11 while week 12 was the presentation session.

Competencies in *mathematising* were largely dependent on task *understanding* and *simplifying* the problem and simplification was interdependent with *mathematising*. Students simplified problems based on the mathematics they wanted to use on the problem. Group improvement in *working mathematically* involved flexibility with decimal numbers, more judicious use of estimating, more negotiation of meaning and measuring more accurately and with better understanding. The mathematical 'toolbox' (Jensen 2007, p. 144) students bring with them and the tools which they use from this toolbox become especially visible during modelling tasks. Students solved all tasks using relatively (surprisingly) simple mathematics even though they have been equipped with more sophisticated mathematics in their classroom experiences. Group competencies in *interpreting* and *validating* were weak throughout the tasks and may be explained by the traditional instruction these students experienced. Their abilities to interpret and validate require teacher support and good task instruction.

Task instruction impacts on modelling competencies. Instructions that require a 'product' such as a letter or report allow students more readily to create a generalisable model and this process relies substantially on interpreting and validating mathematical results. These products also allow the instructor to focus students when they need assistance. Task context must also be considered when discussing competencies and competency development. Some contexts provide supporting platforms for competencies that others do not. Since these tasks required a vast amount of reading, it became evident that student abilities in reading and reading comprehension played a fundamental role in their task understanding and task comprehension Students in the mathematically 'weak' groups also displayed

37 Documenting the Development of Modelling Competencies of Grade

Table 37.2 Indexed competencies

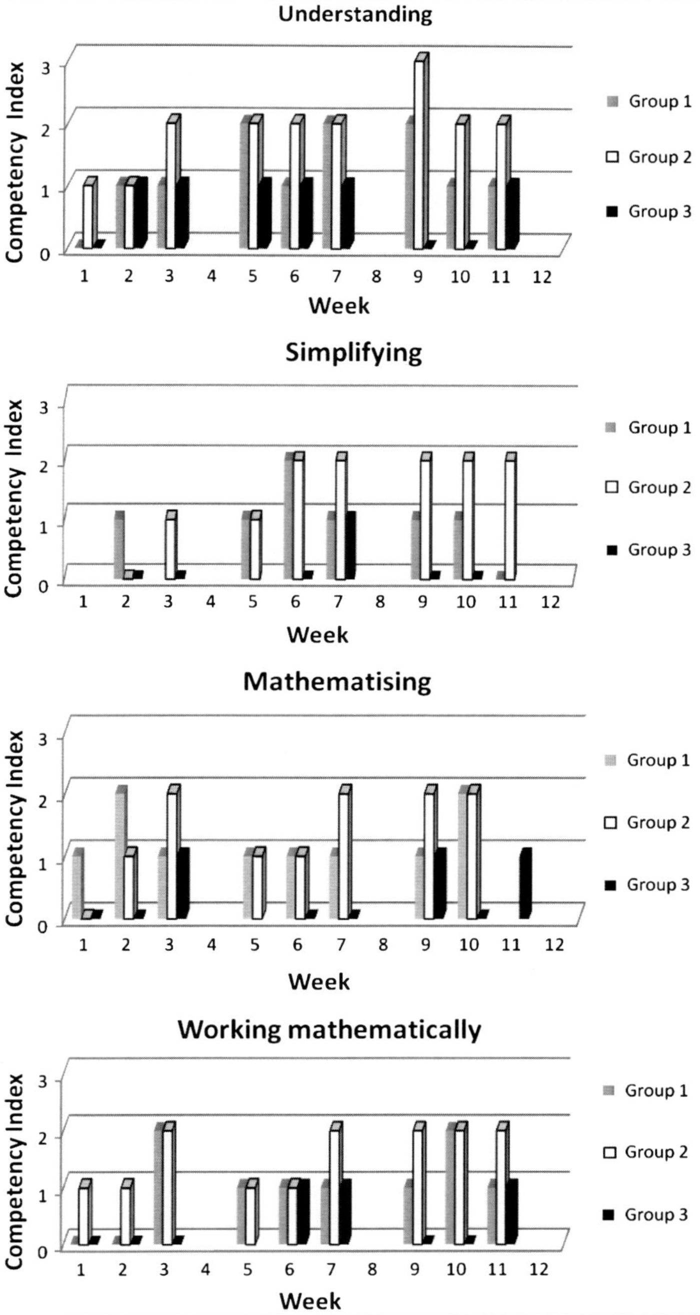

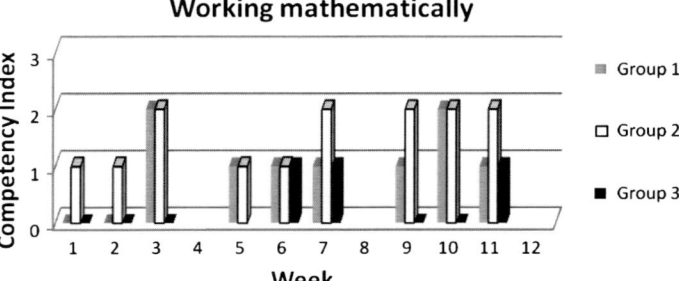

barriers to reading and comprehension. As a starting point in modelling, this competency (*understanding*) was compromised and required extensive assistance from the researcher. The role of language seems to be a dominant factor in competency development.

It would therefore be beneficial to expose students to modelling tasks in a range of mathematical and real life contexts to assist with a wider range of competency development. Students should also be exposed to a broader range of peers in their groups. This will allow for a wider scaffold for interaction, communication and reflection between the group members. This interaction and reflection will support the development of many cognitive and meta-cognitive processes and competencies as well as creating students that are more capable of dealing with future study and career environments.

Acknowledgement The financial assistance of the National Research Foundation (NRF) towards this research is hereby acknowledged. Opinions expressed and conclusions arrived at are those of the authors and are not necessarily to be attributed to the NRF.

References

Biccard, P., & Wessels, D. C. J., (2009). Documenting the development of modelling competencies of Grade 7 mathematics students. Paper presented at the *14th International Conference on the Teaching of Mathematical Modelling and Applications*. 27–31 July 2009. Hamburg: University of Hamburg.

Blomhøj, M. (nd). *Different perspectives on mathematical modelling in mathematics education research – Categorising the TSG 21 papers*. Retrieved from: www.icme11.org/document/get/812. Accessed: 2009-05-12.

Blomhøj, M., & Jensen, T. H. (2007). What's all the fuss about competencies? In W. Blum, P. L. Galbraith, H. W. Henn, & M. Niss (Eds.), *Modelling and applications in mathematics education. 14th ICMI study*. (pp. 45–57). New York: Springer.

Borromeo Ferri, R. (2006). Theoretical and empirical differentiations of phases in the modelling process. *Zentralblatt fur Didaktik der Mathematik (ZDM): The International Journal on Mathematics Education, 38*(2), 86–95.

English, L. D., & Lesh, R. (2003). Ends-in-view-problems. In R. Lesh & H. M. Doerr (Eds.), *Beyond constructivism: Models and modeling perspectives on mathematics problem solving, learning and teaching* (pp. 297–316). Mahwah: Lawrence Erlbaum Associates Publishers.

Galbraith, P. (2007). Beyond the low hanging fruit. In W. Blum, P. L. Galbraith, H. W. Henn, & M. Niss (Eds.), *Modelling and applications in mathematics education. 14th ICMI study* (pp. 79–88). New York: Springer.

Goldin, G. A. (1987). Cognitive representational systems for mathematical problem solving. In C. Janvier (Ed.), *Problems of representation in the teaching and learning of mathematics* (pp. 125–145). Hillside: Lawrence Erlbaum Associates Publishers.

Henning, H., & Keune, M. (2007). Levels of modelling competences. In W. Blum, P. L. Galbraith, H. W. Henn, & M. Niss (Eds.), *Modelling and applications in mathematics education. 14th ICMI study* (pp. 225–232). New York: Springer.

Jensen, T. H. (2007). Assessing mathematical modelling competency. In C. Haines, P. Galbraith, W. Blum, & S. Khan (Eds.), *Mathematical modelling (ICTMA 12) education, engineering and economics* (pp. 141–148). Chichester: Horwood.

Kaiser, G., & Sriraman, B. (2006). A global survey on international perspectives on modelling in mathematics education. *Zentralblatt fur Didaktik der Mathematik (ZDM): International Journal on Mathematics Education, 38*(3), 302–310.

Kuhn, D., & Udell, W. (2003). The development of argument skills. *Child Development, 74*(5), 1245–1260.

Lesh, R., & Doerr, H. M. (2003). Foundations of a models and modelling perspective on mathematics teaching, learning, and problem solving. In R. Lesh, & H. M. Doerr (Eds.), *Beyond constructivism: Models and modeling perspectives on mathematics problem solving, learning and teaching.* (pp. 3–33). Mahwah: Lawrence Erlbaum Associates Publishers.

Lesh, R., Hoover, M., Hole, B., Kelly, A., & Post, T. (2000). Principles for developing thought-revealing activities for students and teachers. In A. Kelly & R. Lesh (Eds.), *Research design in mathematics and science education* (pp. 591–646). Mahwah: Lawrence Erlbaum Associates.

Maaß, K. (2006). What are modelling competencies? *Zentralblatt fur Didaktik der Mathematik (ZDM): The International Journal on Mathematics Education, 38*(2), 113–142.

Mousoulides, N., Sriraman, B., & Christou, C. (2007). From problem solving to modelling: The emergence of models and modelling perspectives. *Nordic Studies in Mathematics Education, 12*(1), 23–47.

Treilibs, V., Burkhardt, H., & Low, B. (1980). *Formulation processes in mathematical modelling.* Nottingham: University of Nottingham Shell Centre for Mathematical Education.

Chapter 38
Students' Reflections in Mathematical Modelling Projects

Morten Blomhøj and Tinne Hoff Kjeldsen

Abstract Students' reflections play an important role in mathematical modelling competency. In this chapter, we argue that there are two kinds of reflections which have to be challenged and supported in different ways. They can be characterised as respectively internal and external with respect to the modelling process. Internal reflections add meaning and quality to the sub-processes involved in a mathematical modelling process, while the external reflections address the role and function of the model in actual or potential applications. If mathematical modelling competency is an educational goal, the teaching needs to provide students with experiences with modelling and applications of models in a variety of authentic contexts in ways that support the students' development of both internal and external reflections. Through analyses of two student projects, we illustrate the two kinds of reflections and discuss how they can be developed in students.

1 Introduction

This chapter is a theoretical one introducing internal and external reflections as two different kinds of reflections related to mathematical modelling. Analyses of two modelling projects from the science bachelor programme at Roskilde University serve to illustrate the two kinds of reflection and as a basis for discussing how to promote such reflection in students. We perceive our work as one important aspect of the more fundamental didactical challenge of developing students' modelling competency in tertiary science education. However, the distinction between internal and external reflections is of relevance for the teaching of mathematical modelling also at other educational levels. Our analyses, argumentations and conclusions are based on the assumption that students develop mathematical modelling competency

---

M. Blomhøj (✉) and T.H. Kjeldsen
IMFUFA, NSM, Roskilde University, Roskilde, Denmark
e-mail: blomhoej@ruc.dk; thk@ruc.dk

through their engagement in modelling activities. Furthermore, if students are to experience that the function and status of modelling and models are context dependent, then they need to be exposed to didactical situations that challenge them to reflect upon and critique the modelling process and the function of models in different contexts.

We begin with some conceptual clarifications of how we view the modelling process and how we define internal and external reflections. Then we present and analyse the project reports from two selected projects. The two projects are different in terms of the subject matter content, of how the students worked with the modelling process and the application of the models. In the first project, an authentic model for city traffic planning and its application in a concrete political decision process is analysed, while in the second project, the role of modelling the hypothalamic-pituitary-adrenal (HPA) axis in physiology to understand the human hormone control system for cortisol, is investigated. Both projects are analysed with the aim of pinpointing the students' internal and external reflections. The chapter ends with a discussion of how to balance these two kinds of reflections with other important elements in modelling competency.

2 Internal and External Reflections in Mathematical Modelling Competency

Our starting point for defining mathematical modelling competency is the general framework of mathematical competency developed in the Danish "KOMpetence project" (Niss and Jensen 2005). Within this framework, modelling competency can be defined as:

> A person's insightful readiness to autonomously carry through all aspects of a mathematical modelling process in a certain context and to reflect on the modelling process and the use of the model (Blomhøj and Jensen 2003, p. 127).

In this definition, mathematical modelling competency refers to *a mathematical modelling process* and therefore, in order to unfold the elements involved in modelling competency, it is relevant to analyse the general structure of a modelling process. The definition also includes "to reflect" on modelling processes and on actual or possible applications of models. In this chapter, we further define two different kinds of reflections: *internal reflections* related to the modelling process and *external reflections* related to the applications or functions of a model. By *reflection* we understand a deliberate act of thinking about some actual or potential action aiming at understanding or improving the action. Reflections take place in the minds of individuals but are strongly influenced by social interactions, and they can only be detected and analysed through communicative acts. The labels *internal* and *external* mirror the reflections' relation to the modelling process and should not be misinterpreted as internal and external with respect to the reflecting subject.

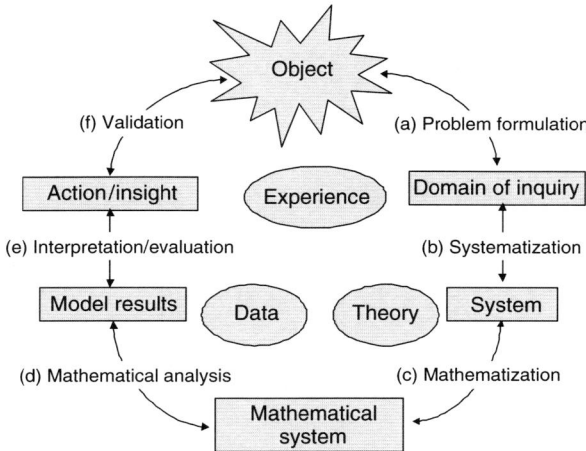

Fig. 38.1 A model of a mathematical modelling process

Our understanding of reflection is close to definitions that can be found in the fast growing literature on the role of reflection in the teaching and learning of mathematics in general (Chamoso and Cáceres 2009, p. 199). However, in educational research on mathematical modelling, the concept of reflection has not jet attracted much attention. We find ideas similar to our work in Greer and Verschaffel (2007) and especially in Henning and Keune (2007) who introduce the notion of 'meta-reflection on modelling' to capture both critical reflections on the modelling process and reflections on the purposes and applications of modelling. However, we have not found research that differentiates between students' reflections related to the modelling process and reflections related to the context of application.

Although, a mathematical modelling process can be conceptualised in different ways for different purposes (Niss et al. 2007, p. 18), it is quite uncontroversial to consider a modelling process as composed of six sub-processes, see Fig. 38.1. In order to create a mathematical model, it is, in principle, necessary to carry through a modelling process. Analytically, we describe a mathematical modelling process as consisting of the six sub-processes (a)–(f) as depicted in Fig. 38.1 (Blomhøj and Jensen 2003).

The modelling process should neither be understood as a linear process nor as an entirely rational process. A modelling process does not always start with the aim of describing or understanding a well-defined object, and very often modelling processes take the form of a cycling process where reflections on the modelling process and the intended applications of the model lead to redefinitions of the object to be modelled. In fact, each of the six sub-processes can lead to changes in the other sub-processes. To indicate these dynamical aspects, the modelling process is depicted in a circular diagram. The diagram may also have a connotation in favour of a sequential understanding of the modelling process, and it may not give full justice to the complexity of the possible connections between the sub-processes nor to the foundations of these sub-processes. However, the important roles of theoretical, empirical and common sense foundations of the modelling processes are indicated by the three ellipses in the middle of the diagram (see Fig. 38.1).

'Theory' here means knowledge about the domain of inquiry used in the modelling process. This knowledge base may have very different epistemological status even within the same modelling process, varying from well-founded theories with built-in mathematisations (often the case in physics) to shared/personal experiences and purely adhoc assumptions. The character of the knowledge base has vital importance for how the model and its possible applications can be validated. Sometimes 'data' exists prior to the modelling process and may then be used to support the processes of systematisation and mathematisation and eventually also as a basis for validating the model. More frequently, however, relevant data have to be collected as part of the modelling process. Here the production of data often presupposes a model. Such data can be used to estimate the model parameters but not as a basis for validating the model. Developing the sensitivity among students for such reflections is an important part of the long-term aim for the teaching of mathematical modelling. Even though this type of high-level reflections concerning the overall epistemological status of a modelling process is an important type of internal reflections, in this chapter, we limit ourselves to analysing reflections connected to particular sub-processes.

In relation to the modelling process, we define *internal reflections* as reflections connected to the sub-processes and their foundations in the modelling process. Students may reflect on the sub-processes in a mathematical modelling process in cases where they are the modellers as well as in cases where they analyse modelling processes behind existing models. The important characteristic is that internal reflections can take place synchronically with the modelling process, and that the object of internal reflections is actions taken within the modelling process. The reflections can be directed towards the foundations of these actions, their consequences for the following steps in the modelling process, for the results of the model, for the possible applications of the model, the model validity or towards bringing possible alternative actions into light. In all cases, the reflections highlight that actions in a modelling process are taken by the modeller(s) for some reasons and with some intentions and consequences. Hence, they are a resource for improving the modelling process.

In opposition to internal reflections, *external reflections* are concerned with actual or possible use of a mathematical model, model results and their role and function in given societal, technical or scientific contexts. This kind of reflection has as its object the role and functions of a model in a particular (type of) application, and therefore necessarily involves the context in which the model is being used or could be used. By nature, external reflections take place diachronically to the modelling process. As a consequence, such reflections cannot be integrated in, or structured by means of, the modelling process.

It is important to stress that this distinction between internal and external reflections is an analytical distinction. In practice, they are intertwined in a dialectic

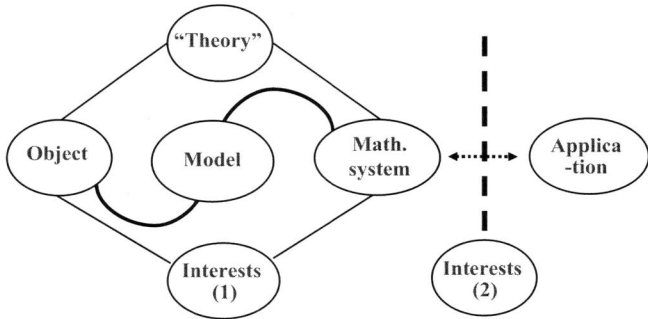

Fig. 38.2 The structural relationship between object and mathematical system in the application process

relationship. In general, when a mathematical model is being applied whether in a societal, technological or scientific context, there are always some interests related to power, political, economical or ethical issues or underlying theories of science involved. Such interests and theories are often (consciously or unconsciously) built into the model through the modelling process, thereby influencing the actual application of the model and its consequences either intended or unintended. The modelling process creates a complex structural relationship between two entities of different epistemological nature, namely, the object that is to be modelled and the mathematical system. Theory, broadly understood as what is known or taken for granted also in the form of available data, is influencing the modelling process in an intricate interplay with the interests behind the model and its intended applications.

The structural relationship between the object and the mathematical system in a process of application is depicted in Fig. 38.2. In the application process, the model is often disconnected from the modelling process. Except for modelling within the framework of a mathematised theory, it is not possible to secure a model's validity theoretically. As a consequence, the validity of a concrete application of a mathematical model has to be judged on the basis of the modelling process behind the model and/or empirical testing.

Application of a mathematical model to a real life problem should be considered as a process of its own, which is separated from and more or less independent of the modelling process. The application process tends to cause changes in the context of the real life problem as well as in the model when it is adjusted to the problem. When a mathematical model is developed, it is influenced by certain interests of the acting subjects. The modelling process that lies behind the model can be theoretically and/or empirically more or less well founded. However, when the model is applied in a political, technological and/or scientific investigation, the model is often separated from the modelling process and the possible critical reflections connected with the process. The model may be used for particular purposes that might be different from what the model was constructed for. Moreover, the application process can be influenced by yet other interests that were not present in the modelling process. Therefore, the process of applying a mathematical model in general tends to cause (a) a reformulation of the problem at hand in order to be adequate for investigation by means of a model, (b) changes in the discourse about the problem towards pro and contra the model and possible adjustments of the model, (c) a limitation of the possible actions taken into consideration to those that can be evaluated in the model, and (d) a delimitation of the group of people that can take part in the discussion and act as a basis of critique (Skovsmose 1990, pp. 129–130). Awareness of, and experiences with, such general phenomena can provide a strong basis for students' external reflections.

Some mathematics teachers may find such reflections to be of no relevance for the teaching of mathematical modelling. However, as will be illustrated in the following examples, how model results are used in a concrete political decision or in a technical or scientific investigation can often only be understood and critiqued on the basis of a mathematical analysis of the modelling process. Therefore, developing mathematical modelling competency includes the obligation to foster among students an ability to reflect critically upon actual and possible applications of mathematical models.

3 Students' Reflections in Modelling Projects

Development of students' mathematical modelling competency is part of the general science education of the bachelor science programme at Roskilde University. This is addressed directly in the students' projects, which is a rather unique way of organising university studies. Therefore, before analysing the two projects, we shall briefly describe the institutional context in which we are working.

3.1 The Institutional Context of the Project Work at Roskilde University

Science students enter the university through a 2-year introductory study programme, which leads to a bachelor degree after two additional semesters of subject specialised studies, and to a master's degree after another six semesters of study. In all programmes, half of the study time in each semester is devoted to group-organised and problem-oriented projects. The other half of the study time is used on subject organised courses.

The students' project work is guided by a problem that needs to be solved or investigated, and not by a curriculum. The students choose the problems they want to work with under some thematic constraints and organise their work with support from a supervisor. In the first three semesters, the problems should be exemplary with respect to (1) the use of science in society, (2) the function of and relationships between models, experiments and theories in the production of scientific knowledge and (3) science as a cultural phenomenon. In the fourth semester, the problem merely has to be exemplary of some aspects of (1)–(3). The project work is organised in houses of 60–80 students and conducted in groups of 4–7 students resulting in 10–12 project groups each semester. A team of 10–12 supervisors representing the various science disciplines (physics, biology, mathematics, environmental science, molecular biology, geography, etc.) are allocated to the house, and each group will be supervised by the same professor during the entire project. See Blomhøj and Kjeldsen (2009) for a more detailed description of the project organised science studies at Roskilde University.

The first project analysed below was made by a group of two fourth semester students and it complies with the first semester requirement, while the second is made by a group of six students in the second semester. Based on the project reports and the groups' final oral presentations, the projects are analysed with the aim of illustrating internal and external reflections and to uncover the students' actual reflections connected to each project.

3.2 The Use of a Traffic Model in the City of Roskilde – The Case of 'Ny Østergade'

The problem that guided the students' work in this project took its point of departure in a contemporary political conflict between the local administration in the city of Roskilde and a group of citizens living very close to a planned new road connection (Jarbøl and Kofoed 1996). The decision to construct this new road was influenced by the use of a mathematical model for the inner city traffic. The model was developed and applied by a consulting firm.

In the modelling process behind this model, the city is divided into 57 zones. Based on interviews and traffic counting, a 57×57 matrix was set up to describe the total traffic flow within and in/out of the city per day. The matrix contains the estimated number of trips between each pair of zones where some of the zones represent 'ports' leading in and out of the city. In the model, these trips are distributed on a road net representing the main roads in the city according to three principles: (a) For all roads, the traffic allocated should be less than the capacity of the road in question. (b) All tours should be placed on the net so that the calculated time of transportation for each tour is minimised. (c) For each road element of every tour, the time of transportation is calculated in minutes according to the formula: $T_i = 60 \cdot \frac{L_i}{v_i} + c_i L_i$, where L_i is the length of the ith element of the considered road measured in kilometres, v_i is the average speed on the ith element in kilometres per hour and c_i signifies that some extra time is used depending on the type of road. The value of c_i in the particular case of the planned new road connection was 0.3 min/km.

In their report, the students analyse the model's behaviour, the empirical basis for the tour matrix and the estimation of the model parameters. Their analysis was structured by a model of a modelling process similar to Fig. 38.1, and in relation to each sub-process the possible reasons for, and effects of, the different assumptions made in the modelling process were discussed. This analysis involves many internal reflections according to our definition, because the students address possible rationales for, assumptions behind, and critique of particular steps in the modelling process. In their work, the students actually reconstructed the entire modelling process based on the official documentation of the model, which in fact was ambiguous and incomplete in many respects. The main focus in the students' project was to analyse the role played by the model and its results in an authentic political decision process and the related public debate. Here the students were challenged to make external reflections. They discovered that the model was used on two different occasions to predict the traffic on the new road to be, in the one case, clearly above 10,000 cars per day and, in the other case, clearly below 10,000 cars per day. The high estimate was used to argue that the new road would reduce the traffic in the centre of the city significantly, and this was an important argument for securing governmental financial support for the construction work. Whereas, the low estimate was used to prevent that the road project should undergo an EU-procedure for evaluation of the effects on the environment. This EU-procedure, which the group of citizens had pleaded for, only applies for roads with a predicted traffic flow above 10,000 cars per day. Based on their analysis of the modelling process, the students realised that the two (contradicting) results were obtained simply by using two different values for the parameter v_i for the new road connection – a change for which there was no valid foundation, since the parameter v_i for a non-existing road cannot be made subject to empirical control. The foundation for the two different model results was not part of the public debate. Based on their own interviews with one of the modellers from the consulting bureau, local politicians and the spokesperson for the group of citizens, the students analysed the discourse concerning the new road connection. They concluded that the model results played a central role in the decision about the new road, and that the results from the model were used beyond its scope of validity in order to serve two different political purposes. The validity of this particular application was not discussed in the public debate. These are clear examples of the importance of external reflections in relation to applications of mathematical models.

3.3 *Modelling in Scientific Investigations: The Project of the HPA-Axis*

In this project, the students decided to investigate a mathematical model presented by Jelic et al. (2005) of the dynamics of the HPA-axis, which is short for the hypothalamic-pituitary-adrenal axis. The HPA-axis is interesting because it is essential in homeostasis in stress-related situations. It controls the secretion of the stress hormone cortisol. Medical companies are interested in gaining a better understanding of the function and dynamics of the HPA-axis, because it is connected to illnesses such as depression. A recent strategy is to use mathematical modelling to achieve this understanding in order to pinpoint biomarkers.

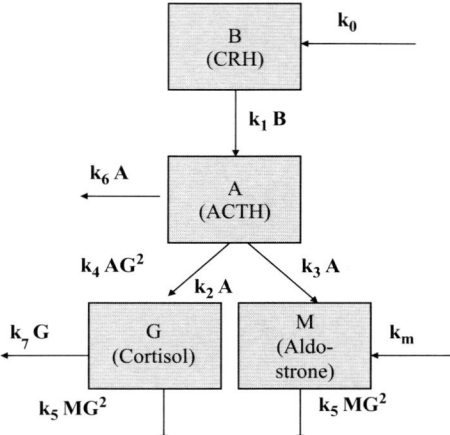

Fig. 38.3 A compartment model of the HPA-axis

The dynamics of the HPA-axis is not fully understood, but the following description is often given: The corticotropin-releasing hormone, CRH, is released from the hypothalamus in response to stress, and in response to CRH, the adrenocorticotropic hormone, ACTH, is secreted from the pituitary. ACTH triggers the secretion of cortisol from the adrenal cortex from where also the hormone aldosterone is secreted. The students represented the structure of the theoretical model of Jelic et al. (2005) by a compartment diagram as shown in Fig. 38.3.

Based on this representation, the students discussed in detail the biological processes of the HPA-axis and how these processes correspond to the individual parts of the model. They realised that the model is based on the assumption that the hormone system can be described by nine biochemical reactions. In their analysis of the modelling process, the students critiqued and evaluated each hypothesis, assumption, and implementation in the model. They became aware that the stimulation of hormone release is implemented in the model as if the hormones are transformed into one another, which is not in accordance with what really happens.

The students also discovered that the feedback mechanism of cortisol is not linked to CRH. In the evaluation of the model, they realised that in the model, the concentrations of ACTH and cortisol are of the same order of magnitude, which contradicts experimental results where they differ by a factor of 10,000–50,000. The model could not be validated by experimental results. These are examples of internal reflections.

Due to the students' external reflections with respect to the function and the status of the model in the application of the model in a scientific investigation, the purpose of which was to gain understanding and knowledge of the dynamics of the HPA-axis, they became aware that the object changed from understanding the dynamics of the HPA-axis to the problem of modelling the dynamics of the concentration of cortisol and ACTH. Through their analysis of the modelling process underneath the model, they experienced a change in discourse from discussions about the dynamics of the HPA-axis to a (model) discussion about whether the model is good or bad, and that this change in discourse was guided by the internal reflections and critique of the modelling process. The students tried to find alternative ideas about the HPA-axis, and they realised that such ideas were reduced to those that can be implemented in the model. The students then revised the model with respect to their internal reflections and critique. As a result of their internal and external reflections, the students experienced a delimitation of the group of people engaged in the discussion which was limited to researchers with sufficient mathematical knowledge to understand the model.

4 Concluding Remarks

In the first project, the students experienced the function of a model in a societal decision-making process while those in the other project learned about the role mathematical modelling can play in scientific investigations and in the production of knowledge. In both cases, internal and external reflections and the interplay between them played an important role. The ability to reflect upon and criticise the modelling process and the function of the model in different contexts as well as awareness of the structural relationship in an application process are essential aspects of modelling competency. The analytical distinction between internal and external reflections is a means to better understand the challenges involved in developing mathematical modelling competency. The analyses of the two student projects show that within a project organised and problem-oriented science programme, it is possible to engage students in both internal and external reflections related to mathematical modelling. Promoting students' internal and external reflections in relation to mathematical modelling is definitely also a relevant didactical challenge in secondary mathematics teaching and this is the area in which we will be focusing our future developmental research.

References

Blomhøj, M., & Jensen, T. H. (2003). Developing mathematical modelling competence: Conceptual clarification and educational planning. *Teaching Mathematics and Its Applications, 22*(3), 123–139.
Blomhøj, M., & Kjeldsen, T. H. (2009). Project organised science studies at university level: Exemplarity and interdisciplinarity. *ZDM – The International Journal on Mathematics Education, 41*(1–2), 183–198.
Chamoso, J. M., & Cáceres, M. J. (2009). Analysis of the reflections of student-teachers of mathematics when working with learning portfolios in Spanish university classrooms. *Teaching and Teacher Education, 25*(1), 198–206.

Greer, B., & Verschaffel, L. (2007). Modelling competencies – Overview. In W. Blum, P. Galbraith, H.-W. Henn, & M. Niss (Eds.), *Modelling and applications in mathematics education: The 14th ICMI-study* (pp. 219–224). New York: Springer-Verlag.

Henning, H., & Keune, M. (2007). Levels of modelling competencies. In W. Blum, P. Galbraith, H.-W. Henn, & M. Niss (Eds.), *Modelling and applications in mathematics education: The 14th ICMI-study* (pp. 225–232). New York: Springer-Verlag.

Jarbøl, M., Kofoed, M. (1996). En trafikmodel i Roskilde kommune – en analyse af modellen bag beslutningsprocessen omkring omfartsvejen Ny Østergade. Group 10, 4. semester, house 13.1, Nat-Bas, Roskilde University.

Jelic, S., Cupic, Z., & Kolar-Anic, L. (2005). Mathematical modeling of the hypothalamic-pituitary-adrenal system activity. *Mathematical Biosciences, 197*, 173–187.

Niss, M., & Jensen, T. H. (Eds.) (2005). Competencies and mathematical learning – Ideas and inspiration for the development of mathematics teaching and learning in Denmark. English translation of part I-VI of Niss & Jensen (2002). Under preparation for publication in the series Tekster fra IMFUFA, Roskilde University, Denmark. To be ordered from imfufa@ruc.dk.

Niss, M., Blum, W., & Galbraith, P. (2007). Introduction. In W. Blum, P. Galbraith, H.-W. Henn, & M. Niss (Eds.), *Modelling and applications in mathematics education: The 14th ICMI-study* (pp. 3–32). New York: Springer-Verlag.

Skovsmose, O. (1990). *Ud over matematikken*. København: Systime.

Chapter 39
From Data to Functions: Connecting Modelling Competencies and Statistical Literacy

Joachim Engel and Sebastian Kuntze

Abstract Over the last two decades, research on learning and teaching mathematical applications greatly advanced our understanding of the processes involved in mathematical modelling. However, the vast majority of examples and concepts developed so far barely include a key source of information: data. Numerical information generated from measurements of the quantities involved is used neither at the validation nor at the modelling step. We adopt a data-oriented approach. In the context of modelling functional relationships, we look at the relationship between modelling competencies and statistical literacy and provide empirical evidence that proficiency in these areas can be jointly improved.

1 Introduction

While the dialectic relationship between context and mathematical model is at the core of any mathematical application, there are also various approaches on how to relate mathematics to extramathematical problems. Some areas of mathematical application are dominated by a structure-oriented approach, in which principal considerations and a structural analysis of the context lead to a mathematical model. For example, in the Luxembourg gas station problem,[1] a straight analysis connects the situation with linear functions, whereas the fire brigade problem[2] is associated with the Pythagorean theorem (for an introduction and discussion of both problems, see Werner Blum's chapter in this volume). On a mathematically more advanced level, many mathematical models are expressed by differential equations, derived on the basis of some structural assumptions modelling local change. Data as a

---

[1] Does it pay to drive across the border to Luxemburg when the gas is cheaper there?
[2] What is the maximum height to save people in a burning house with a new fire ladder?

J. Engel (✉) and S. Kuntze
University of Education, Ludwigsburg, Reuteallee 46, 71634 Ludwigsburg, Germany
e-mail: engel@ph-ludwigsburg.de; kuntze@ph-ludwigsburg.de

check of the obtained mathematical result with reality may then enter at the validation step.

A different modelling approach includes data from the very beginning of the modelling cycle. This idea follows closely the genetic principle (e.g., Safuanov 2004) by studying the phenomena of interest first, beginning with posing good questions, collecting observations, taking measurements, and gradually developing mathematical descriptions of the phenomena. In this approach, components of statistical literacy are inseparably intertwined with modelling competencies.

We take a data-oriented approach and look at how modelling competencies are related to competencies in handling data in the context of modelling functional relationships from a theoretical and an empirical perspective. In Sect. 2 we argue for the merits of taking data into account when modelling and compare the structure-oriented approach with the data-oriented approach from an epistemological perspective. In Sect. 3 we report on empirical evidence that links modelling competencies with statistical literacy.

2 Modelling Competencies and Statistical Literacy

Why use data? Genuine data—as opposed to fabricated or fake data—are a substantial source of information and provide evidence of real problems. They counteract anecdotal evidence, traditional beliefs, prejudice, wishful thinking, or ideology. Data are a most reliable representation of authentic information and the raw material of new knowledge. With looking at data we introduce, in some way or other, statistics and probability into the modelling process. At the core of statistics as the science of modelling, summarizing, and analyzing data is an endeavor to make sense of the data. This goal has led to recent developments in statistics education, which aim at supporting learners' competencies in this area. The past two decades have seen the development of a reform movement in statistics education, emphasizing features such as statistical thinking, active learning, conceptual understanding, genuine data, use of technology, collaborative learning, and communication skills (see, e.g., Moore 1997).

A great deal of research has been initiated (see, e.g., the website of the International Association of Statistical Education[3]) and a wide variety of materials have been developed worldwide to support such instruction. These include:

– Textbooks with emphasis on statistical thinking, conceptual understanding, and the use of genuine data.
– Activity books and lab manuals fostering students' active learning.
– Depositories of genuine datasets, selected and prepared under educational perspectives.
– Java applets and new dynamically linked software, allowing interactive visual explorations of statistical concepts.

---

[3] www.stat.auckland.ac.nz/~iase.

– Assessment tools, such as projects, focusing on students' conceptual understanding and ability to think statistically.

These changes have been institutionalized in curricular documents of many countries (e.g., German Bildungsstandards 2004; NCTM Standards of 1989 and 2000), giving statistics as data science the status of a component of mathematical literacy and linking it to other components of mathematical competency such as modelling.

In particular, when modelling functional relationships, a data-oriented approach means connecting various domains of mathematics such as elementary functions, algebra, and analysis with probability and statistics, and with ideas of modelling. In the following, we provide more detail about the central notions of the theoretical background on which the further analysis will be based. The notion of *modelling proficiency* is based on theoretical considerations by Blum and Kaiser (cited in Maaß 2006), who specify modelling competencies by a detailed list of abilities that are related to the understanding of the modelling process. These abilities include in particular:

– To understand the real problem and set up a model based on reality,
– To set up a mathematical model from the real model,
– To solve mathematical questions within the mathematical model,
– To interpret mathematical results in a real situation, and
– To validate the obtained solution and see the limitations of the model.

Maaß (2006) supplements this list by describing various aspects of these abilities in detail.

A core issue in modelling is that the model is not identical to the situational context. Models, by their nature, are not the real thing, but an oversimplification of the complexity and disorder that reality throws at us. To simplify reality, models sacrifice details. Hence, discrepancies between the model and reality are not necessarily an indication of the model being "wrong," as novices usually suspect when inquiring about the "correct" model. Models let us compare reality to our own ideas about how things work, and allow us to see past that confusing variation and learn to recognize the patterns. They smooth over the natural variation that occurs for all kinds of reasons in order to reveal the underlying pattern. Therefore, they allow us to generalize and to apply our model not only to the reality just observed, but also to similarly structured realities. This serves a number of important purposes: A model describes to us the essential things about a process. It helps us gain insight into the dynamics, to see what is going on. With good models we may be able to forecast the outcome of future events. Finally, models may help us intervene in processes and change things the way we want them to be.

When using the notion of "*statistical literacy*" we follow a comprehensive approach. Despite the intense discussion on the nature of statistical thinking and how it differs from statistical reasoning and statistical literacy (see, e.g., Ben-Zvi and Garfield 2004), for the purpose of our chapter, we rely on the definition by Wild and Pfannkuch (1999, p. 227): "Statistical thinking is concerned with learning and decision making under uncertainty. Much of that uncertainty stems from

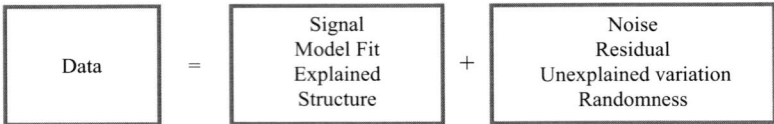

Fig. 39.1 Different versions of the signal–noise representation of data

omnipresent variation. Statistical thinking emphasizes the importance of variation for the purpose of explanation, prediction and control."

Variation is the reason why complex statistical methods were devised in order to filter out signals from noisy data. A core concept in statistical modelling is the signal–noise metaphor. Konold and Pollatsek (2002) characterize data analysis as the search for signals in noisy processes. In their generic form, data are thought of as an additive compound of a structural component plus residuals. This split is our human response to deal with an overwhelming magnitude of relevant and irrelevant information contained in the observed data. Probability hereby plays the role of a heuristic tool to analyze reality. Figure 39.1 shows different versions of expressing the signal–noise idea from various perspectives.

Wild and Pfannkuch (1999) state that the tendency to search for specific causes is very deep-seated and leads people to search for causes even if an individual's data are quite within the bounds of the expected when acknowledging random variation.

In the context of modelling bivariate quantitative data $(x_1, y_1),\ldots,(x_n, y_n)$, the signal–noise idea translates to the formula $y_i = f(x_i) + e_i$, $i = 1,\ldots,n$, where the function f is the structure to be recovered while the e_i's represent the residuals. The y_i's are perceived as a signal f evaluated at x_i, perturbed by a noise e_i. When analyzing bivariate numerical data, the signal–noise metaphor is a very useful concept to bridge the gap between a deterministic view of a function and a statistical perspective that appreciates variation: The signal or structure f captures the explained part of the variation, while the noise comprises the unexplained part of the variation. In the scatter plot of the data it is the unexplained variation that is the reason for several values of y associated with a single x.

We conclude that dealing with statistical data requires modelling competencies, especially when having to describe functional dependencies. Conversely, modelling functional dependency based on bivariate data requires statistical literacy when judging on the discrepancies between model and data. In the following, we will provide empirical evidence supporting these theoretical thoughts.

3 Empirical Evidence on Modelling Competencies and Statistical Thinking

The proximity of statistics and modelling is not surprising. As probability is about modelling random processes, statistics is the science of modelling data. Whereas in approximation theory the focus is on modelling structure, in statistics we model structure *and* residuals, the latter as a random process. The perspective of statistics being a particular type of modelling, that is, data modelling, suggests that instruction emphasizing data-based modelling also improves statistical thinking. Accordingly, in a first empirical study, we investigated the following research question:

- Does statistical thinking improve through a data-based applied mathematics course?

In a second, current study, we investigate the partly complementary question of the role of conceptual knowledge in the areas of probability and functions for the competency of modelling and using representations in statistical contexts (Kuntze et al. 2010). In the following we summarize methods and results of the first study referring for details to Engel et al. (2008), as it pertains to modelling competencies, before giving a short outlook on the second study.

To evaluate the validity of the claim that a data-based course on functional modelling has a positive impact also on more general statistical thinking skills, we conducted a pretest–posttest study with treatment and control groups. Participants were 179 second-year students preparing to be teachers for elementary and secondary schools. During the study they attended one of two different courses in applied mathematics. While the control group of 101 students attended a class with a more traditional syllabus (e.g., elementary functions, linear optimization, in particular no analysis of real data, no residual analysis, and no considerations of variation in data), the class for the treatment group (78 students) followed strictly a data-oriented course of technology-supported modelling functional relationships. Students were instructed about standard functions (e.g., polynomial, exponential, trigonometric, logistic) and learned through projects how to fit them to real data sets, at first by adjusting parameters manually with sliders and then automated by minimizing a least-squares criterion in the case of a linear structure. As this was not an applied stochastics course, the residuals were not modeled as a random process, nor were any concepts used from probability. Throughout the course, however, students were challenged to discuss and interpret deviations of the data from the model and to analyze residual plots, paying increasing attention to the concept "Data = Signal + Noise." Many details of the course material with plenty of examples including student project assignments can be found in Chapters 2 and 4 of Engel (2009).

For measuring statistical thinking skills, a short questionnaire (as an example, see Fig. 39.2.) was given to the 179 participants (78 in treatment group, 101 in control group) at two points in time: the pretest in the first meeting of the class in October 2007 and the posttest during the last course in February 2008 after fourteen 90-minute class meetings. Item 1 was close to the content of the class for the treatment group and requires sketching a free-hand curve onto a scatter plot of data whose context is briefly described. Item 2 is a problem of change point detection based on informal statistical inference. It requires a judgment about a change over time in a system, taking into account some context knowledge and variation in the data. Both items were administered in four different versions and were completely counterbalanced across pretest and posttest, to control for item difficulty. All Item

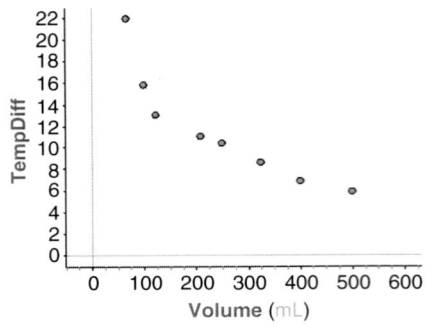

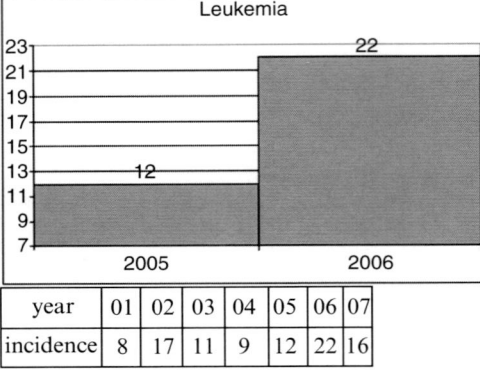

Item 1:
Different amounts of water were heated in the microwave for 30 seconds and the temperatures (before and after) measured and their difference computed. Based on the scatter plot to the left, sketch a free-hand curve describing the relationship between volume and temperature difference.

Item 2:
A member of an anti-nuclear action group presents the following graphics displaying an increase of leukemia within a 50 Km radius of a nuclear power plant. In his plea he argues that the massive increase has to be due to a concealed accident in the power plant in 2005 or 2006.

Inquiring about the incidence of leukemia within the last seven years resulted in the table.

Do you agree with the activist's conclusion? Why or why not?

Fig. 39.2 One version of the questionnaire

2 questions were constructed that at first sight there seemed to be a jump or change point in the data. This impression was aggravated by starting the vertical scale high above the origin. However, when taking into account the variation of measurements over the last several years, which was provided in tabular format, evidence for a change point became very weak.

4 Results

To evaluate the responses we proceeded as follows: Item 1 asked for modelling with functions. Two independently working coders classified each response into one of the following three categories: I for a curve that interpolates, that is, connects all observations in the scatter plot; P for fitting a curve from a chosen parametric class of functions such as a decreasing exponential; and S for data smoothing. While both P and S may be considered as an indication of awareness that a deviation between data and model may be appropriate for the sake of a plausible model, we interpreted interpolation as reflecting a rather deterministic mindset ignoring random variation in real data, that is, attributing every variation in the data to a specific and relevant cause. Results for this coding are presented in Fig. 39.3.

While we observed in the control group only a very modest shift of 8% from the group classified as interpolators (I) to parametric curve fitting (P) and an almost unchanged small percentage of data smoothing, in the treatment group, parametric curve fitting increased between pre- and posttest by more than 27% and the data smoothing gained 5%, and less interpolations. These results were consistent with the expectations, as these types of problems were very close to the course content.

As far as developments of statistical thinking according to the research question are concerned, the analysis focused on a comparison of the Item 2 type problems. Statistical inference had not been taught to either group of students. Scoring of the Item 2 type problems was done according to a scheme that honored recognition of random variation in the past data, enhanced by contextual considerations, whereas attempts to search for or to attribute variations to specific causes led to lower scores. For the Item 2 type problems, Table 39.1 displays the relative scores for

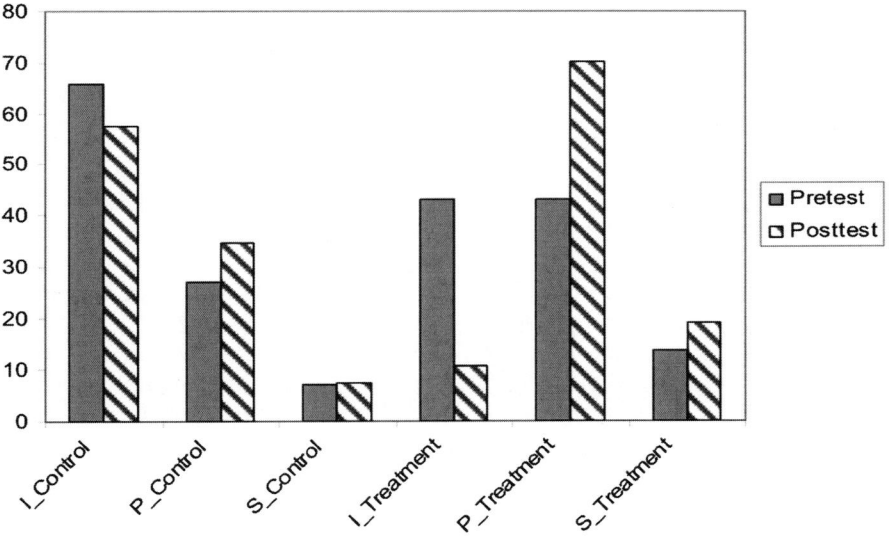

Fig. 39.3 Percentages of students classified as interpolator (*I*), parametric curve fitter (*P*), or smoother (*S*) in pre- and posttest

Table 39.1 Scores on item 2

		Control group (%)	Treatment group (%)
Pretest	Mean	29.6	28.9
	Standard error	2.8	3.3
Posttest	Mean	29.9	43.8
	Standard error	2.9	3.6

pre- and posttests. While in the pretest the two groups barely differed at all, the difference between treatment and control groups in the posttest was highly significant ($p<0.01$; cf. Engel et al. 2008).

4.1 Discussion and Outlook on Further Results

The results of the first study reported earlier indicate a sizeable and highly significant improvement of the treatment group on Item 2 problems and no noticeable changes in the performance of the control group. We interpret this result as an indicator that students were capable of transferring the signal–noise concept from the context of modelling functional relationships to the broader area of dealing with statistical variation. It shows that the data-oriented modelling approach in the applied mathematics course improved statistical thinking skills, even though the course did not explicitly focus on statistics. The results suggest that data-oriented modelling interventions of functional dependence can strengthen learners' knowledge about variation in data and their abilities to deal with variation.

Extensions of this result to a broader scope of domains are the focus of the second study, which is conducted in the framework of the ongoing research project RIKO-STAT (Kuntze et al. 2010). The work of this research project is based on prior work on a competency model for modelling and using representations in statistical contexts (Kuntze et al. 2008). This competency model uses the overarching idea of "data-related reading," which includes modelling activities, when, for example, interpreting information given in a diagram or a table. This competency is a subcomponent of statistical literacy, and a test instrument designed according to the hierarchical competency model has been shown to conform to a one-dimensional Rasch model (Kuntze et al. 2008). These findings suggest that proficiency in the area of statistical literacy is intertwined with abilities of modelling as the competency consisting of both aspects has proved to be one dimensional. However, as various components of conceptual knowledge may substantially contribute to the competency of modelling and using representations in statistical contexts, the project RIKO-STAT investigates several influencing factors such as conceptual knowledge about probability, functional reasoning, and dispositional variables. The complete model of influencing factors is given in Kuntze, Engel, Martignon, and Gundlach (2010). First results of RIKO-STAT indicate that there is a moderate interrelatedness of the competency of modelling and using representations in statistical contexts and the conceptual knowledge in the areas of probability and functional reasoning. For example, for a subsample of 360 university students the competency score correlates with conceptual knowledge in the domains of functional reasoning with $r=0.30$ (Pearson) and probability with $r=0.36$ (correlations significant with $p<0.001$). As the study encompasses further subsamples of more than 450 primary (grade 4) and 600 secondary (grade 9) students, forthcoming analyses in RIKO-STAT will provide us with insights into interdependencies of the competency of modelling and using representations with components of conceptual knowledge across different age groups of learners.

5 Conclusions

Over the last 20 years we have seen great advances in understanding the learning and teaching of mathematical modelling. Yet most of the concepts presented and the examples suggested in the literature may be seen as "data blind" modelling. There have also been great advances in understanding the learning and teaching of probability and statistics, which call for an integration with findings in the area of modelling. It has been a main interest of this chapter to bring forward the mutual exchange between these two mathematics education communities by joining their experience and stimulating synergies. Further detail-oriented empirical research about interdependencies between statistical literacy and modelling competencies and about effective ways of fostering these competencies is needed to advance our understanding, as data-driven modelling is an important area relevant for social participation of responsible citizens.

Acknowledgment The study has been supported by grants of Ludwigsburg University of Education.

References

Ben-Zvi, D., & Garfield, J. B. (Eds.). (2004). *The challenge of developing statistical literacy, reasoning and thinking*. Dordrecht: Kluwer.

Engel, J. (2009). *Anwendungsorientierte Mathematik: Von Daten zur Funktion*. Heidelberg: Springer.

Engel, J., Sedlmeier, P., & Woern, C. (2008). Modeling scatter plot data and the signal-noise metaphor. In C. Batanero, G. Burrill, C. Reading & A. Rossman (Eds.), *Proceedings of the ICMI Study 18 and 2008 IASE Round Table Conference*. www.stat.auckland.ac.nz/~iase/publicatons.

Konold, C., & Pollatsek, A. (2002). Data analysis as the search for signals in noisy processes. *Journal for Research in Mathematics Education, 33*(4), 259–289.

Kuntze, S., Lindmeier, A. & Reiss, K. (2008). Using models and representations in statistical contexts" as a sub-competency of statistical literacy – Results from three empirical studies. In *Proceedings of the 11th International Congress on Mathematical Education (ICME 11)*. http://tsg.icme11.org/document/get/474.

Kuntze, S., Engel, J., Martignon, L., & Gundlach, M. (2010). Aspects of statistical literacy between competency measures and indicators for conceptual knowledge – Empirical research in the Project "RIKO-STAT". In C. Reading (Ed.), *Proceedings of 8th International Conference of Teaching Statistics, Ljubljana, Slovenia*. www.stat.auckland.ac.nz/~iase/publications.

Maaß, K. (2006). What are modeling competencies? *Zentralblatt Didaktik der Mathematik, 38*(2), 113–142.

Moore, D. (1997). New pedagogy and new content: the case of statistics (with discussion). *International Statistical Review, 65*(2), 123–166.
Safuanov, I. (2004). Psychological aspects of genetic approach to teaching mathematics. In M. Johnsen Høines & A. Berit Fuglestad (Eds.), *Proceedings of the 28th Conference of the International Group for the Psychology of Mathematics Education* (Vol. 4, pp. 153–160). http://www.emis.de/proceedings/PME28/RR/RR100_Safuanov.pdf.
Wild, C., & Pfannkuch, M. (1999). Statistical thinking in empirical enquiry. *International Statistical Review, 3*, 223–266.

Chapter 40
First Results from a Study Investigating Swedish Upper Secondary Students' Mathematical Modelling Competencies

Peter Frejd and Jonas Bergman Ärlebäck

Abstract This chapter reports on the first results from a study investigating Swedish upper-secondary students' mathematical modelling competency. Using non-parametric statistical methods the data from 381 12th grade students are analysed, and the students' modelling competency is described in terms of seven sub-competencies. Possible factors affecting the students' mathematical competency such as attitudes towards modelling, previous experiences, last-taken mathematics course, grade, class and gender are also investigated.

1 Introduction and Purpose

What do Swedish upper secondary students know about mathematical modelling, and how capable are they of solving modelling problems? These questions we ask ourselves from the background of the internationally growing interest in the field of educational research in mathematics focused on applications and modelling. This chapter discusses the first results of part of an empirical study conducted to enlighten the present situation at the upper secondary level in Sweden with respect to these issues.

Since 1965 there has been an increasing explicit emphasis on mathematical modelling in the written curricular document governing the Swedish upper secondary mathematics education (Ärlebäck 2009). In the present mathematics curriculum it is stressed that "[a]n important part of solving problems is designing and using mathematical models" and that one of the goals to aim for is to "develop their [the students'] ability to design, fine-tune and use mathematical models, as well as critically assess the conditions, opportunities and limitations of different models" (Skolverket 2000). Indeed, using and working with mathematical models

---

P. Frejd (✉) and J.B. Ärlebäck
Department of Mathematics, Linköping University, Linköping, Sweden
e-mail: peter.frejd@liu.se; jonas.bergman.arleback@liu.se

and modelling, problem solving, communication and the history of mathematical ideas are emphasized as four important aspects of the subject that should permeate *all* mathematics teaching (Skolverket 2000). However, a more explicit definition is not given, and thus this description of mathematical modelling opens up interpretations.

The aim of this chapter is to get an initial indication of the level of the mathematical modelling competency of Swedish upper secondary students. In addition, it investigates if factors such as grade, gender, last-taken mathematics course and different attitudes might affect the level of success of students solving modelling problems.

The research questions we addressed in this research in general terms were:

1. What modelling competency do Swedish upper secondary students in 12th grade display?
2. Are there any connections between the students' modelling competency in relation to their mathematical achievement in general (grade), gender, the students' interest, last-taken mathematical course or their previous experiences?

2 Methodology, Theoretical Considerations and Method

Rather than devising a research instrument of our own we decided to use an already existing and tried tool, and after having scanned the research literature we decided to use the research instrument developed and constructed by Haines et al. (2000). The instrument, also reported on in Haines et al. (2001), originally consisted of 12 multiple-choice questions (five alternative choices), together with a partial credit assessment model assigning a score of 2 to one preferred of the five alternatives in each question; 1 to one or more other choices since "an alternative response could indicate knowledge and understanding in mathematical modelling" (Haines et al. 2000, p. 5); and 0 to the remaining alternatives. Using the wording of Houston and Neill (2003a, pp. 156–159) the 12 questions, grouped in six pairs used in a pre–post test setting, focused on the following aspects of the modelling process (see also Sect. 2.1): making simplifying assumptions; clarifying the goal; formulating the problem; assigning variables, parameters, and constants; formulating mathematical statements; and selecting a model. The instrument is suitable and relevant to use in this study since it was "devised both to address the need for a base level assessment of modelling skills and for application during or on completion of an experience in mathematical modelling" (Haines et al. 2000, p. 2), and the authors argue that using the instrument, "it is possible to obtain a snapshot of students' [modelling] skills at key developmental stages without the student carrying out a complete modelling exercise" (p. 10). The number of test items was extended to 18 by Houston and Neill (2003b), adding one new question to each of the six aspects above. In addition, Haines et al. (2003) extended the numbers of items adding two questions involving *graphical representations* and two questions *exploring real and mathematical world connections*, making a total set of 22 test items covering eight aspects of the modelling process. Besides being used in the research referred to earlier, the research instrument has also been drawn on and used in different settings with a variety of objectives, in Haines and Crouch (2001), Izard et al. (2003), Ikeda et al. (2007), Lingefjärd and Holmquist (2005), and Kaiser (2007), to among other things, investigate the levels of students' modelling competencies.

2.1 Mathematical Modelling and Modelling Competencies

The view of mathematical modelling underlying the construction of the research instrument mentioned in the previous section is represented by the left diagram in Fig. 40.1. Note that the 'content' of the boxes in this diagram are of different types and on different levels; *real world problem* represents the real world situation or phenomena under consideration; *formulating model*, *solving mathematics*, *interpreting outcomes*, *evaluating solution* and *reporting* are all processes, or to use the terminology of Haines et al. (2000) – *skills*, involved in mathematical modelling. *Refining model* is also a process but on another level in the sense that it is often compounded by the other processes just mentioned, meaning that the modeller(s) goes back to the real world problem and possibly *re*-formulates, *re*-solves, *re*-reinterprets and *re*-evaluates her/his(their) work. However, the diagram on the right in Fig. 40.1 makes a more clear distinction between the corresponding processes and *refining model*; in this representation of mathematical modelling it means engaging in another cycle. It also situates the processes in relation to the intra- and extramathematical worlds. In spite of the differences between the two diagrams in Fig. 40.1 we believe that the respective authors as a matter of fact share more or less the same overall view on mathematical modelling, but it must be stressed that both these views of mathematical modelling are highly idealized and schematic representations of the complex processes involved. A similar view is presented by Palm et al. (2004) in their interpretation of the written curriculum documents governing the Swedish upper secondary mathematics education, and this suggests that the research instrument described earlier could adequately be applied to the Swedish context as well.

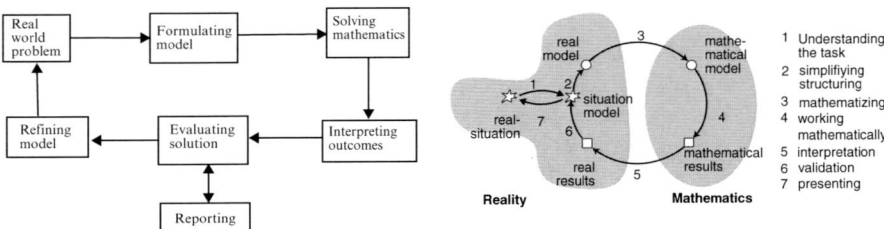

Fig. 40.1 To the *left* are the "[s]tages in the mathematical modelling process" as presented by Haines et al. (2000, p. 3), and to the *right* is the "modelling cycle" of Blum and Leiß (2007) as presented by Borromeo Ferri (2006, p. 92)

Mathematical modelling as presented in Fig. 40.1 is often described using the notion of *modelling competencies*. However, the meaning and content of this concept varies among its users; the Blomhøj and Jensen (2003) definition is "[b]y mathematical modelling competence we mean being able to autonomously and insightfully carry through all aspects of a mathematical modelling process in a certain context" (p. 126), whereas the Maaß (2006) definition is "[m]odelling competencies include skills and abilities to perform modelling processes appropriately and goal-oriented as well as the willingness to put these into action"(p. 117). We find the latter definition ambiguous and problematic for two reasons: first it is not clear what *skills* or *abilities* are, and neither is the relation between these two concepts discussed; and second, the emphasis on *willingness* seems to lack reasonable motivation and has weak grounding, and in addition makes *modelling competencies* a concept hard to operationalize. We believe that the definition suggested by Maaβ is incompatible with the research instrument initiated by Haines et al. (2000) and that the sole use of this definition cannot productively be used to analyse students' mathematical modelling competencies as suggested by Kaiser (2007). Hence, in this chapter we chose to define modelling competence in line with Blomhøj and Jensen (2003) quoted earlier and refer to the processes involved in mathematical modelling, previously described in terms of the eight aspects of the modelling process, as *modelling sub-competencies*.

2.2 Developing an Instrument

All the 22 test items were translated into Swedish, and effort was made to make the translations as true to the original formulations as possible by adjusting only some of the details to the Swedish context where this was appropriate. The translations were checked by a third independent researcher before the items were piloted in a group of 16 students; each individually assigned eight items distributed so that we roughly got the same number of responses on all 22 items. In addition, for each item the students were asked to answer the three questions – (1) Do you think that the problem you just solved is relevant for a mathematics class? (2) Do you think the problem is interesting? (3) Do you think the problem is connected to reality? – by choosing 'yes', 'I don't know' or 'no' and to give a short motivation for their choices. The pilot study served to check how the translated items worked in practice and how much time the students needed to complete the eight items and to give us a first impression of the students' feelings for, and attitudes towards, working on the items. During the time the students worked on the pilot test, which varied between 20 and 30 minutes, the first author surveyed the class and recorded comments.

Taking some of the recorded students' comments from the piloting into account, together with the wish to push the test time down to approximately 20 min, we decided to cut out one of the items. In doing this, we also decided to incorporate the aspects of modelling listed as *graphical representations* with *selecting a model* since both of them focus on selecting a mathematical model, in terms of a graph in the first case and a formula in the second. Hence there are totally four items focusing on the aspect of *selecting a model*. Therefore the view taken on *modelling competency* in this chapter is that it at least constitutes the following sub-competencies: (sC1) *to make simplifying assumptions concerning the real world problem*; (sC2) *to clarify the goal of the real model*; (sC3) *to formulate a precise problem*; (sC4) *to assign variables, parameters, and constants in a model on the basis of sound understanding of model and situation*; (sC5) *to formulate relevant mathematical statements describing the problem addressed*; (sC6) *to select a model*; and (sC7) *to interpret and relate the mathematical solution to the real world context* (cf. Kaiser 2007, pp. 115–116). In addition, the three follow-up questions subsequent to each item were also replaced with the following 7 four-alternative-Likert attitude questions ending the test: *I consider the problems on the test to be* (Q1) *fun*, (Q2) *easy*, (Q3) *interesting*; (Q4) *I think the problems on the test invite you to use mathematics to answer the questions*; (Q5) *I think that problems of this type are well suited for the upper secondary mathematics courses*; (Q6) *In the upper secondary mathematics courses we often work(ed) on similar problems*; and (Q7) *I would like (would have liked) to work more often on similar problems in the upper secondary mathematics courses.*

The pilot study was also used to make a selection of 14 of the original 22 test items, two representing each of the seven sub-competencies. The selection was based on the students' results on the test items, and those items that displayed non-extreme answer distribution, which was not considered to be due to interpretational issues of item formulations, were selected out. This means that the items in which all students got full score was discharged in favour of the items in which they achieved more moderately. After a discussion, the 14 selected items were grouped into two groups, our first-hand choice (FHC) and second-hand choice (SHC), from which four tests consisting of seven items each were constructed; tests T1 and T2 containing solely items from the two respective groups, and tests T3 and T4 containing a mixture of items from the two. The total score a student achieves on either of these four tests is what we take as a measure of the students' modelling competency.

For the final version of our research instrument we decided the first instant to have two quotes from the curriculum guidelines for the Swedish upper secondary mathematics courses (the ones we used in the Introduction and Purpose section) followed up by two questions: *Have you ever encountered the word 'mathematical modelling' during your upper secondary education?* requiring just a 'yes' or a 'no' answer; and the open question *Describe in your own words the meaning you ascribe to the concepts 'mathematical model' and 'modelling'*. Next, the seven items followed and then the attitude questions Q1–Q7. In addition, the students were asked to state their gender, last-taken or ongoing upper secondary mathematics course (in Sweden there are in principle five such courses, one per term, Mathematics A–Mathematics E) and their latest received grade in mathematics (every course is graded as either *IG*=Fail, *G*=Pass, *VG*=Pass with distinction, *MVG*=Pass with special distinction).

In a national science and mathematics teachers' developmental program with participants from all over Sweden, 41 sets of tests were distributed in the spring of

2009, and asked to be brought back to their respective school and given to mathematics teachers teaching 12th graders in the science program (normally, the students are 18 years old and at this time of year the students have completed the Mathematics D course). Each set of tests contained 30 of the tests numbered from 1 to 30, with every fourth test a T1, T2, T3 and T4 test, respectively, starting randomly in that sequence. A letter was also attached to the mathematics teacher informing of the aim of the research and ethical considerations taken as well as practical requests for how to distribute the tests in their classes: to use approximately 20–25 minutes; not to allow the students to use calculators; and to let the students solve the problems individually. In addition, the teachers were asked to fill in a teacher questionnaire, but this will not be reported on in this chapter.

In all, 21 sets of tests (51%) were returned resulting in test scores from a total of 400 students. However, for the statistical analysis, which was made using SPSS, we analysed only the 381 students who answered at least four of the seven items on the tests, since we consider the instrument not to give a reliable measure of the students' modelling competency otherwise. In this chapter we will report only on the quantitative data; the results of the analysis of the students' answers on the open questions in the test will be reported on elsewhere.

2.3 Statistical Analysis

A first statistical analysis using a Kolmogorov–Smirnov test with Lilliefors significance correction showed that the data were not distributed normally, which led us to use non-parametric tests for the continued analysis. The particular tests used were the Mann–Whitney test, the Kruskal–Wallis test and the Kendall's tau. The Mann–Whitney test is the non-parametric equivalent test to the parametric independent t-test, which uses a ranking procedure to compare two independent groups; the Kruskal–Wallis test also uses ranking techniques to compare more than two independent groups; and the Kendell's tau is the non-parametric equivalent to the Pearson's correlation coefficient.

3 Results

With a maximum score of 14 the students scored in average 7.78 (SD 2.26). On the 14 different test items, corresponding to the seven sub-competencies, the students scored on average between 0.33 and 1.67 as illustrated in Fig. 40.2.

Comparing pairs of items in respective sub-competencies (at a level of $p<.05$) indicate a comparability of the pairs only in sC2 ($H(3)=5.868$, $p=.118$) and sC6 ($H(3)=6.397$, $p=.094$). The pair of items in sC5 is a borderline case ($H(3)=7.809$, $p=.050$).

The mean and standard deviation of the students' total score with respect to gender, classes, grades, last-taken course and tests are summarized in Table 40.1.

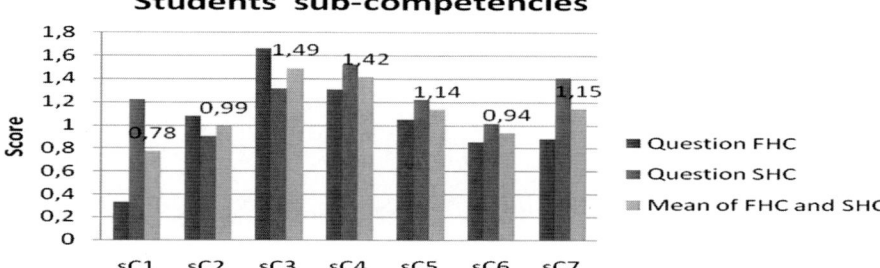

Fig. 40.2 Students' sub-competencies with items relating to FHC and SHC. *The sub-competenices are*: (sC1) to make simplifying assumptions concerning the real world problem; (sC2) to clarify the goal of the real model; (sC3) to formulate a precise problem; (sC4) to assign variables, parameters and constants in a model on the basis of sound understanding of model and situation; (sC5) to formulate relevant mathematical statements describing the problem addressed; (sC6) to select a model; and (sC7) to interpret and relate the mathematical solution to the real world context

Table 40.1 Mean value with respect to gender, classes, grades, courses and tests

Results	Gender		Classes	Grade				Course			Test			
	Female	Male		IG	G	VG	MVG	C	D	E	T1	T2	T3	T4
Mean	7.95	7.76	6.38–9.27	7.50	7.07	7.96	8.43	6.67	7.17	7.99	6.82	8.37	7.63	8.26
Standard Deviation	2.22	2.29	1.57–2.84	2.38	2.16	2.34	2.05	1.73	2.12	2.28	1.93	2.24	2.04	2.48
Number	121	237	8–29	4	130	118	128	9	84	281	95	94	93	99

When analysing the grades and the students' modelling competency the four students with grade IG were excluded, because they scored on average 7.5, which were not considered representative of the IG grade. The grades G, VG and MVG showed a significant effect ($p<.05$) on the students' total scores ($H(2)=27.853$, $p=.000$). However, just considering the grade VG appeared not to have an effect on the students' total scores compared with grade MVG ($U=6720$, $p=.132$).

The students' last-taken mathematics course also had a significant effect ($p<.05$) on the students' modelling competency ($H(2)=10.772$, $p=.005$). A further investigation showed that mathematics course D compared with course E had an effect on the students' total scores ($U=9439$, $p=.005$), but no effect was found between the courses C and D ($U=342.5$, $p=.568$).

Other factors significantly affecting ($p<.05$) the students' total scores were which test they took ($H(3)=27.996$, $p=.000$) and which class the student belonged to ($H(20)=0.437$, $p=.004$). However, no effect was found with respect to gender ($U=13503.5$, $p=.363$).

Table 40.2 summarizes the students' responses to the questions (Q1)–(Q7) and the only attitudes having a significant affect ($p<0.5$) on the students' modelling competency were if the student considered the problems in the test to be (Q2) *easy* ($H(3)=10.912$, $p=.012$) or to be (Q3) *interesting* ($H(3)=18.292$, $p=.000$).

Table 40.2 Mean values of the questions about attitudes and previous experiences (1 = strongly agree; 2 = agree; 3 = disagree; 4 = strongly disagree; mean 2.5)

				Attitudes and previous experiences			
Results	(Q1) Fun	(Q2) Easy	(Q3) Interest	(Q4) Invite math	(Q5) Good mom.	(Q6) Done similar	(Q7) Work more
Mean	2.89	3.25	2.63	2.33	2.41	3.37	2.61
Std.Dev.	0.90	0.70	0.94	0.81	0.93	0.86	0.99
Number	376	374	377	372	374	375	375

Investigating pairwise correlations among the affecting factors on the students' modelling competency, significant correlations ($p<.05$) were found between grades and respective courses ($\tau=-.342$, $p=.000$), easy ($\tau=-.164$ $p=.000$) and interest ($\tau=-.116$, $p=.010$). In addition, interest correlated with course ($\tau=-.109$, $p=.021$) and easy ($\tau=.217$, $p=.000$).

It is notable that only 22.5% of the students have heard/used mathematical models or modelling in school. For these students, it did not show any effect on their total score ($H(1)=.041$, $p=.839$).

4 Discussion

Comparing the sub-competencies of the Swedish upper secondary students' modelling competency, Fig. 40.2 shows that they were most proficient in questions relating to sC3 and sC4, but exhibited more difficulties in questions relating to sC1, sC2 and sC6. The sub-competence sC2 has also been proved to be difficult for the students in previous research (e.g. Houston and Neill 2003a; Kaiser 2007). The notable difference between the two items in sC1 might be an effect due to translation or interpretational problems. Not surprisingly, students' grade and students' last-taken mathematical course have a positive effect on the students' modelling competency. However, due to the found correlation between these two, further analysis is needed.

Looking at the results of the attitude questions Q1–Q7 there are is an overall negative tendency towards working with mathematical modelling as represented in the test items in all answers. In general, the students found the problems very hard (Q2) and did not express any excitement or joy in tackling them (Q1). Neither did the students express that they found the problems especially interesting (Q3), nor that they wanted to (have) work(ed) more on similar problems in their mathematics classes (Q7). However, the students to some extent seemed to recognize the value to use mathematics to solve the problems on the tests (Q4), and in addition they regarded the types of questions asked to be relevant and good to use in mathematics classrooms (Q5). One explanation of these results might be the students expressed that they in principle never worked on similar problems before (Q6). Indeed, such student attitudes may present an obstacle for implementing mathematical modelling at this school level.

In line with Haines et al. (2000) we agree that all the individual stages of mathematical modelling represented in the sub-competencies are part of the modelling process. However, the instrument lacks other aspects of the modelling process such as the use of ICT, the fact that not a 'whole modelling problem' is solved, and collaborative work, which means that the research instrument does not provide a complete picture. Nevertheless, in fulfilling the aims of the study, the test items are adequate in that they allow many students to be tested in a short time and give a first preliminary overview of the present state of the Swedish upper secondary mathematics regarding the students' mathematical modelling competency.

In evaluating their research instrument Haines and Crouch (2001) concluded that "the analogue pairs of items are predicted to perform in a comparable manner" (p. 133), except for the two items used in sC3. Due to our big sample we expected to conclude approximately the same comparability. However, in our study the only comparable pairs of items in respective sub-competencies are sC2 and sC6 (and possibly sC5). Note that the analysis in Haines and Crouch (2001) only investigates 'the original six stages' in the modelling process and that only the first five are comparable to our sC1–sC5, respectively.

The results on the relations between and among the students' modelling competency and their expressed attitudes (Q2) and (Q3), together with the many correlations found between the attitudes, indicate that a more advanced analysis might be fruitful using a more sophisticated statistical model and method. This we plan to do in a forthcoming work.

5 Conclusions

The investigation of the modelling competency of Swedish upper secondary 12th grade students revealed that the students were most proficient in the sub-competencies *to formulate a precise problem* and *to assign variables, parameters, and constants in a model on the basis of sound understanding of model and situation*, and least proficient in the sub-competencies *to clarify the goal of the real model* and *to select a model* (if *to make simplifying assumptions concerning the real world problem* is disregarded). The study also shows that the students' grade, last-taken mathematics course, and if they thought the problems in the tests were easy or interesting were factors positively affecting the students' modelling competency. In addition, only 22.5% of the students stated that they had heard about or used mathematical models or modelling in their education before, and the expressed overall attitudes towards working with mathematical modelling as represented in the test items were negative.

References

Ärlebäck, J. B. (2009). *Mathematical modelling in the Swedish curriculum documents governing the upper secondary mathematics education between the years 1965–2000. [In Swedish]*

(Report No. 2009:8, LiTH-MAT-R-2009). Linköping: Linköpings universitet, Matematiska institutitionen.
Blomhøj, M., & Jensen, T. H. (2003). Developing mathematical modelling competence: Conceptual clarification and educational planning. *Teaching Mathematics and Its Applications, 22*(3), 123–139.
Blum, W., & Leiβ, D. (2007). How do students and teachers deal with modelling problems? In C. Haines, P. Galbraith, W. Blum, & S. Khan (Eds.), *Mathematical modelling: Education, engineering and economics* (pp. 222–231). Chichester: Horwood.
Borromeo Ferri, R. (2006). Theoretical and empirical differentiations of phases in the modelling process. *Zentralblatt für Didaktik der Mathematik, 38*(2), 86–95.
Haines, C., & Crouch, R. (2001). Recognizing constructs within mathematical modelling. *Teaching Mathematics and Its Applications, 20*(3), 129–138.
Haines, C., Crouch, R., & Davis, J. (2000). *Mathematical modelling skills: A research instrument* (Technical Report No. 55). University of Hertfordshire: Dept. of Mathematics.
Haines, C., Crouch, R., & Davis, J. (2001). Understanding students' modelling skills. In J. F. Matos, W. Blum, K. S. Houston, & S. P. Carreira (Eds.), *Modelling and mathematics education: Applications in science and technology* (pp. 366–380). Chichester: Horwood.
Haines, C., Crouch, R., & Fitzharris, A. (2003). Deconstructing mathematical modelling: Approaches to problem solving. In Q. Ye, W. Blum, K. S. Houston, & Q. Jiang (Eds.), *Mathematical modelling in education and culture* (pp. 41–53). Chichester: Horwood.
Houston, K., & Neill, N. (2003a). Assessing modelling skills. In S. J. Lamon, W. A. Parker, & K. Houston (Eds.), *Mathematical modelling: A way of life* (pp. 155–164). Chichester: Horwood.
Houston, K., & Neill, N. (2003b). Investigating students' modelling skills. In Q. Ye, W. Blum, K. S. Houston, & Q. Jiang (Eds.), *Mathematical modelling in education and culture* (pp. 54–66). Chichester: Horwood.
Ikeda, T., Stephens, M., & Matsuzaki, A. (2007). A teaching experiment in mathematical modelling. In *Mathematical modelling: Education, engineering and economics* (pp. 101–109). Chichester: Horwood.
Izard, J., Haines, C., Crouch, R., Houston, K., & Neill, N. (2003). Assessing the impact of teaching mathematical modelling: Some implications. In S. J. Lamon, W. A. Parker, & K. Houston (Eds.), *Mathematical modelling: A way of life* (pp. 165–177). Chichester: Horwood.
Kaiser, G. (2007). Modelling and modelling competencies in school. In C. Haines, P. Galbraith, W. Blum, & S. Khan (Eds.), *Mathematical modelling: Education, engineering and economics* (pp. 110–119). Chichester: Horwood.
Lingefjärd, T., & Holmquist, M. (2005). To assess students' attitudes, skills and competencies in mathematical modeling. *Teaching Mathematics and Its Applications, 24*(2–3), 123–133.
Maaß, K. (2006). What are modelling competencies? *Zentralblatt für Didaktik der Mathematik, 38*(2), 113–142.
Palm, T., Bergqvist, E., Eriksson, I., Hellström, T., & Häggström, C. (2004). *En tolkning av målen med den svenska gymnasiematematiken och tolkningens konsekvenser för uppgiftskonstruktion* (Report No. 199). Umeå: Enheten för pedagogiska mätningar, Umeå universitet.
Skolverket. (2000). *Upper secondary school, syllabuses, mathematics.* Retrieved September 15, 2008, from http://www.skolverket.se/sb/d/190.

Chapter 41
Why Cats Happen to Fall from the Sky or on Good and Bad Models

Hans-Wolfgang Henn

Abstract Teaching students to use mathematical modelling sensibly in realistic context is one general goal of mathematics education in order to educate students to become responsible citizens and future decision makers. Here, I want to discuss three important issues which can – depending on how teaching takes place – either promote or obstruct the development of students' modelling competence. They are:

- Central examinations,
- Use of computers,
- Professional development and motivation of teachers.

Remarks on these issues will be illustrated by examples.

1 The Operation 'Cat Airdrop'

Some years ago the Royal Air Force, on behalf of the World Health Organisation, dropped cats on remote villages in Borneo (Calvin 1986) (Fig. 41.1), where all cats had died and the rat population – potential carriers of dangerous diseases – had increased explosively. What was the reason? Poisonous DDT had been sprayed to kill the malaria-causing flies. The fight against flies and malaria was successful.

However, the poison was also eaten by cockroaches, but in such small doses that it was insufficient to kill them. The cockroaches, with the accumulated DDT, were then eaten by geckos, which accumulated the DDT, again not enough to kill them. The cats of the villages not only fed on rats but also on geckos. Thus, hundreds of cats accumulated the DDT that had been eaten by millions of cockroaches and it was enough to kill the cats. The rats benefited from this! The operation 'cat drop' restored the cat population to its original size and averted the threatening rat plague.

H.-W. Henn (✉)
Faculty of Mathematics, Technische Universität Dortmund, IEEM, Dortmund, Germany
e-mail: wolfgang.henn@tu-dortmund.de

Fig. 41.1 Cat airdrop

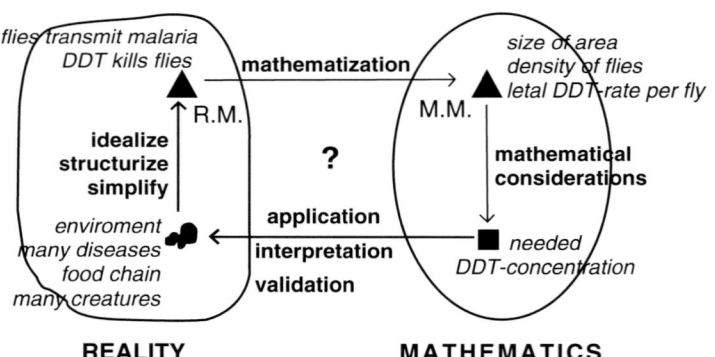

Fig. 41.2 Modelling circuit

The cats example is very convenient to introduce the modelling circuit in school and to discuss what a 'good' and 'bad' model is. In the well-known picture shown in Fig. 41.2 (Blum et al. 2007), we start with a problem of reality. The real problem is idealised to become the real model (RM) and is transferred to the mathematical model (MM). With the help of mathematics we find a mathematical solution, which solves the original problem. Insofar we have developed at the first look a 'good' model. But the model describing reality considers only one aspect of the complex and interconnected situation and does not give credit to the ecological relations. The explosive increase of the rat population, not considered in the model, shows that our first model is not satisfying. By hindsight the model proves 'bad'. As often, we have to again run through the modelling circuit and consider more aspects of reality – in the example, the cats have to be included. Students can understand, discussing this example, that models are simplifying presentations, which consider only certain, somehow objectifiable parts of reality. The purpose of a model is to draw conclusions for reality. Often, it is necessary to go several times through the modelling circuit or to develop a new model. One should never talk about a 'right' model or a 'wrong' model. A model describes the reality better or worse, more or less suitable, regards more or less aspects of reality, and provides more or less sufficient solutions for the problem in question – shortly, one should rather talk about 'good' or 'bad' models. In any case, models will always be of subjective character, owing to the normatively chosen assumptions of the modeller. This aspect also includes the danger of misuse and misinterpretation. It is an important task for school to impart knowledge about these facts.

2   Modelling in School: Chances and Obstacles

Peter Galbraith (2007) formulates a convincing framework about models and modelling. Following his ideas, I will discuss them from a German perspective. Unfortunately, as a rule, reality-oriented teaching on applications outside mathematics is covered only to a limited extent in everyday teaching in Germany although there is a long-standing agreement on the importance of creating relations between realistic situations and mathematics teaching. Heinrich Winter, the well-known German mathematics educator, demands in one of his three 'basic experiences' (Winter 2004) that students should become acquainted with the fundamental contributions of mathematics in acquiring important knowledge about our world. Mathematics proves to be an inexhaustible pool of mathematical models, which allow us to understand better the world around us. Students have to experience true modelling activities in school, have to be involved in the transition from reality to mathematics, mathematical analysis, and the transfer back of the results into the real situation (or a multiple run through of the modelling circuit like the cats example is necessary).

Teaching affects the image that students will take with them into their future life as responsible citizens and future decision makers. I will identify three important factors (of course there are more, see Galbraith (2007)) that can promote the development of modelling competence of students, but can also obstruct them drastically. Unfortunately, the examples in the following three parts are typical for the German school reality. The three factors are:

– The problem area 'central examinations'
  Germany consists of 16 federal states which are responsible for their individual educational policy and planning. That means that we have 16 more or less different school systems, teaching curricula and regulations for the final examinations. Only three of them had central examination at the end of secondary school. Following the TIMSS and PISA shock – Germany reached only a middle ranking – nearly all states changed to written final examinations. Central examinations

have a crucial influence on the content of teaching. The big problem is that real modelling problems and central examinations are difficult to combine. Worse, most of the reality oriented tasks for the written final examinations are counterproductive and create a strange idea of applications and modelling. One possibility we discuss could be a division into two parts: One centrally posed part without formulae, pocket calculators and computers, and one locally posed part. But the administrations oppose it on legal grounds. So up to now modelling tasks posed in examinations are of the very problematic type described in part 3.

– The use of computers
Today's available computer technology can contribute in a special way to aid in the learning process. The computer is a powerful tool to aid in modelling and simulation and can positively influence the generation of adequate basic concepts ('*Grundvorstellungen*') of mathematical ideas – especially through dynamical visualisations. The computer also furthers heuristic-experimental work in problem solving. I have experiences for more than 20 years with many projects using dynamic geometry software (DGS) and computer algebra systems (CAS) in school. One example is the CAS-Project Mobile Classroom (Henn 2001). We experienced how the computer can help to understand mathematical concepts and to apply them in modelling situations. With the help of the new computer tools, open-ended problems lead to individual new solution strategies and motivate creativity. But the problem is that the computer does what you want – sensible or foolish. The example in part 4 is an example for the latter: "Using mathematical modelling as a synonym for curve fitting creates a dangerous aberration of the modelling concept" (Galbraith 2007, p. 49).

– The professional development and motivation of teachers
The not very flattering results of Germany in TIMSS turned out to have the effect of a catalyst inducing a nationwide debate about educational goals and the content of mathematics teaching. The central question is not 'what is to be learned", but 'how should learning take place', 'how can mathematical literacy be promoted', and also 'how can learning processes be measured'. Important is a willingness to question and to rethink current teaching, to change one's own reception and to realise opportunities brought about by new practice and teaching techniques. We had many projects in Germany to change the behaviour of teachers and of students in the wanted direction (Henn 2003). Today, the university education of teachers and the 'classroom culture' has changed. But as shown in part 5, there is a danger to overact.

3 The Problem Field 'Central Examinations

The following example is taken from the central final examination (Year 13) of the German state of Baden-Wuerttemberg, posed in 1998 in the topic Analytic Geometry. It is a typical 'application problem' from a central exam and is only a

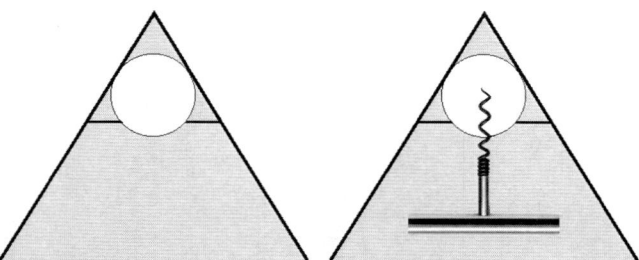

Fig. 41.3 The pyramid

mathematical problem 'in disguise' and not a genuine real life problem. For the students 'uncovering' these problems 'in disguise' is reduced to finding out the algorithms that have been hidden by the teacher, and immediately 'real' mathematics takes over. The problem in question is set in the context of a playground with a wooden pyramid that stands perpendicularly on a square base and is accessible inside. The following text shows part c of the problem (my translation):

> Inside the pyramid a board is fixed parallel to the floor with a circular opening with diameter $d = 2.4$ in its middle. For tidying up, a big foam ball with radius $r = 1.5$ needs to be pushed through the opening towards the upper part of the pyramid. At which height needs the board to be fixed if it is supposed to be as high up as possible with the ball lying loosely in the opening?

The missing measurement units show immediately that the problem poser does not take reality too seriously. Now, let us discuss the task: We assume that the measures are given in metre and make a drawing of the situation (Fig. 41.3, *left*). The board is fixed at a height of 5.6 m, and the ball possesses a volume of 9.4 m³. The Internet states the specific weight of foam: the ball weighs approximately 380 kg.

How should this ball ever be pushed upwards? How should it ever be pulled out again? Maybe the problem poser was thinking about a giant screw pull (Fig. 41.3, *right*) to add the corresponding geometric helix curve to the problem? Anyway, the problem is a typical 'age-of-the-captain' problem (Baruk 1985) and such problems influence teachers in their belief that modelling and applications are meaningless for mathematics teaching.

4 The Use of Computers

Martinez-Cruz and Ratcliff (1998) investigate the men's world record times in 100 m freestyle swimming. Without any mathematics, just using common sense, one would expect qualitatively something like the curve in Fig. 41.4.

This qualitative curve has nothing to do with the modelling assumptions of logistic growth. For the intermediate time there are no reasonable model assumptions pointing at a special curve. The authors use the world record times in Table 41.1

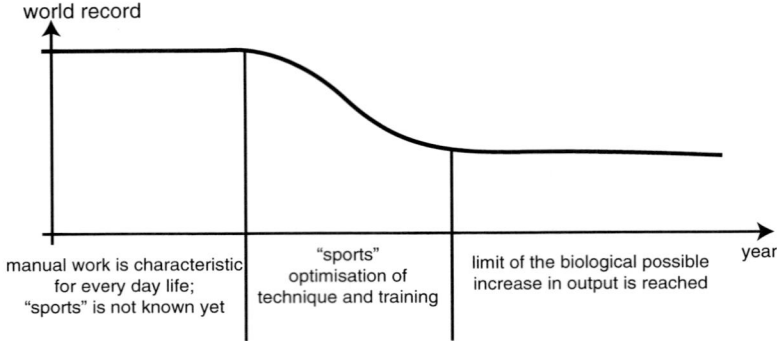

Fig. 41.4 Qualitative model

Table 41.1 World Record Times

Year	Time (s)	Year	Time (s)
1912	61.6	1972	51.22
1924	57.4	1976	49.99
1957	54.6	1988	48.42
1968	52.2	1994	48.21

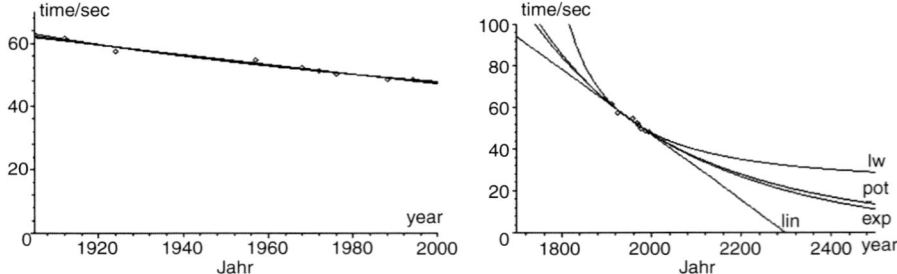

Fig. 41.5 The quantitative models

and fit various curves through the given record data by applying the regression commands available on their calculator. In detail, they fit a linear function, an exponential function, a power function, and a logistic function. Figure 41.5 shows that the choice of the curve is irrelevant for the interval in question. However, extrapolation on both sides shows that all models do not represent the real situation or, in other words, are 'bad' models. The authors favour the logistic model, because its predictions are less meaningless for the future compared to the other predictions! This is nonsense! None of the four models provides a deeper insight or explains the data. This discussion is what Galbraith calls a 'problem of whimsy'.

By the way, the authors do not consider one of the most interesting points of Table 41.1: It is the increase in measurement accuracy from 1968 to 1972. In 1972, the Olympic Games took place in Munich. At first, times were taken with three digits after the decimal point, an accuracy of 1/1,000 s. This can be reconstructed from the results of the 400 m medley swimming contest: At first, times were recorded as 4 min 31.981 s for the swimmer Larsson and 4 min 31.983 s for the swimmer McKee and therefore Larsson was awarded the gold medal. Then, obviously, somebody started thinking: It takes about 50 s to swim 100 m, that means a distance of about 2 mm in 1/1,000 s. Nobody would believe that a 50 m long swimming pool could be constructed so accurately that each swimming lane had an accuracy of less than 2 mm. A little bit more mortar already leads to a larger difference. Therefore, the measurement accuracy was reduced to two digits after the decimal point. But, incomprehensibly no two gold medals were awarded! The same problem but with a different modelling aspect gives new insight in reality and leads to a 'good' model.

5   The Professional Development and Motivation of Teachers

It is an important task to educate teachers to include applications and modelling in their teaching practice. This implies to 'see the world with mathematical eyes', and to find occasions, again and again, to introduce some situations from reality in the mathematics classroom. Lyn English (2003) advocates for these 'rich learning experiences', that is authentic situations, chances for own exploration, multiple possibilities for interpretations, and social competence to take up the responsibility for one's own model up to communicating it to other students. A simple way to do this is to use newspaper clippings. However, caution must be applied not to overshoot the mark. The following text (my translation) shows a newspaper clipping denouncing the often meaningless regulations which are issued in Germany (and elsewhere, too) (from Herget and Scholz 1998).

> **Perfect Official Language**
>
> The perfection of German rule makers has been supported by the Minister of the Interior, Georg Tandler, when he read out a draft for a statutory order concerning calf breeding in the Munich state parliament. There it read, whatever that may mean: "If calves are held in herds each calf has – depending on its height in centimetres – to have a freely usable space in square metres according to the following formula: Minimal space (square cm) equals 0.4 times to the power 2 plus 70 times plus 2,720."

The way how the author describes the functional term shows that he does not understand the meaning of the text. He reads the mathematical symbol x, the variable for the height of the calf, as the symbol for multiplication and gives a totally meaningless text. But even from the correct text it is not easy to develop the correct formula.

I gave the task to grade 9 students who developed about six or seven formulas and wrote them down on the blackboard. Finally, we agreed on the following formula:

$$f(x) = 0.40 \cdot x^2 + 70 \cdot x + 2{,}720$$

After the graph has been drawn (Fig. 41.6) one can reflect on sense and nonsense of this regulation. One of my students immediately argued: It is much simpler to use a straight line!

The thick line would give the same result but would lead to a simpler regulation. So far, so good! But, even this nice problem can be put into bad teaching practice and 'bad' modelling out of sheer enthusiasm about applications and modelling. This is illustrated by the following two examples.

The first example is the schoolbook problem in Fig. 41.7 on the calf regulation (Sigma 1984, my translation). The topic is 'values of polynomial functions'. Without reflection, the regulation is cited, the term '*Widerristhöhe*' (it's an unusual word in German), which means the height of the calves, is explained using a drawing with the variable x, instead of having students search for explanations for themselves – for example by searching in the Internet. Then, without any comment, the formula is given. The task is now to substitute five values in the formula and to add. This is not the way to develop modelling competence, but a typical age-of-the-captain problem.

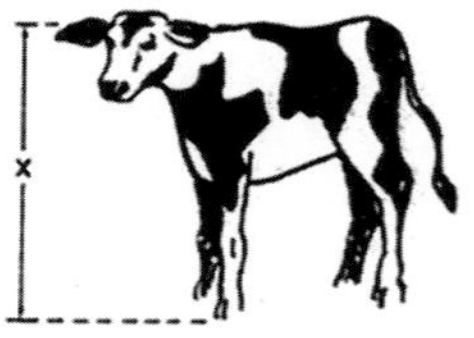

Fig. 41.6 The calves problem

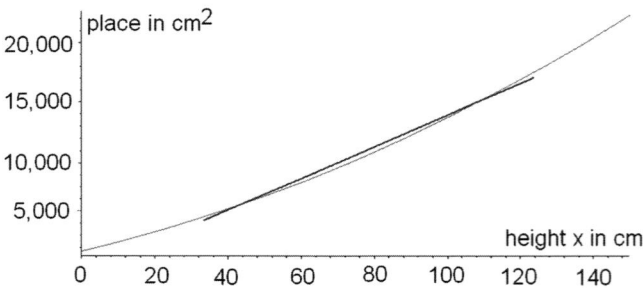

Fig. 41.7 Values of polynomial functions

The second example is the following part of the mathematical diary of a girl. A young teacher, more correctly, a teacher trainee, had covered the calves problem adequately in the classroom and now the girl reports on this (my translation).

I owe the second page to Mrs. Koch, a teacher trainee. The problem was set in the context of calculating the necessary space in a stable for a calf of size x. Maybe this was meant to broaden the students' horizon for the unlimited possibilities to use functions. In this case, the function increased exponentially, which would mean that the farmer needed to apply a straightedge regularly to find out about the growth of each of the calves and then to assign them a new, bigger place in the stable. I would argue in favour of a minimal value that would make any calculation superfluous. However, according to Mrs. Koch, a linear function would turn out to be an indispensable help for the farmer, because he could read off the necessary space comfortably from a graph. I do not agree to this. How would his life be made easier, if his stable needed to look like this?

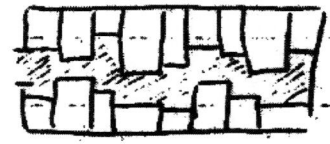

We see, out of pure enthusiasm, the teacher gave the impression that a regulation with a linear formula would be the only reasonable solution. The girl proved to have more common sense than the teacher. And we have an additional example how a 'good' model (the straight line instead of the parabola) can turn into a 'bad' one.

6 Conclusion

The image of mathematics students' experience at school should contain both the beauty *and* the functionality of mathematics. To make orientation in our complex world possible, mathematics lessons must include applications and modelling. It is always pleasant to see how students, who are normally rather uninterested in mathematics, participate critically and actively in the lessons. Of course, it is necessary that the teacher takes seriously both the application treated and the mathematics associated with it. The discussion that takes place must be about the semantic contents and not on the syntactic, formal and algebraic side. Then, students experience applications of mathematics with all their consequences and abundant associations and gain experiences with 'good' and 'bad' models.

References

Baruk, S. (1985). *L'âge du capitaine. De l'erreur en mathematiques*. Paris: Editions du Seuil.
Blum, W., Henn, H. W., Galbraith, P., & Niss, M. (Eds.). (2007). *Modelling and applications in mathematics education*. New York: Springer.

Calvin, W. H. (1986). *The river that flows uphill*. New York: Scribner.
English, L. (2003). Mathematical modelling with young learners. In S. J. Lamon, W. A. Parker, & S. K. Houston (Eds.), *Mathematical modelling: A way of life. ICTMA 11* (pp. 3–17). Chichester: Horwood.
Galbraith, P. (2007). Dreaming a possible dream: More windmills to conquer. In C. R. Haines, P. Galbraith, W. Blum, & S. Khan (Eds.), *Mathematical modelling (ICTMA 12): Education, engineering and economics* (pp. 44–62). Chichester: Horwood.
Henn, H. W. (2001). Mobile classroom – A school project focussing on modelling. In J. F. Matos, W. Blum, S. K. Houston, & S. P. Carreira (Eds.), *Modelling and mathematics education* (pp. 151–160). Chichester: Horwood.
Henn, H. W. (2003). Working and learning in the real world. Early experiences from a mathematics education project in Baden-Wuerttemberg. In S. J. Lamon, W. A. Parker, & S. K. Houston (Eds.), *Mathematical modelling: A way of life. ICTMA 11* (pp. 71–79). Chichester: Horwood.
Herget, W., & Scholz, D. (1998). *Die etwas andere Aufgabe – aus der Zeitung*. Seelze: Kallmeyer.
Martinez-Cruz, A. M., & Ratcliff, M. I. (1998). Beyond modeling world records with a graphing calculator: Assessing the appropriateness of models. *Mathematics and Computer Education, 32*(2), 143–153.
Sigma Mathematik 11. Klasse. (1984). Stuttgart: Klettverlag
Winter, H. (2004). Mathematikunterricht und Allgemeinbildung. In H. W. Henn & K. Maaß (Eds.), *ISTRON Materialien für einen realitätsbezogenen Mathematikunterricht, Band 8* (pp. 6–15). Hildesheim: Franzbecker.

Chapter 42
Assessing Modelling Competencies Using a Multidimensional IRT Approach

Luzia Zöttl, Stefan Ufer, and Kristina Reiss

Abstract We assessed students' modelling competency using a test consisting of different classes of items. Within the first class there are items which cover the whole modelling process, whereas items of the second class focus only on certain parts of this process. To cope with the requirements of the two different classes of items we used a multidimensional Rasch model including subdimensions. In this chapter we describe the structure of the test instrument and compare the subdimensional scaling of the test results with a unidimensional one. The analyses show the superiority of the subdimensional scaling.

1 Introduction

An important field of research within mathematics education is dedicated to the promotion of students' modelling competency. To obtain reliable information, for example, about the effectiveness of new teaching or learning approaches in this field we need appropriate instruments and methods for the assessment of modelling competency. These approaches should take into account the complex structure of modelling competency as this could provide more information about students' problems with modelling tasks and thus propose ways to cope with those difficulties. Furthermore, the scaling of the data obtained from conducting the tests is another challenging task. In this chapter we will use a probabilistic approach, that is item response theory (IRT) and more precisely Rasch modelling (Rasch 1960), instead of classical statistical methods. The advantage of Rasch modelling is, in particular, the fact that despite different test items with different data collection periods, students' progress is estimated and reported on a common scale regardless of fluctuation in test difficulty (Izard 2007). Hence this chapter focuses on an issue

---

L. Zöttl (✉), S. Ufer, and K. Reiss
Institute of Mathematics, Ludwig-Maximilian Universität, Theresienstr. 39,
D-80333 Munich, Germany
e-mail: Luzia.Zoettl@gmx.de

which in this field of research is considered as an important question posed by Blum (2002, p. 276): "What alternative assessment modes are available to teachers, institutions and educational systems that can capture the essential components of modelling competency, and what are obstacles to their implementation?"

2 Modelling Competency

The definition of modelling competency we refer to is the following: "Modelling competencies include skills and abilities to perform modelling processes appropriately and goal-oriented as well as the willingness to put these into action" (Maaß 2006, p. 117). A modelling process is described as a sequence of seven phases which can be summarised in an idealised modelling cycle (see, Blum and Leiss 2006) (Fig. 42.1).

Since this is a rather general definition which does not provide concrete information of how modelling tasks might look like, a further specification is necessary. Therefore we refer to the modelling perspective developed by Blum (1996) as a theoretical basis for our work (see also Zöttl et al. 2010). In this point of view, the most important criterion for a modelling task is not a level of authenticity and complexity that is as high as possible, but its relevance for the students. Thus, modelling tasks assessing modelling competency refer to problems that might be reduced with respect to their complexity and authenticity compared to real problems. Nevertheless adequate modelling tasks should always require the performance of a complete modelling process.

Based on the description of the modelling process subcompetencies can be deduced from the different phases of the modelling process. According to Blum and Kaiser (1997) these subcompetencies encompass understanding of the real problem, setting up a model based on reality, excerpting a mathematical model from the real model, answering mathematical questions within this mathematical model, interpreting mathematical results in a real situation and validating the solution. However, these subcompetencies are necessary but not sufficient to characterise

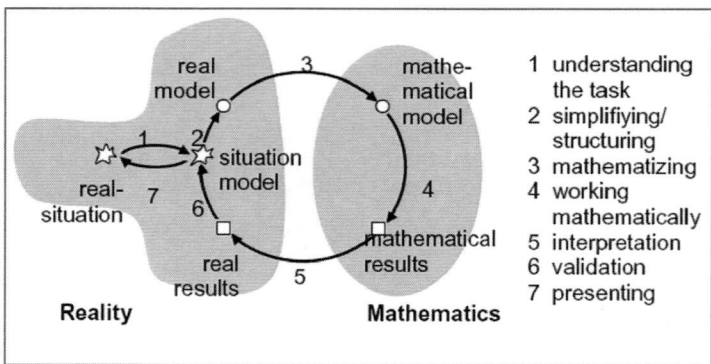

Fig. 42.1 Modelling cycle from Blum and Leiss 2006

modelling competency (Treilibs et al. 1980). Further aspects as for example the coordination of these subcompetencies are relevant as well.

A theoretical framework describing relevant aspects of modelling competency which takes into account the different components as well as their appropriate coordination was developed by Niss and Jensen (Jensen 2007, pp. 143–145). According to their model, the level of someone's modelling competency is determined by three aspects: (1) degree of coverage, (2) radius of action, and (3) technical level.

The *degree of coverage* describes the extent to which a person is able to activate the subcompetencies of modelling competency and to autonomously coordinate them. Accordingly, a person who is able to show all subcompetencies, but only on request, is more competent than someone failing at certain process steps. However, he or she is still less competent than someone performing all process steps without being prompted to do so.

The *radius of action*, in contrast, describes the range of situations in which a person is able to activate his or her modelling competency. Those situations may include different extra-mathematical contexts as well as different mathematical content fields. Thus, a person being able to activate his or her modelling competency within a very wide range of contexts concerning a lot of different mathematical content fields is certainly categorised as more competent compared to someone whose competency is related to only one specific mathematical content field and only certain contexts.

The *technical level*, finally, describes "how conceptually and technically advanced the mathematics is that someone can integrate relevantly in activating the competency" (Jensen 2007, p. 144). Accordingly, a person using only very simple or basic mathematical tools to solve a problem is, hence, less competent than someone using very advanced mathematics. This applies to modelling tasks whose real situation asks for using a more complicated mathematical model.

This theoretical framework which we chose as a theoretical basis for the development of our test instrument induces several consequences with respect to the assessment of modelling competency. These will be described within the next section.

3 Test Instrument

To assess students' modelling competency in a systematic way, that is considering all aspects described in the framework of Niss and Jensen (see above), it is important to provide a wide range of different test items. With respect to the aspect *technical level*, an appropriate test instrument should include tasks allowing to distinguish between different competency levels and thus, asking for the use of mathematical tools at different levels. The aspect *radius of action* however, asks for a variation with respect to the extra-mathematical context of the test items. In a broader sense, this aspect would necessitate also including modelling tasks of various mathematical topics. Nonetheless, a test instrument narrowed to only one specific mathematical topic, as for example circumference and area of certain geometrical figures (circle, triangle, rectangle, and square), can be of specific interest. For example, with the assessment of progress induced by an intervention concerning a specific mathematical content, a modelling test narrowed to that topic might be appropriate, since very far transfer of learning across different mathematical topics is not expected. Another more pragmatic advantage of a narrowed test instrument results from the fact that including very different mathematical topics or content fields in one test might cause difficulties with the scaling of the collected data, since the PISA results showed that the mathematical competency of a person can differ considerably between different content fields. However, one has to be aware that the conclusions drawn from a test narrowed to a specific mathematical topic refer only to modelling competency in this specific topic and thus, consider only a specific level of radius of action.

The third and most important aspect for our work that must be taken into account is the degree of coverage. It implies that a test instrument within the field of modelling should comprise different classes of items, that is items which cover the whole modelling process as well as items focusing on only parts of this process. By use of the latter class of items one can check if the relevant subcompetencies are available – at least on request, whereas the first class of items will assess if the tested person has got an higher degree of coverage and thus is able to perform the whole modelling process on his or her own without being prompted to consider all relevant phases.

Although it has been done for example by Haines and Crouch (2001), nevertheless it refers to the construction of test items which separately assess the subcompetencies deduced from every single process step of the modelling cycle. So, we defined three subprocesses to be assessed (see Fig. 42.2). To get more reliable information about the students' competencies required for the subprocesses, several different tasks had to serve as a scale of items considering different extra-mathematical contexts and requiring different mathematical tools. On the one hand, this guarantees to measure modelling competency also of low-achieving students who are not able to autonomously perform a complete modelling task. On the other hand, these test results might also provide detailed information about students' specific strengths and weaknesses concerning the subprocesses.

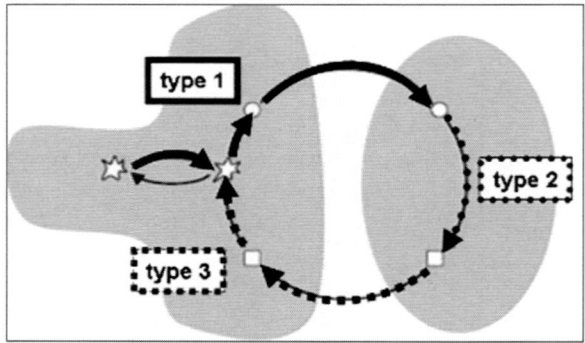

Fig. 42.2 Item types concerning different subprocesses of modelling activity

42 Assessing Modelling Competencies Using a Multidimensional IRT Approach

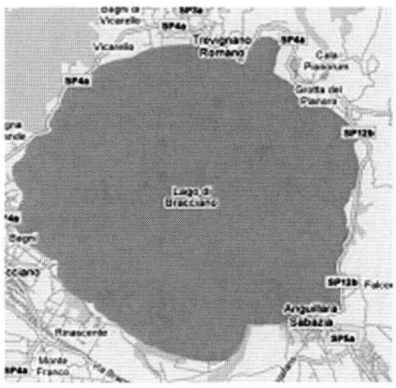

> By using a map, Anke would like to estimate the area covered by this Italian lake. Draw onto the map what Anke has to measure to compute the round area of this lake. **In addition**, write down the formula which is required to do the calculation! (You do not have to perform the calculation!)
>
> **Attention**: Use the same terms in the formula as in the map!

Fig. 42.3 Item "Italian Lake" (type 1)

> Select the correct statement!
>
> "If you triple the side length a of a square, the area of the new square is ...
> ☐ ... three times as big as the area of the primal square."
> ☐ ... nine times as big as the area of the primal square."
> ☐ ... twelve times as big as the area of the primal square."
> ☐ ... not comparable to the area of the primal square."

Fig. 42.4 Item "Variation of a square" (type 2)

We elaborated three different types of items (see Figs. 42.3 to 42.5) which focus on different subprocesses of modelling activity and thus belong to the second mentioned class of items. Items of the first type (type 1) require solely subcompetencies needed to build up a mathematical model (see Fig. 42.3 for an exemplary item). For items of the second type (type 2) intra-mathematical competencies are needed (see Fig. 42.4 for an exemplary item). Items of the third type (type 3) ask for the interpretation of a mathematical result and the validation of a presented problem solution with respect to

In front of the world's biggest Tree (General Sherman Tree) there is an information board shaped like a disk of the tree.

Text on the disk:
Circumference of the trunk (at the base): 31.3 m
Biggest diameter of the trunk (at the base): 11.1 m

Paul wonders: "Using the formula for the circumference of circles and given a circumference of 31.3 m you can compute with that the diameter is about 10.0 m. How is it possible that the biggest diameter of the trunk is 11.1 m.?"
Explain how this is possible!

Fig. 42.5 Item "General Sherman Tree" (type 3)

Fig. 42.6 Item "Spain" (type 4)

the underlying model (see Fig. 42.5 for an exemplary item). Besides those three different item types, we integrated also items which belong to the first class, that is short, but complete modelling tasks, as described above. Thus, we constructed a fourth item type (see Fig. 42.6 for an exemplary item).

Accordingly, a competency model results which focuses on the different components of modelling competency instead of different competency levels. Based on this competency model we developed a modelling test concerning the field of area and circumference of rectangles, triangles and circles, which contains items of all four item types and thus of both classes. The test consists of 36 items evenly distributed over all item types. Thus every subscale consists of nine items. However, we want to stress the fact that those 36 items were used within a multi-matrix design with different testing booklets each consisting of 12 items and thus containing only three items of every item type.

4 Data Scaling

To assess students' modelling competencies on the basis of this test instrument and thus, coping with its subdimensional structure and its different testing booklets we draw on item response theory. As working with partial credits is a potential complicating factor (Izard 2007, p. 160), we scored the students' problem solutions dichotomously. A further reason for doing so relates to the fact that with the subdimensional model, described below, a partial credit model has not been tested so far. On this account it was inevitable to accept this major simplification.

To scale the data it might be convenient to take the following assumption as a basis: All items, notwithstanding the different classes, assess one underlying competency, which is modelling competency (see Fig. 42.7). This would suggest scaling the data by use of a unidimensional Rasch model (Rost 2004, pp. 155 ff). However, this scaling does not consider the different types of items constructed to assess students' competencies within the three different subprocesses (types 1, 2, and 3) and their overall modelling competency concerning the different subcompetencies as well as their coordinate (type 4).

Accordingly, it might be more convenient to use a Rasch model respecting the theoretically evolved subdimensional competency structure the test instrument refers to. This kind of Rasch model would assume that all items measure modelling competency, which consists of the competencies required for the different subprocesses. These subcompetencies, thereby, are assessed by means of the first three

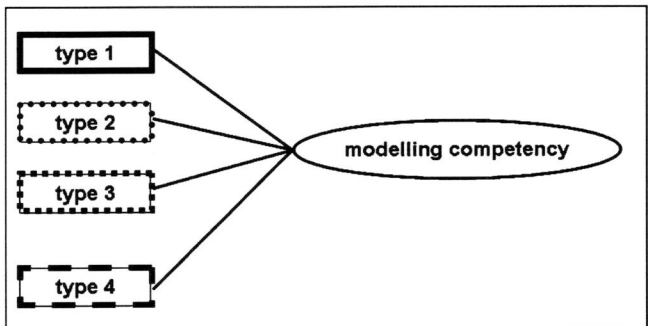

Fig. 42.7 Unidimensional scaling of the test results

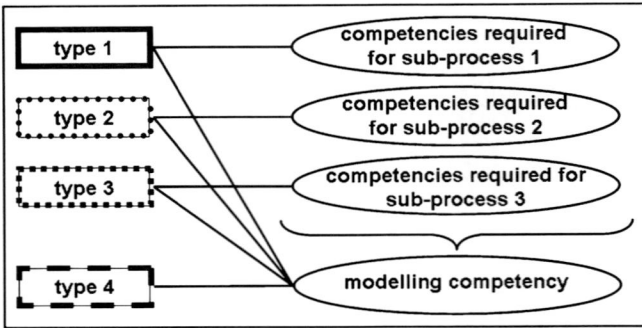

Fig. 42.8 Subdimensional scaling of the test results

item types (see Fig. 42.8). To cope with the requirements resulting from the different classes of items (i.e. items covering the whole modelling process and items focusing on only parts of this process), we used a multidimensional Rasch model including subdimensions (Brandt 2008). This model estimates not on only individual person parameters θ_v, indicating the overall modelling ability of a person, but also corrective parameters γ_{vd} that represent an individual's strengths and weaknesses in the implemented subdimensions, that is subprocesses. Summing up the global person parameter and the corrective parameter one gets a person parameter θ_{vd}, indicating the individual competency for the according subprocess. For more information we refer to the paper of Brandt (2008).

According to the different modes of scaling the data, an interesting matter for the evaluation of the test instrument is the following research question: *Can the subdimensional structure of modelling competency the test refers to be approved on an empirical base? That means: Is there any evidence on a psychometric level that scaling the data by use of a subdimensional Rasch model is more appropriate than scaling the data by use of a unidimensional Rasch model?*

5 Results of the Evaluation

The test instrument was evaluated within the research project KOMMA, which is supported by a grant from the German Federal Ministry of Education and Research (PLI3032). This project mainly aimed at the development and evaluation of a learning environment fostering students' modelling competency (Reiss et al. 2007). In this paper, however, we focus only on the data obtained from conducting the test instrument with 1,657 persons at three data collection periods (pre-, post-, and follow-up test). The testing booklets at each testing period which were linked crosswise via a multi-matrix design consisted of 12 items each. Thus, they comprised three items of every item type. The duration of testing was limited to 30 min per test session.

Table 42.1 Comparison of the uni- and subdimensional data scaling

	Unidim	Subdim	Subdim without item 2
AIC	71,192	66,613	64,687
BIC	71,402	66,910	64,978
CAIC	71,297	66,761	64,833
Reliability	0.57	0.65	0.66
Non-fitting items	7	1	0

To answer the research question we scaled the obtained data using the software ConQuest (Wu et al. 1997) running a unidimensional model as well as the subdimensional model described above. To assess the appropriateness of those two scalings we compared their information indices Akaike Information Criterion (AIC), Bayes Information Criterion (BIC), and Consistent AIC (CAIC). Generally, models with lower values are more likely to be a better means of data description than models with higher such indices, since the former models fit the data better than the latter ones (Bühner 2006, S. 352). These information indices take the number of estimated parameters and thus, the according degrees of freedom of the different models into account. This allows the comparison of the two models with their different numbers of dimensions.

As shown in Table 42.1, the subdimensional model is superior to the unidimensional model with respect to all three information indices. Also with respect to the reliability reported for the main dimension, that is the dimension measuring the overall modelling competency, the subdimensional model with a value of $rel = 0.65$ is better than the unidimensional model ($rel = 0.57$).[1] We analysed also the fit statistic of every single item. Only one of the 36 items did not fit the subdimensional model with respect to a reasonable fit value (meansquare MNSQ ≤ 1.3, Wright and Linacre 1994), whereas with the unidimensional model seven items did not fit the model. After eliminating the non-fitting item the fit values of the subdimensional model even changed for the better. That means, the information indices are smaller, whereas the reliability increases slightly, additionally there remain no items with problematic fit values (see Table 42.1).

6 Discussion

The reliability and the fit values reported with the subdimensional model indicate that the test instrument measures modelling competency the way it is supposed to. Furthermore, the analyses show high evidence that the subdimensional structure of

---

[1] According to Rost (2004, S. 381) the reliability is estimated as follows: $rel = \dfrac{\hat{\sigma}_\theta^2}{var\,\hat{\theta}}$, with $\hat{\sigma}_\theta^2$ as variance of the latent distribution and $var\,\hat{\theta}$ as variance of the estimated person parameters.

modelling competency is reflected also in the data. Thus, the conception of the modelling test and the scaling of the data considering its subdimensional structure show great promise for a reliable assessment of modelling competency. Nevertheless, this approach doesn't cope with all relevant problems either.

For example, although the subdimensional model is detected as more suitable compared to the unidimensional model, there is no global fit index providing information about the general suitability of those models. The reported indices (BIC, AIC, and CAIC) simply detect which of two models is more suitable, and they do not give any evidence whether, for example, neither of the models does fit the data well at all. However, the reliability as well as the item fit values reported with the subdimensional scaling indicate that this model fits well enough. Thus, the subdimensional model seems to be an appropriate method to cope with the requirements of a test instrument considering the specific structure of modelling competency.

A second problematic aspect which is not a specific problem of the subdimensional model either but a general problem concerns the compensatoric assumption underlying the estimation of the persons' modelling competency. A compensatoric approach is based on the assumption that very low competency within one subdimension can be countervailed by high competency in another subdimension. Obviously, this does not correspond to reality because being very competent, e.g. in calculation will not countervail a lack of competency in the third subprocess (interpretation and validation). As mentioned before, this inappropriate assumption is not a specific problem of the subdimensional scaling but it occurs always when a person's modelling competency is estimated by additively reckoning up subcompetencies. In any case it would be more appropriate to use a non-compensatoric, that is a multiplicative approach. However, until now statistical methods to scale the data based on a non-compensatoric approach are not implemented yet in the common software systems. Although the subdimensional model seems to be a very promising way to cope with the subdimensional structure of modelling competency, more research has to be done to optimize this new approach also with respect to statistical methods.

References

Blum, W. (1996). Anwendungsbezüge im Mathematikunterricht: Trends und Perspektiven. In G. Kadunz, H. Kautschitsch, G. Ossimitz, & E. Schneider (Eds.), *Trends und Perspektiven: Beiträge zum 7. Internationalen Symposium zur Didaktik der Mathe-matik* (pp. 15–38). Wien: Hölder-Pichler-Tempsky.

Blum, W. (2002). ICMI Study 14: Application and modelling in mathematics education – Discussion document. *Journal für Mathematikdidaktik, 23*(3/4), 262–280.

Blum, W., & Kaiser, G. (1997). Vergleichende empirische Untersuchungen zu mathematischen Anwendungsfähigkeiten von englischen und deutschen Lernenden (unpublished document). Cited according to K. Maaß (2006). What are modelling competencies? *Zentralblatt für Didaktik der Mathematik, 38*(2), 113–142.

Blum, W., & Leiss, D. (2006). Investigating quality mathematics teaching: The DISUM project. In C. Bergsten & B. Grevsholm, (Eds.). *Developing and researching quality in mathematics teaching and learning. Proceeding of MADIF-5. SMDF*, Linköping, 2007, pp. 3–16.

Brandt, S. (2008). Estimation of a Rasch model including subdimensions. In M. von Davier & D. Hastedt (Eds.), *Issues and methodologies in large-scale assessments, issues and methodologies in large-scale assessments* (pp. 53–71). Hamburg: IEA-ETS Research Institute.

Bühner, M. (2006). *Einführung in die Test- und Fragebogenkonstruktion* (2nd ed.). München: Pearson Studium.

Haines, C. R., & Crouch, R. M. (2001). Recognising constructs within mathematical modelling. *Teaching Mathematics and Its Applications, 20*(3), 129–138.

Izard, J. (2007). Assessing progress in mathematical modelling. In C. Haines, P. Galbraith, W. Blum, & S. Khan (Eds.), *Mathematical modelling (ICTMA 12): Education, engineering and economics* (pp. 158–167). Chichester: Horwood.

Jensen, T. H. (2007). Assessing mathematical modelling competencies. In C. Haines, P. Galbraith, W. Blum, & S. Khan (Eds.), *Mathematical modelling (ICTMA 12): Education, engineering and economics* (pp. 141–148). Chichester: Horwood.

Maaß, K. (2006). What are modelling competencies? *Zentralblatt für Didaktik der Mathematik, 38*(2), 113–142.

Rasch, G. (1960). *Probabilistic models for some intelligence and attainment tests.* Chicago: University of Chicago Press.

Reiss, K., Pekrun, R., Kuntze, S., Lindmeier, A., Nett, U., & Zöttl, L. (2007). KOMMA: Ein Projekt zur Entwicklung und Evaluation einer computergestützten Lernumgebung. *GDM-Mitteilungen, 83*, 16–17.

Rost, J. (2004). *Lehrbuch Testtheorie – Testkonstruktion* (2nd ed.). Bern: Huber.

Treilibs, V., Burkhardt, H., & Low, B. (1980). *Formulation processes in mathematical modelling.* Nottingham: Shell Centre for Mathematical Education.

Wright, B., & Linacre, M. (1994). Reasonable mean-square fit values. *Rasch Measurement Transactions Contents, 8*(3), 370. Retrieved Feb. 11, 2009, from http://www.rasch.org/rmt/rmt83b.htm.

Wu, M., Adams, R., & Wilson, M. (1997). *ConQuest: Generalised item response modelling software*, Draft Release 2. Australian Council for Educational Research, Camberwell.

Zöttl, L., Ufer, S., & Reiss, K. (2010). Modelling with heuristic worked-out examples in the KOMMA project. *Journal für Mathematikdidaktik, 31*(1), 143–165.

Part VI
Modelling in Tertiary Education

Chapter 43
Modelling in Tertiary Education – Overview

Peter Galbraith

Modelling in tertiary education has had a strong tradition within ICTMA conferences and their Proceedings from the beginning. Many ideas developed there have been adapted for teaching at both undergraduate and school levels or have otherwise influenced ways in which the teaching and learning of modelling has been carried out. Examples include group project work (Slater 1986), innovative modelling courses (e.g., Jing et al. 2003), modelling competitions (e.g., Shouting et al. 2003), and the effect of application-based mathematical instruction on achievement and understanding (e.g., Aroshas et al. 2007). Discussions on modelling competencies and their measurement (e.g., Izard et al. 2003) have influenced associated scientific research significantly. The six chapters in this section, representing contributions from seven national contexts continue this tradition, some building upon previous work, while others introduce new emphases. The major common theme among the chapters is a direct focus on modelling issues in undergraduate education, although some make reference to other levels as well.

Alpert reports on a project at a German university that sets out to capture the expertise that a mechanical engineer needs in his or her daily activity. The context is that of two students in the final semester of their course, working collaboratively on an authentic engineering problem in a workplace setting. In addition to the observation of their modelling and related mathematical activity, audio-taped interviews sought information on how the students went about using the resources (mental and physical) available to them for the purposes of addressing the problem and associated decision making. The experiment was carefully planned, thoughtfully conducted, and thoroughly reported. The author provides careful unpacking of essential technical aspects for the interested reader who is not a specialist in the area. The approach is described carefully, with key operational questions, and their implications for practical activity are identified and discussed. The chapter then provides reflective comment on the process and outcomes from a training and

P. Galbraith (✉)
School of Education, University of Queensland, Australia
e-mail: p.galbraith@uq.edu.au

educational perspective. In summary, this chapter provides an excellent example of the integration of demanding mathematical requirements with modelling expertise in an authentic setting, together with thoughtful reflective comment on the educational implications for engineering disciplines as well as more generally.

Deprez outlines a teaching sequence in which a model for the evolution of population in Belgium serves as a natural introduction to the concepts of eigenvalue and eigenvector through the use of Leslie matrices. A need for raising the level of authenticity of problems is given as a motivating force for such an approach, with the need traced back to an excessively 'pure' version of the New Mathematics initiative introduced in Belgium. This promoted a reaction in which so-called applications of mathematics were introduced, which were unrealistic and little more than dressed-up mathematical problems. Some modifications were made to the actual data from official sources, which nevertheless remained realistic and grounded in the Belgian context. Given that the purpose was to enable the concepts of eigenvalue and eigenvector to emerge, some might label the resulting problem as a type of *model eliciting activity* (see Chap. 26). Interesting philosophical questions emerge in a general sense.

For example: To what extent should we sacrifice (amend) reality to achieve a prior mathematical purpose? How different would the outcome otherwise be? What decision criteria are most important?

In fact the author uses simplifications (such as omitting migration) to focus precisely on the role and significance of omitted variables.

In addition to mathematics, some observation of students took place, and some useful student reactions were gathered by means of questionnaires. A negative relationship between attitude and difficulty level resonates with similar findings elsewhere. This suggests that caution is warranted when a small amount of unfamiliar modelling is introduced into an otherwise conventionally taught program.

Gruenwald, Narayanan, Klymchuk, and Zverkova, collaborating across three national contexts, describe a modelling initiative focused around the spread of severe acute respiratory syndrome (SARS) in Hong Kong in 2003. Predictions from three different models based on the data from the World Health Organisation (WHO) were discussed with undergraduates studying engineering or applied mathematics and a group of university staff who teach in the fields of mathematics or mathematical modelling. Basic data and assumptions were discussed, together with the respective model equations, and their predictions compared with the real SARS outcome after a period of 30 days. Questions were asked concerning reasons for differences between predictions and outcomes, what thinking led to the particular reasons given, how the predictions might have been improved, and whether the respondent was interested in learning more about epidemic modelling. Student responses were often speculative, with inferences that could not be inferred from the models – attempts to mix and match between mathematics and common sense could be identified, not all of which were self-consistent. Responses of the lecturers clearly reflected their greater experience and modelling expertise, in all of the question domains. That said, there was a closer match than might have been expected in the first and third questions, with both groups providing a strong majority of appropriate responses. Overall the students were more ambivalent about modelling than the lecturers who were the keener group – this should not occasion surprise.

Heiliø, from the perspective of industrial mathematics in Finland, presents arguments for educational priorities necessary to support future needs and developments in this area. While the main focus is on undergraduate teaching needs, he also offers comment with respect to school level and teacher preparation. He raises the difficult matter of achieving balance between motivational illustrations using real contexts and serious mathematical modelling. Consistent with the thoughts of others in the field, he argues for modelling skills to be developed over time through engaging with a succession of problems, and argues against attempts to teach them using traditional didactical approaches. Then, he lists seven domains of application, with suggestions for examples within each. The examples are confined to topic titles, and no specifics are attempted. As such we recognize material of a type that has found its way into many papers since the early years of ICTMA. What this indicates is that this message still needs emphasizing; for whatever progress has been made, clear areas of need remain. A second aspect that strikes the reader is that many of the suggested problem contexts would never have been suggested for this purpose, even a few years ago – new mathematical and technological developments mean new modelling opportunities. The chapter concludes with a suggestion for a course subject in modelling (containing examples) and raises questions about how much, and when, modelling might be included in school courses with associated questions for teacher education. The author's position is that modelling is unlikely to be effective without a sound base of mathematical knowledge, so such inclusions need to be planned carefully.

Matsuzaki, working in a Japanese context, uses response mapping to display the aspects of progress on a modelling task and illustrates his approach by recording the maps for a graduate school student and an electronics expert. The problem addressed the question 'How much brightness is needed to read a book?' and the data included written material, transcripts of subsequent interviews, and think-aloud protocols. A significant purpose of response maps is to capture and display events that occur 'on the run', as participants progress through a problem. This is a technical paper, in which real world experience and mathematical knowledge and background are connected together, to show their interaction and contribution to the final result. The diagrams are supported by selected excerpts from the commentary of both subjects, with respect to their approach to the aspects of the task. The author's focus was on distinguishing between and describing the contributions of prior knowledge associated with real life experiences and prior knowledge based on previous mathematical experiences.

Dan and Xie report on a study in a Chinese university, designed to investigate the relationships between students' mathematical modelling skills, their creative thinking skills, and their basic knowledge of mathematics. The scene was set, by describing Chinese initiatives during the past decade and identifying areas of education that need addressing to enhance certain abilities of future graduates. The investigation of modelling skills followed the already-mentioned scheme devised by Izard et al. (2003) and used results from an earlier study at the University of

Ulster as a basis for comparison. A high correlation was found between modelling skills and a measure of creative thinking skills obtained using a standard instrument. The tenuous link between basic mathematical knowledge and modelling skills confirms findings from other studies. While this may indicate that different abilities are required for mathematical modelling than suffice for success in basic mathematics, the authors question whether the mathematics tested is too simple to provide a fair test of the importance of mathematical background. The study suggests further questions that invite investigation in pursuing the important goals set out in the introduction.

Collectively the chapters provide an interesting cross section of perceived needs and types of activity that are currently active at the university level in a range of national contexts. Because of this diversity it is inappropriate to look for close linkages or common themes among the chapters, and readers are urged to refer to the respective introductory sections, which establish the purpose and intention of the separate contributions.

References

Aroshas, S., Verner, I., & Berman, A. (2007). Integration of applications in the technion calculus course. In C. Haines, P. Galbraith, W. Blum, & S. Khan (Eds.), *Mathematical modelling in education, engineering and economics: ICTMA12* (pp. 433–442). Chichester: Horwood.

Izard, J., Haines, C., Crouch, R., Houston, K., & Neill, N. (2003). Assessing the impact of teaching mathematical modelling. Some implication. In S. J. Lamon, W. A. Parker, & S. K. Houston (Eds.), *Mathematical modelling: A way of life. ICTMA11* (pp. 165–177). Chichester: Horwood.

Jing, Z., Jihong, J., Qi, D., & Shilu, F. (2003). The knowledge and implementation for the course of mathematical experiment. In Q.-X. Ye, W. Blum, S. K. Houston, & Q.-Y. Jiang (Eds.), *Mathematical modelling in education and culture: ICTMA10* (pp. 225–232). Chichester: Horwood.

Shouting, S., Tong, Z., & Wei, S. (2003). Mathematics contest in modelling: Problems from practice. In Q.-X. Ye, W. Blum, S. K. Houston, & Q.-Y. Jiang (Eds.), *Mathematical modelling in education and culture: ICTMA10* (pp. 93–98). Chichester: Horwood.

Slater, G. L. (1986). Group projects in mathematical modelling. In J. S. Berry, D. N. Burghes, I. D. Huntley, D. J. G. James, & A. O. Moscardini (Eds.), *Mathematical modelling methodology, models and micros* (pp. 90–97). Chichester: Horwood.

Chapter 44
The Mathematical Expertise of Mechanical Engineers: Taking and Processing Measurements

Burkhard Alpers

Abstract This chapter reports about a project that tries to capture the mathematical expertise a mechanical engineer needs in his or her daily work. We study how mechanical engineering students work on typical tasks in their final semester. The task considered in this article is concerned with measuring strain and stress in a critical component of a steering mechanism and processing the measurement data. One major qualification we identified was diligent work in a small algebraic model which has to be interpreted in application terms. Moreover, relating mathematical properties of the measurement curves to behaviour of the steering mechanism is important for making plausibility checks and for drawing conclusions.

1 Introduction

German Universities of Applied Science offer as a distinctive feature a very practice-oriented education. Correspondingly, the mathematical education of engineers should enable students to use mathematical methods for solving practical problems. In order to provide such an education, it is necessary to capture the mathematical expertise a mechanical engineer needs in his or her daily practice. Although mathematics at the workplace has been a topic of research for some time (cf. Bessot and Ridgway 2000), there are just a few studies dealing with presumably 'heavy' users of mathematics such as engineers where it is much harder for a non-professional to understand the work and role of mathematical thinking. Kent and Noss (2002) and Gainsburg (2006) investigated civil engineers and Cardella and Atman (2005) and Cardella (2010) observed mainly industrial engineering students doing their capstone projects. Using an ethnographic qualitative method of

B. Alpers (✉)
Department of Mechanical Engineering, Aalen University - HTW Aalen,
Beethovenstrasse 1, D-73430, Aalen
e-mail: Burkhard.Alpers@htw-aalen.de

research (for engineers and students, respectively), they discovered several aspects and patterns of mathematical thinking. The work described in this chapter is concerned with the mathematical expertise of mechanical engineers, and we study the mathematical skills final year students show when working on 'typical tasks' for a junior engineer. Because it is an in-depth case study performed with two students, we did not systematically investigate real workplaces.

In earlier works (Alpers 2006, 2008, 2010), we investigated tasks which dealt with the construction of a bearing for an ABS box in a car, with the design of a mechanism for a cutting device and with the dimensioning of machine elements in a simple gearing mechanism. This contribution describes the findings concerning a typical measurement task.

The next section describes the method used for the investigation and mechanical engineering task in more detail. Section 3, outlines the approach followed by the students working on the task. Section 4 contains our findings regarding the necessary mathematical qualifications and relates these to the results of the research work mentioned earlier. The final section draws some conclusions with respect to the mathematical education of mechanical engineers.

2 Method of Investigation and Task

Since it is extremely difficult for a mathematician to understand the work of an engineer by simply watching it over a short period of time, we identified a practical measurement task, described below in cooperation with a colleague who worked for several years as an engineer in the car industry. We then hired two students in their final semester to work on the task cooperatively for 100 h. So this was not an educational classroom experiment, but we wanted to study where in the work of the two students mathematics played an essential role. The students were asked to document their working and thinking processes; questions were to be clarified with the colleague who played the role of a group leader. Based on a first understanding obtained by reading their documents and additional background material, the two students were interviewed for further clarification. In these interviews the author particularly asked where the models they used came from (own modelling or choosing existing models known from lectures or literature) and whether decisions such as where to take measurements were based on quantitative or qualitative models. The interviews with the students were audio-taped for later examination, and the screen recording software we used enabled us to let the students point to their documentation during the interview. We then analysed the material for mathematical concepts, models, procedures and how it was used. From this, we identified necessary mathematical qualifications and also examined whether a more mathematical approach might have made work more efficient. The colleague involved and a laboratory engineer who assisted the students were also interviewed in order to check whether or not the students' work resembled real engineering work in industry.

44 The Mathematical Expertise of Mechanical Engineers

Fig. 44.1 Test bench for steering system

In our labs we have a test bench for a steering gear (Fig. 44.1) where a steering wheel can be rotated and via the servo mechanism of the gear the wheels are moved. The students should investigate the most vulnerable components, measure the occurring strain in these components using available measurement technology and a data processing program, and interpret the results.

The test bench already contains a facility for measuring the steering angle and steering moment. An amplifier and a laptop with measurement configuration and processing software were also installed. A student whose diploma thesis was concerned with setting up the test bench and a lab engineer who has a lot of expertise and experience in taking and processing measurements were also available for help.

3 Approach of Students

The major steps for tackling the task were quite clear to the students from their lecture on measurement theory and were performed by them subsequently:

- Analysis of the steering system and identification of critical components
- Identification of load cases for critical components
- Definition of a measurement configuration
- Implementation of measurement configuration and taking measurements
- Processing and interpretation of measurements

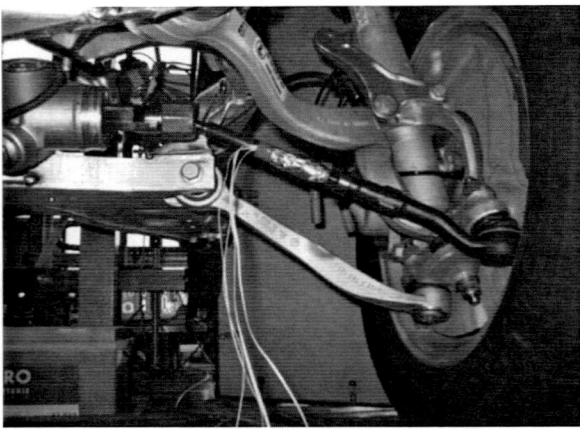

Fig. 44.2 Steering track rod with DMS applied

Using the available test bench one can only consider load cases that resemble the parking situation in which friction forces (between the road and the tyre) during the steering process cause strain within the components. The students identified the steering track rod, which is shown in Fig. 44.2, as the most critical component because of its small cross section, which is a rather coarse qualitative argument. According to the colleague involved, in reality information on critical components comes from analysing the flow of forces in the system, simulation software, or damage reports.

The students were insecure concerning the load case to be considered. There are three major load cases that can occur in combinations: tension/compression, bending, and torsion. Since the rod is connected to other components via ball joints, the only possible load case is tension/compression in the direction of the rod (only if there was considerable friction within the joint could there be bending and/or torsion, but this should not be the case). Nevertheless, the students spent some time thinking about the axis for bending since this would have been important for placing the measurement equipment correctly. According to the laboratory engineer, the biggest mistakes are often made by misjudging the load situation.

Once the load case had been clarified with the colleague involved, the students had to design a proper measurement configuration. An adequate and readily available means to measure strain are strain gauges (for short: DMS), which have to be applied to the track rod. A DMS changes its electrical resistance under strain, and with small elastic strains (i.e., when the load is gone the DMS will again have its original length) there is an approximate linear relation:

$$\frac{\Delta R}{R} = k \cdot \varepsilon \tag{44.1}$$

So the relative change of resistance is proportional to the strain ε with proportionality factor k (the value is approximately 2 but it depends on the material). The changes are normally very small. In order to get a good signal and eliminate other sources of strain than normal forces (e.g. change in temperature), very often a so-called full bridge configuration is used (Wheatstone bridge). There two DMSs are attached in longitudinal direction (marked with '+' in Fig. 44.3) and the other ones in cross direction (marked with '−' in Fig. 44.3). These DMSs are connected as shown in Fig. 44.3, which was drawn by one of the students, and an input voltage U_E is applied. The output voltage U_A is measured between the points shown in the figure (the direction is wrong). If all original resistances are equal (say R_0), then Kirchhoff's laws and a linearization (products of two small changes can be neglected) lead to:

$$U_A \approx \frac{U_E}{4} \cdot \left(\frac{\Delta R_1}{R_0} - \frac{\Delta R_2}{R_0} + \frac{\Delta R_3}{R_0} - \frac{\Delta R_4}{R_0} \right) \qquad (44.2)$$

Using formula (44.1) one gets a relationship between the ratio of voltages and the strains in the four DMSs applied:

$$\frac{U_A}{U_E} \approx \frac{k}{4} \cdot (\varepsilon_1 - \varepsilon_2 + \varepsilon_3 - \varepsilon_4) \qquad (44.3)$$

The strains occurring in the four DMSs might be caused by normal forces (tension/compression: ε_N), by bending ($\varepsilon_{b,x}$), or by temperature (ε_S). The sign of such a partial strain depends on the direction in which the DMS is attached. A strain caused by normal forces is positive when the DMS is attached in longitudinal direction. If it is attached in cross direction, then the DMS is shortened because of lateral contraction, and the magnitude of the strain is to be multiplied by the so-called Poisson ratio μ, so we get $-\mu\varepsilon_N$. Figure 44.4 shows the corresponding algebraic model setup by the students.

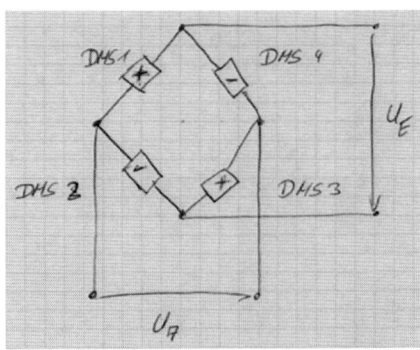

Fig. 44.3 Full bridge configuration

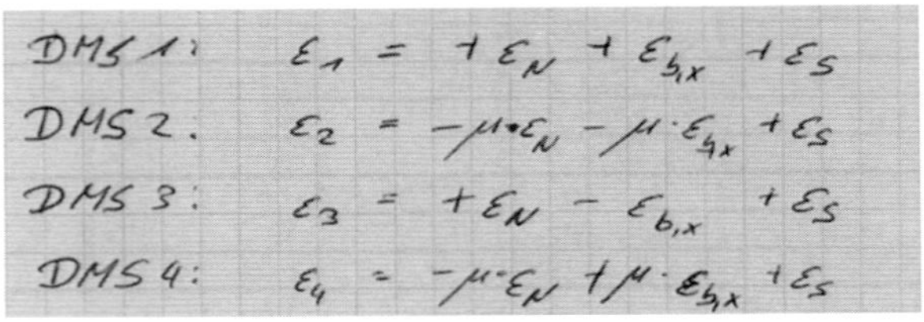

Fig. 44.4 Sign model for strain components

DMS 1: $\varepsilon_1 = +\varepsilon_N + \varepsilon_{b,x} + \varepsilon_S$

DMS 2: $\varepsilon_2 = -\mu \cdot \varepsilon_N - \mu \cdot \varepsilon_{b,x} + \varepsilon_S$

DMS 3: $\varepsilon_3 = +\varepsilon_N - \varepsilon_{b,x} + \varepsilon_S$

DMS 4: $\varepsilon_4 = -\mu \cdot \varepsilon_N + \mu \cdot \varepsilon_{b,x} + \varepsilon_S$

Inserting the right-hand side of the equations in Fig. 44.4 into formula (44.3) yields

$$\frac{U_A}{U_E} \approx \frac{k}{4} \cdot 2\varepsilon_N \cdot (1+\mu) \qquad (44.4)$$

which makes it possible to retrieve the strain caused by normal forces from the ratio between output and input voltages. The small algebraic model in Eq. 44.3 and Fig. 44.4 directly reflects the positioning and the interconnection of the DMS. If there is a mismatch between the model and the real ordering, it leads to totally wrong results. The model is quite standard and is also discussed in the regular lecture on measurement theory, so the students had to recall it and look it up rather than to set it up from scratch. Moreover, they had to transform it correctly into a measurement configuration. According to the experienced laboratory engineer, there are such models for standard situations, but for slightly different situations one has to know how the set up is done using positive and negative signs, in order to isolate the kind of strain one wants to measure.

The ordering is also important when the bridge is connected to the amplifier, which provides the input voltage and amplifies the output voltage. Moreover, the amplifier can be configured such that it directly yields the strain. The necessary data have to be input, and this includes the factor k, the Poisson ratio μ, the resistance of the DMS, and the kind of bridge used (here: full bridge). Although the user in the end gets the strain without any further computation or data processing done by himself or herself, the students had to know the underlying model in order to understand the data needed for amplifier configuration. Therefore, the configuration work of the students was also model-based (the model 'shines through').

The students used the software DIAdem® for configuring the amplifier and sampling the data. As a result they obtained a data table containing the time, the steering angle, and the strain. They copied the table to Excel®, computed stress and force from the strain by simply using proportionality factors, and produced diagrams for visualization of the data. One of the diagrams is shown in Fig. 44.5. It depicts the resulting strain in the rod when the steering wheel is moved to the left for about 550° and then to the right and back to 0° (the sign of the angle is opposite to what is usual in mathematics, i.e., counter-clockwise is negative here).

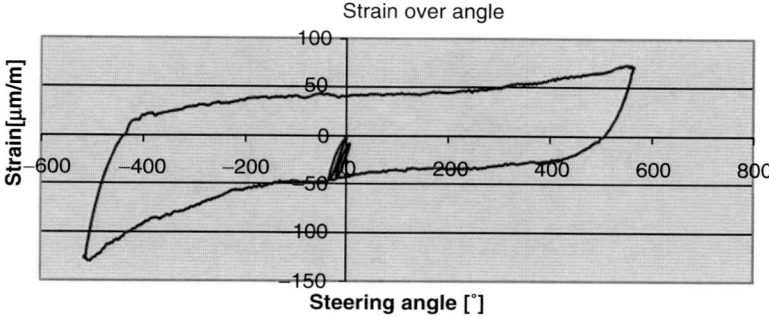

Fig. 44.5 Diagram for strain over steering angle

The students made several experiments with different additional weights (0, 200, 400 kg) and different steering supports (simulating a velocity of 0, 10, 100, and 200 km/h). Having the data, they finally considered two questions:

- Is there a critical load on the track rod?
- How does the load develop with additional weights?

As could be expected, the load is far below the critical load. More interestingly, the students set up the hypothesis that the load develops linearly with the additional weight put on the test bench, which will be discussed in more detail in the next section.

4 Findings and Discussion

4.1 Benefits and Problems of the Method

As in the previous phases of the project, the investigation method applied allowed again a detailed probe into the use of mathematical concepts and procedures during the work on a practical task, because the two students involved were available for in-depth questioning. Since the understanding of the task by the author developed gradually over time, it was important that the students and the colleague were available for clarifications for a longer period of time. Both students were familiar with the basic measurement concepts needed for the task, so regarding their qualifications they were roughly comparable to junior engineers.

There were some problems due to the availability of the laboratory engineer and too little contact between the students and the colleague. So the students were for some time occupied with the question of finding the bending axis of the track rod, and only after a while they contacted the colleague who told them that there should be no bending load case. There were also problems with fixing the DMS, and all these problems delayed the work such that just one component could be investigated. A second critical one, the mounting links of the gear, would have required a slightly more complicated model involving a two-dimensional load case (cf. Issler et al. 1997). In real life, work would have been more goal-oriented, and the data interpretation at the end would have been more intensive. These are typical restrictions of not having a real workplace environment.

4.2 Modelling Qualifications

The work of the students showed that comprehending the load situation is essential for choosing the right measurement configuration. The laboratory engineer emphasized that here the most and worst mistakes are made, which can make a measurement totally useless. For example, when you have a bending load case, the material is prolonged on one 'side' and shortened on the other 'side', and there is no strain 'in the middle' (on the so-called neutral axis), so when you fix the DMSs on the neutral axis you will measure nearly zero strain. In our case, we had just a tension/compression load case. To see this, one needs a rough qualitative mechanical model of forces and interconnections between components: Because the track rod is fixed via ball joints, bending could only occur when there is (unwanted) friction. If the situation is more complicated, one could use simulation results for getting an idea on the most critical place or make measurements at a larger number of places.

A second essential qualification for working on the task consisted of a thorough understanding of the algebraic model underlying the measurement configuration, which is captured in formulae (44.1–44.4) and in Fig. 44.4. The formulae are results of a linearization process. Although knowledge about this is not necessary for successful work with the model, it should at least be clear that in consequence the model is applicable only to small strains. For correctly setting up the equations in Fig. 44.4, the students needed to know the relationship between signs in the model and strain situations for the corresponding DMS. The physical interconnections had to be done exactly according to Fig. 44.3. One of the students made a quick check with the dismantled track rod recognizing that there was nearly no reaction on tension but a 'huge' reaction on bending, although this should be the other way round. Since he knew that for measuring the bending load case the interconnections are made differently (such that ε_N is finally eliminated and $\varepsilon_{b,x}$ remains), he immediately realized that there must have been an interconnection error. This shows that for finding the causes of erroneous configurations the underlying algebraic model is quite important. In the project situation, the check was easy to perform. When the component cannot be dismantled and checked that easily, the work must be done very diligently.

The algebraic model is also important for setting up new configurations when standard configurations do not suffice. Moreover, when setting up the amplifier configuration one has to provide some parameters (e.g., k). It is quite easy to enter wrong input, which can be detected more quickly and hence more efficiently when one understands the meaning of a parameter and its algebraic role in the formulae.

According to the colleague involved, measurement tasks often come from engineers belonging to computational departments who want their simulation results to be checked. In this case measurement and computational engineers need a common model to communicate. The measurement engineers need not know about special numerical procedures, and the computational engineers need not know about peculiarities of attaching DMSs to components, but they both use load cases und models of one-, two-, or three-dimensional strain and stress to feed simulation software or to describe and interpret simulation and measurement data. In our project, we just had to consider the one-dimensional model saying stress = strain*constant, whereas in the two- and three-dimensional situations the model is still linear (for small strains) but more complicated (cf. Issler et al. 1997), involving a so-called stiffness matrix. Both kinds of engineers have to know the model for meaningful communication.

4.3 Data Interpretation and Model Validation Qualifications

When the measurement data have been produced, they must be visualized for interpretation. In our case, we were interested in the development of the strain on the track rod when rotating the steering wheel. The available data table had columns on time, angle, and strain. From this, only angle and strain were interesting, and a curve was produced using Excel® simply by interconnecting the points given in the table. To check for plausibility (model validation) and to get information from the curve, the curve had to be 'read' and curve properties had to be interpreted in application terms. At first one rotates the steering wheel counterclockwise. Correspondingly, one starts at the origin and then traverses the curve clockwise until one reaches the peak in the lower left region. Then the wheel is rotated clockwise until one reaches the upper right peak. Finally, the wheel is rotated counter-clockwise again until one reaches the origin. One student knew this curve form from a lecture on steering systems since she was specialized in automotive engineering. Since the curve looked familiar, the students stopped the validation process here. A closer look at the curve reveals some peculiarities, though. When the steering wheel is rotated fully to the left, which corresponds to the lower left peak of the curve, and is then rotated clockwise again, the wheel is pulled. But still, for more than 50 further degrees, we have a negative strain. The same phenomenon can be observed on the right side of the curve. So there is no abrupt change from negative to positive strain when changing the direction of rotation. The students did not recognize this behaviour because they were content with the rough plausibility and similarity with known curves. A trial at the test bench together with the colleague involved showed that – in addition to the effects of mechanical clearance – the wheel is inclined and lifts the whole test bench slightly. When the direction of rotation is changed, the weight of the test bench still leads to a compression of the track rod for some time (i.e., for some degrees of rotation). So an important qualification here is to detect curve properties and to connect them to application behaviour.

After having taken measurements for different additional weights, the students wanted to check their hypothesis that the strain depends linearly on the weight. They had curves for 0, 200 and 400 kg and different steering gear support scenarios. The first question that comes up is what to compare when you have curves like the one given in Fig. 44.5 for three additional weights. The students chose as criterion the maximum strain occurring along the curve. Since this is important for whether or not the component fails, the criterion seems to be reasonable. One could also compare the relatively stable values in the interval from $-200°$ to $+200°$ (normal load). So the first interesting point here is that it is not obvious how curves for different situations should be compared. Once a decision has been made, the question arises as to how the relationship between weight and strain should be modelled according to the available data. The students called the relationship linear, but what they really meant was monotonous because on questioning in the interview their explanation was that with growing weight the maximum strain also increases. In general, an important qualification here is the knowledge of different mathematical fitting models and the ability to choose an adequate one based on the properties of the models and the data set under consideration.

4.4 Comparison with Other Research

The findings stated earlier are in compliance with the results of practitioners and educational researchers dealing with other engineering professions. Bissell and Dillon (2000) emphasize that control engineers often work within existing models. Besides the mathematical manipulation of these models, they consider the interpretation in application terms and the use of models for prediction and design as the most important activity. They also stress the interpretation of solutions in the language of the field of application. Gainsburg (2006) found all activities related to models (creation, selection, adaptation, application) in her studies. In our project the main activity consisted rather of re-calling and re-understanding the model from the lecture and then using it. Gainsburg also states the importance of 'understanding the phenomenon' and the role 'engineering judgment' plays in this process. In our situation the understanding of the load situation was most crucial at the beginning, and here, judgment grounded in experience is certainly vital. Otherwise, the procedure was pretty clear to the students. Like Bissel and Dillon, Gainsburg points out that an interpretation in application terms is decisive for the acceptance of results. Cardella (2010) looked for mathematical behaviour in categories set up by Schoenfeld. One category deals with mathematical practices including knowledge about when and where to use mathematical models. In our situation, the students knew from their measurement lectures that they needed to use a load model and a further model for the full bridge. Concerning the load model, they experienced the situation of 'uncertainty', which Cardella has often observed. Here, the uncertainty was due to the fact that the students did not see that there was a simple tension/compression load case, but in other measurement situations this might not be so obvious and then an engineer has to deal with such an uncertain situation. Finally, Kent and Noss (2002) emphasize the role of breakdown situations where mathematization is necessary to proceed. We observed a similar situation when the students tested their DMS configuration and realized that it showed bending but not tension/compression. Because of their model knowledge, they were able to localize the error quickly.

5   Conclusions for Education

The work within a small algebraic model of the measuring device was at the heart of the task, and an understanding of the model was essential to arrange the device in a proper way. As a consequence, work within small models (mostly existing, at least as to the way to model) should also be integrated into the mathematical education. Moreover, students should interconnect model properties and application properties and use models to design a configuration with certain properties. The author tries to realize this by setting up first-year mini-projects. In one of these projects, students have to design a truss given a certain load, compute the forces in the bars, build it, and measure the forces. They also design statically over- or under-determined structures resulting in linear systems of equations with no or infinitely many solutions. The projects interconnect mathematical topics and topics from mechanics and stress theory. Our investigation also showed that the detection and interpretation of curve properties in application terms are essential qualifications. These might be obtained in larger mathematical application projects where such curves appear quite naturally.

References

Alpers, B. (2006). Mathematical qualifications for using a CAD program. In S. Hibberd & L. Mustoe (Eds.), *Proceedings of the IMA Conference on Mathematical Education of Engineers, Loughborough*. London: Engineering Council.

Alpers, B. (2008). The mathematical expertise of mechanical engineers – The case of machine element dimensioning. In B. Alpers et al. (Eds.), *Mathematical Education of Engineers, Proceedings of the 14th SEFI (MWG)/IMA Conference*, Loughborough.

Alpers, B. (2010). The mathematical expertise of mechanical engineers – The case of mechanism design. In R. Lesh et al. (Eds.), *Modeling students' mathematical modeling competencies (Proceedings of the ICTMA 13)* (pp. 99–110). New York: Springer.

Bessot, A., & Ridgway, J. (Eds.). (2000). *Education for mathematics in the workplace*. Dordrecht: Kluwer.

Bissell, C., & Dillon, C. (2000). Telling tales: Models, stories, and meanings. *For the Learning of Mathematics, 20*, 3–11.

Cardella, M. (2010). Mathematical modeling in engineering design projects. In R. Lesh et al. (Eds.), *Modeling students' mathematical modeling competencies (Proceedings of the ICTMA 13)* (pp. 87–98). New York: Springer.

Cardella, M., & Atman, C. (2005). A qualitative study of the role of mathematics in engineering capstone projects: Initial insights. In W. Aung et al. (Eds.), *Innovations 2005: World innovations in engineering education and research* (pp. 347–362). Arlington: International Network for Engineering Education and Research.

Gainsburg, J. (2006). The mathematical modeling of structural engineers. *Mathematical Thinking and Learning, 8*, 3–36.

Issler, L., Ruoß, H., & Häfele, P. (1997). *Festigkeitslehre – Grundlagen* (2nd ed.). Berlin: Springer.

Kent, Ph., & Noss, R. (2002). The mathematical components of engineering expertise: The relationship between doing and understanding mathematics. In *Proceedings of the IEE 2nd annual symposium on engineering education*, London.

Chapter 45
Mathematical Modelling Skills and Creative Thinking Levels: An Experimental Study

Qi Dan and Jinxing Xie

Abstract In the last decade, extensive experimental studies were carried out to assess university students' modelling skills in various countries. However, such studies have not been done in China. This chapter tries to fill the gap by introducing the findings from a simple experimental study in a Chinese university. We evaluated 33 engineering students in a class and obtained the distributions of the students' mathematical modelling skills and their creative thinking levels. The data from the experiments show that there is a strong positive correlation between these two kinds of competencies. We also examined the relationship between the students' mathematical modelling skills and the scores they achieved in basic mathematical courses, and found that the correlation between them is insignificant, although some patterns of relationships do exit.

1 Introduction

In the last 30 years China's national economy has achieved much through its Reform-and-Open Policy. However, the reform in China's education system has not progressed rapidly. As a result, the education system is often, especially in recent years, attacked and criticized in China by pointing out that the students graduating from colleges and universities lack creative abilities and are not open to innovative practices and procedures.

For example, in 1999, the Ministry of Education of China and the China Youth League co-sponsored a survey of Chinese students' creative thinking abilities among 19,000 students in 31 provinces (Ban 2001). The survey revealed that only

---

Q. Dan (✉)
Department of Mathematics, Logistic Engineering College, Chongqing, China
e-mail: danqi31@163.com

J. Xie
Department of Mathematics, Tsinghua University, China
e-mail: jxie@math.tsinghua.eud.cn

4.7% of students considered themselves to have curiosity, confidence, perseverance, and imagination. Only 14.9% of students hoped to cultivate their exploring spirits for new things and to enhance their abilities of information collection and imagination. Only 33% of students participated in practical activities during their study life in schools. The proportions of the students with the initial creativity personality and creativity characteristics are as low as being 4.7% and 14.9%, respectively. In addition to this, if a student raised an objection to his/her teacher in the class, 48.1% of students thought that most students would keep silent, and 16.5% of students even thought that most students would criticize the objector.

The above figures from the survey clearly show that most students in China are not open to innovation. The reasons for that are surely complicated. As to our understanding, one of the most important reasons might be that the schools and teachers in China put their attention on teaching the students only about the knowledge and skills, but neglect cultivating the creative thinking ability of students. As a result of this kind of teaching style, the knowledge and skills of the subject concerned are the only focus for the students. The second reason lies in that the evaluation criteria in Chinese schools neglect the student's individuality and personality development. For a long time, we think a good student is the only one who gets very high grades in his class courses, and the students with lower grades but more creative ideas are not valued at all.

In more recent years, it has been widely recognized in China that in order for the country to develop in a sustainable manner, it is crucial to embrace innovation. Since the education system shoulders the special mission of cultivating a national spirit of innovation and fostering creative talents, reforming the teaching styles and the evaluation criteria for students has attracted increasing attention in China. The primary objective of the reform is to regard the cultivation of the innovation spirit and practical ability as the key of the education system.

As one component of the reform in China tertiary education, mathematical modelling courses and related activities are highlighted as the breakthrough of reforming mathematical education in Chinese universities (Jiang et al. 2007a, b; Xiao 2000). The reason behind this is that more and more mathematical teachers in China have recognized the importance and value of the mathematical modelling teaching process and related activities. The Chinese teachers now think that the key to the mathematical modelling teaching process is to create an environment that arouses students' desire to learn and develop their ability of self-study and to enhance their application and innovation ability. In order to improve the students' quality in mathematics, the emphasis is put on the students' ability of acquiring new knowledge and the processes of problem solving, rather than only on knowledge and skills in pure mathematics. Therefore, mathematical modelling is gradually becoming the best bonding point to enhance students' mathematical knowledge and application ability.

In this chapter, we are concerned with a primary question as follows: What is the current status of mathematical modelling skills of the students in Chinese universities? As we know, in the last decade, extensive experimental studies have been carried out to assess students' modelling skills in various countries (e.g., Houston and Neill 2003a, b; Izard et al. 2003; Lingefjärd 2004). Recently, Xu and Ludwig (2007, 2008) and Dan et al. (2007) also carried out experimental analyses on the mathematical modelling ability levels for high school students in China. However, from the authors' knowledge, this kind of experimental studies has never been done in Chinese universities. This chapter tries to fill the gap by introducing the results of a simple experimental study in a Chinese university.

Furthermore, we are more concerned with two relevant questions as follows: What is the relationship between the students' mathematical modelling skills and their creative thinking levels? What is the relationship between their mathematical modelling skills and their basic knowledge of mathematics? The answers to these questions will enhance our understanding about the relationships among the students' mathematical modelling skills, their creative thinking levels, and their achievements in basic mathematical courses, and thus provide evidence why now in China people think mathematical modelling is a vehicle to improve the students' innovation ability. This chapter presents some key findings concerning these questions from our experimental study.

2 Mathematical Modelling Skills

2.1 Test Questions

Multiple-choice questions have been widely used to test students' mathematical modelling skills in the last decade in various countries (Haines and Crouch 2001, 2005; Haines et al. 2001; Houston and Neill 2003a, b; Izard et al. 2003; Lingefjärd 2004; Lingefjärd and Holmquist 2005). It is also reported that the validity, reliability, and stability of the test are very good, thus the approach is adopted as the test instrument in our experiment. Specifically, we used all the 22 multiple-choice questions in our test (the details of the 22 questions may be found in Lingefjärd 2004 and accessed directly on the web from the given reference). The correct answer for each question gains two marks, and a partially correct answer gains one mark. Thus the maximum possible score is 44. Students scoring 29 or more are regarded to have 'strong' mathematical modelling ability; those scoring 21 or less are regarded to have 'poor' mathematical modelling ability; others are considered as 'medium' (Lingefjärd 2004).

2.2 Implementation

We chose to perform the test at Logistics Engineering College (LEC), one of the engineering universities in China with average-level students. All 33 students from Class 2005171, who entered into LEC in September 2005 with a major in engineering (automation), were tested with the previously mentioned questions (in a Chinese version we translated from the original English version) on March 15, 2007. At that time, the students had just completed their courses in higher mathematics (calculus), linear algebra, probability and statistics, and mathematical modelling. The test lasted 40 min, and during the test, the students answered the questions independently without any interruptions from the teacher.

Table 45.1 Basic test results for mathematical modelling skills

	Poor (21− marks)	Medium (21–29 marks)	Strong (29+ marks)
Number of students	4	14	15
Percentage	12.12	42.42	45.45

Table 45.2 The test results by question groups

Modelling skills	Questions	% Agreeing with experts' solutions		% Partially correct	
		LEC	Ulster	LEC	Ulster
Type 1 (simplifying assumptions)	1,2,3	53	47	22	25
Type 2 (clarifying the goal)	4,5,6	37	25	20	35
Type 3 (formulating the problem)	7,8,9	46	64	26	14
Type 4 (assigning variables, parameters, and constants)	10,11,12	52	78	24	12
Type 5 (formulating mathematical statements)	13,14,15	87	77	4	13
Type 6 (selecting a model)	16,17,18	47	28	14	28
Type 7 (graphical representations)	19,20	37	46	56	29
Type 8 (relating back to the real world)	21,22	63	42	22	28

2.3 Results

The basic results are summarized in Table 45.1. The students achieved an average score of 28, with a standard deviation of 6.03 and a range from 17 to 37. The median of the scores is 27, and lower- and upper-quartiles are 23 and 33, respectively. These figures are similar to the test results previously obtained at the University of Ulster, UK (Houston and Neill 2003b, p.160), showing that these students' mathematical modelling skills were satisfactory.

Table 45.2 compares our test results from LEC with those from Ulster by question groups (Houston and Neill 2003b). The students from LEC performed better than the students from Ulster in clarifying the goal, formulating mathematical statements, and relating back to the real world, while they performed worse in formulating the problem; assigning variables, parameters, and constants; and graphical representations.

3 Creative Thinking Levels

Creativity is the sum of a person's mental ability and personality quality in creative activities, and it is the displayed special ability in his or her creative activities (Li and Zhang 1999). When solving problems, a person with a strong creative ability always tends to use a unique way of connecting different concepts and knowledge and makes creative solutions. Students and teachers usually think that mathematical modelling is difficult and it is a creative activity. This section motivates to investigate how the students' creative thinking levels are affected by their mathematical modelling skills.

3.1 Test Questions

Torrance Tests of Creative Thinking (TTCT) is a widely used test instrument for testing one's creative thinking, focusing on one's abilities such as fluency, flexibility, originality, and elaboration (Curtis and Rick 2006; Li and Zhang 1999). Therefore, we adopted TTCT as the test instrument in this study. Specifically, the test questions were from a Chinese book written by Li and Zhang (1999). The test included 20 multiple-choice questions, and each one has a unique correct answer that contributes one mark (the other answers contribute nothing). The test time is 30 min. If one can finish the test in 5 min, he/she can obtain an additional five marks; for 10 min, three marks will be added; and for 20 min, two points will be added. Therefore, the maximum score was 25. Students scoring 14 or more are regarded to be 'strong' in creativity, and those scoring ten or less are regarded as 'poor'. Others are considered as 'medium' (Li and Zhang 1999).

3.2 Implementation

The students tested were from Class 2005171 at LEC, the same as those mentioned previously. The test was carried out for the 33 students on March 20, 2007, just five days later than the test given for their mathematical modelling skills.

3.3 Results

The results are summarized in Table 45.3. The average score of the students' creative thinking levels was 12.3, with a standard deviation of 2.1 and a range from 8 to 16. Overall, the students' creative thinking levels were 'medium'. The students with 'poor' creative thinking levels accounted for 12.12%, which is the same figure we have just observed for the students with 'poor' mathematical modelling skills in Sect. 2. However, the students with 'strong' creative thinking levels accounted for

Table 45.3 Test results for creative thinking levels

	Poor (10– marks)	Medium (10–14 marks)	Strong (14+ marks)
Number of students	4	22	7
Percentage	12.12	66.67	21.21

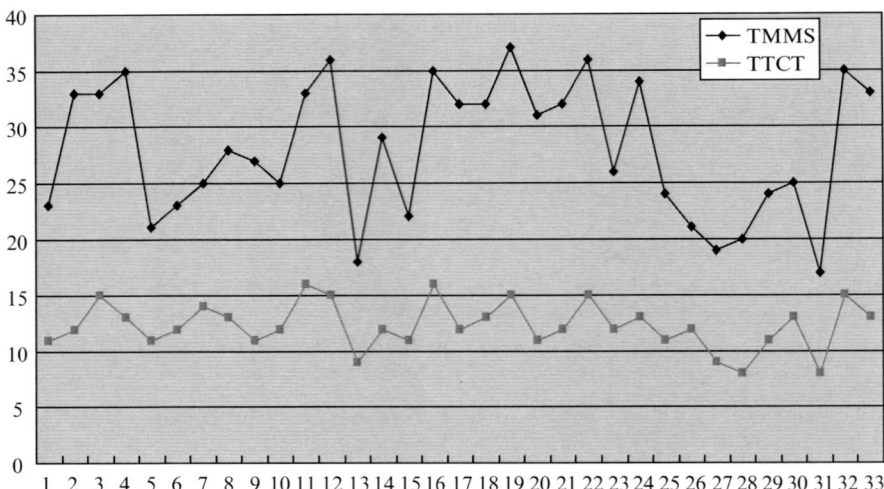

Fig. 45.1 TMMS vs. TTCT

only 21.21%, which is much less than what we have just observed for the students with 'strong' mathematical modelling skills in Sect. 2.

3.4 Relationship with Mathematical Modelling Skills

The detailed data from these two tests (i.e., test for mathematical modelling skills (TMMS) in Sect. 2 and TTCT in this section) are plotted in Fig. 45.1, for all the 33 students tested. The four students who scored less than 9 in TTCT were the same four students who scored less than 21 in TMMS. The seven students who scored over 14 in TTCT were among the fifteen students who scored over 29 in TMMS. This observation clearly indicates that students with 'strong' creative thinking levels were also 'strong' in mathematical modelling skills, and those with 'poor' creative thinking levels were also 'poor' in mathematical modelling skills. The correlation coefficient of mathematical modelling skills and creative thinking levels was 0.815, which indicated that there was a strong positive correlation between these two types of competencies. This observation validated the belief that 'mathematical modelling is necessary for creativity' (D'Ambrosio 1989).

In fact, a careful examination of the TTCT test questions suggests to us that some of the questions were designed to test similar ability as the TMMS questions did. Two example questions of TTCT are given as follows.

Question 12: In Fig. 45.2a, which image is another one's reflection seen in a mirror?

Question 14: For the table in Fig. 45.2b, can you fill a number in the place of "?" by using an arithmetic operation horizontally or vertically?

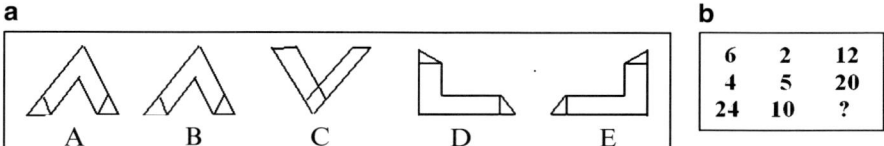

Fig. 45.2 Two example questions of TTCT

Table 45.4 Test results for basic mathematics courses

	Fail (60– marks)	Pass (60–90 marks)	Excellent (90+ marks)
Number of students	2	21	10
Percentage	6.1	63.6	30.3

Clearly, Question 12 tests the students' ability in understanding graphics and their symmetry, which has some links with the ability of graphical representation (Type 7 modelling skills in Table 45.2). Question 14 tests the students' ability in constructing a mathematical relationship between two variables, which has strong links with Type 5 modelling skills in Table 45.2 (formulating mathematical statements). Therefore, it is not surprising that a strong positive correlation exists between these two types of competencies.

4 Knowledge in Basic Mathematics

4.1 Score in Basic Mathematical Courses

The motivation of this section is to investigate how the students' knowledge in basic mathematics is affected by their mathematical modelling skills. The 33 students from Class 2005171 at LEC already had a basic mathematics course test (BMCT) at the end of their first year as freshman (i.e., in July 2006), just after their basic mathematics courses had been completed. The test questions came from a test database developed by Xi'an Jiaotong University (a famous engineering university in China) and published by the Higher Education Press of China. Specifically, the test includes 22 problems in the formats of 'fill-in-the-blank', 'multiple-choice', and 'calculation'. The mathematical content of the test included calculus and linear algebra, and the perfect score was 100. Students with more than 90 marks were regarded as 'excellent'. Those with less than 60 marks were regarded as having 'failed' and 'poor'. The results are summarized in Table 45.4. The average score of the test was 80, with a standard deviation of 16.7 and a range from 25 to 99. Overall, the students performed very well in BMCT.

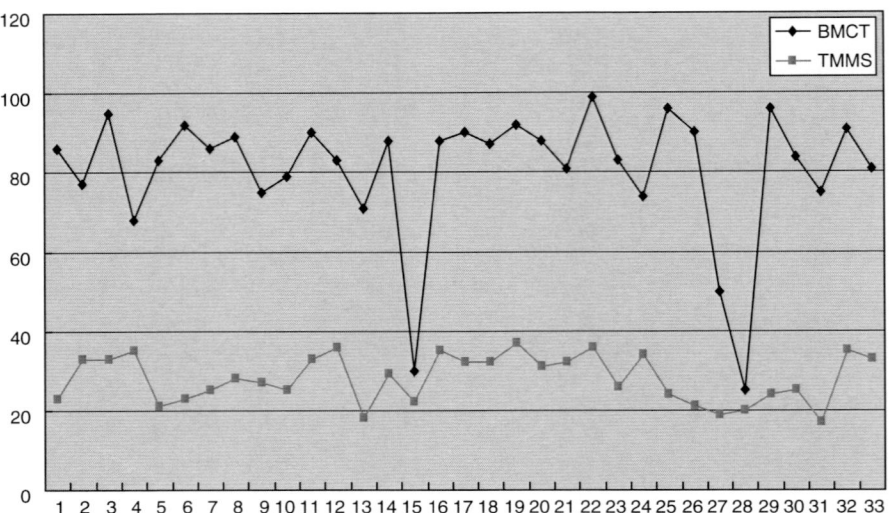

Fig. 45.3 TMMS vs. BMCT

4.2 Relationship with Mathematical Modelling Skills

The detailed data from the two tests (i.e., TMMS and BMCT) are plotted in Fig. 45.3, for all the 33 students tested. The correlation coefficient between the scores of TMMS and BMCT was 0.381, which was very small and thus indicated that the correlation between them was insignificant.

However, some patterns of relationships do exist between these two types of competencies. In more details, their relationship could be summarized as follows:

1. Students with very low scores in BMCT did not have 'strong' mathematical modelling skills (e.g., the students failed in BMCT scored less than 21 in TMMS). In other words, students with 'strong' mathematical modelling skills did not score very low in BMCT.
2. Students with very high scores in BMCT might not have 'strong' mathematical modelling skills. Among the ten excellent students in BMCT, only seven of them scored more than 29 marks in TMMS, and the other three scored only 21, 23, and 24, respectively, in TMMS.
3. Students with strong mathematical modelling skills might not score very high in BMCT. For example, three students who scored over 30 in TMMS scored only 68, 74, and 77, respectively, in BMCT.

For a long time, the teachers and students in China usually think that only those with very high scores in basic mathematics courses can learn and do mathematical modelling well. However, the earlier findings conflict with the traditional thoughts that exist among many teachers and students in China, but they are very consistent with their teaching experience. For example, in mathematical modelling activities such as China's undergraduates mathematical contests in modelling (Jiang et al. 2007b), the students who outperformed the others and were awarded the best achievements were usually not those with very high scores in their basic mathematics courses.

One of the reasons for these observations might be the nature of the TMMS test questions. The mathematical knowledge and skills needed for the test are basically very simple, and advanced mathematical knowledge and skills are not needed for the test. Therefore, mathematical knowledge and skills did not become an obstacle for most of the students when they completed the test. Thus the advantages of these students in mathematical knowledge cannot be reflected in the test.

Another possible reason for these observations might be that if a student gains a very high BMCT score it does not mean that the student really grasps the basic mathematics very well. In other words, the assumptions we make about students regarding their knowledge base and successful completion of earlier modules and/or examinations cannot be relied upon (Anderson et al. 1998; Haines and Crouch 2001).

5 Summary

In this chapter, we introduced some findings from an experimental study concerning mathematical modelling for some engineering students in an average-level university in China. The findings provide strong evidence to support the mathematical education reform in China with regards to mathematical modelling courses and related activities as a vehicle to improve the students' innovation ability. We have a plan to do more experimental studies in China to figure out whether the findings could be extended to other Chinese students as well, since the current study is based only on 33 students. We are also going to investigate how mathematical modelling courses should be taught and how mathematical-modelling-related activities influence the students' creative thinking modes.

References

Anderson, J., Austin, K., Bernard, T., & Jagger, J. (1998). Do third year mathematics undergraduates know what they are supposed to know? *International Journal of Mathematical Education in Science and Technology, 29*, 401–420.

Ban, C. (2001). *Theory and experimental study of mathematical modelling to raise creative thinking of high-school students*. Master's degree thesis, Tianjin Normal University, China (in Chinese).

Curtis, R. F., & Rick, D. R. (2006). Creative thinking and learning styles in undergraduates agriculture students. *Journal of Agricultural Education, 47*(4), 102–111.

D'Ambrosio, U. (1989). Historical and epistemological bases for modeling and implications for the curriculum. In W. Blum, M. Niss, & I. Huntley (Eds.), *Modeling applications and applied problem solving* (pp. 22–27). London: Eillis Horwood.

Dan, Q., Zhu, D., & Song, B. (2007). Impacting factors and training policy for mathematical modelling abilities for high school students. *Journal of Chinese Society of Education, 4*, 61–63 (in Chinese).

Haines, C., & Crouch, R. (2001). Recognizing constructs within mathematical modeling. *Teaching Mathematics and Its Applications, 20*(3), 129–138.

Haines, C., & Crouch, R. (2005). Applying mathematics: Making multiple-choice questions work. *Teaching Mathematics and Its Applications, 24*(2–3), 107–113.

Haines, C., Crouch, R., & Davis, J. (2001). Recognizing students' modeling skills. In J. F. Matos, W. Blum, S. K. Houston, & S. P. Carreira (Eds.), *Modelling and mathematics education – ICTMA 9: Application in science and technology* (pp. 366–380). Chichester: Horwood.

Houston, K., & Neill, N. (2003a). Investigating students' modeling skills. In Q. Ye, W. Blum, S. K. Houston, & Q. Jiang (Eds.), *Mathematical modelling in education and culture: ICTMA 10* (pp. 54–66). Chichester: Horwood.

Houston, K., & Neill, N. (2003b). Assessing modelling skills. In S. J. Lamon, W. A. Parker, & S. K. Houston (Eds.), *Mathematical modelling: A way of life – ICTMA 11* (pp. 155–164). Chichester: Horwood.

Izard, J., Haines, C., Crouch, R., Houston, K., & Neill, N. (2003). Assessing the impact of teachings mathematical modeling: Some implications. In S. J. Lamon, W. A. Parker, & S. K. Houston (Eds.), *Mathematical modelling: A way of life – ICTMA 11* (pp. 165–177). Chichester: Horwood.

Jiang, Q., Xie, J., & Ye, Q. (2007a). Mathematical modeling modules for calculus teaching. In C. Haines, P. Galbraith, W. Blum, & S. Khan (Eds.), *Mathematical modeling – ICTMA12: Education, engineering and economics* (pp. 443–450). Chichester: Horwood.

Jiang, Q., Xie, J., & Ye, Q. (2007b). An introduction to CUMCM. In C. Haines, P. Galbraith, W. Blum, & S. Khan (Eds.), *Mathematical modeling – ICTMA 12: Education, engineering and economics* (pp. 168–175). Chichester: Horwood.

Li, Z., & Zhang, Z. (1999). *Development and cultivation of creativity*. Beijing: Science and Technology Literature Publishing Press (In Chinese).

Lingefjärd, T. (2004). *Assessing engineering student's modeling skills.* http://www.cdio.org/paper/assess_model_skls.pdf. Accessed on 10 Sept 2009.

Lingefjärd, T., & Holmquist, M. (2005). To assess students' attitudes, skill and competencies in mathematical modeling. *Teaching Mathematics and Its Applications, 24*(2–3), 123–133.

Xiao, S. (2000). *Research report on reforms of higher mathematics (for non-mathematical specialties)*. Beijing: Higher Education Press (In Chinese).

Xu, B., & Ludwig, M. (2007). Empirical analysis of mathematical modelling ability levels for middle school students. *Middle School Mathematics Monthly, 11*, 1–2, 30 (In Chinese).

Xu, B., & Ludwig, M. (2008). Comparison of mathematical modelling ability levels between China and Germany students. *Shanghai Research on Education, 8*, 66–69 (In Chinese).

Chapter 46
Modelling the Evolution of the Belgian Population Using Matrices, Eigenvalues and Eigenvectors

Johan Deprez

Abstract We outline a teaching sequence in which a model for the evolution of a population serves as a natural introduction to the mathematical concepts of eigenvalue and eigenvector. Long-term projections using a matrix model are analyzed in terms of a 'long-term growth factor' and a 'long-term age structure'. From a mathematical point of view, these observations can be described by the concepts of eigenvalue and eigenvector. Experiences with students in tertiary education (applied economics, mathematics teacher education) are discussed.

1 Introduction

Matrices are studied in upper secondary and higher education and eigenvalues and eigenvectors in higher education. These topics are interesting from a purely mathematical perspective, but they have numerous interesting and important applications as well. In this chapter we describe a teaching sequence discussing one of these applications: modelling the evolution of a population using a Leslie matrix and using its eigenvalues and eigenvectors to study the long-term evolution of this population. In fact, the teaching sequence uses the application to introduce the mathematical concepts of eigenvalue and eigenvector. Furthermore, in this chapter we discuss experiences with the teaching sequence in three different contexts.

A few decades ago, mathematics education in Flanders (Belgium) was extremely influenced by New Math. Changes in the official curriculum since then have stressed the importance of applications more and more and some applications of mathematics are found in school books and in most classrooms nowadays. However, many of them are just 'dressed up' mathematical problems and not 'really real' situations. One of the reasons to construct the teaching sequence presented in this chapter is to raise the level of authenticity of problems used in the classroom.

J. Deprez (✉)
Hogeschool Universiteit Brussel Katholieke Universiteit Leuven and Universiteit Antwerpen,
Antwerp, Brussels and Leuven, Belgium
e-mail: johan.deprez@ua.ac.be

There are a number of other studies in which population dynamics serves as a context in mathematics education. Niss (2000) discusses in detail the model used by the Danish Bureau of Statistics for the evolution of the Danish population. His model is based on recursive relations between different elements in the system. It aims at good quality projections and, hence, assumes non-constant fertility and death rates. A more simplified model is used by Bradley and Meeks (1986) to let students evaluate implications of the Chinese one-child policy. They assume constant birth and death rates and discard immigration, admitting them to describe the evolution of the Chinese population by a Leslie matrix model. Leslie models were originally introduced for the study of biological populations by P.H. Leslie (1945). Our model for the evolution of the Belgian population is of the same type as the Bradley and Meeks model. However, our teaching sequence differs from theirs in two aspects. Firstly, we start from realistic data of the Belgian Bureau of Statistics and make the simplifications explicit when setting up the matrix model. Secondly, an important part of the teaching sequence is devoted to an analysis of the long-term evolution of the population, which can nicely be described by the dominant eigenvalue and one of its eigenvectors. Our model differs from that of Niss in the sense that, although realistic data are used, the aim is not to give realistic predictions of the population. In fact, we use a simpler and more mathematically structured model in order to be able to provide a motivating context, naturally leading to the mathematical concepts of eigenvalue and eigenvector. Too often, these concepts are introduced rather artificially in tertiary education.

In the terminology of Blum et al. (2002), the teaching sequence focuses more on applications than on modelling, for example, because an existing mathematical model is used. However, certain aspects of the modelling process are present as well, for example when we show to the students the simplifications while setting up the matrix model.

The teaching sequence was used in three different contexts. We briefly report on the first two contexts, from which we do not have data, except for our own observations. In the third context, the main concerns were whether the teaching sequence was not too difficult and/or confusing for the students, whether it made students appreciate the usefulness of mathematical concepts and whether it motivated students. We report on the results of a questionnaire that was administered to 20 randomly chosen students.

2 The Teaching Sequence

2.1 Calculations with Authentic Data

To make students aware of the simplifications that are made when constructing the matrix model in Sect. 2.2, we inserted a phase during which students work with detailed realistic data to answer a number of questions. The data were obtained from the Belgian National Statistical Institute (http://www.statbel.fgov.be). They include population by age and sex on 1 January 2003, life tables for men and women of 2000, 2001, and 2002, and fertility rates for women by age of 1997

Bevolking op 1.1.2003

Bevolking naar burgerlijke staat, geslacht en leeftijd - per provincie

België

		Mannen					Vrouwen				
	Totaal	Totaal	Onge-huwd	Gehuwd	Weduw-staat	Geschei-den	Totaal	Onge-huwd	Gehuwd	Weduw-staat	Geschei-den
Totaal	10.355.844	5.066.885	2.284.072	2.330.799	135.547	316.467	5.288.959	1.990.579	2.336.069	589.901	372.410
Minder dan 1 jaar	111.321	57.076	57.076	0	0	0	54.245	54.245	0	0	0
1 jaar	114.265	58.256	58.256	0	0	0	56.009	56.009	0	0	0
2 jaar	116.079	59.334	59.334	0	0	0	56.745	56.745	0	0	0
3 jaar	114.967	58.558	58.558	0	0	0	56.409	56.409	0	0	0

Fig. 46.1 Population by age and sex

Sterftetafels 2000 - 2002

België - Vrouwen

Leef-tijd (x)	Sterfte-kans (Qx)	Over-levings-kans (Px)	Aantal overleven-den op 1.000.000 geboorten (Lx)	Aantal sterf-gevallen van de ene leeftijd tot de volgende (Dx)	Ver-wachte levens-duur (Ex)
0	0,003375	0,996625	1.000.000	3.375	81,59
1	0,000899	0,999101	996.625	896	80,87
2	0,000333	0,999667	995.728	332	79,94
3	0,000225	0,999775	995.397	224	78,97
4	0,000170	0,999830	995.173	169	77,98

Fig. 46.2 Life table

LEEFTIJD VAN DE VROUWEN	WERKELIJKE VRUCHTBAARHEIDSCIJFERS			
	België			Brussels
	1995	1996	1997	1995
15 jaar	0,0008	0,0008	0,0008	0,0013
16 jaar	0,0027	0,0029	0,0028	0,0050
17 jaar	0,0065	0,0067	0,0070	0,0141
18 jaar	0,0148	0,0150	0,0142	0,0253
19 jaar	0,0244	0,0250	0,0250	0,0386
20 jaar	0,0341	0,0361	0,0359	0,0544

Fig. 46.3 Fertility rates

(the most recent version available!). Figures 46.1–46.3 give an impression of these sets of data (in Dutch).

We chose not to incorporate migration because it makes the matrix model considerably more complicated. Although (or, may be, because) migration is not taken into account, the model will shed light on the role of migration in the evolution of the Belgian population. Using these data, students have to answer questions like: How many men of age 35 were there on 1 January 2003? How many women of age 35 would you predict for 1 January 2010? How many births were there in 2003? How many of those were boys and how many were girls? How many boys of age 3 would you predict for 1 January 2010? While answering these questions, we want students to realize that they have to make assumptions (for example: survival rates and fertility rates are constant) that are subject to criticism. Moreover, calculations tend to become messy and complicated. Students are in the right mood then to be introduced (by the teacher) to a matrix model based on simplified data.

2.2 The Matrix Model

To keep the dimensions of the matrix model within the reach of a graphical calculator, the data are simplified first (by the teacher): the width of the age groups is set to 20 years, the distinction between sexes is neglected, and fertility rates and survival rates for these new age groups are calculated (which is not entirely evident) and rounded to two decimals. Two of these rates have been slightly altered in order to obtain a 'nice' eigenvalue in Sect. 2.3. Table 46.1 shows the simplified data.

Students have to answer a number of questions concerning these data. For example, they have to explain the high fertility rate of the first age group (which is due to the fact that fertility rates and survival rates now refer to periods of 20 years: girls in the first age class have 20 years' time to have their child). Based on these simplified data and assumptions similar to those in the previous section (constant fertility and survival rates, no migration) students make projections for the population in the future, in steps of 20 years, first without mathematical tools. Then the teacher introduces a Leslie matrix model. Let

$$X(0) = \begin{bmatrix} 2\,407\,368 \\ 2\,842\,947 \\ 2\,853\,329 \\ 1\,840\,102 \\ 410\,944 \end{bmatrix} \quad \text{and} \quad L = \begin{array}{c} \text{from} \\ \begin{array}{ccccc} \text{I} & \text{II} & \text{III} & \text{IV} & \text{V} \end{array} \\ \begin{bmatrix} 0.43 & 0.34 & 0.01 & 0 & 0 \\ 0.98 & 0 & 0 & 0 & 0 \\ 0 & 0.96 & 0 & 0 & 0 \\ 0 & 0 & 0.83 & 0 & 0 \\ 0 & 0 & 0 & 0.30 & 0 \end{bmatrix} \begin{array}{c} \text{I} \\ \text{II} \\ \text{III} \\ \text{IV} \\ \text{V} \end{array} \end{array} \quad \text{to}$$

Table 46.1 Simplified data

Age	Label	1 January 2003	Fertility rate	Survival rate
0–19	I	2,407,368	0.43	0.98
20–39	II	2,842,947	0.34	0.96
40–59	III	2,853,329	0.01	0.83
60–79	IV	1,840,102	0	0.30
80–99	V	410,944	0	0
Total		10,354,690		

be, respectively, the initial population and the transition matrix containing the fertility rates (in the first row) and survival rates (below the main diagonal). Using these matrices, a projection for the population in the year $2003 + n \cdot 20$ can recursively be calculated using the relation $X(n) = L \cdot X(n-1)$. An explicit formula is $X(n) = L^n \cdot X(0)$.

Here, it is discussed with the students that the resulting projections do not give realistic predictions, the most important reason being that we discarded migration. However, the projections do make sense. Seemingly contradictory, showing what would happen *without* migration (and if moreover fertility and survival rates would remain constant), they actually draw attention to the role of migration. Stated in other words, the model allows us to simulate a 'reality' that will never exist, but which helps us to understand the 'real' reality. The projections act like a looking glass zooming in on the characteristics of today's population. Doing the same exercise for the Belgian population in, say, the 1950s or for a country in a different part of the world would yield a totally different picture.

2.3 Two Observations Concerning the Long Term Evolution of the Population

Figure 46.4, which is given to the students, shows projections for the five age groups over a period of 240 years.

Firstly, students explore the graph while finding out where they themselves are represented in it (from period to period, they jump from one line to another!). Then the graph is analysed more globally. It shows different behaviours for the (relatively) short term, say the first two or three steps: the numbers in some age groups increase whereas the numbers in other ones decrease. The passage of the baby boom generation and the ageing of the population can be seen very clearly. For the long-term behaviour, however, all the graphs show a common pattern: numbers decrease slower and slower.

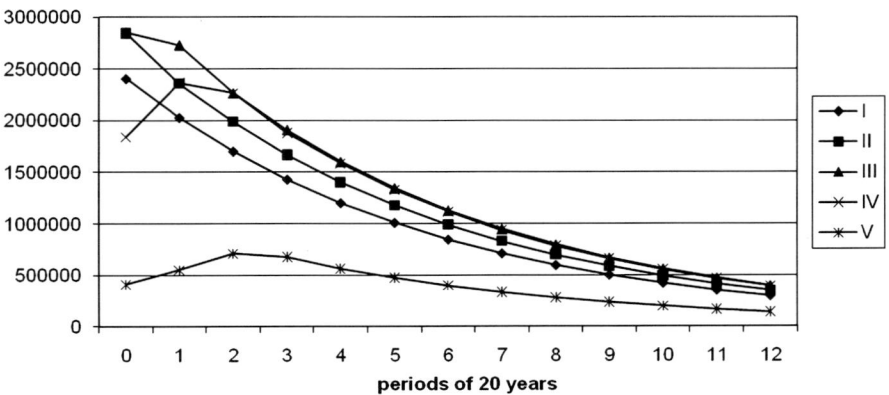

Fig. 46.4 Projections per age group over 12 periods of 20 years

Table 46.2 Growth rates per age group over six periods of 20 years

Number of periods	Growth rates				
	I	II	III	IV	V
1	−15.7%	−17.0%	−4.3%	+28.7%	+34.3%
2	−16.1%	−15.7%	−17.0%	−4.3%	+28.7%
3	−15.9%	−16.1%	−15.7%	−17.0%	−4.3%
4	−16.0%	−15.9%	−16.1%	−15.7%	−17.0%
5	−16.0%	−16.0%	−15.9%	−16.1%	−15.7%
6	−16.0%	−16.0%	−16.0%	−15.9%	−16.1%

Table 46.3 Relative importance of each age group over six periods of 20 years

Number of periods	I	II	III	IV	V
0	23.25%	27.46%	27.56%	17.77%	3.97%
1	20.22%	23.50%	27.19%	23.59%	5.50%
2	19.06%	22.27%	25.35%	25.36%	7.95%
3	18.91%	22.04%	25.24%	24.84%	8.98%
4	18.91%	22.07%	25.20%	24.95%	8.87%
5	18.91%	22.06%	25.22%	24.90%	8.91%
6	18.91%	22.06%	25.21%	24.92%	8.89%

Table 46.2, which is also given to the students, confirms the observations from Fig. 46.4. The numbers in the table are growth rates: they express by how much the number in the age group has increased or decreased over a period of 20 years as a percentage of the number in that age group at the beginning of that period. Now, the pattern in the long-term behaviour can be stated more precisely: in the long run, the number of individuals in each age group decreases by (approximately) 16% per period of 20 years. Students know that this means that the number of individuals in each age group declines exponentially with a growth factor of 0.84 (= 100%−16%). This is the first important observation concerning the long-term evolution of the Belgian population (according to our model). We will call the number 0.84 the *long-term growth factor* of the population.

Table 46.3 leads to a second important observation. Now, the percentages reflect the relative importance of each age group in the total population at a certain time. Students, for example, have to explain how they see the ageing of the population in the table. Moreover, they have to formulate what it means that the percentages no longer change in the long run and they have to explain this phenomenon using the first observation above. We will call the constant distribution of the individuals over the age groups the *long-term age distribution*.

2.4 Mathematical Treatment of the Observations

In the previous section calculations led to two observations that were formulated in terms of the long-term growth factor and the long term age distribution. In this section, the central question is: how can the long-term growth factor and the long-term age distribution be calculated in a more 'mathematical' way? In mathematical terms, the two observations can be formulated as follows:

- We have $X(n+1) = L \cdot X(n) \approx 0.84 \cdot X(n)$ for sufficiently large n. Moreover, if n increases indefinitely, the approximation improves indefinitely.
- The age distribution $X(n)/t(n)$ (where $t(n)$ is the total population at time n) converges to a certain column matrix X.

Combination of these observations leads to the equality $L \cdot X = 0.84 \cdot X$, giving the basis of a more mathematical method to find the long term age distribution if the long term growth factor is known:

The long-term age distribution X is a solution of the system $L \cdot X = 0.84 \cdot X$.

It is important to notice that this (homogeneous) system has an infinite number of solutions, that is, it has non-trivial solutions. As the solution set is one-dimensional, the long-term age distribution is the unique solution satisfying the following supplementary condition:

The sum of the components of the long term age distribution X is equal to 1.

This shows how to calculate the long-term age distribution mathematically. For a mathematical calculation of the long-term growth factor, the starting point is the fact that the system $L \cdot X = 0.84 \cdot X$ has non-trivial solutions. This characterizes the number 0.84, that is, if 0.84 is replaced by another strictly positive number, then the resulting system no longer has non-trivial solutions. Hence, the long-term growth factor can be determined as follows:

The long-term growth factor is the strictly positive number λ for which the system $L \cdot X = \lambda \cdot X$ has non-trivial solutions, that is, $\det(L - \lambda \cdot I_n) = 0$.

2.5 Eigenvalues and Eigenvectors

Finally, the concepts and methods developed in the context of the long-term evolution of the Belgian population are decontextualized, leading to the traditional definitions of eigenvalues and eigenvectors of a matrix in general and to the traditional methods to find eigenvalues and eigenvectors. For Leslie matrices L having two consecutive non-zero fertility rates, the following properties guarantee the existence of a long term growth factor and long term age distribution:

1. L has exactly one strictly positive, real eigenvalue λ_1.
2. One of the eigenvectors of L corresponding to the eigenvalue is a column matrix X consisting of strictly positive numbers adding up to 1.
3. For every $X(0)$ having all elements positive, the sequence $X(n)/t(n)$ (where $t(n)$ is the sum of the components of $X(n)$) converges to X.

3 Experiences

3.1 During the 'Science Week'

The teaching sequence was used during the so-called Science Week in Flanders in 2004 and 2006, during which secondary school pupils visited universities and attended workshops concerning a scientific subject. Students worked in small groups on worksheets based on parts 1, 2, and 3 of the teaching sequence (Sects. 2.1, 2.2 and 2.3). Answers were discussed with the whole class. In addition to the subject matter discussed here, the concept of dependency ratio and its evolution over time were included. This concept is related to financial implications of the ageing of the population and turns up in, for example, newspaper articles. We approximated it by the number of individuals in age groups I, IV, and V, divided by the number of individuals in age groups II and III. The workshop ended with a whole class activity using a spreadsheet centered around the question of how to reach a stable population and/or a socially acceptable dependency ratio by manipulating fertility and survival rates, changing age of retirement, taking migration into account, and so on. Parts 4 and 5 (Sects. 2.4 and 2.5) were not discussed. We have no data from this context, except for our own observations. Experiences are mixed. Although it was announced that a certain level of mathematical ability was expected, in practice, students' mathematical ability and motivation varied considerably. In general, experiences with mathematically sufficiently able students were positive: they managed to understand the model and were interested in the results.

3.2 In Mathematics Teacher Education

Students in mathematics teacher education in Flanders (Belgium) have a masters degree in mathematics or (more and more often) a more or less related subject (like engineering, physics, biology, applied economics, etc.). The whole teaching sequence was used with these students in several years: parts 1, 2, and 3 in the form of a workshop (as described in Sect. 3.1), and parts 4 and 5 as homework (using a text with exercises). We have no data from this context, except for our own observations in class. Experiences were clearly positive. Several students reported that the teaching sequence showed them that eigenvalues and eigenvectors are useful. They added that this was not clear to them before from their courses on linear algebra. Moreover, reactions of students in class showed that the teaching sequence stimulated them to reflect on the subtle relation between mathematics and reality. Finally, the work on the model naturally led some of the students to questions about reality. For example, one of the female students asked: "Does this mean that right-wing parties who promote having more children are right?"

3.3 In an Introductory Mathematics Course for Bachelor Students in Applied Economics

Students in applied economics in Flanders (Belgium) have an introductory mathematics course in their first and second years, covering calculus and some linear algebra. Their mathematical ability and background from secondary education is relatively modest. In this context, parts 2–5 of the teaching sequence were covered, in combination with an application having a simpler long-term behaviour (consumers switching between different brands of a product). The theory of matrices, linear systems, and determinants had been studied in the previous lessons, whereas eigenvalues and eigenvectors had not been dealt with before. Hence, the teaching sequence constituted an application of matrices, on the one hand, and served as an introductory example for the theory of eigenvalues and eigenvectors, on the other hand. The course was teacher-centred with relatively large groups of students (40–60 students) per class. Hence, the teaching methods differed from the previous two contexts. There were strict time constraints as eigenvalues and eigenvectors were the last topic covered in the academic year and not very much time was left. There was no opportunity to work in groups with work sheets. The material was explained by the teacher and student activity in class was limited to answering some questions and doing exercises from time to time. At home, students studied the material and solved problems. The (oral and written) examination was about mathematics and its applications, not about modelling. Lack of time made it impossible to do part 1 of the teaching sequence. So, the modelling started immediately from the simplified data in part 2. However, there were opportunities to draw the attention of the students to the limitations of the model (constant fertility and death rates, no migration).

After the teaching sequence, a questionnaire was filled in by 20 randomly chosen students. The questions intended to find out whether the teaching sequence

- Was not too difficult and/or confusing for the students
- Made students appreciate the usefulness of mathematical concepts
- Actually motivated students
- Made students aware of the simplifications made when describing reality with a mathematical model

Three questions dealt with the level of difficulty of the example. In general students did not find the teaching sequence too difficult. Although they reported that it was more difficult than other parts of the course, they responded neutrally to the statement that the example was difficult to follow during the lesson and positively to the statement that it could be well understood after personal study at home.

Four questions examined whether the students became more aware that mathematics is a useful tool to describe reality. In general, the answers show that the teaching sequence helped students to appreciate the usefulness of mathematics. They responded in the great majority positively to the statement that the example taught them that mathematics can be used to describe phenomena in reality and more, in particular, they agreed that it showed what matrices are used for. Concerning the usefulness of eigenvalues and eigenvectors there were two questions. Students responded neutrally to the statement that the example showed them the usefulness of eigenvalues and eigenvectors. On the other hand, they reacted in the great majority positively to the statement that the example was a good introductory example for the concepts of eigenvalue and eigenvector.

Three questions investigated the motivation of the students. On the one hand, there was a clear general appreciation of the example: students responded in the great majority positively to the statement that more examples of this type should be treated in the course. This appreciation correlated negatively to the experienced rate of difficulty. On the other hand, it is not clear which aspects added to this general appreciation as students reported that they did not find the example of the evolution of the Belgian population more interesting, nor more instructive than other (less authentic) examples in the course.

Finally, the results concerning one question showed that students found the teaching sequence taught them that mathematical models are always a simplification of reality.

There has been no systematic research concerning the results on the examination. The impression was that students performed at the same level as with traditional teaching, which is in agreement with the answers of students to the questions in the questionnaire relating to the experienced level of difficulty of the teaching sequence. Some typical errors, however, appear to be related to the use of the context of growing populations, for example, some students only accept positive numbers as eigenvalues of matrices and sometimes students think that eigenvectors should have components adding up to 1.

4 Conclusion

The outline of a teaching sequence in Sect. 2 shows that it is possible to use the context of age-structured population growth as a natural introduction to the concepts of eigenvalue and eigenvector, which in university teaching are often introduced rather artificially. We discussed experiences with the teaching sequence in three contexts. The most extensive treatment is with students in a tertiary mathematics course in an applied economics degree program. Reactions of students to a questionnaire showed that they experienced the teaching sequence as more difficult than the rest of the course, but not too difficult. They reported that the teaching sequence made them more aware of the usefulness of mathematics. Finally, they found that more examples of this type should be treated in the course. This appreciation correlated negatively with the experienced level of difficulty.

References

Blum, W., et al. (2002). ICMI Study 14: Applications and modelling in mathematics education – Discussion document. *Educational Studies in Mathematics, 51*(1/2), 149–171.

Bradley, I., & Meek, R. L. (1986). *Matrices and society: Matrix algebra and its applications in the social sciences*. Harmondsworth: Penguin.

Leslie, P. H. (1945). On the use of matrices in certain population mathematics. *Biometrika, 33*(3), 183–212.

Niss, M. (2000). Forecasting populations. *The Undergraduate Mathematics and Its Applications Journal, 21*(1), 73–96.

Chapter 47
Modelling and the Educational Challenge in Industrial Mathematics

Matti Heilio

Abstract Modelling is vital for innovation in knowledge-based industries and development of society. Emergence of mathematical technology means a challenge for education. We give an overview and discuss how this should be reflected in educational programs and has implications on the way how modelling should be inserted to the curricula at various levels. The focus of this article is on undergraduate teaching at tertiary level. However, some comments are made concerning teacher training and schools preparing students for universities. Future teachers should be given understanding of the vibrant mathematics-based technology. However, at school level, I suggest certain care regarding how much and when to introduce concepts of modelling into the educational repertoire.

1 Computational Technology

The development of technology has modified in many ways the expectations facing the mathematics education and practices of applied research. Computational tools and modelling competence is needed in many fields and professions. The scope extends to several ordinary professions and science-related fields in engineering, economics and finance, biomedical jobs, not forgetting agriculture and the food chain, health professions and pharmacy, and entertainment and media. Mathematics is a vital resource to promote technical development, innovation, entrepreneurship, structural renovation and public governance. Terms like mathematical technology, industrial mathematics, computational modelling or mathematical simulation are used to describe this active contact zone between technology, computing and mathematics. See a recent Organisation of Economic Cooperation and Development (OECD) report (2008) on *Mathematics in Industry*. Models can be used to perform the following actions:

M. Heilio (✉)
Lappeenranta University of Technology, Lappeenranta, Finland
e-mail: matti.heilio@lut.fi

- Gain understanding of intricate mechanisms by testing assumptions about the systems
- Carry out structural analysis tasks and evaluate the systems performance capabilities
- Replace or enhance experiments or laboratory trials
- Forecast system behaviour and analyse what-if situations
- Perform sensitivity analyses and study the behaviour at exceptional circumstances
- Analyse risk factors and failure mechanisms
- Create virtual and/or visualised images of objects and systems in design processes
- Optimise certain design parameters or the whole shape of a component
- Carry out intelligent analyses on the measurement data through monitoring and experiments
- Manage and control large information systems, networks, data bases.

2 Educational Challenge

There is an obvious need to revise university pedagogy of applied mathematics. Implications for the preparatory levels of high school will be elaborated as well. Regarding the undergraduate programs at universities, the means include: revision of syllabi and curricula, use of computing experiments, data tools and novel teaching methods. We should bring the flavour of a fascinating art to the classroom, to convey the vision about mathematics at work, the diversity of application areas and practical benefits. The challenge is to find ways to make the theoretical content transparent and communicate to the students the end-user perspective of mathematical knowledge.

To become a good applied mathematician, one should be curious about other areas as well, to learn basic facts from a few neighbouring areas outside mathematics. Acquaintance of some application fields, knowledge of physics, engineering or other 'client discipline' of mathematics is important.

Problem-based learning and topical fresh exercises are called for. Mathematics teachers should have interest for different areas of modern professional life. Modelling cases can emerge from grocery packing, ID-card code system, sports betting, laundry machines, brick manufacturing, tram timetables, fermentation processes of food or bioprocesses in gardening. Modelling is a crucial educational challenge; maturing into an expert can only be achieved by 'treating real patients'.

The challenge of exposing students to real world applications is common to school mathematics and undergraduate programs. However, in spite of the importance of modelling in today's world there is a danger of modelling overdrive at school level. Case examples are an excellent means to give meaning and context to mathematical concepts like relation, ratio, equation, pair of equations, solution, multiple solutions, rate of change, derivative, integral, optimum, approximate solution, probability, random variable, etc. The best time to introduce modelling as a serious activity of simulating real world processes and solving problems perhaps comes later. The children however should gain an understanding, interest and curiosity of these technologies already at school.

Mathematics educators have introduced methodical approaches like model-eliciting activities or thought-revealing activities (Chamberlin and Moon 2005; Lesh 2002). Research has also been done to understand the cognitive and metacognitive processes involved in learning of modelling (Stillman and Mevarech 2010). A lot of discussion has appeared around the concept of modelling competencies (Blomhøj and Højgaard Jensen 2007). From the perspective of using mathematical skills in professions, industrial research and development and applied sciences, I present some claims to contribute to the contemporary discussion.

The realm of mathematical models has rich variety in forms, structure and complexity, which is comparable to the diversity of life itself. Any systematisation or classification of the world of models is doomed to be deficient. It may be also impossible to describe the art of modelling as a repertoire of skills and knowledge that can be transferred or 'taught' in the traditional sense. It may be learned over a long process when suitable knowledge base, thought eliciting situations and illuminating examples and problems are presented to a student who has a curious attitude and ability to learn and transfer ideas from one context to another.

A well-known observation is that it is easier to grasp abstract generalisations after one has seen many simple concrete cases. A student should be building mathematics skills, analysing individual functions, curves, equations, data sets, etc., before he is asked to adopt the abstract conceptual scheme of 'mathematical model' much less 'modelling process'. The student should work on rules of manipulation of expressions and learning to understand the basic vocabulary of quantitative thinking (function, equation, constant/variable, solution, rate of change, probability, random variable) for quite awhile. The learning process should couple these concepts into the real world, so case examples (models indeed) are indispensable. However, this activity and individual problem-solving tasks should not be prematurely put into the mathematical modelling paradigm.

My conjecture is that the correct time to introduce the ideas of mathematical model and modelling process and mathematical technology is at the secondary school, in the last two years before entering the university. Excellent ways to organise such modelling activity is the concept of modelling week, workshop or competition (Bracke 2007; Kaiser and Schwarz 2006, 2010; Mathematical Contest in Modelling). These ideas were nicely presented also at ICIAM14.

3 Sphere of Applications

Giving a list of examples, I describe next the increasing sphere of areas where modelling, simulation and intelligent systems are a crucial vehicle of development. The examples illuminate the educational challenge. Students, who are trained to be teachers, should be given the opportunity of obtaining an overall understanding of mathematical technology and industrial mathematics to give them a flavour of the role of mathematics in this modern world. References (Bonilla et al. 2008; Di Bucchianico et al. 2006; European Consortium for Mathematics in Industry) provide more detailed descriptions of mathematical technology and computational engineering. In many cases, mathematics itself is known, but the novelty is implementation of mathematical models and modern computation in ways that were not possible 20 years ago.

Economics and management. The daily functioning of our modern society is based on numerous large-scale systems. Examples are transportation, communication, energy distribution and community service systems. The planning, monitoring and management of these systems offers a lot of opportunities for mathematical approaches.

Corporate management uses methods in which mathematical knowledge is embedded at different levels. Econometric models are used especially in the banking sector to describe the macro level changes and mechanisms in the national economy. Risk analysis, game theory, decision analysis, etc., are used to back up strategic decisions, design a balanced financial strategy or optimise a stock portfolio. The images below represent time series of electricity price variation and the level of water storage in a certain geographic area of an integrated energy market. Modelling attempts are needed to understand the mechanisms leading to these fascinating and intricate stochastic phenomena in energy markets (Fig. 47.1).

Performance analysis: manufacturing systems. The added value in using mathematical methods comes from the possibility of simulating devices, mechanisms, and systems prior to their physical existence. A whole new system – like an elevator system in a high-rise building, a microelectronic circuit containing millions of elements or a high-tech manufacturing system – can be designed and tested for its performance and reliability. Simulation of multibody systems and integrated design of complex mechanisms are examples where computational modelling means real competitive advantage.

Chemical reactions and processes. Chemical processes are being modelled on various scales. In the study of molecular level phenomena, mathematical models are used to describe the spatial structures and dynamical properties of individual molecules to understand the chemical bonding mechanisms etc. Chemical factories use large models to monitor the full-scale production process. Increasingly important are environmental monitoring benefits from models of biochemical processes. The control of microbial processes is quite crucial and adds to the complexity. Food chain security issues and ecological sustainability are becoming major

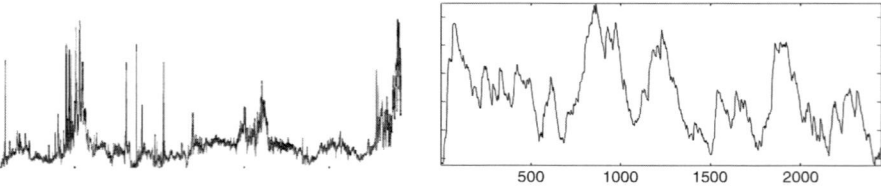

Fig. 47.1 Electricity price variation and the level of water storage

47 Modelling and the Educational Challenge in Industrial Mathematics

Fig. 47.2 River flow

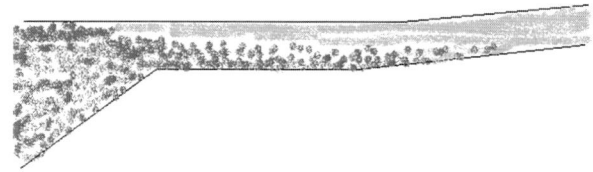

Fig. 47.3 Saline water penetration into river estuary

global problems and there is a demand for sophisticated models. Genomic research and pharmaceutical product development are based on modelling of dynamics of macromolecules, analysis of chemical reaction kinetics and ingenious analysis of astronomic volumes of experimental data.

Flow phenomena. The ability to model sophisticated phenomena, including non-linear effects, the possibility to solve the equations with advanced numerical methods, combined with the latest visualisation tools have created a luxurious environment for mathematical engineering. The computational simulation can be used to support the design of systems from tooth paste tubes, regional heating networks and aircraft fuselage design to ink-bubble printers. Examples are river flow models that are used for flood control and forecasting, planning of hydropower systems and waterways. An estuary model may be needed to understand how the saline water from the sea is penetrating a river estuary. The model should predict the salt concentration and depth of the penetration upstream (Figs. 47.2 and 47.3).

Systems design and control. The design engineers and systems engineers have always been active users of mathematics in their profession. The possibility to set up realistic large-scale system models and the development of modern control

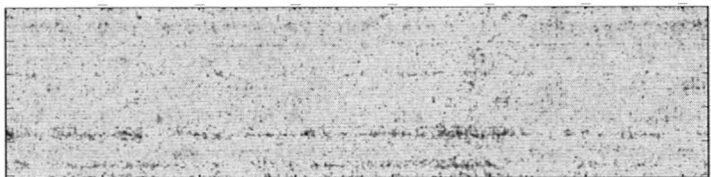

Fig. 47.4 Paper web quality simulation

theory has made the computational platform a powerful tool with new dimensions. Remote control of traffic systems, monitoring and maintenance of power transmission networks, control of windmill farms would be such examples. In traffic systems, the analysis of traffic flow, scheduling, congestion effects, planning of timetables, derivation of operational characteristics, air traffic guidance systems and flight control need sophisticated models. The illustration in Fig. 47.4 describes measurement data from a paper mill where the thickness of a 1,500 m long a paper web has been measured with high precision. The image is actually the product of a simulation model that mimics the actual performance of the paper machine.

The media and entertainment industry is a heavy user of mathematical models. Visualisation techniques, special effects and simulated motion of virtual reality are based on a multidisciplinary approach using mathematics, mechanics and computing power. An example could be the sympathetic character of Gollum from *The Lord of the Rings* movie. The odd and alien skin of the character was created by a technique of simulated subsurface scattering, a combination of modelling, physics of light reflection and computing skills. The image is found at www.ew.com/ew/article/0,,702019,00.html.

Measurement technology, signals and image analysis. The computer and the advanced technologies for measurement, monitoring devices, camera, microphones, etc., produce a flood of digital information. The processing, transfer and analysis of multivariate digital process data have created a need for mathematical theory and new techniques. Examples of advanced measurement technologies are mathematical imaging applications. Applications range from security and surveillance to medical diagnostics, recognition of harmful mould spores in air quality samples or bacteria from virological cell cultures. Modern theory of inverse problems is applied in improving the imaging in dental tomography. So called Bayesian stochastic models are a key to these improvements (Fig. 47.5).

4 Modelling as a Course Subject

Many departments have introduced modelling courses in the curriculum in recent years. A course in modelling may contain study of case examples, reading texts and solving exercises. The actual challenge and fascination is the students' exposure to open problems, addressing questions arising from real contexts. The real world questions may be found from the student's own fields of activity, hobbies, summer

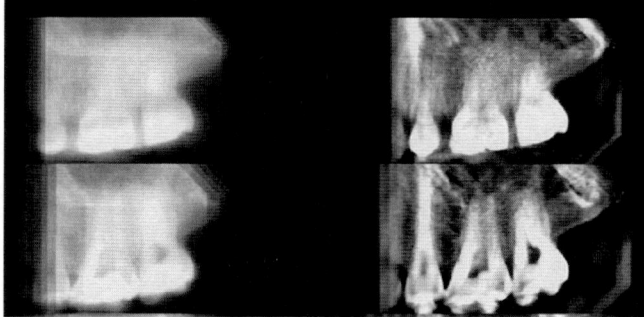

Fig. 47.5 Dental tomography

jobs, from the profession of their parents etc. Reading newspapers and professional magazines with a mathematically curious eye may find an idea for a modelling exercise. A good modelling course should:

(a) Contain an interesting collection of case examples that stir students' curiosity.
(b) Give an indication of the diversity of model types and purposes.
(c) Show the development from simple models to more sophisticated ones.
(d) Stress the interdisciplinary nature, teamwork aspect, communication skills.
(e) Tell about the open nature of the problems and non-existence of 'right' solutions.
(f) Help to understand the practical benefits of the model.
(g) Tie together mathematical ideas from different earlier courses.

One of the innovative educational practices introduced in the recent decades is the 'modelling week', an intensive problem-solving workshop that simulates real life research and development procedures. The cases originate from industry, different organisations or branches of society. The teams are guided by a group of academic staff members who play the role of the problem owners. The students must formulate a model and recognise the typically non-unique mathematical problem. The analysis follows leading to analytical studies and efforts on numerical solutions. Typically the group arrives at an approximate solution. At the end of the week, the student groups present their findings in public. Further, they are assumed to produce a written report to be published in a proceedings booklet.

5 Modelling Problems to Challenge Undergraduates

Figure 47.6 describes a city network of water pipes, water storages and wells. Assume that some harmful contamination enters the network at some point. Describe by a model the spread of the pollution concentration in the network. The next question could be to evaluate certain attempts to clean the system.

Non-trivial modelling examples can be created by unusual pendulum variations like the one illustrated in Fig. 47.7. Describe the system behaviour. Study the system equations limit as the radius of the obstacle goes to zero.

Fig. 47.6 City water network

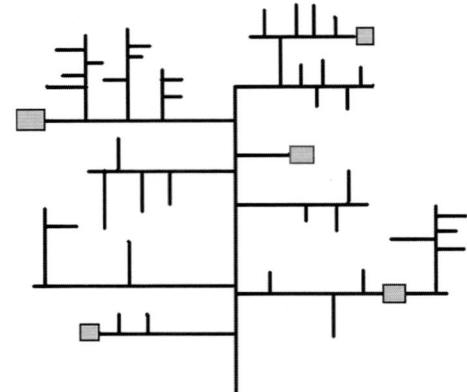

Fig. 47.7 Pendulum with an obstacle

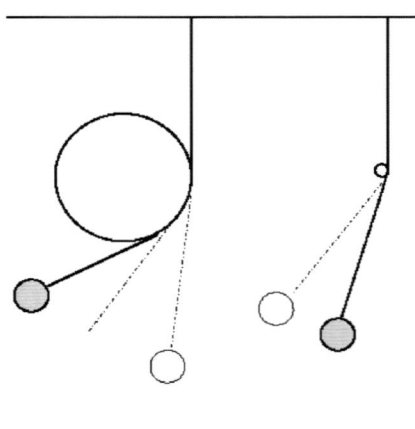

Fig. 47.8 Chain lifting, moving boundary

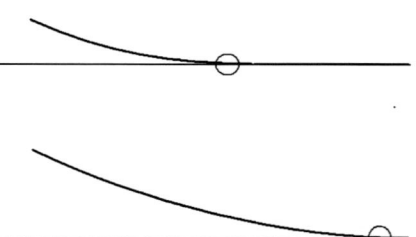

A metal chain or rope is laid on a floor. One end is slowly lifted off the ground. Find the shape of the chain. Try to model the movement of the point where the chain is detached from the floor. This is a so called moving boundary problem, and far from trivial (Fig. 47.8).

Fig. 47.9 Log cutting optimization

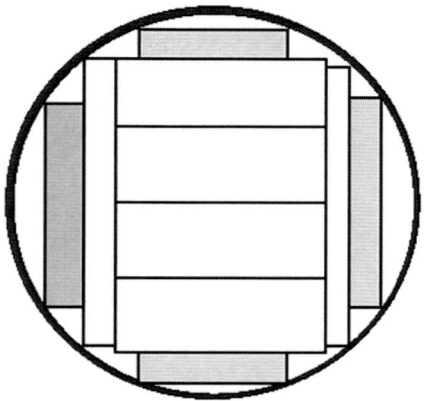

The following example is a true problem in the timber industry. Logs are cut into different wood products of standards sizes. Given the measurements of the various standard products and the diameter of the log, the challenge is the optimal strategy to cut the log (Fig. 47.9).

6 Modelling Education: How Much and When?

Today's technology agenda and the challenge of knowledge-based society indicate the importance of teaching mathematical modelling at various levels of education. Does this support the idea that modelling education should be promoted for all children at all school levels? How much, in what way and when?

The skills of mathematical modelling mean an essential competence which is needed in many science-related fields, technology, engineering, economics, biomedical professions etc. Modelling means a set of specialised science-based skills that can be compared with the expert skills of, let us say, airline pilot and brain surgeon. Our society really needs these skills but we do not arrange minicourses for airline pilots and surgery at primary school levels. This somewhat provocative statement is meant to emphasize the important question of how mathematical modelling should be inserted into the educational system.

The overall understanding of mathematics in today's world should be explained in the mathematics classroom. Examples from applications should be used as fermentation of the learning process. The joy of problem solving and the use of mathematics for real life situations is an ideal way to build interest and enthusiasm. The right timing and pedagogical guidelines are important. The phrase of Prof. Helmut Neunzert at ICTMA14 "…modelling can be learnt but not taught in a usual way" contains an important message. Educational fashions or New Math reforms sometimes tend to drift to overdrive or hype.

Teacher training is a key question. The new generation of mathematics teachers should have a sound overall understanding of the important role of mathematics in today's world so that he or she can bring to the classroom a flavour of the fascinating special skills of mathematics, modelling, simulation and computing. The children hopefully will become aware of several new professional career opportunities in the field of science-based professions where mathematical models are the modern space-age toolkit.

References

Blomhøj, M., & Højgaard Jensen, T. (2007). What's all the fuss about competencies. In W. Blum, P. L. Galbraith, H.-W. Henn, & M. Niss (Eds.), *Modelling and applications in mathematics education: The 14th ICMI study* (pp. 45–56). New York: Springer.

Bonilla, L. L., Moscoso, M., Platero, G., & Vega, J. M. (Eds.). (2008). *Progress in industrial mathematics at ECMI 2006*. Springer Subseries: The European Consortium for Mathematics in Industry, Vol. 12.

Bracke, M. (2007). Turtles in the classroom – Mathematical modeling in modern high school education. site.educ.indiana.edu/Portals/161/Public/Bracke.pdf

Chamberlin, S. A., & Moon, S. M. (2005). Model-eliciting activities as a tool to develop and identify creatively gifted mathematicians. *Journal of Secondary Gifted Education*, September 22, 2005.

Di Bucchianico, A., Mattheij, R. M. M., & Peletier, M. A. (Eds.). (2006). *Progress in industrial mathematics at ECMI 2004*. Springer Subseries: The European Consortium for Mathematics in Industry, Vol. 8.

European Consortium for Mathematics in Industry. www.ecmi-indmath.org

Kaiser, G., & Schwarz, B. (2006). Mathematical modelling as bridge between school and university. *The International Journal on Mathematics Education, 38*(2), 196–208.

Kaiser, G., & Schwarz, B. (2010). Authentic modelling problems in mathematics education – Examples and experiences. *Journal für Mathematik-Didaktik, 31*(1), 51–76.

Lesh, R. (2002). Research design in mathematics education: Focusing on design experiments. In L. English (Ed.), *International handbook of research design in mathematics education, 2002*. Mahwah: Lawrence Erlbaum.

Mathematical Contest in Modelling (MCM). www.comap.com

OECD – Global Science Forum. (2008). Report on "Mathematics in Industry". www.ceremade.dauphine.fr/FLMI/FLMI-frames-index.html

Stillman, G., & Mevarech, Z. (2010). Metacognition research in mathematics education: From hot topic to mature field. *ZDM – The International Journal of Mathematics Education, 42*(2), 145–148. www.citeulike.org/article/6847002

Chapter 48
Modelling of Infectious Disease with Biomathematics: Implications for Teaching and Research

Sergiy Klymchuk, Ajit Narayanan, Norbert Gruenwald, Gabriele Sauerbier, and Tatyana Zverkova

Abstract The chapter compares a variety of models from biomathematics and bioinformatics of the spread of severe acute respiratory syndrome (SARS) that hit dozens of countries worldwide in 2003. It also investigates students' and lecturers' opinions regarding differences in predictions from three different models. All models were based on the real data for Hong Kong published by the World Health Organization (WHO). Although the models were based on the same data, they gave very different predictions of the spread of the disease. The models were discussed with two groups of people: undergraduate students majoring either in engineering or applied mathematics and university lecturers who teach mathematics or mathematical modelling courses. In this chapter we present, analyze, and compare responses to the same questionnaire given to the two groups.

1 Introduction and Framework

Mathematical modelling is a complex process consisting of a number of interrelated steps. Many researchers and practitioners consider skills in mathematical modelling to be different from skills in mathematics. George (1988) states that "model building is an activity which students often find difficult and sometimes rather puzzling. The process of model building requires skills other than simply knowing the appropriate mathematics" (George 1988). Modelling "can be learnt but not taught in a usual way"

S. Klymchuk (✉) and A. Narayanan
Auckland University of Technology, Auckland, New Zealand
e-mail: sergiy.klymchuk@aut.ac.nz

N. Gruenwald and G. Sauerbier
Wismar University of Applied Sciences, Technology, Business and Design, Wismar, Germany

T. Zverkova
Odessa National University, Odessa, Ukraine

(Neunzert and Siddiqi 2000). Relationships between mathematical competencies of students and their skills in modelling were investigated in Galbraith and Haines (1998). Caron and Belair in their exploratory study (2007) examined the phases of the mathematical modelling process that received greater attention from undergraduate students and the competences that were displayed in each phase. They suggested that "some modelling heuristics should be explicitly taught." They also suggested that "more time should be spent discussing the purpose of a model: this would help students clarify the expected outcome and benefits of each stage" (Caron and Belair 2007).

In many cases a major purpose for doing mathematical modelling of a phenomenon is to make predictions. Taking into account uncertainty and a variety of possible models and a number of assumptions in each model, the task of prediction cannot have the 'correct' answer. This fact alone can confuse many students. This chapter investigates students' opinions regarding differences in predictions from three different models based on the same real data. The task given to the students might look very simple. They neither needed to build a model nor solve the given models. All they needed to do was to read the given real life problem, look at the predictions from three different models and give their reasons for the differences in the predictions. We tested one of the modelling competences described by Kaiser in (2007): "Relating back to the real situation and interpreting the solution in a real-world context". We also gave the same task to university lecturers who teach mathematics or mathematical modelling courses. Our idea was to compare the responses of the students and lecturers. The main research question was to investigate possible patterns within each group and also similarities and differences between the two groups when they do the same modelling task. In particular, to which extent the two groups use their intuition, common sense and past experience explaining the differences in predictions from three familiar models.

The theoretical framework of this study was based on the works of Haines and Crouch (2001, 2004). A measure of attainment for stages of modelling has been developed in (Haines and Crouch 2001). The authors expanded their study in (Crouch and Haines 2004) where they compared undergraduates (novices) and engineering research students (experts). They suggested a three-level classification of the developmental processes which the learner passes in moving from novice behaviour to that of an expert. One of the conclusions of that research was that "students are weak in linking mathematical world and the real world, thus supporting a view that students need much stronger experiences in building real world mathematical world connections" (Crouch and Haines 2004). This was consistent with the findings from the study by Klymchuk and Zverkova (2001) on possible practical, not cognitive reasons for students' difficulties linking mathematics and real world. Referring to that study Crouch and Haines wrote: "…students across nine countries all tended to feel that they found moving from the real world to the mathematical world difficult because they lacked such practice in application tasks" (Crouch and Haines 2004).

We believe that doing even simple mathematical modelling activity can be beneficial for students. We agree with Kadjievich who pointed out that "although through solving such … [simple modelling] … tasks students will not realize the examined nature of modelling, it is certain that mathematical knowledge will become alive for them and that they will begin to perceive mathematics as a human enterprise, which improves our lives" (Kadijevich 1999).

2 The Study

2.1 The Models

Infectious diseases hit mankind from time to time on a large scale. In the mid-fourteenth century, the Black Death plague epidemic killed one-third of the population of Europe – 34 million people. In 1563, up to half of the population of London died from the Bubonic plague. In 1918 the Spanish Flu pandemic killed, by different estimations, from 50 to 100 million people worldwide (Patterson and Pyle 1991). An estimated 500 million people were infected. In 1957 the Asian Flu killed up to 4 million people. In 2003 a new highly infectious disease Severe Acute Respiratory Syndrome (SARS) spread rapidly around the world. In March 2003 the WHO, for the first time in history, issued a global warning about the disease. Globally 8,422 people were infected and 916 died. Hong Kong was one of the countries that was hit most by the disease: 1,755 people were infected and 299 died. "Predicting the trend of an epidemic from limited data during early stages of the epidemic is often futile and sometimes misleading. Nevertheless, early prediction of the magnitude of an epidemic outbreak is immeasurably more important than retrospective studies" (Hsieh et al. 2004). Two common types of epidemic models were used to analyze the spread of SARS in Hong Kong. The first type was a long-established susceptible-infected-recovered (SIR) model and its modifications. The model was developed by Kermack and McKerdrick (1927). The fixed population N is divided into three distinct groups: Susceptible (S) (those at risk of the disease), Infected (I) (those that have it), and Removed (R) (those that are quarantined, dead or have acquired immunity). That is: $N = S + I + R$. The model is represented by the following system of differential equations:

$$\begin{cases} \dfrac{dS}{dt} = -rSI \\ \dfrac{dI}{dt} = rSI - aI \\ \dfrac{dR}{dt} = aI \end{cases}$$

where $r > 0$ is the infection rate and $0 < a < 1$ is the removal rate.

A typical prediction from this model based on the first 30 days since WHO started publishing daily reports for Hong Kong was given in Shi and Small (2003). The predicted number of infected people was 1,700 versus 1,755 in reality. So the deviation was only 3.1%.

The second model was a relatively new Small-World (SW) network model. The concept of the SW model was imported from the study of social networks into the natural sciences by Watts and Strogatz (1998). As a full epidemic network in Hong Kong was not available, numerical simulations were applied to construct an epidemic chain based on social contacts. The model was established on a grid network weaved by m parallel and m vertical lines. Every node in the network represented a person. The value of m was 2700 for the population $N = m^2 = 7.29 \times 10^6$. The model predicted 1,830 cases, so the deviation was 4.3% (Shi and Small 2003).

Public measures were very important in controlling the epidemic. If the epidemic was allowed 'to run its natural course,' in other words, to die down by itself, up to several million people would fall victim to SARS in Hong Kong alone. An epidemic will die down only when the basic reproductive number (number of people infected by a patient) is less than 1. This can be achieved only in two ways: when herd immunity is high enough (natural course of events), or when effective public health measures limit the spread of the epidemic.

Apart from complicated mathematical models, three easier models – linear, exponential and logistic – were used for the analysis of the epidemic. These models were offered as a student project in calculus in Hughes-Hallett et al. (2005). Although the models were based on the same data reported in March 2003, they gave very different predictions of the spread of the disease for June 12, 2003 when the last case was reported in Hong Kong. We decided to ask two groups of people – students and lecturers – about their opinions on the differences in predictions from these three familiar models in an unfamiliar (for students) context. Below is the questionnaire given to the participants of the study.

2.2 The Questionnaire

The questionnaire took the following form:

> Please read the case below and answer the questions. You don't need to solve anything.
>
> Models of the Spread of SARS
>
> In 2003 a highly infectious disease SARS spread rapidly around the world. Predicting the course of the disease – how many people would be infected, how long it would last – was important to officials trying to minimize the impact of the disease. A number of mathematical models of the spread of SARS were developed to make the predictions. Below are three simple models of the spread of SARS in Hong Kong. We measure time t, in days since March 17, the date the World Health Organization (WHO) started to publish daily SARS reports. Let $P(t)$ be the total number of cases reported in Hong Kong by day t. On March 17, Hong Kong reported 95 cases. We compare predictions for June 12, the last day a new case was reported in Hong Kong (87 days since March 17). The constants in the differential equations were determined using WHO data from 17 to 31 March (15 days).

(continued)

(continued)

A Linear Model $\frac{dP}{dt} = 30.2$, $P(0)=95$. The prediction for June 12 was 2,722 cases.

An Exponential Model $\frac{dP}{dt} = 0.12P$, $P(0)=95$. The prediction for June 12 was 3,249,000 cases.

A Logistic Model $\frac{dP}{dt} = P(0.19 - 0.0002P)$, $P(0)=95$. The prediction for June 12 was 950 cases.

The actual number of cases on June 12 was 1,755.

Questions:

1. What were possible reasons for the differences in the predictions from the three models above?
2. On what were your reasons from question 1 based (e.g. your experience in modelling, common sense, etc.)?
3. What could make the predictions more accurate?
4. Are you interested in learning more about epidemic modelling and possibly doing research projects in this area (e.g. modelling the spread of swine flu)? Why?

3 The Responses

3.1 Students

The students' group consisted of first-year undergraduate students majoring in engineering from a German university and second and third year students majoring in applied mathematics from a New Zealand university. Ninety questionnaires were distributed over 2009 and 2010. Forty-eight responses were received so the response rate was 53%. It was a self-selected sample. We systematized and grouped students' answers into different categories according to the nature of their responses. We used either the key words or exact quotes to name the categories. Some students gave multiple responses to some of the questions and some students did not answer all the questions. The students' categorized responses are presented below.

1. What were possible reasons for the differences in the predictions from the three models above?
   Different models (16), lack of biological factors (10), different ideas of the speed of spread (8), isolation of infected people (8), population density (6), different assumptions of cases per day, report of cases is not correct (3), different infection rates (3), counter actions, for example pharmaceuticals, different side conditions

(1), different assumptions for each model (1), probability of onset (1), people developed immunity (1), the predictions are theories, which are different from the reality (1), not enough data (1).

2. On what were your reasons from question 1 based (e.g. your experience in modelling, common sense, etc.)?
Common sense (19), mathematical knowledge and experience in modelling (7), both modelling experience and common sense (3), the given information (1), idea of spread of disease (1), I have never seen such problems in [a] mathematical context before, so I don't know exactly, how to solve it (1), reality, never a constant number of persons will be sick (1), my knowledge about curves of elementary functions (1).

3. What could make the predictions more accurate?
Use experiences from studies of other epidemics, in other regions (14), use more data (7), more knowledge of the virus (3), look for preventive steps, compulsory registration (2), improve data collection (1), average value of cases from 7 days (1), a constant showing the rate of infections (1), side effects like number of travellers to and from Hong Kong (1), information of medical doctors or scientists for the course of disease (1), a study of people behaviour and their health state (1), more facts (1), evaluation of the models (1), the logistic model looks more realistic and it could be improved by using more variables (1), set up a limit of resources (1), adjust the models results to the reality all the time (1), compare the first 2–3 days to find the initial condition (1).

4. Are you interested in learning more about epidemic modelling and possibly doing research projects in this area (e.g. modelling of the spread of swine flu)? Why?
Yes – 4. An interesting topic (4), important for the science on viruses (2).

No – 39. Not my area of interest (18), lack of time (8), this is only making panic (3), don't see the point to play with numbers or equations which are not correct (2), it is going to have too many factors and the predictions may not be that reliable (1), not enough knowledge in mathematics (1).

3.2 Lecturers

The lecturers' group consisted of university lecturers from different countries who teach mathematics or mathematical modelling courses. Some of them were involved in research on teaching mathematical modelling and applications. Some of the lecturers were from the same universities as the students participated in the study. Thirty-eight questionnaires were distributed over 2009 and 2010. Twenty-three responses were received so the response rate was 63%. It was a self-selected sample. We systematized and categorized the lecturers' answers in the same way as the students' answers. The lecturers' categorized responses are presented below.

1. What were possible reasons for the differences in the predictions from the three models above?

The models (19), different ideas of the spread of the disease, certain factors were not considered (2), the models were developed for other epidemics, SARS does not fit (1), the assumptions are not the same in all three models (1), did not consider the spread style of the disease (1), infinite number of predictions exist (1).

2. On what were your reasons from question 1 based (e.g. your experience in modelling, common sense, etc.)?
   Experience in modelling (13), common sense (5), both modelling experience and common sense (3).

3. What could make the predictions more accurate?
   More data (6), a better model (3), better parameter estimation (3), knowledge about infection mechanism and other factors e.g. travelling routes, social patterns (2), more accurate analysis of influencing factors (2), a deeper understanding of how infectious disease spreads (1), the parameters in all the models must be the same (1), distribute the observing time in intervals and use different models in different intervals (1), use learning methods (1).

4. Are you interested in learning more about epidemic modelling and possibly doing research projects in this area (e.g. modelling of the spread of swine flu)? Why?
   Yes – 14. Because I work in a similar field, overlapping in research (2), interested in modelling real-life situations to get a grip on it (2), it is important to learn about this modelling since nowadays the disease is increasing (1), interesting subject for my students MS theses (1), modelling of real things is fascinating (1), I am always thrilled how a model helps us in understanding a process and forecasting future data (1), important and relevant area to see maths applied (1), most important in considering today's swine flu, it is a fascinating subject to consider modelling, teachers and students would be motivated by both the mathematics and the consequences for cities and countries to consider (1).
   No – 8. Not my subject (3), no time (2).

3.3 Analysis of the Responses

After consultations with professional mathematicians specialising in epidemic modelling we estimated percentages of appropriate answers to questions 1 and 3 in both groups. The results are presented in the table below. 'CS' means 'common sense' and 'Exp' means 'experience' (Table 48.1).

Table 48.1 Summary of the findings from the questionnaire

	N	Question 1 Appropriate	Question 2 CS	Exp	Both	Other	Question 3 Appropriate	Question 4 Yes	No
Students	48	73%	56%	20%	9%	15%	74%	9%	81%
Lecturers	23	92%	24%	62%	14%	0%	90%	64%	36%

The majority of the students had no or very little experience in mathematical modelling. The closest activity to real mathematical modelling for them was solving application problems. To our surprise the students did quite well in both modelling questions 1 and 3. They were not much behind the lecturers, giving 73% appropriate reasons for the differences in the predictions from the models versus 92% given by the lecturers. They were not much behind the lecturers giving 74% appropriate ways to improve the accuracy of the predictions in the models versus 90% given by the experts. This is consistent with the findings of Haines and Crouch (2001, 2004) where the researchers found that sometimes novices exhibited aspects of expert behaviour although they were not consistent in doing so. In particular, in their study on self-assessment and tutor assessment they found that students were almost as good as tutors in assessing group (project) presentations on modelling and so they could recognize modelling behaviour in others. It is the consistency that demonstrates expert behaviour that perhaps the differentiates lecturers.

In question 2 the reverse polarity on the answers by the students and the lecturers was anticipated: the students relied more on common sense (56%) rather than on experience (20%) compared to the lecturers (24% on common sense and 62% on experience). Apart from lack of modelling experience by the students, one of possible reasons for this reverse polarity might be elements of the lecturers' behaviour where they were reluctant to attribute their responses to common sense, preferring to classify them as experience. After all they have invested a great deal of time in mathematics/modelling.

Based on the participants' comments in the questionnaire and follow-up interviews with some of them, we attempted a comparison of the processes used by the students and the lecturers in terms of links between the mathematical world and the real world in a similar way to that done by Crouch and Haines (2004). We took the first "level (a) where there was clear evidence that the participants took into account the relationship between the mathematical world and the real world" (Crouch and Haines 2004). The students referred explicitly to that relationship in 65% of cases (though not always in a correct way) whereas the lecturers in 20% of cases. The lecturers tended to concentrate more on the mathematical aspects of the models probably implicitly assuming that relationship. One of the possible reasons might be that the lecturers used their experience in modelling and knowledge in mathematics much more than their common sense whereas the students relied more on their common sense and life experiences, lacking the experience in mathematical modelling.

In question 4 very few students (9%) reported that they were interested in doing a research project in epidemic modelling. This was understandable taking into account that the majority were majoring in engineering. The lecturers were more enthusiastic in doing research in epidemic modelling (64%), mostly because of the importance of the topic and/or relevance to their current research. The lecturers also indicated that the topic was very useful for teaching purposes because it was timely and could increase students' motivation.

4 Conclusions

This study indicates that in spite of lack of experience in real mathematical modelling, students can effectively use their common sense and general knowledge of mathematics to evaluate some modelling issues dealing with prediction. The responses at a more general level indicated that both students and lecturers would have preferred to include more parameters in the model to make the modelling more realistic and intuitive, that is, to have a theoretical basis for the modelling that included hypothetical rates of spread, infection mechanisms, etc. It is possible that engineering students, in particular, would have engaged more with the modelling exercise if they saw it as a parameter optimisation problem so that the model was both explanatory and predictive.

We are very aware of the limitations of the study. It was intended as a pilot study to check our assumptions and share the findings with the mathematics education community. Future work should explore students' and lecturers' (i.e., novices and experts according to Haines and Crouch 2004) responses to more sophisticated mathematical models that allow for the adjustment of parameters to optimize the output from the model.

Acknowledgements We would like to express our gratitude to Professor Chris Haines from City University, UK for his help with analyzing participants' responses.

References

Caron, F., & Belair, J. (2007). Exploring university students' competences in modelling. In C. Haines, P. Galbraith, W. Blum, & S. Khan (Eds.), *Mathematical modelling (ICTMA 12): education, engineering and economics* (pp. 120–129). Chichester: Horwood.

Crouch, R., & Haines, C. (2004). Mathematical modelling: Transitions between the real world and the mathematical world. *International Journal on Mathematics Education in Science and Technology, 35*(2), 197–206.

Galbraith, P., & Haines, C. (1998). Some mathematical characteristics of students entering applied mathematics courses. In J. F. Matos et al. (Eds.), *Teaching and learning mathematical modelling* (pp. 77–92). Chichester: Albion.

George, D. A. R. (1988). *Mathematical modelling for economists*. London: Macmillan.

Haines, C., & Crouch, R. (2001). Recognizing constructs within mathematical modelling. *Teaching Mathematics and Its Applications, 20*(3), 129–138.

Hsieh, Y., Lee, J., & Chang, H. (2004). SARS epidemiology modeling. *Emerging Infectious Diseases, 10*(6), 1165–1167.

Hughes-Hallett, D., Gleason, A., McCallum, W., et al. (2005). *Calculus: Single and multivariable* (4th ed.). Hoboken: Wiley.

Kadijevich, D. (1999). What may be neglected by an application-centred approach to mathematics education? *Nordisk Matematikkdidatikk, 1*, 29–39.

Kaiser, G. (2007). Modelling and modelling competences in school. In C. Haines, P. Galbraith, W. Blum, & S. Khan (Eds.), *Mathematical modelling (ICTMA 12): Education, engineering and economics* (pp. 110–119). Chichester: Horwood.

Kermack, W. O., & McKerdrick, A. G. (1927). Contribution to the mathematical theory of epidemics. *Proceedings of the Royal Society of London. Series A, 115*, 700–721.

Klymchuk, S., & Zverkova, T. (2001). Role of mathematical modelling and applications in university service courses: An across countries study. In J. F. Matos, W. Blum, S. K. Houston, & S. P. Carreiara (Eds.), *Modelling and mathematics education: ICTMA-9: Applications in science and technology* (pp. 227–235). Chichester: Horwood.

Neunzert, H., & Siddiqi, A. H. (2000). *Topics in industrial mathematics*. New York: Springer.

Patterson, K. D., & Pyle, G. F. (1991). The geography and mortality of the 1918 influenza pandemic. *Bulletin of the History of Medicine, 65*(1), 4–21.

Shi, P. and Small, M. (2003). Modelling of SARS for Hong Kong. *Populations and Evolution*. http://arxiv.org/abs/q-bio/0312016. Accessed 13 Sep 2009.

Watts, D. J., & Strogatz, S. H. (1998). Collective dynamics of small-world networks. *Nature, 393*, 440–442.

Chapter 49
Using Response Analysis Mapping to Display Modellers' Mathematical Modelling Progress

Akio Matsuzaki

Abstract In this chapter I propose a method to display mathematical modelling progress by using response analysis mapping. I focus on components that are constructed for each model created by the modeller. In addition, I identify components of the modelling based on each modellers' prior experiences; components based on real experiences (CRE) and components based on mathematical experiences (CME). As a case study to illustrate the method, the attempts of two modellers are compared. Links between CRE and CME during modelling have been confirmed by this method.

1 Introduction

The schema of mathematical modelling in modelling literature indicates an ideal modelling cycle, and procedures for progressing through the processes for successful problem-solving. Mathematical modelling attempts are not always successful, so it is important for researchers/teachers to gain insight into actual mathematical modelling progress. It is difficult to capture the essence of such progress because it is different for each modeller. Borromeo Ferri (2007) focussed on the mathematical modelling progress of individual modellers calling their overall attempts "modelling routes". From a cognitive perspective she identified mathematical thinking styles and modelling routes based on the modelling cycle by Blum and Leiß (2007). In this chapter, I focus on components that are constructed in each model created by individual modellers. So I propose to display mathematical modelling progress by using the method based on response analysis mapping (Stillman 1996; Stillman and Galbraith 1998) for viewing the whole progress of an attempt.

A. Matsuzaki (✉)
Saitama University, Saitama, Japan
e-mail: makio@mail.saitama-u.ac.jp

2 Components Based on Experiences

In mathematical modelling, the experiences of one modeller are different from those of another. These can be used to control the solution path as one of the supports to the final solution (Busse and Kaiser 2003; Stillman 2000). For example, when a certain person faces some problem, he or she may recall a situation where he or she has solved a similar mathematics problem. Models constructed based on this situation are related to mathematics and prior knowledge taken into account in the models is based on mathematical experiences. On the other hand, when a certain person faces some problem, he or she may recall a real life situation. Models constructed based on this situation are related to reality, and components constructed in the models are based on real experiences. Thus, components based on the experiences of each modeller can follow two directions. Furthermore, it is possible to investigate modelling progress by visually displaying links between components that are used to construct models using this idea. The aims of this chapter are (1) to identify components based on a modeller's own experiences in order to display details of their mathematical modelling progress and (2) to describe how some components change during the mathematical modelling progress.

3 Applied Response Analysis Mapping as an Analysis Method

Response analysis mapping is one of the methods used for erroneous answer analysis based on answer descriptions, and together with task analysis mapping as a means of the qualitative evaluation of learning (Biggs and Collis 1982; Biggs and Telfer 1987; Stillman 1996; Stillman and Galbraith 1998). A typical map is now described (see Fig. 49.1).

In my use of response analysis mapping, mathematical components (MC) are shown as cues or nodes indicated by ●, and non-mathematical components (RC) are shown as cues or nodes indicated by ○. These components used by modellers in their solution are distinguished by researchers from the written task response after solving ends. The interrelationships between components are shown by arcs. The processes that bind related components together are shown as nodes of

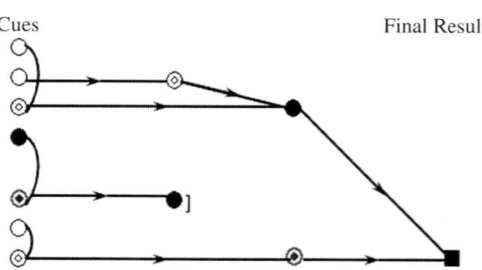

Fig. 49.1 A typical applied response analysis map

intersection of arcs on the way to a final result shown as ■. Thus the relations between components can be followed visually in a diagram. In the case of abandoning the solving process, such nodes are shown with a right side bracket symbol ']'. We can trace modelling progresses from the upper left to the lower right on such a diagram.

In this study I itemize *prior experiences* of modellers and incorporate them into applied response analysis mapping. The objects recalled from prior knowledge are of two different types; firstly based on real world experiences and secondly based on mathematical experiences. In short I itemize all components in the form of prior experiences that each study participant used in his or her solutions that are made clear using the think-aloud method. In applied response analysis mapping, prior knowledge related with reality, that is components based on real experiences (CRE), are shown as cues or nodes indicated by ◎. Prior knowledge related to mathematics, that is components based on mathematical experiences (CME), are shown as cues or nodes indicated by ⊛. The components indicated by ● or ○ are identified from worksheets, whereas the components indicated by ◎ or ⊛ are identified from protocols or responses in interviews. By incorporating CRE and CME into maps, it is possible to describe details of transition components that are not presented in the written response of the modeller.

By using applied response analysis mapping as a method to display mathematical modelling progress, it is possible to display relations or links between components and processes during the construction of models. Modellers will sometimes carry out additional verifying or confirming of the progress of their solutions and this can be shown on the maps after the final node has been connected. An example of this will follow in Sect. 5.1.

4 Research Setting

The participants were asked to tackle an electronics problem using the think-aloud method so the researcher could follow each element of their mathematical modelling progress. The focus cases are a graduate school student, NT, majoring in physical science education and an electronics worker, KN. The data collection involved interviews after responses on worksheets were completed and transcriptions of think-aloud protocols. The investigation involved two stages. The questions and the procedures each time were as follows:

For the first investigation stage, the problem situation 'How much brightness is needed to read a book?' is located in a real situation in accordance with the mathematical modelling cycle. The participants tackled four questions; '(a) What is necessary to solve the problem situation?', '(b) What kind of things did you imagine on thinking about (a)?', '(c) State a problem by using all or some of (a)' and '(d) Solve the problem that you stated for (c)'.

For the second investigation stage, the participants were interviewed by the researcher based on descriptions or protocols from the first stage. Using a lux meter, each participant measured the brightness of the room where the investigation was carried out, and attempted to pose a problem with reference to the measured data. As additional information, they were presented with the sizes of the room and illumination standard for houses before tackling the questions. In this chapter the focus is the first investigation stage. In this stage the participants were required to tackle the problem situation and questions based on their own experiences.

5 Modelling Progress Using Applied Response Analysis Mapping

5.1 The Case of a Graduate School Student

The first participant, NT, was a graduate school student who had studied physical science education. The components of the mathematical modelling progress of NT are displayed by using the method of applied response analysis mapping in Appendix 1.

For questions (a) and (b), NT focused on light and selected components related with light as necessary cues (i.e., first to third in Appendix 1) to solve the original problem situation. Next he recalled two cues: the desk in his own room (i.e., his own house or the laboratory to which he belonged) and the room cue where he tackled the questions. He made the following comment:

04:58   Is it difficult to imagine the thing, which [is] only the thing that I have watched, only a certain thing around me? I think that it is this.

Thus, before posing a problem, NT pointed out several cues that were all related to reality except for the 'numbers of lights' (i.e., 'light' to 'desk in the laboratory' cues in Appendix 1).

For question (c), NT posed the following problem: 'There is a desk right under a light A. Reading was made with less than 1 m distance from light A to the desk. Can you read a book now when you prepare for light B with half the brightness of light A if the distance of light B from the desk is less than what value in metres?'

When NT posed this problem, some *of his prior knowledge* of light was not related with the solving process (shown by use of ']' in Appendix 1), and he made the following comments:

09:07   How much is readable far and wide if I light it more? Is such a problem good?
09:15   It [problem posing] does not come easily.

One problem set by NT was based on mathematics, namely the 'distance between light and desk' shown as ⊙ in Appendix 1. Prior knowledge (from 13th to 15th cues in Appendix 1) is related with reality and so shown as ◎, but these cues are not used for checking the setting of the problem or reflecting on the solution.

For question (d), NT applied his knowledge of physical science (e.g., brightness is inversely proportional to square) to solve the problem set in question (c). These components are CME shown as ◉ and are linked to the final solution shown as ■.

12:22 .... The brightness of the light works for the area ratio of the circumference... It should be inverse proportion to square...

After solving the problem set in question (c) mathematically and finding the final result $r = 1/\sqrt{2}$, that is, 0.7 times, NT recalled prior knowledge of an operating room cue related with reality. He explained this cue (CRE) by the following comments show:

20:52 ... I think that I have heard such a story in physics....
21:09 ... There are ... an operating table ... operating table, and there is a light then. The distance from a light to an operating table assumes that it is 1,000 luxes at the distance of 1 m. It is a case under 2 m. In other words it is just a floor where there is the light of the operating table, and there is an operating table and the floor is said to be 1 m below. During the night, the illumination of the floor was said to decrease considerably.

These components are not used for solving the problem but are used to check whether the answer is right or not. In Appendix 1, the final two cues ('operation table' and 'astral lamp'), shown as ◉, are confirmed as linked to the final result shown as ■.

5.2 The Case of an Electronics Expert

The second participant, KN, was an electronics expert who worked in an electronics job. The components of the mathematical modelling progress of KN are displayed using the method of applied response analysis mapping in Appendix 2.

For question (a), KN divided the initial task into three groups of cues to consider (individual differences, lighting, and books) which are all non-mathematical. He discussed necessary components to solve the original problem situation:

11:18 About an element of the ambiguity that I just said as necessary things [referring to these cues], I think that [a] definite decision is necessary.

For question (b), KN wrote on the worksheet: 'I can't decide whether I apply a problem to general various people or it is not performed under the condition'. His own indecision is evident in the following comment:

19:11 Rather than a problem for calculation, I feel like a problem is demanded experimentally through the piling up of data one by one.

Although the participants had to pose a problem based on the original problem situation for the first investigation stage, the participants would tackle such problems based on experimental data during the second investigation stage (Matsuzaki 2007).

For questions (c) and (d), KN used almost all the components raised in question (b) and set the following problem: 'Find necessary brightness in each age group.'

He made the following observations with regards to solving this problem:

43:02 I make five phases of indexes about brightness and make a questionnaire.
46:26 I handle it using statistical technique…

KN used some ideas based on his own job experiences as the following comments show:

27:01 The case of the general type of house…The distance from roof to desk.
31:43 The illumination is fluorescent, and that is three wavelengths. Because the case used generally is three wavelengths…
32:50 The kind of the illumination is decided in the rating…

KN finished his solutions without coming to a final result, but he indicated the prospect for a mathematical solution; that is a statistical technique be used for a questionnaire, and to indicate this lines from 16th to 18th cues in Appendix 2 are drawn to the node labelled "accumulation of data". After presenting this solution he focused on a large space that was different from the cues used until then.

48:08 When I think about the most suitable illumination, I had better think in space.

These components shown as the last two cues in Appendix 2 are related with mathematics (CME), but they are not used for checking or reflecting on solutions.

6 Discussion

The mathematical modelling progress can be displayed by using the method of applied response analysis mapping, which shows the influences of modellers' own prior knowledge based on their own experiences. This prior knowledge has two aspects (CME and CRE). Especially CRE indicate prior learning experiences or job experiences, and it is shown that these are one of the components that are used to construct some models in modelling.

6.1 Focus on CRE

NT applied physical science knowledge in his mathematical modelling progression; for example, diffusion of light or relationships between distance and brightness. At this time he recalled some cues based on his prior experiences such as lighting instruments in an operating room. These CRE can be based on his current or prior experiences. The former CRE are 'Fluorescents in this room' and 'Lamps in the room' and the latter CRE are 'My own light stand' and 'Desk in the laboratory'. The operating room cue is used for confirming that one of the roles of CRE is as a support of the interpretation of the presented situation (cf., Stillman 2000). Both CRE 'operation table' and 'astral lamp' are used to confirm results. CRE derived from the initial non-mathematical components, 'light' and 'my own desk', are 'How much is readable if one approaches?' and 'How much is readable far and wide?'

In contrast, KN applied his expertise in his electronics job, for example, standards of brightness and wavelengths of fluorescence. We are able to check that CRE are incorporated in lighting cues in his mathematical modelling. In his written response, KN noted that he used these features in his job and tried to solve the problem he posed with statistical knowledge and skills. Additionally, he limited the target RC to 'Each age group from 10 generations to 80 generations' when he attempted to make the questionnaire. He used his own illness, dry eyes, to limit the solution to people with 'not illness of eyes'. One of the characteristics of his modelling progress is to identify three cues, namely, individual differences, lighting and books. In particular, CRE in the lighting group of components (e.g., 'Flicker') are based on his electronics work.

6.2 Focus on CME

Finally I focus on prior knowledge related with mathematics. In the mathematical modelling progress of NT, the 'light A' components (CME) is the starting point for solving mathematically. This component is based on other components, namely two RC and two CRE. At this time, 'light A' is an independent component and 'light B' (MC) is set as a dependent component. Additionally, prior knowledge of 'inverse proportion to square' cue (CME) is derived from 'light' component (RC) via 'area ratio of the circumference' cue (CME). In this way, NT constructed relationships between selected components in his modelling.

The mathematical modelling progress of KN used all components except two (19th cue and 20th cue in Appendix 2). 'Distance between surface of books and a light source' (CME) is a mathematical model that was derived from the connection between lighting cue and book cue, two of the three groups of cues he chose to consider. On the one hand, some CME (e.g., 'size of a pocket edition') are reflected as components in the 'questionnaire', but on the other hand he does not use 'large of space' cue and 'a six-mat room or eight-mat room' cue related with mathematics, that is, these components (CME) are not used for checking or reflecting on solutions. Thus cues related with mathematics might not always function within the solution.

7 Conclusion

In mathematical modelling, the prior experiences of a modeller can be used to control the solution path as one of the supports for the final solution (Busse and Kaiser 2003; Stillman 2000), but these components based on the experiences of each modeller can follow two directions. In this study I distinguished prior knowledge related with reality based on real experiences (CRE) from prior knowledge related with mathematics which is based on mathematical experiences (CME). In addition, I itemized prior experiences of modellers and incorporated them into applied response analysis mapping as an analysis method. By using this method, the influences of CRE and CME on modelling progress can be displayed, including how some of these components are used to construct some models.

Appendix 1: Mathematical Modelling Progress of NT

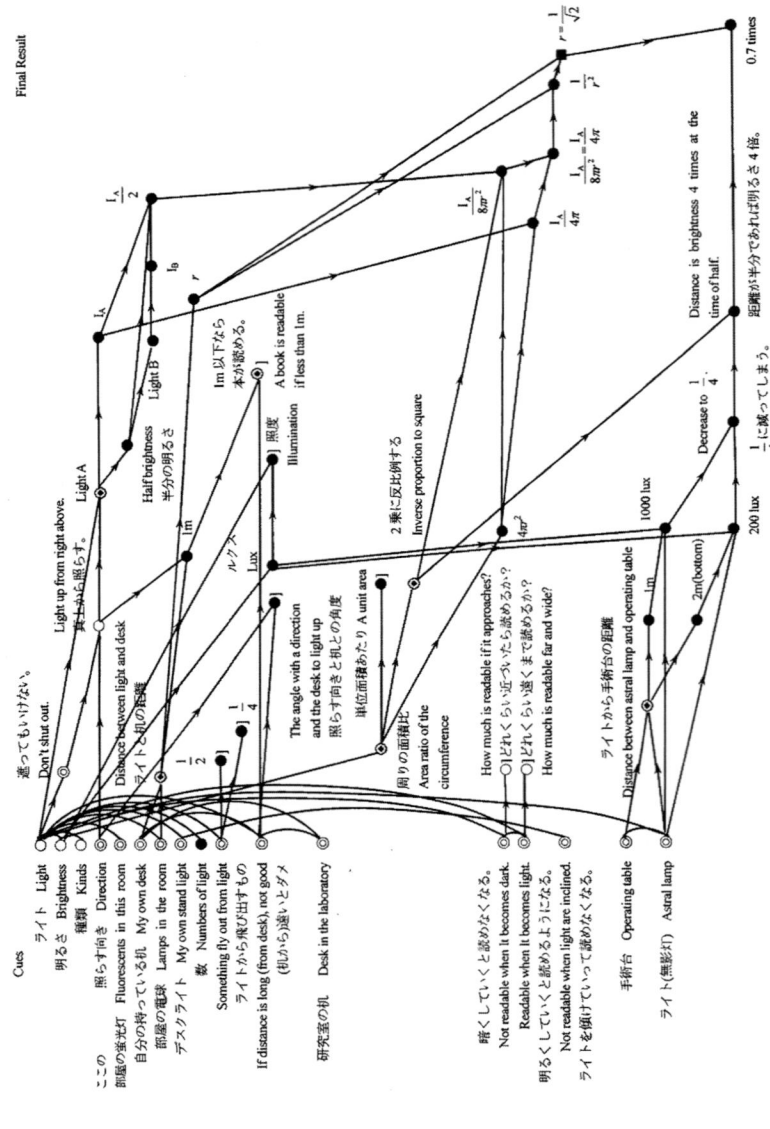

Appendix 2: Mathematical Modelling Progress of KN

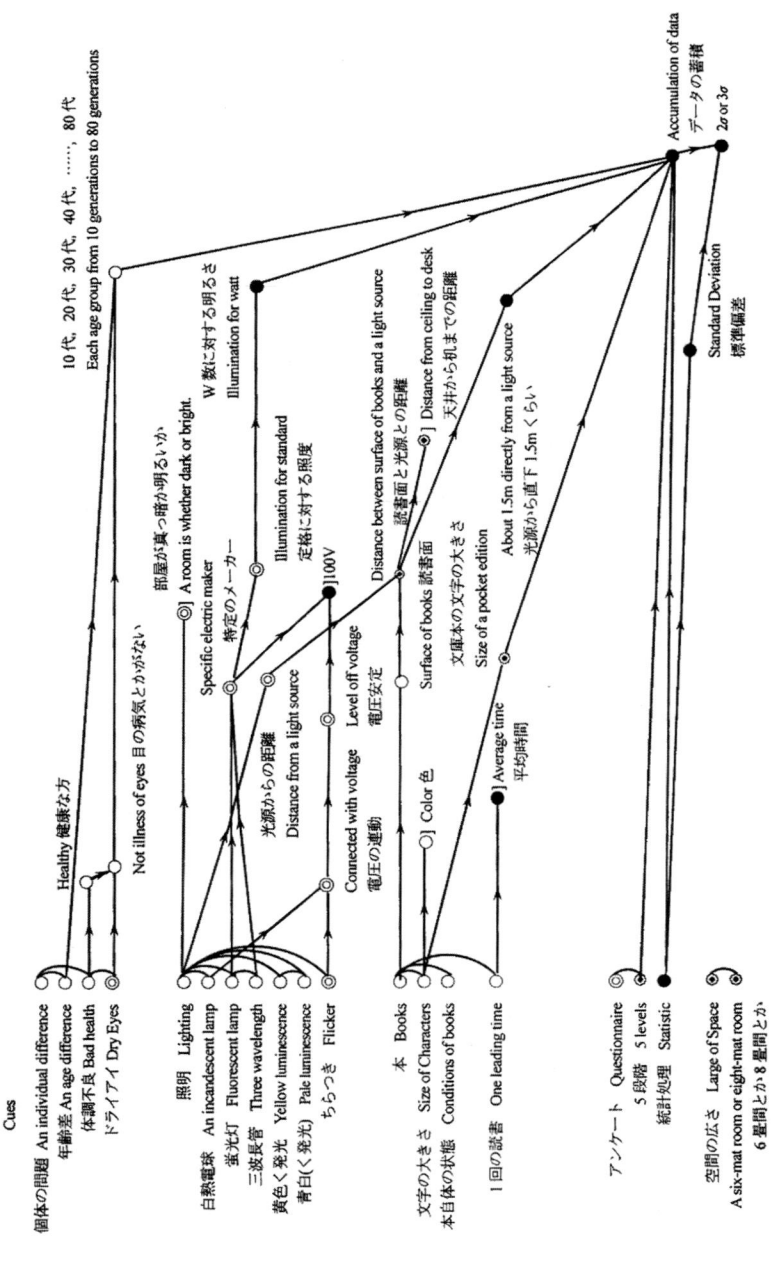

References

Biggs, J., & Collis, K. (1982). *Evaluating the quality of learning: The SOLO taxonomy (structure of the observed learning outcome)*. New York: Academic.

Biggs, J., & Telfer, R. (1987). *The process of learning* (2nd ed.). New South Wales: Prentice-Hall.

Blum, W., & Leiß, D. (2007). How do students and teachers deal with modelling problems? In C. Haines, P. Galbraith, W. Blum, & S. Khan (Eds.), *Mathematical modelling (ICTMA12)* (pp. 222–231). Chichester: Horwood.

Borromeo Ferri, R. (2007). Modelling problems from a cognitive perspective. In C. Haines, P. Galbraith, W. Blum, & S. Khan (Eds.), *Mathematical modelling (ICTMA12)* (pp. 260–270). Chichester: Horwood.

Busse, A., & Kaiser, G. (2003). Context in application and modelling: An empirical approach. In Q.-X. Ye, W. Blum, S. K. Houston, & Q.-Y. Jiang (Eds.), *Mathematical modelling in education and culture: ICTMA10* (pp. 3–15). Chichester: Horwood.

Matsuzaki, A. (2007). How might we share models through cooperative mathematical modelling? Focus on situations based on individual experiences. In W. Blum, P. Galbraith, H.-W. Henn, & M. Niss (Eds.), *Modelling and applications in mathematics education: The 14th ICMI study* (pp. 357–364). New York: Springer.

Stillman, G. (1996). Mathematical processing and cognitive demand in problem solving. *Mathematics Education Research Journal, 8*(2), 174–197.

Stillman, G. (2000). Impact of prior knowledge of task context on approaches to applications tasks. *Journal of Mathematical Behavior, 19*(3), 333–361.

Stillman, G., & Galbraith, P. (1998). Applying mathematics with real world connections: Metacognitive characteristics of secondary students. *Educational Studies in Mathematics, 36*(2), 157–195.

Part VII
Modelling Examples and Modelling Projects: Concrete Cases

Chapter 50
Modelling Examples and Modelling Projects – Overview

Hugh Burkhardt

1 The Challenge

Mathematical modelling as an activity is absent from the implemented curriculum in most classrooms around the world, after the early years where counting is commonplace. Students are taught standard models and expected to apply them. To take some simple examples, proportional models are a major feature of middle-grade mathematics. Students are expected to solve problems like:

> 6 friends bought a six-pack of cola for $3.
>
> How much should each of them pay?

However, in most textbook chapters on proportion, all the situations are indeed proportional. Students are rarely asked to identify whether the situation is one for which a proportional model is appropriate. They do not meet problems like:

> If it takes 40 minutes to bake 5 potatoes in the oven.
>
> How long will it take to bake 1 potato?

where the appropriate model depends on the type of oven[1], let alone a situation like:

> King Henry VIII had 6 wives.
>
> How many wives did King Henry IV have?
>
> – a fact with which modelling cannot help you.

---

[1] Constant, i.e., roughly 40 min, in a regular oven; Proportional, i.e., roughly 8 min, in a microwave oven.

H. Burkhardt (✉)
Shell Centre, University of Nottingham, Nottingham, UK
e-mail: Hugh.Burkhardt@nottingham.ac.uk

Thus modelling presents teachers and students with a new kind of challenge. Students are accustomed to this – school is all about new challenges. However, modelling will take most teachers outside their comfort zone of professional expertise. This chapter is focused on the various pedagogical and mathematical demands of teaching modelling and applications, and how teachers may be helped to respond to them effectively.

1.1 Analysis of the Challenge

In this book there are two papers with the emphasis on analysing the challenge modelling poses on teachers and students.

Barbara Schmidt looks at obstacles from the teacher's perspective. What factors do they say are holding them back from introducing modelling into their classrooms? She reports on a questionnaire-based study of the views of 101 teachers, half of whom take part in training in teaching modelling. There were also in-depth interviews with six teachers. The questionnaire was administered four times over a year – before, during, and after the training period, and 5 months later. It was designed to probe 14 factors. Seven are related to the kinds of obstacle that Blum has described: organizational; pupil-related; teacher-related; and materials related. (The other seven, not analysed here, have a more positive focus – on the affordances that modelling provides).

The teachers' responses identified three major factors that inhibit their teaching of modelling:

- Ninety-seven percent mentioned the time that modelling problems require as an obstacle, a view that was only slightly changed by training. This is a generic problem – the learning of mathematical concepts also requires a process of reflection, building the connections that are essential for robust long-term understanding.
- Before the training, 61% of the teachers said there was too little material available for them to use in the classroom; happily, this concern more or less disappeared during the training, for those who had it but not the control group. (This still leaves the issue of making these materials widely known to teachers, through training and other channels).
- The third obstacle that teachers mentioned was the challenge of assessing performance in modelling. Eighty-four percent of teachers mentioned it, and this view was not affected by the training. While teachers of history or first-language studies are accustomed to assessing essays and other forms of open writing, school mathematics is still dominated by "answers" that are correct or not.

The study notes the importance of teachers' beliefs about mathematics itself, which many felt does not extend to modelling. As with any study of this kind, the overt responses may conceal some deeper discomforts, of the kind that everyone feels when required to extend their professional practice.

Richard Cabassut and Anke Wagner compare French and German primary school textbooks, looking at the role that modelling plays in those curricula from an anthropological perspective. They focus on the use of the term 'modelling', which is rare in the French primary context. They recognize that problem-solving in the primary school includes modelling, but they do not see both of these as strategic/process skills rather than 'knowledge to be taught'. They argue that modelling is perhaps knowledge that is preparatory for the mathematics curriculum. That modelling is not 'knowledge to be taught' presents challenges for teacher education.

1.2 Helping Teachers

Teachers face the biggest challenges when modelling is introduced into the curriculum. All teachers have a well-established pattern of professional practice involving a spectrum of 'moves' in the classroom that covers the situations they face day-by-day. For most teachers of mathematics, this spectrum does not cover all the skills that teaching modelling requires. These include the teacher:

- Moving from a view of mathematics as a large set of separate "things to learn" to seeing it as a well-connected set of concepts and skills that, used flexibly, enable you to solve problems and understand the real world better. (In the teaching of language, this would mean moving from seeing the language as a set of rules for spelling and grammar to a focus on reading and writing substantial pieces – a focus which, of course, all native language teachers have).
- Moving from short item fragments of mathematics to tasks involving longer chains of reasoning – with a teacher's focus on students' reasoning, not just their answers.
- Giving students greater responsibility for their own and each others' learning, moving them into "teacher roles" like explaining their assumptions and assessing each others' reasoning.
- Becoming a diagnostician and adviser rather than a source of answers and summative right/wrong judgements – for modelling, it is an essential part of the students' job to decide if their solutions and reasoning are correct (as in life).

Three of the papers in this chapter address these challenges, and the teacher education programmes that address them. All three relate to learning and education in and through modelling and applications (LEMA), a Europe-wide professional development course focused on modelling.

Katja Maaß and Johannes Gurlitt describe the overall framework of the LEMA project and its evaluation. They outline the theoretical background, the design of the course and its evaluation, and the results.

The design of the course focused on the following theoretical ideas: modelling and its teaching, professional development of teachers, and their beliefs about self-efficacy and about mathematics education – procedures and formalism, processes and application. These ideas were embodied in the five modules of the course.

These addressed, respectively, modelling, tasks, lessons, assessment, and reflection. The course was piloted in six countries.

The cross-country evaluation used a pre–post control group design, investigating the following questions with the differences from the control group briefly noted here:

1. Does the professional development course influence the pedagogical content knowledge of the teachers? Improvement.
2. Does the professional development course influence teachers' beliefs about mathematics education? No change.
3. Does the professional development course influence teachers' self-efficacy regarding modelling? Improvement.
4. Does success in one dimension include success in another dimension (correlational)? Results on 1–3 above correlated at about 0.5.
5. How satisfied were teachers with the professional development course? High level of satisfaction.

The course was appreciated and had a positive effect on the teachers' pedagogical content knowledge and modelling self-efficacy, without affecting their beliefs.

Geoff Wake looks at case-study data generated by five teachers as an 'e-narrative' of their work in their classrooms at the initial stages of introducing modelling activities in the UK context, which is heavily dominated by tests that do not include modelling. He seeks to identify the issues of concern to teachers as they attempt to change their pedagogic practices to include mathematical modelling and to draw conclusions for professional learning both in general, and in relation to modelling in particular. He uses cultural historical activity theory (CHAT), which provides a tool for understanding how the work of a collective, such as teacher and pupils in mathematics classrooms, is mediated by different factors. Success in mathematics (and English) tests of a school's pupils at age 16 is crucial in the school's annual performance measure, an important factor in seeing low-risk strategies in teaching mathematics. This manifests itself in classroom discourse and behaviour that focuses heavily on 'the test', with little in the way of assessing modelling sub-competencies such as 'interpretation'. It is in this culture that the teachers need to find the motivation to adopt new pedagogies that support modelling.

We see, through the lens of CHAT, that expansive professional learning in relation to teacher knowledge cannot be left to chance. The case studies suggest that, in general, professional learning requires the intersection of three important factors:

1. The key personnel involved must have at least approximately aligned long term goals, and a professional expertise and understanding of the context that allows them to work within the rules of the system but adapt these to the benefit of the desired professional learning.
2. A climate in which new or potentially emerging rules appear to mediate an expansion in the object of activity.
3. Networks of personal relationships that facilitate ease of communication and boundary crossing.

In the particular case of teacher development in relation to modelling, the case studies point to the teachers' and students' changing roles in the classroom as being of most concern to teachers as they first use modelling activities. This has important implications for those supporting such changes through facilitating professional development.

Javier García and Luisa Ruiz-Higueras use an anthropological approach to the analysis of teachers' practices to study the changes resulting from the LEMA course in teachers' practices involving modelling and applications. They consider the problem of initial teacher training and professional development (particularly in modelling and applications) as a problem of the teaching profession more than a teacher's problem. They use the anthropological theory of didactics (ATD), which assumes that:

- In order to understand how individuals act, we need to know first how the institutions they belong to *act*.
- Mathematics as a human activity requires the integration of the *practical* with the *theoretical* aspects.

They argue that teaching modelling and applications is more than a problem that teachers face in their classroom; it should be considered a problem that the teaching profession faces due to systemic changes, in the way mathematics is being considered, in the new general aims assigned to schooling, and in the inadequate training of those responsible for developing the curriculum. The paper describes how this generates professional problems like: How can real contexts and situations be used in order to give meaning to mathematics? So, for example, teachers who like a modelling approach face didactical problems like: How can the ideas given in this textbook be restated to encourage students to explore 'variables' in real contexts?

LEMA offers teachers opportunities to develop their practice with a wide range of teaching techniques that support modelling, and to reflect and justify the why of these practices that is to develop their theoretical perspective.

1.3 Situations for Modelling

The final seven papers all present interesting and novel situations that have been used in the teaching and learning of modelling at high school or undergraduate level.

Hans Humenberger's paper is concerned with a problem of interest and concern to most of us – how does Google decide where different websites appear in the list resulting from a search. He shows how this can be tackled at undergraduate level, using a wide range of mathematical techniques including directed graphs and transition matrices that embody the Markov process structure of the problem. The general approach can be realized in a more elementary form with a spreadsheet.

Matthias Brandl describes an internet portal that offers modelling problems aimed at gifted upper high school students. He describes the results of an attitude survey which suggests that mathematically gifted and highly gifted students are (very) interested in applicability and therefore in word problems with connection to real world application. He presents two examples that emphasize modelling processes:

- Using the context[2] of a champagne glass, which cone made from a circle of fixed radius has the greatest volume
- How does the number of winners in a lottery depend on the number of participants?

The first example involves straightforward algebraic modelling and the second more challenging probability theory.

Usha Kotelawala shows how basic models of reliability theory can provide motivating problems for secondary students as they develop skill and understanding in probability and algebra within secondary mathematics. The author gives an outline of reliability theory, presented as a simple application of the probability, and gives simple examples such as:

- Grandma called yesterday to ask you to help her hang holiday lights for the winter season. Her lights are the old type of light strings. If one bulb fails, the whole string fails. The probability of each bulb working is 0.75. If one of the strings has only two bulbs (a short string for illuminating very small areas) what is the probability of the string working?
- FlyCheap Airlines has only one route from London to Hamburg. Recently, the company purchased a used 727 jet airplane having three engines. It can actually fly on just one engine. For each engine, the probability of it working over the course of the trip is 0.98. What is probability that the plane will be able to successfully fly? Find at least two different ways to determine your answer.

before moving onto more complex (both "series" and "parallel") problems. As well as requiring reliable procedural skills (simple and complex substitution, expressing and simplifying polynomials, solving polynomial equations), the problems included opportunities for students to work on inductive reasoning for generalized models and the strategy of simplifying parts of a problem along the path to a larger solution.

Tetsushi Kawasaki and Seiji Moriya address the widespread concern in Japan at the estrangement of science from mathematics in the mathematical curriculum of Japanese senior high schools. For example, in order to connect mathematics to the fields of Newtonian physics through mathematical modelling, they provided concrete examples. The one described in some detail focused on 'Kepler's Laws'. First Kepler's "data", summarized in the first and second laws, is provided and analysed, with IT support. Then Newton's second law and that of gravitation is described. The students focus on numerical solutions using Euler's method,

---

[2] Nice but not directly relevant – they don't make champagne glasses that way.

confronting the error and stability issues. The evaluation suggests that this approach succeeded in increasing students' knowledge about the laws with the simulation of planetary movements and, at the same time, their making of a "mathematical development model". It also suggests that mathematics materials involving physical perspectives are effective for senior high school students.

Mette Andresen and Asbjoern Petersen describe a study of upper secondary students in Denmark using technology in modelling chemical equilibrium, a context that integrates mathematical and chemical modelling.

'Multi-disciplinarity' was prescribed in Danish Upper Secondary Schools' curriculum by governmental regulations, with requirements centred on applications of, and reflections upon each subject. The revision of mathematics teaching is intended to support the students' knowledge about 'how mathematics adds to understanding, formulating and treating problems in different subject areas' and to know about mathematical reasoning. The learning goals served as a basis for the design of multidisciplinary mathematics teaching.

This example is based on dynamic equilibrium in chemistry in which, for example, the two-way reaction comes to equilibrium when the forward reaction rate (proportional to the concentrations of A and B) and backward reaction rate (similarly for the concentrations of C and D) become equal. The students worked on a project that involved both theoretical and experimental work, leading to written reports. These were analysed for their perceptions of modelling in a multi-disciplinary teaching environment. The authors found that there was little evidence of a focus on the modelling[3]. The reason for this are discussed.

Martin Bracke and Andreas Geiger describe in their paper experiences with a long-term teaching experiment, in which they included real-world modelling examples in regular lessons. They describe several challenging realistic tasks such as the design of a track for the German high-speed rail. They evaluated the tasks and showed that students were mainly interested and fascinated by this kind of modelling activities. The students reported that they had learnt modelling and sometimes even changed their attitude towards mathematics.

The paper by Gabriele Kaiser, Björn Schwarz and Nils Buchholtz reports on modelling activities with even more demanding authentic modelling problems in the framework of modelling weeks with students from upper secondary level. Students' solutions to one demanding task, namely the development of infected lady bugs, were described and show that the students reached impressive solutions from a mathematical and real world perspective. Furthermore, the results of the evaluation point out that most of the students were deeply impressed by these kinds of authentic examples and really interested in these activities.

Both approaches show that ambitious modelling examples can be dealt with in mathematics teaching, but they require long-term processes, but then lead to unexpected high results.

---

[3] This reflects earlier work that found that, when actual experiments are involved, making these work took most of the students' attention, leaving little for scientific reasoning.

Chapter 51
Modelling Chemical Equilibrium in School Mathematics with Technology

Mette Andresen and Asbjoern Petersen

Abstract This chapter presents an example of mathematics within a multidisciplinarity context, which took place recently in Danish upper secondary school. The case's topic was modelling a system of chemical equilibrium in a solution of molecules and ions. The teacher deliberately focused on the intertwined mathematical modelling and chemical modelling, as a means to realize a multidisciplinary teaching perspective of the two subjects. The students' written reports were analysed with the aim to study different aspects of their perceptions of modelling, as a result of multidisciplinary teaching. Means and obstacles for support of modelling competency by multidisciplinary teaching are discussed.

1 Introduction

The construct 'multidisciplinarity' was prescribed in Danish Upper Secondary Schools' curriculum by governmental regulations in 2005. The Ministry's intentions and requirements were centred on applications of, and reflections upon, each subject. The revision of mathematics teaching intended to support the students' knowledge about '*important aspects of the interplay between mathematics and culture, science and technology*'. The students were also supposed to acquire knowledge about '*how mathematics adds to understanding, formulating and treating problems in different subject areas*' and to know about mathematical reasoning. The learning goals served as a basis for the design of multidisciplinary mathematics teaching, which was also intended to result in knowledge that enabled the students

---

M. Andresen (✉)
National Knowledge Centre for Mathematics Education, University College Copenhagen, Copenhagen, Denmark
e-mail: mea@ucc.dk

A. Petersen
Avedoere Gymnasium, Hvidovre, Denmark
e-mail: ap@esteban.dk

to competently take a position on the applications of mathematics and to pass further education involving mathematics. The design and potential for the use of multidisciplinary mathematics teaching, as well as the characteristics which differentiate it from interdisciplinary and transdisciplinary teaching, are further discussed in Andresen and Lindenskov (2008).

2 The Case

This chapter reports on the use of multidisciplinary teaching involving the subjects of mathematics and chemistry. The overarching topic was modelling a chemical equilibrium system in a solution of molecules and ions. This topic was chosen as an example of intertwined mathematical modelling and chemical modelling, with the aim to realize a multidisciplinary teaching perspective on modelling in both subjects. The core issue of the case was to study the students' perceptions of models and modelling, developed during a teaching sequence where the teacher deliberately focused on the intertwining of modelling concepts from mathematics and chemistry. The aim of the study was to inquire whether multidisciplinarity as a construct was able to support students' reflections about modelling and to support their modelling competencies in general.

In addition, the topic was also suitable for the study of the impact of technology use on students' learning processes, since it provided the opportunity to combine the chemical and the mathematical modelling in a technological environment. Such a combination provided the opportunity to consider approximations and the use of technology, with elements of each of the four approaches mentioned in (Confrey and Maloney 2007, p. 57):

1. Teach concepts and skills without computers and provide these technological tools as resources after mastery, i.e., to solve the systems of equations by hand and subsequently introduce the discussion of different approximations and the use of MathCad.
2. Introduce technology to make patterns visible more readily, and to support mathematical concepts: in our case, the concept of the sets of solutions and how to choose between them was supported.
3. Teach new content necessitated by a technologically enhanced environment; in our case handling systems of eight or more equations and unknowns.
4. Focus on applications, problem solving and modelling and use the technology as a tool for their solution: this was the very aim of the equilibrium project.

The data for this study consisted of teaching materials (separated into a teacher's part and a student's part) prepared by the second author (Petersen 2009) who taught the class chemistry, the students' written reports, and notes from informal talks with the second author. The teacher's part of the teaching materials is of interest for our analysis because it articulates the intentions and reflections behind the choice of topics and the design of the complete teaching sequence. The students' part of the teaching materials gives information about the expectations for their work, supported by details from the informal interview with the second author.

3 Models of Chemical Equilibria

The teaching sequence studied here deals with chemical reactions in a solution of substances, where each solution simultaneously reaches equilibrium. And, as such, the bidirectional reaction proceeds with the same speed, thereby ensuring that the amount of each substance remains the same. The topic of interest then is to describe the chemical system at equilibrium with regard to the amount of each substance. This chemical system can be described in three, independent but closely related, ways: (1) by experimental modelling in chemistry, (2) by theoretical modelling in chemistry and (3) by theoretical modelling in mathematics.

Modelling the chemical system based on mathematical, chemical equilibrium theory takes as the starting point, that the composition of a system of chemical equilibria in a solution can be described by the actual concentration of molecules and ions contained in it. The relationship between concentrations may be written in a number of equations based on the law of mass action,[1] the law of conservation of matter and energy[2] and conservation of charge. Any mathematical model is intertwined with the chemical model in the sense that the mathematical model of the chemical equilibrium system must consist of a number of equations with the same number of unknowns (the unknown concentrations). The number of equations may correspond to the degree of accuracy of the model, but some approximations are better than others. Traditionally, approximations were needed for technical reasons when these systems of equations had to be solved by hand. Now it is possible to solve any solvable system of equations with the use of technological tools, which gives an opportunity for a discussion with the students about which are the appropriate approximations and why. A number of possible solutions are obtained by solving the system of equations; however, only one of these is acceptable from a chemical point of view. This fact, along with discussions regarding the approximations used, offers the potential for students' to reflect on modelling in chemistry and in mathematics, and on the links between these two branches of modelling. The overall objective then is to support the students' development of modelling competencies in mathematics and in chemistry.

3.1 Chemical Equilibria in Our Case

One example, used in our case, of a system suitable for inquiry is realised in dissolution of silver nitrate in aqueous ammonia. The system may, as described above, be modelled on different levels of complexity, depending on the number

---

[1] The law of mass action: When a reversible reaction has attained equilibrium at a given temperature, the reaction quotient (the product of the molar concentrations of the substances to the right of the arrow divided by the product of the molar concentrations of the substances to the left, with each concentration raised to a power equal to the number of moles of that substance appearing in the equation) is a constant. (Holtzclaw et al. 1984 p. 4).

[2] The law of conservation of matter: During an ordinary chemical change, there is no detectable increase or decrease in the quantity of matter. The law of conservation of energy: During an ordinary chemical change, energy can be neither created nor destroyed, although it can be changed in form. (Holtzclaw et al. 1984 p. 4).

of approximations. In the following, we give one version and outline two other versions (Petersen 2009, first author's translation from Danish).

Version 1. If the law of conservation of matter and the law of equilibrium are applied twice, four equations can be set up to give a medium – complex description involving the four unknown $[NH_3][Ag^+][Ag(NH_3)^+][Ag(NH_3)_2^+]$ concentrations.

Reaction schemes for two steps of the stepwise formation of chemical complexes:

$$Ag^+(aq) + NH_3(aq) \Leftrightarrow Ag(NH_3)^+(aq) \quad \text{(A)}$$

$$Ag(NH_3)^+ + NH_3(aq) \Leftrightarrow Ag(NH_3)_2^+(aq) \quad \text{(B)}$$

The reaction schemes (A) and (B) give the two equations (51.1) and (51.2), respectively, which in total contain the four unknown concentrations (with fixed values of the constants K_1 and K_2, picked from a tabular list).

$$\frac{Ag(NH_3)^+}{[Ag^+][NH_3]} = K_1 \quad K_1 = 2{,}090 \text{ M}^{-1} \quad (51.1)$$

$$\frac{[Ag(NH_3)_2^+]}{[Ag(NH_3)^+][NH_3]} = K_2 \quad K_2 = 8{,}320 \text{ M}^{-1} \quad (51.2)$$

Since no silver ions or ammonia may have disappeared, we find for the (well-known) formal concentration of silver, c_{Ag}, that:

$$c_{Ag} = [Ag^+] + [Ag(NH_3)^+] + [Ag(NH_3)_2^+] \quad (51.3)$$

and for the (well-known) formal concentration of ammonia, c_{NH_3}, that:

$$c_{NH_3} = [NH_3] + [Ag(NH_3)^+] + 2[Ag(NH_3)_2^+] \quad (51.4)$$

Version 2. Chemical arguments may point to decreasing the number of unknown concentrations which will give a model consisting of fewer equations. The system, then, can be reduced to three equations with three unknowns.

Version 3. On the other hand, regarding Ammonium as a base gives rise to a model which includes four additional unknowns and equations, stemming from Ammonia's reaction with water. To these eight unknown concentrations correspond eight equations, some of which are identical to equations in the first system.

4 Aims of the Students' Work

In the students' part of the teaching materials, formulation of the equations and identification of unknowns, etc. is developed as a guided, stepwise modelling process, in parallel with experiments conducted by the students in the chemistry laboratory. The experiments, which are not described in this chapter, involved measuring the concentrations in equilibrium under various conditions. During the modelling process, the students were asked to compare the results of the experiments with their results, obtained by theory under different approximations (as described previously). They were asked to treat the theoretical results mathematically, with the use of appropriate technology. By asking them to evaluate the results of the mathematical modelling, the teacher encouraged the students to reflect upon the process of modelling in mathematics and upon differences and similarities between the processes in mathematics and in chemistry; such as the issue of approximations and the complexity of the models seen from the perspectives of the two subjects, respectively.

The following example illustrates how the teacher intended to establish connections between theory and practice in chemistry by modelling: the students were guided to reach version 2 above, starting with version 1, using the questions provided by the teacher (see below) (Petersen 2009, first author's translation from Danish):

> Try to carry through the chemical argument, that only negligible amount of silver ions are of the form. Write down the three equations, obtained in this case. The one which disappears is an equilibrium equation. Solve this system of equations. Does it fit looser with the experiments' results than the original system of four equations? (…)

The teacher's part of the teaching materials includes the following two excerpts concerning modelling and mathematics, which explains the learning goals of the case (Petersen 2009, first author's translation from Danish):

> In chemistry, students gain experiences in working with calculations of concentrations and with the principle of equilibrium. The goals may be varied, but in this version the aim is to let the students recognise the following:
>
> - Our determinations do not give a perfect model of nature…
> - We may choose to base our calculations on more or less complex models (…), and this choice influences the degree to which the calculations fit with our observations
> - Good simplifications of the mathematical model may be based on chemical arguments (…)

To sum up, the materials were specifically designed to make chemical modelling and mathematical modelling, and the connection between the two, a focus of attention. The idea was to let the students inquire into the connections between chemical theory and practical experiment in the laboratory by comparing the results from both, whilst also building, testing and modifying the mathematical model of the chemical system supported by the use of calculators and/or computers with mathematical software. In their written reports the students were requested to explain and reflect upon the model they used and upon their modifications of it.

5 Methods for Examination of Data

The case was meant to study:

- The students' perception of modelling in mathematics and in chemistry
- The students' understanding of the system of chemical equilibrium
- The students' understanding of connections between theory and practice
- Technology in mathematics as a means to make the modelling process more explicit.

These perspectives were underlined in the teacher part of the teaching material and discussed with the students during the lessons. The analysis of the students' written reports aimed to find signs of the perspectives in the students' own interpretations of the events.

The students prepared written reports on the complete project. Reports from all nine students were read carefully and analysed with the aim to get an impression of the students' perception of models and modelling, as they are revealed in their reports. The excerpts above from the teachers' part of the materials were used to compare the students' outcome, as it was revealed in the reports, with the sequence's learning goals. In the following paragraph, short excerpts from the students' reports are quoted (first author's translation from Danish).

In general, the nine students' reports show a huge variety in quality with regard to language and fluency, thoroughness and reflections. The descriptions of the project's aim in each report, for example, vary from recounting the concentrations of ions to be measured: '*The aim of this exercise is to find the true concentrations of* Ag^+ ...'(Report 8, p. 5), to considering the comparison between theoretical and practical values and reflections upon these: '*The project aims to (...) give an understanding of the interplay between theory and experiment. Besides (...) see if the "laws" in chemistry are in accordance with reality*' (Report 4, p. 4) and '*There is a connection between the worlds of chemistry and of mathematics, since the chemical formulas and expressions are based on mathematics. (...) The aim of this exercise is to measure concentrations of* Ag^+ ...' (Report 5, p. 4).

6 Results

6.1 Students' Perception of Modelling in Mathematics and in Chemistry

Modelling in mathematics and chemistry is most explicitly mentioned in the reports' introduction, where the aims and goals are presented. In the other parts of the individual reports, modelling is for the larger part touched upon via the distinction between mathematical theory and chemical practice, which is a common issue since all the experiments showed huge discrepancies between theoretical and practical results. The reports gave mere technical explanations during the experiments for the discrepancies, with no discussion of the model or its approximations, for example: '*The other possibility for our incorrect result could be that we did not stir careful enough in our vessel with NH_3 while we added HCl*' (Report 9, p. 11). Thus, the validity of the model was taken for granted in all reports, even those which explicitly claimed that the goal was to compare theory with reality, like here: '*Our measurements in the practical and theoretical shows that our experiment failed*' (Report 6, p. 14). Very few reports show examples of comparison between different models as results of different approximations.

6.2 The Students' Understanding of the System of Chemical Equilibrium

In the reports, chemical equilibrium is described using quotations from the text book, not with the students' own words. In our interpretation, the reports did not in general reveal signs indicating a deeper or more profound understanding of chemical equilibrium, than it would usually be the case in any second-year chemistry class in the same school. The reports contained no discussion of the different approximations, as was intended in the teaching materials and during the lessons, according to the chemistry teacher.

6.3 Students' Understanding of Connections Between Theory and Practice

It is remarkable that none of the students discusses the 'useless' solutions to the system of equations. A common argument is the criterion that concentrations must be positive: '*In the result, a lot of numbers appear, but these four are the right ones, since there are no negative results. It is not possible for an actual concentration to be negative*' (Report 3, p. 20). None of the reports refers to the issue of existence and uniqueness of a chemically acceptable set of solutions.

A few students mention 'calculation error' as an error-source to explain the discrepancies. Apart from these few examples, all the discrepancy is dismissed into technical reasons like defective measuring instruments, wrong measurements, contamination of the solutions, etc., for example: '*Suggestion: The voltmeter did not function appropriately*'. (Report 1, p. 13).

6.4 Technology as a Means to Make the Modelling Process More Explicit

This study was not conclusive; the students' use of technology is almost 'invisible' in the reports meaning that the use of MathCad is documented in the reports with excerpts from the calculations but with no discussion of the strategies or other comments, such as: '*MathCad was used to solve these 7 equations with 7 unknowns. More results are obtained for the solution of each substance's concentration, but since a concentration must have a positive, real value, only one set of solutions is useful for each equation*' (Report 4, p. 14).

Focus in the students' work was on the experiments, not on theoretical perspectives or modelling perspectives, except in most of the introductions. Neither was the possibility of treating more complex systems of equations weighed against the chemical validity of different approximations in any of the reports.

7 Conclusion

We have identified a number of 'weak points' in the first trial of this project and these difficulties are listed below.

7.1 Technical Obstacles

There was a lack of time for the students to prepare the written reports. The project lasted 2 weeks including 1 day in the chemistry lab. Consequently, most of the students spent too much time on the experiments and, subsequently, had to write the report without having time for profound discussions in their working groups or for substantial supervision from the teacher during the writing. The lack of time was reflected in the reports' short, summary paragraphs on conclusions and perspectives which did not match the introductions' presentation of the aims and goals. The instruction sheets for the experiments were not tailored for this project. It is a common pedagogical practice in this class that the students must somehow modify or alter their working sheets before the experiments, to ensure that they do not experience only 'cookbook' exercises in a laboratory. When this project is repeated, the teacher will prepare new working sheets, tailored for this experiment. Measuring the potentials of electrodes, which was part of the experiment, was a new method for the students. To reduce the complexity when this experiment is repeated, the students will undertake a small experiment using this method, before the equilibrium project is commenced.

7.2 Few Students' Reflections

To encourage students' reflections upon modelling and strengthen the written conclusions and perspectives, classroom reflection-discussions will be introduced in the next round of the project, as a forerunner and support for the writing of the latter parts of the report. Such reflection-discussions aim to balance the students' 'technical-applications' view by explicitly requiring reflections upon the use of models as well as the modelling process. These discussions can follow the model of combining levels of mathematical activities with levels of reflections as described in Andresen (2009), based on a reflection guide prepared by the teacher in advance. The request for explicated reflections as part of the written reports' conclusions should ensure that more weight will be put on this important section of the report.

7.3 Little Focus on Modelling

The project's intention was to study chemical and mathematical modelling, and the connections between the two, which was not really fulfilled in this case. One reason, apparently, may be the fact that the model of chemical equilibrium is based on fundamental principles like the law of mass action and the laws of conservation. Such fundamental principles are rarely discussed in the classroom; more often, they serve as a prerequisite embedded in the basis for treatment of their consequences in series of concrete or special cases. As a forerunner for the next project, an example of a less-fundamental and trusted scientific model will be a topic for one or two; for example, theories about earth rays or phlogiston. The aim of including this forerunner in the next project is to make the students aware of the role of scientific models to explain observations and establish a shared basis for discussion of criteria for validity of such models. With this background, the possibility to compare different models of chemical equilibrium, resulting from different approximations, will be an issue for discussion in the next round. A comparison between at least two different models (with three, four or eight equations), then, will be requested in the reports.

To sum up, we still find this topic suitable for further investigations with the above-mentioned amendments.

8 Perspectives

This case study leads us to draw an inference in line with the concept of *forced autonomy* introduced by Jeppe Skott in Skott (2004). As in Skott's study, the requirement for multidisciplinary teaching leaves the teacher in a situation, where '*expected classroom practices and learning outcomes* (are) *formulated outside the classroom, but there is no set of well-defined methods for the teacher to carry out and only vague hints as to what kind of practice a certain situation may require*'. Skott argues in his study, that the notion of forced autonomy, based on the conceptions of mathematics and mathematical learning, should be extended to encompass not only the roles of the teacher when supporting students' learning in classrooms, but also the multitude of other obligations that emerge in the course of the classroom interactions. In the case of multidisciplinarity, the teacher's situation appears even more complex when the perspectives from different subjects have to be connected or even intertwined in a challenging teaching task that involves theory as well as practical activities.

An extended notion of forced autonomy may, according to Skott, serve as a better means for researchers to understand the teacher's role for the enacted curriculum. In our case, the complexity of the multidisciplinary teaching sequence may serve to explain why the students' understanding of connections between mathematical modelling and chemical modelling, as it was revealed in their written reports, was rather loose in spite of the deliberate focus.

References

Andresen, M. (2009). Teaching to reinforce the bonds between modelling and reflecting. In M. Blomhøj & S. Carreira (Ed.), *Mathematical applications and modelling in the teaching and learning of mathematics – Proceedings from TSG 21 at the 11th ICME*, Mexico, July 6–13 2008 (p. 73–83). Roskilde, Denmark: IMFUFA.

Andresen, M., & Lindenskov, L. (2008). New roles for mathematics in multi-disciplinary, upper secondary school projects. *ZDM. The International Journal of Mathematics Education*. Berlin: Springer. ISSN 1863-9690 (Print) 1863-9704 (Online) 10.1007/s11858-008-0122-z.

Confrey, J., & Maloney, A. (2007). A theory of mathematical modelling in technological settings. In W. Blum, P. Galbraith, H.-W. Henn, & M. Niss (Eds.), *Modelling and applications in mathematics education – The 14th ICMI study* (pp. 57–68). New York: Springer.

Holtzclaw, H. F., Robinson, W. R., & Nebergall, W. H. (1984). *General chemistry*. Lexington: D. C. Heath and Co.

Petersen, A. (2009). Ligevaegtsprojekt til DASG 1+2 (Equlibrium project for Danish Science Gymnasiums). Available on http://www.navimat.dk/39836/Matematik%20og%20naturvidenskab. 10 Sep 2009.

Skott, J. (2004). The forced autonomy of mathematics teachers. *Educational Studies in Mathematics*, 55, 227–257. Netherlands: Kluwer.

Chapter 52
Real-World Modelling in Regular Lessons: A Long-Term Experiment

Martin Bracke and Andreas Geiger

Abstract This chapter introduces a long-term modelling project that has been conducted at Goethe-Gymnasium-Germersheim with 14-/15-year-old students. The main question of this project was if (and how) it is possible to integrate *real-world modelling tasks* into regular math lessons in a way that demands for the application of knowledge from selected topics of a whole school year. Moreover we wanted to know if the pupils accept the intended frequency of five modelling phases as a convenient diversion or if they consider them not worthwhile (concerning the effort)? Finally, we were curious to learn to which extent we can expect pupils to learn mathematical modelling through frequent repetition. The students had to deal with five realistic modelling tasks and one final comparison task. Solutions developed by the students as well as the concept of the questionnaires used for evaluation are presented.

1 Idea of the Experiment

The Department of Mathematics of TU Kaiserslautern has a long tradition in modelling real-world problems with students and high school students. The following list briefly summarises the important previous steps:

M. Bracke (✉)
Department of Mathematics, University of Kaiserslautern, Kaiserslautern, Germany
e-mail: bracke@mathematik.uni-kl.de

A. Geiger
Leibniz-Gymnasium, Neustadt a.d. Weinstraße, Germany
e-mail: Andreas.Geiger@gmx.de

Introduction of activity	Activity	Educational level of students
Early 1980s	Modelling seminars with industry projects	Graduate
1987	ECMI modelling week	Graduate
1993	Mathematical modelling week for high school students and teachers (annual event)	Secondary level, teacher training
1999	Modelling days in high schools, 1–3 days, compact form	Primary and secondary level (grade 2–13)
2001	Modelling seminars	Undergraduate, teacher students

[1] ECMI = European Consortium for Mathematics in Industry.

It was quite natural to transfer the successful idea of modelling seminars when we started to offer modelling activities for high school students. Hence, the modelling week was planned similar to the ECMI modelling weeks as a compact course with duration of 1 week. Students and teachers work together on real-world problems during those modelling weeks. Thereby, each team has its own academic supervisor (usually researchers from university).

Since the feedback – from both teachers as well as students – was very positive the natural evolution was to develop a concept for the transfer to regular mathematics lessons in schools.[2] But although many teachers wanted to introduce real-world tasks in the style of the modelling weeks into their regular lessons the problem of a chronic lack of time seemed to make this undertaking very hard to realise. Therefore, we started to offer so-called *modelling days*, that is events of 1–3 days duration having a structure very similar to that of the modelling week. After some experiments an appropriate organisational structure was found and today it is a standard task to perform such events for high school students of grade 7–13. Recently, we started to work with even younger students (primary level, grade 2–5) – a task which is more challenging but it is mainly a matter of finding suitable real-world projects and providing appropriate concomitant material, respectively.

In summary, we confidently claim to know how to do modelling projects with students at primary as well as at secondary level *in compact form*. But as mathematical modelling started to play a more prominent role in the math curricula a new question arose: How can mathematical modelling of real-world problems be integrated into mathematics lessons on a regular basis rather than being singular events which take place once a year (or even less frequently)?

There are several studies on this issue (see, e.g. Blum 2006; Kaiser and Maaß 2007; Maaß 2004; Schwarz and Kaiser 2007). Nevertheless, looking at everyday school life we did not find teachers practicing a long-term integration of real-world modelling problems (2–6 lessons duration for each of them) for the period of a whole school year.[3]

---

[2] Another reason was the fact that *mathematical modelling* started to appear more often in the math curricula in Germany.

[3] This is no general statement of course – it relates to our experience with schools in our region, but studies indicate that this phenomenon is not only restricted to Germany (see Kaiser and Maaß 2007).

To summarise, the central questions of our project are:

(Q1) How to integrate *real-world modelling tasks* into regular lessons?
(Q2) Do pupils accept the intended frequency of modelling phases (*attitude*)?
(Q3) To which extent can we expect pupils to learn mathematical modelling throughout frequent repetition (*learning*)?

And there is a fourth question that we would like to have answered but which could not be addressed with the chosen design of the study:

(Q4) Are contents learned more deeply when learned within modelling projects?

2   Design of the Experiment

When we started to think about the implementation of a long-term modelling project the very first question was *How to teach mathematical modelling?* A possible answer to this question comes from *Learning Theory*: Firstly, *our brain cannot be trained in an unspecific way* (Haag and Stern 2003), that is there is no significant learning effect just by working on challenging tasks. And secondly, *strategies for learning and thinking can be learnt but in general they cannot be taught directly. The key to success is Learning by Doing* (Stern 2006)! Our conclusion for the project was that we did not start with a 'lecture' on mathematical modelling but almost immediately started modelling with the students.[4]

Moreover, there is a consensus on the following aspects of a supportive learning environment (Stern 2006):

(S1)   A (challenging) task from the actual topic
(S2)   Application of specific strategies is standing to reason[5]
(S3)   Appropriate material is provided
(S4)   Suitable hints are given by teachers/supervisors.

To fulfil (S1) we had to consider the *Framework Curriculum Mathematics* (MBWJK 2007) which includes two main concepts: On the one hand we have to address the mathematical key competencies (K1) *Mathematical Reasoning*, (K2) *Problem Solving*, (K3) *Mathematical Modelling*, (K4) *Use of Mathematical Representations*, (K5) *Go around with Symbolic, Formal & Technical Aspects of Mathematics* and (K6) *Communication*. On the other hand the central themes (L1) *Numbers and Ranges*, (L2) *Measures and Quantities*, (L3) *Space and Shape*, (L4) *Functional Relations* had to be covered.

---

[4] This concept is not new (see e.g. Kaiser and Maaß 2007; Schwarz and Kaiser 2007) – but it is important!
[5] Specific strategies which have been investigated and trained in preparatory lessons might be helpful during the modelling process.

The next list shows the contents that should be covered in grade 9 mathematics lessons and the corresponding links to the central themes (topics in the order of their appearance).

I   Linear Functions (*L4*)
II Systems of Linear Equations (*L4*)
III Real Numbers, Computing with Square Roots (*L1*)
IV Similarity (Central Dilation, Intercept Theorems) (*L2, L3*)
V   Theorem of Pythagoras (*L3*)
VI Quadratic Functions (*L4*)
VII Quadratic Equations (*L4*)

The next step was to look for five real-world tasks that should be worked on over the whole period of one school year. The next section describes all projects in detail and shows the corresponding central themes that we expected to be addressed. Our goal was to cover as many topics as possible from the list above.

Finally, there have been (anonymous) questionnaires after task numbers 2–5 including control statements for evaluation purpose (see Sect. 5 for details of the evaluation).

3 Choice of Five Real-World Modelling Tasks

The target group was a complete school class (grade 9) with 26 pupils (14 male, 12 female) and the tasks should be integrated into regular lessons. Based on our experience with modelling events in a compact form, we set up the following requests:

- Real-world problems as modelling tasks, questions should be very open (cf. (S1))
- Modelling tasks have to be interesting and challenging (for grade 9 pupils!)
- Rising difficulty from task 1 through task 5
- Modelling teams consist of three to four students
- Intended duration of each task was one to six lessons and every team had to write a report at the end of each task.[6]

The work on the individual tasks was conducted with the greatest possible autonomy but the teams receive hints from the teacher whenever this was necessary (S4). The modelling tasks were chosen in such a way that the students were directed to use methods which have been discussed in the lessons before the start of the respective task in order to meet (S2) and (S3) (see the timeline in Fig. 52.1).

In order to introduce the pupils to the idea of modelling the first task has been conducted as a *guided modelling*. In the remaining four tasks the pupils worked with greatest possible autonomy.

---

[6] 2–4 pages, as a homework.

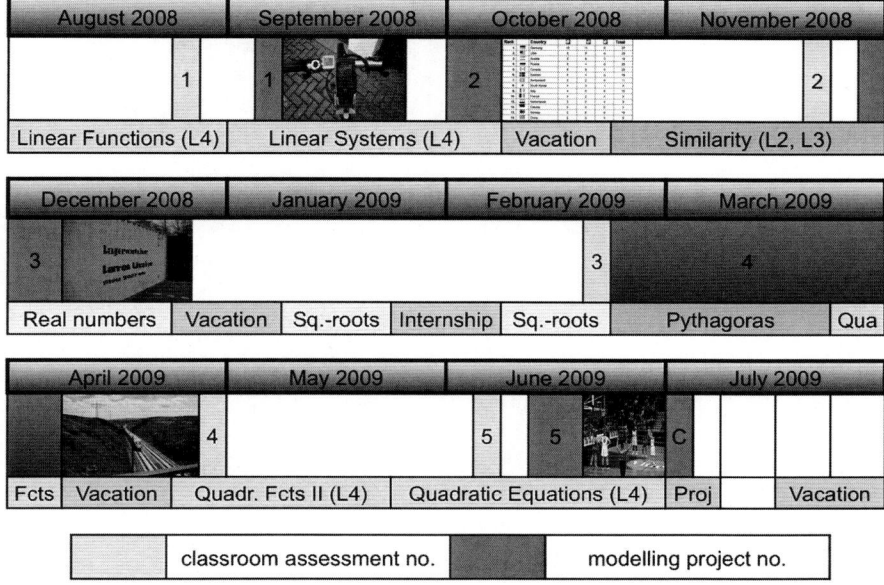

Fig. 52.1 Timeline of the long-term modelling project

3.1 First Task: A Guided Modelling as an Introduction

After just naming some typical questions[7] that could be answered using a suitable mathematical model the main steps of a modelling cycle were discussed by means of a specific example: *How accurate is my bicycle computer?* Of course the first step is to understand the actual problem. Here the simple question is to which extent one can trust the current velocity displayed by a bike computer – and how to verify its accuracy (Fig. 52.2).

The second step is to find an appropriate mathematical model. After some experiments with a real bike one could observe the proportional relation between a *time interval* and the *distance* travelled within that interval *assuming that the velocity is kept constant*. Hence the idea is to collect data and plot distance versus time to obtain the constant velocity as the gradient of the resulting straight line. The third step is to solve the mathematical problem that was set up in the previous step, that is here the question is how to compute the gradient of the straight line plotted from some data points (e.g. using EXCEL). The final step is to interpret the result, that is to compare the gradient that was computed before with the constant velocity that was displayed during the experiment. At this point usually a conversion of units is needed since the distance most probably is measured in

---

[7] How large is the moon? or Will there ever be a man running 100 m below 9 s?

Fig. 52.2 Bicycle computer (From Wikipedia)

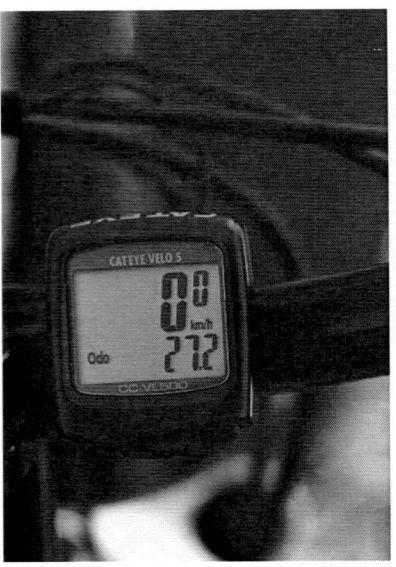

metres and the corresponding time will be some *seconds* while the velocity displayed by the bike computer is in *kilometres per hour*. In most cases the correspondence between the displayed value and the computed velocity will be very good and since the cheap computers display integers only (or maybe velocities in steps of 0.5 km/h) one would conclude that the device's accuracy is almost perfect.

The guided modelling was stopped after shortly mentioning that there is still some room for improvement of the model. Some ideas were collected but there was no time to dive again into the modelling cycle. As we will see later in the discussion of the comparison task[8] it would have been better to take an extra lesson at this time to discuss at least one possible enhancement of the model in detail: The students seemed to have acquired the impression that one can stop modelling after having achieved *any sensible solution* without taking time to think about improvements (see Sect. 7). For this special task one could have asked how sensible the assumption of a constant velocity during data collection really is – and how to change the model if it turns out to be inaccurate (a fact one easily observes when using a bike computer which displays the current velocity up to one decimal place).

Link to curriculum and duration. 1 lesson, L2 (measuring and quantities), L4 (functional relations).

---

[8]The comparison task was presented to the pupils at the end of the project (cf. Sect. 7).

3.2 Second Task: Is the Olympic Medals Table Fair?

After every big sports event like *Olympic Games* or *IAAF World Championships in Athletics* people interested in sports all over the world start discussing the official medals table. Even the ones with lower interest in sports want to see their nation as highly ranked as possible. And the official algorithm to compute the ranking is very simple: The nations are sorted by their number of gold medals (in descending order), if this number is equal for several nations they are ranked by their number of silver medals and the same is done for the bronze medals if again identical numbers appear. In the USA an alternative table is shown quite often which only considers the total number of medals to rank all nations.

Looking at an extract of the 2008 Olympic Summer Games medals table one recognises that there are several configurations which provoke a discussion on the fairness of the official ranking:

Rk	Nation	G	S	B	# Medals	Rk by # medals
1.	China	51	21	28	100	2.
2.	USA	36	38	36	110	1.
3.	Russia	23	21	28	72	3.
4.	Great Britain	19	13	15	47	4.
5.	Germany	16	10	15	41	6.
6.	Australia	14	15	17	46	5.
7.	South Korea	13	10	8	31	8.
8.	Japan	9	6	10	25	11.
9.	Italy	8	10	10	28	9.
10.	France	7	16	17	40	7.
11.	Ukraine	7	5	4	16	10.
12.	The Netherlands	7	5	4	16	15.
13.	Jamaica	6	3	2	11	17.
14.	Spain	5	10	3	18	13.
15.	Kenya	5	5	4	14	16.
16.	White Russia	4	5	10	19	12.
17.	Romania	4	1	3	8	19.
18.	Ethiopia	4	1	2	7	20.
19.	Canada	3	9	6	18	13.
20.	Poland	3	6	1	10	18.

For example, France has won many more medals than Japan but has only seven gold medals (in contrast to nine for Japan) – hence the French team is listed behind the Japanese. A similar situation holds for Jamaica and Spain or the trio White Russia, Ethiopia and Canada. There are a lot of factors, which could be taken into account: The amount of competition for different decisions, number of athletes in the national team, amount of money spent to support sports in different countries or just the population of the country. Or maybe it would be fairer to use a weighted sum of the number of gold, silver and bronze medals?

Fig. 52.3 Text on a freight container – more or less readable

Link to curriculum and duration. 4 lessons, L1 (numbers and ranges), L4 (linear functions).

3.3 Third Task: How to Type on a Container?

The photos in Fig. 52.3 show some advertising that is printed on a freight container. Typically, containers of this type have no flat exterior walls but feature a kind of trapezoidal wall shape (seen in a cross section). As one easily recognises the letters either look strange or even are not readable if one just types on the wall as on a flat plane. Therefore, the question, which immediately arises, is *How to adjust the scripture such that these effects do not appear (if possible at all!)?*

Link to curriculum and duration. 6 lessons, L3 (space and shape)

3.4 Fourth Task: Building of an ICE-Track[9]

This project is described in more detail in Sect. 4 where we exemplarily present some solutions of the pupils. Hence, we just state the main question: *Given the photo of an ICE-track (see Fig. 52.4), can you determine the amount of ground that has to be carried away to build the part of the track depicted in the photo?*

Link to curriculum and duration. 2 (+5)[10] lessons, L3 (similarity, Pythagoras).

---

[9]cf. (Fries et al. 2004).

[10]The pupils worked for approximately 5 weeks on this task in the form of a constant homework with an estimated work load of five lessons; this was in addition to two regular lessons spent on modelling this task.

Fig. 52.4 ICE-track (From Fries et al. 2004)

Fig. 52.5 Free throw by NBA player Dirk Nowitzki (From Wikipedia)

3.5 Fifth Task: How to Do an Optimal Free Throw in Basketball[11]?

In the NBA playoffs of the period 2008/09 the free throw rate ranged from 72% to 80.5%, individual rates of the players were between 30% and over 90%. Of course there are different factors that have an influence on success and failure (e.g. fatigue, psychology). But evidently there exist infinitely many possible trajectories for a successful free throw and the simple question is if there is an optimal one (maybe depending on the height of a player and his/her special skills)(Fig. 52.5).

Link to curriculum and duration. 4 lessons, L4 (quadratic functions, quadratic equations).

---

[11] Idea taken from Gablonsky and Lang (2005).

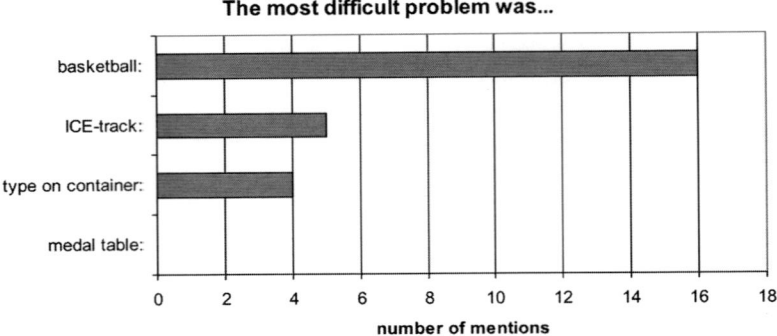

Fig. 52.6 Difficulty felt by the pupils

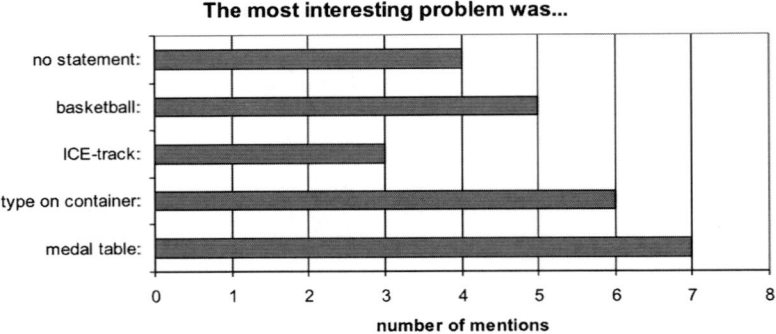

Fig. 52.7 How interesting were the tasks for the pupils?

3.6 Difficulty and Attractiveness of the Projects

From evaluation of the questionnaires it can be seen that our aim of a rising difficulty from task number 2 (medals table) to task number 5 (basketball) was fulfilled quite well (cf. Fig. 52.6) and the second goal of providing interesting tasks (to keep the pupils motivated) was met to a large extent (cf. Fig. 52.7).

Probably it would have been better to allow for multiple choices in the question for the most interesting problem(s) but the answers are still of some value.

4 Detailed Discussion of Task Number 4: ICE-Track

As said before, we now have a closer look at modelling task number 4 which is about the new building of an ICE-track (Fries et al. 2004).

52 Real-World Modelling in Regular Lessons

Fig. 52.8 Photo of an ICE-track

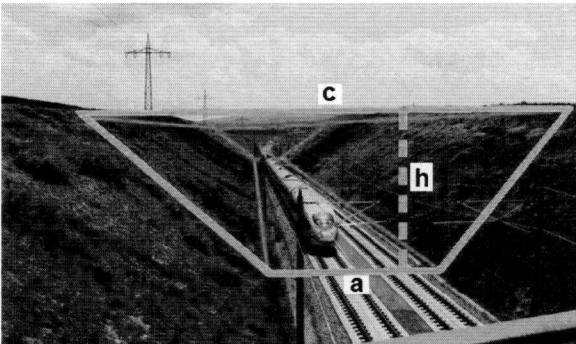

Fig. 52.9 Approximation of the volume by using a prism

At the beginning, the photo shown in Fig. 52.8 was given to the students. The story behind is that for building this new track a lot of ground had to be dug and carried away. In order to coordinate the building project one has to know how many trucks are needed to accomplish this work.

The pupils immediately understood the actual question behind this task as they stated that at first the volume of the ditch must be calculated. But how can this be done since the ditch is of no regular shape like the objects that are normally investigated in school?

After looking at the photo for a while and some team discussions later, the idea of an approximation of the ditch using a prism came up (cf. Fig. 52.9). Since the cross section of a prism is a trapezium, this led to the following result: The volume of the prism V_p can be calculated by multiplying the area of the trapezium A_t with the length l of the ditch which finally leads to the following term:

$$V_p = A_t \cdot l = \frac{1}{2}(a+c) \cdot h \cdot l$$

But now a big problem arose: The pupils wanted us to give them the values of the variables a, c, h and l, otherwise they would not be able to continue. This was something they had never been confronted with before but nevertheless it was part of this particular problem. So we told them that we did not know those quantities and they were forced to think about possibilities of how to obtain the necessary data.

We want to present three different approaches of how the pupils finally managed to come to a result. We try to stick very close to the original solutions – except for the standardised notation – such that the reader gets a genuine impression.

- The *first group* started with writing an e-mail to *Deutsche Bahn AG* in order to get some information about the gauge of an ICE-train. After having received this information (1.435m) they used it to estimate the values of their unknowns to be $a \approx 7\mathrm{m}, c \approx 19\mathrm{m}$ and $h \approx 7\mathrm{m}$. Furthermore, they estimated the length of the ditch at $l \approx 270\mathrm{m}$ without any special source of supply and were finally able to calculate the volume of the prism ($V_p = 24570\mathrm{m}^3$). However, this solution is quite poor as their estimates are all too small and not very elaborate.
- The *second group* tried a form of calculation. First they searched the Internet for the gauge of an ICE-train. Having found this (1.435 m) they used the ratio between the real length and the corresponding length on the photo to compute $a = 10.66\mathrm{m}$, $c = 26.65\mathrm{m}$ and $h = 13.94\mathrm{m}$ (all these values are written down in the same way the pupils did). Additionally the length of the train (which had also been found on the Internet) was used as an estimate for l, ($l = 410.72\mathrm{m}$). This finally allowed for computing the volume of the prism: $V_p = 106808.0235\mathrm{m}^3$ (note the accuracy! The pupils did not see a discrepancy between this value and their assumptions).
- The *third group* followed a similar, but slightly different idea. From the measurement of a real track in their neighbourhood they found out that $a = 9\mathrm{m}$. Then they estimated the values of the other unknowns to be $h = 15\mathrm{m}$ and $b = 22.5\mathrm{m}$ (where b denotes the length of the slope). Making use of the symmetry of the trapezium and Pythagoras' law they were able to calculate $c = 42.5\mathrm{m}$. Finally, taking l as a variable, they arrived at the following functional relationship for the volume of the prism: $V_p(l) \approx 387\mathrm{m}^2 \cdot l$ (leads to $158948.64\mathrm{m}^3$ using $l = 410.72\mathrm{m}$ as group 2 did).

5 Evaluation

In Sect. 1 we presented three questions (Q1)–(Q3) we wanted to answer throughout this long-term project on mathematical modelling. Our first question *How to integrate* real-world modelling tasks *into regular lessons?* has already been answered in the previous sections since we found a way to achieve the integration.

In particular, we wanted to investigate the *attitude* pupils develop towards modelling during a whole school year (Q2) and *learning* to model (Q3). Therefore, we made use of continuous evaluation through *questionnaires*. As the first task was

Fig. 52.10 Part of questionnaire number 4

(1) Gender

(5) I could bring in my ideas to the teamwork

(6) I feel that practice comes with repeated modelling

(8) For dealing with modelling problems I prefer working on my own to teamwork

(9) I have not developed any modellingstrategies yet

(10) Modelling problems should get more space in regular lessons

(14) I think that modelling can be learnt

(15) This problem showed to me where mathematics is applied in real world

(16) By means of modelling you do not learn anything new

(17) I don´t like modelling problems

a guided modelling task, there was no questionnaire at all. After the second, third and fourth tasks questionnaires arranged with respect to the current task were given to the pupils. However, the questionnaire presented after the fifth task was somehow different: Since it was the last task[12] we wanted to have some kind of review on the whole project.

5.1 Concept of Questionnaires

As an example we can have a look at the section questionnaire number 4 (see Fig. 52.10). On the whole this questionnaire consists of 17 items – 10 of them you can see here. Except for the first item they are all formulated as statements to which the pupils can agree or disagree (how this is done in practice will be explained below). The statements can be assigned to different topics. For example statement number 6 clearly corresponds to the question of *learning*. In order to make sure that the pupils' answers are consistent, controlling statements are added. In the case of number 6 the corresponding controlling statement is number 9. Another example can be seen from statement numbers 10 and 17. This time the *attitude* towards modelling is investigated.

Now, in order to start the evaluation process, data has to be generated. A range is added to each statement on which the pupils can agree or disagree continuously.

---

[12] Besides the final comparison task.

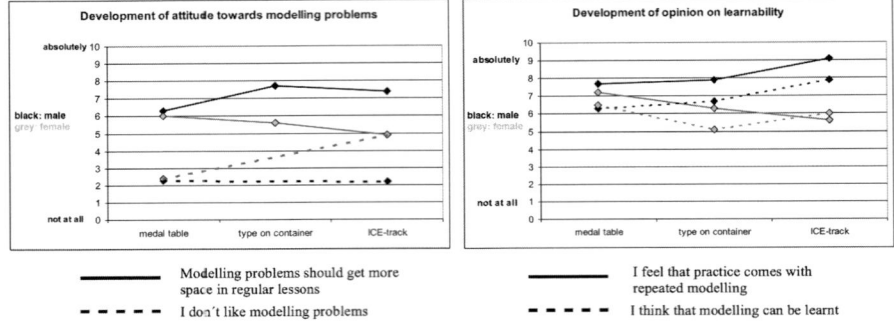

Fig. 52.11 Development of attitude towards modelling problems (*left*) and opinion about learning of modelling (*right*)

After having filled in the questionnaire, a scale is used in order to analyse the judgement. Doing so, we get one value for each pupil and statement. In the next step, this data is analysed by calculating mean values for each statement – distinguishing between male and female students. In order to show these results simultaneously two different colours for boys (black) and girls (grey) will be used in the diagrams. In the next subsection we will look at some selected results with respect to the statements above and particularly how they have developed over time.

5.2 Development from Second to Fourth Task and Final Judgements

First we deal with question (Q2) *attitude* towards modelling problems (cf. Fig. 52.11 (left)). As there is a lot of information contained in this diagram we have to interpret it carefully. First of all, two different statements are analysed – one is formulated in a positive way and the other in a rather negative way. The corresponding scale is on the *y*-axis. The *x*-axis is used for task numbers 2, 3 and 4).[13]

We can see that the boys want to give modelling problems more space in regular lessons on a rather high level; the girls seem to be not that convinced. Except for the second project there is a big difference between their value and the boys' value and this difference even seems to increase with time. This is in clear correlation to the second statement. While the male pupils clearly disagree with this statement constantly the girls' attitude changes throughout time. At the beginning they like modelling as much as the boys do but later they are more reserved about it.

---

[13] Note that the solid and dashed lines shall not indicate a development over time but are there to show the common bonds of the different data sets.

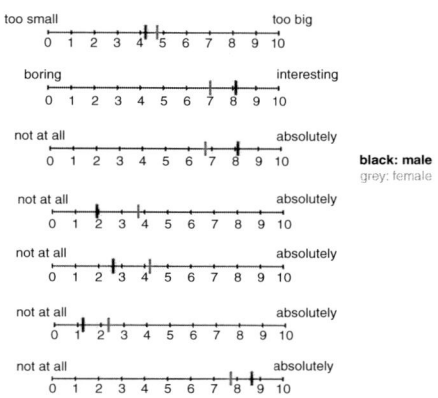

(1) The number of modelling projects during the whole school year was

(2) The subjects which were connected with the modelling problems were altogether rather

(3) **I think that I learnt modelling**

(4) Regular lessons instead of modelling problems would have been more useful to me

(5) I think that the reports we had to hand in are unnecessary

(6) **I do not benefit from working on modelling problems**

(7) **Most of the time I enjoy modelling**

Fig. 52.12 Final judgement (after fifth project)

Almost the same result can be stated for the topic *learning* (see Fig. 52.11 (right)). Note that this time both statements are formulated in a positive way. For the first statement again the assessments of the male pupils are on a high positive level, even higher for the more difficult tasks. Against that the female pupils start at a similar point but the values decrease from task to task. For the second statement we have an almost analogue behaviour except for the fact that this time the difference between male and female students is not as clear as before. But besides this gender-specific discrepancy one should not forget that all the values themselves are in the upper half of the quadrant. Hence, on average they are positive for all students.

In order to get something like a final and general judgement, questionnaire number 5 was designed to be a kind of review – looking back at the whole school year (see Fig. 52.12). Of course to obtain a review the items must be modified but they still aim at the same main questions, such as *learning* (no. 3) or the *attitude* towards modelling (no. 6/7). The statement numbers 1 and 2 were added in order to obtain some information concerning the general framework. These judgements can be useful for a possible repetition of such an experiment.

Having a look at our results we can state the following:

1. The number of modelling tasks during the whole school year was more or less perfect from the pupils' point of view and they dealt with interesting subjects.
2. For all the other statements we have quite positive judgements, for example concerning *learning* (no. 3) or the *attitude* towards modelling (no. 6/7).

But we can make a remarkable observation: There is still a discrepancy between the male and female pupils' judgements and their votes are given in a way that for every single statement girls do not vote as positively as boys do! This does not seem to be a coincidence – there have to be some reasons.

After having found out about this gender-specific discrepancy we looked a bit closer at the filled-in questionnaires. It turned out that there were always two special female students who made negative judgements. Of course this has an effect on the mean values which are used to obtain the graphs. But even if we neglect these two pupils there is still another explanation: Two of the modelling problems had a rather technical background (*type on container/ICE-track*) and, typically, girls are not as interested in technical ideas. This is not an assertion but the result of another survey. Seven out of 11 female students stated that one of these two technical problems was the least interesting one.

6 Comparison Project: Setting and Results

After the fifth task we were almost at the end of the school year and all the results gained so far were based on self-assessment. Fortunately, with the help of other teachers it was finally possible to establish a comparison task in which the trial group and inexperienced pupils of other courses – altogether 95 students – worked on the same two modelling problems:

The first problem is called *Save Teufelstisch* and it deals with a famous and bizarre stone-formation in the *Pfälzer Wald* near Kaiserslautern (see Fig. 52.13). The actual problem now is that Teufelstisch is in danger of breaking down because of erosion. To prevent this happening, the idea is to build an artificial pillar – but for that, one needs to know the mass of the stone-plate.

Thinking of the ICE-track problem some parallels can easily be drawn: Again an irregularly shaped object needs to be approximated by a regular one and again there are no data given. One should now expect that the pupils of the trial group who know the ICE-track problem have a clear advantage.

Fig. 52.13 Teufelstisch, task (A)

52 Real-World Modelling in Regular Lessons

Fig. 52.14 Waiting at the check-out, task (B)

Therefore, we presented a second modelling task which was new to the trial group as well as to the other pupils. This is *Waiting all the time...* (see Fig. 52.14): Imagine you are in the supermarket and want to pay for your purchases – but there are two open check-outs to choose. The first queue is short but all customers have lots of commodities in their shopping trolleys. The second queue is much longer but all customers have only a few things in their trolleys. Now, which check-out would you choose?

There is no similarity between this task and the ones the pupils of the trial group had previously worked on – so this task was new to everybody. Nevertheless it is not too complicated so that 14-year-old students cannot cope with it. What we now want to investigate is the following:

- Did their modelling experiences give the trial group an advantage over the other pupils?
- What kind of attitude do the other pupils develop towards modelling during this single event?

Before we can answer these questions we want to give some information about the general framework of the comparison modelling: As the period of time was only 4 h from 8 o' clock to 12 o' clock the pupils were asked to *work on one of the two problems*.[14] In order to prevent dissemination of ideas the trial group and the other pupils had to work in different classrooms. The comparison event was announced as some kind of mathematical competition: A prize would be awarded to the best

---

[14] Again they were supposed to work in teams of three to four pupils and we had an equal number of teams working on each task.

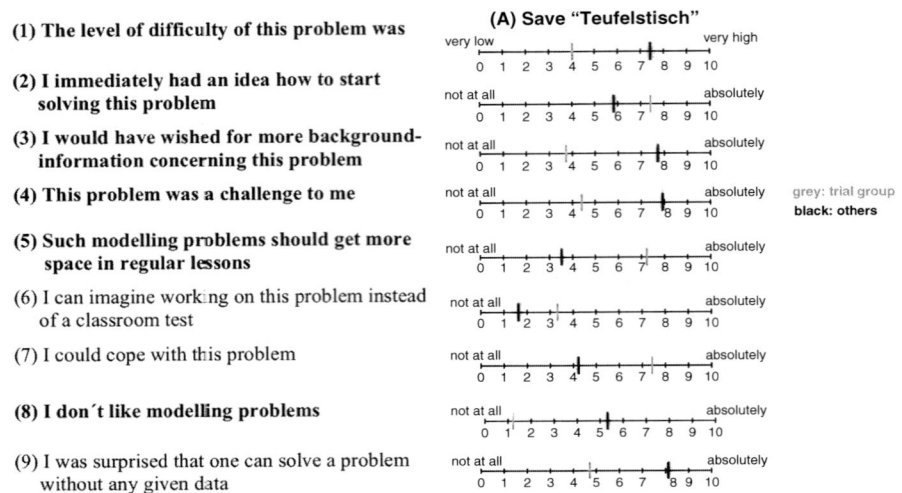

Fig. 52.15 Selected results for task (A) 'Save Teufelstisch'

modelling team of each class. Therefore, it was necessary that *each team documents the solution* carefully and hands this report in at the end. At the end every student (of the trial group and of the other courses) had to *fill in a questionnaire*. This questionnaire together with the personal observations of the supervisors and the assessment of the reports provides information that can be used to answer the central questions stated above.

The structure of this questionnaire is analogue to that of previous projects. Again several statements are given (partly even the same as before) to which the pupils can agree or disagree on a continuous range (see Fig. 52.15). Some of the statements clearly correspond to the question of *learning* (e.g. no. 1–4), others investigate the *attitude* towards modelling problems (such as no. 5/8). When we now present the results, note that this time we do not distinguish between male and female students but between the trial group and the other pupils by using different colours. We start with the results for the first problem: What you can see on the first view is that there is indeed an *advantage* for the trial group over the other pupils, even *in all issues of learning*. For example the level of difficulty of this problem is judged to be much lower by the trial group than by the other pupils. Similar results can be observed by looking at the statements concerning '*background-information*' (no. 3) or the *challenge* (no. 4). In addition to that, the *attitude* towards modelling problems is *a lot more positive* in the trial group than it is among the other pupils as you can see in statement numbers 5 and 8.

Now we compare this with the results of the second modelling task (B) (see Fig. 52.16). One can still observe an *advantage* for the trial group over the other pupils *in almost all issues of learning* but it is *less clear* than in the first task. This is consistent with the fact that the *attitude* towards modelling problems *again* is

52 Real-World Modelling in Regular Lessons

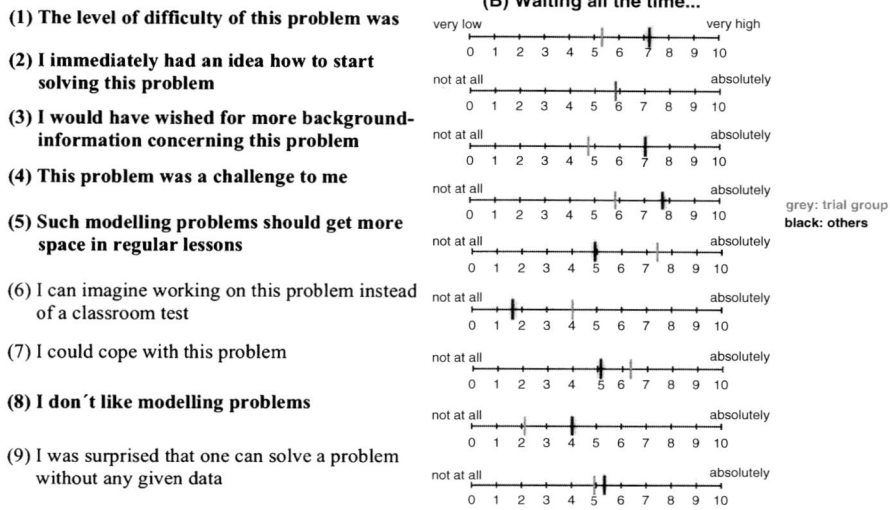

Fig. 52.16 Selected results for task (B) 'Waiting all the time...'

more positive in the trial group than among the other pupils but *again* this is *less clear* than in the case of the first task.

Now we are ready to summarise what we have found in our study[15]:

- There are several indicators for a learning effect – at least the pupils think that they have learned how to do mathematical modelling.
- In case of familiar modelling tasks this is more distinct than in the case of new ones. Nevertheless the effect is still measurable.
- Finally, the attitude towards modelling problems is more positive in the trial group than among the other pupils.

In addition to the self-assessments of the pupils we add a more objective component by looking at the reports that had to be handed in. For assessment of the reports we developed different *categories*[16]:

For the item *Finding a solution* we observed an advantage for the trial group over the other pupils as every team of the trial group managed to develop a solution but only 16 out of 20 teams of the other pupils could cope with their task. This advantage remains when concentrating on the category *Quality of documentation*. Regarding this, the reports of the trial group were a lot better than the other pupils' ones. For the latter, solutions often were not understandable without any deeper knowledge of the

---

[15] Note, that only 95 pupils participated in the comparison project – hence there is no simple generalisation of these findings.

[16] The reports were assessed by members of TU Kaiserslautern who did not know any of the pupils.

underlying task. Regarding *Using appropriate formal notation* there are only slight differences between the trial group and the other pupils. Some of the other pupils' reports were written using prose rather than mathematical terms. Therefore, we claim a slight advantage for the trial group here. Now we come to the most interesting aspect: the *Quality of the solutions* themselves. Comparing solutions of the *same deepness* we note that the pupils of the trial group had finished them earlier. Therefore, we consider the trial group to be faster than the other pupils. But if we investigate the *deepness of solutions on average*, there is a clear advantage for the other pupils over the trial group – to our biggest surprise! For example in the task 'Save Teufelstisch' each team of the trial group used a cuboid as an approximation of the stone-plate, that is the simplest possible model. Against that, some teams of the other pupils used a more realistic and therefore better object, namely, a prism.

We have some ideas of possible reasons for this phenomenon, but since there is no real evidence for one of them we leave these questions without being able to answer them.

7 Summary and Conclusion

In the introduction we formulated certain *aims* that we wanted to achieve by this long-term project: The first one was that we wanted to prove that (against common opinions) *real-world modelling tasks can be integrated into regular lessons of a whole school year*. It can be said that this was successfully done within this project. As a by-product, the pupils have learnt interesting and realistic applications of mathematics during one school year. Secondly, we wanted to collect information about *learning* to model. On the whole the question of learning seems to be answered. At least the pupils claim that they learnt modelling and besides that, some more objective effects could be observed. One consequence of the *unwanted learning effect* we just discussed is the modification of the guided modelling (task no. 1) to include several refinements of the model.

Our third question (Q3) was about the *attitude* pupils develop towards modelling throughout such a long-term project. Especially the comparison tasks showed that this attitude among the pupils of the trial group is not only quite positive on average but in particular is better than among the other students.

Our conclusion at the end of the study is that it is not enough to have a single (and short[17]) modelling event once a school year (or even less frequently). Instead, one should try to create a long-term modelling experience which can be seen as a red thread throughout the whole mathematics education. Ideally, in addition there are modelling days or weeks, which are very intensive experiences and allow for even deeper analysis of the modelling problems.

---

[17] Here it was just 4.5 h for the comparison event – this is definitely too short!

References

Blum, W. (2006). Modellierungsaufgaben im Mathematikunterricht – Herausforderung für Schüler und Lehrer. In A. Büchter et al. (Eds.), *Realitätsnaher Mathematikunterricht – vom Fach aus und für die Praxis* (pp. 8–23). Hildesheim: Franzbecker.

Fries, D., et al. (2004). ICE-Neubaustrecke. In B. Mathea (Ed.), *Mathematik hilft (fast) immer!* (pp. 23–25). Frauen und Jugend Mainz: Ministerium für Bildung.

Gablonsky, J. M., & Lang, A. S. I. D. (2005). Modeling basketball free throws. *SIAM Review, 47*(4), 775–798.

Haag, L., & Stern, E. (2003). In search of the benefits of learning Latin. *Journal of Educational Psychology, 95*, 174–178.

Kaiser, G., & Maaß, K. (2007). Modelling in lower secondary mathematics classrooms – Problems and opportunities. In W. Blum, P. Galbraith, H.-W. Henn, & M. Niss (Eds.), *Applications and modelling in mathematics education. The 14th ICMI study* (pp. 99–108). New York: Springer.

Maaß, K. (2004). *Mathematisches Modellieren im Unterricht*. Hildesheim: Franzbecker.

MBWJK. (2007). *Rahmenlehrplan Mathematik (Klassenstufen 5–9/10)*. Wissenschaft, Jugend und Kultur Rheinland-Pfalz, Mainz: Ministerium für Bildung.

Schwarz, B., & Kaiser, G. (2007). Mathematical modelling in school – Experiences from a project integrating school and university. In D. Pitta-Pantazi & G. Philippou (Eds.), *CERME 5 – Proceedings of the Fourth Congress of the European Society for Research in Mathematics Education* (pp. 2180–2189).

Stern, E. (2006). Lernen – was wissen wir über erfolgreiches Lernen in der Schule? *Pädagogik, 58*(1), 45–49.

Chapter 53
Modelling Tasks at the Internet Portal "Program for Gifted"

Matthias Brandl

Abstract The Internet portal "Program for Gifted" (title in German: "Begabte fördern") was set up by the Chair for the Didactics of Mathematics at the University of Augsburg in the science Year of Mathematics 2008. The published materials are developed for the support of mathematically gifted students both in the context of extracurricular study groups and within-class grouping. There are several topics available already. We present two examples from this variety that arise from application and therefore emphasize the process of mathematical modelling.

1 Fostering of Gifted Students

There are several indications that the fostering of gifted students at German secondary school has been neglected in the last years and decades. First, the Association of Teachers at Higher Secondary Schools in Germany (DPhV) said in the press release DPhV (2008) that in contrast to the fostering of the so-called at-risk students who only achieve competency level 1 in international tests there is too little done in Germany for the gifted or highly gifted students. It is claimed that there are not enough special offerings in the form of enrichment (such as extra courses, competitions, summer schools) or ways of diversification within regular class lessons. Besides this, the missing of a clear top flight within German students at higher secondary level cannot be overlooked (see OECD 2001). Heller (2002a) infers from this that the fostering of gifted or highly gifted students at German higher secondary school is not done properly.

In order to contribute to resolving this situation, the Internet portal "Program for Gifted" was set up by the Chair for the Didactics of Mathematics at the University of Augsburg in the science Year of Mathematics 2008 (see www.lehrer-online.de). The published materials are developed for the support of mathematically gifted students both in the context of extracurricular study groups and within-class grouping by open learning environments as proposed in Heller (2002b), for example.

M. Brandl (✉)
Didactics of Mathematics University of Augsburg, Augsburg, Germany
e-mail: matthias.brandl@math.uni-augsburg.de

2 The Interests of Gifted Students: A Bottom-up Approach

In order to address the needs of students gifted in mathematical thinking we conducted a pre-study for a future questionnaire among 14 (aged 16–18) students from 10th to 12th grade of secondary school who were allowed to study mathematics and informatics at the University of Augsburg (Group 1). This group was followed up by a survey among 26 participants of two intensive mathematics courses from a 12th grade at a higher secondary school (Group 2). Ignoring any gender discrepancy, they gave the following answers that were not pre-classified:

Question 1: "What are you interested or fascinated in concerning mathematics?"

Answer	Group 1	Group 2
Logic (logical), proofs (provable), strength, uniqueness	57%	54%
"Language of nature", applicability	21%	16%
(Surprising) connection between different fields	11%	14%
Demand, difficulty	11%	16%

Question 2: "What is your favourite type of task?"

Answer	Group 1	Group 2
Complex, long non-standard-tasks with surprising solutions and different ways of solving	57%	8%
Riddles, proofs, tasks from competitions	21%	50%
Word problems with connection to (real world) application	11%	42%

So, by the answers to open questions, we can see that mathematically gifted and highly gifted students are (very much) interested in applicability and therefore in word problems with connection to real world application. This context is strongly connected to the process of modelling, which is described in the didactical debate (besides others see Blum 1996; Kaiser 2007). Additionally, designing the learning units in a way that is based on the interests, wishes and needs of the students, promises to be a highly satisfying "bottom-up" approach – in contrast to a "top-down" way inspired by a strongly followed curriculum in mathematics that, according to Burkhardt (2006), is mainly driven by people whose core interest is in mathematics itself, not in its use.

3 Examples Containing Modelling Components

We briefly describe the two units that are designed by the principle given in Heller (2002b), where (in a mathematical sense) rich learning environments are demanded for gifted students. Furthermore, they follow the method of connections as extensively described in Brinkmann (2008), for example, to establish successful learning processes by embedding the new content into a framework of different related mathematical aspects.

All units start with a simple problem taken from reality, whose provisional mathematical content is part of the standard curriculum. By the strategy of "variation" and the corresponding question "What happens when ...?" as proposed in Baptist (2000), the students are led to extra-curricular elements related to their interests.

The modelling aspect shows up in different forms:

(a) Algebraic formulation of a geometric problem (see Sect. 53.3.1)
(b) Algebraic formulation and proof of a kinasectic illustration of a geometric–algebraic transformation or deformation, respectively (see Sect. 53.3.1)
(c) Algebraic formulation and analytic discussion of a stochastic problem (see Sect. 53.3.2)

Furthermore, the units are classified by the addressed fields of mathematical thinking as it was sketched in a simple model for mathematical intelligence in Brandl (2009b) that can be embedded in a more comprehensive model of mathematical giftedness as discussed in Heller and Perleth (2007), for instance. The first unit addresses two- and three-dimensional geometrical thinking as well as numerical, functional, formal and problem-solving thinking. The second topic brings together the fields of stochastics and analysis addressing numerical, functional, formal thinking, problem-solving and reasoning, too. Above all, both units deal with a modelling problem. As to the material there are worksheets guiding the student through the modelling process, additional information sheets for the introduction of new mathematical facts and dynamic software applets for the investigation of certain aspects appropriate for this kind of medium.

3.1 From Cones to Higher Algebraic Curves and Back

This unit strongly connects algebraic and geometric aspects within a modelling context. It therefore concurs with Burkhardt (2006, p. 189), who points out:

> Algebra remains the key to higher performance in modelling as in so much mathematics; however, aspects that are crucial for modelling, particularly the *formulation* of algebraic models, are hardly touched in many current curricula, which focuses elsewhere – mainly on solving given equations. Geometry, too, needs a change of emphasis for modelling – with more emphasis, for example, on design.

The question to be tackled at first view is a very simple one: "Which champagne glass of the form shown in Fig. 53.1 has the biggest capacity?" Hereby the upper section of the glass always is a cone with fixed generator m.

The first approach should be made by building real paper models of possible solutions – quite similar to the one in Affolter et al. (2004) – in order to obtain a first impression of the strongly non-linear aspect. The importance of such an approach is underlined by the experience of Nisawa and Moriya (2011), for instance, where, in the context of two-dimensional functions, paper models are created, too, which were highly recommended by the students. The possible cones made of sectors with fixed radius $m = 10$ cm and sector opening angles ϕ from 30° to 330° in 30° are illustrated in Fig. 53.2.

Fig. 53.1 Cone-shaped glass with fix generator m

Fig. 53.2 Paper models of possible cones with generator $m = 10$ cm

The algebraic modelling uses some formulas linked to the cone or the sector, respectively (see Brandl 2009a). These formulas are well known to a student in the 10th grade, so the mathematical modelling process can be done by the student self-dependently and successfully. With the abbreviations $x := \frac{\varphi}{360°}$ and $k := \frac{\pi}{3} m^3$ we obtain the following "very non-linear" function $V_k : [0;1] \to \mathbb{R}^+$ for the cone volumes

$$V_k(x) = kx^2\sqrt{1-x^2}, \quad k > 0, x > 0.$$

The corresponding graph is shown in Fig. 53.3.

According to Fig. 53.3, the approximate value of the maximal opening angle for the sector can be determined approximately graphically by $x \approx 0.82$ that is $\varphi \approx 295°$. The exact value $x = \sqrt{\frac{2}{3}}$ can be determined either by the use of the calculus (if available) or even without it by methods as illustrated in Schupp (1997), where the inequality between arithmetic and geometric mean is used to detect the maximum of a function in an elementary way. So for the interpretation of the solution in the real world situation: among the cones in Fig. 53.2 the cone we were looking for is the

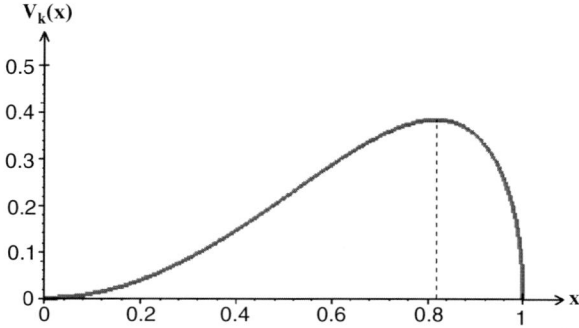

Fig. 53.3 Graph of V_k with maximum at $x \approx 0.82$

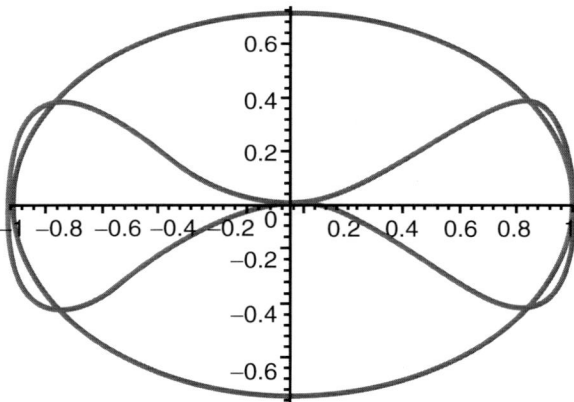

Fig. 53.4 Stimulus to generate "tie" curve

second from the right – a surprising fact. The aspect of validation, which should be carried out in a discussion among the students, leads to the same result as the related optimizing problem asking for the form of a cylindric tin having the smallest surface for a given fixed volume: the optimal glass or tin, respectively, is not very "handy". So aspects of design and ergonomics will succeed over maximal volume.

Up to this point the unit was carried out successfully in a 10th grade. For gifted students there is a second part generalizing the optimizing problem to a discussion of higher algebraic curves (as discussed in Schupp and Dabrock 1995, or Schmidt 1949, for example). Within this context an intra-mathematical modelling problem appears.

Mirroring the graph at both axes results in a smooth "tie" curve (see Fig. 53.4).

Looking at a (suitable scaled) ellipse (see Fig. 53.4), the "tie" curve can be achieved by a clever deformation process. It is motivated by an intuitive kinaesthetic approach suggesting an algebraic model of this deformation that leads to a formal proof of the coincidence with the "tie" curve. For more details and further content of the learning unit see Brandl (2008, 2009a).

3.2 *From the Lottery to the Pascal Triangle*

To quote Burkhardt (2006, p. 189), again: "*Statistics and probability* are essential in thinking sensibly about many problems." This is the context of the second unit to be presented here: "From the lottery to the Pascal triangle – a different kind of curve sketching" (see Brandl 2009c). It is suitable for students of upper secondary level and can be downloaded in German from the Internet portal www.lehrer-online.de.

What's the starting point? It is the question if it is more likely that there are more winners of the (6 out of 49) lottery when there are more participants. And, of course, the intuitive answer is easy: yes, of course. But the question remains, exactly how does this fact depend on the number n of participants. Is it linear? Are there regions of interesting behaviour?

The algebraic formulation of the model turns out to be a combination of the hypergeometric and the binomial distribution:

$$P(n) = P(\text{"At least 2 win the jackpot."})$$
$$= 1 - p_Z(6)^0 \left(1 - p_Z(6)\right)^n - n \cdot p_Z(6)\left(1 - P_Z(6)\right)^{n-1}$$

With $P_Z(r) = \dfrac{P(r) \cdot \binom{1}{1}}{\binom{10}{1}} = \dfrac{1}{10} p(r)$ because of the so-called "Zusatzzahl" (i.e., one additional number out of ten to win the Jackpot) and $p(r) = H(49,6,r) = \dfrac{\binom{6}{r}\binom{43}{6-r}}{\binom{49}{6}}$.

The subsequent curve sketching shows that the probability for n to infinity is 1, of course, and that the graph is strictly monotonically increasing. The high school students need to learn two new easy but powerful mathematical facts besides their curriculum in this context: first the formula for the derivative of a^x, and second the rules of l'Hospital. Both were part of the German mathematics curriculum at upper secondary school previously but they are not today.

The graph of $P(n)$ increases very slowly because of the small value of $p_Z(6)$. But its behaviour can be illustrated for small n if one looks at the chance for just one correct tip, for example, instead of winning the jackpot (see Fig. 53.5). The graph shows that there must be an inflection point somewhere.

So the next question can be: "From what number on does the rising of the graph slow down?" This can be calculated and the answer for one correct tip is three, whereas for the Jackpot it is 280 million, which – as for the interpretation of the result – far exceeds the population of Germany. Hence, one always stays in the

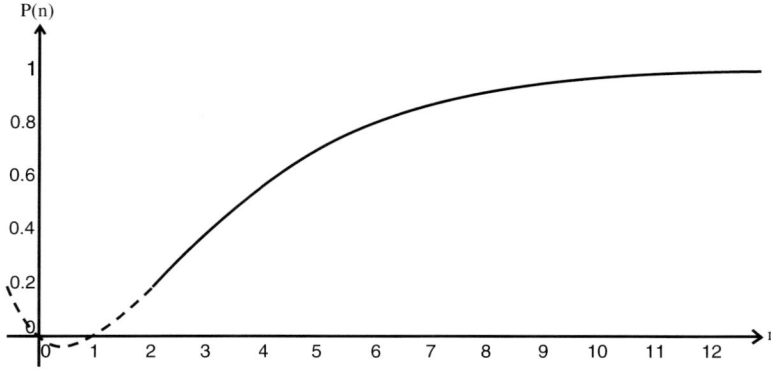

Fig. 53.5 Graph of $P(n)=P$ ("At least to have *one* correct tip.")

unfavourable steeper ascending part if one takes part in the game. However, according to the rush of participants every time the jackpot gets very big, this – just like the fact that with more and more participants the chance for a single jackpot-winner sinks – appears not to be common sense.

In a second, more mathematical part the unit is followed up by a variation that leads to very interesting aspects. The question now is: "How does the possibility vary that k participants have to share a Jackpot with increasing number n of participants?" Hence, the function to discuss is

$$P(n,k,r) = Bin(k \mid n, p(r)) = \binom{n}{k} \cdot p_z(r)^k \cdot (1 - p_z(r))^{n-k},$$

where the odd thing (for the student) is the appearance of the binomial coefficient within this term. By an interpretation of the binomial coefficient as a polynomial of kth degree in n with zeros at $0, 1, 2, \ldots, k-1$, the task can be successfully tackled. Based on the binomial coefficient, the relation $\binom{n}{k} = \binom{n-1}{k-1} + \binom{n-1}{k}$ leads to the Pascal triangle and several other mathematical facts as further items of the unit.

4 Summary

Based on the postulate that there seems to be a lack in the fostering of mathematically gifted students at upper secondary school, we presented two comprehensive learning environments that match the interests of the addressed group as stated in literature hitherto and evaluated in a small pre-study. These units are, or are to be, published at the dedicated established portal "Program for Gifted" at www.lehrer-online.de. Both units start with a real word problem to be solved by mathematical reasoning leading to new aspects of interesting mathematics.

References

Affolter, W., Beerli, G., Hurschler, H., Jaggi, B., Jundt, W., Krummenacher, R., Nydegger, A., Wälti, B., & Wieland, G. (2004). Kegel & Co. In *mathbu.ch – Lernumgebungen 9+* (pp. 30–31). Bern: schulverlag blmv AG, Zug: Klett & Balmer AG.

Baptist, P. (2000). Bausteine für Veränderungen in der Unterrichtskultur. In P. Baptist (Ed.), *Mathematikunterricht im Wandel* (pp. 7–30). Bamberg: C. C. Buchners Verlag.

Blum, W. (1996). Anwendungsbezüge im Mathematikunterricht – Trends und Perspektiven. In G. Kadunz et al. (Eds.), *Trends und Perspektiven* (pp. 15–38). Wien: Hölder-Pichler-Tempsky.

Brandl, M. (2008). *Von Kegeln zu höheren algebraischen Kurven und wieder zurück, große Unterrichtseinheit beim Lehrer-Online-Portal Begabte fördern*. Bonn: Schulen ans Netz e. V.

Brandl, M. (2009a). Kegelvolumen und mehr – Vom Kegel zur Tschirnhaus-Kubik und zurück. In R. vom Hofe and A. Jordan (Eds.), mathematik lehren 154 Themenheft *Wissen vernetzen: Geometrie und Algebra*, pp. 46–49. Seelze: Friedrich Verlag.

Brandl, M. (2009b). Lernumgebungen zur Begabtenförderung am Gymnasium. In M. Neubrand (Ed.), *Beiträge zum Mathematikunterricht 2009*. Münster: WTM-Verlag.

Brandl, M. (2009c). Vom Lotto zum Pascalschen Dreieck – eine etwas andere Kurvendiskussion, große Unterrichtseinheit beim Lehrer-Online-Portal *Begabte fördern*. Köln: lo-net GmbH.

Brinkmann, A. (2008). *Über Vernetzungen im Mathematikunterricht*. Saarbrücken: VDM Verlag.

Burkhardt, H. (2006). Modelling in mathematics classrooms: Reflections on past developments and the future. *Zentralblatt für Didaktik der Mathematik, 38*(2), 178–195.

DPhV (2008). Philologenverband beklagt unzureichende Begabtenförderung in Deutschland. Press release of DPhV on January 8, 2008. Retrieved from http://www.dphv.de/index.php?id=news-archiv-liste&tx_ttnews[pS]=1199142000&tx_ttnews[pL]=31622399&tx_ttnews[arc]=1&tx_ttnews[pointer]=10&tx_ttnews[tt_news]=123&tx_ttnews[backPid]=103&cHash=df3027a047.

Heller, K. (2002a). Bildungsempfehlungen für die Förderung besonders befähigter Gymnasialschüler. In K. Heller (Ed.), *Begabtenförderung im Gymnasium – Ergebnisse einer zehnjährigen Längsschnittstudie* (pp. 235–254). Opladen: Leske+Budrich.

Heller, K. (2002b). Zum Bildungsauftrag des Gymnasiums unter besonderer Berücksichtigung der Begabtenförderung. In K. Heller (Ed.), *Begabtenförderung im Gymnasium – Ergebnisse einer zehnjährigen Längsschnittstudie* (pp. 11–36). Opladen: Leske+Budrich.

Heller, K., & Perleth, C. (2007). Talentförderung und Hochbegabung in Deutschland. In K. Heller & A. Ziegler (Eds.), *Begabt sein in Deutschland* (pp. 139–170). Berlin: LIT Verlag.

Kaiser, G. (2007). Modelling and modelling competencies in school. In C. Haines, P. Galbraith, W. Blum, & S. Khan (Eds.), *Mathematical modelling (ICTMA 12): Education, engineering and economics* (pp. 110–119). Chichester, UK: Horwood.

Nisawa, Y., & Moriya, S. (2011). Evaluation of teaching activities with muli-variable functions in context. In G. Kaiser, W. Blum, R. Borromeo Ferri, & G. Stillman (Eds.), *Trends in teaching and learning of mathematical modelling* (pp. 111–126). New York: Springer.

OECD. (2001). *Knowledge and skills for life. First results from PISA 2000*. Paris: OECD.

Schmidt, H. (1949). *Ausgewählte höhere Kurven: Für Schüler oberer Klassen und Studenten der ersten Semester*. Wiesbaden: Kesselringsche Verlagsbuchhandlung.

Schupp, H. (1997). Optimieren ist fundamental. In H. Schupp (Ed.), *mathematik lehren Heft* (Vol. 81, pp. 4–10). Seelze: Friedrich Verlag.

Schupp, H., & Dabrock, H. (1995). *Höhere Kurven: situative, mathematische, historische und didaktische Aspekte, in der Reihe Lehrbücher und Monographien zur Didaktik der Mathematik, Band 28*. Mannheim/Leipzig/Wien/Zürich: BI-Wiss.-Verl.

Chapter 54
Modelling at Primary School Through a French–German Comparison of Curricula and Textbooks

Richard Cabassut and Anke Wagner

Abstract The teaching of modelling is a place where mathematical knowledge and real world knowledge are transposed in the school institution to become a taught knowledge. We will use a French–German qualitative comparison to propose a theoretical reflection based on Anthropological Theory of Didactic (ATD) to analyse this double transposition at primary school level. The comparison of the curriculum shows the difficulty to designate modelling as knowledge to be taught. The comparison of textbooks illustrates the characteristics of the modelling tasks and the progression of this teaching through the school year. We conclude by pointing the challenges for the production of resources and teacher training.

1 Origin, Method and Theoretical Framework of the Study

The two authors are partners of the European project LEMA[1] proposing a teacher training course on modelling and its application. They have decided to begin a comparison on the teaching of modelling at primary school level. A further study will compare it at secondary school level. For Germany we limit the comparison to Baden-Württemberg where the German author's institution is situated. Primary school is from grade 1–5 in France and from grade 1–4 in Baden-Württemberg. The comparative method through qualitative examples points to similarities and differences. The differences reveal observations that are usual in a country and as a

---

[1] Description of the project LEMA on the site www.lema-project.org

R. Cabassut (✉)
LDAR, Paris 7 University, IUFM, Strasbourg University,
141 avenue de Colmar, 67100 Strasbourg, France
e-mail: richard.cabassut@unistra.fr

A. Wagner
Institut für Mathematik und Informatik, Pädagogische Hochschule
Ludwigsburg, Reuteallee 46, 71634 Ludwigsburg, Germany
e-mail: wagner02@ph-ludwigsburg.de

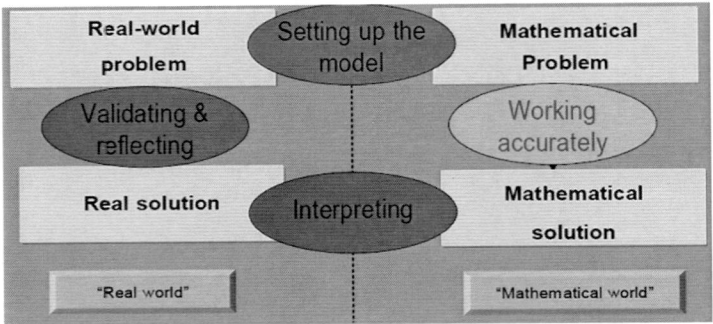

Fig. 54.1 LEMA modelling cycle

consequence less taken into consideration, and that are not present in another country. Comparison helps to break the apparent naturality of the observations showing the role of the institutions. We compare qualitatively curriculum and textbooks. This multiple approach does not intend to define national styles but offers examples in order to analyse questions and problems related to the teaching of modelling. The method is qualitative.

We adopt the definition of modelling considered in the LEMA project, based on the PISA (2006) theoretical framework and summarised in the following modelling cycle (Fig. 54.1).

We shall now use the terms of ATD[2] to point out the role of different institutions. The "scholarly" mathematics institution is composed of institutions (mathematics department of universities, mathematics research centres...) producing and using the "scholarly" mathematical knowledge. The real world institution is composed of different institutions producing extra mathematical knowledge and using it in the real world. For example, everyday life can be considered as one of these institutions[3] producing and using everyday life knowledge.[4] Modelling uses these two types of knowledge. Then other institutions (educational system, noosphere...[5]) decide if modelling will be a knowledge to be taught and explicitly designated in the curriculum. If modelling is a knowledge to be taught, this knowledge will appear as taught knowledge in the classroom and as learned knowledge in the community of study. Modelling that uses mathematical and real-world knowledge produced by different institutions is successively changed into knowledge to be taught, then into taught knowledge and finally learned knowledge. This process is called didactic

---

[2] Anthropological Theory of Didactics.

[3] Different authors use the term "everyday life". For example (Chevallard 1992, p. 88, translation R.C.) asserts "Everyday life is an institution" [La vie quotidienne est une institution]. (Stein 1986, p. 14–15, translation R.C.) considers "everyday life theory" [Alltagstheorie].

[4] Pupils' a class number can be considered as a social knowledge acquired by a child living all along the year in the class. Proportionality between the total amount paid and the number of bread loaves bought at the bakery can be acquired by a child used to buy the bread every day.

[5] Noosphere is the "sphere of those who think about education" (Bosch and Gascon 2006).

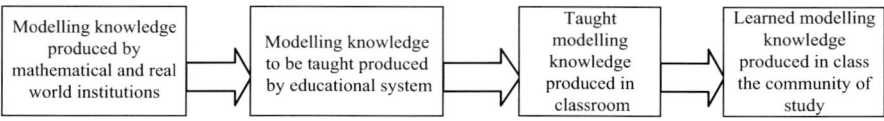

Fig. 54.2 Didactic double transposition process

transposition. Two types of knowledge, the mathematical one and the real world one, are involved in this transposition that we can consider therefore as a double transposition (Fig. 54.2).

2 Comparison of Curricula: Is Modelling a Knowledge to Be Taught?

We shall first compare the curricula in France and in Baden-Württemberg to observe how modelling is designated as knowledge to be taught. In Baden-Württemberg, at secondary school level, curriculum documents for Gymnasium, Realschule and Hauptschule[6] (Ministerium 2004) quotes explicitly "modelling" as a leading idea and all along the curriculum, at every grade, appears a paragraph of contents and competencies allocated to the leading idea "modelling". In this case we can assert that modelling is explicitly a knowledge to be taught. The facts are less clear at primary school level in France and in Baden-Württemberg.

In Baden-Württemberg modelling is quoted only one time[7] related to the ability to transfer a concrete situation in a modelling process using a mathematical model. Modelling is not a leading idea but we find relations with modelling in the leading ideas "measuring and magnitudes",[8] "patterns and structures"[9] and "data and concrete situations"[10] where the mathematisation of concrete situations is mentioned. Problem solving and discovery learning are two important features of mathematics teaching with possible relation with modelling. Modelling is not explicitly a leading ideal as it is in secondary school. We shall consider that, in the Baden-Württemberg curriculum of primary school, modelling is a knowledge to be taught as preparation to modelling as a leading idea at secondary school level.

In France the term "modelling" is not quoted but it is mentioned that mathematic helps to act in everyday life and that problem solving related to everyday life has

---

[6] In Baden-Württemberg, the three main types of secondary school are Gymnasium, Realschule and Hauptschule.
[7] Kennzeichnend für Sachrechenkompetenz ist die Fähigkeit, eine Sachsituation in einem Modellierungsprozess in ein mathematisches Modell zu übertragen, dieses mithilfe des verfügbaren Wissens und Könnens zu bearbeiten und auf dieser Ebene eine Lösung zu finden. Diese Lösung ist dann auf Plausibilität zu prüfen. Translation R.C.
[8] Messen und Grössen (Ministerium 2004).
[9] Muster und Strukturen (Ministerium 2004).
[10] Daten und Sachsituationen (Ministerium 2004).

to be worked for the leading ideas "numbers" and "organization and processing of data" and concrete problems are advocated for "measuring and magnitudes". Problem solving is explicitly designated as knowledge to be taught and two official pedagogical documents were published on "problems to search" and on "problem solving and learning".[11] We shall consider that, in the French primary school curriculum, modelling is a knowledge to be taught as a part of problem solving.

The fact that in both countries modelling does not appear explicitly as a knowledge to be taught has consequences: this knowledge will appear less explicitly, less regularly, less strongly as taught knowledge in the resources and in the class. Why is modelling not explicitly a knowledge to be taught at primary school? Is it because this idea is cognitively too difficult for young pupils? Numerous researchers have shown that modelling can be successful at primary school, even for low achievers (Peter-Koop 2002). Cabassut (2009) has pointed out the difficulties related to the double transposition. But difficulties are not reasons not to teach knowledge. Difficulties could be in the constructivist mind a condition to learn. Help is needed for low achievers and limits to avoid discouragement.

Is it because the beginning of learning in the school has to concentrate on the main leading ideas and so modelling is not one of these? In this case a minor place is allocated to modelling at primary school. This choice has to be argued and the noosphere (those who think about education) has to explain to the teachers the reasons for these choices. Until now, we do not know the answers of the noosphere to the previous questions. Let us consider now textbooks where modelling is taught, in order to describe this taught knowledge.

3 Articulation Between Real and Mathematical World in Textbooks

The importance of textbooks in mathematics lessons was expressed by many researchers.[12] In France the textbook market is a national one because the curriculum is national which enables a big variety in the offerings. In Baden-Württemberg there is a national market with regional editions adapting to the regional curriculum. In both textbooks are often produced by a mixed team with teachers, teacher inspectors, teacher trainers and less often researchers in didactics of mathematics. When comparing French and German textbooks[13] we found no essential differences between the two

---

[11] "Problèmes pour chercher" and "Résolution de problèmes et apprentissage" in (Ministère 2005).
[12] Pepin (2001).
[13] As shown in Cabassut (2007) it is very difficult to get national styles from a comparative study. Here we do not pretend that the textbooks chosen are typical representatives of France and Germany. Our approach is only qualitative. The textbooks show the existence of facts but do not prove how extended among the textbooks and in every country the facts are. In some cases we have illustrated the same facts in both countries. In other examples we illustrate a fact only by a textbook from one country without asserting that it is not possible to find a corresponding example in the other country. The textbooks help to point out facts and problematic questions about them.

54 Modelling at Primary School Through a French–German Comparison

countries concerning the kind of tasks used but we consider that in both countries tasks used in textbooks have special characteristics: the real world domains, the mathematics world domains, and the representations involved in the tasks.

3.1 Real World Knowledge

The first modelling task comes from a German textbook and it shows one example of a real world domain about "consumption of water" in everyday life. The task is illustrated with real photographs. To know how much water a pupil consumes on 1 day we need to know the situations where water is consumed and how to estimate the consumed quantities (Table 54.1).

The second example of modelling task from a grade 2 French book is based on school life: How to calculate the cost of a class trip? In this example we can observe the different register of data representations: drawing of a map (simplified representation of a real map), a flyer, a price table and texts. It is interesting to note that the map brings no information about the cost; the pupils have to sort the useful information. Pupils need real world knowledge to solve these tasks. In German didactical literature (Schipper et al. 2000, p. 208–209) there are proposals to share the real world domain depending on the grade. For grade 1 for example, the following real world domains are proposed: school, classroom, pets, shopping situations, school bus, birthdays and calendar.

In France an editor (Antoine et al. 2007) proposes a mathematics textbook to introduce the contents of the natural sciences syllabus. In this textbook different natural science themes are worked through the different mathematics domains (number, geometry, measuring and magnitudes, problem solving). We show in the following illustration the real world content planned throughout the year (Table 54.2).

In the textbook by Myx et al. (2003) you can find what we could call a real world "theorem": In situations of everyday, prices are generally proportional to the quantities (Table 54.3).

Table 54.1 Example of modelling tasks in textbooks

Germany grade 4 (9–10 years old)	France grade 2 (7–8 years old)
	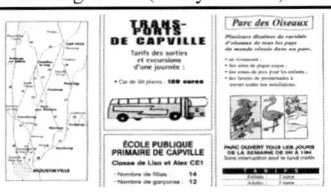
How much water do we consume daily? Estimate it. For what do we need more water? (Eccarius 2002, p. 96)	All pupils, the teachers, two fathers and one mother join the trip. We will help them to calculate the cost of the trip. (Charnay et al. 2001, p. 162)

Table 54.2 Example of real world knowledge planning in a textbook

Real world knowledge	
Grade 4 (9–10 years old)	
Development of insects	
The growth of arthropods and molluscs	
Water in vapour	
Evaporation	
Digestive system	
The rotation of the earth	
Digestion of nutrients	
Attraction of the earth: verticality	

(Antoine et al. 2007)

Table 54.3 Example of real world "theorem"

The price of meat is proportional to its weight.	
The price of a tissue is proportional to its length.	
In situations of everyday, prices are generally proportional to the quantities.	

The questions that we want to pose here are: How is the real world knowledge taught? Is this real world knowledge part of a curriculum? In which subject? Who teaches it (the school, the real life institution, the natural sciences lesson…)? How can teachers obtain this real world knowledge (initial training, in-service training, resources…)?

3.2 Mathematical World Knowledge

Some text books (Charnay et al. 2003, p. 11) try to plan a sharing of mathematical knowledge (Table 54.4).

Another problem is the way to articulate the mathematical knowledge involved in the mathematic problem, modelling the real world problem and the work on this modelling task. Following ATD framework (Artaud 2007) mentions the six moments of the study of a type of mathematics problem: the first encounter, the exploration and the elaboration of a technique to solve this type of problem, the justification of this technique, the technical work (to exercise the technique), the institutionalization and the evaluation. Depending on the moment where the modelling task is worked, the articulation will be different. If a modelling task is used to introduce new mathematics knowledge (first encounter) or to apply available

Table 54.4 Example of mathematics knowledge planning in textbook

	addition soustraction	multiplication division proportionnalité	tableau diagramme graphique	géométrie	mesure	déduction	gestion de contraintes
1. Euros	x	x				x	x
2. Collectionneurs	x	x	x			x	
3. Les fermes d'élevage					x		
4. Fleurs et bouquets	x	x					x
5. Parlons chocolat	x	x					x
6. Le guide-âne				x	x		x
7. La préparation du courrier			x			x	
8. Les chocolats de Nicolas	x	x	x			x	x
9. Le grand prix	x	x	x	x			
10. Énigmes	x	x				x	
11. Le recensement de 1999	x		x				
12. Assemblages de cubes		x			x		x
13. Mise en page	x	x				x	x
14. Surfaces sur quadrillage					x		x
15. Énigmes	x			x		x	x

Mathematics knowledge involved in problem solving (Charnay et al. 2003, p. 11):
Addition and subtraction. Multiplication, division and proportionality. Table, diagram, graph. Geometry. Measuring. Strategies to search Reasoning, deduction. Data processing. Argumentation.
The left table precises the expert knowledge involved in the problems. This knowledge is not necessary the knowledge used by the pupils to solve the problem. They can use personal procedures that use other knowledge.

mathematics knowledge (the technical work), the competencies worked will not be the same. In a similar way, if a modelling task is used to introduce new real world knowledge (first encounter) or to apply available real world knowledge (the technical work), the competencies worked with will not be the same. In the teacher practice, the moments of the mathematical study and the moments of the real world study, where the modelling task is worked, will define the didactic functions of the modelling task. Here is a specific difficulty of modelling with this double transposition articulating the two kinds of moments, especially when the pupils have to switch between the mathematical world and the real world (Cabassut 2009).

3.3 Representation Involved in the Tasks

To illustrate mathematical tasks in textbooks many different representations are used. Realistic representation can be made by photographs, pictures and authentic documents such as bills.

A linguistic representation of reality is usually done by texts that describe a situation. Some representations (photographs) are a kind of bridge between the two worlds. Some searchers (Maaß 2006, p. 115) use an intermediate physical model between real problem and mathematical problem (drawing, simplified problem) that is a kind of representation of the real world problem. In research there exists a long tradition for representations (Bruner 1966; Lompscher 1972) (Table 54.5).

Bruner, for example, distinguishes three modes of representation: enactive, iconic and symbolic. The proximity with the real world means more cognitive complexity, because we have the mathematical world representations, the real world representations, and treatments in and conversions between different registers of representation (Duval 2006). That means that the representation enables focus on specific competencies: to know how to read a table, a graphic, a map and of course to switch between the different kind of representations.

Table 54.5 Example of representations involved in tasks

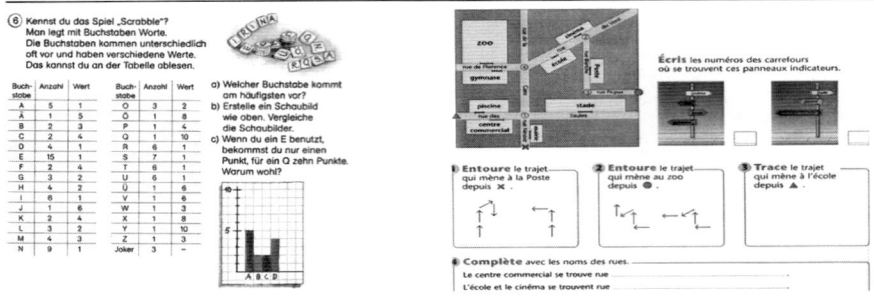

| Treatment and conversion on graph and table representations (Schütte 2004, p. 83) | Treatment and conversion on representations of the space (Blandino and Bourgoint 2006, p. 164) |

3.4 Whole Competencies and Partial Competencies: Didactical Functions of Tasks

In the first presented tasks (water consumption and class trip), the tasks are modelling tasks (opened, complex, authentic and related to real world) and enables use of all the competencies of the PISA modelling cycle (Fig. 54.1). The other previous tasks use only treatment and conversion competencies in order to learn to set the mathematical model.

The example to the right comes from a French teacher guidebook for grade 1 (ERMEL 2000). These guidebooks are proposed all along the primary school curriculum to work problem solving for numerical learning. The attached task proposes a real world context familiar to the pupils: a birthday party The task objectives for grade 1 pupils are to learn to raise questions about a situation and to sort the questions where the answer needs mathematics knowledge (e.g., Are there enough chairs?) and the question you can be answered without using mathematics (e.g., What gift could you offer to the girl?) This task is not a modelling task but a teaching task: it uses prerequisite competences that you need to set a mathematical model.

In the textbook of Table 54.2, the plan of problem solving competencies used throughout the year is the following: to select the pertinent information to answer the questions (3),[14] to analyse solutions: to recognize a false solution and to explain why (2), ask questions: to formulate questions depending on the wording and to answer questions (2), to use or draw a diagram or a graph (6), to solve complex problems with several reasoning and computing steps (3), to solve a research (6).

4 Conclusion: Challenges for Modelling Resources and Teacher Training

In the curriculum of primary school, modelling is not explicit knowledge to be taught but it can be implicitly considered as knowledge to be taught as preparation for the secondary school curriculum (Baden-Württemberg) or as a part of problem solving (France). The consequence is that modelling is not explicitly a study theme in the textbooks. Nevertheless modelling tasks appear in textbooks involving varied domains of the mathematics world and of the real world. Furthermore teaching tasks appear that are not modelling tasks but that support achieving partial competencies as prerequisite of work on modelling tasks, which shows that modelling is a taught object. Some mathematics text books plan, through the school year, the teaching of real world knowledge and mathematical knowledge and their articulation. These different observations show a moving of the textbooks from pupils' practices to teachers' practices. The specific competencies of modelling (in relation to real world and in the role of representations) and the generic competencies of problem solving have to be different enough to justify the interest of modelling in teaching and learning. The double transposition in the teaching of modelling has to be taken into consideration in the resources and in the teacher training in order to support better the learning of modelling. The report (IGEN 2006, p. 66 translation R.C.) on mathematics teaching at French primary school comments: "Some problems are given with a vague objective 'how to play to be a mathematician.' This is against productive when it leads to neglect of the basic math: learning to solve a problem is also a built training, with a progressive and intelligent methodology." Here are challenges for resources and teacher training related to modelling.

References

Antoine, M., Beaufils, G., Burger, O., & Guichard, J. (2007). *Les maths à la découverte des sciences CM1* (pp. 4–5). Paris: Hachette Education.
Artaud, M. (2007). Some conditions for modelling to exist in mathematics classrooms. In W. Blum, P. L. Galbraith, H.-W. Henn, & M. Niss (Eds.), *Modelling and applications in mathematics education* (New ICMI Study Series, Vol. 10). Dordrecht: Kluwer.

---

[14] The number in brackets is the number of citations of the competence in the plan.

Blandino, G., & Bourgoint, P. (2006). *Les Maths à la découverte du monde CP* (p. 164). Paris: Hachette Education.

Bosch, M., & Gascon, J. (2006). *ICMI bulletin 58: 25 years of didactic transposition* (pp. 51–65). Quebec: Université de Laval.

Bruner, J. S. (1966). On cognitive growth: I. In J. S. Bruner et al. (Eds.), *Studies in cognitive growth* (pp. S. 1–29). New York, Wiley.

Cabassut, R. (2007). Examples of comparative methods in the teaching of mathematics in France and in Germany. In *Proceedings of 5th Cerme (Congress of European Society for Research in Mathematics Education)*, Larnaca, Cyprus.

Cabassut, R. (2009). The double transposition in mathematisation at primary school. In *Proceedings of 6th Cerme (Congress of European Society for Research in Mathematics Education)*, Lyon, France.

Charnay, R., Dussuc, M.-P., & Madier, P. (2001). *CE1 cycle des apprentissages fondamentaux. Le matériel photocopiable* (p. 162). Paris: Cap Maths, Hatier.

Charnay, R., Combier, G. & Dussuc, M.-P. (2003). *CM1 Le guide des activités*. Paris: CAP Maths, Hatier.

Chevallard, Y. (1992) Concepts fondamentaux de la didactique: Perspectives apportées par une approche anthropologique, *Recherches en Didactique des Mathématiques, 12*(1), 73–112.

Duval, R. (2006). A cognitive analysis of problems of comprehension in a learning of mathematics. *Educational Studies in Mathematics, 61*(1–2), 103–131.

Eccarius, D. (2002). *Lollipop-Mathematik 4*. Berlin: Cornelsen.

ERMEL. (2000). *Apprentissages numériques et résolution de problèmes Cours préparatoire* (p. 85). Institut National de Recherche Pedagogique. Paris: Hatier.

IGEN Inspection Générale de l'Education Nationale (2006). *L'enseignement des mathématiques au cycle 3 de l'école primaire*. Ministère de l'Education Nationale.

Lompscher, J. (1972). *Theoretische und experimentelle Untersuchungen zur Entwicklung geistiger Fähigkeiten*. Berlin: Volk und Wissen.

Maaß, K. (2006). What are modeling competencies? *Zentralblatt für Didaktik der Mathematik, 38*(2), 113–142.

Ministère (2005). *Documents d'Accompagnement des Programmes Mathématiques Ecole primaire* (pp. 7–19). Ministère de l'Education Nationale. C.N.D.P.

Ministerium. (2004). Bildungsstandards für mathematik. In Ministerium für Kultus, Jugend und Sport Baden-Württemberg (Eds.), *Bildungsplan 2004, allgemein bildendes gymnasium* (pp. 91–102). Stuttgart: Baden-Württemberg.

Myx, A., Dossat, L., Bregeon, J.-P., Vicens, P.-Y., & Poli, B. (2003). *Mathématiques. CM1, collection diagonale* (p. 141). Paris: Nathan.

Pepin, B. (2001). Mathematics textbooks and their use in English, French and German classrooms: A way to understand teaching and learning cultures. *Zentralblatt für Didatik der Mathematik, 33*(5), 158–175.

Peter-Koop, A. (2002). Real-world problem solving in small groups: Interaction patterns of third and fourth graders. In B. Barton, C. Irwin, M. Pfannkuch, & M. O. J. Thomas (Eds.), *Mathematics education in the South Pacific. Proceedings of the 25th Annual Conference of the Mathematics Education Research Group of Australasia, Auckland* (pp. 559–566). Sydney: MERGA.

PISA. (2006). *Assessing scientific, reading and mathematical literacy: A framework for PISA*. Paris: OECD.

Schipper, W., Dröge, R., & Ebeling, A. (2000). *Handbuch für den mathematikunterricht*. Hannover: Schroedel.

Schütte, S. (Ed.). (2004). *Die matheprofis 2*. München: Oldenburg.

Stein, M. (1986). *Beweisen*, Texte zur mathematische-naturwissenschaftlich-technischen Forschung und Lehre, Band 19, Verlag Franzbecker, Bad Salzdetfurth.

Chapter 55
Modifying Teachers' Practices: The Case of a European Training Course on Modelling and Applications

Fco. Javier García and Luisa Ruiz-Higueras

Abstract Many studies consider initial teacher training and professional development a crucial lever for a wider introduction of modelling and applications. Despite the efforts made in recent years, a deeper understanding of teachers' practices in relation to modelling and applications is needed. In this chapter, after some reflections concerning the dimensions of the problem, we introduce a model to describe, at least partially, teachers' practices based on the Anthropological Theory of Didactics. As our interest is modelling, we apply this model in order to analyse to what extent, and in which direction, training materials from a European-funded project promote changes in teachers' practices involving modelling and applications.

1 Introduction

As reported in many studies, despite progress in the research field of modelling and applications, the inclusion of modelling and applications tasks or the adoption of modelling-oriented pedagogies is still marginal in many educational systems (although important differences exist from one country to another). Teachers' initial training and professional development is considered crucial to make this situation evolve; however, the case of modelling and applications seems to be special. According to Burkhardt (2006, p. 191), "in many countries teachers are expected to deliver a curriculum on the basis of the skills they acquired in their pre-service education, consolidated in the early years of classroom practice". On the other hand, for many teachers, modelling and applications did not constitute an important issue either in their past as students, in their training to become teachers, or in their early years in the classroom. As Doerr (2007, p. 69) states: "one reason for the limited use of applications and modelling at the primary and secondary levels of schooling

---

F.J. García (✉) and L. Ruiz-Higueras
Department of Didactics of Science, University of Jaén, Paraje de Las Lagunillas s/n, Building D2, Jaén 23071, Spain
e-mail: fjgarcia@ujaen.es; lruiz@ujaen.es

is the lack of knowledge by those who are expected to teach mathematics through applications and modelling". In addition, as she stresses, beyond the lack of subject knowledge related to modelling, teachers need other kinds of knowledge. The main issue that remains a basic research question is *to determine the knowledge teachers need in order to be effective in using applications and modelling in their practice* (Doerr 2007), which is automatically connected to a new fundamental question: *How should teachers be trained/supported in order to be able to implement modelling-oriented pedagogies effectively?*

Aware of this situation, a group of European researchers applied for a Comenius project (LEMA project) aimed at designing a professional development course on modelling and applications. The materials created during the project are examined although they are not the main focus of this chapter. Particularly, in which direction and to what extent these materials promote changes in teachers' practices are analysed through a specific model of teachers' actions.

2 Teacher Education on Modelling and Applications: From a Teachers' Problem to a Professional Problem

Teacher training in modelling and applications is at the core of tension between two different trends:

- On the one hand, research on modelling and applications concerning teachers' initial training and professional development is a complex domain which is still in its early stages. The situation is not extraordinary. In the extensive review of the state of the art concerning research in mathematics-teacher education carried out by Adler et al. (2005), they conclude that small-scale qualitative research predominates in the field, focused mainly on case studies of a few teachers. The authors consider this an indicator of an emerging research field, where particular issues come first and precede generalisation. Furthermore, they consider this also to be a consequence of a research field that appears to be more complex than others.
- On the other hand, due to the evolution of the educational system in many countries in recent years, teachers' initial training and professional development related to modelling and applications has become a crucial and urgent issue (for instance in Spain, Germany or England, due to the new national curricula, as reported in García et al. 2010). In many countries, increasing pressure is being exerted by society, educational authorities, curriculum, teachers, etc.

The tension is far from being solved but, meanwhile, research and many professional development initiatives are being proposed to encourage teachers' knowledge to evolve.

Following Chevallard's (2006a) distinction, we consider the problem of initial teacher training and professional development (particularly in modelling and applications) as a problem of the teaching profession more than a teacher's problem.

Modelling and applications (as content or as a teaching strategy) is more than a problem that teachers face in their classroom and that they normally express by asking for direct help (e.g., *Where can I find new tasks? What innovative pedagogical approaches exist?*). It should be considered a problem that the teaching profession faces due to profound (intended) changes: in many educational systems, in the way mathematics is being considered, in the new general aims assigned to schooling, and (as reported in many studies) in the inadequate training of those responsible for developing the curriculum.

From this perspective, the focus is not on how teachers are making their personal *models* evolve from reflection concerning their (personal) practice, but rather on *models* that would be useful to the profession in order to deal with the complexity of teaching when modelling and applications are involved. Our approach represents a shift in how the problem is described and undertaken. We might consider a top-down approach to teachers' pedagogical needs concerning modelling against the bottom-up approach from other perspectives. Ultimately, the path is different but the goal is shared – that is, identifying useful models for teachers to deal with the complexity inherent in the teaching and learning of modelling or in the use of modelling-oriented pedagogies.

3 A Theoretical Framework to Describe Teaching Actions

From our approach, any attempt to progress in the profession's needs to make effective use of modelling and applications has to be integrated into a deeper understanding of teachers' actions while using modelling and applications in their teaching (as a content or as a tool). This means that general models or theories to describe teachers' actions need to be considered.

The anthropological theory of didactics (ATD) has been developed over the last 25 years as a comprehensive framework for research in mathematics education. Its origins can be found in the first formulation of the didactic transposition theory, which pointed out that what is being taught at school ('content' or 'knowledge') is, in a certain way, an exogenous production (i.e., something generated outside school that is 'transposed' into school out of a social need of education and dissemination) (Bosch and Gascón 2006). This led to the notion of *institutional relativity* of knowledge and the *ecological metaphor*: mathematical knowledge emerges, evolves, changes, migrates and sometimes dies in institutions and between institutions. Modelling and applications as school content is not exempt from these processes.

In a brief description, ATD is founded on two basic hypotheses. The first one, which distinguishes it from other theories in mathematics education, is the hypothesis that, in order to understand how individuals act, we need to know first how the institutions they belong to *act*:

> Behind the persons, and the knowledge, there appeared the institutions, to be regarded on par with the persons, in the light of a dialectic between persons and institutions. Persons are the makers of institutions which in turn are the makers of persons. Generally, however,

institutions come before those persons – their "subjects" – thanks to whom they will continue to exist and change. So that, in order to understand what persons are made of, we have to understand how institutions live, develop or recede. (Chevallard 2006b, p. 4).

This seems especially crucial both for research in mathematics education as well as in teacher education. Students and teachers are members of various institutions (school, mathematics classroom, the teaching profession, etc.) with explicit and implicit rules, idiosyncrasies, restrictions, connections to other institutions, functions and so forth which largely (although not completely) shape the behaviour of those students and teachers.

A second fundamental hypothesis of the ATD, which is not unique to this framework, is to consider mathematics to be a human activity. This leads to the notion of *praxeology*, in order to describe human mathematical activities and, moreover, any human activity:

> One can analyse any human doing into two main, interrelated components: *praxis*, i.e., the practical part, on the one hand and *logos*, on the other hand. "*Logos*" is a Greek word which, from pre-Socratic times, has been used steadily to refer to human thinking and reasoning – particularly about the cosmos. [...] no human action can exist without being, at least partially, "explained", made "intelligible", "justified", "accounted for", in whatever style of "reasoning" such an explanation or justification may be cast. *Praxis* thus entails *logos* which in turn backs up *praxis*. For *praxis* needs support – just because, in the long run, no human doing goes unquestioned. Of course, a praxeology may be a *bad* one, with its "praxis" part being made of an inefficient technique – "technique" here is the official word for a "way of doing" – and its "logos" component consisting almost entirely of sheer nonsense – at least from the praxeologist's point of view! (Chevallard 2006c, p. 23).

Clearly, the notion evolved as new problems were arising and was later refined in order to produce more precise tools to describe and analyse institutional didactic processes. A distinction between *specific, local, regional and global* mathematical praxeologies was introduced, depending basically on the range of the *practical* and the *theoretical* blocks. This distinction has become critical in order to understand special features of how mathematics is being taught at school as well as to identify didactic phenomena.

3.1 Modelling the Teaching Activity

Not only mathematics can be considered as a human activity and modelled in terms of praxeologies, but also teaching is a human activity and, therefore, a *didactic praxeology* is activated each time a teacher interacts with students in the classroom. Didactic praxeologies are also made up of problematic tasks that teachers face, teaching techniques they use (or could use), and *technologies* and theories that describe, explain and justify their actions. It is also possible to distinguish between *specific, local, regional, and global* forms of didactic praxeologies.

For instance, Fig. 55.1 shows how an English textbook introduces, in a classical way, the notion of variables. Stereotyped and opportunistic relation with reality is used in order to make the situation closer to the students, seeking to stir motivation.

55 Modifying Teachers' Practices

> Write down the short form of these rules.
> Use the red letters and numbers.
> 1 The *t*otal money raised in a sponsored swim at £5 for each *l*ength.
> $t = ... \times ...$
> 2 The *t*otal money raised on a sponsored walk at £4 for each *m*ile.
> $t = ... \times ...$
> 3 The *t*otal cost of a weekly magazine at £2 each *w*eek.
> $t = ... \times ...$

Fig. 55.1 Excerpt from a textbook

> - She proposes a different situation (realistic and open):
>
>    At a petrol station how far apart should the pumps be placed to cater for cars that tow caravans?
>
> - Students work in groups of 4.
> - She circulates and supports students while working (specially, when they are stuck).
> - She says: "at the end, you should produce a poster with your solution".
> - She organizes a plenary: two groups are asked to explain their work.
> - She summarizes the main mathematics involved in the situation.

Fig 55.2 A possible didactic technique

This way of introducing new concepts is quite usual and generates a *professional* problem: *How can real contexts and situations be used in order to give meaning to mathematics?* In our particular example, a didactical problem teachers face could be reformulated as: *How can the ideas given in this textbook be restated to encourage students to explore "variables" in real contexts?*

Any solution to this didactic problem can be considered a didactic technique. While there is no single solution to this problem, Fig. 55.2 offers one possibility.

This approach to the problem is described and justified by (didactic) *technologies* and theories, which constitute the *logos* of this didactic praxis, as shown in Table 55.1.

Compared with describing mathematical praxeologies, it is more difficult to explain how and why teachers act (i.e., their didactic praxeology) because they mix components of mathematical praxeologies with others (e.g., group work, time

Table 55.1 Didactic *logos*

Technological elements	Theoretical elements
Students can develop a more appropriate meaning of "variables" in real situations that offer the possibility of exploring relations between different quantities.	Modelling for learning mathematics
Before being told, students should have the opportunity of exploring the situation.	Collaborative learning
Working in groups is more effective for students to develop their capacities.	Student-centred pedagogies
After working in groups, it is necessary for students to explain and reflect on the work done by certain groups.	ATD: meaningful school mathematics
Before the class ends, it is important for the teacher to summarize the main mathematical ideas embedded in the task.	…

management, assessment, giving feedback to students). Whereas mathematics is characterised as rational structured knowledge, teaching is often viewed as a vague combination of disparate competencies with a strong personal filter.

In contrast with mathematical praxeologies, for which the components are normally not difficult to describe, it is unclear how best to describe teaching practices. That is, what is a teaching technique and what is it made of? What is a teaching technology/theory and what is it made of? Moreover, if teaching processes are not conceived in a normative and prescriptive way, then teaching techniques cannot be considered determined or closed.

To build a framework to model teachers' actions, at least partially, Sensevy et al. (2005) used some general constructs from the theory of didactic transposition. These researchers adopted a non-prescriptive position: "describing the interaction of a teacher and his students in order to improve our understanding while respecting the complexity of the teaching process" (Sensevy et al. 2005, p. 153). Basically, they consider a teacher's work to consist of initiating, establishing and monitoring the students' relationship with knowledge. To do this, teachers' need to employ a wide range of teaching techniques, which comprise their didactic praxeology. Some of these techniques are content-dependent (i.e., different techniques could be observed in an algebra class than, for instance, in a geometry class, and in a teaching process where modelling and applications are involved) and others can be considered general.

Sensevy et al. (2005) considered a triple dimension that describes teachers' work related to starting and maintaining the didactic relationship, giving rise to a classification of teaching techniques in:

- *Mesogenetic techniques*, which involve all the actions teachers take in order to organize the learning situation and its milieu.
- *Topogenetic techniques*, which involve all the actions by teachers in order to divide and orient student–teacher responsibilities during the study process: students' *topos* and teacher's *topos* (a Greek word meaning, in this context, the *place* students and teacher occupy within a teaching process).

- *Chronogenetic techniques*, which involve actions by teachers in order to make the didactic relationship evolve over time.

For instance, in the above example, a *mesogenetic technique* can be identified when the teacher changes the situation (milieu) by introducing the "petrol station" task. By introducing group work, the teacher is performing a *topogenetic technique* (the responsibility has been transferred to the students). After the group work, the teacher introduces the poster to summarize the solution. That technique has mesogenetic (the milieu has changed), chronogenetic (students' activity is extended in time) and topogenetic influences (once again, the responsibility belongs to the students). Finally, when the teacher assumes responsibility for summarizing, this can be considered a topogenetic technique (but in the reverse direction: now responsibility is on the teacher's side).

To what extent these models are useful to explore teaching processes when modelling and applications are involved is a question that we have just started to explore. We use these models in the following section to determine the direction in which training materials from the LEMA project are trying to move teachers' actions related to modelling and applications.

4 LEMA Professional Development: Changing Teachers' Practices

LEMA is the acronym of the Comenius European funded project, *Learning and Education in and through Modelling and Applications*. In this project, mathematics educators from six different countries have being working together on a set of professional development materials for primary and secondary school teachers. The main aim is to change teachers' practices and beliefs towards modelling and applications.

As reported in García et al. (2010), many challenges have been faced: different theoretical backgrounds and research traditions, different school systems in each country, different roles of applications and modelling within each country, different teaching cultures in each country, different teacher professional development systems, etc. All these restrictions have led to a set of training materials organized in modules and sub-modules (see Fig. 55.3) that are mutually dependent but that can be arranged in many different ways in order to be adaptable to each country.

LEMA offers teachers opportunities to develop their *praxis* with a wide range of teaching techniques that support modelling (as content or as vehicle). LEMA offers teachers opportunities to reflect and justify the why of this *praxis*, that is, to develop their *logos*. By bringing both together, LEMA offers the *profession* opportunities to develop current *didactic praxeologies* towards modelling and applications. We extract brief examples from the materials that exemplify these changes.

Example 1: The sub-module "competencies" in the module "lessons" offers different teaching methods aimed at developing both the modelling competency as a whole

Modelling	Tasks	Lessons	Assessment	Reflection
What is modelling?	Exploring	Methods	Formative	Challenges
Why modelling?	Creating	Competencies	Summative	Implementation
	Classifying	Content	Feedback	
	Varying	ICT		

Fig 55.3 LEMA professional development course: modules and sub-modules

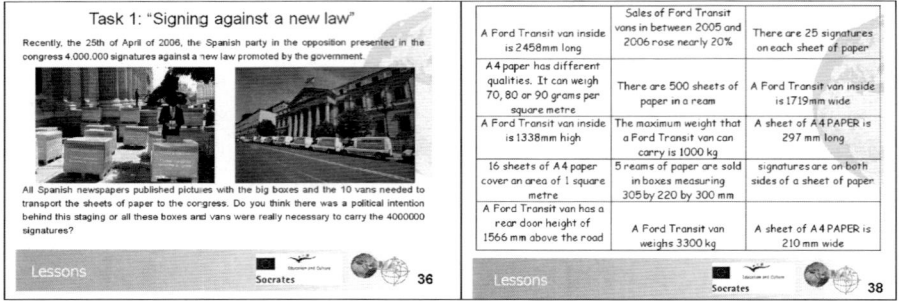

Fig. 55.4 Card game to help set up a model competency

and the associated sub-competencies in particular. For instance, in order to support students' sub-competency of "setting up a model", teachers need to decide, from a set of cards with some hints about the situation, the ones that are useful to arrive at the solution (see Fig. 55.4). Introducing the cards is a way of changing the didactic milieu. Thus, this training activity offers teachers a useful *mesogenetic* technique related to the development of specific modelling sub-competencies. Moreover, it can be considered also to be a *topogenetic* technique because it is meant for assigning the responsibility of establishing the model to students, in contrast with the usual didactic contract where students assume the work *inside the model* (considered to be the authentic mathematics) but they neglect the previous work of structuring the situation, considering that it is teacher's responsibility to offer a situation in which it is clear what type of mathematics should be used.

Example 2: The "assessment" module includes a sub-module dealing with assessment for learning techniques. Teachers explore, among other aspects, how they can encourage students by giving feedback. Following Black and William (1998) *Working inside the Black Box*, questioning techniques are studied and, especially, the fact that many teachers do not give students enough time to look for an answer after a question is given, as reported in this study. Therefore, teachers explore techniques to formulate effective questioning in modelling activities by giving students enough time to construct their own answer. These techniques, which deal with the time during which the teaching and learning processes occur, can be considered *chronogenetic* techniques.

5 Conclusions and Implications

Teacher education related to modelling and applications has become an urgent issue in many countries. Beyond individuals, it can be considered to be a problem the teaching profession faces. At the same time, it is a problem of those institutions in charge of providing initial teacher training and professional development. A deeper understanding of the challenges faced by the profession is needed to provide both the training institutions and the profession with appropriate materials and pedagogical approaches. However, while strong efforts have been made to understand which teaching methods support the best modelling and applications, little is being done to systematize, classify and question them. Frequently, we find lists of teaching methods that support modelling but these are often based on experimental evidence with a weak theoretical background. The framework introduced in this paper attempts to clarify and structure, at least partially, those pedagogical approaches (didactic praxeologies) that support modelling, with the objective of better informing initial training and professional development programmes in modelling.

References

Adler, J., Ball, D., Krainer, K., Lin, F.-L., & Novotna, J. (2005). Reflections on an emerging field: Researching mathematics teacher education. *Educational Studies in Mathematics, 60,* 359–381.

Black, P., & William, D. (1998). *Inside the black box. Raising standards through classroom assessment.* London: Kings College.

Bosch, M., & Gascón, J. (2006). Twenty-five years of the didactic transpositions. *ICMI Bulletin, 58,* 51–63.

Burkhardt, H. (2006). Modelling in mathematics classrooms: Reflections on past developments and the future. *ZDM – The International Journal on Mathematics Education, 38*(2), 178–195.

Chevallard, Y. (2006a). Former des professeurs, construire la profession de professeur. Resource document. Journées scientifiques sur la formation des enseignants du secondaire de la Faculté de psychologie et des sciences de l'éducation (section des sciences de l'éducation) de l'Université de Genève. http://yves.chevallard.free.fr/spip/spip/IMG/pdf/Former_des_professeurs_construire_la_profession.pdf. Accessed 14 Sept 2009.

Chevallard, Y. (2006b). Readjusting didactics to a changing epistemology. Resource document. Invited communication at the *European Conference on Education Research.* http://yves.chevallard.free.fr/spip/spip/IMG/pdf/Readjusting_didactics_to_a_changing_epistemology-2.pdf. Accessed 14 Sept 2009.

Chevallard, Y. (2006c). Steps towards a new epistemology in mathematics education. In M. Bosch (Ed.), *European research in mathematics education* (Vol. IV pp. 21–30). Barcelona: Universidad Ramón Llull.

Doerr, H. (2007). What knowledge do teachers need for teaching mathematics through applications and modelling? In W. Blum, P. Galbraith, H.-W. Henn, & M. Niss (Eds.), *Modelling and applications in mathematics education* (pp. 69–78). New York: Springer.

García, F. J., Maaß, K., & Wake, G. (2010). Theory meets practice: Working pragmatically within different cultures and traditions. In R. Lesh, P. L. Galbraith, C. R. Haines, & A. Hurford (Eds.), *Modeling students' mathematical modeling competencies* (pp. 445–457). New York: Springer.

Sensevy, G., Schubauer-Leoni, M.-L., Mercier, A., Ligozat, F., & Perrot, G. (2005). An attempt to model the teacher's action in the mathematics class. *Educational Studies in Mathematics, 59*, 153–181.

Chapter 56
Google's PageRank: A Present-Day Application of Mathematics in the Classroom*

Hans Humenberger

Abstract Very many people use internet search engines like Google nearly every day, in most cases more than once a day. Here the following question arises: How does Google come to a ranked list, how does Google know which site is an important one and should be placed at the top of the list? It turns out that this question can be dealt with at a very elementary level. Here we show a possible way.

1 Introduction

Very many people use Google; it has become the most used internet search engine all over the world.[1] In most cases the relevant sites (concerning the word we looked up) are more or less on the top, so it is not necessary to have a look at hundreds of sites to read something important and informative. Here the following question arises quite naturally: *How* can Google manage this ranking? How does Google know whether a special site is a relevant one and therefore should be presented quite at the top of the list? It turns out that the answer has to do with the so-called "PageRank". Google's PageRank can be dealt with at school (upper secondary) in two different manners: (1) as another application if "multi stage processes" or "Markov chains" have already been discussed, (2) as an introduction to the mentioned fields.

First a simple possible problem for the introduction: The telephone market of a country is dominated by three companies (A-tel, B-tel and C-tel). They have annual

---

*An extended version with the title "How does Google come to a ranked list – making visible the mathematics of modern society" was first published (online) on May 12, 2011 in Teaching Mathematics and its Applications, doi:10.1093/teamat/hrr007

[1] Market shares (according to television broadcast in June 2009): Google 62%, Yahoo 21%.

H. Humenberger (✉)
University of Vienna, Department of Mathematics, Vienna, Austria
e-mail: hans.humenberger@univie.ac.at

Fig. 56.1 Transition graph

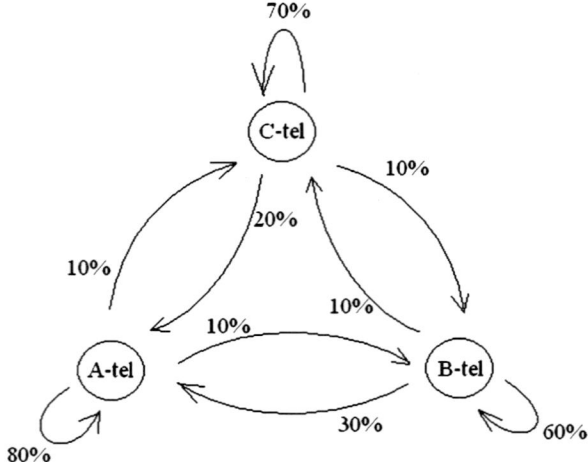

contracts with their customers,[2] the customers stay at their former company to a special percentage and change to other companies, respectively. This situation can easily be described with a so-called *directed graph* (also *transition graph* → Fig. 56.1): This means, for example, for the company C-tel that 70% of their customers stay at C-tel after 1 year, 20% change to A-tel and 10% to B-tel. The other transition rates can be interpreted in a similar way. Let's suppose that these transition rates do not change throughout the next 5 (10; 20) years.

$$0.8A_n + 0.3B_n + 0.2C_n = A_{n+1}$$
$$0.1A_n + 0.6B_n + 0.1C_n = B_{n+1}$$
$$0.1A_n + 0.1B_n + 0.7C_n = C_{n+1}$$

What would be the distribution (fractions, percentages) of the customers to the companies at that time if at the beginning it was $(A_0, B_0, C_0) = (1/3, 1/3, 1/3)$ or $(A_0, B_0, C_0) = (30\%, 50\%, 20\%)$? Even if students have not heard anything about Markov chains or transition matrices they can handle this problem easily by using a spreadsheet programme (e.g. EXCEL). One can establish the associated recursions (→ above) by looking at the transition graph and entering them as a formula. Especially for such iterative situations (problems) spreadsheet programmes are a very useful tool! Using the well-known pulling down method one can easily and quickly see the values after 5, 10, 20 years (using only a calculator would be much more cumbersome here). One will realize that the values quickly tend to be $(A_n, B_n, C_n) = (55\%, 20\%, 25\%)$ independent of the initial distribution (A_0, B_0, C_0).

---

[2] Assumption: these contracts are always made for 1 year, at the end/beginning of a year the customers may possibly change the telephone provider.

For experimentally determining such limit distributions in the case of only a few possible "stations" (above only 3: A-tel, B-tel, C-tel) spreadsheets are a wonderful tool. One does not need matrices or theories behind it, one only needs very elementary knowledge of EXCEL. The process in EXCEL is an iterative one and so is the real determination of the PageRank in the practice of Google. Using spreadsheets here therefore on the one hand is a simple introduction and on the other hand it is not so far away from the procedure in reality (iterative methods are used there too).

For dealing with such "limit distributions" in a more detailed way (especially a few theoretical aspects) EXCEL does not suffice, we need "transition matrices".

2 The WWW as a Directed Graph and the Description by Transition Matrices

Search engines start their procedure by "combing through" the WWW with a so-called *spider* or *webcrawler* (special computer programme): on which pages (sites) is there something written about the word (item) we are interested in and looking for?

The aim of this very large search process is to get a description – as good as possible – of the contents of the pages and the structure of links[3] in the WWW concerning the word (item) looked up. Let us start with a very simple example: A, B, C, D are four different sites that are linked to each other as shown in Fig. 56.2. (e.g., there is a link from site A to B and C, from B there are links to C and D and so on).

Modelling assumption 1: For reasons of simplicity we assume that every link on a page will be used with the same probability.[4] That means if there are two leaving arrows from a site each of them ought to have the number ½, when there are 3 (k) leaving arrows, each arrow ought to have the number 1/3 (1/k). On the basis of this assumption no probabilities are written to the arrows.

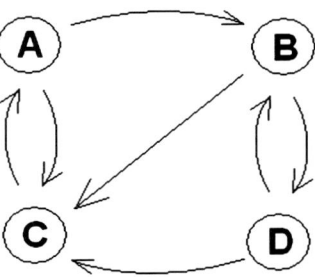

Fig. 56.2 Links of four sites

---

[3] Which site (containing the word looked up) has links to which other one?
[4] Of course in reality this is not exactly the case; a conspicuous link at the top of the page is probably used more often than a "small link" at the bottom. But these kinds of simplifications and idealisations are very typical for mathematical modelling: we have to make such simplifying assumptions in order to be able to use mathematics successfully.

$$C_n = A_{n+1}$$
$$0.5A_n + 0.5D_n = B_{n+1}$$
$$0.5A_n + 0.5B_n + 0.5D_n = C_{n+1}$$
$$0.5B_n = D_{n+1}$$

$$\underbrace{\begin{pmatrix} 0 & 0 & 1 & 0 \\ 0.5 & 0 & 0 & 0.5 \\ 0.5 & 0.5 & 0 & 0.5 \\ 0 & 0.5 & 0 & 0 \end{pmatrix}}_{=:T} \cdot \underbrace{\begin{pmatrix} A_n \\ B_n \\ C_n \\ D_n \end{pmatrix}}_{=:\vec{v}_n} = \underbrace{\begin{pmatrix} A_{n+1} \\ B_{n+1} \\ C_{n+1} \\ D_{n+1} \end{pmatrix}}_{\vec{v}_{n+1}}$$

Now we can imagine – just like in the case with the telephone companies – that many users are in the system of the sites A, B, C, D, at the beginning with the relative frequencies A_0, B_0, C_0, D_0 (fractions, percentages; $A_0 + B_0 + C_0 + D_0 = 1$). The change of a telephone company corresponds here to the change of an internet site. We again think of discrete steps in time: The users change the sites in these time steps (by following the links), so that after n time steps the distribution of the users is A_n, B_n, C_n, D_n. We can again read off the recursions ($\rightarrow$ above) easily from the transition graph (Fig. 56.2). In order to check whether there is a limit distribution $(\overline{A}, \overline{B}, \overline{C}, \overline{D})$ (i.e., a stable distribution in the long run) and if it is the case how does it look like, we could again use EXCEL. But linear equation systems can also be described very comfortably by using matrices and vectors ($\rightarrow$ above).

The vectors $\vec{v}_n$ denote the distribution after n time steps. All transitions $\vec{v}_n \rightarrow \vec{v}_{n+1}$ are given by the same matrix T ("transition matrix"). In column i there are the probabilities that a user on site i changes to site j in the next step by using a link $i \rightarrow j$ ($i, j = 1, \ldots, 4$). Somebody who is at the moment on site C has to go to site A (probability 1) in the next step; this can be seen in the transition graph and in the transition matrix (the probability 1 in column 3 and row 1). For the transitions we get in sequence:

$$T \cdot \vec{v}_0 = \vec{v}_1,\ \underbrace{T \cdot (T \cdot \vec{v}_0)}_{T^2 \cdot \vec{v}_0} = \vec{v}_2,\ T \cdot \overbrace{\left[\underbrace{T \cdot (T \cdot \vec{v}_0)}_{\vec{v}_1} \right]}^{\vec{v}_2} = \vec{v}_3, \ldots, T^n \cdot \vec{v}_0 = \vec{v}_n$$

(under the big bracket: $T^3 \cdot \vec{v}_0$)

When using matrices we have a possibility to get a direct formula for $\vec{v}_n$ (*not* only an *iterative* description like with EXCEL).

A vector that has probabilities (relative frequencies, percentages; sum = 1) as entries is called a "stochastic vector".[5] A square matrix is called "stochastic" if its

---

[5] A vector $\vec{v}$ is called "stochastic", if its components are from the interval [0;1] with sum 1.

column vectors are stochastic. Transition matrices are of course stochastic because they are square matrices and in the first column there are the probabilities for users at A landing at A, B, C, D in the next step. Of course these numbers come from the interval [0; 1] and have sum 1 (analogous with the other columns).

2.1 How Can We Measure the Relevance of a Site?

On relevant sites one expects to read something informative, relevant and worth knowing. Of course, a site s is the more relevant the more sites have a link on s, especially when these links come from relevant sites. But these words do not say how relevance can be measured. Which is the most relevant page in the above graph, which is the next relevant, etc.? How can one determine the relevance of a site within a directed graph? One can think of the following situation:

Many users are in the network (directed graph): What fraction (percentage) of them is at A, B, C, D during their investigations in the long run? If it turns out that a special site attracts 90% of the users, then it is clear that this site is most relevant and must be placed at the top of the list. These fractions in the long run are a possibility to measure the relevance of a site and for these fractions we need "limit distributions".

Let us assume that the users start to surf at the four pages by chance and that the fractions of the users at the beginning are ¼ for A, B, C, D: $\vec{v}_0 = (0.25, 0.25, 0.25, 0.25)'$; if they continue surfing and using the links by chance the distribution in the next step will be: $\vec{v}_1 = T \cdot \vec{v}_0 = (0.25, 0.25, 0.375, 0.125)'$ and in the next but one step $\vec{v}_2 = T \cdot \vec{v}_1 = T^2 \cdot \vec{v}_0 = (0.375, 0.1875, 0.3125, 0.125)'$; the sites A and C seem to have an advantage here. This is also plausible: all sites have a link to C and from there one must go to page A. By multiplying always with T from the left one gets all the following distributions $\vec{v}_n = T^n \cdot \vec{v}_0$; they converge to a "limit distribution" $\vec{v}_n \to \vec{v}$, which is given by $\vec{v} = (3/9, 2/9, 3/9, 1/9)'$. According to this result sites A and C should be placed ex aequo at the first place, followed by B and D.

Such limit distributions can be determined at school in several manners:

1. Repeating the iteration with EXCEL so long until the values do not change anymore.
2. Determining a high power T^n of the matrix with CAS, so that $\vec{v}_n = T^n \cdot \vec{v}_0$ should be near the limit distribution.
3. We are looking for a vector $\vec{v}$ with component sum 1 that does not change under multiplication with T: $T \cdot \vec{v} = \vec{v}$. One has to solve a linear equation system, of course, by CAS.[6]

---

[6] With means of higher mathematics one can speak of an *eigenvector* of T to the *eigenvalue* 1. But in most cases at school students will not know these words and the corresponding ideas. Therefore we do not go into details in this respect.

Problems that could occur:

- Is it possible that there are more such limit distributions $\vec{v}$? If there are more than one, do the vectors (distributions) $\vec{v}_i$ converge sometimes to the one and another time to the other (depending on the start distribution $\vec{v}_0$)? This would be not so good for our purpose because we want to use this limit distribution as a neutral and stable basis for a ranking concerning the relevance. A not unique limit distribution depending on the start distribution would be not a good basis. It would be best if the limit distribution were unique and independent of the start distribution.
- All three possibilities mentioned above to determine the limit distribution $\vec{v}$ only work for relatively low dimensions, as above a 4×4 matrix, eventually also 20×20, but in the case of a $1,000,000 \times 1,000,000$ matrix (or more, as coming up in the practice of Google) other methods are used: *iterative* algorithms that come to an *approximate* solution. They have to be very fast algorithms because there are very many queries given to Google every second. And all users do not want to wait a long time for the result.

Regardless of whether or not Markov chains are dealt with, the following limit theorem is very important because it provides a simple condition to the transition matrix T that guarantees the existence of the limit distribution $\vec{v}$, the uniqueness of it and the independence of the start distribution $\vec{v}_0$ (without proof):

T is stochastic and T^n contains for some $n \geq 1$ (at least) one row with only positive entries $\Rightarrow$ The limit matrix $L := \lim_{n \to \infty} T^n$ exists, is stochastic and has equal columns.[7]

$$\vec{v} = \underbrace{\begin{pmatrix} t_1 & t_1 & t_1 & t_1 \\ t_2 & t_2 & t_2 & t_2 \\ t_3 & t_3 & t_3 & t_3 \\ t_4 & t_4 & t_4 & t_4 \end{pmatrix}}_{L} \cdot \underbrace{\begin{pmatrix} A_0 \\ B_0 \\ C_0 \\ D_0 \end{pmatrix}}_{\vec{v}_0} = \begin{pmatrix} t_1 \\ t_2 \\ t_3 \\ t_4 \end{pmatrix}$$

$$T^{20} = \begin{pmatrix} 0.3333 & 0.3333 & 0.3333 & 0.3333 \\ 0.2222 & 0.2222 & 0.2222 & 0.2222 \\ 0.3333 & 0.3333 & 0.3333 & 0.3333 \\ 0.1111 & 0.1111 & 0.1111 & 0.1111 \end{pmatrix}$$

It is clear that these columns then determine the unique limit distribution $\vec{v}$ independent of the start distribution $\vec{v}_0$: Because of $A_0 + B_0 + C_0 + D_0 = 1$ in case of a 4×4 matrix one gets for the limit distribution $\vec{v}$ with this limit matrix L (independent of the concrete values of A_0, B_0, C_0, D_0).

---

[7] That means the entries are constant in each row. This theorem needs not necessarily be proven at school; one can simply use it for understanding the PageRank algorithm in its basics. Also the other specialties of the theory around it need not be dealt with at school.

Of course we have $\sum t_i = 1$ because L is stochastic. In our example not T itself has such a row with only positive entries but already T^2 has it. So the convergence and the independence of the start distribution is guaranteed by the limit theorem above. In our case we determine, for example, T^{20} and get with a CAS (4 significant digits):

Here $\vec{v} = (3/9, 2/9, 3/9, 1/9)'$ – the limit distribution mentioned already above – can be easily seen.

2.2 Now to an Example Slightly More Complicated

The link structure of a still very small network consisting of 6 internet sites is shown in Fig. 56.3. The transition matrix can be read off easily again ($\rightarrow$ matrix T).

$$T = \begin{pmatrix} 0 & 0 & 1/3 & 0 & 0 & 0 \\ 1/2 & 0 & 1/3 & 0 & 0 & 0 \\ 1/2 & 0 & 0 & 0 & 0 & 0 \\ 0 & 0 & 1/3 & 0 & 0 & 1/2 \\ 0 & 0 & 0 & 1/2 & 0 & 1/2 \\ 0 & 0 & 0 & 1/2 & 1 & 0 \end{pmatrix}$$

This is a new situation: From site ② there is no arrow leaving, there are no links on this page. Within the process of surfing one could call such a situation a "dead end" or "sink". This we can see in the second column of the matrix T; it contains only zeros. This is really bad for our purposes (stochastic matrix, column sum should be 1). What will one do in such a situation if this happens during surfing?

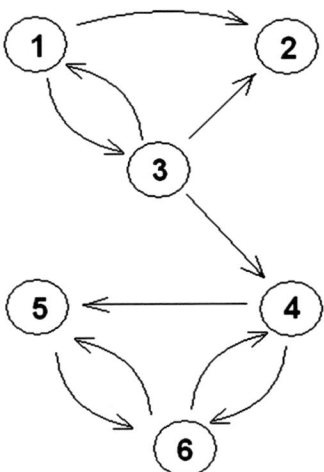

Fig. 56.3 Small network

There are several possibilities:

(a) Stop surfing and stay at site ②; in the matrix this would mean to replace the second zero in the second column by 1, in the directed graph we would have to add an arrow from ② to itself. We will not choose this possibility.
(b) One could go back one step in the browser and then use another link instead of ② (hopefully not again a dead end). One would have to distinguish from what page one came to ② – this would make things rather complicated.
(c) We decide in favour of another alternative: One leaves this site – coming back to the list (which we think as not yet ranked) – and by chance clicks one of the many other sites.

This we want to formulate explicitly as:

Modelling assumption 2: When we come to a dead end during the surfing process we go back to the list and click one of the m possible sites by chance, each with the same probability of $1/m$.

Here we do not consider that one probably won't click at the same page again (if there are really many sites, it will not make a big difference to "take off" the site or not): We replace the entries in the second column (zeroes) by 1/6 (in general: $1/m$ if there are m websites). Instead of the zero column we write the m-dimensional column vector $(1/m,\ldots,1/m)^t$ and get T_1:

$$T_1 = \begin{pmatrix} 0 & 1/6 & 1/3 & 0 & 0 & 0 \\ 1/2 & 1/6 & 1/3 & 0 & 0 & 0 \\ 1/2 & 1/6 & 0 & 0 & 0 & 0 \\ 0 & 1/6 & 1/3 & 0 & 0 & 1/2 \\ 0 & 1/6 & 0 & 1/2 & 0 & 1/2 \\ 0 & 1/6 & 0 & 1/2 & 1 & 0 \end{pmatrix}$$

So we can get a stochastic transition matrix T_1 although there are dead ends in the structure of the network.

Question: What about the situation of a site *to* (instead of *from*) which no link exists? Would this also be so bad?

Modelling assumption 3: From the experiences concerning the dead end situation we can say: Although a page is not a dead end it is very possible that one does not follow the links on a page but comes back to the list and clicks another page (at random). Let us assume that one follows the links on a page with probability α and comes back to the list and takes a new chance with probability $1-\alpha$ (the new chances are taken at random with probability $1/m$). How can this scenario be described mathematically? What does the new transition matrix U look like in this situation?

When following the links of a page the transition matrix is given by T_1.

What must the transition matrix look like in the case of *coming back to the list and taking a new chance* (problem for students)? Because the sites are taken at random with probability $1/m$ in this case the next distribution must be $(1/m,\ldots,1/m)^t$, that means the transition matrix has to be

$$T_2 = \begin{pmatrix} 1/m & \cdots & 1/m \\ \vdots & & \vdots \\ 1/m & \cdots & 1/m \end{pmatrix} \quad \text{because:} \quad \begin{pmatrix} 1/m & \cdots & 1/m \\ \vdots & & \vdots \\ 1/m & \cdots & 1/m \end{pmatrix} \cdot \begin{pmatrix} v_1 \\ \vdots \\ v_m \end{pmatrix}_{\sum v_i = 1} = \begin{pmatrix} 1/m \\ \vdots \\ 1/m \end{pmatrix}.$$

In sum or in combination we get for the new transition matrix U by weighting these two cases with the factors α and $1-\alpha$ respectively:

$$U = \underbrace{\alpha \cdot T_1}_{\text{With probability } \alpha \text{ following the links}} + \underbrace{(1-\alpha) \cdot T_2}_{\text{With probability } (1-\alpha) \text{ "new start"}} \quad (56.1)$$

It is easy to see (problem for students): Because T_1 and T_2 are stochastic matrices U is also stochastic.

2.3 The Crucial Attribute of U

The matrix U has *only positive entries*, no zeroes any more. According to the limit theorem above with this transition matrix we have the wanted and easy case of a unique limit distribution independent of the start distribution. This limit distribution can give us a ranking of the pages concerning their relevance ($\rightarrow$ "PageRank").

Which value should we take for α? It is known that Google has used $\alpha = 0.85$ for a long time. Possibly nowadays Google uses another value. In the example above we get for the solution of the linear equation system $U \cdot \vec{v} = \vec{v}$ ($\alpha = 0.85$; $v_i \geq 0, v_1 + \cdots + v_6 = 1$; CAS; 4 decimal places):

$$(v_1, v_2, v_3, v_4, v_5, v_6)^t = (0.0517, 0.0737, 0.0574, 0.1999, 0.2686, 0.3487)^t.$$

The components v_i of the solution vector $\vec{v}$ determine the relevance of the sites and are called "PageRank values". We get the same result if we determine a high power of U and then take one column of it. Also with EXCEL we would get the same result. According to it the ranked sites (concerning their relevance) would be:

site 6 → site 5 → site 4 → site 2 → site 3 → site 1.

This describes in the general case how the matrix U is created within the PageRank algorithm[8] in an elementary way.

---

[8] For deeper mathematics to this topic see: Chartier (2006), Langville (2006), Wills (2006).

2.4 Explicit Solution (Formula)

When we insert (56.1) in $U \cdot \vec{v} = \vec{v}$ and when we use matrix notation we can manipulate the equation (I hereby denotes the m-dimensional identity matrix). So we get an explicit formula for the limit distribution $\vec{v}$: $\alpha \cdot T_1 \cdot \vec{v} + (1-\alpha) \cdot \underbrace{T_2 \cdot \vec{v}}_{=(1/m,\ldots,1/m)^t} = I \cdot \vec{v} \Rightarrow$

$$(\alpha \cdot T_1 - I) \cdot \vec{v} = (\alpha - 1) \cdot \begin{pmatrix} 1/m \\ \vdots \\ 1/m \end{pmatrix} \Rightarrow \vec{v} = (\alpha - 1) \cdot (\alpha \cdot T_1 - I)^{-1} \cdot \begin{pmatrix} 1/m \\ \vdots \\ 1/m \end{pmatrix}$$

One can show that the matrix $\alpha \cdot T_1 - I$ is not singular so that we always have a unique solution.[9]

However, Google cannot use this explicit solution formula in practice because there we have very big linear equation systems (e.g. $m = 1,000,000$ or more) and in such high dimensions to determine the inverse matrix is a very hard and time consuming job. In practice *iterative* and *approximate* algorithms are used to come to a solution, but we don't want to discuss these numerical problems in detail here.

3 Summary and Reflections

- In this example we can see *the use of mathematics in modern society*. In mathematics education we should have such examples: mathematics is more and more disappearing from societal perception and we should counteract this fact as good as possible.
- Three elementary modelling assumptions (see above) had enormous effects. These ideas are simultaneously elementary and ingenious, they guarantee that the algorithm "always works".[10] Of course this modelling process is not meant to be done by the students themselves (autonomous work) but with teacher's help they get to know a piece of a very up to date application of mathematics (see also Voskoglu 1995).
- All over the world we can detect the so-called relevance paradox of mathematics: Mathematics is used more and more extensively in modern society (mobile phones, internet, electronic cash, cars, computers, insurances, CD players, networks ... one could give hundreds of examples!) and therefore becomes more and more important for us as a society. Mathematics surely is a so-called key technology for our future. But in many cases the mathematics behind these things is very complex and can only be understood by mathematics specialists. And for

---

[9] This follows also from the limit theorem above.

[10] One has to admit: In practice the algorithm is more complicated but its "heart" is a very elementary one – see above – and can be understood by students.

just using these things – we have to admit – understanding is not necessary. This means in some sense mathematics gets less and less important for the individual. This is one reason why many people do not see the importance of mathematics at all. Therefore, in mathematics educating we should come up with examples that show the importance of mathematics in a striking and elementary way. And I think the example of this chapter is one such example.

- *Consequences for teaching*: We should foster examples that show the use of mathematics: for our society and for individuals. Mathematical modelling (with or without autonomous work of students) can be a way of doing that successfully. For teaching in this way it is important that students have basic knowledge of several mathematical fields (above: matrices, vectors, using computers, etc.) and that it is allowed to mention and use *single unproved* mathematical theorems (e.g., the limit theorem from above) in order to come to interesting mathematical phenomena. Of course, this does not mean that reasoning in mathematics education wouldn't be important!
- In many preambles to syllabuses it is stressed that "cross-linking fields" is something desirable. Also most researchers and teachers in the field of mathematics education say that the teaching and learning process should more often give the opportunity for cross-linking mathematical topics. Here we have a very good chance for cross-linking stochastics (probabilities etc.), linear algebra (vectors, matrices, etc.), and analysis (limits, etc.), (see also Wirths 1997). Besides, dealing with this topic provides a possibility for a reasonable use of computers in mathematics education (EXCEL, CAS).
- This example may give reason for motivation and surprise: with how elementary ideas one can establish something world shaking and earn a lot of money.[11] Therefore it may serve as sort of advertisement for mathematics: a great career is possible by cleverly using both elementary and ingenious ideas. We also have an affirmation: Basic ideas are still important!

References

Chartier, T. P. (2006). Googling Markov. *The UMAP Journal, 27*(1), 17–30.
Langville, A. N. (2006). *Google's PageRank and beyond: The science of search engine rankings*. Princeton: Princeton University Press.
Voskoglu, M. G. (1995). Use of absorbing Markov chains to describe the process of mathematical modelling: A classroom experiment. *International Journal of Mathematical Education in Science and Technology, 26*(5), 759–763.
Wills, R. S. (2006). Google's PageRank: The math behind the search engine. *The Mathematical Intelligencer, 28*(4), 6–11.
Wirths, H. (1997). Markow-Ketten – Brücke zwischen Analysis, linearer Algebra und Stochastik. *Mathematik in der Schule, 35*, 601–613.

---

[11] Again: Establishing a company like Google and the real algorithms are not elementary things, but the basic idea of the PageRank is elementary.

Chapter 57
Authentic Modelling Problems in Mathematics Education

Gabriele Kaiser, Björn Schwarz, and Nils Buchholtz

Abstract This chapter presents experiences with modelling activities at the University of Hamburg, in which small groups of students from upper secondary level intensely work for 1 week on selected modelling problems, while their work is supported by pre-service-teachers. The paper presents one authentic solution of a group of students concerning a biological question and describes the approach of the students in detail. The form of an authentic description also includes mathematical errors and thoughts of the students that either have been discounted during the development of a solution or found their way into their ultimate solution. So an insight into the modelling-activities of the students during the modelling week can be gained. Finally some results of an evaluation are presented that has been conducted after the modelling week.

1 Theoretical Framework for Modelling in Mathematics Education

There is a consensus within mathematics education that applications and modelling play an important role. However, how these kinds of examples are implemented, which kinds of examples are used and how the process of modelling can be implemented in school is contentious and a matter for debate. Influenced by the goals, which are connected to the teaching of applications and modelling and mathematics education in general, one can distinguish different perspectives of the modelling debate worldwide: there are perspectives, which emphasise the use of authentic problems – named in a comprehensive framework developed by Kaiser and Sriraman (2006) as realistic or applied modelling. Other positions emphasise more

---

G. Kaiser (✉), B. Schwarz, and N. Buchholtz
Faculty for Education, Psychology, Human Movement, University of Hamburg, Hamburg, Germany
e-mail: gabriele.kaiser@uni-hamburg.de

pedagogical goals such as the development of concepts or the structuring of learning processes. As an overall perspective a meta-perspective is discriminated, called cognitive modelling, which focuses on the cognitive processes taking place during modelling activities. We are not concerned in this paper with the other approaches developed and refer the reader to Kaiser and Sriraman (2006) and the extensive ICMI-study on modelling (see Blum et al. 2007).

The approach described in the following belongs to the so-called realistic modelling perspective. Based on our extensive empirical research (see, e.g. Kaiser-Meßmer 1986) we see the necessity of treating authentic modelling problems, which promote the whole range of modelling competencies and broaden the radius of action of the students. The approach takes its essential starting point that – in order to promote modelling competencies – the students need their own experiences with authentic modelling problems. Similar proposals are developed by Haines and Crouch (2006) or at the beginning of the modelling debate by Pollak (1969).

Authentic problems are defined as problems that are only a little simplified and, in accordance with the definition of authentic problems by Niss (1992), recognised by people working in this field as being a problem they might meet in their daily work. The modelling activities described in this paper – so-called modelling weeks – are developed within a framework based on the realistic or applied modelling approach. A central feature of these activities is the use of authentic modelling problems in order to implement pragmatic-utilitarian educational goals like understanding of the real world or the promotion of modelling competencies. These authentic examples should explain the relevance of mathematics in daily life, the environment and the sciences, and impart competencies to apply mathematics in daily life, environment and sciences. Running modelling weeks is considered to be a powerful and effective way to promote modelling in school on an extensive basis.

2 Framework and Structure of the Modelling Week

Modelling activities have been carried out at the University of Hamburg since 2000 in a cooperation of the department of mathematics (applied mathematics) and the department of education (mathematics education). Since then, many students from the upper secondary level (age 16–18) have taken part in different forms of the project and were supervised by future teachers during their modelling activities. Since 2001, modelling weeks are carried out twice a year with about 200 students from upper secondary level (16–18-years-old students) from schools in Hamburg and its surroundings.

In the following, the characteristics of our modelling activities are described. These are didactical aims, requirements for the examples and didactical reflections about the students' modelling activities especially focusing on a modelling week carried out in spring 2009 with 350 students.

Pragmatic-utilitarian educational goals can be implemented by the students' independent work on modelling examples, precisely in developing mathematical questions from given problems by themselves and developing solutions for real world problems; a comprehensive overview on modelling approaches on the contrary is not taken into focus. In addition, also the students' motivation can be enhanced because at the end of the week, the groups have to present their solution within a presentation to the other participants. A fundamental part of realistic modelling is authenticity; the modelling examples used have to conform to necessary "authentic" requirements. For example, they have to be suggested by applied mathematicians working in industry, and may only be simplified a little, so that the modelling examples are still embedded into reality. This leads to the difficulty that neither the students nor the questioner knows an adequate solution for the modelling examples. Additionally, a problematic situation is described in the example and the students have to determine or develop a question, which they can solve, themselves. Various problem definitions and solutions are possible; they all depend on the norms of the modellers.

But besides the authenticity, it has to be considered whether the examples are compatible with the students' mathematical knowledge, which might not be very high, only covering the beginning of calculus. The modelling competencies differ on a large scale and the authenticity of the given real-world problems often creates difficulties for the students. In order to deal with this heterogeneity, it is necessary to foster independent work within small groups. A fast intervention by the future teachers would hinder the students' independent work. The teacher should rather only support the students in case of lacking mathematical means or when the students are in a cul-de-sac. After all, experiencing helplessness and insecurity is a central aspect and a necessary phase when dealing with mathematical modelling. Modelling is not a "spectator sport", it can only be learnt by own activities.

3 Students Modelling the Spread of Disease in a Population of Ladybirds

Among other examples, the modelling example described in the following has been proposed to the students in the modelling week carried out in March 2009:

How can the spread of a sexually transmitted disease with ladybirds be predicted concerning the development of the population itself?

The problem deals with the reproduction of ladybirds which are affected by sexually transmitted diseases. This problem was developed by Göttlich and Bracke (2009). The source is an article taken from the internet[1] about an Australian study dealing with the reproductive behaviour of ladybirds with two dots (*Adalia bipunctata*). Due to promiscuity great parts of the population of ladybirds are affected by sexually transmitted diseases caused by acarines (mites). Despite these diseases the

---

[1] The article is taken from http://www.wissenschaft.de/wissenschaft/news/258584.html. Bild der Wissenschaft. 25 October 2005. Retrieved 20 April 2010.

size of the population has stayed the same over several years. The students' task was to examine the statements about the long-term development of the population of ladybirds and find out whether the ladybirds might be at risk of extinction. We will restrict our descriptions on the approach of one small group of students in order to have space to show their various ways of tackling the problem, their errors and their final solution. Due to lack of precise information about the long-term development of the population within the article the students[2] started with research in the library and on the internet to find out further information about this species of ladybirds and especially about their reproductive behaviour. To grasp all this information (e.g. about the time to sexual maturity and the incubation period of the sexually transmitted disease, but also climatic information like the temperature profile because the reproductive behaviour of ladybirds depends on the temperature), the students plotted the data in a graph. One can clearly recognize the influence of the context and its complexity on the students' solution approach.

After the research and after getting a first idea the students agreed on factors that might influence the model in their opinion. These were the following:

- Size of the population
- Ratio of bugs affected by the disease
- Reproductive activity
- Temperature
- Mortality rate
- Birth rate
- Time

The students first formulated a hypothesis about the connection between the first derivation of the time-dependent size of the population and the reproductive activity, which was to depend on the time and temperature. They developed as first approach: Let $p(t)$ be the size of the population at the time t and T the temperature as well as $r(T,t)$ the reproductive activity depending on time and temperature (on a scale from 0 to 1). Then the hypothesis was: $p'(t) = r(T,t)$.

The formulation of the hypothesis, which was soon dropped by the students because it did not correspond with the data, clarifies another feature which can often be observed in students' modelling. Usually the students use known concepts, in this case the concept of the derivative rather than developing models which would require further support by the teacher. Furthermore, well-known mathematical concepts are used without further reflection such as steady functions. While discrete models, which are rarely taught in school, would also be possible for population models, the students did not consider this option in the beginning and they formulated a well-known function from their mathematics classes instead.[3]

---

[2] The following descriptions are based on the considerations made by students from the Emil-Krause-Gymnasium in Hamburg.
[3] It has to be added that the students considered this option in the course of the modelling and included it into their model.

57 Authentic Modelling Problems in Mathematics Education

To develop a new hypothesis the students then decided to include the ratio of infected ladybirds as a further factor, which was to be stated on a scale from 0 to 1 just like the reproductive activity. The students called this "infestation" $b(t)$. Furthermore, they decided that the reproductive activity was not to depend on the temperature any more. Instead the temperature dependence was to be included in the time dependence. Motivated by the previous research about the temperature patterns, the students assumed that the temperature mainly depends on the season and they decided that it was therefore sufficient to modify the reproductive activity only in dependence of time. All in all, the students therefore examined the following three factors of which they assumed that they influence the derivative of the size of population as in the previous approach:

- Size of population
- Infestation
- Reproductive activity

The students made all factors depend only on the time. Afterwards the group identified the factors that have to increase or decrease to result in an increase of $p'(t)$ (the students called this "rise of the population") or in a decrease of $p'(t)$ (the students called this "breakdown of the population"). The students presented the results in the above scheme about the connection of the values (Fig. 57.1). An upward arrow represents an increase and a downward arrow represents a decrease of the respective value.

It can be recognized that only the infestation $b(t)$ develops contrarily to the other values. To even this out and to include all the values into one equation, the students examined the value $(1-b(t))$ instead of $b(t)$, which develops like the other values.

Thus, this new equation resulted as hypothesis:

$$p'(t) = r(t)(1-b(t))p(t) \tag{57.1}$$

In the next step, the students added a time-independent mortality rate m to the existing factors, turning the result into the following equation:

$$p'(t) = r(t)(1-b(t))p(t) - m \cdot p(t) \tag{57.2}$$

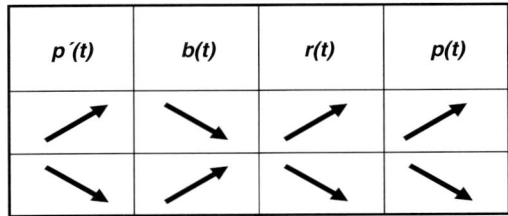

Fig. 57.1 Scheme about the interrelation of the changing of the different factors

This formula matches at least in parts structurally those population models that do not include a limit of capacities of the system. Notable is the way the students chose to arrive at this formula. They multiplied the previous equation (57.1) with $(1-m)$ to adjust also the new factor to the existing ones just like they did with the infestation. This resulted in:

$$p'(t) = r(t)(1-b(t))p(t)(1-m)$$

The students expanded this formula with $(1-m)$. And they called the first addend of the right side of the expanded equation "birth", the second addend "death" to differentiate between the different causes influencing the development of the population (in this case the derivative of the population size). This resulted in the following equation:

$$p'(t) = \underbrace{r(t)(1-b(t))p(t)}_{birth} - \underbrace{m \cdot r(t)(1-b(t))p(t)}_{death}$$

Based on this equation the students stated the following: "We were able to cross out $r(t) \cdot m$ and $(1-b(t)) \cdot m$ because the mortality rate only refers to the population". This led to Eq. 57.2. Even if the mortality rate really only refers to the population, the students of course did not cross out the factors $r(t)$ and $(1-b(t))$, but equated them with 1.

As a last factor the students added the offspring per female bug in the first addend as a time-independent value. They called this "birth rate" X. The students estimated the value of X on the basis of their previously researched information. For this estimation they multiplied the amount of eggs a female bug produces in her life (400[4]) with the chance of survival of an egg (0.38%) and the ratio of female bugs in the bug population (9/10[5]). This resulted into $X = 1.4$ approximately and as formula:

$$p'(t) = X \cdot r(t)(1-b(t))p(t) - m \cdot p(t)$$

Subsequently the group took the value 1,000 for the initial size of the population $p(0)$ and 10% for the mortality rate.

To be able to calculate the population's development using the presented differential calculus the students would have had to identify a closed form for $p(t)$ besides developing the functions for $b(t)$ and $r(t)$. That means that they would have had to solve the differential equation or alternatively they would have had to calculate it numerically with a computer. The group did not use any of these

---

[4] The values concerning this matter vary in different sources. In the text we strictly use the values used by the students independent of their biological correctness.

[5] Independent of its exact value the ratio of female bugs indeed is very high for two-dotted ladybugs. This is due to symbiotic bacteria in the ovule and explains the promiscuous lifestyle of these bugs.

57 Authentic Modelling Problems in Mathematics Education

methods possibly due to insufficient knowledge about these means. The group therefore did not decide to ask the teacher for assistance to solve the mathematical problem, but decided to use a discrete model with a recursive formula instead of a steady model. The decision was therefore not based on considerations with regard to content, but rather on considerations about available mathematical methods, which enabled them to work on a related discrete model. To transform the steady model into a discrete one, the students decided to keep the original form and the previous considerations and only add them in the sense of a recursive procedure to the size of the population in the previous month. The recursive formula was therefore:

$$p(m+1) = X \cdot r(m)(1 - b(m))p(m) - m \cdot p(m) + p(m)$$

Furthermore, the students calculated graphs for the development of $b(t)$ and $r(t)$, which allocates one value for $b(t)$ and one for $r(t)$ for each month of the year, based on information they had looked up and on the data from the text as well as on estimations. The resulting graphs of $b(t)$ and $r(t)$ are depicted in Figs. 57.2 and 57.3.

With this formula the students were able to calculate the development of the ladybirds' population over a year according to their model. Figure 57.4 depicts the development of the population on the basis of their calculations:

Although the composition of the formula and its calculated values are disputable, the students succeeded in developing a model which displays the stability of the population over the years. In their model the population size varies throughout the year but it reaches the initial size again at the end of the year. The important step of validation of the model was unfortunately omitted due to missing comparative data and time spent on the preparation of the presentation of the solution, but it would have confronted the students with their own estimation of adequacy of

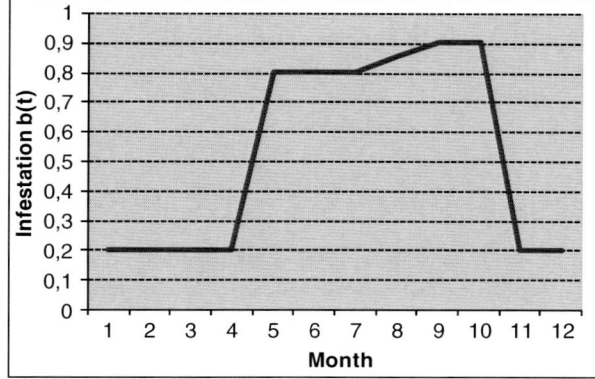

Fig. 57.2 Students' graph of the infestation $b(t)$

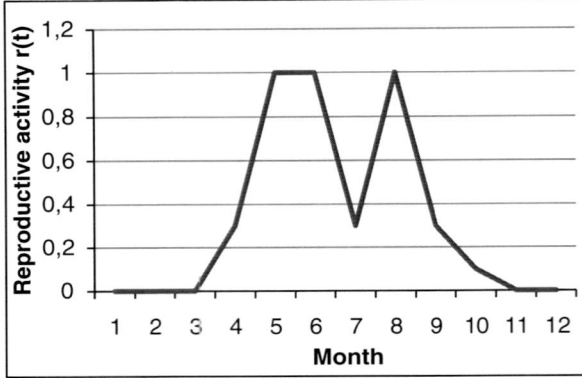

Fig. 57.3 Students' graph of the reproductive activity $r(t)$

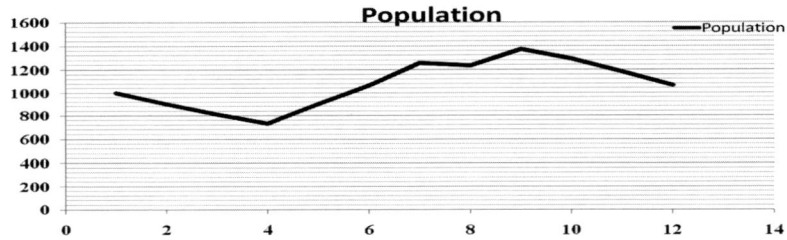

Fig. 57.4 Development of the population calculated by the students

their model and they would possibly have revised or corrected the model. But yet after this first result, without validation, the students thought of further modifications of the model, such as a stronger temperature dependence of the development of the population. This was especially motivated by the wording of the problem, in which the author speculated about a breakdown of the population due to a rise of temperature because of climate change. Furthermore, they discussed the possibility of looking at the infestation $b(t)$ not only as a value depending on time, but also as a value $b(t, r(t))$ depending on the reproductive ratio. Other ideas concerned including predator–prey relationships into the mortality rate or refining the time intervals and considering the fact that a bug is sexually mature only after 4–5 weeks. Even if these modifications are comprehensible, the validation of the model is missing and this could indicate that the developed model is arbitrary to the students, as long as they get to a model, so further didactical insistence on the validation of solutions should be taken into focus when teaching mathematical modelling.

4 Evaluation of the Modelling Week

For the evaluation of the modelling week in March 2009 we used a questionnaire with four mainly open questions on the beliefs of students about mathematics teaching and three open questions on the appreciation of the modelling examples tackled in the modelling week and five closed questions, which had to be answered on a five-point-Likert-scale. The questionnaire was filled out at the end of the modelling week by 289 students from 19 schools from Hamburg and its surroundings; originally 350 students participated, but not all of them were present on the last day or were willing to fill out the questionnaire. Based on methods of Grounded Theory (see Strauss and Corbin 1998) we used in-vivo-codes for the open questions, that is codes extracted from the students' answers, written as verbatim quotations, and grouped them to similar quotations under a theoretical perspective. In order to analyse the students' answers in the open questions we transformed the grouped in-vivo-codes into theoretical codes. We used methods of consensual coding for reasons of quality assurance, which means that a coding team consisted of two coders, who conducted all steps described above together. According to Steinke (2000) interpretations in groups are a "discursive way to create inter-subjectivity and comprehensibility by means of handling data and the respective interpretations explicitly" (p. 326, translation by Karen Stadtlander) and thus serve to assure quality.

Within this paper, we concentrate on questions about the students' perception of their learning outcomes. The following question was asked: "From your point of view, what did you learn when dealing with the modelling example?". Two hundred and six (about 71%) of the students gave a detailed answer that could be coded; only 49 did not see any learning outcomes from the modelling week. The answers demonstrated that the students perceived the main factors of their learning progress to be insights into the applications of mathematics and working techniques. Other factors like learning formal mathematics, the (formal) application of computers or the comprehension of mathematics or social aspects were also named, but more seldom. The students described that they now were aware of the relevance of mathematics in their environment and able to use their mathematical knowledge to apply mathematics in daily life.

> ...that mathematics not only can be applied in school, but also to specific examples in real life. (Female student, 19 years old)
>
> ...to solve arising problems autonomously. (Female student, 16 years old)

Even though most of the reactions and answers of the students to the learning outcome of the modelling week were positive, 82 of 260 students (about 32%) gave a negative answer to the question: "Should these examples be increasingly dealt with as part of regular math classes or would you reject this?" Apparently, these students are not willing to deal with modelling, when it comes to modelling in their own mathematics education.

Taking a look at the negative answers, time pressure was the most often named factor against including mathematical modelling in ordinary mathematics teaching.

Other factors like disinterest or insignificance of modelling problems, difficulty of modelling problems or the impreciseness of mathematical solutions were also named, but again, more rarely. This clearly shows that the students are aware of the time pressure in their mathematics curriculum, which conflicts with the time-consuming character of mathematical modelling experienced in the modelling week. Thus, even though they may have liked mathematical modelling, they were not in favour of including modelling in their ordinary mathematics classes. On the other hand, it shows, that still today there are only few opportunities to integrate mathematical modelling in school.

> To my mind this would be too time-consuming. It is more important to master the subject itself. (Female student, 18 years old)

> It would not make sense to use this in class, because it has little relation to maths. It is more like something you could puzzle over. (Male student, 17 years old)

> No, because the tasks were not narrowed down and there were too many possible solutions to choose from. (Female student, 17 years old)

Concerning the positive answers for the inclusion of mathematical modelling in ordinary teaching, the most named reason here is the relation of mathematical modelling to reality. Other arguments for the inclusion are the promotion of working techniques or variation in mathematics classrooms. Some students proposed to change the modelling problems before working on them in regular mathematics classes. Finally, the positive answers of the students demonstrate that dealing with mathematical modelling enables the students to understand the utility and necessity of mathematics in real life.

> I think that these examples should DEFINITELY be dealt with in maths classes. Because of these examples one will only realise what mathematics is needed for. (Male student, 16 years old)
> Such problems SHOULD be dealt with in class, because they will improve the so-called 'competence in problem solving'. Furthermore, it will train the knowledge gained in previous grades. (Male student, 19 years old)

Our evaluation demonstrates that these kinds of examples can be tackled successfully by ordinary students in upper secondary level. The students describe high learning outcomes reflecting all the goals of modelling in mathematics education, ranging from psychological goals, such as motivation, meta-aspects, such as promoting working attitudes, to pedagogical goals, namely enhancing the understanding of the world around us. The strong plea of the students for the inclusion of these kinds of examples in usual mathematics lessons support our position that it is appropriate to include these kinds of problems in ordinary mathematics lessons, clearly not everyday, but on a regular basis.

References

Blum, W., Galbraith, P. L., Henn, H.-W., & Niss, M. (Eds.). (2007). *Modelling and applications in mathematics education. The 14th ICMI study*. New York: Springer.

Göttlich, S., & Bracke, M. (2009). Eine Modellierungsaufgabe zum Thema "Munterer Partnertausch beim Marienkäfer". In M. Neubrand (Ed.), *Beiträge zum Mathematikunterricht* (pp. 93–96). Münster: WTM-Verlag Stein.

Haines, C. R. & Crouch, R. M. (2006). Getting to grips with real world contexts: Developing research in mathematical modelling. In M. Bosch (Ed.), *Proceedings of the Fourth Congress of the European Society for Research in Mathematics Education* (pp. 1655–1665). Barcelona: Fundemi IQS - Universitat.

Kaiser, G., & Sriraman, B. (2006). A global survey of international perspectives on modelling in mathematics education. *ZDM – The International Journal on Mathematics Education, 38*(3), 302–310.

Kaiser-Meßmer, G. (1986). *Anwendungen im Mathematikunterricht. Band 1: Theoretische Konzeption. Band 2: Empirische Untersuchungen*. Bad Salzdetfurth: Franzbecker.

Niss, M. (1992). *Applications and modelling in school mathematics – Directions for future development*. Roskilde: IMFUFA Roskilde Universitetscenter.

Pollak, H. (1969). How can we teach applications of mathematics. *Educational Studies in Mathematics, 2*, 393–404.

Steinke, I. (2000). Gütekriterien Qualitativer Forschung. In U. Flick & E. V. Kardorff (Eds.), *Qualitative Forschung – Ein Handbuch* (pp. 319–331). Reinbek: Rowohlt.

Strauss, A., & Corbin, J. (1998). *Basics of qualitative research*. Newbury Park: Sage.

Chapter 58
Using Modelling Experiences to Develop Japanese Senior High School Students' Awareness of the Interrelations between Mathematics and Science

Tetsushi Kawasaki and Seiji Moriya

Abstract The estrangement of science from mathematics in the mathematical curriculum of Japanese senior high schools is a subject of widespread and grave concern. Thus, in order to connect mathematics to the fields of Newton's science, for example, through mathematical modelling, it is suggested that it will be necessary to provide concrete examples such as "Kepler's Laws". This approach succeeded in increasing a class of Year 12 students' knowledge about the laws with the simulation of planetary movements and the making of a "Mathematical development model" occurring together. Consequently students will realize the necessity of differential equations in order to analyze actual phenomena. This empirical research suggests that mathematics materials involving physical perspectives are effective for senior high school students.

1 Introduction

In school education in Japan, it is thought that it is enough for mathematics teachers to teach only theory. It is believed that the science teachers practice the application and the use of mathematics; but it has not been actually achieved. Mathematics is removed from the phenomena of daily life and science. Students will understand the ideas through the formal theory of mathematics; but they have not gleaned the scientific spirit. Additionally, it is difficult for them to understand the concept of mathematics theory. The reason why students study mathematics is "to notice its necessity and to learn the thought" (cf. Stephens and Yanagimoto 2001). The solution of problems by mathematical modelling is necessary to achieve these two purposes.

T. Kawasaki (✉)
Kyoto Prefectural Sagano Senior High School, Kyoto, Japan
e-mail: tetsushi@kyokyo-u.ac.jp

S. Moriya
Tamagawa University, Tokyo, Japan
e-mail: smoriya@edu.tamagawa.ac.jp

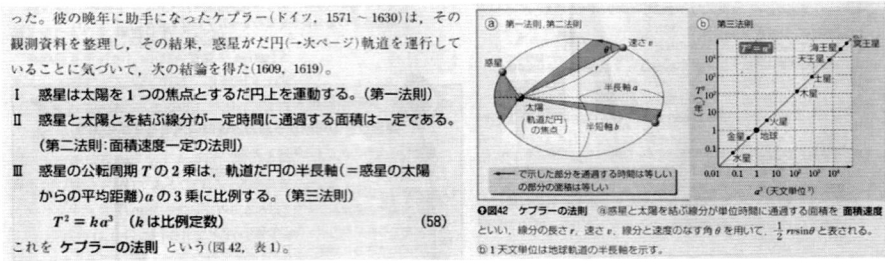

Fig. 58.1 Formal knowledge and illustration of Kepler's Laws in physics textbook (Suken Syuppan 2003)

Fujii (1987) pointed out the following problems about mathematics that is aimed at preparation for science: "The calculus education in Japanese senior high schools includes a unit in which students must learn a theory of Newton's science. However, senior high school mathematics does not aim at the dynamics. An educational practice to determine the elliptic orbit of a planet using Newton's law of gravitation is necessary" (p. 86). The differential equation of the law of universal gravitation (Newton's Law) is necessary to prove Kepler's Laws. However, high school physics does not prove Kepler's Laws, they are only presented (Fig. 58.1). High school mathematics mainly treats the ellipse in orthogonal axes coordinates. High school physics mainly treats uniform circular motion. Both high school mathematics and physics only show formal knowledge. Students will not be able to notice the necessity of mathematics to solve problems in science or as it applies to their lives. They will also be unable to learn correct thinking about applying mathematics. The result is uncertainty as to whether the students can gain the spirit of modern science well by the current situation in school education in Japan. As an alternative, Kawano (2001) conducted a class using mathematical modelling but there were no data from the students about the recognition of change of understanding and satisfaction.

We carried out a survey of student knowledge concerning Kepler's laws. The low percentage of students in a Year 12 class (N = 39) with the mathematical background theory used for Kepler's Laws is shown in Fig. 58.2. They did not know the deep meaning though they knew the formal theory. The students knew that a planet orbits the sun. They learned Kepler's law in Year 11. However, they learned eccentricity in their mathematics class, but they did not remember the word or its meaning. In addition, they had not used a computer for such a class. Perhaps it is thought that the students would not understand "the common point of the focus and the sun" or "the planet's motion".

When we analyze a real event that needs mathematics, it is necessary for us to apply mathematics (Applied Problem) and it is necessary for us to think about the solution by understanding the meaning of the problem. We practice this, and it becomes necessary to look back at the results as in the modelling cycle process (Borromeo Ferri 2006) as in Fig. 58.3a.

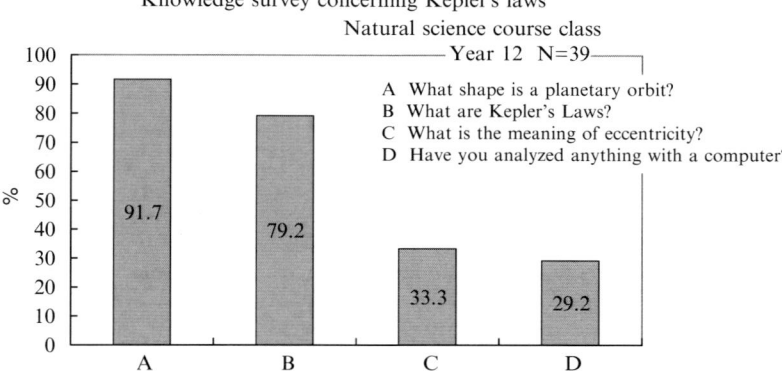

Fig. 58.2 Knowledge survey concerning Kepler's Laws (Kawasaki 2007)

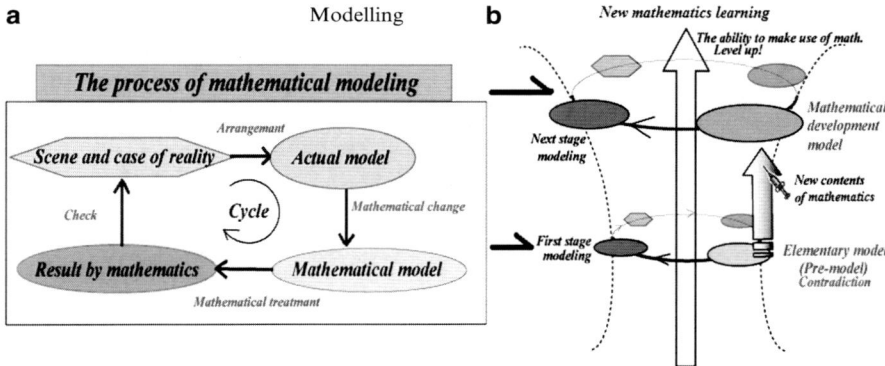

Fig. 58.3 (a) Kaiser's Mathematical modelling cycle (b) The development of modelling image

Then, when the actual model describes the scene and the case of reality, we call it an "Elementary Model" (see Fig. 58.3b). The model described by using mathematics that the students have studied is called a "Pre-model", because it becomes a part of, and a connection to, the development model though it is unsatisfactory as a model. The model which guides students to the new mathematics is called a "Mathematical development model". These models provide the basis for introductory teaching materials of the new mathematics unit where students develop an image of modelling (see Fig. 58.3b).

When the model changes from simplicity into complexity, students repeat this modelling. However, it is advisable that the student do not think about this process alone, because they cannot find the new model from only their experiences. A minimum of support by the teachers is necessary (cf. Blum and Ließ 2007).

An example of this modelling is the teaching materials using "Kepler's Law" and the law of universal gravitation which will be explained in the next section.

To make high school students consider how mathematics is connected with science is very difficult. Why is it difficult?

The following educational situations have always existed in Japan. According to the course of study (Ministry of Education, Culture, Sports, Science and Technology 2005) in Japan, all teachers must comply with curriculum guidelines that are managed by the commonwealth institution. Teachers should use the textbooks that follow these guidelines. The guidelines have not treated the approach of mathematical modelling until now. So in mathematics and science education in Japan, a curriculum for modelling activities does not exist. The school system in Japan does not permit mathematics teachers to teach other subjects such as physics or information processing. Also, educational materials beyond the guidelines are not test requirements for university entrance examinations. Given the circumstances, teachers are presently unable to further develop teaching materials. So teachers depend on the textbook. In addition, the students' classes emphasize mathematics to gain entrance into university as demanded by their parents or guardians. Consequently, the reason students study mathematics is for university entrance examinations. A lot of high school students do not have basic knowledge of the connections between mathematics and other areas (Fig. 58.2). If this continues, not only students and citizens but also teachers will eventually come to lack recognition of the necessity of mathematics. To address such situations, new teaching materials need to be developed and evaluated in a teaching experiment.

2 An Example Using Modelling: Teaching Materials for Kepler's Law for High School Students Becoming Scientists

This example is not only to foster modelling. It is also intended we should address the problems of school education outlined in the previous section. We want to recommend the proof of the elliptical orbit of the planet as mathematics material with physical viewpoints, because the essence of modern science is connected to Keplerian and Newtonian science.

2.1 Premodel

First of all, students will practice using the observational data of Mercury. This will be called the elementary model to understand Kepler's Law (Table 58.1, Fig. 58.4).

The elliptical orbit of Mercury appears as the envelope of the tangential lines, "Kepler's 1st Law" (Fig. 58.5). And, when the length of observational periods are equal, the sectoral areas caused by the segment that connects the sun with Mercury are equal, "Kepler's 2nd Law".

Usually only this model must be satisfactory, and as a result, we think that students will confirm the image of Kepler's Law.

Table. 58.1 The observational data of mercury (cf Kyoto Chigaku Kyoiku Kenkyukai 1993)

Year	Eastern max. angle (E)		Western max. angle (W)	
1990	April 14	20*	February 1	25*
	August 12	27*	May 31	25*
	December 6	21*	September 24	18*
1991	March 27	27*	January 14	24*
	July 25	27*	May 13	26
	November 19	22*	September 8	18*
			December 28	22
1992	March 10	18*	April 23	27*
	July 6	26*	August 21	19*
	November 1	24*	December 9	21*

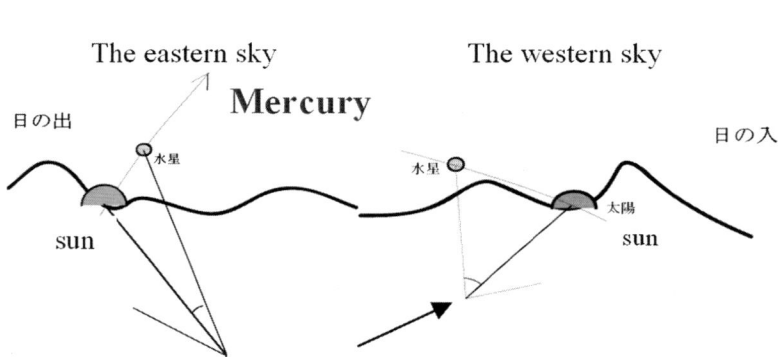

Fig. 58.4 The maximum angle (cf Kawasaki 2007)

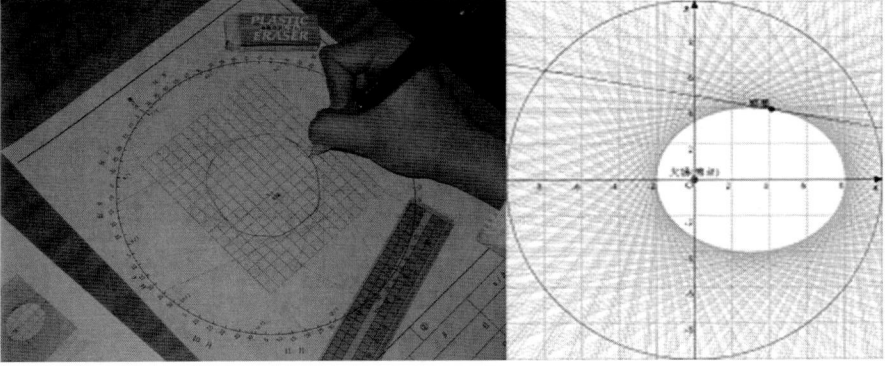

Fig. 58.5 Drawing in mercury orbit using tangential lines: (*right*) elementary model, (*left*) premodel (cf Kawasaki 2007)

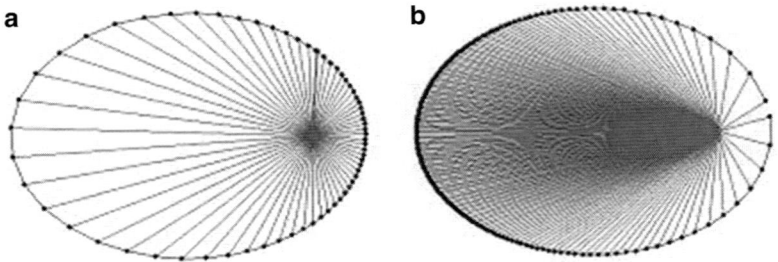

Fig. 58.6 (a) Uniform motion (b) Planetary motion (cf Kawasaki 2007)

However, it is doubtful whether the students understand these correctly. In fact we showed them a simulation of two planetary movements (Fig. 58.6), and we made them judge which was correct. At once students could not judge, and selected Fig. 58.6a later.

This elementary model is not enough for students' knowledge to mature, but it is a good result as a model. Before long they will demand a new model. Therefore, we call it the premodel that prepares for a new model. It is necessary to prepare the new content of mathematics. The new model is shown in the next section. It becomes one composition by these two models. The new model is also a means to lead to a new mathematics unit. Students will make the best use of the new model for the next practice.

2.2 Mathematical Development Model

It is necessary to introduce the differential equation. The equation of Kepler's Laws by Newton's Law is a second order linear differential equation with scalar constant. This is very difficult and the students cannot find the new model from only their own investigations. A minimum of support by the teacher is necessary, and the first author tried to help the students gain understanding, using the analysis method of numerical values by computer programming. They will be able to have full realization of Kepler's Laws and Newton's Law by drawing the solution curve using Euler's method (Fig. 58.7).

The solution curve by Euler's method has the feature that the error grows when the input value increases, and the graphs separate (Fig. 58.8). When teachers give the student the base program, regard for error is necessary. Figure 58.9 is the system diagram showing how differential equations are introduced. A feature of this is the treatment of simple harmonic motion in high school physics by the differential equation. This second order differential equation relates to the law of universal gravitation. Other necessary mathematical content includes polar coordinates and linear transformations (see Fig. 58.10).

Fig. 58.7 Euler method

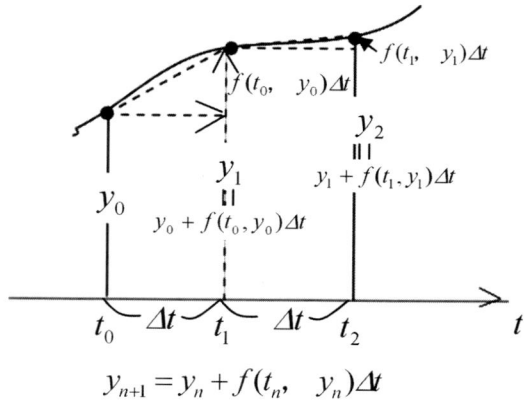

$$y_{n+1} = y_n + f(t_n, y_n)\Delta t$$

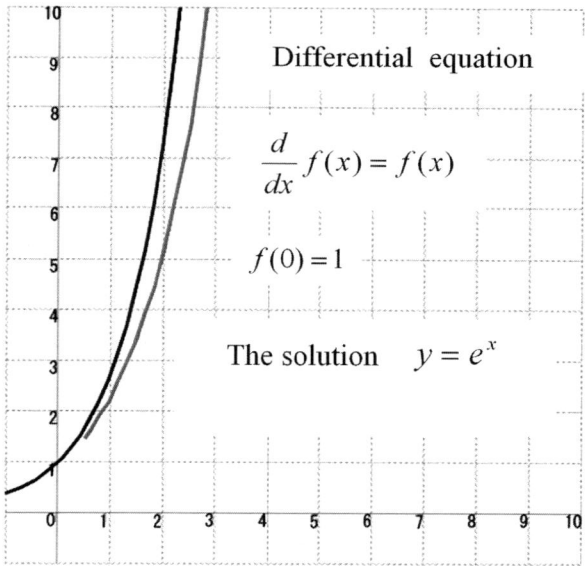

Differential equation

$$\frac{d}{dx}f(x) = f(x)$$

$$f(0) = 1$$

The solution $y = e^x$

Fig. 58.8 By Euler method

Students were able to smoothly make the mathematical development model using a supplied worksheet and computer programming (Fig. 58.11). In addition, this mathematical development model means the polar equation of ellipse is, $r = \dfrac{\ell}{1 - e\cos\theta}$ ($\ell > 0, e > 0$). Students confirmed whether this solution accorded with the drawing of the solution curve (Fig. 58.12).

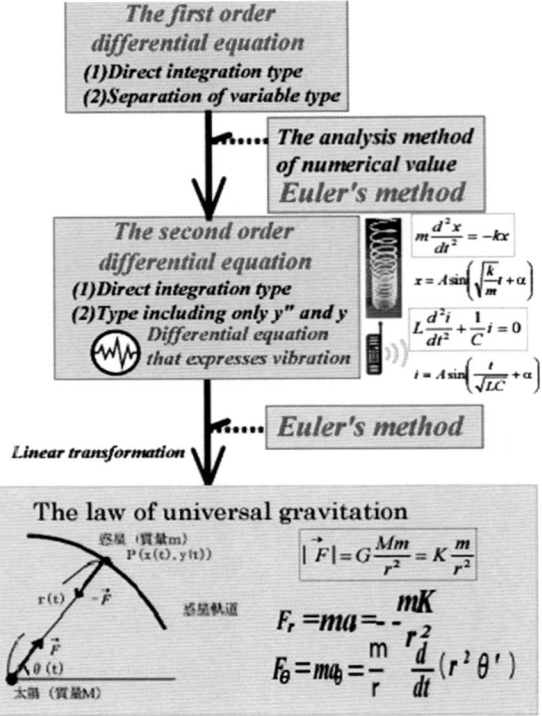

Fig. 58.9 System diagram to introduce differential equations (Kawasaki 2007)

3 Students' Evaluation and Impressions

The whole Year 12 class (N=39) participated in the teaching experiment up to the premodel stage of the teaching materials and were given a questionnaire where items were rated from 1 (lowest) to 5 (highest) evaluating their understanding of the modelling lessons. A further treatment of the mathematical development model was conducted with only three students from the class and this was also evaluated in a similar fashion.

3.1 The Class Treated the Premodel

The students understood the mathematics background to Kepler's Law well. Understanding and interest show high scores (Fig. 58.13). It appeared they liked the work on the planetary orbit. They noticed that Kepler's 2nd law used mathematics they had not learned. However, they could not interpret the new mathematics very

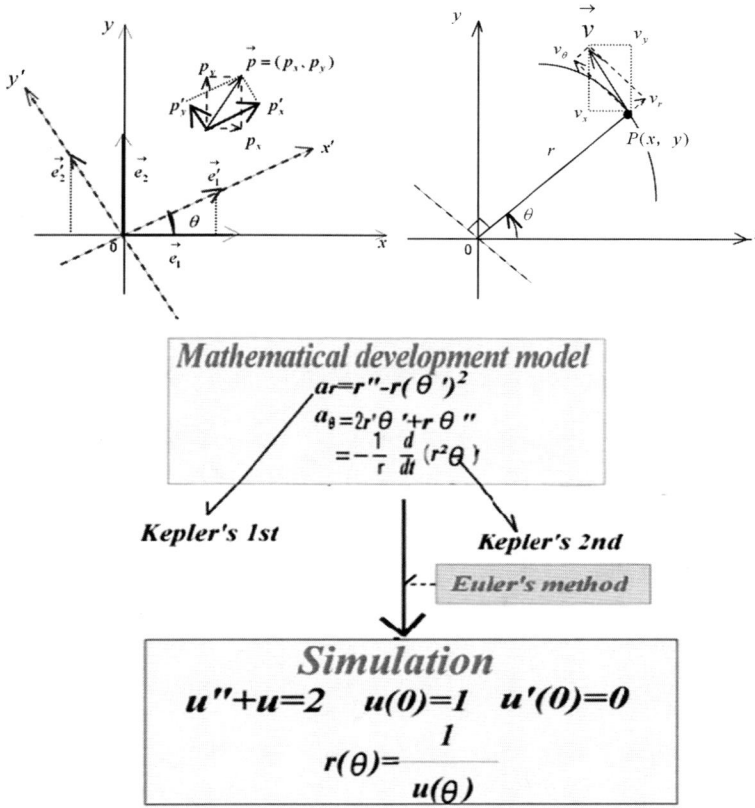

Fig. 58.10 Transform of coordinates and mathematical development model (Kawasaki 2007)

well, because they had trouble choosing the correct answer from two planetary movements. They seemed to be unable to understand, but had the will to learn new mathematics (Fig. 58.13 (6)).

There were some students who gave low scores for Fig. 58.13 (6) and (7) as they were not accustomed to the calculator, and also the classes advanced quickly.

3.2 The Class Treated the Mathematical Development Model

At this time, students were busy for the university examination, so this next lesson was carried out for only three students. They entered the science faculty of Kyoto University. Almost all of the survey answers were composed of high scores. One student's impression was "This lesson was a valuable experience of learning the interest of natural science before I get into college". Analysis by programming

Fig. 58.11 Computer programming

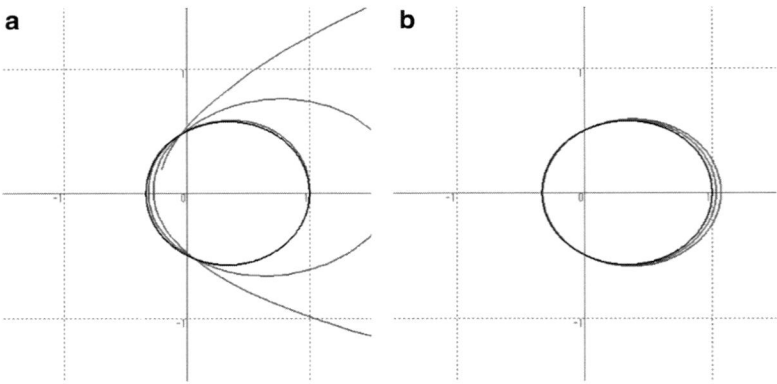

Fig. 58.12 Simulation result (**a**) $d\theta = 0.1$ (**b**) $d\theta = 0.01$

proved to be very effective in increasing the students' understanding (Fig. 58.14 (11) (12) (14)). However, there was a lot of content covered in only a few classes. It seemed that students had difficulty understanding the content though they were interested in it. After all, it was difficult for students to make the model by themselves. Moreover, it was also difficult for the teacher not to provide assistance and allow them to work unaided (Fig. 58.14 (9)). Another student's opinion was that,

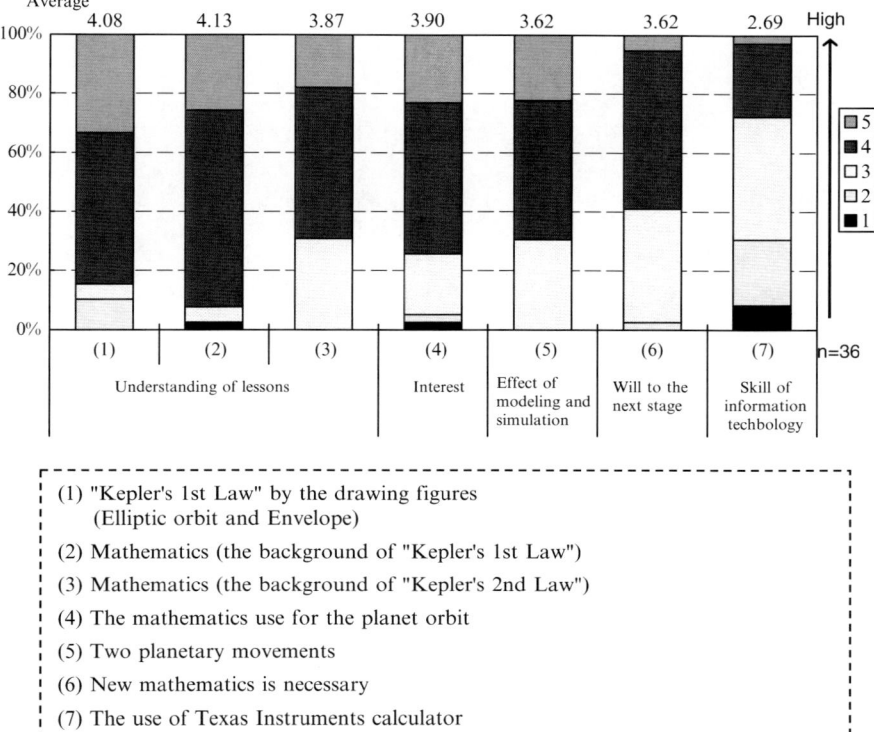

Fig. 58.13 "Premodel" – from survey of students

"Both content of modelling and guidance to Kepler's motion is too greedy". A little more time might have been necessary so that the students might master the knowledge. However, this modelling aims at the introduction into a new mathematics unit such as differential equations by making a mathematical development model.

The students would have achieved this step with further study and repeated modelling tasks. In schooling in Japan, there is no custom of using information technology effectively and the strong and weak points of the programming were caused by the students' ability. This is difficult if there is no educational environment that treats information technology properly.

4 Conclusion and Future Subjects

Both models showed utility as educational content raising awareness of interrelations between science and mathematics. This example resulted in students understanding the true concept of the mathematical theory against the background of a great scientific

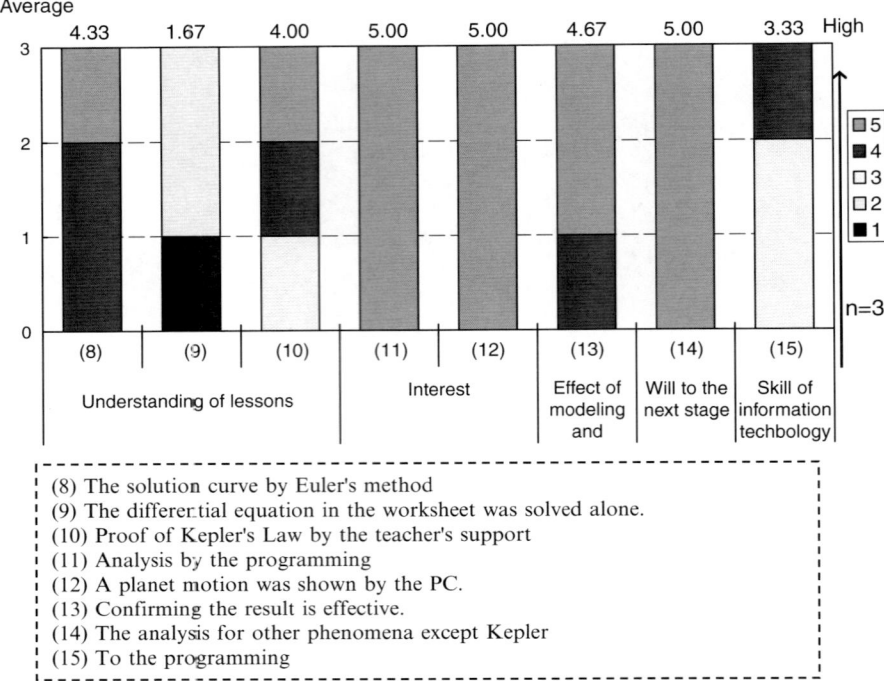

Fig. 58.14 "Mathematical development model" – students' evaluation

discovery. Students also learnt some more mathematics. Incomplete models might not be helpful when trying to solve a problem by applying mathematics knowledge. It is also important to give students new mathematics knowledge, if the model has contradictions or limits. Modelling practice in more advanced stages is also necessary so teachers should prepare new content. Students will grow if preparation is well founded, and the modelling can bridge to a new mathematics unit. When students must make a previously unseen mathematical model, teachers should assist when the application of learnt mathematics or new mathematics is needed. High school students in Japan are not accustomed to problem solving by modelling. There are few chances for them to appreciate the necessity of mathematics, when they study it. Even if it is late when students treat modelling, it is necessary to improve teacher's guidance when applying such modelling. This time, students made the mathematical model with teacher support. It is difficult for a teacher to judge whether it was the minimum support needed, but in the future students might have to research in teams, or solve problems alone. In order to do this, students need as much such self-help as possible at early stages of their growth (cf. Blum and Ließ 2007). This teaching experiment showed mathematics was useful for understanding science. Students could glimpse the scientific spirit. Though not all students achieved, they experienced and understood the mathematical theory. Such modelling practice is important for students who aim to become scientists and engineers and for students to become wise citizens.

References

Blum, W., & Leiß, D. (2007). How do students and teachers deal with modelling problems? In C. Haines, P. Galbraith, W. Blum, & S. Khan (Eds.), *Mathematical modelling (ICTMA12)* (pp. 222–231). Chichester: Horwood.
Borromeo Ferri, R. (2006). Theoretical and empirical differentiations of phases in the modelling process. *The International Journal on Mathematics Education, 38*(2), 86–95.
Fujii, J. (1987). Kansu Kyoiku No Konpan Mondai. *Sugaku Kyoiku Kenkyu, 16* (pp. 75–88). Osaka: Osaka Kyoiku Daigaku.
Kawano, Y. (2001). Sougou kyoiku No Ici Kyozai Kepler No Housoku No Sugakuteki Syomei. *Kenkyu Kiyo 48* (pp. 53–66). Hiroshima: Hiroshima Daigaku Fuzoku Chu Koutou Gakkou.
Kawasaki, T. (2007). The development of teaching materials connected with Newton science in the senior high school. In *Proceedings of the Exchange of Mathematics Studies beiween Japan and China* (pp. 96–100). Osaka: HANKAI.
Kyoto Chigaku Kyoiku Kenkyukai. (1993). 23 Kepler No Housoku. *Chigaku Jissyucho* (pp. 28–109). Kyoto: Kyoto Chigaku Kyoiku Kenkyukai.
Ministry of Education, Culture, Sports, Science and Technology. (2005). *Kotogakko Gakusyu Shido Yoryo Kaisetsu Sugaku Hen* (pp. 1–123). Tokyo: Jikkyo Syuppan.
Stephens, M., & Yanagimoto, A. (2001). Sugaku no ouyo no atsukai. *Sogogakusyu ni ikiru sugaku kyoiku* (pp. 27–28). Tokyo: Meiji Tosyo Syuppan.
Suken Syuppan. (2003). Banyu Inryoku. *Kotogako Butsuri, 2* (pp. 58–67). Tokyo: Suken Syuppan.

Chapter 59
Stochastic Case Problems for the Secondary Classroom with Reliability Theory

Usha Kotelawala

Abstract Basic models of reliability theory can provide relevant and motivating problems for secondary students as they develop skill and understanding in probability and algebra. This paper introduces the stochastic measurement of a system's reliability. It then presents problems which can be used in secondary mathematics classrooms discussing the prerequisite mathematics and the variation in the types of problems which can be posed within the framework of reliability theory. This includes providing an example of an open-ended project with an assessment rubric. Finally, it summarizes the mathematical residue as a rationale for secondary teachers to consider incorporating interesting applied stochastic problems within their curricula.

1 Introduction

While examples of modelling have been developing rapidly over the past few decades, the proportion of teachers using these materials remains low (Blum et al. 2002). To encourage teachers to incorporate modelling within their curricula, they must see the tasks accessible. This chapter provides examples demonstrating the utility of stochastic case problems in motivating and engaging secondary students in developing modelling skills. The problems presented demonstrate contextual problems that can be accessed by students in a range of varying skill levels.

This report introduces a case of mathematical modelling and mathematical applications for the secondary classroom taken from reliability theory. In light of the continuing challenge of getting teachers to consider areas which they are not familiar with, the following two questions guided the development of this report:

1. Do applications of reliability theory provide accessible, engaging, and motivating problems for secondary students?

U. Kotelawala (✉)
Graduate School of Education, Fordham University, New York, USA
e-mail: kotelawala@fordham.edu

2. Can the mathematics from reliability theory be seen as accessible and useful to broad audiences of secondary mathematics teachers?

First, this paper summarizes the basic concepts of reliability theory and its history. The following section presents sample reliability tasks for secondary classrooms. This section includes discussions of prerequisites and a variety of questions which can be posed within the framework of reliability theory. This includes providing an example of an open-ended project with an assessment rubric. Finally, the paper closes by summarizing the mathematical residue which can be extracted through the use of reliability theory problems in the classroom. This provides a discussion of characteristics which can persuade teachers to consider varying their traditional curricula to sample modelling in their classrooms.

2 Reliability Theory

Utilizing ideas from reliability theory provides a platform for presenting interesting problems while demonstrating a need and use for basic algebraic skills and concepts in probability. Examples of contexts approachable by secondary school students include:

1. How many engines does a 747 need to fly? What's the probability of all engines failing at the same time?
2. Christmas lights go dark when just one fails. What is it worth to get better bulbs?

2.1 Brief History

Building blocks of reliability theory were developed by German rocket engineers in Peenemunde, Germany during World War II and then in Huntsville, Alabama following the war (Kececioglu 2002). In both cases collaboration was within the teams created by Wernher von Braun. The two early pioneers of the field were Eric Pieruschka and Robert Lusser who initially worked on trying to improve a failing missile by improving the weakest link. They later recognized that a rocket's reliability was equal to the product of the reliabilities of component parts and this concept became known as Lusser's law.

It may be its utility in the aviation industry which elevated reliability theory as its own field. It is argued that Z.W. Birnbaum, a professor at the University of Washington and a consultant to the mathematics division at the Boeing Scientific Research Laboratories, led formalized development of the field in a 1961 issue of Technometrics, "Multi-Component Systems and Structures and Their Reliability[1]"

---

[1] The term "reliability" has meanings in common language and with slight variances in different regions of the world. For the purpose of this paper, the author is using the specific stochastic meaning initiated in Mathematical Theory of Reliability by Barlow and Proschan in 1965.

(Barlow 2002). This was significantly motivated by industry needing and wanting to increase the safety of flying. Through a set of simplifying assumptions, such as dichotomy (systems either perform or fail), and components with common reliabilities, they examined requirements for building multicomponent systems more reliable than individual components.

2.2 A Summary of the Reliabilities of Simple Fundamental Systems

Reliability is the probability that a multicomponent system performs adequately over an interval of time (Barlow and Proschan 1965).

Systems can often be modeled using series systems, parallel systems, or combinations of both. In these models, a system functions successfully as long as there is a path through working components from the start (generally the left-most point) to the finish.

2.2.1 Series Systems

In Fig. 59.1, components 1 through n must all work for the system to work.

$$R = p_1 \times p_2 \times \cdots \times p_n = \prod_{i=1}^{n} p_i$$

The probability that the i-th component will work is denoted as p_i. In a system of components in a series, the reliability of the system is the probability that all components work. The reliability of the whole system is labeled as R, the probability that all components work.

2.2.2 Parallel Systems

In a parallel system, the system will work as long as at least one of the components is working. In Fig. 59.2, the system works as long as at least one of the components is working and it fails if and only if all components are failing.

The reliability of the system of n parallel components is,

$$R = 1 - \left[(1 - p_1)(1 - p_2)\ldots(1 - p_n)\right] = 1 - \prod_{i=1}^{n}(1 - p_i).$$

Fig. 59.1 n components in series

Fig. 59.2 n components in parallel

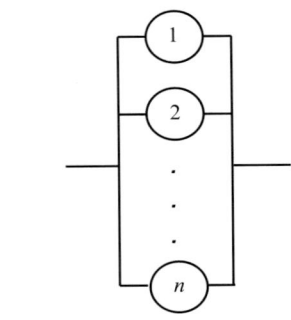

Fig. 59.3 A combined system

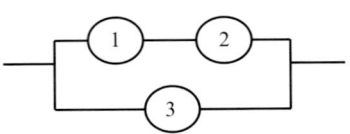

2.2.3 Combined Systems

For systems composed of both series and parallel components, the whole system's reliability can be found by finding the reliabilities of subsystems and then again applying the routines for series or parallel sets of components. For example Fig. 59.3 uses components in series and parallel.

By finding the reliability of the subsystem of components 1 and 2 in series, what will be labeled as R_1, I can then treat the new reliability as a single component in parallel with component 3 as is shown in Fig. 59.4. The reliability of R_1 is $p_1 \times p_2$. Thus the whole system's reliability is,

$$R = 1-(1-R_1)(1-p_3) = 1-(1-p_1 \times p_2)(1-p_3)$$

This method of dealing with combined systems provides an explicit and visual case for students to see how simplifying a problem becomes a critical strategy in modelling and determining the reliability of a system. This process can be repeated with any system to obtain the system reliability. This paper focuses on this solution method[2].

---

[2] The purpose of this paper is to demonstrate how an extracted type of problem from a field generally taught only in tertiary programs can be used to generate interesting and educationally useful problems. The methods demonstrated thus far only present a small portion of the beginning ideas of reliability theory. It is common for next steps to be consideration of *minimal paths* and *minimal cuts* to generate an algorithm for determining the reliability of a system. While this too may have potential with secondary students, it is outside the scope of this paper.

Methods which build the reliability function from minimal paths or minimal cuts can frequently be found in textbooks for Operations Research (Hillier and Lieberman 2010) and in the work of Birnbaum et al. (1961).

Fig. 59.4 Simplified modification of Figure 59.3

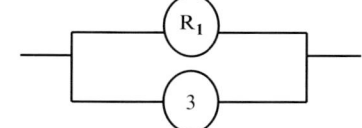

3 Reliability Theory Problems for the Secondary Classroom

3.1 Prerequisites for Reliability Theory

Problems using reliability theory require understanding of a few fundamental concepts of probability, namely, independent events and the consequential multiplication law of joint independent events. While other ideas are also critical to utilizing reliability theory, such as the probabilities of complementary events, they can be taught through the use of reliability theory in considering a single component and generalizing.

The level of algebraic proficiency required varies from none to advanced depending on the design of the problems used, as is evident in the problems which follow.

3.2 Examples of Problems for Secondary School

A variety of problematic contextual tasks can be created for students to explore and solve. These have been separated into Levels I and II to show both simple and complex problems within the realm of secondary students. These are followed by a sample project giving students a chance to create their own designs while maximizing reliability.

3.2.1 Level I Problems

In the Level I problems below, the systems are either exclusively series or exclusively parallel. Questions have been designed to push students to see the patterns emerging from multiple components either in series or in parallel. A choice was made to select simpler component reliability values over the more realistic true reliabilities. For example, in problem B below, the real reliability of a jet engine is much greater than the 0.98 given. If a teacher wanted to focus on concepts around decimals and scientific notation, they might choose to use more realistic numbers. These numbers were chosen to prioritize the objective of examining the probabilities while lowering anticipated student struggles with decimals and significant digits.

Problems A and B ask students to find more than one way of answering a problem. This allows different methods of solution to arise which can be discussed to compare the efficiency of different methods. Problem A asks students to get to the general case for components in series, which is Lusser's Law for n component parts in series:

$$R = R_1 \times R_2 \times \cdots \times R_n.$$

Problem B asks the student to list all possible cases of success and failure for a system. Additional problems could be added to confirm that students realize the role of multiplication in generating all possible cases for a set of components.

Problem A. Grandma called yesterday to ask you to help her hang holiday lights for the winter season. Her lights are the old type of light strings. If one bulb fails, the whole string fails. The probability of each bulb working is 0.75. If one of the strings has only two bulbs (a short string for illuminating very small areas) what's the probability of the string working?

   (a) Draw a picture of this situation.
   (b) Find two different ways to solve the problem.
   (c) What is the probability of a string of 5 lights working? 13 lights? N lights?

Problem B. FlyCheap Airlines has only one route from London to Hamburg. Recently, the company purchased a used 727 jet airplane having three engines. It can actually fly on just one engine. For each engine, the probability of it working over the course of the trip is 0.98.

   (a) Draw a picture to represent this scenario.
   (b) What is the probability that the plane will be able to successfully fly? Find at least two different ways to determine your answer.
   (c) There are various different possibilities which could occur such as: Engine 1 works, Engine 2 works, Engine 3 fails. We could represent this particular case as (1, 1, 0) where a 1 indicates an engine working, a 0 indicates an engine failing and its place in the order indicates which engine it is.

      i. List all possible cases where the system would work.
      ii. List all possible cases where the system would fail.

   (d) If the reliability of an engine was unknown, call it x, how could you represent the system's reliability?
   (e) How much would the reliability improve if 4 engines were used?

3.2.2 Level II Problems

Each of these problems includes systems in which both series components and parallel components are both used.

Problem C provides a contextual description and leaves the task of creating the diagram to the student. Using this exercise in pairs with discussion allows the teacher to check student understanding and interpretation of parallel and series systems which connects and supports understanding the importance of the words "and" and "or" in various areas of mathematical logic and problem solving.

Problem D asks for finding the system's reliability as a function of one of the component reliabilities. It provides an example of how reliability problems can be used as simple applied practice problems of basic algebra. The reliability model can be expressed as $R = 1 - [(1 - 0.5x)(1 - 0.7)]$. Teachers can also use this in the study of functions and different problems can be used to generate more complicated polynomial functions.

Problem C. Consider a stereo system consisting of the following components:

1. A radio
2. A CD player
3. An iPod connection device
4. An amplifier
5. Two speakers

Assume the system works only if the amplifier, at least one speaker, and either the radio, CD player, or iPod connector works.

With this stereo system, we can ask what components must work for the system to produce music[3]. There are a number of possibilities that would allow the system to work. For example if 1, 4, and 5 work we can hear music from the radio. If the amplifier, component 4, is down the system cannot function.

(a) Create a diagram to model the stereo.
(b) What are other combinations of working components that would allow the system to work at least minimally? List as many possibilities as you can.

Problem D. In the system below, the reliabilities of two components are fixed.

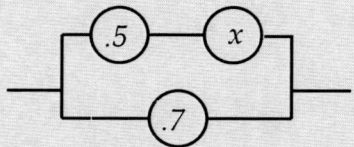

(a) What is the maximum reliability of the system?
(b) What must the reliability of the third component be in order for the system to have a reliability of 0.8?
(c) Express the system's reliability as a function of x.
(d) Graph the function and determine the maximum and minimum reliabilities of the overall system.

---

[3]This question could become an entry point into the varied methods of solution involving minimal paths and minimal cuts.

Problems E and F get the student to compare parallel components over components in series as they consider cost and varying reliabilities of components in the same system. They push student thinking with a simple three component system to determine where weaker components do the most harm to a system. Problem E puts the question into a realistic context where increasing a component's reliability is associated with a cost. Problem F lets student consider changing arrangements with given reliabilities.

Problem G leads to examination of higher-degree polynomial functions. It provides an example of how more complex problems can be utilized to differentiate work and determine how well students can organize and track their solutions. Question c is a problem where technology can be used to solve the resulting reliability polynomial.

Problem E. The reliabilities for each component in this system are given.

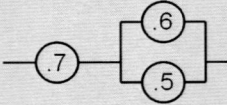

(a) The cost of improving a component is $10,000 to obtain an increase of 0.1 in the component reliability.
(b) What would be the minimum cost of raising the system's reliability to 0.95?

Problem F. How do you arrange three components with reliabilities 0.5, 0.6, and 0.7 in each system below to maximize the system reliability? Is there a different arrangement of three components which would generate an even higher reliability?

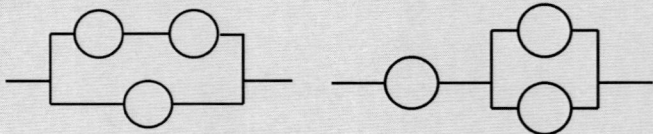

Problem G.

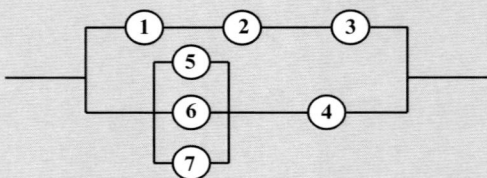

(a) Create the reliability function for the diagram above.
(b) Suppose the reliability of each component has the same value, x. Determine the resulting reliability function.
(c) If each of the components has the same reliability, what must that reliability be for the whole system to have a reliability of 0.95?

3.2.3 Sample Project

The Dormitory Electrical Wiring project asks students to design and assess the reliability of the college dormitory electrical system. This provides opportunities to consider how various designs increase or decrease overall reliabilities with an open-ended task which would be appropriate for multiple-student teams. This project accomplishes several of the critical design principles recognized by practitioners and researchers in mathematical modelling (Lesh and Kelly 2000), such

---

Dormitory Electrical Wiring: Project in Reliability Theory

You are an electrical contractor bidding for a job with Fordham University. You need to find an appropriate model for wiring a college dormitory. Your goal is to maximize reliability while minimizing wiring costs.

The new dormitory will be a 12 story building. Each floor has a hallway, two mini-kitchens, and eight clusters. A cluster consists of 4 two person dorm rooms, bathroom, and a small common area. Here are the SPECS (specifications):

Hall: 6 ceiling lights
Mini-kitchen: 1 ceiling light, 1 double plug-in outlets, one oven/cook-top plug-in
Dorm room: 1 ceiling light, 2 desk lights, 4 double plug-in outlets
Bathroom: 2 central ceiling lights, a row of 6 lights, 3 double plug-in outlets
Common Area: 1 ceiling light, 2 double plug-in outlets

You also know the following (reliabilities are given for successful use over a 10 year period)

$2 for a 95% reliable plug-in outlet
$4 for a 98% reliable plug-in outlet
Reliability of ceiling light hook-up and row lighting: 0.95
Reliability of oven/cook-top plug-in: 0.97

1. Draw a separate diagram for each of the following:
   - One diagram representing wiring to one dorm room
   - One diagram representing wiring to one cluster
   - One diagram representing wiring on one floor
   - One diagram representing the wiring to each floor

2. For each of the following systems create the model that allows you to calculate the reliability of that system based on variable reliability values for the components:
   - One dorm room
   - One cluster
   - One Floor
   - The whole building

> 3. Using the information given in the SPECS, calculate the reliability of each of the systems listed in two.
> 4. Discuss the difference in reliability that results for using the more expensive parts. Which parts would you recommend?
>
> Your project must include each of the following:
>
> (A) A written proposal to submit to the Fordham University
> (B) A table of components, the reliability, and their failure probability
> (C) Work showing how you found the reliability for each part of the system
> (D) Research on a manufacturing/reliability issue.

as requiring construction of a model, connecting to a meaningful context (Kaiser and Schwarz 2006), encouraging self-assessment in the process, and promoting seeking reusable subsystems.

To focus on greater student independent thinking, the teacher can make a pedagogical choice to begin with the larger case problem such as the dormitory wiring problem. Then problems such as those presented at Level I and Level II can be used as needed to help students consider simpler problems such as the sound system problem.

A teacher may also develop a mini-unit letting problems grow gradually from the more simple series, parallel, and combined problems of Level I to gradually arrive at the point where students are ready to work on the more complicated project problem. There can also be opportunities for students to compete in generating more reliable systems.

4 Conclusion: The Mathematical Residue of Reliability Tasks (RT)

The work of Maaβ (2006) resulted in a set of questions to use in guiding the development of modelling tasks as presented by Mousoulides in "Mathematical Modelling for Elementary and Secondary School Teachers" (2009). In conclusion this paper focuses on two of these central questions: (1) How relevant and authentic is the task? and (2) How does the task fit into the curriculum and general framework?

Contexts such as sound systems and wiring provide a way to present topics with real world relevance for students. Authenticity and relevance are often subjective and often arguable. However, the problems also demonstrated how variations of problem types are possible and this can also provide materials for differentiation within a classroom. The strength of these problems is in addressing relevant topics, understandings, and skills which are extendable to other problems with contexts which can capture student interest.

The mathematical residue of simple reliability problems fit into curriculums by focusing on central concepts of secondary mathematics. The primary topics utilized and potentially developed through use of the problems focus on probability and algebra, central topics to any secondary curriculum. Problems utilized procedural skills such as simple and complex substitution, expressing and simplifying polynomials, and solving polynomial equations. The problems also included opportunities for students to work on inductive reasoning for generalized models and the strategy of simplifying parts of a problem along the path to a larger solution.

In conclusion, the problems presented provide relevant contextual motivating problems which secondary teachers with basic understandings of probability can consider using in their classrooms. The mathematics of these problems provides a rationale for their use, which may then become a path toward greater incorporation of modelling in the classroom.

References

Barlow, R. E., (2002, June). *Mathematical reliability theory: From the beginning to the present time*. Invited paper at the Third International Conference on Mathematical Methods in Reliability, Trondheim, Norway.

Barlow, R. E., & Proshan, F. (1965). *Mathematical theory of reliability*. New York: Wiley.

Birnbaum, Z. W., Esary, J. D., & Saunders, S. C. (1961). Multi-component systems and structures and their reliability. *Technometrics, 3*(1), 55–77.

Blum, W., et al. (2002). ICMI study 14: Applications and modelling in mathematics education—Discussion document. *Educational Studies in Mathematics, 51*(1/2), 149–171.

Hillier, F. S., & Lieberman, G. J. *Introduction to operations research*, (9th ed.) New York: McGraw-Hill.

Kaiser, G., & Schwarz, B. (2006). Mathematical modelling as bridge between school and university. *Zentralblatt für Didaktik der Mathematik, 38*(2), 196–208.

Kececioglu, D. (2002). *Reliability and life testing handbook*. Lancaster: DEStech Publications, Inc.

Lesh, R. A., & Kelly, A. E. (2000). Multi-tiered teaching experiments. In R. A. Lesh & A. Kelly (Eds.), *Handbook of research design in mathematics and science education* (pp. 197–231). Mahwah: Lawrence Erlbaum Associates.

Maaß, K. (2006). What are modelling competencies? *Zentralblatt für Didaktik der Mathematik, 38*(1), 113–142.

Mousoulides, N. G. (2009). Mathematical modeling for elementary and secondary school teachers. In A. Kontakos (Ed.), *Research and theories in teacher education*. Rhodes: University of the Aegean.

Chapter 60
LEMA – Professional Development of Teachers in Relation to Mathematical Modelling*

Katja Maaß and Johannes Gurlitt

Abstract LEMA was an international project which aimed at designing a professional development course for modelling. Materials for professional development which were to be used in different national contexts were designed, piloted and evaluated. In this chapter, we present the overall framework of the project and its evaluation by outlining the theoretical background, the design of the professional development course, the design of the evaluation, and summative results. In brief, the summative results of the evaluation showed that the professional development course had no effect on teachers' beliefs but a strong positive effect on their pedagogical content knowledge and self-efficacy in terms of modelling, as well as a high degree of satisfaction among participants regarding their professional development.

1 Theoretical Background

Researchers, practitioners and policy makers in mathematics education agree that an educational goal should be to enable students to apply mathematics to their everyday lives and contribute to the development of active citizenship. Nevertheless, throughout Europe modelling is still rare in day-to-day teaching (Blum et al. 2002). LEMA (Learning and Education in and through Modelling and Applications) was

---

*The material for the professional development course is freely available at www.lema-project.org (in the six partner languages). If you want to use it, please refer to LEMA. Please also send a brief e-mail to katja.maass@ph-freiburg.de to acknowledge the use of the materials.

K. Maaß (✉)
University of Education Freiburg, Freiburg, Germany
e-mail: katja.maass@ph-freiburg.de

J. Gurlitt
University of Freiburg, Freiburg, Germany
e-mail: johannes.gurlitt@ezw.uni-freiburg.de

a transnational European Project[1] (2006–2009) that attempted to tackle this problem at teacher level by designing a common course of professional development in mathematical modelling. To this end, LEMA was a design research project (Burkhardt 2006). Given the different types of schools and educational structures in partnership countries and the different theoretical backgrounds of partners situated in various cultural contexts, the main challenge within this transnational project was to agree on essential concepts and to design materials which can be used across Europe within a wide range of different national contexts.

Mathematical modelling means applying mathematics to realistic, open problems. There are many descriptions of modelling processes (Kaiser-Meßmer 1986) that vary according to the described modelling cycle, the relevance given to the context, and the justifications seen for modelling in mathematics lessons (Kaiser and Sriraman 2006). In this study, we followed the description of the modelling process in PISA (Baumert et al. 2001), although we restricted it to context-related problems. There are also a large variety of types of *modelling tasks* which can be differentiated according to a number of features, such as the modelling activity carried out and its relation to reality or data given (e.g., Burkhardt 1989; Galbraith and Stillman 2001). *Modelling competency* is the ability to carry out modelling processes independently. It comprises competencies to carry out the steps of the modelling process, competencies in mathematical reasoning and metacognitive modelling (see e.g., Haines and Izard 1995; Maaß 2006).

Teaching modelling: Empirical educational research has provided insights into the basic dimensions of good teaching quality which supports insightful learning (Prenzel et al. 2004). These are among others: clarity, comprehensibility and structure in the subject and tasks presented, promoting insightful learning by taking metacognition into account (Weinert 1998), and dealing with heterogeneity by offering different tasks and by varying methods. Formative assessment which gives constructive, motivating feedback is seen as a method that supports students in developing their competencies (Black and Williams 1998). In relation to modelling, a variety of studies have shown that working in small groups, and students working independently can support the development of modelling competencies (see e.g., Galbraith and Clatworthy 1990; Ikeda and Stephens 2001).

Professional development of teachers: When considering teachers' competencies in teaching, we follow Krauss et al. (2004) and Shulman (1986) by distinguishing professional knowledge (content knowledge, pedagogical content knowledge, pedagogical knowledge), beliefs, motivational orientation, self-efficacy and competencies

---

[1] Partners of LEMA: Katja Maaß (Coordinator) University of Education Freiburg, Geoff Wake, University of Manchester, Fco. Javier Garcia Garcia, University of Jaen, Nicholas Mousoulides, University of Cyprus, Ödon Vancso & Gabriella Ambrus, Eötvös Lóránd University Budapest, Anke Wagner, University of Education Ludwigsburg, Richard Cabassut, IUFM Strasbourg. This project has been funded with support from the European Commission. This publication reflects the views only of the authors, and the Commission cannot be held responsible for any use which may be made of the information contained therein.

in reflection. Empirical studies of teachers' professional development (Tirosh and Graeber 2003; Wilson and Cooney 2002) have shown that professional development interventions lead to changes if the courses are long term, with embedded phases of teaching and reflection, and take into consideration both the context in which teachers work (e.g., the school director, parents etc.) and teachers' own beliefs.

Teachers' beliefs about mathematics and its education are thought to have a major impact on if and how a teacher employs innovation in everyday teaching. According to Pehkonen and Törner (1996), beliefs are composed of a relatively long-standing subjective knowledge of certain objects as well as the attitudes linked to that knowledge. Kaiser (2006) showed that innovations required by the curriculum are interpreted by the teacher in such a way that they fit into his or her existing belief system. Grigutsch et al. (1998) classified beliefs about mathematics into the aspect of scheme (fixed set of rules), the aspect of process (problems are solved), the aspect of formalism (logical, deductive science), and the aspect of application (important for life and society). These aspects of beliefs can also be found in teachers' beliefs about effective teaching (Maaß 2009).

Teachers' self-efficacy beliefs in this context can be described as teachers' ability to believe in their own capabilities to organize and execute mathematical modelling activities in their planning and classroom practice (see Bandura 2006). Based on Bandura's social cognitive theory and theoretical models of behaviour prediction (e.g., theory of planned behaviour from Ajzen 1985), self-efficacy about modelling should be a valuable predictor for the intention to use modelling and (although beyond the scope of this study) for the actual use of modelling in the classroom.

In conclusion, teachers' knowledge and beliefs about the nature of the subject, their views on how to teach the subject and their self-efficacy concerning modelling all influence how they design or select tasks, plan, implement and evaluate their lessons (e.g., Brickhouse 1990).

2 Design of the Course of Professional Development

In order to design the course for professional development, the development team tried to answer the question of what knowledge a teacher needs in order to teach modelling. Based on our theoretical backgrounds, we came up with a theoretical model of the pedagogical content knowledge needed for modelling. We distinguish between four main categories, which are further divided into sub-categories.

1. *Modelling*: To implement modelling in lessons, teachers need background information about this concept (Sub-categories: What is modelling? Why use it?).
2. *Tasks*: When it comes to planning lessons, teachers need to learn how to select appropriate tasks for their students and anticipate the modelling outcomes. In line with our assumptions on how to teach modelling, a variety of tasks should be chosen. (Sub-categories: Exploring tasks, Creating tasks, Classification of

tasks, e.g. according to area and context, and Variation of tasks, e.g. in order to adapt them to the specific needs of a class).
3. *Lessons*: Teachers need information about how to design lessons appropriate for modelling and how to act in the classroom (Sub-categories: Teaching methods, Using ICT, Supporting the development of modelling competencies, Exercising mathematical content through modelling).
4. *Assessment*: If modelling is implemented in lessons, it also has to be evaluated. Assessment should be used not only for grading but also for supporting learning through feedback (Sub-categories: Formative Assessment, Summative Assessment, Feedback).

Incidentally, this model has been validated by the fact that Borromeo Ferri and Blum (2009) came up with a similar model for pedagogical content knowledge with reference to prospective teachers.

As the designed course for professional development should not only be based on theory but also meet teachers' needs, an analysis of needs in relation to teachers' beliefs about mathematics, its teaching, modelling and the actual modelling tasks themselves, was carried out with 561 voluntary teachers from all participating partner countries (Maaß and Gurlitt 2009). The results revealed a discrepancy between teachers' beliefs in mathematics as an important method for problem solving in everyday life and negative perceptions of open, complex, tasks related to everyday life.

Based on the needs, analysis and our theoretical background, we developed a professional development course consisting of five key modules, of which four mirrored the four categories named above. Further, we added a fifth module on reflection. As outlined in the theoretical background, reflection on implementation in lessons and dealing with challenges is crucial for the success of professional development courses. The course was designed for use with primary and lower secondary teachers, and it was piloted in 2008 over a period of 5 days and evaluated in all six partnership countries. Implementation, however, was quite different in each country, as it had to meet the different statutory requirements and contexts of every country. Further, every partner had to draw on given opportunities to implement the course. For example, in France the training was given as a one-block course in January 2008, addressing teachers of students aged 6–8 years. In Spain, the course contained two blocks in April and May. In Germany the course consisted of five separate days from January to November.

3 Design of the Evaluation

Given the different national contexts and the various piloting approaches (see above), one may question whether it makes sense to evaluate the course as it took place in all countries together. As the course materials, which are designed to be used in different countries, are detailed and as all countries followed the materials closely, we felt that a cross-country evaluation would mirror the quality of the materials while depending neither on specific implementation nor target group. For the evaluation, we did not consider the students because this seemed to be almost impossible, given the variety of pupils involved (age 6–16) and the given national contexts. Thus, we focused on teachers and used questionnaires.

Therefore, first and foremost the goal of the evaluation was to assess whether, and in which dimensions, the professional development course had an impact on teachers. This included teachers' pedagogical content knowledge about modelling, beliefs about mathematics education and teachers' self-efficacy regarding implementing mathematical modelling. We were also interested in determining whether teachers who scored highly in one dimension also succeeded in the other dimensions and vice versa. Finally, the evaluation also investigated the degree of satisfaction teachers felt with the professional development course and their intention to implement modelling. We will give reasons for this selection later. Subsequently, our evaluation focused on the following research questions: (1) Does the professional development course influence the pedagogical content knowledge of the teachers? (2) Does the professional development course influence teachers' beliefs about mathematics education? (3) Does the professional development course influence teachers' self-efficacy regarding modelling? (4) Does success in one dimension include success in another dimension (correlational)? (5) How satisfied were teachers with the professional development course?

Design and participants: In order to evaluate the course, we used a pre-post-control group design. This design allowed us to evaluate changes over time (before and after the course of professional development) and compare these changes within the intervention group and with a control group. This seemed necessary in order to make sure that changes were not due to any other influencing factors such as media coverage or new curricula. The participants were 155 teachers (124 females, 31 males) with a mean age of 41 years (SD=9.86) and a mean teaching experience of 16 years (SD=10.35). More than half of these teachers, specifically 89, taught primary school, 63 teachers taught in secondary schools and 3 did not provide an answer. One hundred and four teachers were assigned to the intervention group and 51 teachers were assigned to the control group. Whenever possible, the assignment was randomized; however, in some instances this was not possible and thus it was necessary to find similar teachers who filled out the questionnaire during similar time intervals. Teachers not selected for the intervention group were offered the chance to receive the training at a later time. All teachers completed the pre-questionnaire and post-questionnaire. Participants in the teacher training also filled in a so-called "optimization questionnaire" after each of the five training days (investigating satisfaction with the course and suggestions for future improvements of the course).

Instruments and measures: The design and implementation of the questionnaire conformed to the following principles. First, the questionnaire was designed to mirror the theoretical background and key aspects of the modules of the professional development. Second, considering the target group and their understandable preference for a short questionnaire, our target was a balance between a reasonable length and what would still provide a reliable assessment. Subsequently, we provided no questions about pedagogical knowledge or motivational aspects. Third, careful guidelines were developed to ensure uniformity (as far as possible) when asking teachers to complete the questionnaire. Fourth, to further build on previous research, the scale construction was based on established scales wherever possible (see below). Fifth, the questionnaire was designed in several steps to improve its validity and reliability (for details see Maaß and Gurlitt 2009). The questionnaire comprised four parts: (i) pedagogical content knowledge, (ii) beliefs, (iii) self-efficacy and (iv) modelling intention. We did not insert any questions relating to content knowledge, as we did not want the teachers to feel as though their subject knowledge was being tested. Due to time limits, when completing the questionnaire we restricted our questions about beliefs to those concerning mathematics education, as we felt that they would be relevant for all teachers. For the pedagogical content knowledge we used open items, as we reasoned that this would be less suggestive in eliciting teacher responses. Three out of the four questions related to one particular modelling task, following on from the results of the analysis of needs, which had shown that the more concrete an item was the more evident the objections became. Identical questionnaires were used before and after the intervention. Among other things, teachers had to answer to the following items:

1. What characteristics do modelling tasks have?
2. Imagine you are teaching children whom you regard the right age for this task. The following five questions are all related to the task below and all connected with each other.

Task: It is the start of the summer holidays and there are many traffic jams. Chris is on holiday in Germany and has been stuck in a 20-km traffic jam for 6 h. It is hot and she is longing for a drink. Although there are rumours that the Red Cross is coming around with a small lorry distributing water, she has received nothing so far. How long will the Red Cross need to provide everyone with water?

2a. Give as many reasons as possible (pros and/or cons) to use the task and mark them as such (+/−)
2b. Imagine you are planning your lesson and you decide to use this task the next day. You want students to work on this task in groups and afterwards to present their results. How will you organize the presentation phase?

The scale of belief items about the nature of mathematics education was based on Grigutsch et al. (1998). Teachers rated their beliefs on a five-point scale, ranging from strongly disagree to strongly agree. The following items are an extract from the scale:

	School mathematics in my lessons from my point of view as a teacher	Strongly disagree				Strongly agree
5.1.1	School mathematics is a collection of procedures and rules which determines precisely how a task is solved.	○	○	○	○	○
5.1.2	School mathematics is very important for the students later in life.	○	○	○	○	○
5.1.3	Central aspects of school mathematics are flawless formalism and formal logic.	○	○	○	○	○

Based on Bandura's method for measuring self-efficacy beliefs (Bandura 2006), we designed a self-efficacy scale assessing efficacy beliefs on a 100-point scale, ranging in 10-unit intervals from 0% ("cannot do at all"), to 100% ("highly certain can do") (see examples).

7.1.1	I feel able to adapt tasks and situations in text books to provide realistic open problems.	–
7.1.2	I feel able to distinguish between modelling tasks and other reality-based tasks.	–

To measure the future modelling intention we used the following question:

	Not very likely				Very likely
In future I will use the modelling approach in my teaching.	○	○	○	○	○

4 Results and Discussion

We used a two-factorial repeated measure ANOVA, to analyze the data with group (intervention vs. control) and time (pre-test vs. post-test) as factors. The effect under consideration for the research questions of the current study was whether the intervention group outperformed the control group over time, which is the interaction effect between group and time, thus only this effect will be reported below (neglecting main effects indicative of mean differences between the groups or the two points in time). As an effect size measure, we used partial η^2, qualifying values <0.06 as small effects, values in the range between 0.06 and 0.13 as medium effects, and values >0.13 as large effects (Cohen 1988). Table 60.1 provides an overview of the results.

Table 60.1 Means and standard deviations (in parentheses) of dependent variables

		Pre-test	Post-test	p
Pedagogical content knowledge	Control	0.47 (0.60)	0.69 (0.48)	<0.01
	Intervention	0.69 (0.52)	1.27 (0.47)	
Beliefs	Control	3.19 (0.40)	3.22 (0.40)	0.84
	Intervention	3.43 (0.46)	3.45 (0.47)	
Self-efficacy	Control	45.31 (22.73)	45.15 (23.75)	<0.01
	Intervention	49.47 (18.88)	70.92 (11.55)	

The two groups (control vs. intervention) were analyzed in respect of the aspects below at two points in time (pre and post-intervention in terms of the intervention group).

Pedagogical Content Knowledge (PCK): This yielded the effect that the participants of the intervention group improved their pedagogical content knowledge in relation to modelling while the control group did not change. This was indicated by an analysis of variance, more precisely the large effect of the interaction between group and time $F(1,153)=41.52, p<0.01, \eta^2=0.21$. Analyzing the crucial interaction between group and time further, post hoc tests showed that participants in the intervention group improved their pedagogical content knowledge during the intervention $F(1,103)=123.49, p<0.01$ while participants in the control group did not change their pedagogical content knowledge over time $F(1,50)=1.55, p=0.22$.

Beliefs about Mathematics Education: There was no difference in change over time, $F(1,152)=0.04, p=0.84$, and no main effect for time $F(1,152)=0.50, p=0.48$.

Self-efficacy for modelling: The intervention group showed a greater improvement in self-efficacy than the control group. This was indicated by a large effect of the interaction between group and time $F(1,148)=65.25, p<0.01, \eta^2=0.31$. Analyzing the crucial interaction between group and time further, post hoc tests showed that participants in the intervention group improved their self-efficacy during the intervention $F(1,101)=149.49, p<0.01$ while participants in the control group did not change their self-efficacy regarding modelling over time $F(1,48)=0.02, p=0.89$.

Satisfaction: Evaluating satisfaction with the intervention, we assessed satisfaction ratings after each respective module. The mean satisfaction was 4.25 (SD=0.47); the satisfaction ratings of participants ranged from 3.28 to 5.00. As the rating scale had a scale from 1 to 5, a mean of 4.25 can be interpreted as a rather high level of satisfaction with the intervention.

Correlations between post-test measures: The correlations between the post-test measures (PCK, beliefs, self-efficacy) were all positive and significant and ranged from $r(152)=0.27$ to $r(152)=0.54$, indicating that high scores in one dimension were related to high scores in the other dimensions measured. In detail, the lowest correlation was between beliefs and the self-efficacy measures in the post-test with $r(152)=0.27, p<0.01$; the highest correlation was between self-efficacy and pedagogical content knowledge with $r(154)=0.54, p<0.01$.

The mean *intention to use modelling* in the classroom was 3.91 on a scale from 1 to 5. Theoretically plausible is the high correlation between pedagogical content knowledge in the post-test and the intention to use modelling r (148)=0.50, $p<0.01$, and especially self-efficacy and this intention r (148)=0.60, $p<0.01$. Without knowledge of modelling it would not be very likely that teachers will use modelling in their classes. Similarly, according to Banduras social cognitive theory (Bandura 1997, 2006) and theoretical models of behaviour prediction (e.g., theory of planned behaviour from Ajzen 1985), self-efficacy is an important predictor for the intention to conduct the specific behaviour. Taken together, these relationships support the validity of the conducted evaluation. Last, the beliefs measured before and after the intervention showed a negative correlation with age r_{pre} (155)=−0.31, $p<0.05$; r_{post} (154)=−0.16, $p<0.05$, and years of teaching r_{pre} (155)=−0.37, $p<0.01$ and r_{post} (154)=−0.24, $p<0.01$, indicating that older teachers with more years of teaching experience had less favourable beliefs about modelling.

In conclusion, the evaluation of the intervention showed strong positive effects on pedagogical content knowledge and self-efficacy ratings and received high satisfaction ratings from the participants. Teachers showed a high intention to integrate modelling into their lessons. These results can be regarded as important necessary preconditions for the implementation of modelling. Beliefs across the different countries and target group were surprisingly similar. However, not surprisingly, beliefs about modelling did not change as a result of the intervention. On the one hand, beliefs are well established and deeply entrenched frameworks and are probably very difficult to change through an intervention – even one that took several weeks. On the other hand, the questionnaire may not have been sensitive to any changes that took place. Looking at the changes, the high level of satisfaction with the course hints to the fact that the design of the course – based on prior empirical findings (about modelling and professional development, e.g. including phases of reflection on practice) and the analysis of needs – indeed met teachers' needs and supported them in the development of both self-efficacy and pedagogical content knowledge.

The correlations between the post-test measures show that success in one domain was linked to success in another domain. This relationship between pedagogical content knowledge, beliefs and self-efficacy may provide an important basis for a successful implementation. For example, it may not be helpful if the PCK is high but the self-efficacy is low. The correlations between age and beliefs may be due to new modes of teacher education which have already started, perhaps giving us hope that 1 day modelling will be found more and more in day-to-day teaching.

So far, we cannot assess to what extent the project contributed to a more widespread implementation of modelling into day-to-day teaching. However, the empirical results indicate that the intervention made the use of modelling in day-to-day teaching more likely. Within LEMA, as a European project, it was necessary to analyze the given national contexts carefully and to strive for common ground in relation to modelling in order to design a course for *use in* the different countries. We used the different perspectives and experiences to learn from each other (Garcia et al. 2010) so as to improve teachers' pedagogical content knowledge and self-efficacy about modelling. LEMA thus contributes to the discussion about modelling by providing an internationally usable course and by providing empirical evidence that the course improved teachers' self-efficacy and pedagogical content knowledge of modelling in different European countries.

References

Ajzen, I. (1985). From intentions to actions: A theory of planned behavior. In J. Kuhl & J. Beckmann (Eds.), *Action control: From cognition to behavior*. Berlin/Heidelberg/New York: Springer.
Bandura, A. (1997). *Self-efficacy: The exercise of control*. New York: Freeman and Company.
Bandura, A. (2006). Guide for constructing self-efficacy scales. In F. Parjares & T. Urdan (Eds.), *Self-efficacy beliefs of adolescents* (Adolescence and education, Vol. 4, pp. 307–337). Greenwich: Information Age Publishing.
Baumert, J., Klieme, E., Neubrand, M., Prenzel, M., Schiefele, U., Schneider, W., et al. (Eds.). (2001). *PISA 2000: Basiskompetenzen von Schülerinnen und Schülern im internationalen Vergleich*. Opladen: Leske+Budrich.
Black, P., & Williams, D. (1998). Assessment and classroom learning. *Assessment in Education: Principles, Policy and Practice, 5*(1), 7–68.
Blum, W., et al. (2002). ICMI study 14: Applications and modelling in mathematics education – Discussion document. *Educational Studies in Mathematics, 51*, 149–171.
Borromeo Ferri, R., & Blum, W. (2009). Mathematical modelling in teacher education – Experiences from a modelling seminar. In *Proceedings of CERME 6*. Available from http://www.inrp.fr/editions/editions-electroniques/cerme6/.
Brickhouse, N. W. (1990). Teachers' beliefs about the nature of science and their relationship to classroom practice. *Journal of Teacher Education, 41*, 53.
Burkhardt, H. (1989). Mathematical modelling in the curriculum. In W. Blum, J. S. Berry, R. Biehler, I. Huntley, G. Kaiser-Meßmer, & L. Profke (Eds.), *Applications and modelling in learning and teaching mathematics* (pp. 1–11). Chichester: Horwood Publishing.
Burkhardt, H. (2006). From design research to large-scale impact: Engineering research in education. In J. Van den Akker, K. Gravemeijer, S. McKenney, & N. Nieveen (Eds.), *Educational design research* (pp. 121–150). London: Routledge.
Cohen, J. (1988). *Statistical power analysis for the behavioral sciences*. Hillsdale: Lawrence Erlbaum Associates.
Galbraith, P., & Clatworthy, N. J. (1990). Beyond standard models – Meeting the challenge of modelling. *Educational Studies in Mathematics, 21*(2), 137–163.
Galbraith, P., & Stillman, G. (2001). Assumptions and context: Pursuing their role in modelling activity. In J. F. Matos, W. Blum, K. Houston, & S. P. Carreira (Eds.), *Modelling and mathematics education, ICTMA 9: Applications in science and technology* (pp. 300–310). Chichester: Horwood Publishing.
Garcia, F. J., Maaß, K., & Wake, G. (2010). Theory meets practice – Working pragmatically within different cultures and traditions. In R. Lesh, P. Galbraith, C. Haines, & A. Hurford (Eds.), *Modelling students' mathematics modeling competencies – ICTMA 13*. New York: Springer.
Grigutsch, S., Raatz, U., & Törner, G. (1998). Einstellungen gegenüber Mathematik bei Mathematiklehrern. *Journal für Mathematik-Didaktik, 19*, 3–45.
Haines, C., & Izard, J. (1995). Assessment in context for mathematical modelling. In C. Sloyer, W. Blum, & I. Huntley (Eds.), *Advances and perspectives in the teaching of mathematical modelling and applications* (pp. 131–150). Yorklyn: Waterstreet Mathematics.
Ikeda, T., & Stephens, M. (2001). The effects of students' discussion in mathematical modelling. In J. F. Matos, W. Blum, K. Houston, & S. P. Carreira (Eds.), *Modelling and mathematics*

education: ICTMA 9: Applications in science and technology (pp. 381–390). Chichester: Horwood Publishing.

Kaiser, G. (2006). The mathematical beliefs of teachers about applications and modelling – results of an empirical study. In: J. Novotná et al. (Eds.): Mathematics in the centre. *Proceedings of the 30th Conference of the International Group for the Psychology of Mathematics Education.* (Vol. 3, pp. 393–400). Prague: Charles University.

Kaiser, G., & Sriraman, B. (2006). A global survey of international perspectives on modelling in mathematics education. *Zentralblatt für Didaktik der Mathematik, 38*(3), 302–310.

Kaiser-Meßmer, G. (1986). *Anwendungen im Mathematikunterricht.* Bad Salzdetfurth: Franzbecker.

Krauss, S., Kunter, M., Brunner, M., Baumert, J., Blum, W., Neubrand, M., et al. (2004). COACTIV: Professionswissen von Lehrkräften, kognitiv aktivierender Mathematikunterricht und die Entwicklung von mathematischer Kompetenz. In J. Doll & M. Prenzel (Eds.), *Bildungsqualität von Schule: Lehrerprofessionalisierung, Unterrichtsentwicklung und Schülerförderung als Strategien der Qualitätsentwicklung* (pp. 31–53). Münster: Waxmann.

Maaß, K. (2006). What do we mean by modelling competencies? *Zentralblatt für Didaktik der Mathematik, 38*(2), 113–142.

Maaß, K. (2009). What are German teachers' beliefs about effective mathematics teaching? In J. Cai, G. Kaiser, B. Perry, & N. Y. Wong (Eds.), *Effective mathematics teaching from teachers' perspectives: National and cross-national studies.* New York: Sense Publisher.

Maaß, K., & Gurlitt, J. (2009). Designing a teacher questionnaire to evaluate professional development in modelling. In *Proceedings of CERME 6.* Available from http://www.inrp.fr/editions/editions-electroniques/cerme6/

Pehkonen, E., & Törner, G. (1996). Mathematical beliefs and different aspects of their meaning. *Zentralblatt für Didaktik der Mathematik, 28*(4), 101–108.

Prenzel, M., Baumert, J., Blum, W., Lehmann, R., Leutner, D., Neubrand, M., et al. (2004). *PISA 2003 – Der Bildungstand der Jugendlichen in Deutschland – Ergebnisse des zweiten internationalen Vergleichs.* Münster: Waxmann.

Shulman, L. S. (1986). Paradigms and research programs in the study of teaching: A contemporary perspective. In M. C. Wittrock (Ed.), *Handbook of research on teaching* (pp. 3–36). New York: Macmillan.

Tirosh, D., & Graeber, A. O. (2003). Challenging and changing mathematics teaching practices. In A. Bishop, M. A. Clements, C. Keitel, J. Kilpatrick, & F. Leung (Eds.), *Second international handbook of mathematics education* (pp. 643–688). Dordrecht/Boston/London: Kluwer Academic Publishers.

Weinert, F. E. (1998). Guter Unterricht ist ein Unterricht, bei dem mehr gelernt als gelehrt wird. In J. Freund, H. Gruber, & W. Weidinger (Eds.), *Guter Unterricht – was ist das? Aspekte von Unterrichtsqualität* (pp. 7–18). Wien: ÖBV.

Wilson, M., & Cooney, T. J. (2002). Mathematics teacher change and development. The role of beliefs. In G. C. Leder, E. Pehkonen, & G. Törner (Eds.), *Beliefs: A hidden variable in mathematics education?* (pp. 127–148). London: Kluwer Academic Publishers.

Chapter 61
Modelling in the Classroom: Obstacles from the Teacher's Perspective

Barbara Schmidt

Abstract Modelling is not only written into educational standards throughout Germany; other European countries also demand the integration of reality-based, problem-solving tasks into school mathematics. In reality, however, things look quite different: in many places mathematics lessons are still dominated by exercises in simple calculation. So why? What is stopping teachers from introducing modelling? In order to explore this issue in depth, an empirical study was conducted. A 55-item questionnaire to investigate teachers' perspectives on mathematical modelling in classrooms was developed and refined. A sample of 101 teachers, 52 of whom had undergone a training programme in modelling, completed the questionnaire. Major factors perceived by teachers as hindrances to using modelling were lack of time, assessment of performance and lack of materials.

1 Basic Theory

1.1 Mathematical Modelling

Mathematical modelling generally refers to using mathematics to solve realistic and open problems. At the same time, the exact definition varies depending on the aims, which model of the modelling process is being used, and the nature of the context assigned to a modelling task (Kaiser-Meßmer 1986; Kaiser and Sriraman 2006).

1.2 Obstacles to the Integration of Modelling

In day-to-day school life, modelling still plays a much smaller role than one would wish (Burkhardt 2006; Maaß 2004). It appears that at the moment teachers

B. Schmidt (✉)
Tobelstraße 10, 72379 Hechingen, Germany
e-mail: Schmidt.Barbara@gmx.de

see more obstacles to using modelling than advantages. Blum (1996) divided these obstacles into four categories: organisational, pupil-related, teacher-related and material-related.

Organisational obstacles refer mainly to the short amount of time – 45-min – for lessons.

Pupil-related obstacles: Modelling makes the lesson too difficult and less predictable for pupils (e.g., see Blum and Niss 1991). Pupils can have difficulties carrying out individual steps or even the whole modelling process (Maaß 2004). Standard calculating tasks are more popular with some pupils because they are easier to understand and to solve the problem one simply has to apply a particular formula making it easier for pupils to obtain good grades in mathematics.

Teacher-related obstacles: There appear to be a variety of obstacles for teachers. The literature on this issue refers repeatedly to the time aspect. Teachers need more time to update tasks, to adapt them to the needs of the respective class, and to prepare them in detail (Blum and Niss 1991). In addition, there are obstacles in relation to the actual lessons: teaching becomes more demanding and more difficult to predict. Furthermore, a teacher requires other skills and competencies in order to be able to deal with a changed approach to teaching. The most recent literature also refers to teachers' beliefs about – or attitudes to – mathematics teaching as being an obstacle to innovation in the classroom (Pehkonen 1999; Törner 2002). These studies point to the fact that, in general, teachers do not view modelling as mathematics. Moreover, some teachers do not consider themselves competent enough to carry out modelling tasks when the context is taken from a subject area they did not study. In addition, a significant aspect of the perceived obstacles is the question of how to assess performance, as teachers feel overwhelmed by the increasing complexity of this process.

Material-related obstacles: Often teachers simply do not know of enough modelling examples which they feel would be suitable for their lessons, or they select excessively detailed materials.

1.3 Research Questions

The previous section identified some of the obstacles to the introduction of modelling in mathematics classrooms. However, these have not been subjected to empirical analysis. This suggests the need for some kind of instrument with which to measure or assess empirically the arguments against modelling. In order to ensure the resulting point of view is not one-sided, this instrument should also analyse the arguments for modelling. This has the additional advantage that not only are the deficiencies revealed, but also solutions may be revealed. Therefore, the central questions to be explored are:

1. What are the obstacles which teachers perceive with regards to modelling?
2. Which obstacles appear meaningful when putting modelling into practice?
3. Which changes in the obstacles can be identified during a teacher training course?

2 Methodology

2.1 Instruments for the Study

To find out which aspects teachers view as obstacles to modelling, quantitative and qualitative methods were used. Among other instruments, a questionnaire was designed with the aim of ascertaining the obstacles. In addition, guided interviews were conducted.

2.2 Study Design

The questionnaire was implemented at three points in time: at the beginning of the study before any training was undertaken; immediately following the completion of five training modules (12 months later); and 5 months after completing the training (see Fig. 61.1). The interviews were conducted at the same points of time. However, one additional elicitation of the interviews was of interest in order to register possible changes during the teacher training course (see case study in Schmidt 2010).

2.3 Sample

The study took place in Germany in the Bundesland Baden-Württemberg (south west Germany) with teachers from primary and secondary school. From secondary school there were teachers from the low achievement school (Hauptschule) and the

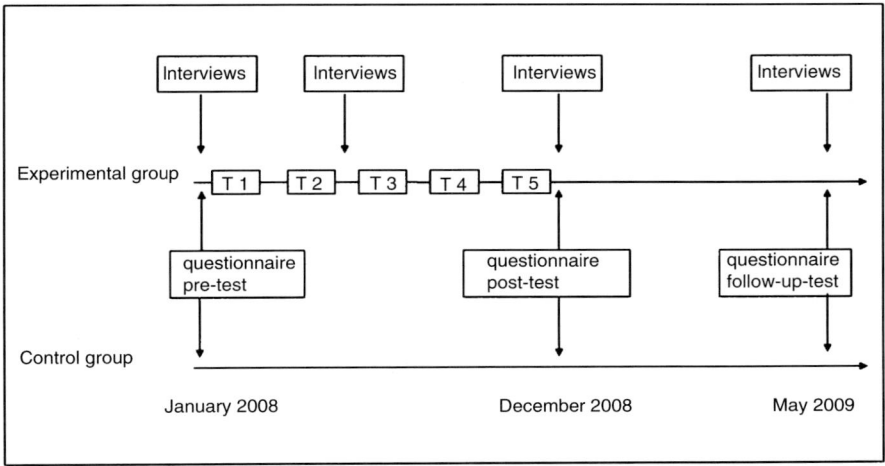

Fig. 61.1 Study schedule

middle school (Realschule). One hundred and five teachers[1] registered interest in the training course. From these, participants were selected at random for the experimental group ($n=52$) who would take part in the training with the remainder being the control group ($n=49$), who would not participate in the course.

The interviews were not conducted with all the teachers of the experimental group, but with a selection of six. The selection was based on the results of the pre-tests with three teachers being selected who saw many obstacles to introducing modelling and three who instead saw many affordances.

3 Questionnaire Development

To lay the foundations for the study and to answer the first research question, a questionnaire was developed whose purpose was to throw light on the constraints and affordances for the teacher regarding modelling in mathematics lessons. To be able to guarantee this, a four-stage design was developed.

3.1 Questionnaire Development

The first items were developed from the subjective theories of researchers (i.e., using deductive item construction). For this, the obstacles described above were restated as items. To guarantee the authenticity of the items, the "natural" polarity of the obstacles was retained. The result was a preliminary questionnaire that included a total of 65 items. The response format corresponded to a five-level Likert scale (Rost 1996), ranging from "strongly disagree" to "strongly agree". As the items were not expected to provide complete insight into the issues, additional open questions were integrated allowing the teachers to identify any further obstacles to the use of modelling. With the help of these open items, together with the evaluation and optimisation of the closed items, the aim was to create a second and third test version of the questionnaire. This was necessary in order to be able to change the phrasing of items with ceiling effects, thereby minimising the effect. At the same time, it was important to check the changed items once again in another test version in order to ensure that all ceiling effects were eliminated. The initial open items also generated new items that were checked in subsequent test versions for ceiling effects.

The questionnaire was piloted on three occasions on 240 mathematics teachers from south-west Germany.

The fourth step was to organise the 55 items into scales. To do this, the scales should be formed from the items (inductive categorisation). The first indications for

---

[1] Drop out rate: four teachers

scales were provided by Blum's classification (1996) as illustrated above. In addition, the items, however, were repeatedly analysed together as a whole, so as to check for more possible scale indicators with the help of the qualitative content analysis (Mayring 2007). Finally, the 55 items generated 14 scales each of which have internal consistency of content.

3.2 Format of Questionnaire

In Fig. 61.2 there is an example of the items for the scale: "time". What was of interest was whether teachers were being discouraged from modelling because modelling tasks take too much time. But first of all it was important to find out *if* teachers do actually think that modelling tasks take too much time. Therefore, there was a five-point scale on which teachers could mark the extent to which they agreed with an item (see Fig. 61.2). However, agreement does not actually indicate if the teacher then feels deterred from or motivated into carrying out modelling tasks. So there was a second column with a scale on which teachers could express this more exactly. This approach corresponds to Vroom's expectancy-times-valence model (1964). This states that expectancy multiplied by valence produces motivation.

As you can see in the questionnaire not only obstacles/constraints were referred to but also motivations/affordances for modelling were referred to (see second scale). That was important to not only know what hinders teachers from modelling but also to know factors that make them more likely to engage in modelling. However in this chapter I only focus on the obstacles.

	strongly disagree				strongly agree	obstructive	often obstructs	irrelevant	often motivates	motivates me
	−2	−1	0	1	2	−2	−1	0	1	2
Working through the tasks in the classroom is very time consuming.	○	○	○	○	○	□	□	□	□	□
It takes a lot of time to find the approach to the solution.	○	○	○	○	○	□	□	□	□	□
A lot of time is needed to pay attention to, and consider the pupils' various solution possibilities.	○	○	○	○	○	□	□	□	□	□

Fig. 61.2 Scale: time

4 Results

In the following the main aspects which teachers named as obstacles in the questionnaire are explored. Two analyses were made: (i) the first column of the questionnaire was evaluated to find out to what extent teachers agree with particular obstacles to modelling (ii) the extent to which this particular obstacle constrains their teaching was evaluated. The scale values of both columns were multiplied according to Vrooms model mentioned above.[2]

In addition, the established results from the questionnaire were triangulated with statements from the interviews.

Seven of the 14 scales were supposed to become a hindering factor for the teachers.[3] However, most of the teachers saw mainly obstacles in three aspects: the time, the performance assessment of modelling tasks and the material. They are described more in detail in the following.

4.1 I Have Too Little Material

Sixty one percent of the questioned teachers complained about too little material and considered this as an obstacle ($M=-1.67$[4]; $SD=3.34$). With the control group this remained constant over time; with the experimental group, on the other hand, after the training course this no longer represented an obstacle (see Fig. 61.3 and Table 61.1). The teachers received lots of materials in the training course and also learned how to develop tasks. This may be a reason why they no longer claimed to have too little material to support their teaching after the training course. This difference is of statistical significance and had a medium effect. $F(1,97)=6.79, p<0.011$, $\eta^2=0.07$ (medium effect). This effect remained stable in the follow-up period.

The interviews gave a profound insight concerning the material. One teacher, for example, was of the opinion that modelling tasks had to fit to the particular mathematical topic she was teaching at that moment. She wanted to have five tasks for every topic that she could use. As she did not have such a collection, she complained about having too little material. Another teacher had little problem finding modelling tasks for pupils at grades 5–7 (age 10–13) but did for pupils at grades 8–10 (age 13–16). This was because she could not find a realistic context for a modelling task which fitted with the mathematical topics that were taught in these grades.

Another teacher found the content of the modelling tasks too far-fetched. In her opinion the tasks were not interesting for the pupils and the questions were too unrealistic and not relevant. As a result she complained about having too little *good*

---

[2] The numbers of the first column have been transformed into 0–4 for the further calculations.
[3] The other 7 scales were supposed to become a motivation factor for the teachers.
[4] A negative mean stands for a hindering factor; a positive mean stands for a motivation factor regarding modelling.

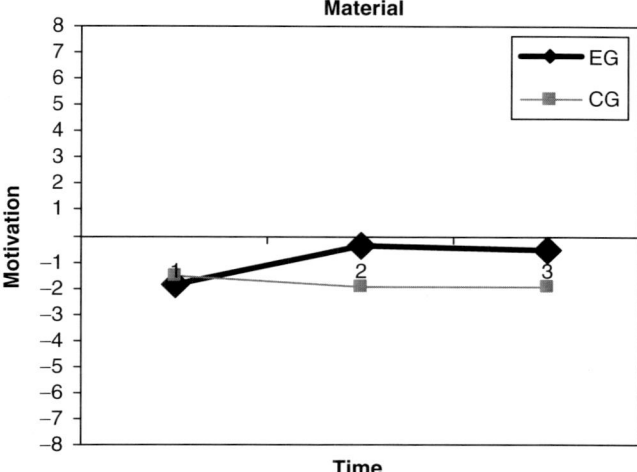

Fig. 61.3 Material

Table 61.1 Material

		M	SD	n
Pre-test	EG	−1.83	3.13	51
	CG	−1.51	3.57	48
	Total	−1.67	3.34	99
Post-test	EG	−0.39	1.33	51
	CG	−1.86	3.91	48
	Total	−1.10	2.96	99
Follow-up-test	EG	−0.58	2.14	50
	CG	−1.86	2.92	48
	Total	−1.20	2.62	98

material. When asked for an example of a good context for a task which pupils might be interested in, she had no idea at all.

4.2 Performance Assessment is Too Complex

Another aspect was the assessment of pupils' performance which 84% of the teachers thought to be complex. The teachers named this aspect as an obstacle ($M=-2.03$; $SD=2.70$). There was no significant change during training ($F(1,98)=0.46, p=0.5$) (see Table 61.2). It appears that this aspect has to be considered further in future training courses.

Table 61.2 Performance assessment

		M	SD	n
Pre-test	EG	−1.17	2.65	51
	CG	−2.93	2.49	49
	Total	−2.03	2.70	100
Post-test	EG	−1.41	2.99	51
	CG	−2.74	3.26	49
	Total	−2.07	3.18	100
Follow-up-test	EG	−1.51	2.89	51
	CG	−2.65	3.43	49
	Total	−2.07	3.20	100

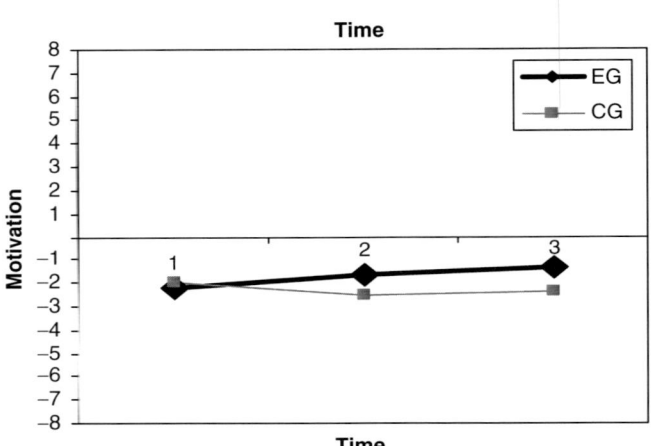

Fig. 61.4 Time

The interviewed teachers reflected this thinking with some of them seeing no reason in assessing tasks as long as the final examination does not include modelling tasks. The fact that the examination does not yet include modelling tasks was also a reason for some teachers not to do modelling tasks at all.

4.3 I Don't Have Enough Time for Modelling

The most important reason for not using modelling was the time aspect. Ninety seven percent of the teachers agreed that modelling tasks take up too much time in class. The teachers named this as an obstacle ($M=-2.11$; $SD=2.75$). The experimental group changed their opinion marginally (see Fig. 61.4 and Table 61.3) within the period of the training course, but this change was not significant ($F(1,99)=3.1$, $p=0.08$).

Table 61.3 Time

		M	SD	n
Pre-test	EG	−2.19	2.78	52
	CG	−2.02	2.73	49
	Total	−2.11	2.75	101
Post-test	EG	−1.65	2.92	52
	CG	−2.55	3.04	49
	Total	−2.08	3.00	101
Follow-up-test	EG	−1.33	3.13	52
	CG	−2.36	3.37	49
	Total	−1.83	3.28	101

Teachers spoke a lot about the time in the interviews drawing attention to this acting as an obstacle to using modelling. Beside the time taken for pupils to work on modelling tasks, teachers also referred to other areas where time is an issue. For example, one teacher expressed having too little time to prepare modelling tasks in advance. Another was of the opinion that the presentation of the results absorbs too much time. The same teacher was of the opinion that pupils do not understand modelling tasks as quickly as the usual standard tasks which has a time problem as consequence. Similar to this, some teachers expressed that the curriculum contains already too many topics so that it seems impossible for them to fulfil all of their obligations in covering the curriculum. The time pressure becomes even greater again for these teachers when using modelling tasks. In line with this argument what another teacher said was modelling is "too little mathematics in too much time". Another argument of a teacher was the insufficient availability of computers for use by pupils and internet connection which wastes too much time as well.

These statements from the interviews point to how complex the issue of time is. It is not surprising therefore that this aspect demonstrated only a marginal improvement during training.

5 Discussion

Three obstacles were identified which teachers see as the main factors hindering their use of modelling. From these, the aspect of the lack of material can be relatively easily addressed. The study showed that most teachers who received material in the teacher training course changed their opinion and saw this no longer as an obstacle. However, the other two aspects, the time problem and the assessment of performance, seem to be very resistant to change. Even a 1-year teacher training course was not able to change beliefs in relation to these significantly.

These three aspects attracted attention in the categories mentioned by Blum (1996) (see above) and could be confirmed through this empirical study. Additionally, the study gave more background regarding each aspect.

It seems that the lack of material might not be a general problem for all teachers but might be a special problem for teachers who teach older pupils (from class 8 (age 13) and above). Teachers had no problem in finding modelling tasks for lower classes, but had no idea for real contexts for the curriculum of the upper classes. This suggests that many mathematics teachers are teaching topics every day without knowing of applications in the real world.

Teachers found it consistently difficult to assess modelling tasks. Even the teacher training course was not able to change this opinion. Consequently further thought must be applied to considering how these issues might be addressed in the future.

Not only the problem of the time used by modelling in class was mentioned but also the time required to prepare modelling tasks was an important issue. Another new aspect concerning the lack of time was identified: that associated with the availability of equipment and internet connection associated with using computer technology. Other teachers complained about the enormous demands of the curriculum. As teachers found it difficult to meet these needs in a short amount of time they wanted to concentrate on fulfilling the curriculum and not "wasting time" in doing modelling. This statement is in conflict with reality, because modelling has been part of the curriculum in Baden-Württemberg since 2004. This means that if these teachers want to fulfil the curriculum they have to include modelling.

In conclusion, this study was able to empirically confirm in which areas there are problems that hinder teachers from using modelling in their teaching and point to which of these obstacles might be reduced through teacher engagement with an appropriate training course.

References

Blum, W. (1996). Anwendungsbezüge im Mathematikunterricht – Trends und Perspektiven. In G. Kadunz et al. (Eds.), *Trends und Perspektiven, Schriftenreihe Didaktik der Mathematik, Band 23* (pp. 15–38). Wien: Hölder-Pichler-Tempsky.

Blum, W., & Niss, M. (1991). Applied mathematical problem solving, modelling, applications, and links to other subjects – State, trends and issues in mathematics instruction. *Educational Studies in Mathematics, 22*(1), 37–68.

Burkhardt, H. (2006). Modelling in mathematics classrooms: Reflections on past developments and the future. *Zentralblatt für Didaktik der Mathematik, 38*(2), 178–195.

Kaiser, G., & Sriraman, B. (2006). A global survey of international perspectives on modelling in mathematics education. *Zentralblatt für Didaktik der Mathematik, 38*(3), 302–310.

Kaiser-Meßmer, G. (1986). *Anwendungen im Mathematikunterricht* (Vol. 2). Bad Salzdetfurth: Franzbecker.

Maaß, K. (2004). *Mathematisches Modellieren im Unterricht: Ergebnisse einer empirischen Studie*. Hildesheim: Franzbecker.

Mayring, P. (2007). *Qualitative Inhaltsanalyse. Grundlagen und Techniken* (9th ed.). Weinheim/ Basel: Beltz.

Pehkonen, E. (1999). Beliefs as obstacles for implementing an educational change in problem solving. In E. Pehkonen, & G. Törner (Eds.), *Mathematical beliefs and their impact on teaching and learning of mathematics. Proceedings of the Workshop in Oberwolfach* (pp. 109–117). Gerhard-Mercator-University, Duisburg.

Rost, J. (1996). *Lehrbuch Testtheorie, Testkonstruktion*. Bern: Huber.

Schmidt, B. (2010). *Modelling in mathematics classrooms – Motives and obstacles from the teachers' perspective*. Hildesheim: Franzbecker.

Törner, G. (2002). Mathematical beliefs – A search for a common ground: Some theoretical considerations on structuring beliefs, some research questions, and some phenomenological observations. In G. Leder, E. Pehkonen, & G. Törner (Eds.), *Mathematical beliefs: A hidden variable in mathematics education?* (pp. 73–94). Dordrecht: Kluwer.

Vroom, V. H. (1964). *Work and motivation*. New York: Wiley.

Chapter 62
Teachers' Professional Learning: Modelling at the Boundaries

Geoff D. Wake

Abstract This paper explores teachers' professional learning in mathematical modelling using a range of theoretical tools. The study on which it is based gives a snapshot of the work of a development group of teachers near the outset of their journey into modelling. Narrative accounts of their development in terms of both their teaching and students' learning are analysed using an instrument developed for this purpose. The results provide insight into important issues to consider when supporting professional learning in general and modelling in particular. This small scale study points to the importance for teachers of renegotiating the didactical contract of their classrooms when introducing modelling and consequently the need for professional learning that expands their repertoires in relation to both subject knowledge and particularly pedagogy more generally.

1 Introduction

This paper explores the work of a teacher development group consisting of five teachers and two local authority mathematics specialists from two different local authorities of the conurbation of Greater Manchester, England. Their aim was to build on the work of the European project LEMA (Learning and Education in and through Modelling and Applications[1]) that had stimulated the desire of two of the group, one a teacher and the other one of the local authority specialists, to ensure that more pupils had experience of modelling in their mathematics lessons. The group was funded by a research grant from the English National Centre for Excellence in

---

[1] http://www.lema-project.org/web.lemaproject/web/eu/tout.php

G.D. Wake (✉)
Centre for Research in Mathematics Education, School of Education, University of Nottingham, Nottingham NG8 1BB, UK
e-mail: geoffrey.wake@nottingham.ac.uk

Teaching Mathematics (NCETM). Throughout the work was supported by me as mathematics education researcher and partner in the LEMA project.

Here I seek to identify the issues of concern to teachers as they attempt to change their pedagogic practices to include mathematical modelling and to draw conclusions for professional learning both in general, and in relation to modelling in particular. In doing so I draw on case study data produced by the five teachers themselves as they developed an 'e-narrative' of their work in their classrooms at the initial stages of introducing modelling activities. The majority of the case study data on which the analysis is based is publicly available at a Web site[2] where text, images and video clips document the teachers' and local authority specialists' journeys into mathematical modelling. This site served multiple purposes including the archiving of case study data as well as the provision of a resource for teachers who wish to know how mathematical modelling can become part of teaching and learning in the context of urban schools. However, it was found that the development of the Web site in itself, as a product, also gave a clear focus and purpose to teacher engagement with the project and consequently their professional development.

2 Classroom Practice and Its Transformation

In England in mathematics classrooms mathematical modelling is rarely seen as teachers appear constrained to using pedagogies that are low risk and that in the main conform to a normative cultural 'script' (Wierzbicka 1999) that sees periods of transmissionist exposition by teachers (Askew et al. 1997) interspersed with periods in which students practise rules and procedures that are in the main situated in contexts that favour mathematics itself above any other form of reality (Ofsted 2008). It should not be assumed that such lessons are necessarily in line with what teachers might favour or believe should be the case, but it does seem that such practice is widespread and forms the dominant style of didactical contract (Brousseau 1997).

In this paper, therefore, I consider professional development of the individual teacher as being culturally and historically situated and in order to take account of this draw on theoretic analytical tools provided by Cultural Historical Activity Theory (CHAT) (for example see Engestrom and Cole 1997). Such an approach conceptualises the work of a collective such as teachers and pupils in mathematics classrooms as an activity system focused on an object such as the learning of mathematics with their activity mediated by different factors including explicit and implicit rules and how the division of labour is organised by the community.

Engestrom (2001) in third generation activity theory re-conceptualises how expansive learning may occur in workplace situations. In doing so he points to how historically accumulating tensions within and between activity systems, perhaps

---

[2] http://www.education.manchester.ac.uk/research/centres/lta/LTAResearch/lema/

brought about by a change in the rules that 'govern' the system or the introduction of a new technology, can lead to contradictions which may give rise to conflicts that may be personal to the members of the community, and ultimately innovative attempts to change the nature of the collective's activity. In the study here we have multiple activity systems (namely, the school classroom, the development group, my university research group and the local authority support teams) that have the same object, student learning of mathematics, that we consider as a 'boundary object' (see for example, Engestrom 2001). In each of these communities the individual key players (subjects in CHAT terms) have shorter-term goals and actions. For example, (1) I, as researcher have goals in relation to fulfilling obligations relating to the NCETM funding such as writing a report of the work (Wake 2009a), whereas these are not the goals of the individual teachers; (2) one of the teachers (E) operating in both the development group and her school, and therefore in CHAT terms a boundary crosser, has needs that include supporting her goal of informing curriculum development and specification in her school (see Wake 2009b). Engestrom (2001) suggests that for expansive learning to take place individuals need to question and deviate from established norms and that a collaborative and deliberate effort towards change that can support a collective journey through the Vygotskyian Zone of Proximal Development (ZPD) can result in a full cycle of expansive transformation. In such cases this may lead to changes in actions of individuals as they seek to ensure continued successful outcomes but in new and enhanced activity.

In the case explored here we see the teachers seeking to take part in such an expansive transformation as they attempt to introduce modelling although the rules of the system (including the implicit didactical contract) are such that, on the whole, they discourage this. We do, however, detect potentially supportive structural contradictions in the school activity system where performance is measured not only by pupil attainment but also by government inspectorial judgment of teaching. This is highly critical of the type of low-risk transmissionist teaching which we have seen dominating in response to national assessment (Ofsted 2008) and favours pupils engaging in a range of mathematical activity that includes modelling. A problem for the mathematics teacher, therefore, is how to meet the demands of ensuring that pupils continue to obtain high grades in procedurally orientated assessment whilst also ensuring that their teaching also promotes the learning of process skills and modelling. It is with this backdrop that this paper explores the concerns of the development group's teachers as they attempt to bring about change by introducing modelling in their classrooms.

3 Knowledge for Teaching and Learning in Modelling Classrooms

In an attempt to identify teachers' professional learning in relation to their initial steps in modelling I draw on Ball et al.'s (2008) recent attempts to build on Shulman's construct of pedagogic content knowledge (Shulman 1986) to gain empirical insight into knowledge for teaching. In doing so their analysis of many mathematics lessons leads them to propose that 'content knowledge for teaching' may be categorised using two overarching categories, subject knowledge and pedagogic content knowledge, each of which may be subdivided into three further sub-categories that I summarise in Table 62.1.

These categories were used in a systematic analysis of the teachers' accounts that also categorised these in a second dimension in relation to their pupils' learning. This was informed by Cardella (2008) who, building on the work of Schoenfeld (1992), identifies five key aspects relating to mathematical thinking of learners as summarised in Table 62.2.

This gives a two dimensional framework (with six sub-domains for knowledge for teaching and five in relation to student learning), Table 62.3.

Before analysing the experiences of the teachers as they went about introducing modelling into their teaching I briefly consider their potential reflections in relation

Table 62.1 Ball et al.'s (2008) categorisation of knowledge for teaching

Subject knowledge	
Specialised content knowledge	Mathematical knowledge and skills unique to teaching – "unpacking" mathematics – understanding "why?" – explaining so that important ideas are made visible to learners
Common content knowledge	Mathematical knowledge and skills that others have as well – not special to the work of teaching
Horizon content knowledge	Understanding how mathematics is connected both internally and externally (to other subjects/disciplines, its use in the world of work etc.)
	Understanding how mathematical ideas develop and build on earlier concepts and form the foundations for later concepts/topics
Pedagogic content knowledge	
Content of knowledge and students	Understanding the way that different students and groups of students talk about mathematics
	Anticipating what students
	• Are likely to think and what they will find confusing
	• Will find interesting/motivating
	• Will find hard/easy
Content of knowledge and teaching	Understanding of the design of teaching tasks/sequences of instruction
	In the classroom knowing when and how to interact with students' responses. Evaluating which materials to use and in what sequence
Content of knowledge and curriculum	Understanding the place of particular mathematics in the prescribed curriculum – how this relates to expectations and assessment
	Understanding of both content and process skills and how these are developed in the curriculum – for example, in relation to modelling and applications

Table 62.2 Cardella's (2008) categorisation of key aspects of learning mathematics

Knowledge base	Specialised knowledge of school mathematics content
Problem solving strategies	Global or local strategies learned from mathematics courses, e.g. (local) considering special cases, (global) refining a model by revisiting assumptions
Mathematical practices	Activities or actions that people engage in and with mathematics, i.e. what it is to use mathematics (e.g. as a mathematician, as a scientist …)
Use of resources	Including social resources, time and metacognitive processes such as planning and monitoring work
Beliefs and effects	Beliefs and emotions about mathematics and one's mathematical ability

Table 62.3 Two dimensional framework used to analyse teachers' concerns in early teaching of modelling

		Student related				
		Knowledge base	Problem solving strategies	Mathematical practices	Use of resources	Beliefs and effects
Teacher related	Specialised CK					
	Common CK					
	Horizon CK					
	PCK content and students					
	PCK content and teaching					
	PCK content and curriculum					

to this framework in light of the literature regarding the knowledge of teachers about mathematical modelling. Much of this has in effect focused on the difference in mathematical activity that results for students due to the introduction of mathematical modelling tasks and teacher knowledge appropriate to this. Central to the concerns of researchers has been teacher knowledge of modelling competencies and meta-cognition of the modelling cycle as a pre-requisite for working effectively with students (see, e.g. Kaiser et al. (2010)). Clear understanding of such matters is considered as a pre-requisite for competent teaching using modelling approaches and suggests that in the framework proposed the *horizon content knowledge* domain will be of particularly increased importance as teachers consider how mathematics might be used effectively to mathematise and describe 'reality' (or the non-mathematical world). It appears that the *content knowledge and curriculum* domain would be important in this regard as teachers seek to understand how the process skills and competencies associated with modelling might best be developed by students. From the point of view of the student, the changing nature of the mathematical activity in which they engage might be expected to result in concerns about the *problem solving strategies* and *mathematical practices* they employ and may therefore result in an increased focus by the teachers in the appropriate domains of the framework.

Research in relation to teacher knowledge for mathematical modelling recognises the challenge that the introduction of new pedagogic practices poses for teachers and the development of professional knowledge that this entails, but also recognises an underlying requirement that teachers often need to make a change in their attitudes and beliefs in relation to their classroom practice (see for example, Kaiser and Maaß (2006)). With respect to this the didactic contract which results in the enactment of long-held expectations about the nature of mathematics teaching and learning in our classrooms by teachers is likely to prove difficult to modify by both teachers and students alike. However, it is also the case that much research suggests that the attitudes and beliefs that are brought to bear by teachers in lesson implementation are deep seated (Pehkonen and Torner 1996) and difficult to modify.

The self reports of the teachers as they worked with their classes on a series of initial lessons in mathematical modelling were written for sharing within the group and more widely as web-based accounts for the project Web site. The framework (Table 62.3) was used to analyse for each teacher to situate each of their statements in relation to their early experiences of using modelling. For example, statements such as 'a resources table with calculators and metre rules was set up for pupils to use as they wanted' and 'each group was given mini white boards to use for planning as well as paper and markers for the presentations of their solutions' were located as being concerned with students' use of resources and teachers' PCK: content and teaching. Table 62.4 shows the results of this analysis for each of the five teachers' case studies of their classrooms.

Whilst each of the sub-domains of teacher knowledge and student learning will have a role to play in modelling classrooms it seems clear as I suggest above that we might expect concerns about pedagogic content knowledge relating to curriculum (teacher related) and problem solving strategies (student related) to have some priority. However, Table 62.4 shows that the main areas of teachers' concerns are related to pedagogic content knowledge in relation to teaching and students and the intersection of this with students' practices and use of resources. In general teachers are concerned about:

1. Learning to employ group work effectively (e.g. 'Before I attempted a modelling task I used some other closed group work tasks to get classes used to working in groups and to producing a presentation of their solutions'. Teacher C)
2. Use and adaptation of tasks (e.g. 'I have started with some modelling tasks that have sporting connections … The sporting link seems to have motivated the pupils well…' Teacher B)

Table 62.4 Analysis of teacher statements in relation to teacher knowledge and aspects of student learning (*Note*: Teachers have been anonymised as A, B, C, D and E)

		Student related				
		Knowledge base	Problem solving strategies	Mathematical practices	Use of resources	Beliefs and effects
Teacher related	Specialised CK	B				
	Common CK	A B C	A B	B		
	Horizon CK	B				
	PCK content and students		C C E E E E	A B C E E E E E E E E	A B B E E E E	A B B B B D D D D C C C C C C C C C E E
	PCK content and teaching		A A C E	A A B B C C C D D D D D D D D D D	A A A B B B B B B C C C C C C D D D D D D D D E E E E	
	PCK content and curriculum		B B B	A A B B D	D	

3. Perceptions of pupils' increased motivation (e.g. 'Feedback from the students about the tasks has been extremely positive'. Teacher E)

In telling of their initial steps in using mathematical modelling with their classes, therefore, the teachers point to their concerns in relation to managing changes in their and students' practices and in particular, using resources:

> For the modelling task I allowed pupils to sit at one of four tables in groups of up to five. Two groups worked really hard, one was divided where two people did all of the work and one group did virtually no work. The disengaged pupils were the same each lesson and so in future I need to not only look at group structuring but at social skills for learning. (Teacher C).

> My preparation before the lesson was different to that for a 'traditional lesson' as my main concerns became how to set up the classroom and how to organise pupils into groups. (Teacher E).

The general lack of concern over issues of student knowledge in relation to mathematics and modelling is perhaps just as telling. It seems that overwhelmingly the concerns of teachers focus on changing roles and relationships both socially and with regard to mathematical activity as teachers consider how they might change the didactical contract in their classrooms. It is important not to interpret this finding as meaning that issues of content knowledge are not important, but rather that in the initial stages of introducing modelling approaches in mathematics lessons teachers are more concerned about general pedagogic approaches that fall outside of their usual pedagogic repertoires.

4 Further Theoretical Reflections

As I have noted earlier Engestrom's analysis, using third generation Activity Theory, would suggest that in this case we appear to observe the start of expansive learning of the Activity System of the teacher related to their classroom practice. Importantly the learning that we detect might be considered as 'horizontal' as the teachers seek to bring about the changes they pragmatically need to instigate modelling lessons in their classrooms. This might be considered 'horizontal' transformation following the analysis of Engestrom (2001) who after drawing attention to the classical understanding of concept formation being at the intersection of everyday and scientific understanding (Vygotsky 1987) and the vertical direction of this, suggests that in his analysis of the professional learning of health workers he detected 'sideways' moves in their re-conceptualisations of their practice. In the case I describe here, I suggest we also see a horizontal shift in the teachers' expanded view of their classrooms as they focus on re-negotiating the didactical contract and associated pedagogies with their pupils.

For the individual teachers this development may, therefore, be conceptualised as a consequential transition, in the sense of Beach (1999), as it sees a developmental change in the teacher in relation to their professional activity that is consciously reflected on, and stimulated within the development group, particularly due to the teacher's engagement with the production of their e-narrative or case study. Whilst ultimately the developmental progress in relation to modelling will occur within the school, for the individual teachers at the key moment of kick-starting this development it is their engagement in both classroom and development group that is crucial. This suggests that for these particular teachers their transition may be considered *collateral* as it requires their to-and-fro movement between activity systems as boundary crossers. Ultimately, it is to be hoped that other teachers in their schools will be able to undergo *encompassing* transitions, as stimulated by the development group teachers, their departments attempt to shift practice within the school. Therefore, for the teachers of the development group we again need to consider the idea of horizontal learning, development or progress. Beach suggests that such horizontal development might be more difficult to understand because of the multi-directionality of movement of the individuals which is not necessarily aligned with their development. Here, therefore, it is the teachers' reflections on their professional practice, afforded within the development group, that allows a space in which they can meet with new scientific concepts (for example, the modelling cycle) and with colleagues consider how they might develop new conceptualisations of teaching and learning.

We should note the similarities (and of course differences) we have here with Engestrom's boundary crossing laboratory where CHAT itself provides a mediational tool in the professional workers' reflections on their practice. Here the meetings of the development group provide a similar reflective space with the tool of the production of teachers' e-narratives important in supporting their reflections and horizontal re-conceptualisations of their practice.

Crucially we note that expansive professional learning in relation to teacher knowledge for modelling cannot be left to chance. The case studies reflected upon here suggest that in general professional learning requires the intersection of three important factors:

1. The key personnel involved must have at least approximately aligned long-term goals in relation to the object of the activity systems in which they operate and a professional expertise and understanding of the context that allows them to work within the rules of the system but adapt these to the benefit of the desired professional learning.
2. A climate in which new or potentially emerging rules appear to mediate an expansion in the object of activity (importantly in this case, we draw attention to the contradiction that occurs in the different measures used to judge schools as being potentially significant in this respect).
3. Networks of personal relationships that facilitate ease of communication and boundary crossing (for example, so that teachers can comfortably operate in both their school classrooms and the post-LEMA development group).

In the particular case of teacher development *in relation to modelling* the case studies here point to the teachers' and students' changing roles in the classroom as being of most concern to teachers as they first use modelling activities. This has important implications for those supporting such changes through facilitating professional development.

With regard to the tools we might use to support such professional learning *in general* I draw attention to the potential of the production of e-narratives by the teachers as providing an important reflective instrument for them as well as providing important data for researchers. Additionally, in supporting professional development *in modelling* in particular, the analysis tool developed and used here might prove useful in stimulating teachers' reflections, particularly as it allows individuals and groups opportunities to conceive of, and understand their practice in relation to both important aspects of content knowledge in relation to modelling such as modelling competencies and students' learning more widely than a focus on modelling per se. The initial findings here suggest that in the early stages of introducing modelling in classrooms teachers appear to need to focus on the development of general pedagogic practices such as managing group work rather than specific content knowledge. The use of an analytic tool such as that developed here might allow teachers to monitor their changing practices over time and in reflective discussion of this bring to the focus of attention their attitudes and beliefs and any changes they might recognise or negotiate in these. This seems particularly important in the case of modelling, as certainly in the English context, it struggles to find space in the curriculum.

References

Askew, M., Brown, M., Rhodes, V., Johnson, D., & William, D. (1997). *Effective teachers of numeracy – Final report*. London: King's College.

Ball, D. L., Hoover Thames, M., & Phelps, G. (2008). Content knowledge for teaching: What makes it special? *Journal of Teacher Education, 59*(5), 389–407.

Beach, K. D. (1999). Consequential transitions: A sociocultural expedition beyond transfer in education. *Review of Research in Education, 24*, 124–149.

Brousseau, G. (1997). *Theory of didactical situations in mathematics*. Dordrecht: Kluwer.

Cardella, M. E. (2008). Which mathematics should we teach engineering students? An empirically grounded case for a broad notion of mathematical thinking. *Teaching Mathematics and Its Applications, 27*(3), 150–159.

Engestrom, Y. (2001). Expansive learning at work: Toward an activity theoretical reconceptualisation. *Journal of Education and Work, 14*(1), 133–156.

Engestrom, Y., & Cole, M. (1997). Situated cognition in search of an agenda. In J. A. Whitson & D. Kirshner (Eds.), *Situated cognition: Social, semiotic, and psychological perspectives* (pp. 301–309). Hillsdale: Lawrence Erlbaum Associates.

Kaiser, G., & Maaß, K. (2006). Modelling in lower secondary mathematics classrooms – Problems and opportunities. In W. Blum, P. Galbraith, H.-W. Henn, & M. Niss (Eds.), *Modelling and applications in mathematics education*. New York: Springer.

Kaiser, G., Schwarz, B., & Tiedemann, S. (2010). Future teachers' professional knowledge on modeling. In R. Lesh, P. Galbraith, C. Haines, & A. Hurford (Eds.), *Modeling students mathematical competencies: ICTMA 13* (pp. 445–457). New York: Springer.

Ofsted (2008). *Mathematics: Understanding the score*. London: Office for Standards in Education.

Pehkonen, E., & Torner, G. (1996). Mathematical beliefs and different aspects of their meaning. *Zentralblatt fur Didaktikder der Mathematik, 28*, 101–108.

Schoenfeld, A. H. (1992). Learning to think mathematically: Problem solving, metacognition and sense – Making in mathematics. In D. Grouws (Ed.), *Handbook for research on mathematics teaching and learning*. New York: Macmillan.

Shulman, L. S. (1986). Those who understand: Knowledge growth in teaching. *Educational Researcher, 15*(2), 4–14.

Wake, G. (2009a). Learning through mathematical modelling. Report to NCETM. May be downloaded at http://www.ncetm.org.uk/files/504526/. Final_Report_G080713_University_Manchester.pdf.

Wake, G. (2009b). Conceptualising and implementing teacher professional development in mathematics in England. Paper presented at *European Conference on Educational Research*, Vienna, Sep 2009.

Vygotsky, L. S. (1987). *Thinking and speech*. New York: Plenum.

Wierzbicka, A. (1999). German "cultural scripts": Public signs as a key to social attitudes and cultural values. *Discourse and Society, 9*(2), 241–282.

Part VIII
Theoretical and Curricular Reflections on Mathematical Modelling

Chapter 63
Theoretical and Curricular Reflections on Mathematical Modelling – Overview

Pauline Vos

Over the years we have seen many mathematical modelling projects being carried out and described in past ICTMA proceedings. In these projects, aims and objectives of mathematical modelling, varying between pragmatic or scientific-humanistic (Kaiser-Messmer 1986), were brought into practice in classrooms. Often these projects depended on personal initiatives of passionate teachers and dedicated researchers. In those projects, modelling education was often small scale, temporary, and detached from the mathematics curriculum. Formalization and sustainability of modelling education were not easily achieved due to lack of institutionalization, funds, or encouragement.

However, since the turn of the millennium in the ICTMA proceedings we see descriptions of mathematical modelling being shaped into curricula, whereby formal documents frame the intentions for modelling education. Guided by curriculum documents and teaching manuals, teachers implement these intentions in the classroom, whether trained or untrained. In the ICMI-14 Study on Modelling and Applications by Blum et al. (2007) there are descriptions of large-scale curriculum innovations in Ontario (Suurtamm and Roulet 2007), Queensland and Victoria (Stillman 2007) and South Africa (Julie and Mudaly 2007). In the ICTMA-13 proceedings Vos (2009) added another description of a large-scale curriculum innovation favoring modelling in the Netherlands. These country case studies show that modelling can take up different positions with regard to the mathematics curriculum. While in traditional mathematics curricula modelling is at best a fragment of the mathematics curriculum, being used as a marginal illustration that comes after the core, the new curricula put modelling to the fore as a basis for learning mathematics (modelling as a vehicle for mathematics) or as the central focus of the mathematics curriculum in itself.

P. Vos (✉)
University of Amsterdam, Amsterdam, The Netherlands
e-mail: fpvos@hotmail.com

When studying curricula, many educational researchers use a framework to distinguish aspects of curricula at different levels, going back to a framework established by Goodlad and Richter (1966):

- The intended curriculum at the macrolevel consists of what society at large prescribes students to learn; the intended curriculum is often described in formal documents containing aims and exemplary tasks; an intended curriculum may be synonymous to, or approximate, the planned and the formal curriculum.
- The implemented curriculum at the mesolevel is what happens at classroom level; the implemented curriculum includes teacher's instruction and students' activities; the implemented curriculum contains the enacted and the experienced curriculum.
- The attained curriculum at the microlevel is the final result at the level of the students: it consists of what is actually learnt by students, but also includes beliefs, perspectives, values, and motivation that emerge from the implemented curriculum.

The papers in this chapter mainly deal with aspects of the implemented curriculum and thus extend prior research by our ICTMA colleagues Burkhardt and Pollak (2006), who have written about barriers and levers for implementing intentions of mathematical modelling. The papers contribute to further understanding of curricular aspects of mathematical modelling, in particular with respect to the implementation of mathematical modelling curricula.

1 Paper Summaries

This chapter contains five papers that deal with the implementation of mathematical modelling into education. Two papers (the joint paper by Ikeda and Stephens and the paper by Vos) are literature studies, while the studies described in the other papers contain evidence from surveys, interviews, and observations. One paper studies the implementation through the eyes of instructional actors: teachers (by Villa-Ochoa and Jaramillo). Another paper also includes teachers' views, but additionally has data about the views of curriculum experts (the paper by Stillman and Galbraith). Finally, one paper studies the implementation of modelling through the eyes of academic actors: mathematical modelling experts at a university on aims and pedagogical practices in their modelling courses (the paper by Spandaw).

The paper by Ikeda and Stephens studies a medium for implementing mathematical modelling into education: a textbook. Textbooks are the embodiment of curricular intentions and guide and shape classroom implementation. Object of Ikeda and Stephen's study is a series of textbooks, published in 1943–1944, which covered the grades 7–11. The textbooks contain an abundance of real world examples, and these were used for mathematical modelling, both as a basis for learning mathematics and as a validation of pure mathematics. Ikeda and Stephens point at two interesting design principles applied in the textbooks. The first is the recurrence of one and the same real world situation in subsequent textbooks, with tasks that increase in difficulty and complexity. The second is the use of real world situations for modelling tasks in which a variety of mathematical concepts is applied.

The paper by Vos studies the use of the term 'authentic' as a qualification of the correspondence between reality and mathematical modelling. The term 'authentic' is used at the level of the intended curriculum, when designers describe 'authentic learning environments' or 'authentic contexts' for mathematical modelling. Vos criticizes this holistic use of the term 'authenticity', in particular for simulations, replicas, or copies which are not 'authentic' by definition. She explains that 'authenticity' is a social construct, which is actor-independent (unlike 'relevance'). It requires a direct and certifiable origin in real life. At the level of the implemented curriculum, for example, a resource should only be characterized as 'authentic' if it truly originates from out-of-school. Or, an 'authentic' model is one that is used in real-life modelling research, and which can be certified by scientific researchers in person.

The paper by Spandaw studies the knowledge of authentic mathematical modelers (i.e. experts at universities who use modelling in their research) on the teaching and learning of modelling. At the level of the implemented curriculum, the participants mention the difficulties of supervising modelling. At the level of the intended curriculum, they strongly advocated an approach in which mathematical concepts are taught first, and therefore, they recommend only limited modelling at secondary schools, ignoring motivational aspects of modelling. They were horrified by numerical methods. Yet, Spandaw offers an interesting example of how the attained curriculum can be a reversal of the intended curriculum: for civil engineering students a theoretical course was identified as a necessary prerequisite for a computer modelling course, but most students only managed to pass the theoretical course after seeing applications of the theory in the computer modelling course.

The paper by Villa-Ochoa and Jaramillo starts from the intended mathematics curriculum in Columbia, which includes modelling. Within this context, the researchers interviewed and observed four teachers in light of their beliefs and their abilities to implement the curriculum, using a framework termed as 'sense of reality', which includes the capacity to detect modelling opportunities suitable for students. The teachers indicate that many modelling situations are technological and inaccessible to students. Villa-Ochoa and Jaramillo conclude that implementing a modelling curriculum requires teachers to unveil mathematics from situations close to students' experiences, which is a complex competence.

Finally, the paper by Stillman and Galbraith reports on the evaluation of a state-wide and longitudinal curriculum intervention in Queensland (Australia), in which modelling was included into the intended curriculum as a focus of mathematics education. Their study deals with the implementation thereof, containing data from interviews with both teachers and curriculum experts. The study shows that even after 20 years of implementation, the views differ, for example, when asked about the robustness of modelling as an established practice. Although the teachers expressed satisfaction with their skills with respect to designing modelling tasks, the designed tasks showed a large variety from minimalist to very rich approaches. It is interesting that the respondents experience applications and modelling as promoting mathematical thinking and understanding, as demonstrating utility and relevance, and as engaging students.

References

Blum, W., Galbraith, P. L., Henn, H.-W., & Niss, M. (Eds.). (2007). *Modelling and applications in mathematics education: The 14th ICMI study.* New York: Springer.

Burkhardt, H., & Pollak, H. (2006). Modelling in mathematics classrooms: Reflections on past developments and the future. *Zentralblatt für Didaktik der Mathematik, 38*(2), 178–195.

Goodlad, J. I., & Richter, M. N. (1966). *The development of a conceptual system for dealing with problems of curriculum and instruction.* Los Angeles: UCLA.

Julie, C., & Mudaly, V. (2007). Mathematical modelling of social issues in school mathematics in South Africa. In W. Blum, P. L. Galbraith, H.-W. Henn, & M. Niss (Eds.), *Modelling and applications in mathematics education: The 14th ICMI study* (pp. 503–510). New York: Springer.

Kaiser-Messmer, G. (1986). *Anwendungen im Mathematikunterricht (Applications in mathematics education).* Bad Salzdetfurth: Franzbecker.

Stillman, G. (2007). Implementation case study: Sustaining curriculum change. In W. Blum, P. Galbraith, H.-W. Henn, & M. Niss (Eds.), *Applications and modelling in mathematics education: The 14th ICMI study* (pp. 497–502). New York: Springer.

Suurtamm, C., & Roulet, G. (2007). Modelling in Ontario: Success in moving along the continuum. In W. Blum, P. Galbraith, H.-W. Henn, & M. Niss (Eds.), *Applications and modelling in mathematics education: The 14th ICMI study* (pp. 491–496). New York: Springer.

Vos, P. (2009). The Dutch maths curriculum: 25 years of modelling. In R. Lesh, C. R. Haines, P. L. Galbraith, & A. Hurford (Eds.), *Modelling students' mathematical modeling competencies (ICTMA 13)* (pp. 610–618). New York: Springer.

Chapter 64
Making Connections Between Modelling and Constructing Mathematics Knowledge: An Historical Perspective

Toshikazu Ikeda and Max Stephens

Abstract This study will look at a surprising resolution of the tension that arises in trying to strike a balance between modelling and pure mathematics by examining Japanese textbooks for the junior high school nearly 70 years ago. Three characteristics are found: (1) two distinct roles –first as objects to mathematize in order to solve real world problems and second as evidence by which to test the validity of mathematical concepts; (2) repeated instances of the same contexts through which new phases of mathematization could be developed; and (3) a series of real world questions focussed on the reason for solving a real world problem.

1 Aims in This Study

One of the unresolved tensions in the teaching and learning of mathematics is how to strike a balance between modelling and pure mathematics. At the Rome 2008 Centennial of ICMI, Niss (2008) reiterated the importance of such a balance, challenging today's accepted opinion that instruction in mathematical ideas and techniques should come *first*, and *only then* might it be possible or desirable for students to apply those ideas in modelling activities. This study will look at a surprising resolution of that tension by examining Japanese textbooks for the junior high school nearly 70 years ago – before and during World War II (Ikeda 2008).

T. Ikeda (✉)
Yokohama National University, Yokohama, Japan
e-mail: ikeda@edhs.ynu.ac.jp

M. Stephens
The University of Melbourne, Victoria, 3010, Australia

The aims of this study are to clarify:

1. The roles of real world problems in acquiring mathematics knowledge in mathematics textbooks before World War II in Japan.
2. What kinds of teaching sequences can connect solving real world problems with acquiring mathematics knowledge?

To attain these aims, the following procedure is applied:
Focussing on 7th to 9th Grade mathematics textbooks of that era,

(a) To extract the contents concerned with real world problems and examine the role of these problems in helping students to acquire mathematical knowledge.
(b) To examine what kinds of teaching sequences can be seen to combine or connect solving real world problems with acquiring mathematical knowledge.

2 Mathematics Textbooks Before World War II in Japan

The "Midori Hyoshi" textbooks in elementary school, which had already focussed on mathematization from a real world, created a necessity of improving the national curriculum in junior high school mathematics for those who were entering to junior high school from 1941. In 1942, the "Examining Committee of Secondary Mathematics Curriculum" was set up under the Ministry of Education. In the same year, a "Revised Curriculum of Mathematics and Sciences in Junior High School" was announced. In this document, the following aim was pointed out (Nagasaki 1990):

> It is required for the teacher to cultivate students' ability to mathematize a real world phenomenon based on number, quantity and space, treat the result mathematically and apply it into national life. Through these activities, it is expected for students to contribute to society as a nation. (p. 93)

In 1943, the textbooks for 7th, 8th and 9th Grades were published (Tyuto Gakko Kyokasho Kabushiki Kaisya 1943), and in 1944 those for 10th and 11th Grades. These were the most common textbooks which targeted all junior high school students in Japan. Nagasaki (1990) outlined the five aims of this series of mathematics textbooks:

1. The contents of mathematics are divided into two parts: content concerned with number and quantity, and one concerned with space. These parts should be connected to attain the aims of mathematics.
2. In lower levels, the ability to apply basic mathematical methods should be fostered by treating concrete activities, and gradually more accurate methods should be introduced.
3. Connections between mathematics with science should be emphasized.
4. Mathematical methods should be applied in other areas based on intuition about number, quantity and space, avoiding concepts that are too abstract or placing too much emphasis on logic.

5. Practical content, such as measurement, statistics and descriptive geometry, should be enlarged and concepts of limit should be emphasized. (p. 95)

Consequently, real world situations were incorporated into mathematical textbooks with a view to helping students to construct mathematical knowledge.

3 Roles of Real World Situations

Real world situations in these textbooks serve two distinct roles in their relation to students' mathematical thinking. These two roles are not stated explicitly by the authors and are distinct from the five curriculum principles outlined above. They are:

[Role 1]: As objects to mathematize in order to solve real world problems.
[Role 2]: As evidence by which to test the validity of mathematical concepts.

This is an interesting reversal of accepted opinion where mathematical ideas are simply assumed to possess *for students* their own validity. We pick up the example concerning role 2. Here, a real world situation is used as evidence of how to expand number concepts, in other words, how to define the rule of multiplication of negative numbers. At the 1st Grade in Ichirui, the following explanation is described.

We are not suggesting that it is easy or even possible to construct mathematics knowledge simply by focussing on the direction from a real world situation into a mathematical model. The reverse direction which connects mathematical knowledge to real world situations is, however, very important. Rarely do teachers employ concretized models to show how existing mathematical concepts can be validated in such concretized models. If mathematics education is to enable students to form mathematics concepts or methods in their internal world, not simply having them injected from an external world of a teacher or textbook, then it is important to introduce students to concretized models that give these mathematical concepts personal meaning and validity. This is shown in the above example (see Fig. 64.2), where they can see that moving in a negative direction for x on a negative slope results in a positive increase in height.

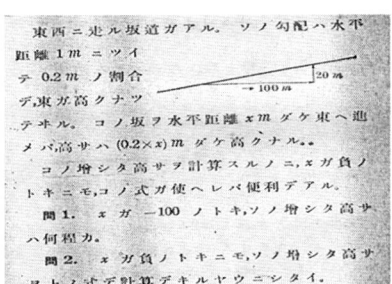

Let's consider how to determine the code (plus or minus) if you multiply a positive or negative number by negative number. There is a slope in the direction from East to West (Fig. 64.1). When we go east, the height is increased "0.2 m" per "1 m". If we go East "x m" in the horizontal direction, the height is increased "$(0.2 \times x)$ m".

Fig. 64.1 Concretized Model 1

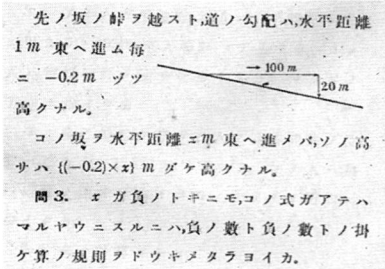

In the next example there is a different slope (Fig. 64.2), when we go East, the height is increased "−0.2 m" per "1 m". If we go East "x m" in the horizontal direction, the height is increased "$(−0.2 \times x)$ m". We would like to calculate the increased height by the previous formula when horizontal distance, x, is a negative number. How can we define the rule of multiplication, namely negative number times negative number?

Fig. 64.2 Concretized Model 2

4 Repeated Instances of the Same Contexts

As students' grasp of mathematical knowledge and techniques becomes more sophisticated, it is often thought desirable to seek out and use different contexts and situations to illustrate the usefulness of those ideas and techniques. By contrast, these Japanese textbook authors tended to use repeated instances of the same contexts through which new phases of mathematization could be developed.

This characteristic has been pointed out by Sato (2001) and Tanaka (2008). Following Sato, we argue that, even though they may lack the authenticity of real world situations, some problems strongly appeal to the usefulness of mathematics for students, for example, the problem of making a square prism box by cutting same-sized squares from the four corners of a square. These kinds of problem situations are repeatedly located in the textbooks at several grades. Students can re-encounter these situations, by applying new mathematical concepts or methods. Further, students see how their solution becomes more precise by applying new mathematical concepts or methods.

Actually, the problem making a square prism box by cutting same-sized squares from the four edges of square is treated 3 times as shown in Figs 64.3–64.5.

Tanaka (2008) follows the analysis of Sato in arguing that similar and related real world problems can be used by students:

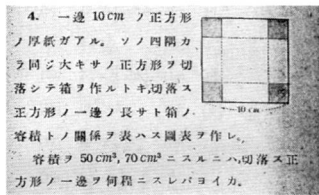

Here is a square in which the one side length is 10cm. When making a prism box by cutting the same size of squares at the four corners of it, make a table regarding the relation between the one side length of each cut square and the volume. Find out the one side length of each cut square when the volume is 50cm^2 and 70cm^2.

Fig. 64.3 Unit "Figure, table and formula" at Grade 7

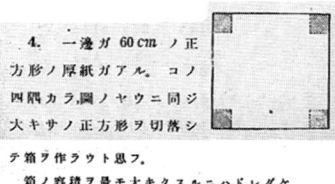
Here is a square in which the one side length is 60cm. I will make a prism box by cutting the same size of squares at the four corners of it. What is the one side length of each cut square so that the volume will become the maximum?

Fig. 64.4 Unit "Quadratic equation" at Grade 8

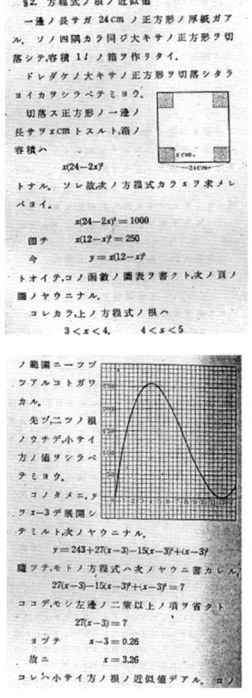

From a square in which the one side length is 24cm. I will make a prism box so that the volume becomes 1L by cutting the same size of squares at the four corners of it. When the one side length is x cm, the formula is $x(24-2x)^2$ and the equation is as follows
$$x(24-2x)^2 = 1000$$
$$x(12-x)^2 = 250$$
When we set $y = x(12-x)^2$, the graph of this function is shown as the following figure.

By graph, we can find (or have found) that the two required values of x are $3 < x < 4$, and $4 < x < 5$.
Let's examine the value of x which is close to 3. We can expand the function with $(x-3)$, the function becomes as follows:
$$y = 243 + 27(x-3) - 15(x-3)^2 + (x-3)^3$$
The equation is expressed as follows.
$$27(x-3) - 15(x-3)^2 + (x-3)^3 = 7$$
Neglecting $(x-3)^2$, $(x-3)^3$ $\quad 27(x-3) = 7$
giving $x = 3.26$
This is one of the approximate values of x. [Students are expected to understand and explain why the square and cube terms can be neglected in this approximate solution.]

Fig. 64.5 Unit "Quadratic equation" at Grade 8

... to mathematize the phenomenon according to their mathematical knowledge and skill. Therefore, the phase of mathematization should be changed according to (students') mathematical knowledge and skill by treating the same problem situations repeatedly. (p.13)

In the following geometrical example, an open-ended problem in a unit on Measurement in Grade 7 (see Fig. 64.6) is posed for students at first before learning the particular mathematical concepts and methods. Two years later, more complex problems are posed for students in an introduction to the unit "Triangles and Trigonometric function" unit at Grade 9 as shown Fig. 64.7.

2. 山ノ影ガ湖ニ映ツテヰル，山ノ頂ノ仰角ト影ノ俯角トカラ山ノ高サヲ測ル方法ヲ考ヘヨ。

Let's devise methods for how to measure the height of a mountain by using the angle between a line from my eye to the top of a mountain and a horizontal line, and the angle between a line from my eye to the shadow of the top of a mountain on the pond and a horizontal line.

Fig. 64.6 Unit "Measurement" at Grade 7

2. 池ノ岸ニ立ツテ,対岸ノ塔ト水面ニ映ル影トヲ見タラ,塔ノ尖ハ30°ノ仰角ニ見エ,ソノ影ハ40°ノ俯角ニ見エタ。
眼ハ水面カラ 3m ノ高サニアルトシテ,塔ノ水面カラノ高サヲ作図デ求メヨ。

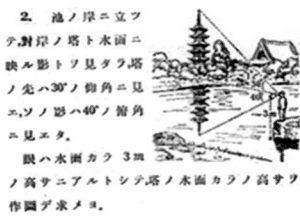

I am standing on the ground and looking at the top of the tower and a shadow of it on the pond. There is a 30 degree angle between a line from my eye to the top of the tower and a horizontal line, and there is a 40 degree angle between a line from my eye to the shadow of the top of the tower on the pond and a horizontal line. The height of the location of my eye from the surface of the water is 3m. Find the height of the tower from the surface of the water by constructing the geometric figure.

Fig. 64.7 Introduction of unit "Triangle and Trigonometric function" at grade 9

Thus, new and more sophisticated mathematical knowledge and techniques are applied to familiar contexts and familiar results.

5 Making Connections Between Real World Problems

Several different ways of connecting real world problems can be observed from the textbooks. The first uses a *series of real world questions from general to specific, or from specific to general.* To illustrate this first idea, students are introduced to the general (basic) problem of how to measure a distance where for some reason direct measurement is not practicable. Then students are expected to apply the techniques they have learned to similar situations.

For example, the "Measurement" unit at 7th Grade is sequenced based on a series of real world questions in which students are taken through several distinct stages as follows.

(a) *Introducing the question in a basic form*: In Sub-unit 1 called "Measuring the distance", the reason why students need to solve the problem is explained in Fig. 64.8.
(b) *Widening the original question:* In the same unit, the explanation is noted to connect problems tackled before with a new question such as "We have thought how to measure the distance between one point and another point where it is

It is impossible to measure the distance directly from here to the island across the sea even though we would like to know it. Of course it is possible to measure the distance with the eye; *b*ut when we would like to know it precisely, it is necessary to devise a solution.

Fig. 64.8 Island problem

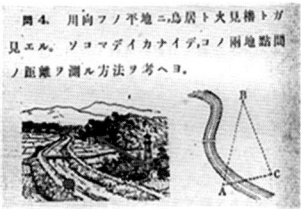
Across the river, we can see both the gateway at the entrance to a Shinto shrine and a fire tower. Let's consider methods to measure the distance between them, keeping away from them.

Fig. 64.9 Gateway and fire tower problem

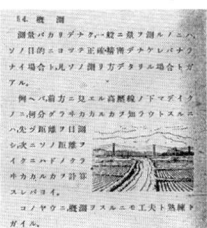
There are two general cases when we measure according to aims. The first case is to measure precisely, and the other is to measure approximately. For example, when we would like to know the time taken from here to below the high-voltage cable in front of us, it is possible for us to measure the distance by eye at first, then calculate the time. Like this, it is required for us to make a device and to get in practice.

Fig. 64.10 High-voltage cable problem

impossible to measure. From now on, let's consider how to measure the distance between two points which are both far from here". The following problem is posed for students (see Fig. 64.9).

(c) *Posing a different but related question:* In sub-unit 4 called "Approximate measurement", students here must select an appropriate method according to the aims in a real world situation shown in Fig. 64.10.

At other times, the textbook writers use a sequence of real world questions using related or similar kinds of scientific ideas. For example, a sub-unit called "Contour lines" is located in the unit "Chart and formula" at 7th Grade. In this sub-unit, a series of real world situations such as isotherms, isobars, contour lines and lines of inclination are treated successively. Different mathematical concepts or methods are used to solve these real world problems. We cannot show these problems in detail. In a problem, based here on lines of inclination, the concept of cosine is introduced and applied as follows:

> When we go up 100 m in the slope, the height is increased by 20 m. We say this slope is of "inclination 20" (Fig. 64.11 upper). [Problem] If we climb it at an angle of 60° to the straight line we create a different (easier) slope (this path is not straight up but is a succession of zigzags.) (Fig. 64.11 lower). What is the new inclination in this climbing?" (See Fig. 64.12 for solution).

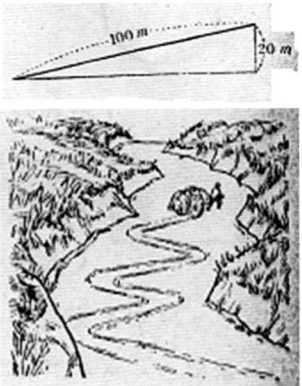

Fig. 64.11 Slope of inclination 20 and new inclinaticn in the climbing

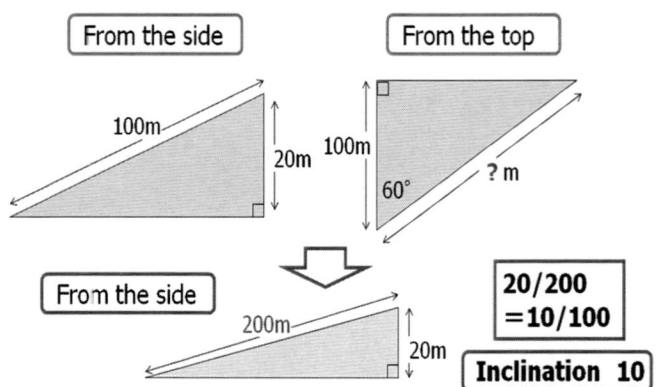

Fig. 64.12 Solution of inclination problem

6 Implications for Teaching of Mathematics Today Including Modelling

Three points are suggested for teaching of mathematics today including modelling:

Students do not construct mathematics knowledge by only focussing on the direction from a real world situation into a mathematical model. The reverse direction from mathematical knowledge into a real world situation is also important. Especially, when students consider how to expand the existing mathematical concepts, it is crucial to build up a concretized model in a concrete world (especially, real world) and consider it under the concretized model. In these historical Japanese textbooks, these two distinct roles are assumed rather than stated as key organizing principles: first as objects to mathematize in order to solve real world problems; and second as evidence by which to test the *validity or reasonableness* of mathematical concepts. Both abstracted models (real world $\rightarrow$ mathematics) and concretized models (mathematics $\rightarrow$ real world) are treated complementarily to strike a balance between modelling and constructing mathematics concepts.

As students' grasp of mathematical knowledge and techniques becomes more sophisticated, it is now thought desirable to seek out different contexts and situations to illustrate the usefulness of those ideas and techniques. By contrast, the Japanese textbook authors tended to use repeated instances of the same contexts through which new phases of mathematization could be developed. Thus, new and more sophisticated mathematical knowledge and techniques were applied to familiar contexts and familiar results. This helps students to see that phases of mathematization can be changed according to the growth of their mathematical knowledge.

Several sequences were explicitly based on a series of real world questions that focus on the reason why students have to solve it. Sometimes, these sequences moved from specific real questions to discuss general principles; at other times they moved from the general to the specific; and made sure that students could see mathematical connections between the different contexts used. Knowing why to solve a real world problem is important for both teachers and students, more important than knowing how, and very important for fostering in students a sense of the power and utility of mathematics.

7 Conclusion

While the context of early secondary education in Japan has changed greatly over the past 70 years, the guiding principles of curriculum design which informed these earlier textbooks continues to provide a helpful reference point – and a point of challenge to those assumptions which are too readily made today – in deciding how to balance modelling and the construction of mathematical knowledge in the teaching and learning of mathematics and in the writing of school textbooks.

References

Ikeda, T. (2008). Reaction to M. Niss's plenary talk – Perspectives on the balance between applications and modelling and 'pure' mathematics in the teaching and learning of mathematics. In M. Menghini, F. Furinghetti, L. Giacardi, & F. Arzarello (Eds.), *The first century of the International Commission on Mathematical Instruction (1908–2008) reflecting and shaping the world of mathematics education* (pp. 85–90). Rome: Enciclopedia Italiana.

Nagasaki, E. (1990). Reconstructing movement of mathematics education and the arrival of "Mathematics Dai Ichirui and Nirui". *Journal of National Institute for Educational Policy Research, 20*, 85–102 (in Japanese).

Niss, M. (2008). Perspectives on the balance between applications and modelling and 'pure' mathematics in the teaching and learning of mathematics. In M. Menghini, F. Furinghetti, L. Giacardi, & F. Arzarello (Eds.), *The first century of the International Commission on Mathematical Instruction (1908–2008) reflecting and shaping the world of mathematics education* (pp. 69–84). Rome: Enciclopedia Italiana.

Sato, E. (2001). Mathematics education during World War 2. *The Japanese Journal of Curriculum Studies, 10*, 17–29 (in Japanese).

Tanaka, Y. (2008). Analysis of teaching materials with common situations in the course, mathematics category I. focusing on "Mathematization of Phenomenon". *Journal of Japan Society of Mathematics Education, 90*(1), 12–25 (in Japanese).

Tyuto Gakko Kyokasho Kabushiki Kaisya. (1943). *Mathematics 1–3 (Textbooks at Grade 7-9) Dai Ichirui and Nirui*, Sanseido (in Japanese).

Chapter 65
Practical Knowledge of Research Mathematicians, Scientists, and Engineers About the Teaching of Modelling

Jeroen Spandaw

Abstract This chapter describes the results of interviews with 12 research mathematicians, scientists, and engineers exploring their professional knowledge about modelling and the teaching of mathematical modelling. The interviews deal with the issues of goals, competencies, meta-cognition, beliefs, epistemology, computers, and implementation. Differences and similarities between the interviewees' views and mathematics education literature on mathematical modelling are discussed. Some suggestions regarding how mathematical modelling in secondary education might profit from the interviewees' experience are provided.

1 Introduction

In the Netherlands, as elsewhere, there have been several attempts to strengthen the role of mathematical modelling in secondary mathematics and science education (Lijnse 2006). Lijnse attributes this focus on modelling to attention for students' preconceptions (mental models), attention for the nature of scientific knowledge, and the availability of computers for doing numerical simulations. In fact, already two decades ago, Hestenes claimed that "mathematical modelling should be the central theme of physics instruction" (1987, p. 25) and he has repeated that ever since (e.g., Hestenes 2006). Researchers in mathematics education also advocate modelling, even for young students, not only to learn to apply mathematics, but also to learn mathematical concepts (see e.g., Blum et al. 2007). However, this popularity of modelling in secondary mathematics education is not ubiquitous. In the Netherlands, there has been strong opposition from mathematics professors against the joint efforts of mathematics and science educators to develop interdisciplinary modelling teaching materials. As a result, the mathematicians backed out of the project, effectively leaving modelling to science education.

J. Spandaw (✉)
Delft University of Technology, Delft, The Netherlands
e-mail: j.g.spandaw@tudelft.nl

In this chapter, we investigate the practical knowledge about mathematical modelling of several mathematicians, scientists, and engineers, who use modelling in their research and teach modelling at university level. Although they are not experts on secondary education, this is interesting since (a) one might profit from their expertise in modelling and teaching modelling, and (b) their opinions have an impact on secondary education curricula. Duffee and Aikenhead (1992) identified teachers' practical knowledge as a major factor in their response to curriculum change. Practical knowledge is defined as the integrated set of knowledge, beliefs, intentions, and attitudes teachers develop with respect to their teaching practice (Grimmett and MacKinnon 1992). The literature on mathematics education (e.g., Blum et al. 2007) has also identified beliefs as an important factor in the educational debate.

We have the following research questions: What do the interviewees have to say about modelling and modelling education? What are the similarities and differences between interviewees and mathematics education researchers? What is the range of opinions about modelling within the group of interviewees?

To answer these questions the author conducted semistructured in-depth interviews with 12 scientists: four mathematicians (researching and teaching differential equations, numerical mathematics, and mathematical physics), two statisticians, one physicist, two system biologists, one geo-physicist, and two civil engineers. The interview findings are summarized after presenting the theoretical framework and the design of the case study. In the final section we discuss the findings and try to answer the research questions.

2 The Case Study

In this section the theoretical framework and the design of the case study are presented.

2.1 *Theoretical Framework*

As in part 1 of Blum et al. (2007) I take mathematical modelling to be "the entire process consisting of structuring, generating real world facts and data, mathematizing, working mathematically, and interpreting/validating (perhaps several times around the loop)" (pp. 9–10). In particular, it differs from applied mathematics and applied problem solving. I distinguish two directions: using known mathematics to solve a non-mathematical problem using modelling and conversely, using mathematical modelling to develop mathematical concepts ("modelling for mathematics"). Mathematical modelling concerns many related issues. In part 3 of the volume cited above the most important ones are grouped as follows: epistemology, authenticity and goals, competencies, applications and modelling for mathematics, pedagogy, implementation, and assessment. To this I add mathematical knowledge and domain knowledge.

Mathematical modelling from the point of view of physics educators is described in Hestenes (1987, 2006), and Lijnse (2006). Some interesting descriptions of implementations of modelling in secondary science education can be found in Vollebregt (1998) on the use of modelling in scientific theory, and in Löhner (2005) on computer modelling. Ormel (2010) describes an interesting recent experiment in complex dynamic "modelling for physics" using *Powersim*, a user friendly general modelling tool with graphic interface. Modelling goals are described in Blum and Niss (1991), who distinguish formative, critical, practical, cultural, and instrumental goals. The modelling process is usually described using a modelling cycle. For my purposes, the simple version in Maaß (2006) suffices. I interpret the mathematical step "working mathematically" liberally: it may include substantial use of software. I adopt Maaß' view that modelling competencies consist not only of the competencies of performing the steps of the modelling cycle, but also of meta-cognitive competencies such as monitoring the modelling process, and attitudes (Maaß 2006).

2.2 Design of the Case Study

As mentioned above, the interviewees use mathematical modelling in their scientific research. They also teach modelling at different universities. The interview questions were divided into six groups, covering all major issues from the literature mentioned above: (1) goals of modelling and education in modelling; (2) implementation and assessment of modelling education; (3) students' competencies, meta-cognition, beliefs, and epistemological understanding; (4) use of computers; (5) mathematical and context knowledge; (6) opinions about mathematical modelling in secondary education. The first 5 groups concerned the interviewees' research and teaching experience at tertiary level. Using my notes I summarized the interviews and submitted the summaries for verification to the interviewee to guarantee their correctness. This led to some minor modifications. The corrected summaries, together with the lecture notes used in the modelling courses for bachelor degree mathematicians and civil engineers, as well as the 2009 report "Biomathematics – A vision for success" by the Royal Dutch Academy of Sciences form the input of my analysis below, based on the theoretical framework described in Sect. 2.1.

3 Summary of Interview Findings

The most important interview findings are now overviewed: first, goals of teaching modelling; second, implementation of teaching modelling, including authenticity and the use of computers; third, students' modelling competencies, including epistemological understanding, beliefs, and meta-cognition; and finally opinions about modelling in secondary education.

3.1 Teaching Goals of Modelling in Tertiary Education

The main goals mentioned by most interviewees were (1) becoming acquainted with, and learning how, to apply standard models and standard techniques such as balance equations and dimensional analysis, (2) identifying the appropriate tools such as the relevant concepts or standard models, and (3) the use of common sense and rough estimates. Also widely mentioned were awareness that models have restricted applicability, the relation with the research question, validation, and the iterative nature of modelling, and learning to use the appropriate software.

Another goal was the critical use or evaluation of models made by fellow engineers or scientists. Biology and geophysics masters students had to learn how to deal with mathematical models found in scientific journals. For example, they had to learn to uncover hidden assumptions in scientific papers (cf. Blomhøj and Kjeldsen, this volume).

The main goal for most mathematicians was to connect mathematics with reality, to teach students how to apply mathematics which they were supposed to be familiar with. Similarly, the physics professor wanted to teach how to apply basic physical knowledge to understand complex physical processes. To him the *process* of analyzing the problem conceptually using modelling was more important than the resulting model itself.

The steps of the modelling cycle played a prominent role in the bachelor courses in modelling for mathematicians and civil engineers. Both courses dealt with the use of models to make predictions, or as a substitute for experiments. The professor who designed the latter course also mentioned development of "a critical modelling attitude," the courage to simplify, and understanding the necessity to do so.

The masters courses for biologists and geophysicists aimed at computer-aided modelling for *quantitative* analysis of complex systems. One biology professor distinguished "top-down" and "bottom-up" modelling. In top-down modelling one tries to find patterns in experimental data to formulate new research questions. Bottom-up modelling on the other hand aims at understanding complex systems starting from the basic theory.

3.2 Modelling Courses in Tertiary Education

In all cases modelling was preceded by teaching basic mathematics (e.g., differential equations) and context knowledge (e.g., heat transport). Modelling for mathematics was almost completely absent: Only one professor mentioned the possibility of emergent modelling in a stochastics course for engineers. The following example shows how persistent belief in "skills first" can be. A masters program for civil engineers consisted of two theoretical courses on partial differential equations followed by a course on computer modelling, where students learned how to analyze engineering problems using a dedicated software package.

Most students only managed to pass the theoretical exams *after* seeing applications of the theory in the computer modelling course. As a result, most students now postpone the theoretical part until after computer modelling. The faculty, however, still insists that the theoretical course is a *necessary prerequisite* for computer modelling!

Epistemological aspects of modelling and the process of modelling were usually treated in passing, often reduced to heuristic rules such as "simplify," "proceed conceptually," and "use physical dimensions." A bachelor course for engineers which *did* pay extensive attention to different types of models (conceptual and phenomenological, quantitative and qualitative, deterministic and stochastic), the nature of modelling, and the steps of the modelling cycle was abolished after a few years. A drastically shortened version is now used for bachelor degree students in applied mathematics, with much emphasis on numerical issues.

Authenticity of the modelling problems varied among the courses. The bachelor courses for engineers, mathematicians, and physicists mentioned above use everyday contexts, such as waiting queues or gas bubbles in soft drinks. Information could be deliberately inadequate, so students had to decide whether they needed more data, and which data they should ignore. In more advanced courses, on the other hand, modelling problems were "authentic" in the very different sense of "standard models used by scientists."

Supervising open modelling problems is difficult, especially the first steps. One of the mathematics professors confessed that supervising new problems with unfamiliar contexts is "impossible" even for highly educated university staff! To prevent students getting completely stuck and frustrated with open modelling problems, one mathematics department introduced the following interesting educational design: They gave students a simple mathematical model and a research question which was just too subtle to be answered using the model, so the students had to refine the model.

In most cases, the degree of idealization decreased during the course. For example, the bachelor course for physicists started with simple standard models, which were later combined to model more complex situations. A masters course for biologists, on the other hand, started with concrete experiments and experimental data from scientific papers which students had to analyze (top-down modelling). Next, the course turned to bottom-up modelling. The course then gradually moved away from experiments toward abstract mathematical models.

Computers were widely used. The bachelor course for civil engineers used *Powersim*. The bachelor course for applied mathematicians used Matlab. Since this is an important tool for masters students and professional engineers, designing and programming numerical methods in Matlab is an important part of the course. Some masters courses used more specialized software, such as *FlexPDE* in the case of geophysics and *DIANA* in the case of civil engineers. An important goal of these courses is to learn how to transform the problem context into input of these software packages: partial differential equations with boundary conditions for *FlexPDE,* and basic engineering elements such as beams for *DIANA*.

3.3 Students' Modelling Competencies

Each step of the modelling cycle was reported to present difficulties to students. All students had great difficulties getting started in open modelling problems. They had no recipes; at most some heuristic rules how to proceed. They needed help with dealing with ill-posed problems (leading to complaints about unclear assignments), simplification, conceptual analysis (e.g., understanding and choice of appropriate theory, introduction of relevant variables, feedback loops), and mathematization. The mathematical step of solving the mathematical equations was difficult for most students in biology and geophysics. Physics students had difficulties working with general variables and parameters, rather than numerical values. Mathematics students had difficulties with interpreting mathematical equations and programming in *Matlab*. Students were reported as omitting interpretation and validation, leading to absurd model predictions, which were not identified as such by the students. Students often did not validate their models using dimensional analysis, common sense, or rough estimates. Losing touch with the research question sometimes led to unnecessary refinements of models.

Problems about the epistemology of modelling were also mentioned. Some students invented arbitrary new "laws of nature" to rescue their flawed models. Masters students in civil engineering dismissed a rather realistic 3D model of H-beams simply because it differed from a more familiar, but less realistic 1D model! A third example concerns biology masters, who were so impressed by mathematical modelling that they forgot about the simplifications and restricted applicability of models. They did not realize that by tuning the many parameters in complex models you can "explain" almost anything!

According to the interviewees the most important factors contributing to successful modelling are good command of both mathematics and context knowledge which can be applied effortlessly, the courage to simplify the problem, common sense, and interest in the mathematical and nonmathematical aspects. The latter was seen as problematic, especially for mathematicians and biologists. It was suggested that only masters students are "mature" enough for modelling.

3.4 Opinions about Modelling in Secondary Education

In the previous subsections we discussed the professional knowledge of the interviewees in university education. Now we consider their opinions about secondary education.

Most interviewees were skeptical about mathematical modelling in secondary education. In their opinion even many bachelor degree students were "too immature" for modelling! (According to the 2009 report by the Royal Dutch Academy of Sciences about mathematical modelling in biology, modelling should be restricted to masters students.) It was fine to show simple applications if possible, but they abhorred "quasi-real applications riddled with numerical coefficients," rather than general parameters. Modelling in secondary education should be restricted to teaching simple, classical examples, such as exponential decay or Hooke's law. Problems and contexts should be cleanly idealized, rather than "messy and blurred." Modelling is time consuming, and that time should be spent on basic mathematical skills (algebra, analysis) and basic physics. Early use of computers was seen as a threat for developing and maintaining algebraic skills. Finally, supervision of modelling was said to be too difficult for teachers.

4 Discussion and Conclusions

We now turn to the remaining research questions.

4.1 Similarities Between Interviewees and Education Researchers

Being expert modelers, the interviewees knew, of course, that modelling is more than "applying mathematics"; all steps of the modelling cycle were mentioned. Some steps such as mathematical analysis and validation were worked out in more detail than in Maaß (2006). They also realized that basic mathematical and context knowledge do not lead automatically to successful modelling. They mentioned cognitive and affective barriers, as well as meta-cognitive aspects such as monitoring of the modelling process. They considered students' epistemological understanding of models important and not unproblematic. Nonmathematical competencies such as communication, cooperation, gathering, and judging information, and common sense were also considered important. Modelling open problems was said to be time consuming and very difficult to supervise.

4.2 Differences Between Interviewees and Education Researchers

The main sources of students' difficulties with modelling were reported to be, first, lack of interest, and, second, lack of basic skills and knowledge in mathematics, physics, and biology. This lack of interest was attributed to lack of maturity. Furthermore, the interviewees believed strongly in "skills first." As a result, they concluded that mathematical modelling at school should be restricted to teaching a few clean standard models, classics such as exponential decay or harmonic oscillators. Modelling of fuzzy real world problems was said to be too time consuming, time better spent on basic mathematics and science. Besides, in their opinion, supervision of such modelling tasks is very difficult, and cannot reasonably be expected from teachers in secondary education.

The interviewees had a strong preference for working qualitatively with general (dimensionful) parameters, rather than (dimensionless) numerical coefficients. For example, high school students should not solve $y' = ay$ numerically for different values of a using software; instead they should investigate the dependence of y on the parameter a using mathematical means only.

Since the interviewees are not experts on secondary education, and often implicitly reduce secondary education to the upper secondary science stream, it is not surprising that they have blind spots. The interviewees did not know that software can be used to develop and maintain mathematical skills, and can greatly enhance students' motivation. As mentioned in Sect. 3.2, belief that theory should precede use of computers was very persistent, even when it obviously failed in practice. Most interviewees did not realize the possibilities to use modelling to *develop* mathematical concepts, not to mention meta-cognitive skills. They did not know how important applications are to motivate high school students (cf. e.g., Muller and Burkhardt 2007). They attributed students' difficulties to blend mathematical and nonmathematical skills to "lack of maturity," rather than to their lack of experience with mathematical modelling and their misguided beliefs about mathematics and science, which are the result of current secondary education. Kaiser and Maaß (2007) report how modelling helps to improve these issues. The respondents underestimated the importance of applications and modelling (everyday and scientific) to motivate high school students, to teach them how mathematics is used in science and society through mathematical modelling, and to develop meta-cognitive skills. Making estimates and using common sense were identified as important for university students, but not for high school students. The same holds for the formative, critical, practical, and instrumental goals of modelling (Blum and Niss 1991). Finally, the interviewees were unaware of the possibilities of open modelling tasks even for very young students (Kaiser and Maaß 2007).

4.3 *Differences Among Interviewees*

The interviewees had different notions of "authenticity." For some "authentic" means "meaningful for students." For others it means "models used in science." This is related to the use of models: engineers or applied mathematicians solve a (complex) everyday life problem, whereas (geo-)physicists model idealized situations to gain scientific understanding of natural phenomena.

Although all modelling courses used computers, the opinions about the use of software in modelling education varied from strong reservations ("Early use of software is harmful for acquiring basic mathematical skills") via acceptance of its necessity ("Modelling complex phenomena is impossible without computers") to enthusiasm ("It's silly not to use computers" and "If only I'd had this software when I was a student!").

Finally, although all respondents were rather wary about modelling in secondary education, some did mention the importance of illustrating the use of mathematics in other subjects. One even mentioned "understanding restricted applicability of models" as a goal.

4.4 Possible Implications for Secondary Education

The interviewees are expert modelers and experienced teachers in tertiary education. Their opinions do affect the design of new curricula for secondary education, so they cannot be ignored. In my opinion they are right in emphasizing the difficulty of supervising open modelling tasks, the necessity of basic skills, and the merits of working conceptually. Modelling in (upper) secondary education could pay more attention to balance equations, as a rather general method to construct models. Physical units should not be considered as a (nonmathematical) nuisance, but rather as a welcome source of information. Dimensional analysis is a purely mathematical (algebraic!) technique, and is extremely valuable for understanding and checking modelling equations and the behavior of their solutions, and for estimating the characteristic quantities of a modelling problem. This necessitates the use of abstract parameters instead of numerical coefficients. The mathematical study of the behavior of solutions is indeed important, but contrary to some interviewees' beliefs this study can be preceded and supported by computer experiments.

The resistance to messy models and the insistence on classics receives unintended experimental support from Ormel (2010): During an emergent modelling teaching experiment in high school the author had to retreat twice from "the realistic constraints of climate modelling" to classic contexts: a double pendulum for chaos and radioactive decay for differential equations. He concluded that it is preferable to teach differential equations and basic concepts from thermodynamics first, using simple standard models, *before* embarking upon more complex modelling tasks.

I conclude that, while modelling in secondary mathematics education should profit from experts' experience, it should not be abolished in favor of "pure" mathematics. We need a "diverse curriculum featuring both abstract and applied mathematics [to] enable more students to achieve higher levels of mathematical competence" (Muller and Burkhardt, p. 270). High school students must learn what mathematics is actually good for and how they can use it.

References

Blum, W., & Niss, M. (1991). Applied mathematical problem solving, modelling, applications, and links to other subjects – State, trends and issues in mathematics instruction. *Educational Studies in Mathematics, 22*, 37–68.

Blum, W., Galbraith, P. L., Henn, H.-W., & Niss, M. (Eds.). (2007). *Modelling and applications in mathematics education: The 14th ICMI study*. New York: Springer.

Duffee, L., & Aikenhead, G. (1992). Curriculum change, student evaluation, and teacher practical knowledge. *Science Education, 76*, 493–506.

Grimmett, P. P., & MacKinnon, A. M. (1992). Craft knowledge and the education of teachers. *Review of Research in Education, 18*, 385–456.

Hestenes, D. (1987). Toward a modelling theory of physics instruction. *American Journal of Physics, 55*(5), 440–454.

Hestenes, D. (2006). Notes for a modelling theory of science, cognition and instruction. In E. van den Berg, T. Ellermeijer, O. Slooten (Eds.), *Modelling in physics and in physics education* (pp. 34–65). *Proceedings Groupe Internationale de Recherche sur l'Enseignement de la Physique 2006*, University of Amsterdam, Amsterdam.

Kaiser, G., & Maaß, K. (2007). Modelling in lower secondary mathematics classroom – Problems and opportunities. In W. Blum, P. L. Galbraith, H.-W. Henn, & M. Niss (Eds.), *Modelling and applications in mathematics education: The 14th ICMI study* (pp. 99–108). New York: Springer.

Lijnse, P. (2006). Models of/for teaching modelling. In E. van den Berg, T. Ellermeijer, O. Slooten (Eds.), *Modelling in physics and in physics education* (pp. 20-33). *Proceedings Groupe Internationale de Recherche sur l'Enseignement de la Physique 2006*, University of Amsterdam, Amsterdam.

Löhner, S. (2005). *Computer based modelling tasks. The role of external representation*. Amsterdam: Universiteit van Amsterdam.

Maaß, K. (2006). What are modelling competencies? *Zeitschrift für Didaktik der Mathematik, 38*(2), 113–142.

Muller, E., & Burkhardt, H. (2007). Applications and modelling for mathematics – Overview. In W. Blum, P. L. Galbraith, H.-W. Henn, & M. Niss (Eds.), *Modelling and applications in mathematics education: The 14th ICMI study* (pp. 267–274). New York: Springer.

Ormel, B. J. B. (2010). *Scientific modelling of dynamical systems. Towards a pedagogical theory for secondary education*. Utrecht: Centre for Science and Mathematics Education.

Vollebregt, M. J. (1998). *A problem posing approach to teaching an initial particle model*. Utrecht: Centre for Science and Mathematics Education.

Chapter 66
Evolution of Applications and Modelling in a Senior Secondary Curriculum

Gloria Stillman and Peter Galbraith

Abstract In Queensland Australia, mathematical modelling and applications have featured in senior secondary mathematics curricula for two decades. Part of a longitudinal study of the implementation of this initiative, as seen through the eyes of selected teachers and administrators who have been centrally involved in its development and on-going practice, is reported. The data consist of responses to structured and open interview questions, syllabus documents, and application and modelling tasks designed and implemented by teachers. Perceptions of why modelling and applications are valuable at this level of schooling, the distinction between applications and modelling, how established applications and modelling are in the curriculum, the sources of such tasks, and the sufficiency of support for the development of these tasks by teachers are presented.

1 Introduction

One of the challenging educational issues with regard to the teaching of applications and modelling in schools identified by Niss et al. (2007) was curricular implementation in practice. Three questions relevant to this issue are the following:

1. How are different versions of the modelling process being applied within particular contexts and levels of education, and what are their achievements and ongoing challenges?

G. Stillman (✉)
School of Education (Victoria), Australian Catholic University (Ballarat),
P.O. Box 650, Victoria, Australia
e-mail: gloria.stillman@acu.edu.au

P. Galbraith
The University of Queensland, Brisbane, Australia
e-mail: p.galbraith@uq.edu.au

2. What pressures do teachers indicate are significant in deciding when and how they will incorporate applications and modelling tasks in their teaching? What purposes do they have in using such tasks?
3. How have efforts at incorporating modelling at a curriculum level unfolded, and what impediments have impacted on the outcomes achieved?

Such questions are not answered from single shot designs; but rather they are informed through accumulated evidence and data from studying a variety of initiatives. In this chapter we provide such a contribution by examining one substantial initiative covering 20 years – mathematical modelling in Queensland senior secondary mathematics courses (i.e., Years 11 and 12).

2 Mathematical Modelling and Applications in Queensland

Presently school education in Australia is the responsibility of the respective states, which means that syllabus content and objectives, and assessment procedures, are designed and implemented at state level. Unlike other states, assessment at the upper secondary (pretertiary) level in Queensland is entirely school based, with panels at district and state levels performing critical reviewing roles to assure comparability of outcomes across schools and districts. There are no common external leaving examinations across the system as such. Applications and mathematical modelling were first introduced in 1990–1991 within senior mathematics curricula in Queensland in a limited number of schools. A particular emphasis in the chapter is the development of the initiative from the perspective of implementing teachers and key curriculum figures responsible for its introduction and evolution. For this purpose structured interviews were conducted with 23 individuals (9 at the end of 2005 and 14 at the end of 2007) as part of an on-going study of *Curriculum Change in Secondary Mathematics* (CCiSM).

2.1 Syllabus Objectives

The objectives of the 1989 Trial/Pilot Syllabuses (e.g., Queensland Board of Senior Secondary School Studies QBSSSS 1989) included the identification of the assumptions and variables of a mathematical model, formulation of a model, derivation of results from a model, and interpretation of these in terms of the given situation. Once statewide implementation occurred, despite syllabus refinements over the years (e.g., QBSSSS 1992, 2000; Queensland Studies Authority QSA 2009), the mathematical modelling component remained essentially stable although different emphases brought to the fore different aspects in implementation at different times.

Four approaches to mathematical modelling within curricular implementations have been identified in the Australasian context (Stillman et al. 2008). An approach using *real world examples to motivate the study of mathematics* is employed by some teachers to stimulate positive affect rather than promote mathematics learning (Pierce and Stacey 2006). Modelling as *curve fitting* reduces model development to the exercise of regression options on technological devices, without regard to the mathematical structures underlying the data – effectively distorting the modelling process by overemphasizing one (albeit important) possibility within the solution process. At times mathematical modelling is used as a *vehicle* for teaching mathematical content, emphasizing the provision of "an alternative – and supposedly engaging – setting in which students learn mathematics without the primary goal of becoming proficient modelers" (Zbiek and Conner 2006, p. 89). In contrast, treating modelling as *content* (e.g., Ikeda et al. 2007) aims to equip students with skills that enable them to apply and communicate mathematics in relation to solving problems in their world. This approach enables students to use their mathematical knowledge to solve a range of real world-related problems and simultaneously scaffolds their development as modelers. The generic objectives of the Queensland syllabuses provide for both these purposes where the intention is for the continuing presence of modelling across the four semesters of Year 11 and 12 mathematics subjects.

2.2 Implementation in Schools

Since, as noted above, assessment at upper secondary level in Queensland (years 11 and 12) is entirely school based, individual schools and teachers design specific work programs (including assessment tasks) within syllabus expectations. Hence, a key element is the translation of general objectives into specific criteria for teaching, learning, and assessment of school-based activity. In alignment with syllabus objectives, assessment criteria such as those for a task based around koala populations in Australia using information from the Australian Koala Foundation website (https://www.savethekoala.com/) (see Fig. 66.1) are developed by teachers in schools. To evaluate the student modelling efforts, the following criteria were used as a basis for awarding credit:

- Appropriateness of interpretation of data
- Reasonableness of assumptions
- Quality of mathematical model for population changes

The enclosed pages (data from the Koala Facts pages of the Australian Koala Foundation website Media Centre and a newspaper article by Harbutt, 2004) provide information about koala populations in Australia. You are to develop a mathematical model based on Leslie matrices to represent the population changes of the koala population. Choose some region in which the population might be increasing, investigate the growth over at least 10 years, and then use this to discuss ways to maintain a population that the environment can sustain.

Fig. 66.1 Koala task

- Justification of choice of values for model parameters
- Discussion of strengths and weaknesses of model
- Evaluation of model and recommended changes to maintain a suitable population

This exemplifies how a school has turned the syllabus general objectives into criteria for assessing student work. For this to be successful, a viable pathway between syllabus statements and school implementations must exist and be able to be validated by moderating panels. The extent to which schools and teachers in general find this essential process workable and valid in terms of syllabus expectations, given other administrative and curricular pressures, acts as a litmus test for evaluating the continuing implementation of this initiative of mandating modelling in syllabus requirements. To pursue this goal more broadly an on-going longitudinal study is being conducted into this curriculum initiative.

3 Research Methods

Syllabus and review documents from the late 1980s up to the latest syllabuses implemented in 2009 were examined. In addition purposeful samples of 5 *key curriculum figures* (e.g., nonteacher members of expert advisory committees or statutory authority officers overseeing syllabus implementation), 6 *key teachers* (i.e., secondary mathematics teachers in *key* implementation roles such as members of state review panels or district review panel chairs), and 12 secondary *mathematics classroom teachers* were selected. The teachers were representative of several school districts and of state, Catholic and independent schools systems. A series of questions covering the period of introduction and later periods of widespread implementation and modification were asked of these 23 interviewees. In addition, practising teachers provided artefacts, usually in the form of tasks that typified their use of real world applications and modelling in teaching and assessment, and their use of technology in these contexts. In order to identify emergent themes within the interview responses, and the teaching and assessment artefacts, these data were entered into an NVivo 8 database (QSR 2008) and analysed through intensive scrutiny of the data to develop and refine categories related to these themes. This chapter addresses emergent themes related to interviewee responses to the following questions which link to the three general questions articulated in the introduction:

1. Why was the introduction of applications and modelling considered to be a valuable initiative in senior secondary mathematics?
2. While the general objectives of the 1992 syllabuses were for mathematical modelling, it could be argued that the implemented syllabuses were mainly concerned with mathematical applications. Does that distinction have meaning for you?
3. Do you believe that mathematical modelling is an established part of upper secondary school practice in Queensland? Why?

4. The courses require a balanced assessment plan that includes a variety of techniques such as extended modelling and problem-solving tasks and reports. What is your source of tasks? Do you believe you have sufficient support to develop tasks, or have you gone past that?

4 Findings

4.1 Reasons Applications and Modelling Valuable Initiative at Senior Secondary

At the upper secondary level applications and mathematical modelling are not always seen as relevant let alone central to the mathematics curriculum. The interviewees in this study, however, having experienced the realities of curriculum implementation at this level of schooling had little trouble providing arguments for why modelling is seen as valuable in the context of the mainstream upper secondary mathematics curriculum. The spectrum of responses is shown in Table 66.1.

Most were highly positive as typified by the following response:

The content is important and you need the content but it is the application of that content in a way that is meaningful and the kids can make sense of that so there is a purpose for it. Nowadays, unless the kids can see a reason for doing something they just park up [refuse to budge]. You will get the small handful of kids who will just jump through any hoop ... but the vast majority of kids, unless there is a reason for doing it, they just won't engage. [Key Teacher 3, Nov., 2005 interview].

Two interviewees expressed reservations as exemplified here by a classroom teacher:

I think there is a huge divergence of opinion still...people who say that students are not getting enough practice at the nitty gritty, and trying to solve problems...is a bit

Table 66.1 Frequency of arguments pro modelling in upper secondary curriculum

Arguments	Frequency
Promotes mathematical understanding/ thinking rather than regurgitation	9
Makes mathematics applicable/demonstrates utility of mathematics	9
Demonstrates relevance of mathematics	7
Engenders student interest in mathematics promoting engagement	6
Promotes variety in teaching strategies	3
Promotes enjoyment in mathematics	2
Promotes success leading to retention of students in mathematics	2
Develops a world view of modelling to solve problems	2
Counterbalances abstract mathematics	1
Enhances recall of mathematical concepts and processes	1
Appeases employer demands	1

like putting the cart before the horse. They really ought to be spending more time understanding the concepts and looking at application later on. [Classroom Teacher 1, Nov., 2005 interview].

4.2 Distinction Between Applications and Modelling

The objectives of the 1992 syllabuses were meant to encourage development of problem solving and mathematical modelling skills. The latter focus was generally not evident in implementation (Stillman 1998) and this was confirmed by most interviewees reflecting back on this syllabus implementation. The main reason for this was seen as mathematical applications being much closer to what teachers were already doing whilst mathematical modelling was not well understood.

They didn't understand what mathematical modelling was about basically—the whole idea of building a model, use a set of data, build a model, then generalising that model. They probably tended to look at an application where the kids had a model or something like that and the kids had to do something with it…so I would say it was probably lack of understanding of what was expected. (Key Teacher 3, Nov., 2005 interview).

Other reasons given for modelling being less likely to be taken up were that it was perceived as difficult and time consuming resulting in it often being confined to alternative assessment.

Applications could easily descend into teaching this application, and that application, and that application. So today we will do the application of this Calculus we have been doing to this, this biological situation here and that is a lot easier to do I think than teaching the other more messy business with the modelling. … teachers generally felt that they didn't have much time and therefore it is the modelling sorts of things that were really time consuming. (Key Curriculum Figure 3, Dec., 2007, interview).

There was, however, a group of enthusiasts who pursued modelling and often these teachers were also technology innovators and this resulted in an alternative route for modelling to grow in prominence through the period of the 1992 syllabuses. "*Graphing calculators … were valuable tools in any modelling exercise…that removed the grinding mathematics which was the aim of a lot of people and allowed people to discuss applications without being bogged down in the arithmetic of doing it*" (Classroom Teacher 7, Oct., 2007 interview). Furthermore, when in-services were conducted for new technologies, "*the material being used was modelling material … giving teachers some idea which way to go, how they could use the technology and develop modelling*" (Key Teacher 5, Dec., 2007 interview).

Gradually monitoring panels applied more pressure to make sure all general objectives with respect to modelling were being fully addressed. Some teachers embraced the new changes after an initial period of caution as expressed by the following teacher:

I was very nervous at first but I really like the approach … From my point of view I think it gives us, with the technology that we have got on hand, a chance to do

*mathematics that has more meaning for the students so you can use real life data....
It has given me great insight into how students actually think. (Key Teacher 5, Dec.,
2007 interview).*

Sometimes efforts by panels to assure better teaching of the general objectives as a whole had unexpected outcomes. Ensuring students address particular elements such as "extending and generalising from solutions" in the standard for, say, an A in *Modelling and Problem Solving* resulted in some mechanistic approaches as the following demonstrates.

Our kids will ask us, 'Have I got enough G's? Have I generalised enough?' because we actually count how many times. I don't suppose a lot of schools do that. And we think. 'Oh, we might have to put a G on the next test' or we might have to put a 'refine the model', something which gives them the opportunity to demonstrate what we call A attributes in that criteria [sic]. So I just feel sometimes that they are jumping through hoops. (Classroom teacher 6, Nov., 2007 interview).

In some cases this could be construed as a response by schools to a perceived desire by the panels for schools to show explicit evidence of developing particular competencies in tasks targeting these specifically (see Fig. 66.2), or alternatively, it could be due to the continued persistence of a lack of differentiation between applications and modelling.

One of the interviewees, for example, had been involved in the implementation as a classroom teacher and mathematics department head since the early 1990s but still in 2007 did not distinguish between applications and modelling despite this long involvement. Not surprisingly, she did not recognise the essential elements in the assessment criteria distinguishing modelling from applications. She described the panel's focus on such elements of the criteria as "nit picky" continuing: *"I would like to see this trimmed down. I am happy with things like synthesis of procedures and strategies, I think that is very important in the problem solving; selecting appropriate procedures for the particular modelling question involved"* (Classroom teacher 5, Nov., 2007 interview). However, when it came to exploring the strengths and limitations of a model, extending and generalising from solutions, recognition of the effects of assumptions used, evaluation of the validity of arguments, these were all considered superfluous.

A streamlined object falls from a satellite towards earth. One model to represent the motion of this object is to consider there is no air resistance. Considering downwards as positive, the velocity at any time will be given by $v = gt$. How long will it take for the object to be travelling at $40 ms^{-1}$? Another model to represent the motion of this object is to consider there is air resistance to the motion producing a retardation of $0.2v$, so that the acceleration of the object is given by $a = g - 0.2v$, where downwards is considered positive. Solve the differential equation $\frac{dv}{dt} = g - 0.2v$ to find an expression for velocity at any time t. How long will it take now for the object to be travelling at $40 ms^{-1}$? Compare the two answers in light of the assumptions made and the limitations of each model.

Fig. 66.2 Task exploring strengths and limitations of models and effects of assumptions

4.3 Embedding Applications and Modelling in Current Practice

Opinions about the robustness of modelling as an established practice in Queensland upper secondary schools were almost equally divided between those who agreed it definitely was and those who saw a continuum from minimalist approaches to very rich. In the main, classroom teachers agreed modelling was established often influenced by what was happening in their local context. Teachers in key implementation roles gave a broader view. "*I still think there are schools who teach all the purely mathematical aspects of it and they'll throw in the pseudo-real world problem to satisfy us.*" (Key Teacher 5, Dec., 2007, interview). In contrast, two curriculum figures most remote from classroom practice thought there was little modelling occurring. The most positive interviewees saw the establishment of modelling in classroom practice as a "*slow evolutionary process*" (Key Teacher 4, Dec. 2005) strongly supported by the panel system and the insistence on alternative assessment. A tolerance for diversity in uptake of modelling has led to the situation where "*everybody has moved a certain amount, some people have moved a large amount, others are still being pushed*" (Key Teacher 4).

4.4 Designing Tasks

Questions relating to task design were asked of only teachers from the classroom and key teacher groups (i.e., $n=18$). All teachers interviewed expressed satisfaction with their own or their mathematics department's capability of designing suitable tasks seeing this as a creative, enjoyable and interesting task.

I think we are now getting deluged with applications... Teachers are thinking of them themselves now. It's been a wonderful development in the 15 years since that 92 syllabus came in. (Classroom Teacher 9, Nov., 2007, interview).

It can take you several hours to start this little idea off and then get the things together and then write it and trial it and share it with someone else and get the criteria sheet right but I like doing it. I enjoy doing that. (Key Teacher 5, Dec., 2007, interview).

Teachers interviewed were generally satisfied that they had sufficient support to produce or source tasks by themselves through colleagues, the Board or professional bodies as indicated by this classroom teacher.

I am lucky in a sense that I've been involved, so that I've got people that I could contact for help. Personally, yes, I feel I would have the support. (Classroom Teacher 12, Nov., 2007, interview).

Some reservations were expressed regarding other schools especially from interviewees on local monitoring panels and with respect to teachers experiencing geographical or professional isolation as Queensland is a large state with a diversity of school sizes.

Table 66.2 Most frequently mentioned sources for applications and modelling tasks

Sources	Frequency
Teaching colleagues	12
Personal experiences	9
Books	8
Professional journals	4
Newspapers	4
Monitoring panel meetings	4
Presentations at professional gatherings (national, regional, local)	3

I think I am pretty right myself and the staff here are pretty good. They are coming up with some good stuff too…other schools you see on panels (like 1 in 5) are having problems I think. (Classroom Teacher 11, Nov., 2007, interview).

You need to contact other schools out west though, because they do struggle. (Classroom Teacher 10, Nov., 2007, interview).

The main sources of ideas mentioned by interviewees for tasks are given in Table 66.2. Colleagues, personal experiences and print sources figured strongly, but the internet, television and the commercial world infrequently.

5 Discussion and Conclusion

With respect to the responses of participants, we note that arguments for inclusion of modelling as a valuable component of the upper secondary mathematics curriculum were, in the main, predominately in service of the promotion of mathematics learning by students at this level along the lines of modelling being one of the "tactical devices to improve the situation for traditional mathematics instruction" (Blum & Niss, 1991, p. 47). Despite being emphasised in the rationale of the syllabuses, there was no mention of critical competence, that is, "preparing students to live and act with integrity as private and social citizens" (p. 43). Opportunities provided by modelling were taken up by those teachers welcoming a syllabus supporting a desire for change. Others wanted to remain within familiar territory by interpreting 'applications' as little different from previous activity as noted by Stillman (1998) in the early implementation. Several teachers embraced the mutually supportive relationship between modelling and technology use.

The teachers felt they had sufficient support to produce appropriate tasks but raised concerns on behalf of more isolated colleagues. Suggestions from colleagues and personal experiences were the most frequent sources of tasks but internet sources were not amongst these. The tasks in Figs. 66.1 and 66.2 illustrate the variety exhibited in this task development. The former is much more open, with the making of assumptions, choice of mathematics, and interpretation in context essential aspects of this modelling problem. The latter is an application where essential activities central to modelling are absent. The elaboration of modelling assessment criteria in a later syllabus revision was generally viewed as helpful in providing enhanced guidelines for task production and assessment of performance. However, there was evidence of some teachers continuing with minimalist approaches, attempting to assimilate challenging new requirements into traditional conservative practices. If review panels are viewed as agents for change and guardians of comparability, evidence of implementations along the continuum from minimalist to very rich suggests that these functions require further work. The assessment by interviewees of current practice reflected this variety. A deliberate tolerance for diversity in uptake of modelling has ensured some progress in all schools, perhaps more than if full immediate compliance was required.

To conclude, we reflect back on the three general questions listed in our introduction with regard to curriculum implementation in practice. For the first, we have obtained a cross-sectional view of representative implementations at senior secondary level in a state context. There is evidence that implementations include opportunities for students to use their mathematical knowledge to solve a range of real world-related problems and at the same time scaffold their development as modellers but this is by no means universal. For the second, our interviewees have indicated pressures that apply to teachers in meeting syllabus requirements, and the challenges they feel these impose on what can be achieved. The main purpose for using such tasks was to enhance the uptake and learning of mathematics. Finally, useful insights have been obtained into ways in which changing focus on assessment criteria have been used (or not) to facilitate the teaching of modelling at senior secondary.

References

Blum, W., & Niss, M. (1991). Applied mathematical problem solving, modelling, applications, and links to other subjects: State, trends and issues in mathematics instruction. *Educational Studies in Mathematics, 22*(1), 37–68.

Harbutt, K. (2004, May 8–9). How much can a koala bear? *The Weekend Australian Magazine*, pp. 18–21.

Ikeda, T., Stephens, M., & Matsuzaki, A. (2007). A teaching experiment in mathematical modelling. In C. Haines, P. Galbraith, W. Blum, & S. Khan (Eds.), *Mathematical modelling (ICTMA12)* (pp. 101–109). Chichester: Horwood.

Niss, M., Blum, W., & Galbraith, P. (2007). Introduction. In W. Blum, P. Galbraith, M. Niss, & H.-W. Henn (Eds.), *Modelling and applications in mathematics education* (pp. 3–32). New York: Springer.

Pierce, R., & Stacey, K. (2006). Enhancing the image of mathematics by association with simple pleasures from real world contexts. *ZDM, 38*(3), 214–225.

QBSSSS. (1989). *Trial/pilot senior syllabus in mathematics C*. Brisbane: Author.

QBSSSS. (1992). *Senior mathematics B*. Brisbane: Author.

QBSSSS. (2000). *Mathematics B senior syllabus 2001*. Brisbane: Author.

QSA. (2009). *Mathematics B senior syllabus 2008*. Brisbane: The State of Queensland.

QSR. (2008). *NVivo v.8 [Computer software]*. Melbourne: QSR.

Stillman, G. A. (1998). The emperor's new clothes? Teaching and assessment of mathematical applications at the senior secondary level. In P. Galbraith, W. Blum, G. Booker, & I. Huntley

(Eds.), *Mathematical modelling: Teaching and assessment in a technology rich world* (pp. 243–253). Chichester: Horwood.

Stillman, G. A., Brown, J. P., & Galbraith, P. L. (2008). Research into the teaching and learning of applications and modelling in Australasia. In H. Forgasz et al. (Eds.), *Research in mathematics education in Australasia 2004–2007* (pp. 141–164). Roterdam: Sense Publishers.

Zbiek, R. M., & Conner, A. (2006). Beyond motivation: Exploring mathematical modeling as a context for deepening students' understandings of curricular mathematics. *Educational Studies in Mathematics, 63*(1), 89–112.

Chapter 67
Sense of Reality Through Mathematical Modelling*

Jhony Alexander Villa-Ochoa and Carlos Mario Jaramillo López

Abstract We present the results of a research project which deals with a qualitative case study in the field of mathematical modelling. The study investigates the role of "real" life modelling situations of the learner in the construction of mathematical knowledge at school. The study of episodes, interviews, questionnaires, and direct observations allowed analysis on how teachers describe their teaching performance when approaching the content of school mathematics. However, the most important aspect was to be able to detect the necessity of a "sense of reality", which is characterized in this paper. The development of this research project shows that it is still necessary to develop a sense of reality as a tool to facilitate interaction between the sociocultural context and school mathematics, all through modelling.

1 Introduction

Modelling and model application have their roots in the study of *real world* problems, which have served both the mathematician applied to the development of theories that explain phenomena, and the mathematics education teachers and researchers who approach the study of such reality, connect it to mathematical knowledge, and use it as a teaching resource for the learning of mathematics. Mathematics teachers who use modelling as a resource to introduce mathematics concepts become engaged in reflection processes about real problems, their relevance, and a way of approaching them in their classrooms. This paper describes the way *sense of reality* is understood as the knowledge that mathematics teachers have.

---

*This research was funded by the Committee for the Development of Research – CODI and the Regionalization Direction of the University of Antioquia in Medellín-Colombia.

J.A. Villa-Ochoa (✉) and C.M.J. López
Research group of Mathematics Education and History, University of Antioquia,
Calle 67 No 53 108 of 4-108 Medellín, Colombia
e-mails: javo@une.net.co; cama@matematicas.udea.edu.co

2 Brief Description of Modelling in Colombian Educational Regulations

Colombia, since the publication of the document *Lineamientos Curriculares* (curricular guidelines) in 1998, has given mathematics a wider scope, which makes it possible for the students to use their knowledge outside the classrooms in settings where they can formulate a hypothesis and make decisions to face new situations and adapt to them. In this sense, Ministry of National Education [MEN] (1998, p. 35) states that it is necessary to relate the content of learning to students' daily life, presenting them in the context of problem situations and exchange of points of view. According to this perspective, one of the purposes of teaching mathematics at school is to develop mathematical thinking and therefore, modelling and problem solving are fundamental to achieve this goal and at the same time overcome the perspective of "concept transmission" that is sometimes held at school regarding mathematics teaching.

Among the arguments that support the importance of modelling in Colombian classrooms, MEN (1998, p. 101) states that modelling allows students to observe, reflect, discuss, explain, predict and revise, and thus build mathematical concepts based on meaning. Therefore, it is considered that all students need to experiment with mathematization processes that lead them to discover, create, and establish models at all levels. Villa-Ochoa and Ruiz (2009) claim that both MEN (1998) and MEN (2006) suggest incorporating modelling and problem solving processes, but do not make reference to the elements on which such processes converge or differentiate, which is indispensable at the moment of classroom implementation, since the classroom is where "reality " makes sense.

In spite of the fact that modelling was incorporated as a process in the curricular guidelines of the Ministry of National Education (MEN), we have not found enough evidence to observe an important development of this process in the classroom, which is in keeping with the abundant empirical data reported in international literature (Kaiser and Maaβ 2007).

3 The Project

3.1 The Context

This research counted on the participation of four teachers who worked in various levels of public schools. The four teachers hold bachelor degrees in Mathematics and one of them holds a specialization degree in mathematics teaching. They worked in a Colombian subregion 3.5 h away by road from Medellín, one of the main cities in Colombia. The four teachers volunteered their participation in the research project after working as *teaching internship* counselors with four students in the undergraduate program Bachelor of Education in Mathematics in the same subregion. The teachers were observed by the researchers while they were engaged in their teaching practice from April to June 2008. Later on they answered a questionnaire, and held a 5-hour meeting in which they discussed episodes related to modelling and finally had an interview. Based on the results, we established some beliefs that teachers have in the presence of elements corresponding to the modelling process (reality), which determines certain practices in the classroom.

3.2 Methodological Approach

The project paid special attention to the ways teachers acknowledge the importance of modelling in the classroom and the practical work used to implement mathematics at school. This way, the main questions of the project were: (a) What beliefs do teachers have regarding the mathematical modelling process in the classroom? And (b) What factors enable or not enable the implementation of mathematical modelling as a process in the classroom? In this chapter we present some results of the second question, in particular some aspects of sense of reality. In that sense, we adopted case study as our research method. Yin (2009) establishes that a case study is an empirical inquiry that investigates a contemporary phenomenon within its real existence context, when the boundaries between the phenomenon and the context are not clearly evident. We recorded information gathered through various means (audio, video, field diaries, and other written records) to later on organize and analyze it through a data triangulation process. According to Yin (2009, p. 115) "the most important advantage presented by using multiple sources of evidence is the development of converging lines of inquiry, a process of triangulation and corroboration". In Table 67.1, we present other details of instruments and some results.

4 Results

We return to the concept of *sense of reality* presented in Villa-Ochoa et al. (2009) and discuss how such meaning, regarding modelling and applications, is influenced by the relationship between the subject's (teacher) academic training and his/her interaction with sociocultural contexts. In this sense, we report the cases of Alberto and Alexander[1]; Alberto is a teacher who had about 13 years of experience, and Alexander had 5 years of experience. Both Alberto and Alexander have been mathematics teachers in several grades of elementary and high school (students from 11 to 18 years of age). These cases were chosen due to the close link between

---

[1] Alberto and Alexander are not their real names.

Table 67.1 Some instruments and results in this research

Instruments/records/dates	Purpose	Achievements
Classroom observation/ Field diary Data: from April to June, 2008	To identify the main trends, methodologies, activities, and tasks used by teachers	It was found that teachers still use explanatory strategies in sequences such as: introduction → concept definition → explanation → examples → exercises and/or application → evaluation
Episode analysis/audio and video Data: collected on the second week of November, 2008	To identify some beliefs about school mathematics and contribute to the reflection and interpretation of several modelling situations	There was reflection about the role of mathematics when solving real everyday problems. Also, about the need to develop a sense of reality that makes it possible to link the real world and school mathematics.
Questionnaire/written information Data: collected in the first week of December, 2008	To collect information about the *why* and the *how of* mathematics in Secondary School (11–15 years of age) and to identify some beliefs about school mathematics	Three types of situations used in the classroom were found: (1) Situations or exercises that favor made-up reality (Alsina 2007), (2) prototype exercises within mathematics and (3) exercises related to the themes studied, applied to generally artificial contexts
Interview/audio and video Data: collected on the second week of February, 2009	To collect information about beliefs and the why of real problems in school mathematics	Three beliefs were found: (1) Mathematical concepts are first taught and then applied, (2) constructing a model is a matter of finding the representation of the situation, and (3) a real problem is any situation that uses every day words and that shows the applications of mathematics

the concepts the teacher has about *reality* and the way he is observed in his teaching practice regarding modelling, so we need to study not only the conceptions teachers have about mathematics and mathematics teaching, but also the ways they observe their own teaching practice.

4.1 What is Sense of Reality?

The concept ***sense of reality*** was present by Villa-Ochoa et al. (2009) as *sensitivity that a teacher should have towards reality, that also includes intuition and the* *capacity to detect situations and opportunities in the sociocultural context towards which students' knowledge can be geared. Such sense includes a good dose of imagination and creativity.* A *sense of reality*, more than a rational component of teacher's knowledge, is a subjective component that metaphorically acts as a magnifying glass with which the teacher observes *objective reality* and facilitates the (re)significance of such reality as of a mathematical modelling process. The concept appeared as a need observed in a teacher, in whom we identified certain positive attitudes towards mathematics and their role in teaching. This teacher made evident a search for situations that would establish relationships between school mathematics and the real world. However, such a search was unattainable, since in most cases, they were situations characterized as made-up reality, manipulated, faked, or far fetched (Alsina 2007).

4.2 What is Reality for Alberto and Alexander?

Alberto	Alexander
Episode No 2, shows the following word problem: "*A family of four (4) people has invited three (3) friends to eat at home. How many places will be set at the table?*" (MEN, 1998: p 78). The teacher said that:	In episode No 1 "The cartoon" (see Attachment 67.1) Alexander said: We have seen that question many times. You get there and tell them [the students] what they are good for [mathematics]; you even make diagrams for them to show them how they can use math. However, the kid [student] does not want to look at it, uh, it is not part of mathematics, but the procedural part, the part where it is used. But the kid does not want to learn the concepts and I think that if something is missing in mathematics, it is learning concepts.
I would think that it is not [real], it may not happen [...] my kids would say "we don't have food for one, let alone to invite three" What is real, the language used? what is common in reality is what I live and what may actually happen [...]. For him a real situation is not something that exists per se, but something that exists in the ambit of the language he knows, he sees, he has touched, he can work with, but not those real situations that are real, but out of context, or concealed and actually all the problems posed are of the same kind: cross a river, throw a liana, estimate an angle, measure the distance. Anyway [...]. To ask us to use real situations to solve problems, no. They are rather common situations, and the language; it wouldn't justify [mathematics] usefulness either.	I study auto mechanics, an associate degree, and one of the electronics professors commented that electronics would be practically impossible without mathematics because everything that has to do with circuits and everything else, is actually mathematics. So you start to observe that and feel motivated, but with the kids, you can show them how to do it, how to take it to real life. However you are followed by the same two or three kids that are always interested, and the others [say] "*It is not that I am not interested in the device, I just want to know what it is good for*" that's then, the great difficulty.

We also observed how for Alberto, *reality* is a characteristic that situations should have per se, that have meaning for students, which becomes a tool, and when implemented in the classroom, helps the students see the importance of mathematics; it is a way to answer the everyday question *what is mathematics good for?* However, the incorporation of such *reality* through situations in the classroom is a dream that for him is still difficult to achieve. Alexander identifies himself with the answer given by the teacher in the cartoon, to his students, and considers that students are not interested in such reality. In that sense, it is observed that the list of applications that can be presented to the students, makes part of a *far fetched, not really attainable reality*, since even though mathematics originated such applications, it is also true that such reality is not part of school discourse, as it only enumerates a series of applications through which the students cannot visualize why and how mathematics played an important role in the development of those applications. It is observed, then, that while for Alberto reality is closer to situations lived by the students in their context, for Alexander reality is given by a series of technological and/or scientific developments that play an informational role, but do not manage to make the students participate in their construction.

Attachment. Episode N° 1. The cartoon

Attachment 67.1 Episode No 1. The cartoon

4.3 Sense of Reality in School Mathematics

Alberto	Alexander
In episode No 1, "The cartoon". The teacher said that:	In episode No 1, "The cartoon". The teacher said that:
I say that it depends on the grade you are going to; I say, if you are going to sixth grade, I think that, if, if mathematics didn't have clothes. […] but anyway, I don't have enough [enough clear arguments]. What is it good for? […] it is in college where you are gonna need it buddy, it's going to be tough [you'll have a lot of difficulty] but I know that it is an answer that comes out of my lack of capacity to tell them: uh that trigonometric identities, uh; I don't know; of course. When I don't know, I have said I don't know, but in this particular case, I haven't' been able to […] We have to tell the students. ¡You have to study it because you have to!	Regarding this theme, you say, "if the binary system didn't exist, the airplane wouldn't fly". How do you link it [relate] to electronics. But how do you take the kid there, if you are looking at, for example, the binary system, out of the binary system? That problem has a lot of mathematics. So, you show the students a minimum part and it is not enough for the kid, because the kid says "show me everything"

Both for Alberto and Alexander, the question, what is mathematics good for? presented in the cartoon, appears in many different ways in the classroom, usually in the local contexts of mathematics, which is to say, in a specific theme or concept. However, the way of facing it is different in both teachers. Alberto acknowledges that the training he received in college was not enough to unveil the relationship existing between mathematics and the real world, and his relationship with the world has not been wide enough to develop a sense of reality that allows the identification of such relationship in order to design situations that help him shape them into classroom situations. On the other hand, for Alexander, his knowledge of Electronics has helped him use this discipline to "mention" some applications of mathematics, without making it an object to design situations through which the students face serious modelling processes for experiments, search for data, variable identification, regularities, abstractions, and possible simplifications; all of them elements of modelling as a classroom process. It is observed, then, how Alexander's experience with academic contexts in electronics has made it possible for him to expand his relationship with mathematics and the real world. However, it has not been enough to develop a *sense of reality* that involves the modelling of situations in a sociocultural context.

4.4 A First Approach

Seeking teachers' reflection about the role of real situation modelling problems in the classroom, a series of situations were proposed for teachers in order to analyze aspects like: appropriateness in the classroom, reality within the situation, appropriateness of the situation in the particular classroom context, the mathematical concept involved in the situation, and the specific school grade in which the situation is developed. In that respect, the following comments appeared:

Alberto	Alexander
In episode No 3 (see Villa-Ochoa et al. 2009). The teacher said:	In episode No 2, 3, 4. The teacher said:
I think this situation can be used with the students in ninth grade. Even though it is not a problem per se in their reality, I think it can be adapted to utilities service or Internet service, but within the companies in the township where I work. Also, for example, the bills in the hotels or banks can be used to work mathematics in class and ask questions they can start working on.	They are real situations that students might like, but the problem is, they would say they want the formula to solve the questions, so they won't find the answer. However these are situations that require the kind of mathematics they are being taught; the problems are not so big that the students don't have enough mathematical knowledge to solve them.

It can be observed that the discussion about some episodes (modelling situations created in other contexts) presented elements that the teachers could assume in a critical way, claiming that they wouldn't be appropriate in the students' context. Throughout this discussion, the teachers make some reflections on the subject of their own contexts, and establish elements for the design of situations that are adequate for their context. Though new reality views were developed, for the teacher, it is not enough to affirm that sense of reality was developed by these teachers.

5 Discussion

The question *what is Mathematics good for?* appears very often in mathematics classes and brings along an invitation for the student to understand, somehow, the mathematical concepts locally[2] addressed from a more realistic and useful perspective. However, it was found that such a question puts the participants in this project in deep trouble, since even though they showed command of global concepts in which mathematics is "applicable" they had difficulty posing particular situations for local concepts. The teachers agreed that *mathematics is everywhere!* but, like the teacher in the cartoon, it poses examples that relate to scientific discoveries restricted to some kind of reality that is not really accessible to the students, thus preventing them from assuming the challenge of implementing modelling processes.

---

[2] When we say locally, we refer to the mathematical concepts presented in a particular class or a segment of a class.

The statements above confirm the need to develop a *Sense of Reality,* with which the teacher can use modelling as a tool to answer questions at a local level, instead of just giving a general list of themes and applications that give the students a passive role when considering mathematical concepts. Through a s*ense of reality,* teachers can recognize situations from their sociocultural context to use with the students, letting the role of modelling and applications of school mathematics overcome the concept that mathematics is just "information or data" modeled by "others", to acquire a more meaningful role where the students become part of knowledge development.

Based on the above, we consider that when facing real situations in the sociocultural context of the classroom, modelling becomes a tool that allows the (re) significance of such contexts. But in addition to this, we think that modelling must advance toward the notion of *practice that includes the (re)elaboration and interpretation of models already developed.* Therefore, the *problems* must be assumed as *real context problems;* understanding *real context as* daily, social, cultural, consumption, and other science contexts, where students are required to identify and manipulate data, simplify and make abstraction of amounts and variables in order to build a model to solve the problem.

6 Conclusions

The cases reported in this article are evidence that there are teachers who have learned that mathematics is everywhere in nature, and in all other sciences, and in the context; but they haven't learned how to unveil it yet. S*ense of reality* built through mathematical modelling outlines the need to unveil mathematics within the sociocultural context. It must transcend beliefs about reality as something artificial, made up, dressed, to put ourselves into a kind of reality closer to the needs of the students' contexts.

In the case of Alexander, he points out the fact that *sense of reality* is strongly influenced by the academic situations lived by the subject regarding mathematics, even though in this case "technical" knowledge of mathematics can be observed. However, contact with modelling situations belonging in the sociocultural context seems to propitiate a certain degree of familiarization and a new look at reality out of school, thus promoting alternatives to seek new contexts to work modelling in the classroom. This way, *sense of reality* may permeate the development of mathematical knowledge in school, since according to D'Ambrósio (2005) the development of mathematical knowledge in school is mainly built as of the way the individual acknowledges reality and its various manifestations (individual, social, planetary, and cosmic reality).

The *sense of reality* must make possible the appraisal of *school realities* (Alsina 2007) and promote the implementation of other realities closer to daily life in the students' sociocultural contexts. This way, *close or tangible reality* (see, Villa-Ochoa et al. 2009) can be privileged in the first instance, since being part of the students' contexts, transcends the idea of *possible contexts* to set up real problems that make it possible for students to have a critical perspective of some social demands (such as consumption). This is a way to comply with one of the social duties of modelling in school mathematics, since through a *sense of reality*, modelling not only is in charge of the interpretation and solution of real life problems, but also promotes its transformation (subjective reality) by (re)signifying such (objective) reality.

These case studies have created certain implications for teaching undergraduate and graduate programs, since, based on Alberto's case, the need for reflection about the sociocultural context to develop a *sense of reality,* is clear. It demands serious research processes that should inquire into various aspects, among them:

- The ways teachers interpret reality.
- The ways teachers think that a situation "is in harmony" with the textbook.
- The way teachers consider that a situation "accommodates" or adjusts to school reality.
- The level of comfort and appropriation that the teachers have about the sociocultural context.
- The "rationality" that teachers assign the phenomenon related to the concept to be built.
- The level of "training" that teachers have regarding the modelling process.

Finally, we consider that just awakening a *sense of reality* does not transform conditions in the classroom, but without such a sense, there is a risk of remaining in the knowledge transmission model, which does not know about the tools offered by the sociocultural context to build mathematical knowledge at school.

The results of this research raise others new questions for future inquiry:

- Are there relationships between *sense of reality* and teacher's mathematical knowledge?
- How can a *sense of reality* be developed?
- What are the relationships between *sense of reality* and competences in mathematical modelling?

References

Alsina, C. (2007). Si Enrique VIII tuvo 6 esposas, ¿cuántas tuvo Enrique IV? *Revista Iberoamericana de Educación* (43), 85–101.

D'Ambrósio, U. (2005). Sociedade, cultura, matemática e seu ensino. *Educação e Pesquisa, 31*(1), 99–120.

Kaiser, G., & Maaβ, K. (2007). Modelling in lower secondary mathematics classroom – Problems and opportunities. In W. Blum, P. Galbraith, H.-W. Henn, & M. Niss (Eds.), *Modelling and applications in mathematics education. The 14th ICMI Study* (pp. 275–284). New York: Springer.

Ministerio de Educación Nacional [MEN]. (1998). *Lineamientos Curriculares: Matemáticas.* Bogotá: Magisterio.

Ministerio de Educación Nacional [MEN]. (2006). *Estándares básicos de competencias.* Bogotá: Magisterio.

Villa-Ochoa, J.A., & Ruiz, M. (2009). Modelación en Educación Matemática. Una mirada desde los Lineamientos y Estándares Curriculares Colombianos. *Revista Virtual-Universidad Católica del Norte* (27), 1–21. With access by http://revistavirtual.ucn.edu.co/index.php?option=com_docman&task=doc_download&gid=57&Itemid=21.

Villa-Ochoa, J. A., Bustamante, C. A., Berrio, M., Osorio, J. A., & Ocampo, D. A. (2009). Sentido de realidad y modelación matemática. El caso de Alberto. ALEXANDRIA. *Revista de Educação em Ciência e Tecnologia, 2*(2), 159–180. With access by http://www.ppgect.ufsc.br/alexandriarevista/numero_2_2009/jhony.pdf.

Yin, R. (2009). *Case study research, design and methods.* Thousand Oaks: Sage.

Chapter 68
What Is 'Authentic' in the Teaching and Learning of Mathematical Modelling?

Pauline Vos

Abstract There are different perspectives on the use of the adjective 'authentic' (c.q. the noun 'authenticity') in the teaching and learning of mathematics, and in particular in mathematical modelling. Researchers use a variety of meanings, describing authentic tasks, authentic situations, authentic learning environments, authentic models, and so forth. On the one hand, authenticity refers to being genuine (true, honest); on the other hand, authenticity refers to properties of simulations (copies) of out-of-school aspects. I use a framework from sociology to describe authenticity as a social construct, pointing at a number of definition problems. I will propose a pragmatic resolution to these problems with a pragmatic definition of authenticity for separate aspects in tasks (themes, resources, activities) if these are "clearly not created for educational purposes".

1 Introduction

The terms *authentic* and *authenticity* are frequently used in research on the teaching and learning of mathematical modelling. Researchers report on authentic contexts (situations), authentic problems (tasks, assessment), authentic learning environments (learning situations), authentic models, and so forth. The use of the terms goes back to the 1990s, when the term authentic became functional to criticize the multiple-choice format in assessment, and in mathematics education to criticize stereotype word problems. For example, Kaiser (2002) writes on German mathematics classes: "… *it is typical for German mathematics teaching that real-world examples discussed in lessons are not authentic real-world problems, but made to illustrate mathematical contents. Therefore, these examples give a quite artificial and far from reality impression*" (p 253). In mathematics education research, *not*

P. Vos (✉)
University of Amsterdam, Amsterdam, The Netherlands
e-mail: fpvos@hotmail.com

authentic is synonymous to: pseudorealistic, artificial, constructed, deviating from and distorting out-of-school practices, and concealing underlying mathematical problems. Still, we remain with the question: what can be labelled as authentic in mathematical modelling education?

According to several dictionaries (e.g. Webster's Online Dictionary) authentic means: genuine, known to be true, being of undisputed origin. The term is illustrated with the discipline Archaeology, where found artifacts are considered authentic when they truthfully originate from human activities in the past. In this discipline, the term authentic is used as a contrast to *being a copy*, such as imitations and forgeries. The qualification is binary: an artifact is either authentic or not; one cannot say that an artifact is 'more authentic than another'. It is possible that an artifact receives the tag: 'authenticity cannot be established', but this doubt does not render the artifact 'a little bit authentic'. In Archaeology, only acknowledged experts can give an artifact the classification of authenticity; if an outsider finds an object in his/her backyard, she/he will need an expert to confirm the authenticity.

In this exemplary discipline, we observe that authenticity is constructed in interplay between objects and actors. The objects are: (1) an artifact as object of study that may be qualified as authentic or not; (2) an origin that has produced the object. The actors are: (3) the experts who attest the qualification of authenticity; and (4) the public who observe the artifact as outsiders and trust the expert's authority, see Fig. 68.1.

In this paper, I draw on sociological studies of social constructs (Berger and Luckmann 1966). Social constructs are agreements pertaining to perceptions, norms and values, and these are developed and sustained in relations between actors and objects. I will analyze authenticity as a social construct with respect to task design for mathematical modelling.

My personal interest in task design and authenticity goes back to a series of design studies in Mozambican mathematics classes. We designed geometry lessons starting from reed baskets, fish traps and drums, and we designed statistics lessons starting from genuine newspaper articles (Vos et al. 2007). We termed the artifacts as authentic resources, we brought these physically into the classroom, and the students appreciated how these embodied a connection of mathematics to out-of-school life. However, we noted that the artifacts were not educational in themselves.

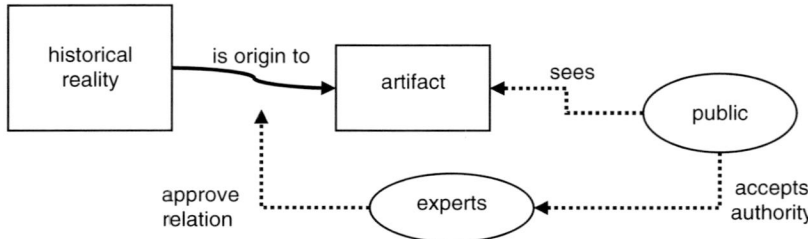

Fig. 68.1 Authenticity in archaeology as a social construct

For example, baskets are created to contain objects and this container quality can be connected to geometrical calculations of area and volume. Nevertheless, neither the basket nor the basket maker directly generated mathematical questions suitable for classroom tasks. The task designer creates these with educational aims in mind, and in the design process the tasks may become quite *inauthentic*, such as: what is the circumference of the basket in cm? Thus, we observed the combination of authentic resources and inauthentic activities and operationalized the term authentic for separate aspects in tasks, if these are "*clearly not created for educational purposes*" (pp 53, 64). According to this definition, authenticity requires an aspect to originate from an out-of-school environment, and authenticity can be agreed in consensus regardless of the actor: whether one is student, teacher, or fisherman, all can agree that a reed basket is an out-of-school object with an out-of-school purpose.

In the ensuing text I discuss definitions of authenticity, compare between definitions and validate these with examples from education, in particular on mathematical modelling.

2 Variations of Tasks in Mathematics Education

In the definitions of *authenticity*, terms such as *situation, realism* and *relevance* reappear. In this section, I offer a number of exemplary tasks to illustrate these terms.

In mathematics education, the abstract discipline of mathematics can be connected to out-of-school reality through 'situated problems', 'context problems', 'real-world problems', or 'work-related problems'. For example, the abstract exercise $3\frac{1}{2} \div \frac{1}{4} = ...$ can be adapted into "*how many quarters of an hour go into three and a half hours?*" In this reformulation the mathematical problem is situated (contextualized). The exercise can also be set into a money situation or a pizza situation, and such situations can allow students to apply common sense knowledge for solving the problem. In a situated problem all numbers have a meaning (e.g. in time units, in money value, or in pizza slices), but this does not make the exercise *meaningful* (interesting) to all students. Task designers can make the problem *realistic* (as if from real-life, but clearly not real) by setting this problem on the division of fractions into the office of a doctor during a morning (8.30–12.00 h), where patients have visits of a ¼ h, and the question is: how many patients can the doctor see?[1]

In this last example, the doctor's practice is a *situation* (figurative context) for a task on the division of fractions and one may question the authenticity of the described situation. Thus, authenticity can be a qualification of a task situation. However, a task is carried out within a social context (learning environment) through task *activities*. These task activities can also be rendered authentic, for

---

[1] Adding reality implies adding *complexity*. The doctor needs a coffee break and the schedule requires flexibility to fit in emergencies or late comers.

example, by the response format of the task. Palm (2002, 2007) studied achievement results on two variations of the bus problem: *how many buses are needed for a school excursion with 360 children, while each bus has 48 seats?* In this study, Palm offered students a variation of this traditional, pseudorealistic word problem, asking students to fill in an order sheet from a bus company: on the sheet they had to put the school name, date of the excursion, the number of buses needed, and additional remarks. Palm found that the group in the sheet-condition outperformed the control group with the traditional word problem. Nevertheless, this variation on the bus problem was constructed by the researcher, the bus company was imaginary and the numbers in the task were invented. Thus, the problem situation was not authentic, but the activity was special: to fill in an order form was a simulation of real-life activities and this activity rendered the task more *worthwhile* (inviting to spend effort) or *relevant* (probably useful in the future).

With Palm's study, we see that one mathematical task (the bus problem) can be designed in task variations. We also see task variations in the Giant Shoe problem, a case encountered in a number of studies on mathematical modelling. Blum (2011) presented a giant shoe sculpted in bronze, depicted by a photograph, and the students were asked to calculate the height of a person who fits the shoe. This Giant Shoe problem matches the curricular concept of similarity and proportionality, and this makes the exercise valuable to mathematics teachers. However, depending on one's background, one may raise other questions. A shoe maker may ask how much leather is needed for a leather copy of the bronze shoe. A bronze thief may ask how much bronze the statue contains. An art student may ask into what artistic tradition the statue fits. Therefore, the task resources may be authentic, but the posed mathematical question may lack relevance, both to people working with bronze statues and to students. In an interesting adaptation of the Giant Shoe problem, Biccard and Wessels (2011) designed a task variation, asking students to assist the police in relating foot prints found at crime scenes to the possible size of suspects, a question truly asked in crime scene investigation. In this variation, the Giant Shoe problem did not merely ask for a number related to the size of an imaginary giant, but the question asked was identical to a question that professional crime scene investigators also set, and thus: in this variation the question was authentic. Thus, variations of tasks and differences in the correspondence of task aspects to reality indicate that separate task aspects can be considered for authenticity.

3 Definitions of 'Authenticity' in Mathematics Education

In this section I offer a cross-section of definitions of 'authenticity' in mathematics education. For example, Gulikers et al. (2005) write:

> An authentic learning environment provides a context that reflects *the way knowledge and skills will be used in real life. This includes a physical or virtual environment that* resembles *the real world with real-world complexity and limitations, and provides options and possibilities that are also present in real life.* (p 509) [emphasis added by PV].

68 What Is 'Authentic' in the Teaching and Learning of Mathematical Modelling?

Here are a number of definitions of an 'authentic task':

The task asks students to address a concept, problem, or issue that is similar *to one they have encountered or are likely to encounter in life beyond the classroom.* (Newmann et al. 1995, emphasis added by PV).

authentic tasks (are) based on situations which, while sometimes fictional, represent *the kinds of problem encountered in real life.* (OECD 2001:p 23, emphasis added by PV).

the term authentic task refers to a task in which the situation described in the task (…) is a situation from real life outside mathematics itself that has occurred or that might very well happen. In addition, the task situation is truthfully described and the conditions under which the task solving takes place in the real situation are simulated *with some reasonable fidelity in the school situation.* (Palm 2002:p IV-7, emphasis added by PV).

In the above definitions, authenticity means that the object (a task/an environment) is a copy that honestly simulates reality. Thus, the object does not originate from reality, but it has been designed to mirror reality. In the diagram, the arrow between object and reality is reversed, see Fig. 68.2. There is no authorization on the *honesty* of the simulation, thus, the simulation may be quite fake. According to this definition, the student may not even recognize the reality that is simulated by the task. Of course, the task presentation can includes media (e.g. photographs) as a window on reality. Also, the task designer/teacher may be an expert of the problem area (e.g. the task designer has a background in scientific research). In these cases, authenticity can be added to the task to assist students to imagine that the copy simulates reality.

In the above definitions authenticity is connected to simulations (copies) of reality, while the dictionary definition states that authenticity is *opposite to being a copy*. Somehow, some educators detached the term authentic from the meaning in the dictionary and started to use it synonymously to *realistic*, to *worthwhile* or to *relevant*. Used in this way, authenticity can be relative to the actor's view, and then a task can be less authentic to some and more to others.

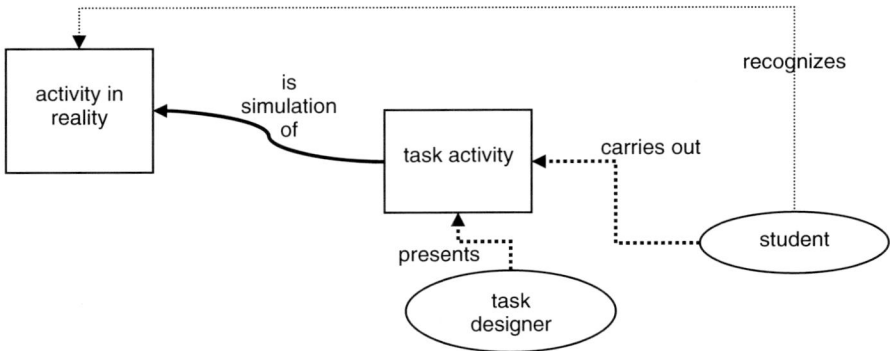

Fig. 68.2 Simulation tasks constructed as being 'authentic'

Below are definitions of authenticity that differ from the simulation definitions above:

> *Authentic mathematical activities are* actual work samples *taken from a representative collection of activities that are meaningful and important in their own right. They are not just surrogates for mathematical activities that are important in "real-life" situations.* (Lesh and Lamon 1992:p 17, emphasis by L&L).

> *We define an authentic extra-mathematical situation as one which is* embedded *in a true existing practice or subject area outside mathematics, and which deals with objects, phenomena, issues, or problems that are genuine to that area and* recognized *as such by people working in it.* (Niss 1992, cited in Palm 2002:p I-20, emphasis added by PV).

In these definitions, the object (task, situation, activity) originates from an out-of-school reality, just like original Mozambican reed baskets for geometry lessons. The task designer uses true out-of-school reality for a task. These definitions align with the dictionary definition of 'not being a copy'. Thus, authenticity cannot be added, because the authenticity was already present before the task designer created the task.

Also, the above definitions mention a social agreement. The Niss-definition includes an actor who authorizes the qualification of authenticity: people working in a true existing practice or subject area outside mathematics. By including experts, the Niss-definition converges with the archeological construct of authenticity, in which experts authorize the authenticity. However, the Niss-definition has its limitations, with experts being equated to 'people working in that practice or area'. With this formulation, Niss excludes other stakeholders of problem situations. For example, consumer problems (finding the best price) or environmental problems (optimizing CO_2 reduction) can be recognized as being important, not only by people working in these areas, but also by out-of-school stakeholders, such as consumers, environmentally engaged citizens, or the students themselves, and other 'not formally' working people who take up responsibilities.

4 Problems with Defining 'Authenticity'

Below, I will present a number of definition problems, which emanate from the fact that authenticity is a social construct and needs to be agreed upon in different communities (educators, students). In the first place, the term has been used by many different educational specialists as shown in the previous paragraph. Besides being frequently used, the term is used in a variety of interpretations, and sometimes not clearly defined at all. I agree with Palm (2002), who in his study on 'authentic assessment', observes: *"authentic assessment can mean almost anything"* (p I-10). In the same way, authenticity in mathematical modelling education can mean almost anything. For example, authenticity can both refer to not being a copy, and to being a copy. Thus, contradictory and unclear definitions are being used, and this calls for a clarification of the differences.

A second problem emanates from what quality we want 'authenticity' to have: should it be a binary qualification, or do we accept it to have an ordinal scale? Do we accept it as an absolute qualification, which does not depend on the viewpoint of the actor, or do we accept that we may have different opinions and that 'something can be authentic to some students, but not to others'?

A third problem arises from whether we want authenticity to apply holistically to a complete task (or to a complete learning environment), or only to partial aspects in a task. For those who use holistic terminology such as 'authentic student projects' or 'authentic enquiry', the question then is: how much correspondence with reality is sufficient to qualify for authenticity? Or in other words: what essential aspects can, and what cannot be cut out for the sake of education?

I will illustrate the last problem with an example from vocational education. In the training of pilots flight simulators are crucial training tools. Externally, they are containers standing on poles that enable the simulator to shake. Inside the container is a partial cockpit with one or more pilot's seats, steering rudders, engine throttles, and all around and overhead are panels with instruments, keyboards, displays and signals, radio and navigation equipment, and many other details that make the simulator resemble an original cockpit. When a candidate sits there, he/she sees through the windows a number of panels on which a realistic horizon is projected. To 'add authenticity', a candidate is requested to dress in full pilot gear. According to some definitions (e.g. Gulikers et al. 2005), a flight simulator is an 'authentic learning environment' because it simulates a real plane with sufficient accuracy. Nevertheless, flight simulators do not fly and there is not one account of a fatal accident with flight simulators. What clearly distinguishes flight simulators from real planes is that a candidate has no responsibility over lives or material. Thus, if essential aspects of the origin have been cut out for educational purposes, can we still speak of an authentic task, of authentic activities, or of an authentic learning environment? Defining authenticity holistically for complex units (tasks, environments) creates a problem of delineation: what are the essential aspects of the original that need to be taken into the definition, and what can be deleted without losing the qualification of authenticity? The example of the flight simulator shows us that deleting authenticity does not downplay the value of an effective learning environment: deleting responsibility over material and lives offers students a safe environment to learn.

5 Pragmatically Constructing Authenticity in Mathematical Modelling

As described before, there are several aspects that require attention in defining authenticity. First, one needs to consider whether the term covers copies and simulations, or whether the qualification is only applicable for true originals. As I have argued above, applying the term 'authenticity' for simulations, such as a flight simulator, will only obscure the term and raise questions on how much authenticity must be added or can be deleted to allow for the qualification of authenticity.

Second, one needs to consider whether it has a binary (yes/no) or an ordinal (more/less) scale. In my view, the definition should align with the Archeology-construct of authenticity, with the binary scale to be used. As a consequence, the definition can then only apply to aspects within a task and not to complete tasks. I therefore criticize terms such as 'authentic tasks', because the educational setting will require adaptations in the same way as with the flight simulator: it is not an authentic airplane (but an effective learning environment); the horizon seen inside is not authentic (but realistic). Still, a flight simulator can contain authentic instruments (originating from a real plane), candidates wear their authentic pilot's uniforms, and a flight simulator can contain authentic software equal to the one used in reality. Also, the training activities can be authentic, being exactly identical to activities in reality. Similarly, in mathematical modelling a task can ask students to work with authentic data, with authentic modelling software (the original software that modelling researchers use), while the activity to use the software is adapted for training purposes. Some aspects in a task/learning environment can be authentic while others aren't.

Besides simply defining authenticity, it is a social construct on which a community agrees on its qualification. For this, the term cannot be dependent on an actor's viewpoint. I claim that the term authenticity can be a qualification clear to all actors, even if the aspect has no meaning or relevance to them. This implies that the link between the object (aspects of the task) and its out-of-school origin must be direct and certifiable. With such an origin, the term authentic differs from actor-dependent terms such as *relevance* and *meaning*.

The construct of authenticity is certifiable if experts authorize the qualification. In fact, out-of-school experts embody the authenticity. In the Niss-definition experts were 'people working in the area', but in my view the actors with expertise can also be stakeholders linked to the area by concerns and responsibilities. In the field of mathematical modelling education, there are two out-of-school areas for experts. First, we have the area of the problem situation with stakeholders for workplace problems, consumer problems, environmental problems, scientific research problems, and so forth. Second, we have mathematical modelling researchers, that is: people who apply mathematical techniques to resolve the aforementioned problems. The modelling researchers can testify to the authenticity of the problem-solving approaches, to the authenticity of used mathematical models (including the symbols used in reality), to the authenticity of computer software (software used by professional modelling researchers). By truthfully incorporating authentic aspects of mathematical modellers' work, we can *enculturate* students into the research field of mathematical modelling. Thus, authenticity is not only a qualification for aspects in the nonmathematical side of the modelling cycle; also the mathematical side of the modelling process can contain authentic aspects (see Fig. 68.3).

Asserting that the qualification of authenticity should only apply to genuine originals, be binary, apply to separate task aspects, and be independent of the actor, I now

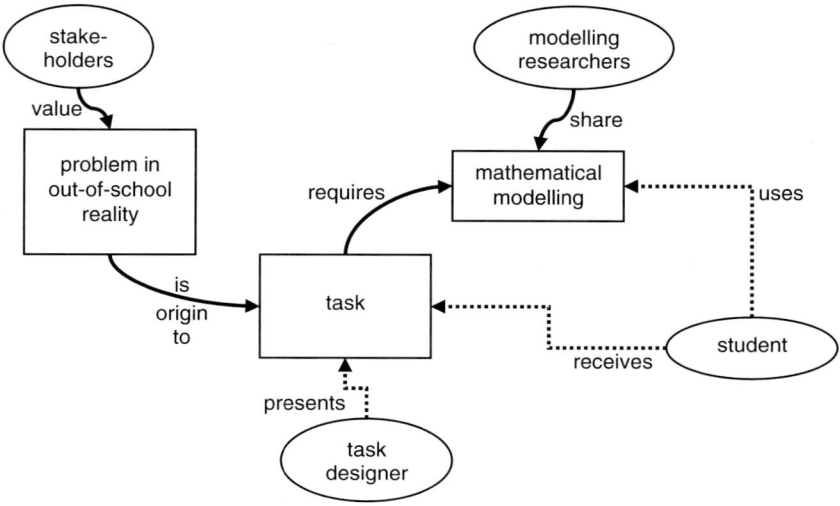

Fig. 68.3 The social construct of 'authentic' aspects in a mathematical modelling task

return to the pragmatic definition of being authentic in mathematical modelling education for objects that are *"clearly not created for educational purposes"*. This operationalisation has a number of advantages: (1) it connects to an out-of-school origin, (2) it is binary, (3) it can be applied to separate task aspects, and (4) it is actor-independent ('clear to anybody'), which can be certified by actors (stakeholders, modelling researchers). If objects serve out-of-school purposes, there are out-of-school actors by or for whom the object was created.

According to the above, pragmatic definition, a task can have some authentic aspects, while other aspects are included for educational purposes. I shall illustrate this with a final example of a modelling task: in the *Porsche Task* (Maaß 2006) students are asked to calculate the amount of paint needed for a Porsche 911. This task contains aspects, which originate from reality and have clearly no educational purpose: a Porsche 911 (purpose: driving fast), paint (purpose: protect against corrosion), the need to know the amount of paint (purpose: factory stock). Even the calculation method used by the students (segmenting the car's surface into triangles) was identical to the method of the mathematical modelers in the factory and thus, even the calculation method served out-of-school purposes. On the other hand, the Porsche Task also contained aspects with an educational purpose: for example, there was no absolute urge for correctness of the calculation. The students weren't in an internship in the factory and they weren't responsible for correct answers. If that would be the case, an error has repercussions in the factory (wrong stock) or with the consumers (rusting cars). However, in education we allow students to commit errors. Fortunately, not all aspects in a task need to be authentic, but tasks are more engaging if a number of aspects are.

References

Berger, P. L., & Luckmann, T. (1966). *The social construction of reality: A treatise in the sociology of knowledge.* Garden City: Anchor Books.

Biccard, P., & Wessels, D. (2011). Documenting the development of modelling competencies of Grade 7 mathematics students. In G. Kaiser, W. Blum, R. Borromeo Ferri, & G. Stillman (Eds.), *Trends in teaching and learning of mathematical modelling* (pp. 375–383). New York: Springer.

Blum, W. (2011). Can modelling be taught and learnt? Some answers from empirical research. In G. Kaiser, W. Blum, R. Borromeo Ferri, & G. Stillman (Eds.), *Trends in teaching and learning of mathematical modelling* (pp. 15–30). New York: Springer.

Gulikers, J. T. M., Bastiaens, T. J., & Martens, R. L. (2005). The surplus of an authentic learning environment. *Computers in Human Behavior, 21,* 509–521.

Kaiser, G. (2002). Educational philosophies and their influence on mathematics education – An ethnographic study in English and German mathematics classrooms. *ZDM – The International Journal on Mathematics Education, 34*(6), 241–257.

Lesh, R. A., & Lamon, S. J. (1992). *Assessment of authentic performance in school mathematics.* Washington, DC: AAAS.

Maaß, K. (2006). What are modelling competencies? *ZDM – The International Journal on Mathematics Education, 38*(2), 113–142.

Newmann, F. M., Secada, W. G., & Wehlage, G. G. (1995). *A guide to authentic instruction and assessment: Vision, standards, and scoring.* Madison: WCER.

OECD. (2001). *Knowledge and skills for life. First results from the OECD Programme for International Student Assessment (PISA) 2000.* Paris: OECD.

Palm, T. (2002). *The realism of mathematical schools tasks – Features and consequences.* Ph.D. thesis, Umeå University, Umeå, Sweden.

Palm, T. (2007). Features and impact of the authenticity of applied mathematical school tasks. In W. Blum et al. (Eds.), *Applications and modelling in mathematics education* (ICMI studies series no. 10, pp. 201–208). New York: Springer.

Vos, P., Devesse, T. G., & Rassul Pinto, A. A. (2007). Educational design research in Mozambique: Starting mathematics from authentic resources. *African Journal for Research in Mathematics, Science and Technology Education, 11*(2), 51–66.

Corresponding Authors

Burkhard Alpers HTW – Hochschule Aalen für Technik und Wirtschaft, Aalen, Germany, Burkhard.Alpers@htw-aalen.de

Mette Andresen NAVIMAT – National Knowledge Centre for Mathematics Education, UCC – University College Copenhagen. Denmark, mea@ucc.dk

Piera Biccard University of Stellenbosch, Stellenbosch, South-Africa, pbiccard@yahoo.com

Maria Salett Biembengut Universidade Regional de Blumenau, Department of Mathematics, Blumenau, Brazil, salett@furb.br

Morten Blomhøj IMFUFA, NSM, Roskilde University, Roskilde, Denmark, blomhoej@ruc.dk

Werner Blum Department of Mathematics, University of Kassel, Kassel, Germany, blum@mathematik.uni-kassel.de

Marcelo C. Borba UNESP – Universidade Estadual Paulista Júlio de Mesquita Filho, São Paulo, Brazil, mborba@rc.unesp.br

Rita Borromeo Ferri Department of Mathematics, University of Kassel, Kassel, borromeo@mathematik.uni-kassel.de

Martin Bracke Department of Mathematics, University of Kaiserslautern, Kaiserslautern, Germany, bracke@mathematik.uni-kl.de

Matthias Brandl Didactics of Mathematics, University of Augsburg, Augsburg. Germany, matthias.brandl@math.uni-augsburg.de

Jill P. Brown Australian Catholic University, Melbourne, Australia, jill.brown@acu.edu.au

Hugh Burkhardt Shell Center, Nottingham University, Nottingham, UK Hugh.Burkhardt@nottingham.ac.uk

Andreas Busse Faculty for Education, Psychology, Human Movement, University of Hamburg, Hamburg, Germany, andreas.busse@uni-hamburg.de

Richard Cabassut Laboratoire André Revuz- Paris 7 University, Paris, France, richard.cabassut@unistra.fr

Susana Carreira Universidade do Algarve & CIEFCUL, Portugal, scarrei@ualg.pt

Ching-Kuch Chang Graduate Institute of Science Education, National Changhua University of Education, Changhua City, Taiwan, ckuchang@gmail.com

Qi Dan Shanghai Maritime University, Logistic Engineering College, Shanghai, China, danqi31@163.com

Johan Deprez Universiteit Antwerpen, Hogeschool Universiteit Brussel and Katholieke Universiteit Leuven (Belgium), johan.deprez@ua.ac.be

Dirk de Bock Hogeschool-Universiteit Brussel, Belgium, dirk.debock@hubrussel.be

Helen Doerr Syracuse University, Syracuse, N.Y., USA, hmdoerr@syr.edu

Andreas Eichler University of Education Freiburg, Germany, andreas.eichler@ph-freiburg.de

George Ekol Simon Fraser University, Canada, george_ekol@sfu.ca

Joachim Engel Ludwigsburg University of Education, Ludwigsburg, Germany, engel@ph-ludwigsburg.de

Frank Förster Technische Universität Braunschweig, Braunschweig, Germany, f.foerster@tu-braunschweig.de

Peter Frejd Department of Mathematics, Linköping University, Linköping, Sweden, peter.frejd@liu.se

Peter Galbraith The University of Queensland, School of Education, Brisbane, p.galbraith@uq.edu.au

Javier García University of Jaén, Jaén, Spain, fjgarcia@ujaen.es

Andreas Geiger Hohenstaufen-Gymnasium Kaiserslautern, Germany, Andreas.Geiger@gmx.de

Vince Geiger Australian Catholic University, Brisbane, Australia, vincent.geiger@acu.edu.au

Boris Girnat University of Education Freiburg, Freiburg, Germany, boris.girnat@ph-freiburg.de

Gilbert Greefrath University of Cologne, Cologne, Germany, g.greefrath@uni-koeln.de

Roxana Grigoras University of Bremen, Bremen, Germany, roxana@math.uni-bremen.de

Corresponding Authors

Christopher Haines City University, London, UK, C.R.Haines@city.ac.uk

Matti Heilio Lappeenranta University of Technology, Lappeenranta, Finland, matti.heilio@lut.fi

Hans-Wolfgang Henn Technische Universität Dortmund, Faculty of Mathematics, Dortmund, IEEM, wolfgang.henn@tu-dortmund.de

Hans Humenberger University of Vienna, Vienna, Austria, hans.humenberger@univie.ac.at

Toshikazu Ikeda Yokohama National University, Yokohama, Japan, ikeda@edhs.ynu.ac.jp

Gabriele Kaiser Faculty for Education, Psychology, Human Movement, University of Hamburg, Hamburg, Germany, gabriele.kaiser@uni-hamburg.de

Tetsushi Kawasaki Kyoto Prefectural Sagano Senior High School, Kyoto, Japan, tetsushi@kyokyo-u.ac.jp

Sergiy Klymchuk Auckland University of Technology, Auckland, New Zealand, sergiy.klymchuk@aut.ac.nz

Usha Kotelawala Fordham University, Graduate School of Education, New York, USA, kotelawala@fordham.edu

Sebastian Kuntze Ludwigsburg University of Education, Ludwigsburg, Germany, kuntze@ph-ludwigsburg.de

Richard Lesh Indiana University, School of Education, Bloomington, USA, ralesh@indiana.edu

Thomas Lingefjärd University of Gothenburg, Gothenburg, Sweden, Thomas.Lingefjard@ped.gu.se

Katja Maaß University of Education Freiburg, Freiburg, Germany, katja.maass@ph-freiburg.de,

Akio Matsuzaki Saitama University, Saitama, Japan, makio@mail.saitama-u.ac.jp

Nicolas G. Mousoulides Cyprus University of Technology, Limassol, Cyprus, nicholas.mousoulides@cut.ac.cy

Rui Gomes Neves Unidade de Investigação Educação e Desenvolvimento (UIED) e Departamento de Ciências Sociais Aplicadas (DCSA), Faculdade de Ciências e Tecnologia (FCT), Universidade Nova de Lisboa (UNL), Monte da, Portugal, rgn@fct.unl.pt,

Kit Ee Dawn Ng National Institute of Education, Nanyang Technological University, Singapore, Singapore dawn.ng@nie.edu.sg

Yoshiki Nisawa Kyoto Prefectural Rakuhoku Senior High School, Japan, Nisawa y-nisawa@kyoto-be.ne.jp

José Ortiz University of Carabobo, Venezuela, ortizjo@cantv.net

Sanne Schaap University of Groningen/University of Amsterdam, Groningen/Amsterdam, Netherlands, s.schaap@uva.nl

Barbara Schmidt Realschule Hechingen, Hechingen, Germany, Schmidt.Barbara@gmx.de

Manuel Sol Vilatzara School, Vilassar de Mar, Spain, msol@xtec.cat

Jeroen Spandaw Delft University of Technology, Netherlands, j.g.spandaw@tudelft.nl

Gloria Stillman Australian Catholic University (Ballarat), Victoria, Australia, gloria.stillman@acu.edu.au

Jhony A. Villa-Ochoa Research Group of Mathematics Education and History-University of Antioquia, Colombia, javo@une.net.co

Pauline Vos University of Amsterdam, Netherlands, f.p.vos@uva.nl

Geoff Wake University of Nottingham, Nottingham, UK, geoffrey.wake@nottingham.ac.uk

Jinxing Xie Tsinghua University, Beijing, China, jxie@math.tsinghua.eud.cn

Shih-Yi Yu Graduate Institute of Science Education, National Changhua University of Education, Taiwan, sheree318@yahoo.com.tw

Luzia Zöttl Institute of Mathematics, Ludwig-Maximilian Universität, Luzia.Zoettl@gmx.de

Index

A

Affinity, 289–298
Algebraic curves, 553–556
Alpers, B., 445–455
Amado, N., 199–209
Andresen, M., 519–528
Anthropologic theory of didactic (ATD), 86, 87, 89, 91–94, 515, 560, 564, 571, 574
Applications, 65–73, 107–116, 165–179, 431–432, 552, 569–577, 579, 631, 689–698
    of models, 386–390, 392, 393
Applied perspective, 57–64
Ärlebäck, J.B., 407–415
Assessment, 251, 262, 362, 363, 458, 576, 618, 630, 632, 655, 690–692, 694, 695, 697, 698
    examination, 108, 153, 419
    examination tasks, 325–329, 648
    of performance, 642, 646–648
    test instrument, 429–436
Attitudes, 306, 408, 410, 411, 413–414, 600, 631, 658
Authenticity, 467, 592–600, 672, 683, 686, 713–721
    authentic tasks, 23

B

Baioa, A.M., 211–220
Beliefs, 23, 65–73, 75–84, 145, 151, 281, 289–292, 295–298, 575, 630–633, 657, 658, 703
    about nature of mathematics, 295
    about teaching mathematics, 295–297, 635–638
    self-efficacy, 631, 635–638
Biccard, P., 375–382

Biembengut Faria, T.M., 269–278
Biembengut M.S., 269–278
Blockages, 137–145
Blomhøj, M., 343–347, 385–394
Blum, W., 1–5, 15–27
Borba, M., 31–35
Borromeo Ferri, R., 1–5, 181–185
Boundary, 208
    crosser, 655
    crossing, 202
    objects, 654
Bracke, M., 529–548
Brandl, M., 551–557
Brown, J.P., 187–197, 243–246, 289–298
Buchholtz, N., 591–600
Burkhardt, H., 511–517, 552, 553, 569
Busse, A., 37–45

C

Cabassut, R., 559–567
Carreira, S., 159–162, 199–209, 211–220
Case study, 128, 191, 203, 309
    ethnographic, 234
    qualitative, 65–73, 234
Chang, C.-K., 147–156
Chemistry, 520, 521, 523–526
Classroom, 511–515
Cognition, 22, 33, 200
    cognitive, 31, 86, 94, 128, 232, 233, 565
    cognitive apprenticeship, 26
    cognitive activation, 22
    cognitive barriers, 18, 20
    cognitive complexity, 20
    cognitive demands, 23, 369
    cognitive levels, 21, 24
Comparison, 559–567

Competencies, 19, 21, 47, 62, 63, 139, 145, 203, 292–294, 297, 375–382, 386–390, 394, 397–405, 407–415, 427–436, 565–567, 575, 591–593, 600, 630, 685
  metacognitive, 145, 377, 379, 681
Computing, 479, 480. 484, 488
Conceptions
  of mathematical, 271, 273
Content, 83, 281–282, 286
  authentic, 467–470
Content knowledge, 108
  mathematical (MCK), 290, 630, 659
  nature of, 291, 295–297
  pedagogical (PCK), 281, 290, 297, 630–632, 636–638, 655–658
Context, 134, 159–162, 200, 594
  real, 233, 574
  real world, 37–45, 107, 108, 114, 349–352, 362, 566
  situation, 200, 201
  task, 38, 200
Cultural Historical Activity Theory (CHAT), 654, 660
Curriculum
  guidelines, 200
  individual curricula, 75–76, 82
  primary school, 559–567
  secondary school, 562, 567, 670, 687, 689–698
Curved surface model, 120–123, 125

D
Dan, Q., 457–465
Data
  processing, 447, 450
  representations, 563
De Bock, D., 47–54
Deprez, J., 467–476
Development, 247–259, 261–267
Differential equation, 604, 608, 610, 613
Difficulties
  authenticity, 593
  double transposition, 565
  lack of materials, 642, 646, 650
  lack of time, 67, 646, 648–649, 685, 694
  students, 16, 19, 20, 23, 25, 108, 233, 354, 686
  supervision of modelling, 686
DISUM project, 16, 17, 20, 22, 24, 26
Doerr, H.M., 247–267

E
Education
  adult, 199–209
  engineering, 221–229
  realistic mathematics, 214–216, 220
  science and mathematics, 331–338
  secondary, 684, 687
  tertiary, 73, 468, 476
Educational practice, 604
Edwards, I., 187–197
Eichler, A., 75–84
Eigenvalues, 467–476
Eigenvector, 467–476
Ekol, G., 57–64
Empirical research, 15–27, 407
Enculturation, 720
Engel, J., 397–405
English, L.D., 221–229
Environments, 212, 213
Epidemic models, 491–496
Evaluation, 599–600
Experience, 42, 111, 200, 201, 206, 208
Experimental study, 457–465

F
Förster, F., 65–73
Framing, 101–106
Frejd, P., 407–415
Function, 117–126, 188, 190, 194, 195
  multi-variable, 117–126
Functional reasoning, 404

G
Galbraith, P., 441–444, 689–698
Garcia, F.J., 85–94, 569–577
Geiger, A., 528–548
Geiger, V., 305–313
Geometry, 75, 76, 78–81, 83, 84, 119
Giménez, J., 231–239
Girnat, B., 75–84
Goedhart, M., 137–145
Greefrath, G., 301–304, 315–329
Grigoras, R., 85–94
Gruenwald, N., 489–497
Gurlitt, J., 629–638

H
Haines, C., 349–364
Halverscheid, S., 85–94

Index

Heilio, M., 479–488
Henn, H.-W., 417–425
Humenberger, H., 579–589

I
Ikeda, T., 669–677
Industry, 479–488
Interactive learning environments, 332, 337
Interdisciplanary, 107–116
Interests, 44, 270–272, 552, 557

K
Kaiser, G., 1–5, 591–600
Kawasaki, T., 603–615
Kepler's Laws, 603–613
Kjeldsen, T.H., 385–394
Klymchuk, S., 489–497
Knowledge, 199–209
　diffusion, 89
　life, 209
　linking mathematical, 352–353
　mathematical world, 564–565, 567
　meta, 145
　prior, 109, 112, 159, 160, 162, 271
　real world, 114, 115, 563–565, 567, 570
　teacher, 256–267, 655–659
Kotelawala, U., 617–627
Kuntze, S., 279–287, 397–405

L
Learning
　interactive environments, 332, 337
　situated, 44
　student, 247–267
Learning and Education in and through Modelling and Applications (LEMA), 515, 559, 569, 575–577, 630–638, 653
Lecoq, F., 199–209
Lesh, R., 247–267
Linear algebra, 589
Linear process, 319–320
Lingefjärd, T., 9–14, 97–106
Long term experience, 548
López, C.M.J., 701–710

M
Maaß K., 367–371, 629–638
Markov chains, 579, 581, 584
Mathematical
　components, 500–502, 505
　connections, 267, 677
　content knowledge, 290, 630, 660
　creativity, 212
　development model, 605, 608–613
　experiences, 93, 500–502, 504, 505
　model, 88, 560
Mathematics, 519–528, 574
　experimental, 212
　human activity, 86, 214, 572
　in a modern society, 588
　nature of, 290, 295–297
Mathematising, 18, 21, 85, 94, 95, 139, 214, 379, 669, 671, 689–698
　activities, 85, 139–140
Mathematisation, 18, 43, 80, 86, 132, 139, 214, 219, 231, 235, 561, 670, 672–674, 677
Matrix model, 467–476
　Leslie, 467, 468
　transition, 581
Matsuzaki, A., 499–507
Measurements, 112, 114, 445–455
Mechanical engineering, 467–476
Meier, S., 97–106
Metacognition, 165–174, 630
　teacher, 161, 162, 658
Metacognitive, 21, 22, 24, 26, 165–179, 248, 256, 657
　activities, 139
　blindness, 172, 174, 176
　mirage, 172, 173
　misdirection, 172, 174
　monitoring, 685
　productive acts, 169–174, 179
　red flags, 172–175
Meta-metacognitive, 169, 174–179
Model-eliciting activity (MEA), 23, 147–156, 250–256, 260, 261, 375
Modelling, 85–94, 97–106, 127–134, 140, 269–270, 305–313, 315–329, 561–562
　application of models, 385–395
　applications *versus* modelling, 694–695
　competencies, 26, 31, 47, 57–64, 85, 375–382, 385–392, 394, 397–405, 407–415, 427–436
　computational, 331–338
　conceptual, 223, 224, 226–229
　cycle, 17, 21, 24, 76, 77, 79, 80, 85, 86, 94, 128, 138, 202, 212, 232–233, 238, 259–260, 559, 560, 566, 604, 605, 682
　distance course, 269–278
　drivers for, 349–364
　education, 127–134, 679–687

Modelling (*cont.*)
    experimental, 211–220
    explorative, 138
    expressive, 138
    group, 376, 379
    mathematical, 47, 48, 50, 51, 53, 92, 120, 407–415, 519–528, 533, 540, 546, 679–687, 701–710
    metacognitive, 630
    model of the HPA-axis, 385, 392–393
    obstacles, 641–650
    pedagogies, 565, 571, 577
    practices, 542
    process, 18, 85, 87, 92, 127, 133, 138, 232, 385–390, 392, 394, 428–429, 433
    routes, 20, 24, 86, 139, 211, 215, 231–239, 352
    schemes, 128, 133
    skills, 457–465
    subject-centred, 162
    teaching, 147–156
    traffic model, 385, 390–394
Modelling tasks, 279–287
Models and modelling perspective (MMP), 147, 214, 215, 219, 247–267
Moriya, S., 117–126, 503–615
Motivation, 43, 71, 72, 272, 572, 588, 600, 660, 686
Mousoulides, N.G., 221–229
Multidisciplinary, 63, 519, 520, 527

N
Narayanan, A., 489–497
Neves, R.G., 331–338
Ng, K.E.D., 107–116
Nisawa, Y., 117–126
Non-mathematical components, 500, 504

O
Object manipulation, 211–220
Opportunities and blockages, 137, 138, 140, 143
Ortiz, J., 127–134

P
Pedagogy, 569–571, 574, 577, 654–659, 661
Perspective, 19, 33, 34, 79, 86, 212–215, 219, 232, 247, 264, 376
    model-eliciting, 214
Petersen, A., 517, 519–528
Physics, 332–334, 337

Population dynamics, 468
Pragmatism, 349–364
Praxeology, 87–88, 90, 94, 572–574
Predictions, 490, 492–496
Primary school, 48, 49, 222, 239, 559–567
Probability, 556, 581, 619, 622
Problems
    content-related, 631
    missing-value, 54
    with modelling requirement, 47–54
    real life/world, 140, 530–532, 672, 674
    reliability theory, 621–626
    solving strategies, 21, 148
    sound propagation, 129
    vehicle stopping distance, 121
Professional development, 420, 423–425, 513–515, 569, 570, 575–577, 654
    course, 629–634, 637
    design of, 621–625629–635, 637
    effects of, 635, 636
Project, 33
    COACTIV, 23
    design based, 107–116
    DISUM, 16, 17, 25, 26
    dormitory electrical wiring, 625, 626
    interdisciplinary, 107–116
    LEMA, 515, 559, 569, 575, 629–638, 653
    modelling, 385–394
    modelling route, 233
    realistic mathematical project, 232, 233
    SINUS, 25
    sun hour, 97–106
    work, 232, 234, 390, 592
Proportional
    methods, 54
    reasoning, 49, 110, 111, 115, 208, 259
Proxy for real world, 353, 359, 362

Q
Qualitative case study, 65–73, 234

R
Real experiences, 500, 501, 505
Reality, 701–710
Real-world context, 37–45
Real world problems, 529, 530, 532, 557, 630, 661
Reflection, 23–25, 41, 139, 256, 293, 385–394, 571, 576
Regular lessons, 529–548
Reiss, K., 434
Relevance, 715, 716, 720

Representations, 127, 128, 131–133, 233, 250, 259, 261, 563, 565–567
Research
  empirical, 15–27, 83, 86, 147, 407
  research methods, 248, 254
Response analysis mapping, 499–507
Rich situations, 513, 515–517
Rosich, N., 231–239
Ruiz-Higueras, L., 569–577

S
Santos, A.D., 127–134
Sauerbier, G., 489–497
Scale drawings, 109, 112, 115
Schaap, S., 137–145
Schmidt, B., 641–650
Schwarz, B., 591–600
Secondary School, 38, 99, 127–134, 307–309
Self-efficacy, 630–632, 634–638
Sense making/made sense, 23, 108, 162, 202, 203, 232, 258
Sense of reality, 701–710
Signal-noise metaphor, 400, 401
Siller, H.S., 315–329
Silva, J.C., 331–338
Simulation, 252, 479, 481–484, 487
Situated meanings, 200, 201
Socio-mathematical norms, 44, 45
Sol, M., 231–239
Spandaw, J., 679–687
Statistical Literacy, 397–405
Stephens, M., 669–677
Spandaw, J., 679–687
Stillman, G., 1–5, 165–179, 289–298, 689–698
Stochastics, 75, 82–83, 583, 589, 617, 682
Students, 489, 490, 492–496
  gifted, 551, 552, 555, 557
  secondary, 127–134, 167–169, 171, 174–176, 178, 188, 189
Support, 514–517

T
Tasks, 22, 23, 40–42, 44, 47–54, 82, 84–94, 159–162, 187–197, 208, 562–566, 573, 575, 617, 630, 631, 641, 642, 644–649, 658
  characteristics, 261, 279, 566
  context, 38, 187–190, 192–197, 280, 647, 654
  design, 696
  variations of, 715
Teacher
  competencies, 24
  in-service, 279–287
  meta-knowledge about modelling, 281, 284, 285, 287
  perception, 150
  pre-service/prospective, 88, 279–287, 289–298
  professional, 277
  role, 239
  secondary, 65–73, 75–84, 258, 264, 306, 307, 309, 575
  training, 132, 280, 488, 564, 567, 570, 577
Teaching
  application oriented, 420, 424
  challenges, 511–513
  competencies, 591–593
  examples, 591–593, 599, 600
  experiment, 215
  mathematics, 147–156
  quality, 22, 23, 630
  style, 25, 26
Technology, 33, 87, 315–329, 519–528
  computer algebra system/CAS, 100, 303–313, 583, 584, 587, 589
  computer use, 100, 101, 105, 106, 121, 122, 124–125, 420, 588, 589, 599, 604, 608
  Dynamic Geometry Sketchpad (DGS), 59
  mathematical, 479, 481
  spreadsheet, 580
Technology-rich and learning environment (TRTLE), 188, 190, 197
Tendencies
  in distance course, 273
Teodoro, V.D., 331–338
Textbook, 559–567, 606, 669–677
Theory, 87, 90, 91, 641–642
Thinking
  creative, 457–465
  higher order, 188, 190, 194–197
  mathematical, 254
  mathematically, 48, 214
  styles, 21, 86
Torrance test, 460
Triangulation, 39, 40, 44

U
Ufer, S., 427–436
Upper secondary level, 408, 519

V

Validate, 21, 132, 239, 322, 324–325, 329, 380, 393, 598. 671, 682, 684, 685
Validation, 18, 133, 214, 233, 239
Van Dooren, W., 47–54
Variation, 262, 266, 552, 557, 600, 632, 716
Verschaffel, L., 47–54
Views
   of modelling, 32
   socio-cultural, 200
   specific professional knowledge, 279–281, 285–287
Villa-Ochoa, J.A., 701–710
Vleugels, K., 47–54
Vos, P., 137–145, 665–667, 713–721

W

Wagner, A., 559–567
Wake, G.D., 653–661
Weitendorf, J., 315–329
Wessels, D.C.J., 375–382

Word problems, 19, 38, 47, 72, 713, 715
Workplace mathematics, 445
Workplace study, 445
World
   figured, 200, 201, 208
   lived-in, 200, 201, 208
   mathematical, 88, 91
   rest of the world, 88
   real, 91, 94

X

Xie, J., 457–465

Y

Yu, S.-Y., 147–156

Z

Zöttl, L., 427–436
Zverkova, T., 489–497

Printed by Books on Demand, Germany